Contemporary Mathematicians

Gian-Carlo Rota
Editor

Hassler Whitney in the 1970s
(Photograph by Sally Whitney)

Hassler Whitney at age 14 in the Swiss Alps

Hassler Whitney
Collected Papers

Volume II

James Eells
Domingo Toledo
Editors

Birkhäuser
Boston · Basel · Berlin
1992

James Eells
Department of Mathematics
University of Warwick
Coventry CV 4 7AL
England

Domingo Toledo
Department of Mathematics
University of Utah
Salt Lake City, Utah 84112
U.S.A.

Library of Congress Cataloging-in-Publication Data

Whitney, Hassler.
The collected papers of Hassler Whitney / edited by
James Eells, Domingo Toledo
p. cm. – (Contemporary mathematicians)
Includes bibliographical references.

ISBN-13: 978-1-4612-7741-5

1. Topology. 2. Geometry, Differential. 3. Combinatorial analysis. I. Eells, James, 1926- II. Toledo, Domingo.
III. Title. IV. Series.
QA611.W4854 1992
514–dc20 91-9885
CIP

Printed on acid-free paper.

Softcover reprint of the hardcover 1st edition 1992

ISBN-13: 978-1-4612-7741-5 e-ISBN-13: 978-1-4612-2974-2
DOI: 10.1007/978-1-4612-2974-2

9 8 7 6 5 4 3 2 1

The Collected Papers of Hassler Whitney

Contents — Volume 2

(Bracketed numbers are from the Bibliography)

Chapter 3

Chapter 4

Chapter 5

Permissions

The Collected Papers of Hassler Whitney

Contents — Volume 1
(Bracketed numbers are from the Bibliography)

Chapter 2
Differentiable Functions and Singularities 227

Chapter 3
Analytic Spaces 455

Permissions

Preface

We present here the mathematical papers of Hassler Whitney. This collection contains all the published papers, with the exception of some short announcements that Whitney did not wish to be included. We also include the introduction to his book *Geometric Integration Theory*, and one previously unpublished manuscript on the four-color problem. The papers are presented under some broad categories: graphs and combinatorics, differentiable functions and singularities, analytic spaces, manifolds, bundles and characteristic classes, topology and algebraic topology, geometric integration theory.

Whitney intended to write an introduction to this collection. Unfortunately he left us no manuscript at the time of his death, May 10, 1989. We had discussed the possibility of using his paper "Moscow 1935 – Topology moving toward America," written for the Centennial of the American Mathematical Society, as part of his introduction to this collection, an idea which he much liked. We therefore include this paper, which contains personal information as well as mathematical reflections, as Whitney's own introduction to these volumes.

Whitney's mathematical style, like his personal style, was that of an explorer and pioneer. One of the pictures included in these volumes shows him as a mountain climber. In mathematics, he preferred to work on undeveloped areas: break new ground and build foundations. During the last twenty years of his life he concentrated his efforts on developing an educational system that builds on the natural tendency in children to be explorers.

We find that his papers are as fresh today as when they were written, and feel that any comments from us on their contents could only detract from their clarity. The influence of his work is too vast to review meaningfully in a few pages, event if we felt competent to do so.

We do find it appropriate to comment on his unpublished manuscript (dating from around 1940) of a book that contained the details of all the results announced in "On the theory of sphere bundles," [42]. It contains a lengthy proof of his duality theorem (the formula for the Stiefel–Whitney classes of the direct sum of two vector bundles) obtained by developing an obstruction theory for direct sums, and computations with cochains. Whitney has referred to that proof as the hardest thing he ever did. In an unguarded moment he at once remarked that he felt that the duality theorem was his most original contribution to mathematics.

J. Eells
D. Toledo

Academic Appointments and Awards

Born March 23, 1907, in New York City
Yale University, Ph. B., 1928; Mus. B., 1929; honorary Sc. D., 1947
Harvard University, Ph. D. 1932
National Research Council Fellow in Mathematics, 1931–33
Instructor to Professor of Mathematics, Harvard University, 1933–52
Professor of Mathematics, Institute for Advanced Study, 1952–77;
Professor Emeritus, 1977–89

Member: American Mathematical Society, Mathematical Association of America, National Council of Teachers of Mathematics, National Academy of Sciences, American Philosophical Society, Swiss Mathematical Society (honorary)

American Mathematical Society:

Colloquium Lecturer, 1946
Vice President, 1948–50
Editor, American Journal of Mathematics, 1944–49
Editor, Mathematical Reviews, 1949–54
Committee on Visiting Lectureship (chairman), 1946–51
Committee for Summer Institutes (chairman), 1953–54

National Science Foundation: Chairman of Panel on Mathematics, 1953–56
Researcher for Applied Mathematics Panel of NDRC of OSRD, 1943–45
Exchange Professor to France from Harvard, 1951–52
Exchange Professor to College de France (Fulbright), 1957
Committee on Support of Research in the Mathematical Sciences of the
National Research Council, 1966–67
L. R. Ford Award for paper, "The Mathematics of Physical Quantities," 1969
National Medal of Science, 1967
Wolf Prize in Mathematics (shared) for 1982
Steele Prize, American Mathematical Society, 1985
Consultant to School Mathematics Study Group, Cambridge Conference in School Mathematics, Education Development

Center, and other groups International Commission on Mathematical Instruction, President, 1979–82

Bibliography of Hassler Whitney

[1] The coloring of graphs, NAS Proc. v. 17, 1931, 122–125 .

[2] Non-separable and planar graphs, NAS Proc. v. 17, 1931, 125–127 .

[3] A theorem on graphs, Annals of Math. (2) v. 32, 1931, 378–390 .

[4] Note on Perron's solution of the Dirichlet problem, NAS Proc., v. 18, 1932, 68–70.

[5] Non-separable and planar graphs, AMS Transac. v. 34, 1932, 339–362.

[6] Congruent graphs and the connectivity of graphs, Am. Jour. Math. v. 54, 1932, 150–168.

[7] Regular families of curves I, NAS Proc., v. 18, 1932, 275–278.

[8] Regular families of curves II, NAS Proc., v. 18, 1932, 340–342.

[9] A logical expansion in mathematics, AMS Bull., v. 38, 1932, 572–579.

[10] The coloring of graphs, Annals of Math., (2) v. 33, 1932, 688–718.

[11] A characterization of the closed 2-cell, AMS Transac., v. 35, 1933, 261–273.

[12] A set of topological invariants for graphs, Am. Jour. Math., v. 55, 1933, 231–235.

[13] On the classification of graphs, Am. Jour. Math., v. 55, 1933, 236–244.

[14] 2-Isomorphic graphs, Am. Jour. Math., v. 55, 1933, 245–254.

[15] Regular families of curves, Annals of Math. (2) v. 34, 1933, 244–270.

[16] Characteristic functions and the algebra of logic, Annals of Math. (2) v. 34, 1933, 405–414.

[17] Planar graphs, Fundamenta Math., v. 21, 1933, 73–84.

[18] Analytic extensions of differentiable functions defined in closed sets, AMS Transac., v. 36, 1934, 63–89.

[19] Derivatives, difference quotients and Taylor's formula, AMS Bull., v. 40, 1934, 89–94.

[20] Differentiable functions defined in closed sets I, AMS Transac., v. 36, 1934, 369–387.

[21] Derivatives, difference quotients and Taylor's formula II, Annals of Math. (2) v. 35, 1934, 476–481.

[22] Functions differentiable on the boundaries of regions, Annals of Math. (2) v. 35, 1934, 482–485.

[23] On the abstract properties of linear dependence, Am. Jour. Math., v. 57, 1935, 509–533.

[24] Differentiable manifolds in Euclidean space, NAS Proc., v. 21, 1935, 462–464, reprinted in Receuil Mathematique. T. 1(43) N. 5 (1936) 783–786.

[25] Sphere-spaces, NAS Proc., v. 21, 1935, 464–468, reprinted in Receuil Mathematique. T. 1(43) N. 5 (1936) 787–791.

[26] A function not constant on a connected set of critical points, Duke Math. J., v. 1, 1935, 514–517.

[27] Differentiable functions defined in arbitrary subsets of Euclidean space, AMS Transac., v. 40, 1936, 309–317.

[28] Differentiable manifolds, Annals of Math. (2) v. 37, 1936, 645–680.

[29] The imbedding of manifolds in families of analytic manifolds, Annals of Math. (2) v. 37, 1936, 865–878.

[30] On regular closed curves in the plane, Compositio Math. 4, 1937, 276–284.

[31] On matrices of integers and combinatorial topology, Duke Math. J., 3, 1937, 35–45.

[32] On the maps of an n-sphere into another n-sphere, Duke Math. J., 3, 1937, 46–50.

[33] The maps of an n-complex into an n-sphere, Duke Math. J., 3, 1937, 51–55.

[34] On products in a complex, NAS Proc., v. 23, 1937, 285–291.

[35] Analytic coordinate systems and arcs in a manifold, Annals of Math. (2) 38, 1937, 809–818.

[36] Topological properties of differentiable manifolds, AMS Bull. 43, 1937, 785–805.

[37] A numerical equivalent of the four color problem, Monatshefte fur Math. und Phys. 3, 1937, 207–213.

[38] Cross sections of curves in 3-space, Duke Math. J., 4, 1938, 222–226.

[39] On products in a complex, Annals of Math. (2) 39, 1938, 397–432.

[40] Tensor products of abelian groups, Duke Math. J., 4, 1938, 495–528.

[41] Some combinatorial properties of complexes, NAS Proc., 26, 43–148.

[42] On the theory of sphere-bundles, NAS Proc., 26, 1940, 148–153.

[43] On regular families of curves, AMS Bull., 47, 1941, 145–147.

[44] On the topology of differentiable manifolds, Lectures in Topology, U. of Michigan Press, 1941, 101–141.

[45] Differentiability of the remainder term in Taylor's formula, Duke Math. J., 10, 1943, 153–158.

[46] Differentiable even functions, Duke Math. J., 10, 1943, 159–160.

[47] The general type of singularity of a set of $2n - 1$ smooth functions of n-variables, Duke Math. J., 10, 1943, 161–172.

[48] Topics in the theory of Abelian groups, I. Divisibility of Homomorphisms, AMS Bull., 50, 1944, 129–134.

[49] On the extension of differentiable functions, AMS Bull., 50, 1944, 76–81.

[50] The self-intersections of a smooth n-manifold in $2n$-space, Annals of Math. (2), 45, 1944, 220–246.

[51] The singularities of a smooth n-manifold in ($2n$-1)-space, Annals of Math. (2), 45, 1944, 247–293.

[52] Algebraic topology and integration theory, NAS Proc., v. 33, 1947, 1–6.

[53] Geometric methods in cohomology theory, NAS Proc., v. 33, 1947, 7–9.

[54] Complexes of manifolds, NAS Proc., v. 33, 1947, 10–11.

[55] On ideals of differentiable functions, Am. Jour. Math. 70, 1948, 635–658.

[56] Relations between the second and third homotopy groups of a simply-connected space, Annals of Math. (2) 50, 1949, 180–202.

[57] Classification of the mappings of a 3-complex into simply-connected spaces, Annals of Math. (2) 50, 1949, 270–284.

[58] An extension theorem for mappings into simply-connected spaces, Annals of Math. (2) 50, 1949, 285–296.

[59] La topologie algebrique et la theorie de l'integration, Colloques Internationaux du CNRS XII, Topologie Algebrique, 1947, 107–113, published by CNRS, Paris, 1949.

[60] (With L. H. Loomis) An inequality related to the isoperimetric inequality, AMS Bull., 55, 1949, 961–962.

[61] On totally differentiable and smooth functions, Pacific J. Math. 1, 1951, 143–159.

[62] r-dimensional integration in n-space, Proc. Int. Cong. Math., 1950, vol. 1, 245–256, Amer. Math. Soc. 1952.

[63] On singularities of mappings of Euclidean spaces, I. Mappings of the plabe into the plane, Annals of Math. (2) 62, 1955, 374–410.

[64] On functions with bounded n-th differences, J. de Maths. Pres et Appl. 36, 1957, 67–95.

[65] Geometric Integration Theory, Princeton University Press, 1957, Princeton, NJ, xv + 397 pages. (Princeton Math. Series 21) Translated into Russian [book].

[66] Elementary structure of real algebraic varieties, Annals of Math. (2) 66, 1957, 545–556.

[67] Singualrities of mappings of Euclidean spaces, Symposium Internacional de Topologia Algebraica, Mexico, 1956, 285–301, Mexico, La Universidad Nacional Autonoma.

[68] (With F. Bruhat) Quelques proprietes fondamentales des ensembles analytiques-reels, Comm. Math. Helv. 33, 1959, 132–160.

[69] (With A. Dold) Classification of oriented sphere bundles over a 4-complex, Annals of Math. (2) 69, 1959, 667–677.

[70] On bounded functions with bounded n-th differences, AMS Proc. 10, 1959, 480–481.

[71] (With A. M. Gleason) The extension of linear functionals defined on H-infinity, Pacific J. Math. 12, 1962, 163–182.

[72] The work of John W. Milnor, Proceedings ICM 1962m Institut Mittag–Leffler Djursholm, Sweden, x1viii-1.

[73] Local properties of analytic varieties, in: differential and combinatorial topology (Symposium in Honor of Marston Morse), Princeton, NJ., Princeton University Press.

[74] Tangents to an analytic variety, Annals of Math (2) 81, 1965, 496–549.

[75] The mathematics of physical quantities. Part I, Mathematical Models for measurement, Am. Math. Monthly 75 (1968), 115–138, Part II, Quantity structures and dimensional analysis, ibid., 237–256.

[76] Logic Fad or tool? Nico 4, 1969, Revue per du Centre Belge de Pedagogie de la Mathematique, 2–14.

[77] On reducibility in the four color problem, unpublished manuscript, 1971.

[78] (With W. T. Tutte) Kempe chains and the four color problem, Utilitas Mathematica 2(1972), 241–281.

[79] Complex Analytic Varieties, Addison–Wesley Pub. Co., Reading, MA 1972, xii +399 pp. [book].

[80] Math Activities, multilithed, Institute for Adv. Study, 1974 [book].

[81] Comment on the division of the plane by lines, Am. Math. Monthly 86(1979), p. 700.

[82] Moscow 1935: Topology moving toward America, in *A Century of Mathematics in America*, Part I, P. L. Duren Ed., 96–117, Amer. Math.

Chapter 1

Manifolds

Annals of Mathematics
Vol. 37, No. 3, July, 1936

DIFFERENTIABLE MANIFOLDS[1]

By Hassler Whitney

(Received February 10, 1936)

Introduction

The main purpose of this paper is to provide tools of a purely analytic character for a general study of the topology of differentiable manifolds, and maps of them into other manifolds. A differentiable manifold is generally defined in one of two ways; as a point set with neighborhoods homeomorphic with Euclidean space E_n, coördinates in overlapping neighborhoods being related by a differentiable transformation,[2] or as a subset of E_n, defined near each point by expressing some of the coördinates in terms of the others by differentiable functions.[3]

The first fundamental theorem is that the first definition is no more general than the second; any differentiable manifold may be imbedded in Euclidean space. In fact, it may be made into an analytic manifold in some E_n. As a corollary, it may be given an analytic Riemannian metric. The second fundamental theorem (when combined with the first) deals with the smoothing out of a manifold. Let f be a map of any character (continuous or differentiable, without an inverse) of a differentiable manifold M of dimension m into another, N, of dimension n. (Either manifold might be an open subset of Euclidean space.) Then if $n \geqq 2m$, we may alter f as little as we please, forming a regular map F. (A map is *regular* if, near each point, it is differentiable and has a differentiable inverse.) Moreover, if $n \geqq 2m + 1$, F may be made (1-1). We show in Theorem 6 that if $n \geqq 2m + 2$, then any two regular maps f_0, f_1 of M into E_n are equivalent, in the following sense. $f_0(M)$ may be deformed into $f_1(M)$ by maps $f_t(0 \leqq t \leqq 1)$ so that the path crossed by the manifold is the regular map of an $(m + 1)$-dimensional manifold. Moreover, if $n \geqq 2m + 3$, and $f_0(M)$ and $f_1(M)$ are non-singular, so is the $(m + 1)$-manifold.

A fundamental unsolved problem is the following: *Can any analytic manifold be mapped in an analytic manner into Euclidean space?*[4]

---

[1] Presented to the Am. Math Soc. Sept. 1935. An outline of the paper will be found in Proc. Nat. Ac. of Sci., vol. 21 (1935), pp. 462–463.

[2] Differentiable manifolds have been studied for instance by O. Veblen and J. H. C. Whitehead, *The foundations of differential geometry*, Cambridge Tracts, 1932. An example of a differentiable (in fact, analytic) manifold is the manifold of k-planes through a point in n-space. See §24.

[3] Manifolds in E_n which are defined by the vanishing of a set of differentiable functions are of a special character; see H. Whitney, *The imbedding of manifolds* $\cdots$, in the October 1936 issue of these Annals.

[4] This seems quite probable. It is proved for some special analytic manifolds in §§23–24.

Theorem 1 shows only that there is a differentiable map (with all derivatives), such that the resulting point set forms an analytic manifold.

Many portions of the proofs are based on the Weierstrass approximation theorem, if the manifolds are closed; if they are open, this theorem must be replaced by a corresponding theorem on functions defined in open sets. This and other theorems which will be useful may be found in a previous paper.[5] In proving both fundamental theorems, the following method is used continually. Let f be a differentiable map of M into E_n, and let U be a small portion of M. We consider a class S of maps f' of U into E_n which approximate to f in U: S forms a part of a Euclidean space. The maps f' we do not wish are characterized by subsets of S whose dimensions may be learned,—we use here the notion of "k-extent" of a set similar to a definition of Carathéodory.[6] We find a desirable map f' in U, and do the same in other neighborhoods until we have found F in the whole manifold M.

The arrangement of the paper is indicated by the sentences introductory to each part.

Contents

I. Definitions and Preliminary Results

In this part we collect definitions and facts which will be used constantly in what follows. In one section, §4, we assume a knowledge of the imbedding theorem and of Lemma 23.

1. Manifolds of class C^r. By an *m-dimensional manifold M of class C^r*, or, a *C^r-m-manifold* (r finite or $r = \infty$),[7] we shall mean a system composed of a set of points, which we shall also call M, and certain maps, as follows: Let $Q = Q_m$ be the interior of the unit $(m-1)$-sphere in Euclidean space E_m.[8] Let $\theta_1, \theta_2, \cdots$ be a finite or denumerable number of (1-1) maps of Q into M. Define the sets of points

$$(1.1) \qquad U_i = \theta_i(Q), \qquad U_{ij} = U_{ji} = U_i \cdot U_j, \qquad Q_{ij} = \theta_i^{-1}(U_{ij}),$$

[5] H. Whitney, *Analytic extensions of differentiable functions defined in closed sets*, Trans. Am. Math. Soc., vol. 36 (1934), pp. 63–89. We refer to this paper as AE.

[6] C. Carathéodory, *Über das lineare Mass von Punktmengen*, Gött. Nachr., 1914, p. 426.

[7] We always suppose r is finite and > 0 unless otherwise stated. However, the results of §§1–4 all hold for $r = \infty$.

[8] That is, the space of all ordered sets of n real numbers.

and the (1-1) maps

$$h_{ij}(x) = \theta_j^{-1}(\theta_i(x)) \text{ in } Q_{ij}. \tag{1.2}$$

We make four assumptions:

(α) The maps θ_i cover M: For each p in M there is an i and an x in Q such that $\theta_i(x) = p$.

(β) The Q_{ij} are open.

(γ)[9] There are no i, j, and sequence of points $\{x^k\}$ such that

$$x^k \text{ is in } Q_{ij}, \qquad x^k \to x \text{ in } Q - Q_{ij}, \qquad h_{ij}(x^k) \to x' \text{ in } Q - Q_{ji}.$$

(δ) $h_{ij}(x)$ (if defined) is of class C^r (see §2), and if $r > 0$, it has a non-vanishing Jacobian.

If, further, the h_{ij} are analytic, we say M is *analytic*. A 0-manifold consists of a finite or denumerable number of isolated points. A k-manifold ($k < 0$) contains no points.

We call the θ_i the *maps defining* M, and the U_i, *neighborhoods* in M. Note that M *need not be connected.*

We define *limit points* in M as follows: $p_k \to p$ if and only if there are an i, an s, a point x of Q and a sequence $\{x^k\}$ of points of Q such that

$$\theta_i(x^k) = p_{s+k}, \qquad \theta_i(x) = p, \qquad x^k \to x. \tag{1.3}$$

We prove two facts. *If $p_k \to p$ and p is in U_j, then for some s, p_{s+k} is in U_j and $\theta_j^{-1}(p_{s+k}) \to \theta_j^{-1}(p)$.* For, say (1.3) holds. Then p is in U_{ij}, and hence $x = \theta_i^{-1}(p)$ is in Q_{ij}. As Q_{ij} is open and $x^k \to x$, there is an s such that x^{s+k} is in Q_{ij}. Hence $h_{ij}(x^{s+k}) = \theta_j^{-1}(p_{s+k})$ is in Q_{ji}, and as h_{ij} is continuous, $\theta_j^{-1}(p_{s+k}) \to \theta_j^{-1}(p)$. *If $p_k \to p$ and $p_k \to p'$, then $p = p'$.* For suppose $p \neq p'$, (1.3) holds, and similar relations hold with i, p, x^k, x replaced by j, p', x'^k, x'. (We may evidently take $s' = s$.) Then $h_{ij}(x^k) = x'^k$. x is not in Q_{ij}; for if it were, then $h_{ij}(x) = x'$ as h_{ij} is continuous, and

$$p' = \theta_j(x') = \theta_j(h_{ij}(x)) = \theta_i(x) = p.$$

Similarly x' is not in Q_{ji}. But this contradicts (γ).

These two facts show that the obvious criteria for p_k not $\to p$ hold: If p is in U_i and there are no $\{x^k\}$, x such that (1.3) hold, then p_k does not $\to p$; if $\theta_i(x^k) = p_k$, $x^k \to x$ in Q, and $\theta_i(x) \neq p$, then p_k does not $\to p$. We can now define *open*, *closed*, *compact* sets, etc. as usual. A manifold M is closed or open according as the set of points M is compact or not.

Note that *any manifold of class C^r is of class C^s for $s < r$; any open subset of a manifold is a manifold* (with suitably chosen maps); *a finite or denumerable number of manifolds together form a manifold. E_n is a manifold with the obvious maps.*

Given two sets of maps in M, we say they define *C^r-equivalent manifolds* if not only each set separately, but also the two sets together, satisfy the above condi-

[9] (β) and (γ) correspond to (C^1) and (C^2) in Veblen and Whitehead, loc. cit.

tions. We may speak of a point being the identical point on both manifolds, of a function being identical on them, etc. Two manifolds are *C^r-homeomorphic* if there is a (1-1) correspondence between their points such that, on identifying corresponding points, the two sets of maps define C^r-equivalent manifolds. In other words there is a (proper) (1-1) regular C^r-map of either one into the other (see §2). It is often convenient not to distinguish between two C^r-equivalent manifolds. We may then say for instance "choose maps defining M such that $\cdots$." The number r should then be definitely associated with the manifold; for a C^r-manifold may be C^r-equivalent to a C^s-manifold with $s > r$ (see Theorem 1).

2. **Functions defined in manifolds.** We shall use the words "function" and "map" interchangeably. Let R be an open set in E_m, and let f be a real-valued function defined in R. It is *of class C^r* if it has continuous partial derivatives through the rth order. If R is any subset of E_m,[10] we say f is *of class C^r in the subset R of E_m* if its definition can be extended through an open set containing R so that it is of class C^r there (see AE). It is sufficient that the extension be possible separately about each point of R. If the values of f are points of E_n, we say it is of class C^r (or of class C^r in a subset) if each of its coördinates is.

Suppose f is a function defined in a subset R of a C^r-manifold M, with values in a C^s-manifold N. Let θ_i, χ_j; U_i, V_j; Q_m, Q_n be the maps etc. defining M and N respectively.[11] Take any p_0 in R, and say p_0 is in U_i, $q_0 = f(p_0)$ is in V_j. The function $f_i(x) = f(\theta_i(x))$ is defined in the set $R_i = \theta_i^{-1}(R \cdot U_i)$. Suppose that, for some neighborhood U of $x^0 = \theta_i^{-1}(p_0)$ in Q_m, x in $R_i \cdot U$ implies $f_i(x)$ in V_j; set $f_{ij}(x) = \chi_j^{-1}(f_i(x))$, and suppose that f_{ij}, defined in $R_i \cdot U$ and with values in Q_n, is of class C^t ($t \leqq r, s$). If this is true of each p_0 and each corresponding i, j, we say f is *of class C^t in R*, or in the subset R of M, if R is not open. (If the condition is satisfied at p_0 for one pair (i, j), it is satisfied at p_0 for each such pair (i, j), on account of (δ).) If M and N are analytic, f may be analytic.

Suppose M is of class C^r, $r \geqq 1$. Let x be a point of Q, and let $C_1, \cdots, C_\mu$ be differentiable curves ending at x, whose tangents at x form a set of independent vectors. If x is in Q_{ij}, then parts of these curves, and the vectors, transform under h_{ij} into other such curves and vectors in Q; by (δ), the new vectors are independent. The corresponding curves in M we shall say define a set of *independent directions* at $\theta_i(x)$. If f is a C^1-map of M into N, these curves, and hence "directions," go into curves and directions in N. We define: f is a *regular*[12] map of M into N if it is of class C^1, and any set of independent directions at a point in M goes into such a set in N. If f is defined in a subset of M, we say it is regular if its definition can be extended through an open subset of M so that it is regular there.

[10] This case does not occur in the fundamental theorems. In this connection, see also H. Whitney, *Differentiable functions* $\cdots$, Trans. Am. Math. Soc., vol. 40 (1936).

[11] We shall always use these symbols in this manner.

[12] In Veblen and Whitehead, loc. cit., it is also required that the map be (1-1).

The map f of M into N is *completely regular* if it is regular, and has the following property: at most two points of M go into any single point of N; if $f(p_1) = f(p_2)$, $p_1 \neq p_2$, then a set of m independent directions at p_1 together with such a set at p_2 go into a set of $2m$ independent directions at $f(p_1)$ in N. This is of course only possible (if f is not (1-1)) if $n \geqq 2m$.

Given the map f of M into N, we define the *limit set* Lf as follows: A point q of N is in Lf if there exist sequences $\{p_k\}$ in M and $\{q_k\}$ in N such that $q_k \to q$, $f(p_k) = q_k$, and the sequence $\{p_k\}$ has no limiting point in M. The map f is *proper* if $f(M)$ does not intersect its limit set: $f(M) \cdot Lf = 0$. If M and N are C^r-m- and C^r-n-manifolds, and f is a (1-1) regular proper C^r-map of M into N, we shall say f *C^r-imbeds* M in N. $f(M)$ is then a C^r-m-manifold in N (see §3).

3. **Manifolds in manifolds.** If f is a regular C^r-map of M into N, we shall call the combination (M, f) a *local C^r-manifold* in N. Each point p of M is then in a neighborhood U in which f is (1-1). In general, the nature of M is determined by the nature of the point set $f(M)$; but this is not necessarily the case. We shall commonly speak of $f(M)$ as a local manifold in N, keeping in mind that M and f must both be given. (But see below.) The *limit set* $Lf(M)$ of $f(M)$ is the limit set Lf. The local manifold is *proper* if f is; has at most *regular singularities* if f is completely regular; is *non-singular* if f is (1-1). (M, f), or the resulting point set $f(M)$, is a *C^r-manifold in N* if it is non-singular and proper. If $f(M)$ is a local manifold in N, and f is of class C^r, we shall say M is *C^r-homeomorphic* with $f(M)$.

We shall show now that if (M, f) is a C^r-manifold in N, then, using the point set $f(M)$ alone, we may determine a C^r-manifold M', which is necessarily C^r-homeomorphic with M. This justifies calling the point set $f(M)$ a manifold in N. Also, setting $N = E_n$, we justify our original definition of a manifold. (See also Theorem 1.) Moreover, as M and $f(M)$ are C^r-homeomorphic, there is in general no harm in identifying them. This justifies the phrase "the manifold M in N."

LEMMA 1. *Let (M, f) be a C^r-m-manifold in the C^r-n-manifold N. Then the subset $f(M)$ of N has the following property. Any q_0 in $f(M)$ is in a neighborhood U in $f(M)$*[13] *such that U is in some V_j, and the points $\chi_j^{-1}(U)$ are given by*

$$y_{m+k} = \psi_k(y_1, \cdots, y_m) \qquad (k = 1, \cdots, n-m) \tag{3.1}$$

(if $y_1, \cdots, y_m$ are suitably chosen rectangular coördinates in E_n), where $(y_1, \cdots, y_m)$ runs through an open set in the $(y_1, \cdots, y_m)$-plane. Moreover, if M' is any subset of N with the above property, then maps in M' determined by (3.1) *make M' a C^r-m-manifold; if, further, $M' = f(M)$, then M is C^r-homeomorphic with M'.*[14]

Given M, f and q_0, say $q_0 = f(p_0)$; we may take a neighborhood U^* of p_0 in M such that U^* is in some U_i and $U = f(U^*)$ is in some V_j (as f is continuous).

[13] That is, q_0 is in U which is in $f(M)$, and U is open in $f(M)$, i.e., no q in U is the limit (in N) of a sequence of points of $f(M) - U$.

[14] It is easily seen that the lemma holds if f is completely regular and proper.

Let $(x_1, \cdots, x_m)$ be rectangular coördinates in E_m. $f_{ij} = \chi_j^{-1} f \theta_i$ maps the open subset $\theta_i^{-1}(U^*)$ of Q_m into Q_n. As f is regular, the matrix $\| \partial y_k / \partial x_l \|$ of partial derivatives of f_{ij} is of rank m at $\theta_i^{-1}(p_0)$; we may suppose that the first determinant is $\neq 0$. Then, taking U^* small enough, (3.1) holds. To show that U is a neighborhood of q_0 in $f(M)$, take any q in U, and suppose there is a sequence $\{q_k\}$ in $f(M) - U$, $q_k \to q$. Say $q = f(p)$, $q_k = f(p_k)$. Suppose there were a subsequence $\{p_{\nu_k}\}$ of $\{p_k\}$ such that $p_{\nu_k} \to p'$ in M. Then $f(p_{\nu_k}) = q_{\nu_k} \to f(p')$, hence $f(p') = q$, and $p' = p$, as f is (1-1). As $p_{\nu_k} \to p$ in U^*, there is an s such that $p_{\nu_{s+k}}$ is in U^* (see §1). But then $q_{\nu_{s+k}}$ is in U, a contradiction. Therefore $\{p_k\}$ has no limit in M. But then f is not proper, again a contradiction.

Next suppose that M' is a subset of N defined by equations (3.1). In each such equation, the domain of definition $R = (y_1, \cdots, y_m)$ is open; we may cover R by spheres, and map Q_m into each sphere and hence into M', defining maps in M'. Using the fact that the h_{ij} in N are of class C^r with non-vanishing Jacobian, it is easily shown that the same is true for the maps in M'. Suppose further that $M' = f(M)$. The equations $y = f_{ij}(x)$ previously considered, when solved as before, give a further set of equations (3.1) and thus another set of maps defining M'; but this set is clearly C^r-equivalent to the other, and thus defines a C^r-homeomorphism between M and M'.

LEMMA 2. *Let M be a C^r-m-manifold in N, and N, a C^r-n-manifold in the C^r-n'-manifold N'. Then M is a C^r-m-manifold in N'. We may replace "manifold" by "local manifold."*

Say f maps M into N and g maps N into N'; then $f' = gf$ maps M into N'. As f and g are regular and of class C^r, so is f'. If f and g are (1-1), so is f'; the same is easily seen to be true with "(1-1)" replaced by "proper."

4. Functions defined in submanifolds of a manifold.

Let M be a submanifold of N, and let f be defined in one of the manifolds; we propose to study the relation of f to the other manifold. The values of f may be points of another manifold. The results of this section will be used only occasionally.

LEMMA 3. *Let M be a local C^r-m-manifold in the C^r-n-manifold N, let R be an open subset of N, and let f be of class C^r in R. Then f is of class C^r in the subset $R \cdot M$ of M.*

By this we mean, if g is the map of M into N, that $f' = fg$ is of class C^r in that open subset R' of M for which $g(R')$ is in R. Take q_0 in $R \cdot M$, and say $q_0 = g(p_0)$, p_0 in U_i, q_0 in V_j. The hypothesis is that $f^*(y) = f\chi_j(y)$ is of class C^r in a neighborhood of $y^0 = \chi_j^{-1}(q_0)$ in Q_n. As θ_i, g, and χ_j^{-1} are of class C^r, so is

$$f'(\theta_i(x)) = f^* \chi_j^{-1} g \theta_i(x)$$

in a neighborhood of $x^0 = \theta_i^{-1}(p_0)$, as required.

A converse of this lemma is

LEMMA 4. *Let M be a C^r-m-manifold in the C^r-n-manifold N, let R' be an open subset of M, and let f be of class C^r in R'. Then its definition may be* *extended throughout an open subset R of N containing R′ so that it is of class C^r there.*

As R' is a C^r-m-manifold in N, we may suppose without loss of generality that $R' = M$. By Theorem 1, we may C^r-imbed N in E_ν ($\nu = 2n + 1$); then, by Lemma 2, M becomes a C^r-m-manifold in E_ν. Define f in $R(M)$ by setting $f(p) = f(H(p))$ (Lemma 23). f is now of class C^r in $R(M)$. For, let U be a neighborhood of a point p_0 in M, and let S be the product of $\theta_i^{-1}(U)$ and $E_{\nu-m}$: $z = (x, y)$, x in $\theta_i^{-1}(U)$, y in $E_{\nu-m}$, z in S. S may be considered as a subset of E_ν. Setting $f'(z) = f(\theta_i(x))$, f' is obviously of class C^r in S. Let ϕ be a congruent map of $E_{\nu-m}$ into $P(p_0)$, and for any p in U, let $T_p = T_{p,P(p)}$ be the map of $P(p_0)$ into $P(p)$ of §19. Set

$$q = \psi(x, y) = T_{\theta_i(x)}\phi(y).$$

This is a C^r-map of S into E_ν. If $S' = (x, y)$ for $\| y \| <$ some α, and $\phi(O) = p_0$, ψ maps S' into part of $R(M)$. Moreover,

$$x = \theta_i^{-1}(H(q)), \qquad y = \phi^{-1}T_{H(q)}^{-1}\, q \text{ in } \psi(S');$$

hence ψ has an inverse of class C^r in $\psi(S')$. But in $\psi(S')$, by the definitions of f and f', $f(q) = f'(\psi^{-1}(q))$, which shows that f is of class C^r in $R(M)$. Set $R = R(M) \cdot N$; by Lemma 3, f is of class C^r in R.

We remark that the lemmas hold if we replace everywhere "of class C^r" by "analytic." The lemmas show that if M and N are as given, and f is defined in M, then *"f is of class C^r in M" is the same as "f is of class C^r in the subset M of N."*

5. **Admissible sets of maps in a manifold.** Let M be a C^r-m-manifold with maps θ_i. If (a) each θ_i is of class C^r in $\bar{Q}$,[14a] and (b) any compact subset of M has points in common with but a finite number of the U_i, we say the maps form an *admissible set.* If, further, Q' is the sphere concentric with Q and of half the radius, $U_i' = \theta_i(Q')$, and the U_i' cover M, we say the maps form a *completely admissible set.* Any manifold may be defined by a completely admissible set of arbitrarily small maps:

LEMMA 5. *Let M be a C^r-m-manifold, and let $R_1, R_2, \cdots$ be a set of open sets covering M. Then M is C^r-equivalent to a manifold with completely admissible maps θ_i such that each $\bar{U}_i$ is in some R_j.*

Let θ_i^* be the given maps in M. Let $Q_1, Q_2, \cdots$ be the spheres of rational center and rational radius such that each $\bar{Q}_k$ is in Q. Q may be mapped into Q_k by a linear transformation ϕ_k. Set $\theta_{ki}(x) = \theta_i^*(\phi_k(x))$; then the θ_{ki} are defined in $\bar{Q}$, and obviously define a manifold C^r-equivalent to M. Call these maps χ_i, and the corresponding neighborhoods, V_i.

Set $W_i' = V_1 + \cdots + V_i$; then $\bar{W}_i'$ is compact and closed. Each point of $\bar{W}_i'$ is in some W_j', and hence $\bar{W}_i'$ is in the sum of a finite number of the W_j' and is thus in some W_k'. Hence we may pick out a finite or infinite subsequence $W_1, W_2, \cdots$ of the W_i' such that $\bar{W}_i$ is in W_{i+1} and the W_i cover M. Each

[14a] $\bar{Q} = Q$ plus limit points.

point of $\overline{W_i - W_{i-1}}$ is in a $V'_j = \chi_j(Q')$ such that $\bar{V}_j$ lies in some R_k and has no points in W_{i-2}; a finite number of these V'_j may be chosen such that they cover $\overline{W_i - W_{i-1}}$. Choose such neighborhoods for each i; the corresponding maps, arranged in a sequence $\theta_1, \theta_2, \cdots$, obviously have the required properties.

6. **Approximations to functions defined in manifolds.** Let M and N be C^r-m- and C^r-n-manifolds with admissible maps θ_i, χ_j, and let $f(p)$ be a C^r-map of M into N. Take any p_0 in M, and say p_0 is in U_i, $f(p_0)$ is in V_j. Then $f_{ij}(x) = \chi_j^{-1} f \theta_i(x)$ is a C^r-map of part of Q_m into Q_n, with derivatives

$$D_k f_{ij}(x) = \frac{\partial^{k_1 + \cdots + k_m}}{\partial x_1^{k_1} \cdots \partial x_n^{k_n}} f_{ij}(x) \qquad (\sigma_k \leqq r),$$

where $\sigma_k = k_1 + \cdots + k_m$. Each $D_k f_{ij}(x)$ is a vector function defined in part of Q_m. Let $\eta(p)$ be a positive continuous function in M, and let $F(p)$ be another C^r-map of M into N. We shall say *F approximates to f in M through the sth order with an error* $< \eta$, or, *F approximates* (f, M, s, η), if the following is true: For any point p in any U_i there is a j such that $f(p)$ and $F(p)$ are in V_j and

$$\| D_k F_{ij}(p) - D_k f_{ij}(p) \| < \eta(p) \qquad (\sigma_k \leqq s).$$

If $M = E_m$ and $N = E_n$, this reduces to the ordinary definition. For $s = 0$, this inequality is independent of the maps defining M.

Lemma 6. *Given two sets of admissible maps θ_i, θ_i^* in M and two sets χ_j, χ_j^* in N, and given $f(p)$ and $\eta(p)$ as above, there is a positive continuous function $\zeta(p)$ in M such that if F approximates (f, M, s, ζ) in terms of the θ_i^* and χ_j^*, then F approximates (f, M, s, η) in terms of the θ_i and χ_j.*

Let f_{ij}, f_{ij}^*, F_{ij}, F_{ij}^* be the corresponding maps of Q_m into Q_n, and set $u_{ik}(x) = \theta_k^{*-1}(\theta_i(x))$, $v_{jl}(y) = \chi_l^{*-1}(\chi_j(y))$ where defined. Now given i, j, there are numbers k, l such that near a given point

$$f_{kl}^* = \chi_l^{*-1} f \theta_k^* = v_{jl} \chi_j^{-1} f \theta_i u_{ik}^{-1} = v_{jl} f_{ij} u_{ik}^{-1};$$

hence the derivatives of f_{kl}^* of order $\leqq s$ are polynomials in those of v_{jl}, f_{ij}, and u_{ik}^{-1}, of order $\leqq s$. It is sufficient to show that all such derivatives are bounded in the neighborhood of any point p_0 of M. p_0 is in a finite number of U_i and U_k^*, and $f(p_0)$ is in a finite number of V_j and V_l^*. Each θ_i etc. is defined in a bounded closed set, and hence its derivatives are bounded; therefore the same is true of the derivatives of the above functions.

We remark that *if f is regular, completely regular, proper, or* (1-1) *regular and proper, then the same is true of any function which approximates to f through the first order closely enough.*

II. The Theorems

We collect here the principal results of the paper (apart from Lemma 23). Using the first two theorems and the results of §9, the remaining theorems may be proved with little difficulty; hence we give these proofs practically in full in this part.

7. **Two types of properties of maps.** The purpose of this section is to explain (e) and (f) of Theorem 2. See also §9. Let M and N be C^r-m- and C^r-n-manifolds, and let f be a fixed C^r-map of M into N ($r \geqq 0$). Let $\eta(p)$ be a positive continuous function in M. We shall say a property Ω of maps of M into N is an *(f, r, η)-property* if the following is true:

(a) Ω is defined for all maps f' which approximate (f, M, r, η).

(b) There is a compact open subset W' of M such that whether any f' has the property Ω depends only on the values of f' in $\overline{W}'$.

(c) If f' approximates (f, M, r, η) and has the property Ω, then for some continuous $\eta'(p)$, $0 < \eta'(p) < \eta(p)$, any map f'' which approximates (f', M, r, η') has the property Ω.

(d) If f' approximates (f, M, r, η), then for an arbitrary continuous $\eta'(p)$, $0 < \eta'(p) < \eta(p)$, there is a map f'' which approximates (f', M, r, η') and has the property Ω.[15]

Before defining the second type of property, we shall consider certain functionals. We suppose now that M and N are analytic manifolds in E_μ and E_ν respectively.

Let $\Delta(p, q)$ be a continuous function of the pair of points p, q of M such that $\Delta(p, p) = 0$ and $\Delta(p, q) > 0$ for $p \neq q$. Let $\zeta(p)$ be a positive continuous function in M. To each C^r-map f of M into E_ν such that

$$\| D_k f(q) - D_k f(p) \| \leqq \Delta(p, q) \qquad (\sigma_k \leqq r), \tag{7.1}$$

(see §6),[16] let there correspond a map $\mathfrak{L}f$ of M into E_ν. We shall say $\mathfrak{L}$ is an *analytic linear $(M, E_\nu, r, \Delta, \zeta)$-functional* if $\mathfrak{L}f$ is analytic, and approximates (f, M, r, ζ), and, if f_1, f_2 and $f_1 + f_2$ satisfy (7.1), then[17]

$$\mathfrak{L}(f_1 + f_2) = \mathfrak{L}f_1 + \mathfrak{L}f_2.$$

The existence of such a functional is given by Lemma 27. Note that if $\mathfrak{L}$ is an analytic linear $(M, E_\nu, r, \Delta', \zeta')$-functional, then it is an analytic linear $(M, E_\nu, r, \Delta, \zeta)$-functional for $\Delta \leqq \Delta'$, $\zeta \geqq \zeta'$.

We shall say a property Ω is an *$[f, r, \eta, \Delta, \zeta]$-property* if (a), (b) and (c) hold, and also:

(d′) There is a compact open subset W of M containing $\overline{W}'$, and there are C^r-maps $G_1, \cdots, G_h$ of M into E_ν such that $G_i(p) = O$ for p in $M - W$, with the following property. If $\mathfrak{L}$ is any analytic linear $(M, E_\nu, r, \Delta, \zeta)$-functional and if f' approximates (f, M, r, η) and satisfies (7.1), then there is an arbitrarily

[15] (c) and (d) may be phrased as follows: If $\mathfrak{S}$ is the space of maps of M into N which approximate (f, M, r, η), using a very strong topology, then those maps which have the property Ω form an everywhere dense open subset of $\mathfrak{S}$.

[16] $D_k f(p)$ shall mean some $D_k f_i(x)$, where $p = \theta_i(x)$. If p and q are both in some U_i, we shall use the same i in defining $D_k f(p)$ and $D_k f(q)$.

[17] We add points, etc., by considering them as vectors from the origin O. If $p = \Sigma\alpha_i p_i$ with $\Sigma\alpha_i = 1$, then p is independent of the choice of O.

small $\alpha = (\alpha_1, \cdots, \alpha_h)$ such that

$$f_\alpha(p) = H\mathfrak{L}[f'(p) + \sum \alpha_i G_i(p)]$$

has the property Ω. (We suppose η and ζ are so small that f_α is in $R(N)$ for $|\alpha_i| <$ some $\bar{\alpha} > 0$, and thus H is defined; see Lemma 23.)

Note that any $[f, r, \eta, \Delta, \zeta]$-property is an $[f, r, \eta, \Delta', \zeta']$-property for $\Delta' \geqq \Delta$, $\zeta' \leqq \zeta$; also, for large enough Δ, it is an (f, r, η)-property.

8. **The fundamental theorems.** We state here the two theorems on which most of the other results of the paper are based.

THEOREM 1. *Any C^r-m-manifold ($r \geqq 1$ finite or infinite) is C^r-homeomorphic with an analytic manifold in Euclidean space E_{2m+1}.*

See also Theorem 3 and footnote [32].

THEOREM 2. *Let M and N be analytic m- and n-manifolds in Euclidean spaces E_μ and E_ν respectively. Let f be a C^r-map of M into N ($r \geqq 0$ finite). Let η be a positive continuous function in M. Then there is a C^r-map F of M into N with the following properties:*

(a) *F approximates (f, M, r, η).*

(b) *If $n \geqq 2m$, then F is completely regular.*

(c) *If $n \geqq 2m + 1$, then F is* (1-1).

(d) *F is analytic.*

(e) *Let $\Omega_1, \Omega_2, \cdots$ be (f, r, η)-properties, let $W'_1, W'_2, \cdots$ be the corresponding subsets of M* (§7, (b)), *and suppose any compact subset of M has points in common with at most a finite number of the W'_i. Then F has the properties $\Omega_1, \Omega_2, \cdots$.*

(f) *For some functions $\Delta(p, q)$, $\zeta(p)$ as in* §7, *let $\Omega'_1, \Omega'_2, \cdots$* be $[f, r, \eta, \Delta, \zeta]$-*properties. Then F has these properties.*

In place of (d), (e) *and* (f) *we may have if we choose* (e'): (e) *holds without the finiteness restriction.*

If we are satisfied with a function F of class C^r, we would naturally use (e') in place of (d), (e) and (f). Note that if f is proper in M, then we can insure that F be proper in M by taking η small enough; then, if $n \geqq 2m + 1$, (b) and (c) show that F is a homeomorphism, and thus $F(M)$ is a C^r- (or analytic) manifold in N. We might generalize the theorem by making $F = f$ at certain isolated points of M, or, if r is replaced by ∞, letting $F(p)$ approximate to $f(p)$ together with higher and higher partial derivatives as p approaches the limit set LM. (Compare AE, Theorem III.) We could also consider manifolds of different classes in different subsets; an example is given in Theorem 5.

9. **Consequences of** (e) **and** (f) **of Theorem 2.** We may give the function F various properties, either because these are of one of the two types, or because they are the logical sums of a denumerable number of such properties. We give some examples below; for the proofs that they are of the required nature, see §§34–35.

(A) and (B) are (b) and (c) of the theorem.

(C) If K is a subset of N which is the sum of at most a denumerable number of sets of zero $(n - m)$-extent (see §17), then $F(M)$ does not intersect K.

(D) If $f(M)$ and N' are local C^1-m- and C^1-n'-manifolds in N, then if $m + n' < n$, the local manifold $F(M)$ does not intersect N', while if $m + n' \geqq n$, $N^* = F(M) \cdot N'$ is a local $(m + n' - n)$-manifold in N. At each point p of N^*, there are n independent directions in N, each being parallel to $F(M)$ or to N'. If $f(M)$ and N' are non-singular, or proper, so is N^*.

10. **A further imbedding theorem; Riemannian manifolds.** We may replace E_{2m+1} by E_{2m} in Theorem 1 as follows:

THEOREM 3. *Any C^r-m-manifold M ($r \geqq 1$ finite or infinite) is C^r-homeomorphic with a proper analytic local manifold with at most regular singularities in E_{2m}.*

To prove this, let M' be a C^r-homeomorphic analytic manifold in E_{2m+1}. Let $\rho(p)(p$ in $M')$ be the smaller of (a) 1, (b) the reciprocal of the distance from p to a fixed point of E_{2m+1}, (c) the distance from p to the limit set LM' if $LM' \neq 0$. Let q_0 be a fixed point distinct from the origin O in E_{2m}, and set $f(p) = \rho(p)q_0$ in M'. This is a continuous map of M' into E_{2m} such that $f(p) \neq O$, and either $Lf(M')$ is void or $Lf(M') = O$; hence f is proper. Let F be the analytic completely regular proper map given by Theorem 2 with $N = E_{2m}$; $F(M')$ is the required local manifold.

THEOREM 4. *Any C^r-manifold M ($r \geqq 1$ finite or infinite) may be given an analytic Riemannian metric, the coefficients of the fundamental quadratic form being of class C^r in terms of the original maps defining M.*

This follows from Theorem 1 or 3 on using the ds^2 of E_{2m+1} or E_{2m}.

11. **An extension theorem.** Suppose a closed subset of a manifold M is mapped into a manifold N; can the map be extended over the rest of M so as to be differentiable? Or, M might be replaced by a manifold M with boundary B, and the closed subset, by B. An answer is given by the following theorem.

THEOREM 5. *Let A be a separable metric space, let B be a closed subset of A, and let $M = A - B$ be a C^r-m-manifold ($r \geqq 1$ finite or infinite).*[18] *Let N be a C^s-n-manifold ($s \geqq r$), and let f be a continuous map of B into N. Suppose f can be extended so as to be continuous throughout M.*[19] *Then there is a map F of A into N with the following properties:*

(a) *F is continuous in A and of class C^r in M; $F = f$ in B.*

(b) *If $n \geqq 2m$, F is completely regular in M.*

(c) *If $n \geqq 2m + 1$, F is (1-1) in M.*

Suppose, in addition, that A is a $C^{r'}$-m-manifold,[20] *$1 \leqq r' \leqq r$; f is of class C^t in the subset B of A (see §2), $1 \leqq t \leqq r'$. Then we have also*

[18] We suppose that continuity in M agrees with continuity in A.

[19] This is always possible if $N = E_n$. See for instance Kuratowski, *Topologie* I, p. 211, or Alexandroff-Hopf, *Topologie* I, p. 76.

[20] We suppose that the maps defining M are a subset of those defining A.

(d) *F is of class C^t in A.*

(e) *If $n \geqq 2m$ and f is regular [completely regular] in B, then F is regular [completely regular] in A.*

(f) *If $n \geqq 2m + 1$ and f is regular and (1-1) in B, then F is regular and (1-1) in A.*

We might also apply (e) and (f) of Theorem 2. If the extension of f over M is proper in M [in A], we can make F proper in M [in A]. If N is analytic, we may make $f(M)$ analytic, etc., as in Theorem 2.

By Theorem 1, there are analytic manifolds M' and N' in E_μ and E_ν, C^r- and C^s-homeomorphic with M and N respectively. f[20a] gives a map f' of M' into N'. Choose $\eta(p)$ positive and continuous in M so that $\eta(p) \to 0$ as $p \to B$. Applying Theorem 2 with its r replaced by 0, we replace the extension f' over M' by a function F': the resulting map F of M into N is of class C^r and has the properties (a), (b) and (c) (setting $F = f$ in B).

Now suppose that A is a $C^{r'}$-manifold, $1 \leqq r' \leqq r$. Let A_1' be a $C^{r'}$-homeomorphic analytic manifold in E_μ; then M is $C^{r'}$-homeomorphic with the corresponding subset M_1' of A_1'. Let M'' be an analytic manifold in E_μ', C^r-homeomorphic with M. The map g of M'' into M_1' thus defined is of class $C^{r'}$. We may choose $\eta(p)$ positive and continuous in M so that $\eta(p) \to 0$ as $p \to B$, and so that, considering $\eta(p)$ in M'', if g'' approximates (g, M'', r', η), then g'' is a homeomorphism and $g''(M'')$ does not intersect $B' = A_1' - M_1'$. Let g'' be such a function which is analytic (Theorem 2); the resulting map g' of M into $g'(M)$ is a C^r-homeomorphism. Letting g', in B, be the map already given, g' is (1-1) in A. From Lemma 10 below it is seen that g' is of class $C^{r'}$ in A. It is regular, and taking $\eta(p)$ small enough insures that it is proper: hence *g' is a $C^{r'}$-homeomorphism in A and a C^r-homeomorphism in M.* Let $A' = g'(A)$, $M' = g'(M)$. M' (but not A') is analytic. Let N' and f' be as before.

f' is continuous in A'; considering the values of f' in B' alone, it is of class C^t in the subset B' of E_μ (Lemma 4). Suppose that we have proved Lemma 7; then there is a function F'' of class C^t in $R(A')$ (see Lemma 23) which equals f' together with derivatives of order $\leqq t$ in B', and such that $F''(M')$ is in $R(N')$.[21] Then $F''' = HF''$ (see Lemma 23) is of class C^t in A', and maps A' into N'. Define $\eta(p)$ in M' as before, and let F' be the approximation to F''' in M' given by Theorem 2. Set $F' = F''$ in B'; then F' is of class C^t in A', as is easily seen from Lemma 10. The resulting map F of A into N is of class C^t in A and of class C^r in M, and has the properties (a) through (d). The regularity condition in (e) is satisfied automatically; we obtain complete regularity if f is completely regular in B by applying the method of proof of (A) of §9 in using Theorem 2. Suppose $n \geqq 2m + 1$ and f is (1-1) in B. Then $F(B)$ is the sum of a denumerable number of sets of zero $(m + 1)$-extent, by Lemmas 13, 14 and 15, and hence we may make $F(M)$ avoid $F(B)$ by applying (C) of §9.

There remains to prove

[20a] I.e. the extension of f over M.

[21] If $N = E_\nu$, we could obviously avoid Lemma 7.

LEMMA 7. *Let A^* be a subset of the open set R in E_μ, $\bar{A}^* \cdot R = A^* \cdot R$, and let B^* be a subset of A^*, $\bar{B}^* \cdot A^* = B^* \cdot A^*$. Let $\zeta(p)$ be a positive continuous function in A^*. Let f' be a continuous map of A^* into E_ν, and let f', considered in B^* alone, be of class C^t in B^*. Then there is a C^t-map F of A^* into E_ν which equals f' in B^*, together with partial derivatives of order $\leqq t$, and approximates $(f', A^* - B^*, 0, \zeta)$.*

We may suppose that $E_\nu = E_1$, the space of real numbers. By a direct application of the method of proof of AE Lemma 3,[22] we find a function f which is continuous in R, is of class C^t in a neighborhood R' of B^* (R' in R), $= f'$ in B^*, and approximates $(f', A^* - B^*, 0, \zeta)$. Let R'' be a neighborhood of B^* such that $S = \bar{R}'' \cdot A^*$ is in R'. Set

$$A_s = E_\mu - R \qquad (-1 \leqq s \leqq t - 1),$$

$$A = A' = A_s = A_0 + B^* \qquad (s \geqq t),$$

$$B_s = A^* - S \qquad (0 \leqq s \leqq t - 1),$$

$$B = B_s = A^* - B^* \qquad (s \geqq t).$$

The conditions of AE Theorem III are seen to be satisfied; the function F given by the theorem (if $\epsilon(x)$ is small enough) has the required properties.

12. **A deformation theorem.** We first introduce some definitions. Let M and N be C^r-m- and C^r-n-manifolds, let I be the closed interval (0, 1), and let I' be an open interval containing I. Let $M' = M \times I'$ be the product of M and I'; this is a C^r-$(m + 1)$-manifold with easily defined maps. Let f be a regular C^r-map of M' into N. For each p in M and t in I', set $\phi_t(p) = f(p \times t)$. Each ϕ_t is a regular C^r-map of M into N. Set $M_0 = \phi_0(M)$, $M_1 = \phi_1(M)$; these are local C^r-manifolds in N. If, given M_0 and M_1, there exist such M, I' and f, we shall say the set of maps ϕ_t, t in I, forms a *regular C^r-deformation* of M_0 into M_1. A regular deformation such that M' has at most regular singularities, we shall call *completely regular*; one in which each ϕ_t is (1-1), we call *topological*; one in which f is (1-1), we call *completely topological*. If there is such a map f which is merely continuous, it defines a *deformation* of M_0 into M_1.

THEOREM 6. *Let M_0 and M_1 be C^r-homeomorphic local C^r-m-manifolds in the C^r-n-manifold N ($r \geqq 1$ finite or infinite, $n \geqq 2m + 2$). Suppose there exists a*

[22] Let f_1 be a continuous extension of f' throughout R. Take a subdivision of $R - B^*$ as in AE §8, and define an extension f_2 of class C^t of f', considered in B^* alone, throughout R. (In using AE §9, we take for x^ν a point of B^* whose distance from y^ν is less than twice the distance from y^ν to B^*). Define the ϕ_k in $R - B^*$ as in AE. Given a neighborhood R^* of B^*, R^* in R, let ϕ_{i_k} be those functions which are $\neq 0$ somewhere in $R - R^*$, and set

$$f = f_2 + \sum \phi_{i_k}(f_1 - f_2).$$

As $\sum \phi_{i_k} = 1$ in $R - R^*$, $f = f_1$ in $R - R^*$; also $f = f_2$ in a neighborhood of B^*. For R^* small enough, f evidently approximates to f' as required.

deformation of M_0 into M_1 in N.[23] *Then there is a regular C^r-deformation of M_0 into M_1. If M_0 and M_1 are completely regular, so is the deformation. If M_0 and M_1 are non-singular, it is topological. If, further, $n \geqq 2m + 3$ and M_0 and M_1 do not intersect, it is completely topological.*[24]

We shall not consider the question of proper maps, but merely note that any map of a closed manifold is proper. The following lemma is necessary.

LEMMA 8. *Let M be a local C^r-m-manifold in the C^r-n-manifold N, $n \geqq 2m + 1$. Then there is a vector function $v(p)$ of class C^r in M such that for each p, $v(p)$ is independent of the directions in M at p.*[25]

The lemma may be proved most simply by imbedding N in E_ν (Lemma 19), triangulating M, and defining $v(p)$ successively over the 0-cells, 1-cells, $\cdots$ of M. It may also be proved easily by the methods in this paper.

Let $v_i(p)$ be a vector function in M_i as in the lemma, $i = 0, 1$. Let us C^r-imbed N in some E_ν (Lemma 19); then $v_0(p)$ becomes a vector $v_0'(p)$ in E_ν parallel to N but independent of M at p. Set

$$g_0(p \times t) = p + tv_0'(p);$$

this gives a C^r-map of $M' = M \times I'$ into E_ν. For $t <$ some $\zeta(p)$, $g_0(p \times t)$ lies in $R(N)$ (see Lemma 23); set $g_0'(p \times t) = Hg_0(p \times t)$. Let $f = g_0'$ in M_0; g_0' also defines the derivatives of f of order $\leqq r$ in the subset M_0 of M'. By the choice of v_0, f is a regular map of the subset M_0 of M' into N (see §2). Define f similarly in M_1.

By hypothesis, there is a continuous map f' of M' into N which agrees with f in M_0 and in M_1. We now apply Theorem 5 with A, B, t replaced by M', $M_0 + M_1$, r. This gives a regular C^r-map of M' into N, and hence a regular C^r-deformation of M_0 into M_1. The other statements in the theorem follow at once from Theorem 5, except for the statement on topological deformations; we leave the proof of this to §36.

13. **Spheres bounding differentiable cells.** Let S be the unit m-sphere in E_{m+1}, let Q' be the interior of S, and let Q be the interior of a larger concentric m-sphere. Let f be a regular C^r-map of S into N; we shall call $S_1 = f(S)$ *a local C^r-m-sphere* in N; we leave out "local" if f is (1-1). If f may be extended throughout Q so as to be regular and of class C^r, we say S_1 *C^r-bounds regularly* the $(m + 1)$-cell $f(Q')$. If f is completely regular, or (1-1), we call the bounding *completely regular*, or *topological.* If there exists an f which is merely continuous, we say S *bounds* a cell in N.

THEOREM 7. *Any local C^r-m-sphere S in the C^r-n-manifold N ($r \geqq 1$, $n \geqq 2m + 2$) which bounds a cell in N, C^r-bounds regularly a cell in N. If S has at*

[23] This is of course always the case if $N = E_n$.

[24] If N is an analytic manifold in E_ν, we may make the map of M' analytic except over M_0 and M_1 (see Theorem 2). We shall strengthen the theorem in §36.

[25] The lemma holds if $N = E_n$ and $n \geqq 2m$, as will be shown in a paper on "sphere-spaces." "Vector function" means here "direction function."

most regular singularities, the bounding is completely regular; if $n \geqq 2m + 3$ *and* S *is non-singular, the bounding is topological.*[26]

The proof runs almost exactly like that of Theorem 6.

14. **Examples** (A) Let M be the interval $(-\infty, +\infty)$. Let $a_0, a_1, \cdots$ be a sequence of points dense in E_n, $n \geqq 2$. For each integer $i \geqq 0$, set $f(i) = a_i$, and let f map the interval $(i, i + 1)$ into the segment $\overline{a_i a_{i+1}}$. Set $f(-t) = f(t)$. This is a continuous map of M into a subset of E_n. Set $\eta(t) = 1/(1 + |t|)$. Theorem 2 then gives, if $n = 2$, an analytic curve everywhere dense in the plane, and if $n \geqq 3$, a non-singular analytic curve dense in E_n. On applying (C) of §9, we may (say in E_3) make the curve avoid all rational lines; we may also replace the curve by a denumerable number of such curves which are non-intersecting.

(B) Let B be the unit circle in the plane, with interior M, and set $A = B + M$. Map B into the whole of the sphere S_n of dimension $n \geqq 4$. Any continuous map of B into S_n may be extended continuously over M; hence, by Theorem 2, there is an analytic regular map F of M into S taking on the given boundary values; if $n \geqq 5$, $F(M)$ is non-singular. As in (A), we may make $F(M)$ avoid sets of $(n - m)$-extent zero, may find a denumerable number of non-intersecting surfaces of this sort if $n \geqq 5$, etc.

III. The Imbedding Theorem

In this part we shall prove Lemma 19; this is Theorem 1, except for the analyticity condition. The proof of Theorem 1 will be completed in Part V. The present proof falls into two parts. We first show, in Lemma 12, practically, that Theorem 2 with the conclusions (a) and (e′) alone holds; we then show that (B) and part of (A), §9, hold. This, with the lemma, gives the imbedding theorem. We first give some lemmas of a general nature.

15. **Some general lemmas.** The lemmas which follow are mostly simple extensions of results from AE.

Lemma 9.[27] *Let* $f(p)$ *be a* C^r*-map* $(r \geqq 0)$ *of the open set* R *in* E_m *into* E_n, *and let* $\eta(p)$ *be positive and continuous in* R. *Then there is an analytic map* $F(p)$ *in* R *which approximates* (f, R, r, η).

If $n = 1$, this follows at once from AE Lemma 6. We define open sets $R_1, R_2, \cdots$ as in that lemma, and let ϵ_i be the lower bound of $\eta(p)$ for p in R_{i+1}. For the general case, we apply the lemma separately for each coördinate in E_n.

Lemma 10. *Let* A *be a closed subset of the* C^r*-manifold* M, *let* $\eta(p)$ *be positive and continuous in* $M - A$, *let* $\eta(p) \to 0$ *as* p *approaches any point of* A, *and let*

[26] Compare footnote[24].

[27] See also Lemmas 22 and 26. We may make $F(p)$ approximate to $f(p)$ to higher and higher orders as p approaches the boundary of R; see AE Lemma 6.

$f(p)$ be a C^r-map of M into the C^r-manifold N. If F approximates $(f, M - A, r, \eta)$ and $F = f$ in A, then F is of class C^r in M.

Suppose first $M = E_m$, $N = E_1$; the theorem then follows from AE Lemma 1 (see AE, end of §11). The general case is an immediate consequence of this.

LEMMA 11.[28] *Let A be a subset of the open set R in E_m, with $\bar{A} \cdot R = A \cdot R$, and let $f(p)$ be of class C^r $(r \geqq 0)$ in the subset A of R. Then the definition of f may be extended throughout R so it is of class C^r there. The values of $f(p)$ may be points of E_n.*

If $R = E_m$ and f is real-valued, the proof is given by AE Lemma 2. If $R \neq E_m$, the proof needs only a slight alteration; or we may use AE Theorem III. If the values of f are points of E_n, we apply the lemma to each coördinate separately.

16. **Maps of a manifold with given properties.** The first of the three lemmas giving the imbedding theorem is the following.

LEMMA 12. *Let f be a C^r-map of the C^r-m-manifold M into the C^r-n-manifold N, let $\eta(p)$ be positive and continuous in M, and let $\Omega_1, \Omega_2, \cdots$ be (f, r, η)-properties. Then there is a C^r-map F of M into N which approximates (f, M, r, η) and has these properties.*

It is clear that there is a sequence of positive continuous functions $\eta_i''(p)$ such that if $f_0 = f$ and f_i approximates $(f_{i-1}, M, r, \eta_i'')$, then $F = \lim f_i$ exists and approximates (f, M, r, η). We shall choose functions $\eta_{1j}(j \geqq 1)$, f_1, $\eta_{2j}(j \geqq 2)$, $f_2, \cdots$ in that order so that if $\eta_i(p)$ for each p is the smallest of $\eta_{1i}(p), \cdots, \eta_{ii}(p), \eta_i''(p)$, then f_i approximates (f_{i-1}, M, r, η_i), and $f_j (j \geqq i)$ and F have the property $\Omega_i (i = 1, 2, \cdots)$. Suppose we have found these functions through f_{i-1}. For each p in M, let $\eta_i'(p)$ be the smallest of the numbers $\eta'(p)$ of (c), §7, for the properties $\Omega_1, \cdots, \Omega_{i-1}$; then if f' approximates (f_{i-1}, M, r, η_i'), f' will have the properties $\Omega_1, \cdots, \Omega_{i-1}$. Choose $\eta_{ij}(p) (j \geqq i)$ so that if $f_{i-1}' = f_{i-1}$ and f_j' approximates $(f_{j-1}', M, r, \eta_{ij})(j \geqq i)$, then $F' = \lim f_j'$ approximates (f_{i-1}, M, r, η_i'). Define η_i, and let $f_i(p)$ approximate (f_{i-1}, M, r, η_i) and have the property Ω_i, by (d), §7. We thus find all the above functions, and the function F has the required properties.

17. **The k-extent of a set.** Let A be a subset of E_m. We shall say the *k-extent of A is finite* if there is a number G such that if $0 < \epsilon < 1$, then there are sets $A_1, \cdots, A_\nu$ of diameter $< \epsilon$ which cover A, and $\nu\epsilon^k < G$. If M is a C^1-m-manifold with admissible maps θ_i etc., and A is a subset of M, we say the *k-extent of A is finite* if A is compact (and hence is in a finite number of the U_i), and each $\theta_i^{-1}(A \cdot U_i)$ is of finite k-extent in E_m. The subset A of E_m is of *k-extent zero* if for any $\epsilon' > 0$ there is a $\delta > 0$ such that if $0 < \epsilon < \delta$, then there are sets $A_1, \cdots, A_\nu$ of diameter $< \epsilon$ covering A, and $\nu\epsilon^k < \epsilon'$. Similarly

[28] This lemma is not needed in the proof of the fundamental theorems, though it is useful in Lemma 17. When we have proved the imbedding theorem, we may show easily that the lemma holds with E_m and E_n replaced by a C^r-manifolds (see §4).

if A is a subset of M. A set of zero k-extent ($k \leqq 0$) or of finite l-extent ($l < 0$) is vacuous. These definitions (in M) are independent of the admissible set of maps defining M (see Lemma 15). The sum of a finite number of sets of finite [zero] k-extent is of finite [zero] k-extent.

LEMMA 13. *A bounded subset A of E_k has finite k-extent; if A contains inner points, then it has not zero k-extent.*

This is obvious on cutting up E_k into equal cubes of any desired size. Equally obvious is

LEMMA 14. *If A has finite k-extent, it has zero $(k + 1)$-extent.*

LEMMA 15. *If M and N are C^1-m- and C^1-n-manifolds, f is a C^1-map of M into N, and A is a subset of M[29] of finite [zero] k-extent, then $f(A)$ is of finite [zero] k-extent.*

Consider first the case of finite k-extent. A is in a finite number of the U_i, and $f(A)$ is in a finite number of the V_j; hence we can put $A = A_1 + \cdots + A_t$, A_s in $U_{i(s)}$, $f(A_s)$ in $V_{j(s)}$. Set $B_s = \theta_{i(s)}^{-1}(A_s)$; then B_s is of finite k-extent in E_m. It is sufficient to show that if $g_s = \chi_{j(s)}^{-1} f \theta_{i(s)}$, then $g_s(B_s)$ is of finite k-extent in E_n. g_s is of class C^1 in the compact set B_s; hence, for some number μ, any subset B' of B_s is mapped by g_s into a set $g_s(B')$ whose diameter is at most μ times that of B'. Set $G^* = \mu^k G$. (G corresponds to B_s.) Given ϵ, $0 < \epsilon < 1$, set $\epsilon_1 = \epsilon/\mu^k$, and divide B_s into sets $B_{s1}, \cdots, B_{s\nu}$ of diameter $< \epsilon_1$, so that $\nu\epsilon_1^k < G$. Then $g(B_{s1}), \cdots g(B_{s\nu})$ cover $g(B_s)$, these sets are of diameter $< \epsilon$, and $\nu\epsilon^k < G^*$, as required.

Consider now the case of zero k-extent. Define the B_s etc. as before. Given $\epsilon' > 0$, set $\epsilon_1' = \epsilon'/\mu^k$, and choose $\delta > 0$ so that if $0 < \epsilon_1 < \delta$, then there are sets $B_{s1}, \cdots, B_{s\nu}$ of diameter $< \epsilon_1$ covering B_s, and $\nu\epsilon_1^k < \epsilon_1'$. Now take any ϵ, $0 < \epsilon < \delta$, and set $\epsilon_1 = \epsilon/\mu$. Define the sets B_{si}; then $g(B_{s1}), \cdots, g(B_{s\nu})$ have the required properties.

18. **Transformations of one set away from another.** Let R and R' be open subsets of E_m and E_h, and for each $\alpha = (\alpha_1, \cdots, \alpha_h)$ in R' let T_α be a C^r-map of R into E_n. Let $x' = T_\alpha(x)$ be of class C^1 in terms of the $n + h$ variables (x, α) (x in R). If for each x in R and α in R' the vectors

$$\partial x'/\partial\alpha_1, \cdots, \partial x'/\partial\alpha_h$$

are independent, we shall say the T_α form an *h-parameter family* of C^r-maps. We may also define such families of maps of one manifold into another in an obvious manner; Lemma 16 holds for such families also.

LEMMA 16. *Let $T_\alpha(x)$ be an h-parameter family of C^1-maps of R into E_n, and let A and B be closed sets in R and E_n of finite k-extent and zero $(h - k)$-extent respectively. Then for some α in R', $T_\alpha(A)$ does not intersect B.*[29a]

We may suppose that $0 \leqq k \leqq h$. For some α^0 in R' and $\eta > 0$, all points α

[29] Note that M may be replaced by any open subset R of M which contains A. It is evidently sufficient that the map satisfy a Lipschitz condition.

[29a] We may evidently interchange "finite" and "zero".

within η of α^0 are in R'. Take $\xi \geqq 1$ so that if A^*, in A, is of diameter $< \epsilon$, then $T_{\alpha^0}(A^*)$ is of diameter $< \xi\epsilon$. The condition that the $\partial x'/\partial\alpha_i$ are independent for x in A and $\alpha = \alpha^0$ is equivalent to the condition that for some $\beta > 0$, any directional derivative for x in A and $\alpha = \alpha^0$ is a vector of length $\geqq \beta$. Hence we may take ζ, $0 < \zeta < \eta$, such that for any $\epsilon > 0$, if A^* and B^* are subsets of A and B of diameter $< \epsilon$, α and α' are points of R' within ζ of α^0, and the distance from α to α' is $> 4\xi\epsilon/\beta$, then either $T_\alpha(A^*)\cdot B^* = 0$ or $T_{\alpha'}(A^*)\cdot B^* = 0$.

Let G be the number corresponding to A, and take $\epsilon' \leqq (\beta\zeta)^h/[2^h(4\xi)^h G]$. Find $\delta < 1$ corresponding to B and ϵ'. Choose an ϵ, $0 < \epsilon < \delta$. Then we may set

$$A = A_1 + \cdots + A_\nu, \qquad B = B_1 + \cdots + B_\sigma,$$

so that the A_i cover A, the B_j cover B, the diameter of each A_i and each B_j is $< \epsilon$, and so that

$$\nu\epsilon^k < G, \qquad \sigma\epsilon^{h-k} < \epsilon'.$$

Let Z_{ij} be the set of all points α within ζ of α^0 such that $T_\alpha(A_i)\cdot B_j \neq 0$; then, by the choice of ζ, the diameter of Z_{ij} is $< 4\xi\epsilon/\beta$, and hence its ordinary h-volume is $< (4\xi\epsilon/\beta)^h$. Therefore the ordinary h-volume of all the Z_{ij} is less than

$$\nu\sigma(4\xi\epsilon/\beta)^h = (4\xi)^h(\nu\epsilon^k)(\sigma\epsilon^{h-k})/\beta^h < (4\xi)^h G\epsilon'/\beta^h \leqq \zeta^h/2^h.$$

Hence there is an α within ζ of α^0 which is in no Z_{ij}; for this α, $T_\alpha(A)$ does not intersect B.

19. **The transformation $T_{p,P}$.** Let P_0 be a fixed k-plane through a fixed point p_0 in E_n. Corresponding to any k-plane P not perpendicular to P_0 and any point p of P we may let correspond a non-homogeneous orthogonal transformation of space $T_{p,P}$ which carries p_0 and P_0 into p and P, and is analytic in p and P (see §24), for instance as follows. Let P_0' and P' be the parallel planes through the origin. Let $v_1, \cdots, v_n$ be mutually orthogonal unit vectors such that $v_1, \cdots, v_k$ are in P_0'. The following rules determine Tv_i. The Tv_i are mutually orthogonal unit vectors. If $1 \leqq i \leqq k$ and P_i' is the plane in P' determined by the projections $v_1', \cdots, v_i'$ of $v_1, \cdots, v_i$ into P', then $Tv_1, \cdots, Tv_i$ are in P_i' and determine the same orientation in P_i' as $v_1', \cdots, v_i'$; for each j ($j = k + 1, \cdots, n$), if P_j' is the plane determined by P' and $v_{k+1}, \cdots, v_j$, then $Tv_{k+1}, \cdots, Tv_j$ are parallel to this plane, and $Tv_1, \cdots, Tv_j$ determine the same orientation in it as $Tv_1, \cdots, Tv_k, v_{k+1}, \cdots, v_j$. $T_{p,P}$ is the unique linear transformation carrying p_0 into p and v_i into Tv_i.

20. **(1-1) maps and properties.** We can now prove

LEMMA 17. *If f is a regular C^r-map ($r \geqq 1$) of the C^r-m-manifold M into E_n, $n \geqq 2m + 1$, and η is a sufficiently small positive continuous function in M, then the property of maps f' which approximate (f, M, r, η) of being (1-1) and avoiding* *a fixed point b_0 is the logical sum of a denumerable number of (f, r, η)-properties.*[30]

Let $\theta_1, \theta_2, \cdots$ be a completely admissible set of maps defining M; we choose them so that if $\bar{U}_i$ and $\bar{U}_j$ have common points, then f is (1-1) in $\bar{U}_i + \bar{U}_j$. (This is simple; see §5.) Consider all sets $U_i + U_j$ such that $\bar{U}_i \cdot \bar{U}_j = 0$; we arrange these in a sequence $W_1, W_2, \cdots$. Take $\eta(p)$ so small that any map f' which approximates (f, M, r, η) is regular and (1-1) in all $\bar{U}_i + \bar{U}_j$ for which $\bar{U}_i \cdot \bar{U}_j \neq 0$. Take any k, and say $W_k = U_i + U_j$; let Ω_k hold for the map f' if $f'(\bar{U}'_i) \cdot f'(\bar{U}'_j) = 0$ and b_0 is not in $f'(\bar{U}'_i)$. As the U'_i cover M, the property of f' being (1-1) and avoiding b_0 is the sum of the properties $\Omega_1, \Omega_2, \cdots$.

It remains to show that Ω_k is an (f, r, η)-property. (a) of §7 holds; (b) holds with W' replaced by $W'_k = U'_i + U'_j$. As $f'(\bar{U}'_i)$ and $f'(\bar{U}'_j) + b_0$ are bounded closed subsets of E_n, (c) is obvious; it remains to prove (d). Let f' approximate (f, M, r, η). Let $\lambda'(x)$ be a function of class C^r in E_m which $= 1$ in $\bar{Q}'_m$ and $= 0$ in $E_m - Q_m$. (Such a function is given by Lemma 11, replacing A by $\bar{Q}'_m + (E_m - Q_m)$; or the function may be constructed directly without great difficulty.) Then $\lambda(p) = \lambda'(\theta_i^{-1}(p))$ in U_i and $= 0$ in $M - U_i$, is of class C^r in M, and $= 1$ in $\bar{U}'_i$. For any vector v in E_n, set

$$f(p, v) = f'(p) + \lambda(p)v.$$

Given an arbitrary $\eta'(p)$, we may choose $\beta > 0$ so that if[31] $\| v \| < \beta$, then $f(p, v)$ approximates (f', M, r, η'). By Lemmas 13, 14 and 15, $f'(\bar{U}'_j) + b_0$ is of zero $(m + 1)$-extent and $f'(\bar{U}'_i)$ is of finite m-extent. The transformations $T_v q = q + v$ with $\| v \| < \beta$ form an n-parameter family in E_n; by Lemma 16, there is a v_0 such that $T_{v_0} f'(\bar{U}'_i)$ does not intersect $f'(\bar{U}'_j) + b_0$. $f'' = f(p, v_0)$ with this v_0 is then the required approximation; for $f'' = f' + v_0$ in $\bar{U}'_i$ and $= f'$ in $\bar{U}'_j$.

21. **Regular maps and properties.** The final lemma is

LEMMA 18. *If f is a C^r-map $(r \geqq 1)$ of the C^r-m-manifold M into E_n, $n \geqq 2m$, and η is a positive continuous function in M, then the property of maps f' which approximate (f, M, r, η) of being regular is the logical sum of a denumerable number of (f, r, η)-properties.*

Let $\theta_1, \theta_2, \cdots$ be completely admissible maps defining M, and let Ω_i hold for f' if f' is regular in $\bar{U}'_i$; then if $\Omega_1, \Omega_2, \cdots$ hold, f' is regular in M. Each Ω_i satisfies the conditions (a), (b) and (c) of §7; to show that it is an (f, r, η)-property, we must show that it satisfies (d). Set $g(x) = f'(\theta_i(x))$ in Q_m; this is a C^r-map of Q_m into E_n. As in the last §, it is sufficient to show that for an arbitrary $\zeta > 0$ there is a map $g'(x)$ of Q_m into E_n which approximates (g, Q_m, r, ζ) and is regular in Q'_m. We then set

$$f''(p) = f'(p) + \lambda(p)[g'(\theta_i^{-1}(p)) - f'(p)] \text{ in } U_i,$$

and $f''(p) = f'(p)$ in $M - U_i$; as $f''(p) = g'(\theta_i^{-1}(p))$ in $\bar{U}'_i$, it is regular in $\bar{U}'_i$.

[30] In §34 we shall express the property as a sum of $[f, r, \eta, \Delta, \zeta]$-properties.

[31] $\| v \|$ is the length of v.

If $r = 1$, let $g_0(x)$ be a function of class C^2 which approximates (g, Q_m, r, ζ') (Lemma 9); otherwise, set $g_0(x) = g(x)$. We shall find functions $g_1(x), \cdots, g_m(x) = g'(x)$ such that $g_i(x)$ approximates $(g_{i-1}, Q_m, r, \zeta')$, and so that the vectors $\partial g_i/\partial x_1, \cdots, \partial g_i/\partial x_i$ are independent in $\bar{Q}_m'$; if ζ' is small enough, g' is the required function.

Suppose we have found g_{i-1}. For any vector v in E_n, set

$$g_i(x, v) = g_{i-1}(x) + x_i v;$$

then

$$\frac{\partial g_i(x, v)}{\partial x_i} = \frac{\partial g_{i-1}(x)}{\partial x_i} + v, \qquad \frac{\partial g_i(x, v)}{\partial x_j} = \frac{\partial g_{i-1}(x)}{\partial x_j} \qquad (j \neq i).$$

As the $\partial g_{i-1}(x)/\partial x_j$, $j = 1, \cdots, i-1$, are independent in $\bar{Q}_m'$, we need merely show that there is an arbitrarily small v_0 such that $\partial g_i(x, v_0)/\partial x_i$ is independent of these vectors at each point of $\bar{Q}_m'$; we then set $g_i(x) = g_i(x, v_0)$. By Lemma 13, it is sufficient to show that the vectors v, $\| v \| \leqq 1$, which do not have the required property, form a set of zero n-extent in E_n. Given the point x^0 of $\bar{Q}_m'$, it is sufficient to show that the vectors not having the required property in a closed neighborhood S of x^0 are of zero n-extent; for a finite number of such sets S cover $\bar{Q}_m'$.

Let $P(x)$ be the $(i-1)$-plane through the origin O in E_n determined by $\partial g_{i-1}(x)/\partial x_1, \cdots, \partial g_{i-1}(x)/\partial x_{i-1}$, and set $P_0 = P(x^0)$. (If $i = 1$, $P(x) = O$.) Choose S so that $P(x)$ is not perpendicular to P_0 for x in S. Let $y_1, \cdots, y_n$ be rectangular coördinates (with origin O) in E_n such that P_0 is the $(y_1, \cdots, y_{i-1})$-plane. Let $E = E_{m-i-1}$ be the space with coördinates $(x_1, \cdots, x_m, y_1, \cdots, y_{i-1})$. Set

$$v(x) = \frac{\partial g_{i-1}(x)}{\partial x_i}, \qquad K = \max \| v(x) \| \ (x \text{ in } S).$$

Let D be the subset of E with x in S, $\| \bar{y} \| \leqq K + 1$, where $\bar{y} = (y_1, \cdots, y_{i-1})$. Let $T_x = T_{O,P(x)}$ be the transformation of §19 leaving O fixed and carrying P_0 into $P(x)$. Given $\bar{y}$, set $y = (y_1, \cdots, y_{i-1}, 0, \cdots, 0)$ and

$$w(x, \bar{y}) = T_x(y) - v(x).$$

as g_{i-1} is of class C^2, this is a C^1-map of D into E_n. By Lemma 15, $w(D)$ is of finite $(m + i - 1)$-extent, and hence of zero n-extent. Now let v be any vector, $\| v \| \leqq 1$, such that for some x in S, $v(x) + v$ is in $P(x)$; we shall show that v is in $w(D)$. As $P(x) = T_x(P_0)$, there is a $\bar{y}$ such that

$$v(x) + v = T_x(y), \qquad v = w(x, \bar{y});$$

as $\| v(x) + v \| \leqq K + 1$, $\| y \| \leqq K + 1$, and v is in $w(D)$. Hence there is an arbitrarily small v_0 such that no $v(x) + v_0$ is in $P(x)$; $v(x) + v_0 = \partial g_i(x, v_0)/\partial x_i$ is independent of $\partial g_{i-1}/\partial x_1, \cdots, \partial g_{i-1}/\partial x_{i-1}$ in S, and the lemma is proved.

22. **Proof of the imbedding theorem.** The last three lemmas lead at once to

LEMMA 19. *Any C^r-m-manifold M may be C^r-imbedded in E_{2m+1}.*[32]

Let $\eta(p)$ be a positive continuous function in M such that if $p_1, p_2, \cdots$ is a sequence of points of M with no limit in M, then $\lim \eta(p_i) = 0$.[33] Let f map M into the origin O in E_{2m+1}; f is of class C^r. Applying Lemma 12 with the properties of Lemma 18 gives a regular C^r-map f' approximating (f, M, r, η). We now apply Lemma 12 again, this time with the properties of Lemma 17, setting $b_0 = O$. (The new η may have to be smaller than the last.) The resulting function F is (1-1) regular and of class C^r in M, and $F(p) \neq O$ in M. By the choice of η, the limit set $LF(M)$ either is void or equals O; hence F is proper. Therefore $F(M)$ is a C^r-manifold in E_{2m+1} C^r-homeomorphic with M, and the proof is complete.

IV. THE NEIGHBORHOOD OF A MANIFOLD IN E_n

Suppose M is a C^r-m-manifold in E_n, $r \geqq 2$, $n > m$. To each point p of M there is a normal plane $P'(p)$; this family of planes fills out a neighborhood of M in E_n in a (1-1) way. If $r = 1$, this may not hold; for the $P'(p)$ depend on the first derivatives of functions defining M. Our object in this part is to find an approximating family of planes $P(p)$ of class C^r. §24 is necessary in this proof, and also directly in the next part.

23. **Projective spaces in Euclidean spaces.** The following lemma will be needed in the next §.

LEMMA 20. *Projective n-space E_n^* may be imbedded analytically in Euclidean space E_{2n+1}.*

The points of E_n^* are the sets of numbers $(x_1, \cdots, x_{n+1}) \neq (0, \cdots, 0)$, proportional sets being the same point. Let S_n be the unit n-sphere $\Sigma x_i^2 = 1$ in E_{n+1}. To each pair of "opposite" points p, $-p$ of S_n corresponds a point $\gamma(p) = \gamma(-p)$ of E_n^*. S_n is an analytic manifold with obvious neighborhoods; mapping these neighborhoods into E_n^* under γ defines E_n^* as an analytic manifold. By Lemma 19, E_n^* is C^1-homeomorphic with a manifold M' in E_{2n+1}. Let ϕ denote this homeomorphism. Then $\psi(p) = \phi(\gamma(p))$ is a (2-1) regular map of S_n into E_{2n-1}.

For any map g of S_n into E_{2n+1}, set[34]

$$Ag(p) = [g(p) + g(-p)]/2; \text{ then } A\psi(p) = \psi(p).$$

It is easily seen that if g approximates $(\psi, S_n, 1, \delta)$, then the same is true of Ag. As $Ag(p) = Ag(-p)$, we may let correspond to g a map f of E_n^* into E_{2n+1}; this

[32] Note that we may let $F(M)$ have no (finite) limit set in E_{2m+1}, by applying a transformation with reciprocal radii at the end of the proof.

[33] Let θ_i be a set of admissible maps in M. Let $\eta'(x)$ be continuous in E_m, > 0 in Q_m, and $= 0$ in $E_m - Q_m$. Set $\eta_i(p) = \eta'(\theta_i^{-1}(p))$ in U_i and $= 0$ in $M - U_i$. If $\alpha_1, \alpha_2, \cdots$ are small enough, we may set $\eta(p) = \Sigma \alpha_i \eta_i(p)$.

[34] Compare footnote[17].

might be written $f(q) = Ag(\gamma^{-1}(q))$. Take $\epsilon > 0$ such that any map approximating $(\phi, E_n^*, 1, \epsilon)$ is (1-1) regular. Choose $\delta > 0$ so that if g approximates $(\psi, S_n, 1, \delta)$, then the corresponding f approximates $(\phi, E_n^*, 1, \epsilon)$. Extend ψ through a neighborhood R of S_n in E_{n+1}, say by letting it be constant on any half ray from the origin; it is then of class C^1 in R. By Lemma 9 (or the Weierstrass approximation theorem) there is an analytic map g approximating $(\psi, R, 1, \delta)$; considering g on S_n alone, the corresponding function f has then the required properties.

24. k-planes in n-space. Let $\mathfrak{S}$ be the space whose points are the k-planes in n-space through the origin. We shall express this space as an analytic manifold $M(n, k)$ in a Euclidean space $E(n, k)$. Given the plane P, let $v_1, \cdots, v_k$ be a set of independent vectors in P; their coördinates form a matrix, with k-rowed determinants $D_{i_1 \cdots i_k}(P)$. These determinants, arranged in a sequence $D_1^*(P), \cdots, D_{\gamma_{nk}}^*(P)$, $\gamma_{nk} = \binom{n}{k}$, form the homogeneous coördinates of a point $D^*(P)$ in projective space $E_{\gamma_{nk}-1}^*$. $D^*(P)$ is independent of the vectors $v_1, \cdots, v_k$, and $D^*(P) \neq D^*(P')$ if $P \neq P'$; thus we have a (1-1) map of $\mathfrak{S}$ into a subset $\mathfrak{S}'$ of $E_{\gamma_{nk}-1}^*$. By Lemma 20, we may imbed $E_{\gamma_{nk}-1}^*$ analytically in Euclidean space $E(n, k)$; this carries $\mathfrak{S}'$ into a subset $M(n, k)$ of $E(n, k)$.

We may show that $\mathfrak{S}'$ and hence $M(n, k)$ is an analytic manifold by taking any P_0, choosing a determinant, say $D_{1 \cdots k}(P_0)$, which is $\neq 0$, and expressing each $D_{i_1 \cdots i_k}$ in terms of the determinants $D_{1, \cdots, s-1, s+1, \cdots, k, t}$ by Vahlen's relations, which are analytic; this determines maps of the required nature in $\mathfrak{S}'$. An analytically equivalent set of maps may be given as follows: Given P_0, let $p_1^0, \cdots, p_k^0$ be points of P_0 which form linearly independent vectors from the origin, and let $L_1, \cdots, L_k$ be $(n - k)$-planes (or analytic $(n - k)$-manifolds) through $p_1^0, \cdots, p_k^0$ orthogonal to P_0. If P (through the origin) is near P_0, it intersects each L_i in a point p_i; the positions of the p_i determine a map of part of $E_{k(n-k)} = E_{n-k} \times \cdots \times E_{n-k}$ (k factors) into part of $M(n, k)$.

Another important space is the space $\mathfrak{S}^*$ of all k-planes in E_n; this also forms an analytic manifold. $M(n, k)$ is closed; the present manifold is open. We may map $\mathfrak{S}^*$ into $M(n, k)$ by letting any plane correspond to the parallel plane through the origin. We shall use the symbol P for points of either space; it will always be clear which space is meant.

By an *analytic function of k-planes* we shall mean a function which, when considered in $M(n, k)$, is analytic.

We shall say two planes of any dimensions (in either space) are *orthogonal* if any vector in (or perhaps better, parallel to) one is orthogonal to any vector in the other; *independent* if they have no common vector $\neq 0$; *perpendicular* if some vector $\neq 0$ in one is orthogonal to each vector of the other. If P, P' are points of $\mathfrak{S}$, we may let $\| P' - P \|$ be the distance between the corresponding points of $M(n, k)$.

25. The neighborhood of a manifold in space. Two more lemmas lead up to the main result of this part, Lemma 23.

LEMMA 21. *Let M be a C^r-m-manifold in E_n ($r \geqq 1$ finite or infinite), and let $P(p)$ be a function of class C^r in M satisfying* (a) *of Lemma* 23. *Then there is a function $\xi(p)$ in M satisfying the remaining conditions.*

Take any p_0 in M. A neighborhood U of p_0 may be determined by functions (3.1) of class C^r. Define the transformation $T_{p,P}$ in terms of p_0 and $P(p_0)$ as in §19. Set $w_i(p) = T_{p,P(p)}(v_i)$; then the points of $P(p)$ for p in U are given by

$$q = p + \sum_{i=1}^{n-m} \alpha_i w_i(p). \tag{25.1}$$

Using (3.1), we may express p in terms of $y_1, \cdots, y_m$: $p = \psi(y_1, \cdots, y_m)$. Putting in (25.1) gives q as a function of $y_1, \cdots, y_m, \alpha_1, \cdots, \alpha_{n-m}$:

$$\begin{aligned} q &= \psi(y_1, \cdots, y_m) + \Sigma \alpha_i w_i(\psi(y_1, \cdots, y_m)) \\ &= g(y_1, \cdots, y_m, \alpha_1, \cdots, \alpha_{n-m}). \end{aligned} \tag{25.2}$$

g is of class C^r. Consider the vectors $\partial g/\partial y_i$, $\partial g/\partial \alpha_j$ at $q = p_0$ in U. The $\partial g/\partial y_i$ are independent vectors in the tangent plane T to M at p_0, as the α_i vanish there, and the $\partial g/\partial \alpha_j = w_j(p_0)$ are independent vectors in $P(p_0)$; as $P(p_0)$ is independent of T, the whole set of vectors is independent. In other words, the Jacobian of (25.2) is $\neq 0$ at p_0, and hence in a neighborhood R' of p_0. Solving for $y_1, \cdots, y_m$ gives p in terms of q in R': $p = H(q)$. H is of class C^r.

We may cover M by such neighborhoods R' so that any bounded closed subset of M has points in but a finite number of the R'. It is easy to construct a positive function $\xi(p)$ in M such that if $R(p)$ is that part of $P(p)$ within $\xi(p)$ of p, then $R(p)$ lies in some R'. (c) and hence (b) of Lemma 23 now hold.

LEMMA 22. *Lemma* 9 *holds with R and E_n replaced by analytic manifolds M and N in E_μ and E_ν respectively.*[35]

Let $P(p)$ be the normal plane to p in M. Then $P(p)$ is analytic, and hence we may define $H(p)$ by the last lemma. H is analytic. Similarly we define P' and H' for N. Set $f(q) = f(H(q))$ in $R(M)$; this is a C^r-map of $R(M)$ into the subset N of E_ν (see Lemma 4). If the analytic function F' approximates to f closely enough and $R'(M)$ is a small enough neighborhood of M in $R(M)$, then F' maps $R'(M)$ into $R(N)$, and $F = HF'$ is an analytic map of $R'(M)$ into N. F, considered on M alone, is analytic (see Lemma 3).

LEMMA 23. *Let M be a C^r-m-manifold in E_n($r \geqq 1$ finite or infinite). Then there is a positive continuous function $\xi(p)$ and a function $P(p)$ of class C^r in M, such that:* (a) *$P(p)$ is an $(n - m)$-plane through p independent of the tangent plane to M at p.* (b) *If $R(p)$ is that part of $P(p)$ within $\xi(p)$ of p, then the $R(p)$ fill out a neighborhood $R(M)$ of M in a* (1-1) *way.* (c) *If $H(q) = p$ for q in $R(p)$, then H is of class C^r in $R(M)$. Moreover, if M is analytic, so are $P(p)$ and $H(p)$.*

[35] See also Lemma 27.

We have just considered the analytic case. If $n = m$, the lemma is trivial (then H is the identity); suppose $n > m$. Let $P'(p)$ be the normal plane to M at p; there is a corresponding point $D'(p)$ in $M(n, n - m)$. D' is of class C^{r-1} and is thus continuous in M. Extended D' so as to be continuous throughout a neighborhood R of M in E_n. (Almost any method in use will do this; or we may use Lemma 11.) If R is small enough, $D'(R)$ is in $R(M(n, n - m))$; then $D'' = H'D'$ is a continuous map of R into $M(n, n - m)$, and $D'' = D'$ in M. (H' is defined in $R(M(n, n - m))$ as in the last lemma.) By Lemma 22 we may approximate D'' in R by an analytic function D so closely that if $P(p)$ is the plane through p in M parallel to the plane defined by $D(p)$, then $P(p)$ is independent of the tangent plane to M at p. $P(p)$, considered in M alone, is of class C^r, by Lemma 3. The lemma now follows from Lemma 21.

V. Analytic Manifolds

26. The lemma and method of proof. Our object in this part is to prove

LEMMA 24. *Let M be a C^r-m-manifold in E_n ($r \geqq 1$ finite or infinite). Then there is a C^r-homeomorphic analytic manifold M^* in E_n.*

This, together with Lemma 19, completes the proof of Theorem I. Actually, M^*, as constructed, will approximate to M to any desired degree, but it is easier to find an approximating analytic manifold after a homeomorphic analytic one is found. (See Lemma 22.) We may suppose that $n > m$; if $n = m$, then M is analytic.

To prove the lemma, we first construct an analytic $(n - 1)$-manifold S "surrounding" M, and then find in an analytic fashion a "center" M^* of S. The proof is most easily visualized for $n = 3$, $m = 1$. The construction of S is straightforward. We determine a function positive and analytic near M and vanishing in M, subtract a very small positive analytic function, and let S be the set of points where the resulting function vanishes. The inside of S is filled up by $(n - m)$-planes $P(p)$ approximately normal to M (see Lemma 23). The resulting function $D(P)$ with values in $M(n, n - m)$ (see §24) is of class C^r inside of S. We approximate this function by an analytic function, and thus determine an analytic family of planes $P^*(p)$. (These planes, unlike the $P(p)$, intersect each other inside S.) A point p inside S is in M^* if and only if p is at the center of mass of that connected part of $P^*(p)$ inside S which contains p.

The following lemma is necessary.[36]

LEMMA 25. *Given an open set R in E_n, a positive continuous function $\eta(p)$ in R, and $r \geqq 0$, there is an analytic function $\omega(p)$ in R such that*

$$(26.1) \qquad \omega(p) > 0, \qquad |D_k\omega(p)| < \eta(p) \text{ in } R \qquad (\sigma_k \leqq r).$$

Let $C_1, C_2, \cdots$ be a denumerable set of overlapping cubes covering R, and let

[36] This lemma, except for analyticity, is practically equivalent to a theorem of Ostrowski, Bull. des Sciences Math., 1934, pp. 64–72. See also our lemma 10. The theorem was known to the author in 1933. Note that we may make $r = \infty$ in a manner similar to that in AE Lemma 6.

$\phi_i(x)$ be a function of class C^∞ in R which is > 0 within C_i and $= 0$ in $R - C_i$ (see for instance AE, §9). If the a_i are small enough and positive, then $\phi(x) = \Sigma a_i \phi_i(x)$ is a positive function of class C^∞ in R satisfying the inequality (26.1). If $\omega(p)$ is an analytic function approximating (ϕ, R, r, ζ) for small enough $\zeta(p) > 0$ (see Lemma 9), then (26.1) holds.

27. **The manifold** S **and spheres** $S^*(p, P)$. In this section we shall find S, and shall show that certain $(n - m)$-planes P through points p near M intersect S in $(n - m - 1)$-spheres $S^*(p, P)$, which are analytic and vary analytically with p and P. In the next section we shall find the analytic manifold M^*.

Define the planes $P(p)$ and the projection $H(p)$ in the neighborhood $R(m)$ as in Lemma 23. We extend the definitions of $P(p)$ and $\xi(p)$ through $R(M)$ by setting

$$P(p) = P(H(p)), \qquad \xi(p) = \xi(H(p)).$$

Define the function $\Phi(p)$ in $R(M)$ by

$$\Phi(p) = \| p - H(p) \|. \tag{27.1}$$

As $H(p)$ is of class C^r in $R(M)$, $\Phi(p)$ is of class C^r in $R(M) - M$. By Lemmas 9 and 10 there is a function $\Phi'(p)$ continuous in $R(M)$ and analytic in $R(M) - M$ such that $\Phi' = 0$ in M, and it and its gradient satisfy

$$| \Phi'(p) - \Phi(p) | < \tfrac{1}{3} \xi(p), \qquad \| \nabla\Phi'(p) - \nabla\Phi(p) \| < \tfrac{1}{3} \tag{27.2}$$

in $R(M) - M$. By Lemma 25, there is a positive analytic function $\omega(p)$ in $R(M)$ such that

$$| \omega(p) | < \tfrac{1}{3} \xi(p), \qquad \| \nabla\omega(p) \| < \tfrac{1}{3}. \tag{27.3}$$

Set

$$\Phi^*(p) = \Phi'(p) - \omega(p); \tag{27.4}$$

then $\Phi^*(p)$ is continuous in $R(M)$ and is analytic in $R(M) - M$, and $\Phi^* < 0$ in M. S is determined by the vanishing of Φ^*.

To prove the existence of and properties of S, we shall introduce some auxiliary functions. Let p_0 be any point of M. Some neighborhood U of p_0 in M is defined by equations (3.1). Given any subset K of M, let $R(K)$ be the set of all points p of $R(M)$ such that $H(p)$ is in K. $P(p)$ is independent of T (see Lemma 23); hence there is a neighborhood U' of p_0 in M, U' in U, and a $\delta > 0$, such that if P is an $(n - m)$-plane through a point p of $R(U')$ and

$$\| P - P(p) \| < \delta, \tag{27.5}$$

then P is independent of T and hence intersects T in a unique point $H^*(P)$. (T is the tangent plane to M at p_0.) H^* is analytic. Set

$$H'(p) = H^*(P(p)), \qquad u(p) = \frac{p - H(p)}{\| p - H(p) \|}, \qquad u'(p) = \frac{p - H'(p)}{\| p - H'(p) \|}. \tag{27.6}$$

We may choose U' and δ so small that for any p in $R(U')$ and any P through p satisfying (27.5),

$$\Phi^*(H^*(P)) < 0, \qquad \| H'(p) - H(p) \| < \xi(p)/6, \tag{27.7}$$

and

$$\text{if } \Phi^*(p) \geqq 0, \quad \text{then} \quad \| u'(p) - u(p) \| < \tfrac{1}{3}. \tag{27.8}$$

For any P satisfying (27.5), let T_P be the transformation $T_{H^*(P),P}$ of §19, using the fixed point p_0 and plane $P(p_0)$, and let $S(P)$ be the unit $(n - m - 1)$-sphere in P about $H^*(P)$. Given any point $\bar{q}$ of $S_0 = S(P(p_0))$, there is a corresponding point

$$q' = T_P(\bar{q}) = \mu(P, \bar{q}) \text{ in } S(P); \tag{27.9}$$

μ is analytic. To each P satisfying (27.5), each $\bar{q}$ in S_0, and any $\alpha > 0$, let correspond points p', q', q by

$$q = p' + \alpha(q' - p') = H^*(P) + \alpha[\mu(P, \bar{q}) - H^*(P)]. \tag{27.10}$$

For such values of $\alpha > 0$ which make q lie in $R(U) - U$ we define the analytic function

$$\sigma(P, \bar{q}, \alpha) = \Phi^*(q). \tag{27.11}$$

We shall show next that for some γ, $0 < \gamma < \delta$, if P is a plane through a point p of $R(U')$, $\| P - P(p) \| < \gamma$, and $\bar{q}$ is in S_0, then there is unique number

$$\alpha = \rho(P, \bar{q}) > 0 \tag{27.12}$$

which, put in (27.10) and (27.11), makes σ vanish (with q in $R(U)$); moreover, ρ is analytic. Set

$$\sigma'(p, \bar{q}, \alpha) = \sigma(P(p), \bar{q}, \alpha); \tag{27.13}$$

it is sufficient to show that, using $P(p)$, there is a unique point q of the line segment $\overline{p'q'}$ in $R(U')$ such that $\Phi^*(q) = 0$, and $\partial\sigma'/\partial\alpha > 0$ at this point.

By definition of $R(M)$, $R(U')$ contains all points of $P(p)$ within $\xi(p)$ of $H(p)$. As $p' = H'(p)$ for $P = P(p)$, (27.7) gives

$$\| p' - H(p) \| < \xi(p)/6. \tag{27.14}$$

Hence, if q'' is the point q for which $\alpha = 5\xi(p)/6$ (keeping $\bar{q}$ fixed), all of $\overline{p'q''}$ lies in $R(U')$. Moreover, as $H(q'') = H(p)$, (27.1) through (27.4) with (27.14) give

$$\Phi^*(q'') > \Phi(q'') - \tfrac{2}{3}\, \xi(p) > 0. \tag{27.15}$$

By (27.7), $\Phi^*(p') < 0$; hence there is a point of $\overline{p'q''}$ for which $\Phi^* = 0$.

Now take any q on $\overline{p'q''}$ such that $\Phi^*(q) \geqq 0$, keeping $P = P(p)$. As

$$\| q' - p' \| = 1, \qquad p' = H'(q'),$$

and $u'(q') = u'(q)$, differentiating (27.13) and using (27.10) gives

$$\frac{\partial \sigma'}{\partial \alpha} = |\nabla \Phi^*(q) \cdot (q' - p')|_{P=P(p)} = \nabla \Phi^*(q) \cdot u'(q). \tag{27.16}$$

The projection of $\nabla \Phi$ into any plane P equals the gradient of Φ as a function defined in P; hence, by (27.1) and (27.6),

$$\text{Proj}_{P(p)} \nabla \Phi(q) = u(q),$$

and if u' is any vector parallel to $P(p)$, then

$$\nabla \Phi(q) \cdot u' = u(q) \cdot u'.$$

Hence, by (27.2) through (27.4) and (27.8),

$$\left| \frac{\partial \sigma'}{\partial \alpha} \right| = | [\nabla \Phi^*(q) - \nabla \Phi(q)] \cdot u'(q) + u(q) \cdot [u'(q) - u(q)] + 1 | \tag{27.17}$$
$$> - \tfrac{2}{3} - \tfrac{1}{3} + 1 = 0.$$

This shows that Φ^* vanishes at a unique point q of $\overline{p'q''}$, and the existence and analyticity of ρ is proved.

Take any P satisfying (27.5) with δ replaced by γ; putting (27.12) in (27.10) gives q as a function of $\bar{q}$. As $\bar{q}$ ranges over S_0, q ranges over an analytic $(n - m - 1)$-sphere $S^*(P)$; this sphere varies analytically with P. It is the intersection of P and S. A finite or denumerable number of neighborhoods U' cover M; for each there is a corresponding $\gamma > 0$. Let $\gamma(p)$ be a positive continuous function in $R(M)$ such that $\gamma(p) = \gamma(H(p))$, and if p is in any U', then $\gamma(p)$ is less than the corresponding γ. Now if p is any point of $R(M)$, P contains p, and

$$\| P - P(p) \| < \gamma(p), \tag{27.18}$$

then $R(p)$ intersects S in an analytic sphere $S^*(p, P)$ which varies analytically with p and P.

28. **The analytic manifold M^*.** For any p in $R(M)$ and any plane P through p satisfying (27.18), let $Q^*(p, P)$ be that part of P inside $S^*(p, P)$. Let $g(p, P)$ be the center of mass of $Q^*(p, P)$. We shall show that if $P^*(p)$ is any analytic function in $R(M)$ approximating to $P(p)$ closely enough in $R(M)$, then the set M^* of points in $R(M)$ satisfying

$$g(p, P^*(p)) = p \tag{28.1}$$

is an analytic manifold in $R(M)$, C^r-homeomorphic with M.

We shall first show that $g(p, P)$ is analytic. Consider a point p_0 of M and a neighborhood U of p_0 in M etc. as before. If $V(p, P)$ is the $(n - m)$-volume of $Q^*(p, P)$ and $|dp|$ denotes the volume element, then for p in $R(U')$ and any P

through p within $\gamma(p)$ of $P(p)$,[37]

$$g(p, P) = \frac{1}{V(p, P)} \int_{Q^*(p, P)} q \mid dq \mid. \tag{28.2}$$

We shall express this integral in a different form. The points of $Q^*(p, P)$ are given by the pairs

$$(\bar{q}, \alpha); \qquad \bar{q} \text{ in } S_0, \qquad 0 \leqq \alpha \leqq \rho(P, \bar{q}).$$

Letting W_0 be the $(n - m - 1)$-volume of S_0 and noting that T_P preserves volume, (28.2) may be written

$$g(p, P) = \frac{1}{W_0} \int_{S_0} \left[\frac{n - m}{\rho^{n-m}} \int_0^\rho \alpha^{n-m-1} \{p' + \alpha(q' - p')\} \, d\alpha \right] \mid d\bar{q} \mid, \tag{28.3}$$

where $\rho = \rho(P, \bar{q})$. This expression is easily seen to be analytic.

Let M' be the set of points in $R(M)$ satisfying $g(p, P(p)) = p$. Each $Q^*(p, P(p))$ has exactly one point in M', namely, its center of mass. Taking p_0 etc. again, set

$$\tau(p, P) = T_P^{-1}(p) - T_P^{-1}(g(p, P)), \qquad \tau(p) = \tau(p, P(p)). \tag{28.4}$$

$\tau(p, P)$ is a point (or vector) of $E_{n-m} = P(p_0)$. $\tau(p) = 0$ if and only if p is in M'. We shall show that if $P^*(p)$ is an analytic function approximating $P(p)$ closely enough in $R(U')$ through the first order, and

$$\tau^*(p) = \tau(p, P^*(p)), \tag{28.5}$$

then the vanishing of $\tau^*(p)$ determines an analytic manifold through p_0. To this end, let $\tau_1(p), \cdots, \tau_{n-m}(p)$ and $\tau_1^*(p), \cdots, \tau_{n-m}^*(p)$ be the components of $\tau(p)$ and $\tau^*(p)$ in the directions of fixed mutually orthogonal vectors in $P(p_0)$; then $\tau(p) = 0$ if and only if the $\tau_i(p) = 0$, and similarly for $\tau^*(p)$. The $\tau_i(p)$ vanish at a unique point of each $Q^*(p_1, P(p_1))$, and the $\nabla\tau_i(p)$ are independent as functions in $P(p_1)$; hence the same is true of the $\tau_i^*(p)$ and the $\nabla\tau_i^*(p)$, if the approximation of $P^*(p)$ is close enough. Therefore $\tau^*(p) = 0$ defines an analytic manifold M_1^* in $R(U')$, which cuts each $Q^*(p, P(p))$ in a unique point p_1', and such that the tangent plane to M_1^* at p_1' is independent of $P(p_1)$.

If $P^*(p)$ approximates to $P(p)$ closely enough in $R(M)$, then the above will hold near each point of M, and the vanishing of $\tau^*(p)$ will determine an analytic manifold M^* cutting each $Q^*(p, P(p))$ as noted. As is seen from (28.4) and (28.5), the points of M^* satisfy (28.1). The map $p' = H(p)$ of M^* into M is (1-1) and of class C^r. As the tangent plane to M^* at p is independent of $P(p)$, the inverse is also of class C^r (see Lemma 21); hence the map is a C^r-homeomorphism and the proof is complete.

[37] Compare footnote[17].

VI. PROOF OF THEOREM 2

We shall first prove the existence of analytic linear functionals as defined in §7; we will then be able to prove Theorem 2 and the properties in §9. Finally, we shall prove the unproved statement in Theorem 6.

29. Real-valued analytic linear functionals. We shall generalize AE Lemma 7 as follows:

LEMMA 26. *Let R be an open set in E, let $\Delta(p, q)$ and $\zeta(p)$ be as in §7 in R, and let r and s be finite, $s \leqq r$. Then there is an analytic linear $(R, E_1, r, \Delta, \zeta)$-functional $\mathfrak{L}$; moreover, $\mathfrak{L}$ is defined for any polynomial P of degree $\leqq s$, and $\mathfrak{L}P = P$.*

Let $R_1, R_2, \cdots$ be bounded open subsets of R such that $\bar{R}_i$ is in R_{i+1} and $R_1 + R_2 + \cdots = R$, and let ϵ_i be the minimum of $\zeta(x)$ for x in $\bar{R}_{i+1}$. Let $\Delta_i'(\rho)$ be the maximum of $\Delta(x, x')$ for points x and x' of $\bar{R}_i$ whose distance apart is ρ; these functions are easily seen to satisfy the requirements in AE Lemma 7. Let a be a fixed point of R. Given any function f of class C^r in R, set

$$\mathfrak{L}^*f(x) = \sum_{\sigma_k \leqq s} \frac{D_k f(a)}{k\,!}(x - a)^k; \tag{29.1}$$

This is the polynomial of degree $\leqq s$ approximating to f most closely at a. (See AE for the notation.) $\mathfrak{L}^*$ is a linear functional, and for any polynomial P of degree $\leqq s$, $\mathfrak{L}^*P = P$. As seen in AE, footnote on p. 78, for each i there is a number K_i such that if f satisfies (7.1) and hence

$$| D_k f(x') - D_k f(x) | \leqq \Delta_i'(\| x' - x \|) \text{ in } \bar{R}_i \qquad (\sigma_k \leqq s), \tag{29.2}$$

then

$$| D_k f(a) | < K_i \qquad (0 < \sigma_k \leqq s). \tag{29.3}$$

Let $\Delta_i''(\rho)$ be the maximum in $\bar{R}_i$ of

$$| D_k P(x') - D_k P(x) | \quad \text{for} \quad \| x' - x \| = \rho, \qquad \sigma_k \leqq r,$$

for polynomials $P(x)$ of degree $\leqq s$ whose derivatives at a are $\leqq K_i$, and set $\Delta_i(\rho) = \Delta_i'(\rho) + \Delta_i''(\rho)$. Now if f is any function of class C^r in R satisfying (29.2), then $f - \mathfrak{L}^*f$ satisfies the same equation with Δ_i' replaced by Δ_i.

Let $\mathfrak{L}'$ be the linear functional given by AE Lemma 7 with $M = 0$, and set

$$\mathfrak{L}f = \mathfrak{L}'(f - \mathfrak{L}^*f) + \mathfrak{L}^*f. \tag{29.4}$$

$\mathfrak{L}$ is defined for all $f = f' + P$, where f' satisfies (7.1) and hence (29.2), and P is a polynomial of degree $\leqq s$; for

$$f - \mathfrak{L}^*f = f' - \mathfrak{L}^*f',$$

and this function satisfies (29.2) with Δ_i, and is 0 at a. As both $\mathfrak{L}^*$ and $\mathfrak{L}'$ are linear, $\mathfrak{L}$ is linear. As $\mathfrak{L}'$ is analytic and $\mathfrak{L}^*f$ is a polynomial, $\mathfrak{L}$ is analytic.

Obviously $\mathfrak{L}P = P$ for polynomials P of degree $\leqq s$. Finally, $\mathfrak{L}f$ approximates (f, R, r, ζ), and the proof is complete.

30. Analytic linear functionals. We replace R and E_1 in the last lemma by M and E_ν, as follows.

LEMMA 27. *Let M be an analytic m-manifold in E_μ, let $\Delta(p, q)$ and $\zeta(p)$ be as in §7, and let r be finite. Then there is an analytic linear $(M, E_\nu, r, \Delta, \zeta)$-functional.*[38]

It is sufficient to prove this for real-valued functions; the general case then follows on applying it to each coördinate separately. Given any f in M, define f' in $R(M)$ by $f'(p) = f(H(p))$. We shall show that the functional of the last lemma, which we now call $\mathfrak{L}'$, may be applied to f'; then $\mathfrak{L}f$ is $\mathfrak{L}'f'$ considered in M alone.

Suppose f is of class C^r in M; then if $f'(p') = f(H(p'))$, f' is of class C^r in $R(M)$ (see the proof of Lemma 4); we let $P(p)$ be the normal plane to M at p. Differentiating $f'(p') = f'(H(p'))$ shows that $D_kf'(p')$ is a polynomial of degree $\leqq \sigma_k$ in the derivatives of order $\leqq \sigma_k$ of f' at $p = H(p')$ and of $H(p')$ at p'. Say $p = \theta_i(x)$. Then $D_kf'(p)$ is determined by the $D_sf_i(x) = D_sf(\theta_i(x))$ and the $D_t\theta_i(x)$. (The latter determine $P(p)$.) Hence

$$D_kf'(p') = \Phi[D_sf_i(x), D_t\theta_i(x), D_uH(p')] \qquad (\sigma_s, \sigma_t, \sigma_u \leqq \sigma_k), \tag{30.1}$$

for $\sigma_k \leqq r$. (k, s, t and u have respectively ν, m, m and ν components.) As the θ_i are admissible, there are but a finite number of such expressions for $D_kf'(p')$. Let a be a fixed point of M. Given any f in M or $R(M)$, set $\bar{f}(p) = f(p) - f(a)$. For any compact subset A of M there is a number K such that if f satisfies (7.1) in M, then

$$| D_k\bar{f}(p) | < K \text{ in } A \qquad (\sigma_k \leqq r), \tag{30.2}$$

(see AE, footnote on p. 78). Hence, by (30.1), for any two points p' and q' of $R(M)$ there is a number Δ such that for any such f, $(\bar{f})' = \bar{f}'$ and hence f' satisfies

$$| D_kf'(q') - D_kf'(p') | \leqq \Delta \qquad (\sigma_k \leqq r). \tag{30.3}$$

Let $\Delta^*(p', q')$ be the minimum of such numbers Δ. There are several (but a finite number of) choices for $D_kf(p)$ in (7.1); we take $\Delta^*(p', q')$ large enough for all these.

We show now that if $p'_h \to p'_0$ and $q'_h \to p'_0$, then $\Delta^*(p'_h, q'_h) \to 0$. Suppose not. Then there are a k $(\sigma_k \leqq r)$, sequences $\{p'_h\}$ and $\{q'_h\}$ approaching p'_0, and functions f_h in M satisfying (7.1), such that

$$| D_kf'_h(q'_h) - D_kf'_h(p'_h) | > \alpha > 0. \tag{30.4}$$

We may suppose that $p_0 = H(p'_0)$, $p_h = H(p'_h)$, and $q_h = H(q'_h)$ are in some U_i; (30.1) then applies. Replace f_h, f'_h by $\bar{f}_h$, $\bar{f}'_h$ as before. Then the $D_s\bar{f}_{hi}(x_h)$ are

[38] If $M = R$, we may have $\mathfrak{L}P = P$ as in the last lemma; P is a polynomial with values in E_ν if each of its coördinates is.

bounded ($x_h = \theta_i^{-1}(p_h)$), and we may suppose $D_s f_{hi}(x_h) \to D_{is}$. As $y_h = \theta_i^{-1}(q_h) \to \lim x_h$, (7.1) shows that $D_s f_{hi}(y_h) \to D_{is}$ also. Therefore all the variables in (30.1) approach the same limit when p is replaced by p_h as when it is replaced by q_h. As Φ is continuous, $D_k f'_h(p_h)$ and $D_k f'_h(q_h)$ approach the same limit; but this contradicts (30.4). It is now easy to construct a continuous function $\Delta'(p', q')$ for p', q' in $R(M)$ of the required nature, such that $\Delta'(p', q') \geqq \Delta^*(p', q')$.[39] Now if f is any function of class C^r in M satisfying (7.1), and $f'(p) = f'(H(p))$ in $R(M)$, then f' satisfies

$$(30.5) \qquad | D_k f'(q) - D_k f'(p) | \leqq \Delta'(p, q) \qquad (\sigma_k \leqq r).$$

Applying Lemma 26 to f' with $s = 0$ gives an analytic function $\mathfrak{L}'f'$ approximating $(f', R(M), r, \zeta')$; then $\mathfrak{L}f = \mathfrak{L}'f'$ in M approximates (f, M, r, ζ), if ζ' is sufficiently small.

31. **Proof of Theorem 2 with** (b) **and** (c) **omitted.** The proof of Theorem 2 with just (a) and (e') is given by Lemma 12. We shall prove it with (a), (d), (e) and (f); the proof will be complete when we have proved (A) and (B) of §9.

We first apply Lemma 12 to find a function F' of class C^r which approximates (f, M, r, η) and has the (f, r, η)-properties $\Omega_1, \Omega_2, \cdots$. If $n \geqq 2m$, we include in these properties those of §21, to make F' regular. This is permissible, as the finiteness condition of (e) of the theorem is satisfied for these properties. Only a slight change in the proof of Lemma 18 is necessary because of E_n being replaced by N. Let W_i^* and η_i^* be the neighborhoods and functions of §7(b) and (c) corresponding to Ω_i and F'. Because of the finiteness condition, there is a positive continuous function ζ in M such that if F approximates (F', M, r, ζ), then it approximates (f, M, r, η), and for each i, it approximates $(F', \overline{W}_i^*, r, \eta_i^*)$; F then has the properties $\Omega_1, \Omega_2, \cdots$. It remains to show that the analytic function F may be chosen so as to approximate (F', M, r, ζ) and have the properties $\Omega'_1, \Omega'_2, \cdots$.[40]

Replace the $\Delta(p, q)$ of the theorem if necessary by a larger Δ so that (7.1) is satisfied with f' and Δ replaced by F' and $\frac{1}{2}\Delta$. Let $\mathfrak{L}$ be the analytic linear $(M, E_\nu, r, \Delta, \frac{1}{2}\zeta)$-functional given by Lemma 27; we suppose ζ is so small that if F'' approximates (F', M, r, ζ), then $F''(M)$ is in $R(N)$. For some ζ', if F'' satisfies (7.1) and approximates (F', M, r, ζ'), then $\mathfrak{L}F''$ approximates (F', M, r, ζ). We must now choose F'' so that it approximates (F', M, r, ζ') and satisfies (7.1), and so that $F = \mathfrak{L}F''$ has the properties $\Omega'_1, \Omega'_2, \cdots$.

[39] Let $\rho(p)$ be the smaller of 1 and half the distance from p to $E_\nu - R(M)$, and let $\rho(p, q)$ be the smaller of $\rho(p)$, $\rho(q)$, $\| q - p \|$. Take p and q in $R(M)$, let $\delta_\alpha(p, q)$ be the upper bound of $\Delta^*(p', q')$ for $\| p' - p \| \leqq \alpha$, $\| q' - q \| \leqq \alpha$, and set

$$\Delta'(p, q) = \| q - p \| + \frac{1}{\rho(p, q)} \int_0^{\rho(p, q)} \delta_\alpha(p, q)\, d\alpha.$$

[40] *If $n \geqq 2m + 1$ and f is proper*, we may find an F which is analytic, regular, (1-1) and proper, and has the properties $\Omega_1, \Omega_2, \cdots$, by including in these properties those of §§20 and 21, and then applying Lemma 22 with its $\eta(p)$ sufficiently small. (See the end of §6.)

Let W_i, W'_i, G^i_j be the open sets and functions corresponding to Ω'_i ($i = 1, 2, \cdots$). We shall choose sets of numbers α^i so that if

$$f_0 = F', \qquad f_i = f_{i-1} + \sum_j \alpha^i_j G^i_j, \qquad F'' = \lim f_i,$$

then F'' is the required function. As the W'_i are bounded, if f' is a function satisfying an inequality of the nature of (7.1), and α is small enough, then $f' + \sum_{i,j} \alpha^i_j G^i_j$ approximates to f' as closely as we please and satisfies (7.1) with a new Δ as near the old as we wish. Hence we may choose numbers $\bar{\alpha}^1, \bar{\alpha}^2, \cdots$ such that if $\alpha^1, \alpha^2, \cdots$ is any sequence with $|\alpha^i_j| \leqq \bar{\alpha}^i$, then F'', using these α^i, satisfies (7.1) and approximates (F', M, r, ζ'). The proof now runs exactly like that of Lemma 12, except that $H\mathfrak{L}f_j$ ($j \geqq i$) and $H\mathfrak{L}F''$ will have the property Ω'_i. The existence of α^i at each step such that $H\mathfrak{L}f_i$ has the property Ω_i is given by (d′) of §7.

32. **Certain maps of M into N like translations near a point.** Let f be a C^r-map of M into $N (r \geqq 0)$, take p_0 in M, and set $q_0 = f(p_0)$. Given any $\delta > 0$, we shall find neighborhoods W', W of p_0 in M such that $\overline{W}'$ is in W and $\overline{W}$ is within δ of p_0 (measuring in E_μ), and we shall find C^r-maps $G_1, \cdots, G_n$ of M into E_ν such that $G_i(p) = O$ for p in $M - W$, and such that

(32.1) $$f_\alpha(p) = H\mathfrak{L}[f(p) + \sum \alpha_i G_i(p)] = H[\mathfrak{L}f(p) + \sum \alpha_i \mathfrak{L}G_i(p)]$$

for $|\alpha_i| \leqq 1$ is an n-parameter family of C^r-maps in $\overline{W}'$. $\mathfrak{L}$ is any $(M, E_\nu, r, \Delta, \zeta)$-functional for Δ large enough and ζ small enough near p_0.

Let θ be a C^r-map of Q_m into a neighborhood W of p_0; we will determine the size of W later. Set $W' = \theta(Q'_m)$. Let $\lambda(p)$ be of class C^r in M, $= 1$ in $\overline{W}'$, and $= 0$ in $M - W$ (see §20). Let P be the tangent plane to N at q_0, let $y_1, \cdots, y_\nu$ be rectangular axes in E_ν such that P is the $(y_1, \cdots, y_n)$-plane, and let v_i be the unit vector in the direction of y_i. Set

(32.2) $$G_i(p) = \lambda(p) v_i \text{ in } M \qquad (i = 1, \cdots, n).$$

In using Lemma 23, let $P(q)$ be the normal plane to q in N. Then obviously

(32.3) $$\frac{\partial H(q_0)}{\partial y_j} = v_j \ (j = 1, \cdots, n), \text{ and } = 0 \qquad (j = n+1, \cdots, \nu).$$

Putting (32.2) in (32.1) and differentiating gives therefore

(32.4) $$\frac{\partial f_\alpha(p_0)}{\partial \alpha_i} = \sum_{j=1}^{n} [\mathfrak{L}G_i(p_0)]_j v_j = \text{Proj } \mathfrak{L}G_i(p_0),$$

where Proj v is the projection of a vector v in E_ν into P. If we leave out $\mathfrak{L}$ in (32.4), we may choose W so that the resulting vectors $\partial f_\alpha / \partial \alpha_i$ are independent in $\overline{W}'$. We then choose Δ and ζ so that $\mathfrak{L}$ is defined for the terms in (32.1) with $|\alpha_i| \leqq 1$, and so that the vectors $\partial f_\alpha / \partial \alpha_i$ are independent in $\overline{W}'$; then $f_\alpha(p)$ *is an n-parameter family in $\overline{W}'$.*

33. **Certain maps of M into N like rotations plus translations near a point.** Let f be a C^r-map ($r \geqq 1$) of M into N. Take p_0, q_0, etc. as before; we shall choose W, W', and C^r-maps G_{ij}. Let $\bar{E} = E_{(m+1)n}$ be a Euclidean space with coördinates

$$z_{ij} \qquad (i = 1, \cdots, m+1; j = 1, \cdots, n).$$

The subscripts i, j will, in this section, always range over the values shown, unless otherwise stated. Let $\bar{v}_{ij}$ be the unit vector in the direction of z_{ij}. Let θ map Q_m into W. We may consider $(x_1, \cdots, x_m)$ (in Q_m) as coördinates in W, and write

$$x \text{ for } p = \theta(x), \qquad \frac{\partial g(p)}{\partial x_i} \text{ for } \left.\frac{\partial g(\theta(x))}{\partial x_i}\right|_{x=\theta^{-1}(p)} \text{ in } W.$$

Corresponding to any C^r-map g of W into E_ν, define the C^{r-1}-map $\bar{g}$ of W into $\bar{E}$ by

$$\bar{g}(p) = \sum_{j;i \leqq m} \frac{\partial g_j(p)}{\partial x_i} \bar{v}_{ij} + \sum_j g_j(p)\bar{v}_{m+1,j}\,; \tag{33.1}$$

g_j is the jth component of g in E_ν. (Note that j runs to n only.) Define

$$G'_{ij}(p) = \lambda(p)x_i v_j \ (i \leqq m), \qquad G'_{m+1,j}(p) = \lambda(p)v_j\,. \tag{33.2}$$

For these G, (33.1) gives in $\overline{W}'$

$$\bar{G}'_{ij}(p) = \bar{v}_{ij} + x_i\bar{v}_{m+1,j} \ (i \leqq m), \qquad \bar{G}'_{m+1,j}(p) = \bar{v}_{m+1,j}\,. \tag{33.3}$$

These vectors are obviously independent for p in $\overline{W}'$.

The family of maps $f_\beta(p)$ will be defined by

$$f_\beta(p) = H\mathfrak{L}[f(p) + \sum_{i,j} \beta_{ij}G'_{ij}(p)]. \tag{33.4}$$

As before, we find

$$\frac{\partial}{\partial \beta_{ij}} f_\beta(p_0) = \text{Proj } \mathfrak{L}G'_{ij}(p_0).$$

As Proj $G'_{ij}(p) = G'_{ij}(p)$, the vectors $\overline{\text{Proj } \mathfrak{L}G'_{ij}(p_0)}$ are linearly independent for small enough ζ. It is easily seen that the operations of passing from g to $\bar{g}$ and of differentiating are permutable; hence, as before, we may take W and W', Δ, and ζ so that the vectors $\partial \bar{f}_\beta(p)/\partial\beta_{ij}$ are linearly independent in $\overline{W}'$. Hence $\bar{f}_\beta$ *is an* $(m+1)n$*-parameter family of* C^{r-1}*-maps of* $\overline{W}'$ *into* $\bar{E}$.

34. **Proof of** (B) **and** (C), §9. In the hypothesis of (B), $n \geqq 2m+1$. As regularity is taken care of by (e) of the theorem, we may first replace the given map by a regular C^r-map; let the new map be f. As f is locally (1-1), we may find a positive continuous function $\delta(p)$ in M such that $f(p)$ is (1-1) for p within $\delta(p_0)$ of p_0, for any p_0. For each p_0 in M there is a neighborhood W as in §32; moreover, these may be taken arbitrarily small. Hence we may choose such

neighborhoods $W_1, W_2, \cdots$ such that $W_1' + W_2' + \cdots$ covers M, any compact subset of M has points in common with but a finite number of the $\bar{W}_s$, and if p is in $\bar{W}_s' + \bar{W}_t'$, $\bar{W}_s \cdot \bar{W}_t \neq 0$, then $\bar{W}_s + \bar{W}_t$ lies within $\delta(p)$ of p. Next define the maps $G_1^s, \cdots, G_n^s$ $(s = 1, 2, \cdots)$ as in §32. We may suppose that the η of the theorem is so small that any f' approximating $(f, M, 1, \eta)$ is (1-1) in all $\bar{W}_s + \bar{W}_t$ with $\bar{W}_s \cdot \bar{W}_t \neq 0$.

Arrange the pairs of numbers (s, t) for which $\bar{W}_s \cdot \bar{W}_t = 0$ in a sequence. For any k, let (s, t) be the kth member of the sequence, and let Ω_k be the property of maps f' which holds if $f'(\bar{W}_s')$ does not intersect $f'(\bar{W}_t')$. The property of f' of being (1-1) is the sum of these properties. We shall show that Ω_k is an $[f, r, \eta, \Delta, \zeta]$-property. (a), (b) and (c) of §7 hold; we shall prove (d′), with W, W', G_i replaced by $W_s + W_t$, $W_s' + W_t'$, G_i^s. Given f' and $\mathfrak{K}$, set

$$(34.1) \qquad f_{\alpha^s}'(p) = H\mathfrak{K}\left[f'(p) + \sum_{i=1}^{n} \alpha_i^s G_i^s(p)\right].$$

Applying §32, we now see that f_{α^s}' is an n-parameter family of C^r-maps $(r \geqq 1)$ in $\bar{W}_s'$. As $f'(\bar{W}_t')$ is of zero $(m + 1)$-extent in N, $\bar{W}_s'$ is of finite m-extent in M, $(m + 1) + m \leqq n$, and $f_{\alpha^s}' = f'$ in W_t, there is an arbitrarily small α^s such that $f_{\alpha^s}(\bar{W}_s') \cdot f_{\alpha^s}(\bar{W}_t') = 0$ (Lemma 16), and (d′) is proved.

To prove (C), §9 we proceed as above. Let $K = K_1 + K_2 + \cdots$ be the subset of N, each K_s being of zero $(n - m)$-extent. Arrange the pairs (s, t) in a sequence, and let Ω_k hold if $f'(\bar{W}_s')$ does not intersect K_t. The proof runs now exactly as above.

35. **Proof of** (A) **and** (D), §9. As before, we may suppose that the given map is regular. If $r = 0$ or 1, we may at the beginning replace the map f by a C^2-map (Lemma 22). Hence we suppose that $r \geqq 2$. Define $\delta(p)$ and the W_s, W_s' exactly as in §34, and define the G_{ij}' as in §33. Again, let k correspond to (s, t), and let Ω_k be the property of maps f' which holds if f' has at most regular singularities in $\bar{W}_s' + \bar{W}_t'$; complete regularity is the sum of these properties. We must prove (d′) for Ω_k.

Before proceeding, consider (D). Let $V_1, V_2, \cdots$ be admissible neighborhoods in N', and let Ω_k^* hold if $\bar{W}_s'$ intersects $\bar{V}_t$ only in the proscribed manner. If $n' = m$, this is the same as stating that $f'(\bar{W}_s') + \bar{V}_t$ has at most regular singularities. If we show how to transform $\bar{W}_s'$ with reference to $\bar{V}_t$, the same process transforms $\bar{W}_s'$ with reference to $\bar{W}_t'$; hence we need merely prove (d′) for (D).

Given f' and $\mathfrak{K}$, set

$$(35.1) \qquad f_{\beta^s}'(p) = H\mathfrak{K}\left[f'(p) + \sum_{i,j} \beta_{ij}^s G_{ij}'^s(p)\right],$$

To each f' corresponds an $(m + 1)n$-parameter family of maps $\bar{f}_{\beta^s}'$ of $\bar{W}_s'$ into $\bar{E}$, by §33. We shall show that if $\bar{f}_{\beta^s}'(\bar{W}_s')$ avoids a certain set S, then $f_{\beta^s}'(\bar{W}_s')$ intersects $\bar{V}_t$ in the proper manner. S will be the sum of a denumerable number of sets of finite $(m + 1)(n - 1)$-extent. As the number of parameters in $\bar{f}'_{\beta^*}$ is $(m + 1)n$ and the dimension of W_s is m, we may apply Lemma 16 and make $\bar{W}'_s$ avoid any of these. Applying the process in the proof of Lemma 12, we may make $\bar{W}'_s$ avoid S, and thus complete the proof.

We shall leave out the indices s, t in what follows. The vector $\partial'\bar{f}'_\beta(p)/\partial x_i$ in P (see §32) with components $\partial f'_{\beta j}(p)/\partial x_i$ $(j = 1, \cdots, n)$, by definition, (see §33) is the projection of the vector $\partial f'_\beta(p)/\partial x_i$ in E_ν into P. As f'_β is regular (for small β^s_{ij}), the latter vectors $(i = 1, \cdots, m)$ are independent. As the W'_s may be taken arbitrarily small, we may suppose that the $\partial'\bar{f}'_\beta(p)/\partial x_i$ are independent in $\bar{W}'$. As N' may be cut into arbitrarily small pieces, we may suppose that $\bar{V}$ projects in a (1-1) regular manner into P (if $\bar{V}$ intersects $f'_\beta(\bar{W}')$ for any β). As $f'_\beta(\bar{W}')$ and $\bar{V}$ are both in N, and the projection of N into P is regular near q_0, they intersect in an allowable manner in N if and only if their projections do in P. Let $P\bar{V}$, Pq etc. denote the projections of $\bar{V}$, q, etc. into P.

Let p be a point of $\bar{W}'$, and q, a point of $\bar{V}$; we shall consider under what conditions $Pf'_\beta(p) = Pq$, the intersection being of an unallowable character. This is so if the vectors $\partial'\bar{f}'_\beta(p)/\partial x_i$ determine a plane P_m in P which has a plane P_h of dimension $h > k = m + n' - n$ in common with the plane $P_{n'}$ tangent to $P\bar{V}$ at $q' = Pq$. Hence the set S in $\bar{E}$ which $\bar{f}'_\beta(\bar{W}')$ must avoid is the set of points z with the following property. For some q in $\bar{V}$, some $h > k$ and plane P_h in the tangent plane $P_{n'}$ to $P\bar{V}$ at $q' = Pq$, and some plane P_m which contains q' and has exactly P_h in common with $P_{n'}$, the last n coördinates of z determine (in P) the point q' and the first mn coördinates determine the direction of P_m. Let z_0 be that point of S we have just described; we shall consider that part of S_h near z_0, S_h being those points of S with this corresponding h.

A point z of S_h is determined by the set

$$(q, P_h, P_m, v), \quad \text{where} \quad v = (v_1, \cdots, v_m), \qquad v_i = \frac{\partial'\bar{f}'_\beta(p)}{\partial x_i},$$

q, P_h, P_m, v being chosen in the order given. (The last n coördinates of z are then determined.) q runs over a set of dimension n'. Now keep q fixed, and vary P_h. P_h lies in $P_{n'}$ and contains q'; the dimension of such a set of planes is $h(n' - h)$. Next vary P_m. It is determined by naming a plane P_{m-h} through q' in the plane P_{n-h} through q' orthogonal to P_h; hence the set of planes P_m is of dimension $(m - h)(n - m)$. Finally, vary the vectors v_i. Each one may vary freely as long as it remains in P_m; hence the dimension of this set of positions of the set of vectors is m^2. Consequently, if we set $h = k + h'$ and note that $n' - h = n - m - h'$, the above set runs over a part of a Euclidean space of dimension

$$\begin{aligned} d_h &= n' + h(n' - h) + (m - h)(n - m) + m^2 \\ &= n' - hh' + (h + m - h)(n - m) + m^2 \\ &= mn + n' - hh' \leqq mn + n' - h \leqq mn + n - m - 1. \end{aligned}$$

The map of this set into the corresponding part of S_h is of class C^1, and hence the $(m + 1)(n - 1)$-extent of the latter is finite, as required.

36. **Completion of the proof of Theorem 6.** Let $\theta_1, \theta_2, \cdots$ be a completely admissible set of maps in that part M^* of M' for which $0 < t < 1$. We take them so that if $\bar{U}_i \cdot \bar{U}_j \neq 0$, then f is (1-1) in $\bar{U}_i + \bar{U}_j$. In applying Theorem 2 in the proof of Theorem 6, let us introduce (f, r, η)-properties as follows. Arrange the pairs of numbers (i, j) in a sequence. Let the kth member be (i, j). Then Ω_k holds for $f(p \times t) = \phi_t(p)$ if the following is true. If $0 < t < 1$, $p \times t$ is in $\bar{U}'_i$, and $q \times t$ is in $\bar{U}'_j$, then $f(p \times t) \neq f(q \times t)$. We shall show that each property is an (f, r, η)-property (for small enough η). It will follow that *each* $\phi_t(0 < t < 1)$ *may be made* (1-1), *irrespective of* M_0 *and* M_1. The new map of M' into N is of class C^r, by Lemma 10.

We may suppose the U_i are so small that if $\bar{U}_i \cdot \bar{U}_j \neq 0$, then $\bar{U}_i + \bar{U}_j$ is in some V_h in N, and so that vectors $v_1, \cdots, v_{n-m}$ may be chosen in E_n with the following property. Set $f'_h(p \times t) = \chi_h^{-1} f(p \times t)$, and let $U'_j(t)$ be the set of points $p \times t$ of U'_j whose second coördinate is t. If P and P' are the m- and $(n - m)$-planes through the origin in E_n, the first being orthogonal to and the second parallel to $v_1, \cdots, v_{n-m}$, then any $f'_h(\bar{U}_j(t))$ projects into a subset of P so that both the projection and its inverse are of class C^r. The points

$$q = p + \sum_{s=1}^{n-m} \alpha_s v_s, \qquad p \text{ in } f'_h(\bar{U}_j(t)),$$

fill out an open set R_t in E_n, and if solving this (see §3) gives

$$p = H(q, t), \qquad \alpha_s = \Phi_s(q, t),$$

then H and the Φ_i are of class C^r. To any C^r-map g of U_i into Q_n let $\bar{g}$ be the map of U_i into E_{n-m} whose s^{th} component is

$$\bar{g}_s(p \times t) = \Phi_s(g(p \times t), t).$$

Define the family of maps

$$g_\beta(p \times t) = f'_h(p \times t) + \sum_{s=1}^{n-m} \beta_s v_s \text{ in } \bar{U}_i;$$

then the $\bar{g}_\beta$ have the property that

$$\frac{\partial}{\partial \beta_s} \bar{g}_\beta(p \times t) = v_s .$$

Hence the $\bar{g}_\beta$ form an $(n - m)$-parameter family in E_{n-m}. As $\bar{U}_i$ is of dimension $m + 1 < n - m$, there is an arbitrarily small β such that $\bar{g}_\beta(p \times t) \neq 0$ in $\bar{U}_i$ (see Lemmas 14–16). For this β, no $g_\beta(\bar{U}'_i(t))$ intersects any $f'_h(\bar{U}'_j(t))$ (same t), for no $\Phi_s(g_\beta(p \times t), t) = 0$. Consequently, using $\lambda(p)$ etc. as in §20, we prove (d) of §7. The other properties are obvious, and the statement is proved.

HARVARD UNIVERSITY.

ANNALS OF MATHEMATICS
Vol. 37, No. 4, October, 1936

THE IMBEDDING OF MANIFOLDS IN FAMILIES OF ANALYTIC MANIFOLDS[1]

BY HASSLER WHITNEY

(Received February 26, 1935, revised February 10, 1936)

I. THE THEOREM

1. **Introduction.** Let $f_1, \cdots, f_{n-m}$ be differentiable functions defined in an open subset of Euclidean n-space E_n. At each point p at which all the f_i vanish, let the gradients $\nabla f_1, \cdots, \nabla f_{n-m}$ be independent. Then the vanishing of the f_i determines a differentiable manifold M of dimension m. Any such manifold we shall say is in "regular position" in E_n. Only certain manifolds are in regular position; see below. The purpose of the paper is to show that any m-manifold M in regular position in E_n may be imbedded in an $(n - m)$-parameter family of homeomorphic analytic manifolds; these fill out a neighborhood of M in E_n.

We may extend the above definition as follows: M is in regular position if, roughly, there exist $n - m$ continuous vector functions in M which, at each point p of M, are independent and independent of vectors determined by pairs of points of M near p. If M is differentiable, the two definitions agree; the ∇f_i are the required vectors. The theorem holds also for this more general class of manifolds.

Some manifolds which are necessarily in regular position if they are differentiable in E_n are:[2]

(a) Any curve; it may have corners.
(b) Any orientable surface.
(c) Any orientable $(n - 1)$-manifold.
(d) Any closed orientable $(n - 2)$-manifold.
(e) Any manifold which is an m-cell.

No non-orientable manifold can be in regular position, and no such manifold can be imbedded in an $(n - m)$-parameter family of manifolds. For examples of orientable manifolds not in regular position, see Sphere-spaces, 8(e) and (f).

Some definitions and lemmas from a paper by the author, Differentiable Manifolds, these Annals, vol. 37 (1936), pp. 645–680, will be used. We refer to this paper as DM. AE will mean the paper referred to in footnote 5 of DM.

---

[1] Presented in part to the Am. Math. Soc., Sept. 4, 1934. A statement of the theorem is given in the Proc. Nat. Ac. of Sci., vol. 21 (1935), p. 463.

[2] See H. Whitney, *Sphere-spaces*, Proc. Nat. Ac. of Sci., vol. 21 (1935), p. 467, 8 (d). (e) has been proved by W. Wazewski, Compositio Math., vol. 2 (1935), pp. 63–68. (d) has been proved also by H. Seifert, Math. Zeitschr., vol. 41 (1936), pp. 1–17. He proves here the existence of an analytic (and hence algebraic) approximation for the case of a closed manifold of class C^2.

2. The independence of a set of vectors. This section will be used extensively in part III. With any point p is associated a vector, drawn from the origin to p; with this in mind, we shall use the terms point and vector interchangeably (as long as we do not translate axes). Let $\| v \|$ denote the length of v; then $\| q - p \|$ is the distance from p to q. If v_i is a vector with components $v_{i1}, \cdots, v_{in}$, then from $v_1, \cdots, v_s$ we form the matrix

$$\mathbf{M} = \left\| \begin{matrix} v_{11} & \cdots & v_{1n} \\ \cdots & \cdots & \cdots \\ v_{s1} & \cdots & v_{sn} \end{matrix} \right\|. \tag{2.1}$$

The square root of the sum of the squares of all s-rowed determinants $D_{i_1 \cdots i_s}$ from $\mathbf{M}$ we call the *independence* of $v_1, \cdots, v_s$; this is also the square root of the determinant of the scalar products of the v's, by the general Lagrange identity:[3]

$$I(v_1, \cdots, v_s) = [\sum D^2_{i_1 \cdots i_s}]^{\frac{1}{2}} = \begin{vmatrix} v_1 \cdot v_1 & \cdots & v_1 \cdot v_s \\ \cdots & \cdots & \cdots \\ v_s \cdot v_1 & \cdots & v_s \cdot v_s \end{vmatrix}^{\frac{1}{2}}. \tag{2.2}$$

$I(v_1, \cdots, v_s)$ is a continuous and symmetric function of $v_1, \cdots, v_s$; it is independent of the (rectangular) coördinate system employed, as each $v_i \cdot v_j$ is. Geometrically, it is the s-dimensional volume of the parallelopiped determined by $v_1, \cdots, v_s$. It vanishes if and only if $v_1, \cdots, v_s$ are dependent.

Some properties of I following at once from (2.2) are:

$$I(v) = \| v \|, \tag{2.3}$$

$$I(v_1 + cv_2, v_2, \cdots) = I(v_1, v_2, \cdots), \tag{2.4}$$

$$I(cv_1, v_2, \cdots) = | c | \, I(v_1, v_2, \cdots). \tag{2.5}$$

Choose the $x_2, \cdots, x_s$-axes so that $v_2, \cdots, v_s$ lie in the $x_2, \cdots, x_s$-plane, and choose the x_1-axis so that v_1 lies in the $x_1, \cdots, x_s$-plane. Writing out the resulting matrix (2.1) and using the first expression for I in (2.2) gives, as

$$| v_{11} | \leqq \| v_1 \|,$$

$$I(v_1, v_2, \cdots, v_s) \leqq \| v_1 \| \, I(v_2, \cdots, v_s). \tag{2.6}$$

Repeated use of this inequality gives

$$I(v_1, \cdots, v_s) \leqq \| v_1 \| \cdots \| v_s \|, \tag{2.7}$$

an extension of Hadamard's inequality on determinants. A similar choice of axes with the addition theorem for determinants gives

$$I(v_1' + v_1'', v_2, \cdots) \leqq I(v_1', v_2, \cdots) + I(v_1'', v_2, \cdots). \tag{2.8}$$

Suppose

$$w = \alpha_1 v_1 + \cdots + \alpha_s v_s. \tag{2.9}$$

[3] See O. Schreier and E. Sperner, *Analytische Geometrie* I, Leipzig 1931, pp. 113–114.

By (2.4) and (2.5),

$$I(w, v_2, \cdots, v_s) = I(\alpha_1 v_1, v_2, \cdots, v_s) = |\alpha_1| \, I(v_1, \cdots, v_s).$$

Hence, using (2.6), *if* $I(v_1, \cdots, v_s) \neq 0$,

$$(2.10) \qquad |\alpha_1| \leqq \| w \| \frac{I(v_2, \cdots, v_s)}{I(v_1, \cdots, v_s)},$$

with a similar inequality for each other $|\alpha_i|$. In (2.8), replace v_1' and $v_1' + v_1''$ by v_1 and w_1; we obtain with (2.6)

$$I(w_1, v_2, \cdots) \leqq I(v_1, v_2, \cdots) + \| w_1 - v_1 \| \, I(v_2, \cdots).$$

A similar inequality with v_1 and w_1 interchanged gives, with this,

$$(2.11) \qquad | I(w_1, v_2, \cdots) - I(v_1, v_2, \cdots) | \leqq \| w_1 - v_1 \| \, I(v_2, \cdots).$$

An s-fold application gives, using (2.7), *if the* v_i *and* w_i *are unit vectors,*

$$(2.12) \qquad | I(w_1, \cdots, w_s) - I(v_1, \cdots, v_s) | \leqq \| w_1 - v_1 \| + \cdots + \| w_s - v_s \|.$$

A vector function $v(p)$ defined in a point set M is of *Class* C^r or *analytic* if each of its components is. It satisfies a *local Lipschitz condition* in M if there exists a positive continuous function $K(p)$ in M such that if p and q are in M, then

$$(2.13) \qquad \| v(q) - v(p) \| \leqq K(p) \| q - p \| \qquad (\| q - p \| < 1/K(p)).$$

For this to hold, it is evidently necessary and sufficient that each component of $v(p)$ satisfy a local Lipschitz condition.

3. **Manifolds in regular position in E_n.** Let M be a point set in Euclidean n-space E_n. We shall say M is an *m-manifold in regular position in E_n* if the following is true:

(1) Each point p of M is in a neighborhood U of p in M which is an m-cell.[4]

(2) There are $n - m$ continuous unit vector functions $v_1(p), \cdots, v_{n-m}(p)$ and a positive continuous function $\rho(p) \leqq 1$ defined in M, such that if p and q are in M and $\| q - p \| < \rho(p)$, then

$$(3.1) \qquad I[v_1(p), \cdots, v_{n-m}(p), q - p] \geqq \rho(p) \| q - p \|.$$

If there is a limit set LM (see DM §2), we take $\rho(p)$ less than the distance from p to LM.

(3.1) says that for a certain $\epsilon > 0$, the vector $q - p$ makes an angle $\geqq \epsilon$ with the $(n - m)$-plane determined by $v_1(p), \cdots, v_{n-m}(p)$. From (3.1) we find, using (2.6),

$$(3.2) \qquad I[v_1(p), \cdots, v_{n-m}(p)] \geqq \rho(p),$$

and thus $v_1(p), \cdots, v_{n-m}(p)$ are linearly independent.

[4] That is, U is homeomorphic with the set of points $\sum_{i=1}^{m} x_i^2 < 1$ in m-space.

If M is of class $C^r(r \geqq 1)$, an equivalent formulation of (2) is evidently: If v is any unit vector parallel to M at p, then

$$I[v_1(p), \cdots, v_{n-m}(p), v] \geqq \rho(p). \tag{3.3}$$

In other words, the tangent plane to M at p and the plane of $v_1(p), \cdots, v_{n-m}(p)$ are independent.

Let M and M^* be manifolds of class $C^r(r \geqq 0)$, and let $\eta(p)$ be a positive continuous function defined in M. We say *M^* approximates to M through the rth order with an error* $< \eta(p)$ if M and M^* are homeomorphic and the following is true:

(A) If p and p^* are corresponding points of M and M^*, then $\| p^* - p \| < \eta(p)$.

(B) If $r > 0$, let T be the m-plane tangent to M at p, and let $x_1', \cdots, x_n'$ be a set of rectangular axes such that T is the $x_1', \cdots, x_m'$-plane. Let

$$x_i' = f_i(x_1', \cdots, x_m'), \qquad x_i' = f_i^*(x_1', \cdots, x_m') \qquad (i = m+1, \cdots, n)$$

be functions (of class C^r) determining a neighborhood of p in M and a neighborhood of the corresponding point p^* of M^*. Then if q and q^* are the projections of p and p^* into T,

$$| D_k f_i^*(q^*) - D_k f_i(q) | < \eta(p) \qquad (\sigma_k \leqq r;\, i = m+1, \cdots, n). \tag{3.4}$$

For the notation, see DM §6, or AE.

THEOREM I. *Let M be an m-manifold of class $C^r(r \geqq 0$ and finite) in regular position in E_n, and let $\eta(p)$ be any positive continuous function defined in M. Then M can be imbedded in an $(n-m)$-parameter family of manifolds*

$$M(c_1, \cdots, c_{n-m}), \qquad | c_i | < 1 \qquad (i = 1, \cdots, n-m) \tag{3.5}$$

with the following properties:

(1) *$M(0, \cdots, 0) = M$.*

(2) *$M(c_1, \cdots, c_{n-m})$ is analytic if $(c_1, \cdots, c_{n-m}) \neq (0, \cdots, 0)$.*

(3) *Each manifold $M(c_1, \cdots, c_{n-m})$ approximates to M through the rth order with an error $< \eta(p)$.*

(4) *If*

$$p^* = Z(p; c_1, \cdots, c_{n-m}) \tag{3.6}$$

is the point of $M(c_1, \cdots, c_{n-m})$ corresponding to the point p of M, then Z is a continuous function of its arguments, and the points p^* fill out an open set R containing M in a one-one way; hence no two of the manifolds intersect. The points p^* for p fixed fill out a portion of an $(n-m)$-plane through p.

(5) *If the sets $(c_1^{(i)}, \cdots, c_{n-m}^{(i)})(i = 1, 2, \cdots)$ approach the set $(c_1, \cdots, c_{n-m})$, then the manifolds $M(c_1^{(i)}, \cdots, c_{n-m}^{(i)})$ approach the manifold $M(c_1, \cdots, c_{n-m})$, together with partial derivatives of order $\leqq r$.*

(6) *If M is of class C^∞ and has a boundary*[4a] *M', then we can make higher and*

[4a] Boundary shall throughout mean limit set.

higher partial derivatives of $M(c_1, \cdots, c_{n-m})$ *approximate to the corresponding derivatives of* M *as we approach* M'.

II. The Proof, M Differentiable

4. The vector functions $w_i(p)$. Set $s = n - m$. The vector functions $v_1(p), \cdots, v_s(p)$ defined in M may not satisfy a local Lipschitz condition; we shall replace them by unit vector functions $w_1(p), \cdots, w_s(p)$ which do, and will in a certain sense be analytic.

Let R_1 be an open set containing M but no boundary points of M. (For instance, R_1 = all points whose distances from M are less than half their distances to the boundary M' of M if M' exists; otherwise, $R_1 = E_n$.) Extend each component $v_{ij}(p)$ of each $v_i(p)$ so it will be continuous in R_1.[5] Set $\zeta(p) = \rho(p)/(4ns)$ and extend it so it will be positive and continuous in R_1. Define the analytic functions $w'_{ij}(p)$ in R_1 by AE Lemma 6 (DM Lemma 9) 1 so that

$$(4.1) \qquad | w'_{ij}(p) - v_{ij}(p) | < \zeta(p) \text{ in } R_1.$$

Now for any p in M,

$$| \, \| w'_i(p) \| - 1 \, | = | \, \| w'_i(p) \| - \| v_i(p) \| \, | \leqq \| w'_i(p) - v_i(p) \|$$
$$\leqq \sum_j | w'_{ij}(p) - v_{ij}(p) | < n\zeta(p) < 1,$$

and hence $\| w'_i(p) \| \neq 0$ in a neighborhood R_2 of M. Define the unit vector function $w_i(p) = w'_i(p)/\| w'_i(p) \|$. Then

$$w_i - v_i = \frac{w'_i}{\| w'_i \|}(1 - \| w'_i \|) + (w'_i - v_i),$$

$$\| w_i(p) - v_i(p) \| \leqq n\zeta(p) + n\zeta(p) = \rho(p)/(2s) \text{ in } M.$$

Hence, using (2.12), if u is a unit vector,

$$(4.2) \qquad | I[w_1(p), \cdots, w_s(p), u] - I[v_1(p), \cdots, v_s(p), u] | \leqq \tfrac{1}{2}\rho(p),$$

which shows that (3.3) holds with $\rho(p)$ replaced by $\frac{1}{2}\rho(p)$. Therefore M is in regular position in terms of $w_1, \cdots, w_s$. As the first partial derivatives of the w_{ij} are bounded in a neighborhood of any point of R_2, the law of the mean shows that each w_{ij} and hence each w_i satisfies a local Lipschitz condition in R_2 and hence in M: There is a continuous function $K(p) \geqq 1$ in M such that

$$(4.3) \qquad \| w_i(q) - w_i(p) \| \leqq K(p) \| q - p \| \qquad (\| q - p \| < 1/K(p)).$$

5. The functions $\phi_i(p)$. Given any point p of M and any set of numbers $\alpha_1, \cdots, \alpha_s (s = n - m)$, define the point p^* by

$$(5.1) \qquad p^* = p + \sum_{i=1}^{s} \alpha_i w_i(p) = G(p, \alpha_1, \cdots, \alpha_s).$$

We wish to solve this equation for p and the α_i in terms of p^*.

[5] Almost any method of extending continuous functions can be used to do this.

LEMMA 1. *There is an open set R_3 containing M, and a vector function H and functions ϕ_i, all of class C^r,*

$$p = H(p^*), \qquad \alpha_i = \phi_i(p^*) \qquad (i = 1, \cdots, s) \tag{5.2}$$

defined in R_3, such that

$$H[p + \sum \alpha_i w_i(p)] = p, \qquad \phi_k[p + \sum \alpha_i w_i(p)] = \alpha_k \tag{5.3}$$

if the point in brackets is in R_3, and conversely, if p^* is in R_3,

$$H(p^*) + \sum \phi_i(p^*) w_i[H(p^*)] = p^*. \tag{5.4}$$

This follows at once from the proof of DM Lemma 21.

LEMMA 2. *Let $T'(p)$ be the s-plane through p in M determined by $w_1(p), \cdots, w_s(p)$, let $x_1', \cdots, x_n'$ be rectangular axes chosen so that $T'(p)$ is the $x_1', \cdots, x_s'$-plane with p at the origin, and with x_1' in the $w_1(p)$-direction, x_2' in the plane of $w_1(p)$ and $w_2(p)$, etc. Then if $\phi_i(x_1, \cdots, x_n) = \phi_i'(x_1', \cdots, x_n'; p)$,*

$$J' = \frac{\partial(\phi_1', \cdots, \phi_s')}{\partial(x_1', \cdots, x_s')} \neq 0. \tag{5.5}$$

If we add $hw_k(p)$ to both sides of (5.1), we obtain

$$p^* + hw_k(p) = p + \sum_{i=1}^{s} (\alpha_i + \delta_{ik} h) w_i(p).$$

This, with (5.3), gives

$$\phi_i[p^* + hw_k(p)] = \phi_i(p^*) + \delta_{ik} h, \qquad \text{if } p = H(p^*). \tag{5.6}$$

From this it follows that

$$\nabla\phi_i \cdot w_k(p) = \delta_{ik}, \tag{5.7}$$

$$\frac{\partial\phi_i'}{\partial x_k'} = 0 \ (k < i) \text{ and } \neq 0 \ (k = i),$$

and hence $J' \neq 0$.

LEMMA 3. *Let $T(p)$ be the tangent plane at p in M, and let $\bar{x}_1, \cdots, \bar{x}_n$ be rectangular axes chosen so that $T(p)$ is the $\bar{x}_1, \cdots, \bar{x}_m$-plane, with p at the origin. Then if $\phi_i(x_1, \cdots, x_n) = \bar{\phi}_i(\bar{x}_1, \cdots, \bar{x}_n; p)$,*

$$\bar{J}\big|_p = \frac{\partial(\bar{\phi}_1, \cdots, \bar{\phi}_s)}{\partial(\bar{x}_{m+1}, \cdots, \bar{x}_n)}\bigg|_p \neq 0. \tag{5.8}$$

By the last lemma (using $x_1', \cdots, x_n'$ in place of $x_1, \cdots, x_n$)

$$I(\nabla\phi_1, \cdots, \nabla\phi_s) \neq 0 \text{ on } M. \tag{5.9}$$

Now express I in terms of the $\bar{x}_1, \cdots, \bar{x}_n$ coördinate system. As $\partial\bar{\phi}_i/\partial\bar{x}_j = 0 (j \leqq m)$ at p, all determinants in the new expression except possibly the determinant in (5.8) vanish; hence this determinant in absolute value equals I, and (5.8) is proved.

6. **Two lemmas on implicit functions.** We shall use here the symbols σ'_k, D'_k if n is replaced by m.

LEMMA 4. *Let the functions*

$$\Phi_i(x_1, \cdots, x_s) \qquad (i = 1, \cdots, s)$$

be defined and of class C^1 in a neighborhood[5a] *V of the origin q_0, let them vanish at q_0, and let the Jacobian be $\neq 0$ there. Then for any $\epsilon > 0$ there is a $\delta > 0$ with the following property: If $\Theta_1, \cdots, \Theta_s$ are defined and of class C^1 in V, and*

$$(6.1) \qquad |\Theta_i - \Phi_i| < \delta, \qquad \left|\frac{\partial\Theta_i}{\partial x_j} - \frac{\partial\Phi_i}{\partial x_j}\right| < \delta \qquad (i, j = 1, \cdots, s)$$

in V, then the Jacobian of the Θ_i is $\neq 0$, and there is a unique point q^* in V such that

$$(6.2) \qquad \| q^* - q_0 \| < \epsilon, \qquad \Theta_i(q^*) = 0 \qquad ((i = 1, \cdots, s).$$

Moreover, δ can be made continuous in the Φ_i, their derivatives, and ϵ; the point q^* is continuous in the Θ_i.

LEMMA 5. *Set $s = n - m$. Let the functions*

$$\Phi_i(x_1, \cdots, x_n) \qquad (i = 1, \cdots, s)$$

be defined and of class C^r $(r \geqq 1)$ in V, the set of all points within a distance ρ of the origin p_0, let them vanish at p_0, and let

$$J_{p_0} = \left.\frac{\partial(\Phi_1, \cdots, \Phi_s)}{\partial(x_{m+1}, \cdots, x_n)}\right|_{p_0} \neq 0.$$

Then for every $\epsilon > 0$ there is a $\delta > 0$ with the following property: If $\Theta_1, \cdots, \Theta_s$ are defined and of class C^r in V, if

$$(6.3) \qquad | D_k\Theta_i(p) - D_k\Phi_i(p) | < \delta \qquad (\sigma_k \leqq r; i = 1, \cdots, s)$$

in V, if $p^* = (x_1^*, \cdots, x_n^*)$ and

$$(6.4) \qquad \| p^* - p_0 \| < \delta, \qquad \Theta_i(p^*) = 0 \qquad (i = 1, \cdots, s),$$

and if

$$\psi_i(x_1, \cdots, x_m), \qquad f_i(x_1, \cdots, x_m) \qquad (i = 1, \cdots, s)$$

are continuous functions defined for $x_1^2 + \cdots + x_m^2 < \delta^2$ such that

$$(6.5) \qquad \psi_i(0, \cdots, 0) = 0, \qquad f_i(x_1^*, \cdots, x_m^*) = x_{m+i}^*,$$

$$(6.6) \qquad \Phi_i(x_1, \cdots, x_m, \psi_1, \cdots, \psi_s) \equiv 0,$$

$$(6.7) \qquad \Theta_i(x_1, \cdots, x_m, f_1, \cdots, f_s) \equiv 0,$$

in $x_1, \cdots, x_m$ for $i = 1, \cdots, s$, then the ψ_i and the f_i are of class C^r and

[5a] Which we take sufficiently small.

$$|D'_k f_i(x_1^*, \cdots, x_m^*) - D'_k \psi_i(0, \cdots, 0)| < \epsilon \qquad (\sigma'_k \leqq r) \tag{6.8}$$

for $i = 1, \cdots, s$. *Moreover,* δ *is continuous in* ρ, ϵ, *and the* Φ_i *and their derivatives.*

These lemmas are forms of the implicit function theorem, with uniformity and continuity conditions. They may be proved by standard methods.

7. **Proof of** Theorem I, $r \geqq 1$. Given any point p_0 of M, we may choose axes and define the functions $\bar{\phi}_i(\bar{x}_1, \cdots, \bar{x}_n; p_0)$ as in Lemma 3. Moreover, we may evidently choose such axes in a continuous manner for each point p of a whole neighborhood U of p_0 in M; the $\bar{\phi}_i$ will then be continuous in $x_1, \cdots, x_n$ and p for $(x_1, \cdots, x_n)$ near $(0, \cdots, 0)$ and p in U. Lemma 3 shows that the conditions of Lemma 5 are satisfied; the number ρ may obviously be taken continuous in U. Hence, covering M by a set of neighborhoods U, we see that there is a positive continuous function $\Omega_1(p) < \eta(p)$ in M such that if (6.3) through (6.7) hold with Φ_i and δ replaced by $\bar{\phi}_i$ and $\Omega_1(p_0)$, then (6.8) holds with ϵ replaced by $\eta(p_0)$.

Next, define $T'(p_0)$ and axes $x'_1, \cdots, x'_n$ in a continuous manner for any p in a neighborhood U of p_0 as in Lemma 2. For points of $T'(p_0)$, the functions ϕ'_i are then

$$\phi'_i(x'_1, \cdots, x'_s, 0, \cdots, 0; p_0) = \Phi_i(x'_1, \cdots, x'_s; p_0), \tag{7.1}$$

are defined in a neighborhood of the origin, and are continuous in $x'_1, \cdots, x'_s, p_0$. As $J' \neq 0$, the conditions of Lemma 4 are satisfied. Hence there is a positive continuous function $\Omega_2(p) < \Omega_1(p)$ in M such that if (6.1) holds for axes and functions corresponding to p_0 in M and with δ replaced by $\Omega_2(p_0)$, then (6.2) holds with ϵ replaced by $\Omega_1(p_0)$.

Next, set $\Omega_2(p) = \Omega_2(H(p))$ in R_3. Set

$$\Omega(p) = \Omega_2(p)/(2\, N_r), \tag{7.2}$$

where N_r is a number such that if F and F' are the expressions for any function of class C^r in terms of any two sets of rectangular axes, and

$$|D_k F| \leqq a \ (\sigma_k \leqq r), \qquad \text{then } |D_k F'| \leqq N_r a \ (\sigma_k \leqq r). \tag{7.3}$$

Let $\Omega'(p)$ be the minimum of $\Omega(p)$ and the distance from p to M. Then $\Omega'(p)$ is continuous in R_3, zero in M, and positive in $R_3 - M$.

By DM Lemma 9 there are functions $\phi''_i(p)$ analytic in $R_3 - M$ and such that

$$|D_k\phi''_i(p) - D_k\phi_i(p)| \leqq \Omega'(p) \text{ in } R_3 \qquad (\sigma_k \leqq r;\ i = 1, \cdots, s), \tag{7.4}$$

or at least in $R_3 - M$. If we set $\phi''_i(p) = \phi_i(p)$ in M, then $\phi''_i(p)$ is of class C^r in R_3, and (7.4) holds in M also, by DM Lemma 10. By DM Lemma 25 there is a positive analytic function $\omega(p)$ in R_3 such that

$$|D_k\omega(p)| < \Omega(p) \text{ in } R_3 \qquad (\sigma_k \leqq r). \tag{7.5}$$

Take any set of numbers

(7.6) $$c = (c_1, \cdots, c_s), \qquad |c_i| < 1 \qquad (i = 1, \cdots, s).$$

Corresponding to this set we define in R_3

(7.7) $$\theta_i(p; c) = \phi_i''(p) - c_i\omega(p) \qquad (i = 1, \cdots, s);$$

these functions are analytic in $R_3 - M$. (7.4), (7.5) and (7.6) show that

(7.8) $$| D_k\theta_i(p; c) - D_k\phi_i(p) | \leqq 2\,\Omega(p) \qquad (\sigma_k \leqq r; i = 1, \cdots, s)$$

in R_3. The manifold $M(c)$ is defined as the set of points p in R_3 for which

(7.9) $$\theta_i(p; c) = 0 \qquad (i = 1, \cdots, s).$$

Given p in M and $T'(p)$, choose axes $x_1', \cdots, x_n'$ as before, and consider the functions $\Theta_i(x_1', \cdots, x_s')$, the values of the $\theta_i(p; c)$ on $T'(p)$. (7.2), (7.3) and (7.8) show that (6.1) is satisfied with δ replaced by $\Omega_2(p)$; hence there is a unique point

(7.10) $$p^* = Z(p; c)$$

in $T'(p)$ within $\Omega_1(p)$ of p at which (7.8) is satisfied. Lemma 7 shows also that p^* moves continuously as we move p in M. If we let p^* in $M(c)$ correspond to p in M, Lemma 1 shows that $M(c)$ and M are homeomorphic.

Evidently the function Z is continuous in both p and c. No two of the manifolds $M(c)$, $M(c')(c \neq c')$ have a common point, as follows at once from their definition. Given any point (7.10), the corresponding c is given by:

$$c_i = \phi_i''(p^*)/\omega(p^*).$$

As ϕ_i'' and ω are continuous and R_3 is open, to each p' near p^* there corresponds a c', $|c_i'| < 1$. Hence the points (7.10) fill out a neighborhood R of M.

Take p_0 in M, a small neighborhood U of p_0 in M, and a fixed c. For each p in U there is a $p^* = Z(p; c)$; using the axes of Lemma 3, these points p^* are given by functions f_i as in (6.5). As $\Theta_i(p^*) = \theta_i(p^*; c) = 0$ at these points, (6.7) is satisfied. Setting $c = 0$ shows that (6.6) is also. By (7.8) and (7.3), and as $\| Z(p; c) - p \| < \Omega_1(p)$, we may apply Lemma 5; (6.8) now completes the proof of (3) in Theorem I. Finally, the $M(c)$ are analytic for $c \neq 0$, as the $\theta_i(p; c)$ are analytic in $R_3 - M$. (5) is obviously true. If $r = \infty$ and we wish to have (6) satisfied, we use AE Lemma 6 with $m = \infty$. Theorem I for $r \geqq 1$ is now completely proved.

III. The Proof, M not Differentiable

If M is not differentiable, the functions ϕ_i will be continuous, but not differentiable. However, they will have directional derivatives in the directions of the w_i; we may approximate to the ϕ_i by analytic functions having similar directional derivatives. To show this, we first need some preliminary lemmas.

8. **Averages of a function.** Let $S(p, \lambda)$ denote the set of points q within λ of p, let $S'(p, \lambda)$ denote its boundary, let $V(\lambda)$ be the volume of $S(p, \lambda)$, and let

$V'(\lambda)$ be the $(n-1)$-volume of $S'(p,\lambda)$. If $\phi(p)$ is a function continuous in the open set R, we may define the average

$$A(p,\lambda) = \frac{1}{V(\lambda)} \int_{S(p,\lambda)} \phi(q)\, dq. \tag{8.1}$$

Similarly we define $B(p,\lambda)$, the average of ϕ over $S'(p,\lambda)$. A simple computation (noting that $V'(\lambda) = \partial V(\lambda)/\partial\lambda$) shows that $A(p,\lambda)$ has a continuous derivative with respect to λ, given by

$$\frac{\partial}{\partial\lambda} A(p,\lambda) = \frac{n}{\lambda}[B(p,\lambda) - A(p,\lambda)]. \tag{8.2}$$

Also, $A(p,\lambda)$, as a function of p, has continuous first partial derivatives. (In fact, the directional derivative of $A(p,\lambda)$ in the direction of the vector v is given by taking the formula for $B(p,\lambda)$, and putting in the integrand the factor $\cos\theta$, where θ is the angle between $q-p$ and v.)

9. **Approximation to a certain continuous function by an analytic function.** We prove here

LEMMA 6. *Let $\phi(p)$ be a continuous function defined in the open set R, let $\psi_1(p), \cdots, \psi_s(p)$ be continuous in R, and let $w_1(p), \cdots, w_s(p)$ be continuous unit vector functions in R. Let there be a continuous function $\mu(p) \geqq 1$ in R such that if p and q are in R and*

$$\begin{gathered} q = p + h[w_i(p) + \eta], \\ \| q - p \| < 1/\mu(p), \qquad | h | < 2/\mu(p), \end{gathered} \tag{9.1}$$

then

$$\left| \frac{\phi(q) - \phi(p)}{h} - \psi_i(p) \right| \leqq \mu(p) \, \| \eta \|, \tag{9.2}$$

for $i = 1, \cdots, s$. Then for any positive continuous function $\zeta(p)$ in R there is an analytic function $\phi''(p)$ in R such that

$$| \phi''(p) - \phi(p) | < \zeta(p), \qquad | \nabla\phi''(p)\cdot w_i(p) - \psi_i(p) | < \zeta(p) \tag{9.3}$$

for any p in R and $i = 1, \cdots, s$.[6]

It is sufficient to find a differentiable function $\phi^*(p)$ such that (9.3) holds with $\zeta(p)$ replaced by $\frac{1}{2}\zeta(p)$; we can then at once find the required analytic function by DM Lemma 9. It may be shown that the analytic approximation to $\phi(p)$ given in the proof of AE Lemma 6 has the required property, but this is not very simple.

We show first that $\phi(p)$ satisfies a local Lipschitz condition. Let $L_1(p) \geqq 1$ be a positive continuous function in R such that

$$| \psi_i(p) | < L_1(p) \qquad (i = 1, \cdots, s). \tag{9.4}$$

[6] The lemma evidently holds still if we add $\frac{1}{2}\zeta(p)$ to the right hand side of (9.2).

Take p and q in R, $\|q - p\| < 1/\mu(p)$. Set

$$h = 2\|q - p\|, \qquad p^* = p + hw_1(p), \qquad \eta = \frac{q - p^*}{h};$$

then

$$q = p^* + h\eta = p + h[w_1(p) + \eta].$$

Also

$$\|\eta\| \leqq \frac{1}{h}[\|q - p\| + \|p^* - p\|] = \frac{3}{2}.$$

Hence, by (9.2), *if* $\|q - p\| < 1/\mu(p)$, *then*

(9.5) $$|\phi(q) - \phi(p)| < 2[L_1(p) + 3\mu(p)/2]\,\|q - p\| = L_2(p)\,\|q - p\|.$$

Take p in R and $\lambda < 1/\mu(p)$, and define $A(p, \lambda)$ by (8.1). (9.5) gives

(9.6) $$|A(p, \lambda) - \phi(p)| < \frac{1}{V(\lambda)}\int_{S(p,\lambda)} \lambda L_2(p)\, dq = \lambda L_2(p),$$

with a similar inequality for $|B(p, \lambda) - \phi(p)|$. Hence, by (8.2), *if we keep* $\lambda < 1/\mu(p)$, *then*

(9.7) $$\left|\frac{\partial}{\partial\lambda} A(p, \lambda)\right| < 2nL_2(p).$$

Let $\zeta'(p)$ be a positive continuous function in R so small that

(9.8) $$\zeta'(p) < 1/\mu(p), \qquad \zeta'(p) < \zeta(p)/[8n^2 L_2(p)],$$

and if $\|q - p\| < \zeta'(p)$, then

(9.9) $$\|w_i(q) - w_i(p)\| < \zeta(p)/[16\,\mu(p)],$$
$$|\psi_i(q) - \psi_i(p)| < \zeta(p)/8, \qquad \mu(q) < 2\mu(p),$$

for $i = 1, \cdots, s$, and if $\|q' - p\| < 2\,\zeta'(p)$, then q' is in R. Let $\lambda(p)$ be any positive differentiable function in R such that

(9.10) $$|\lambda(p)|, \left|\frac{\partial\lambda(p)}{\partial x_j}\right| < \zeta'(p) \qquad (j = 1, \cdots, n)$$

(see DM Lemma 25). We define $\phi^*(p)$ by

(9.11) $$\phi^*(p) = A(p, \lambda(p)).$$

To prove (9.3) with $\zeta(p)$ replaced by $\frac{1}{2}\,\zeta(p)$, take any i and any p in R. Take any h such that $|h| < \zeta'(q)$ for $\|q - p\| < \lambda(p)$, and set $\xi = hw_i(p)$. Take any q, $\|q - p\| \leqq \lambda(p)$, and set $\eta = w_i(p) - w_i(q)$; then q and $q + \xi$ are in R, and

$$q + \xi = q + h[w_i(q) + \eta].$$

Applying (9.2) and (9.9) gives

$$\left| \frac{\phi(q+\xi) - \phi(q)}{h} - \psi_i(p) \right| < \frac{\mu(q)\,\zeta(p)}{16\,\mu(p)} + \frac{\zeta(p)}{8} < \frac{\zeta(p)}{4}.$$

Choose axes $x'_1, \cdots, x'_n$ so that $w_i(p)$ lies in the x'_1-direction. Knowing that $A(p, \lambda)$ is differentiable, the last inequality gives (keeping λ fixed)

(9.12) $$\left| \frac{\partial}{\partial x'_1} A(p, \lambda) - \psi_i(p) \right| \leqq \frac{\zeta(p)}{4}.$$

Hence

$$\left| \frac{\partial \phi^*(p)}{\partial x'_1} - \psi_i(p) \right| < 2n\, L_2(p)\, \frac{n\,\zeta(p)}{8n^2\, L_2(p)} + \frac{\zeta(p)}{4} = \frac{\zeta(p)}{2},$$

proving the second inequality of (9.3). The first follows from (9.6), (9.10) and (9.8). The lemma is now proved.

10. **The functions $\phi_i(p)$.** Given p in M, let $\rho'(p)$ be the smaller of the two numbers $\rho(p)$, $1/K(p)$ (see (4.3)), and take $\rho''(p) \leqq \frac{1}{2}\rho'(p)$ positive and continuous in M so that if $\| q - p \| < \rho''(p)$, then

(10.1) $$\tfrac{1}{2}\rho'(p) < \rho'(q) < 2\rho'(p), \qquad \tfrac{1}{2}\rho'(p) < \rho(q) < 2\rho(p).$$

Let $T^*(p)$ be the set of points (5.1), with

(10.2) $$|\alpha_i| < \rho''(p)/[3\, s\, K(p)].$$

LEMMA 7. *Set $K'(p) = 12/\rho(p)$ in M. If p^* and q^* are in $T^*(p)$ and $T^*(q)$ respectively and $\| q^* - p^* \| < \frac{1}{3}\rho''(p)$, then*

(10.3) $$\| q - p \| \leqq K'(p)\, \| q^* - p^* \|.$$

Say $q^* = q + \Sigma\, \beta_i w_i(q)$. Subtracting (5.1) from this and using (10.2) gives

$$\| q - p \| < \frac{\rho''(p)}{3} + \frac{s\rho''(p)}{3sK(p)} + \frac{s\rho''(q)}{3sK(q)}.$$

We show now that

(10.4) $$\| q - p \| < \min\,[\rho'(p), \rho'(q)].$$

According as $\rho''(q) \leqq \rho''(p)$ or $> \rho''(p)$, we find

$$\| q - p \| < \rho''(p) \leqq \tfrac{1}{2}\rho'(p) < \rho'(q),$$
$$\| q - p \| < \rho''(q) \leqq \tfrac{1}{2}\rho'(q) < \rho'(p).$$

Set

(10.5) $$\Delta_i = w_i(q) - w_i(p);$$

then

$$q - p = q^* - p^* + \sum (\alpha_i - \beta_i) w_i(q) - \sum \alpha_i \Delta_i.$$

Applying (2.4), (2.7), (4.3) and (10.2) gives

$$\begin{aligned} I &= I[w_1(q), \cdots, w_s(q), q - p] \\ &= I[w_1(q), \cdots, w_s(q), q^* - p^* - \sum \alpha_i\Delta_i] \\ &\leqq \| q^* - p^* \| + \frac{\rho(p)}{6} \| q - p \|. \end{aligned}$$

Also, by (3.1) with v_i and ρ replaced by w_i and $\frac{1}{2}\rho$, and (10.1),

$$I \geqq \| q - p \| \tfrac{1}{2}\rho(q) > \| q - p \| \tfrac{1}{4}\rho(p).$$

That we may apply (4.3) and (3.1) follows from (10.4). The last two inequalities give (10.3).

LEMMA 8. *The sets $T^*(p)$ fill out a neighborhood R' of M in a 1-1 way.*

By the last lemma, no two sets $T^*(p)$, $T^*(q)$ $(q \neq p)$ have a common point. To show that R' is an open set, take p_0 in M, and let U be a neighborhood of p_0 in M which is an m-cell. Map U into a portion of E_n with its last $n - m$ coördinates zero. If p in U is mapped into p' in E_n, take any p^* of (5.1) and map it into the point with the same first m coördinates as p' and with $\alpha_2, \cdots, \alpha_s$ as its last coördinates. The resulting region S' of E_n is homeomorphic with the set S of points on $T^*(p)$, p in U. By the theorem on invariance of regions, S is a region. R' is the sum of such sets S, and hence is open.

We may now define the functions ϕ_i and H as in Lemma 1, and they will be single valued and continuous in R'.

11. **The difference quotients of the $\phi_i(p)$.** We first extend the definition of the $w_k(p)$ over R' by setting $w_k(p) = w_k(H(p))$.

LEMMA 9. *The functions $\phi(p) = \phi_k(p)$, $\psi_i(p) = \delta_{ik}$, $w_i(p)$ $(i = 1, \cdots, s)$ satisfy the conditions of Lemma 6 for each $k = 1, \cdots, s$ and an open subset R of R' containing M.*

Let R be the set formed by keeping α_i less than one half the right hand side of (10.2). Let $\mu(p)$ be a positive continuous function in M so large that

$$\mu(p) > 9/\rho''(p), \qquad 3[K'(p)\,K(p) + K'(p) + 1]/\rho(p), \tag{11.1}$$

and if $\mu(p^*) = \mu(H(p^*))$ in R and $\| p' - p^* \| < 2/\mu(p^*)$, p^* in R, then p' is in R'. Now take any p^* and q^* in R and any h so that

$$\| q^* - p^* \| < 1/\mu(p^*), \qquad | h | < 2/\mu(p^*). \tag{11.2}$$

Take a fixed i, set $p = H(p^*)$, $q = H(q^*)$, and define α_j, β_j, η, p' by

$$\begin{aligned} p^* &= p + \sum \alpha_j w_j(p), & q^* &= q + \sum (\alpha_j + \beta_j) w_j(q), \\ q^* &= p^* + h[w_i(p) + \eta], & p' &= p^* + h w_i(p). \end{aligned} \tag{11.3}$$

As $\| p' - p^* \| < 2/\mu(p^*)$, p' is in R'. By (11.3) and (11.1),

(11.4) $\qquad \| q^* - p' \| \leqq \| q^* - p^* \| + | h | < 3/\mu(p^*) < \frac{1}{3} \rho''(p);$

hence we may apply (10.3), giving, with (11.1),

(11.5) $\qquad \| q - p \| < K'(p) \| q^* - p' \| < 3 K'(p)/\mu(p) < 1/K(p).$

This shows that we may apply (4.3), giving with (10.5) and (11.5),

(11.6) $\qquad \| \Delta_j \| \leqq K(p) K'(p) \| q^* - p' \| = K(p) K'(p) | h | \| \eta \|,$

as $q^* - p' = h\eta$. From (11.3) we find

$$q - p = hw_i(p) + h\eta - \sum (\alpha_j + \beta_j)\Delta_j - \sum \beta_j w_j(p).$$

Hence, by (11.5) and (11.6),

$$\| \sum (\beta_j - \delta_{ij} h) w_j(p) \| < [K'(p) + 1 + K(p) K'(p)] | h | \| \eta \|.$$

Dividing by $| h |$ and applying (2.10), (2.7), and (3.2) with v_j and ρ replaced by w_j and $\frac{1}{2} \rho$ gives, with (11.1),

(11.7) $\qquad \| \beta_k/h - \delta_{ik} \| < \mu(p) \| \eta \| \qquad (k = 1, \cdots, s).$

As $\beta_k = \phi_k(q^*) - \phi_k(p^*)$ and $\mu(p) = \mu(p^*)$, (9.2) is proved.

12. **Proof of** Theorem I, $r = 0$. The proof follows the lines of §7, except that we do not use Lemmas 3 and 5. Given a point p_0 of M, we may obviously take a neighborhood U of p_0 and choose axes $x_1', \cdots, x_n'$ as in Lemma 2 in a continuous manner for p in U. For each p in U we have functions (7.1) satisfying the conditions of Lemma 4, (by Lemma 2, which holds even if $r = 0$). Take $\Omega_2(p) < \eta(p)$ so that if (6.1) holds with δ replaced by $\Omega_2(p)$, then (6.2) holds with ϵ replaced by $\eta(p)$. Set $\Omega(p) = \Omega_2(p)/(2n)$, and define $\Omega'(p)$ as before.

By lemmas 9 and 6 we may define the functions $\phi_i''(x_1, \cdots, x_n)$ in $R_3 - M$ so that, setting $\phi_i'' = \phi_i = 0$ in M,

(12.1) $\qquad | \phi_i''(p) - \phi_i(p) |, \qquad | \nabla\phi_i''(p) \cdot w_k(p) - \delta_{ik} | \leqq \Omega'(p)$

in R_3; the ϕ_i'' are continuous in R_3. We now define $\omega(p)$ and the $\theta_i(p; c)$ by (7.5) and (7.7). (7.8) is replaced by

(12.2) $\qquad | \theta_i(p; c) - \phi_i(p) |, \qquad | \nabla\theta_i(p; c) \cdot w_k(p) - \delta_{ik} | \leqq 2 \Omega(p)$

in R_3. $M(c)$ is defined by (7.9). The rest of the proof goes as before.

Harvard University.

On regular closed curves in the plane [1])

by

Hassler Whitney

Cambridge, Mass.

We consider in this note closed curves with continuously turning tangent, with any singularities. To each such curve may be assigned a „rotation number" γ, the total angle through which the tangent turns while traversing the curve. (For a simple closed curve, $\gamma = \pm 2\pi$.) Our object is two-fold; to show that two curves with the same rotation number may be deformed into each other, [4]) and to give a method of determining the rotation number by counting the algebraic number of times that the curve cuts itself (if the curve has only simple singularities, — see Lemma 2).

This paper may be considered as a continuation of a paper of H. Hopf [2]); we assume a knowledge of the first part of his paper.

1. *Regular closed curves.*

Ordinarily, a curve in the plane is defined as a point set with certain properties; but when we allow singularities, this mode of definition cannot be used. (See footnote [3]).) Our first purpose is therefore to define a regular closed curve.

Let E be the Euclidean plane. Let E' be the vector plane (which we might let coincide with E), with origin O. Let I be the closed interval $(0, 1)$. Any differentiable function $f(t)$ with values in E has, as its derivative, a function $f'(t) = \frac{df(t)}{dt}$ with values in E'. By a *parametrized regular closed curve*, or parametrized curve for short, we shall mean a differentiable function $f(t)$ defined in I and with values in E, such that

$$(1) \qquad f(1) = f(0), \quad f'(1) = f'(0), \quad f'(t) \neq 0 \text{ for } t \text{ in } I.$$

---

[1]) Presented to the American Mathematical Society, Sept., 1936.

[2]) Heinz Hopf, Über die Drehung der Tangenten und Sehnen ebener Kurven [Compositio Math. 2 (1935), 50—62].

The first two conditions are the conditions for the curve to be closed; the last condition makes t a „regular parameter". To any such f there corresponds a unique differentiable function $\bar{f}$ defined in $(-\infty, \infty)$, such that

$$(2) \qquad \bar{f}(t) = f(t) \text{ in } I,\ \bar{f}(t+1) = \bar{f}(t),\ \bar{f}'(t) \neq 0,$$

and conversely.

It is natural to call two parametrized curves equivalent if one can be obtained from the other by a change of parameter (preserving orientation). The exact definition is: f and g are *equivalent* ($f \sim g$) if there exists a function $\eta(t)$ in $(-\infty, \infty)$ whose first derivative is continuous and positive, and is such that

$$(3) \qquad \eta(t+1) = \eta(t) + 1,\ \bar{g}(t) = \bar{f}(\eta(t)).$$

Obviously $f \sim f$, $f \sim g$ implies $g \sim f$, and $f \sim g$ and $g \sim h$ imply $f \sim h$. Hence the parametrized curves fall into classes; we call each of these a *regular closed curve*, or curve for short. With any curve C is associated many (equivalent) *parametrizations* f. Let $\bar{C}$ be the corresponding set of points in the plane E (all points $f(t)$). C is by no means determined by $\bar{C}$[3]).

Given any C, *a parametrization* g *may be chosen so that* $|g'(t)|$ *is constant,* that is, so that the parameter is a constant times the arc length.

To prove this, set

$$(4) \qquad L(t) = \int_0^t |\bar{f}'(s)|\, ds, \quad L = L(1).$$

$L = L(C)$ is the length of C. As $\bar{f}'(t) \neq 0$, $L(t)$ is a differentiable increasing function; hence we may solve $L \cdot s = L(t)$ for t, giving $t = \eta(s)$. The derivative $\eta'(s)$ is continuous and positive. As $\bar{f}$ is periodic,

$$L(t+1) - L(t) = \int_t^{t+1} |\bar{f}'(s)|\, ds = \int_0^1 |\bar{f}'(s)|\, ds = L;$$

hence $\eta(s+1) = \eta(s) + 1$. Therefore

$$\bar{g}(t) = \bar{f}(\eta(t))$$

[3]) Let $\bar{C}$ be the unit circle in E; then for each integer $n \neq 0$ there is a corresponding curve C_n with $\bar{C}_n = \bar{C}$, determined by letting $f(t)$ traverse $\bar{C}$ in the positive sense n times while t runs over I. Again, if we take an ellipse and pull the ends of the minor axis together till they are tangent, then there are four corresponding curves, in each of which the corresponding $f(t)$ traverses each point but one of the ellipse only once.

is a parametrization of C. Moreover,

$$\bar{g}'(t) = \bar{f}'(\eta(t)) \frac{L}{L'(\eta(t))}, \qquad |\bar{g}'(t)| = L. \tag{5}$$

If h is any parametrization with $|h'(t)| = k$, then $k = L$ and $\bar{h}(t) = \bar{g}(t+a)$ for some constant a.

First, as $h \sim g$, there is an η such that $\bar{h}(t) = \bar{g}(\eta(t))$. As

$$h'(t) = g'(\eta)\eta'(t), \text{ hence } k = L\eta'(t),$$

we have

$$1 = \eta(1) - \eta(0) = \int_0^1 \eta'(t)dt = \int_0^1 \frac{k}{L} dt = \frac{k}{L},$$

and $k = L$. Hence $\eta'(t) = 1$, and $\eta(t) = t + a$.

Let f_0 and f_1 be parametrized curves. We say one may be *deformed* into the other if $f_u(t)$ may be defined for $0 < u < 1$ such that it is continuous in both variables for $0 \leqq t \leqq 1$, $0 \leqq u \leqq 1$, and each f_u is a parametrized curve.

If f_0 and f_1 are parametrizations of C, then one may be deformed into the other within C, that is, we can make each f_u a parametrization of C.

To prove this, say $f_1(t) = f_0(\eta(t))$. Set

$$\eta_u(t) = u\eta(t) + (1-u)t, \quad f_u(t) = \bar{f}_0(\eta_u(t)), \tag{6}$$

for $0 \leqq u \leqq 1$. Then $\eta_0(t) = t$, $\eta_1(t) = \eta(t)$, so that $\bar{f}_0$ and $\bar{f}_1$ bear the proper relation to f_0 and f_1. As

$$\eta_u(t+1) = u[\eta(t)+1] + (1-u)(t+1) = \eta_u(t) + 1,$$

$$\frac{d\eta_u(t)}{dt} = u\frac{d\eta(t)}{dt} + (1-u) > 0 \text{ for } 0 \leqq u \leqq 1,$$

each f_u is a parametrized curve equivalent to f_0.

We say C may be deformed into C' if some parametrization of C may be deformed into one of C'. By the above statement, this is independent of the parametrizations chosen.

2. *The deformation theorem.*

The following lemma is fundamental in this section.

LEMMA 1. *Let $f'(t)$ be a continuous vector function in I, such that $f'(t) \neq 0$. If p is a point of E, then*

$$f(t) = p + \int_0^t f'(s)ds \tag{7}$$

is a parametrized curve if and only if

$$f'(1) = f'(0), \quad \int_0^1 f'(s)ds = 0. \tag{8}$$

This is obvious. The last relation may be stated as follows: The average value of $f'(s)$ is O.

Given any parametrized curve f, we define its rotation number $\gamma(f)$ as the total angle through which $f'(t)$ turns as t traverses I. The function $f^*(t) = \frac{f'(t)}{|f'(t)|}$ is a map of I into the unit circle; $\gamma(f)$ is 2π times the degree of this map. (See Hopf. loc. cit., 1c, and our equation (11).)

If f may be deformed into g, then $\gamma(f) = \gamma(g)$.

For $\gamma(f_u)$ is continuous in u, and is an integral multiple of 2π; hence it is constant. Hence, by **1**, we may define $\gamma(C)$ for a curve C as $\gamma(f)$ for any parametrization f of C.

THEOREM 1[4]). *The curves C_0 and C_1 may be deformed into each other if and only if $\gamma(C_0) = \gamma(C_1)$.*

One half of the theorem was proved above. Suppose now that $\gamma(C_0) = \gamma(C_1) = \gamma$. Let g_0 and f_1 be parametrizations of C_0 and C_1 such that

$$|g_0'| \equiv L(C_0) = L_0, \quad |f_1'| \equiv L(C_1) = L_1.$$

Set

$$g_u(t) = g_0(0) + \left[u\frac{L_1}{L_0} + (1-u)\right][g_0(t) - g_0(0)];$$

this deforms the parametrized curve g_0 into one g_1. Set $f_0 = g_1$; then $|f_0'| \equiv |g_1'| \equiv L_1$. We must deform f_0 into f_1.

The proof runs as follows. We consider the maps f_0' and f_1' of I into the circle K of radius L_1. They are both of degree $\frac{\gamma}{2\pi}$; hence one map may be deformed into the other, say by the maps h_u. We alter each h_u by a translation to obtain a map f_u' whose average lies at O; these functions then define the required deformation, at least if $\gamma \neq 0$.

We begin by defining the vector function

$$\theta(t) = (L_1 \cos t,\ L_1 \sin t); \tag{9}$$

this gives an angular coordinate t in K. Suppose first that $\gamma \neq 0$. By rotations in the plane E we may alter f_0 and f_1 so that

[4]) This theorem, together with a straightforward proof, was suggested to me by W. C. GRAUSTEIN.

$f'_0(0) = f'_1(0) = \theta(0)$. As $f'_i(t)$ lies on K, we may give it an angular measure $F_i(t)$:

(10) $$f'_i(t) = \theta(F_i(t)), \text{ with } F_i(0) = 0 \qquad (i = 0, 1).$$

(See Hopf., loc. cit., 1a.) Then, by definition of γ,

(11) $$F_i(1) = \gamma \qquad (i = 0, 1).$$

Set

(12) $$\begin{aligned} F_u(t) &= uF_1(t) + (1-u)F_0(t), \\ h_u(t) &= \theta(F_u(t)) \end{aligned} \qquad (0 \leqq t \leqq 1),$$

(13) $$\begin{aligned} f'_u(t) &= h_u(t) - \int_0^1 h_u(s)\,ds, \\ f_u(t) &= f_0(0) + u[f_1(0) - f_0(0)] + \int_0^t f'_u(s)\,ds. \end{aligned}$$

It is clear that $\int_0^1 f'_u(t)\,dt = 0$. As $F_u(0) = 0$, $F_u(1) = \gamma$, and γ is an integral multiple of 2π,

$$f'_u(1) - f'_u(0) = \theta(F_u(1)) - \theta(F_u(0)) = \theta(\gamma) - \theta(0) = 0.$$

Finally, as $\gamma \neq 0$ and hence $h_u(t)$ passes over all of K, its average value lies interior to K; therefore for no t does $h_u(t)$ equal the average, and $f'(t) \neq O$. This proves that each f_u is a regular closed curve. As $f_u(t)$ is continuous in both variables, and it reduces to f_0 and f_1 for $u = 0$ and $u = 1$, it is a deformation of f_0 into f_1, as required.

Suppose now that $\gamma = 0$. If we alter $F_u(t)$ so that it is constant for no u, then again $f'(t) \neq 0$, and the above proof will hold. Choose a t_0 for which $F_1(t_0) \neq 0$, and deform $F_0(t)$ in a small neighborhood of t_0 into $F_1(t)$ in this neighborhood; now deform the new F_0 into F_1 by the process given above. Then (as $F_u(0) = 0$) no F_u is constant.

3. *Crossing points of curves.*

Let $f(t)$ be a parametrized curve. Let p be a point of the plane. If there are exactly two numbers t_1, t_2, such that

$$0 \leqq t_1 < t_2 < 1, \quad f(t_1) = f(t_2) = p,$$

and if $f'(t_1)$ and $f'(t_2)$ are independent vectors, we call p a (*simple*) *crossing point* of the curve. This is evidently independent of the parametrization. If the curve has no singularities other than a finite number of simple crossing points, we say the curve is *normal.*

LEMMA 2. *Any curve may be made normal by an arbitrarily small deformation.*

Given $\varepsilon > 0$, cut I into intervals $I_1, \ldots, I_\nu$ so small that each corresponding arc $A_i = f(I_i)$ is of diameter $< \varepsilon$, and the tangents at different points of A_i differ by at most ε. By a small deformation we may clearly obtain arcs A'_i such that neither end of any A'_i touches other points of the curve. Now for any i and j, it is easy to replace A'_j by an arc A''_j arbitrarily near it and with the same ends so that A''_j cuts A'_i in simple crossing points only[5]). Alter thus A'_2 in relation to A'_1; then A'_3 in relation to A'_1; then A'_3 in relation to A'_2, altering it so slightly that its relation to A'_1 is not impaired, etc.

Let f be a parametrized curve, and let $\overline{C}$ be the corresponding set of points $f(t)$ in the plane. We say f has an *outside starting point* if there is a line of support to $\overline{C}$[6]) containing $f(0)$.

Let $f(t_1) = f(t_2)$, $t_1 < t_2$, be a crossing point. If the vectors $f'(t_1)$ and $f'(t_2)$ are oriented relative to each other in the *opposite* manner to the (fixed) x- and y-axes, we say the crossing point is *positive*; otherwise, *negative*[7]). If we set $\bar{g}(t) = \bar{f}(t+\tau)$ with $t_1 < \tau \leqq t_2$, then the above crossing point changes its type. Corresponding to any normal parametrized curve are the numbers

$$(14) \qquad \left.\begin{matrix} N^+ \\ N^- \end{matrix}\right\} \text{ of crossing points of } \left\{\begin{matrix} \text{positive} \\ \text{negative} \end{matrix}\right\} \text{type.}$$

These may be found by following the curve from its starting point, and watching the intersections with the part of the curve already traversed.

THEOREM 2. *If f is a normal parametrized curve with an outside starting point, then*

$$(15) \qquad \gamma(f) = 2\pi[\mu + (N^+ - N^-)], \qquad \mu = \pm 1.$$

If the axes are moved so that the x-axis is the line of support at

---

5) The proof is simplified by first replacing $f(t)$ by a function $g(t)$ with continuous second derivatives. The lemma is contained in Theorem 2 of H. WHITNEY, Differentiable manifolds [Annals of Math. 37 (1936)]. (We replace I by the unit circle M and use (b) of the theorem.)

6) That is, a straight line touching $\overline{C}$ and having each point of $\overline{C}$ on it or on a single side of it.

7) An example of a positive crossing point is given in Fig. 2.

$f(0)$ and the curve is on the same side of this line as the positive y-axis, then $\mu = +1$ or -1 according as $f'(0)$ is in the positive or negative x-direction.

In particular, if the curve has no singularities, then $\gamma = \pm 2\pi$, which is the „Umlaufsatz".

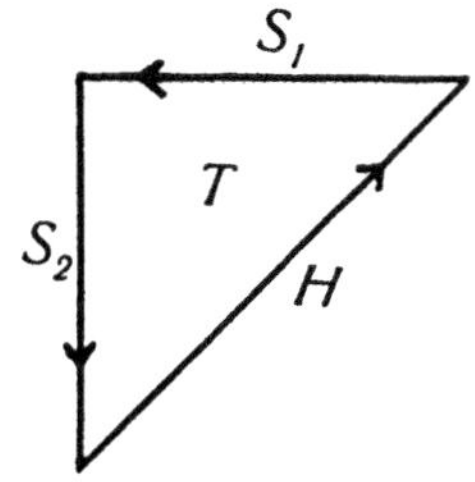

Fig. 1.

Let T be the triangle of all pairs of numbers

$$(t_1, t_2),\ 0 \leqq t_1 \leqq t_2 \leqq 1.$$

Let $l(t_1, t_2)$ be the smaller of $t_2 - t_1$ and $(1+t_1) - t_2$. Set

$$(16)\qquad \begin{aligned} &\psi(t_1, t_2) = \frac{f(t_2) - f(t_1)}{l(t_1, t_2)} \text{ if } l(t_1, t_2) \neq 0\,, \\ &\psi(t, t) = f'(t), \qquad \psi(0, 1) = -f'(0)\,. \end{aligned}$$

ψ is continuous in T, and is 0 at (t_1, t_2) if and only if $t_1 < t_2$ (but not $t_1 = 0$, $t_2 = 1$), and $f(t_1) = f(t_2)$, i.e. if and only if $f(t_1)$ is a crossing point [8]).

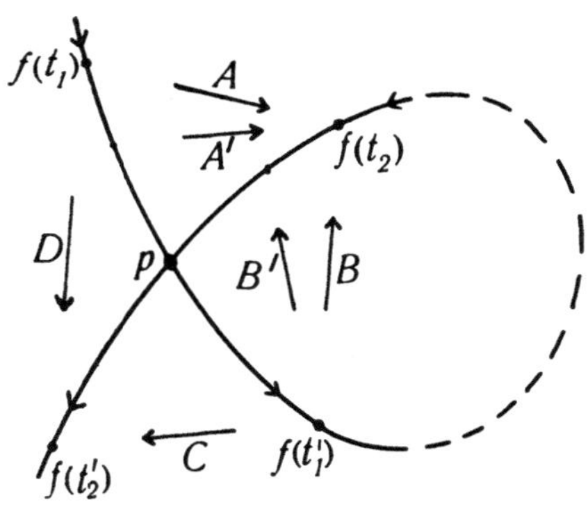

Fig. 2.

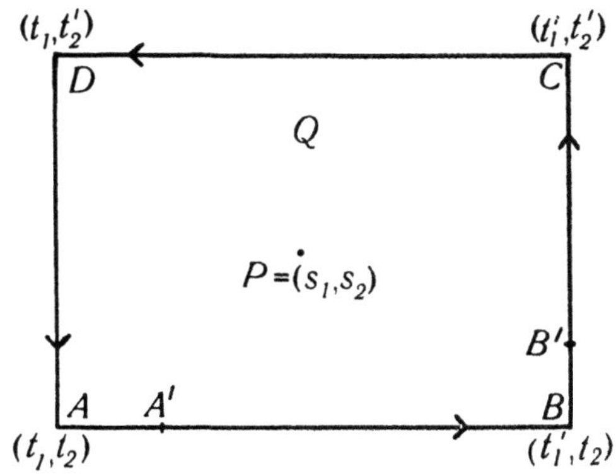

Fig. 3.

Take any crossing point $p = f(s_1) = f(s_2)$; suppose it is positive. As $s_1 < s_2$, $P = (s_1, s_2)$ is not on the hypothenuse of T. As $f(0)$ is an outside point, it is obviously not a crossing point; hence $f(0) \neq f(t)$ for $0 < t < 1$, and P is on neither side of T. As $P \neq (0,1)$, it follows that P is interior to T. Choose numbers t_1, t_1' very close to s_1, and t_2, t_2' very close to s_2, so that

[8]) This function replaces the function $f(s_1, s_2)$ of Hopf (p. 54). It will be seen that $N = N^+ - N^-$ is the algebraic number of times that T covers 0 under ψ (see footnote [10])).

$$t_1 < s_1 < t_1', \quad t_2 < s_2 < t_2',$$

and

$$f(t_1),\ f(t_1'),\ f(t_2),\ f(t_2') \text{ are equidistant from } p.$$

Let Q be the rectangle in T containing P, with coordinates (t_1, t_2), etc. It is easily seen that if we run around Q once in the positive sense, the corresponding ψ runs around O once in the positive sense. For running around each side of Q turns the vector ψ through an angle of approximately $\frac{\pi}{2}$ (see the diagram); hence it turns, in all, approximately 2π; but it turns an integral multiple of 2π, and hence exactly 2π [9]). If the crossing point is negative, the result is obviously -2π.

Let $P_1, \ldots, P_m$ be the points of T corresponding to crossing points, and let $Q_1, \ldots, Q_m$ be corresponding rectangles enclosing them, no two of which have common points. Cut the rest of T into triangles $Q_{m+i}, \ldots, Q_r$. If we run around the boundary of any Q_{m+i}, ψ runs around 0 zero times [10]). To show this, consider the vector $\psi^* = \frac{\psi}{|\psi|}$. This is defined throughout Q_{m+i}, and its values are on the unit circle. Hence an angular coordinate may be defined, giving the position of ψ^* throughout Q_{m+i} (see Hopf, loc. cit., 1b). If we run around the boundary of Q_{m+i}, the angular coordinate comes back to its original value, and hence ψ has turned around zero times.

Let $\alpha_1, \ldots, \alpha_l$ be all sides of triangles or rectangles in T. Let $\alpha_1, \ldots, \alpha_k$ be those lying on the boundary B of T, oriented the same as T; the remaining α_i are oriented arbitrarily. With each α_i we associate a number $\varphi(\alpha_i)$, the angle through which ψ turns when α_i is traversed in the positive direction. Let $\varphi(Q_i)$ be the angle through which ψ turns when the boundary of Q_i is traversed in the positive direction; similarly for $\varphi(T)$. Now

$$\sum_{i=1}^{r} \varphi(Q_i) = \varphi(T). \tag{17}$$

For each $\varphi(Q_i)$ may be expressed as a sum $\Sigma^{(i)} \pm \varphi(\alpha_j)$, summing over the boundary lines of Q_i; when these sums are added, the two terms corresponding to each α_j interior to T cancel, and we are left with the sum over the α_j on the boundary of T.

---

9) By choosing the proper degree of approximation, it is easy to make this reasoning rigorous.

10) Hence, in all cases, $\varphi(Q_j)$ (see below) is the algebraic number of times that the map ψ of Q_j covers 0.

We have seen above that

$$\sum_{i=1}^{r} \varphi(Q_i) = \sum_{i=1}^{m} \varphi(Q_i) = 2\pi(N^+ - N^-) = 2\pi N. \tag{18}$$

Suppose $\mu = 1$. If S_1, S_2 and H are the positively oriented sides and hypotenuse of T (see Fig. 1), it is easily seen that

$$\varphi(S_1) = \varphi(S_2) = -\pi. \tag{19}$$

(See Hopf, pp. 54—55. The change in sign is caused by the difference in orientation of S_1 and S_2 from that used by Hopf.) Hence, using (17) and (18),

$$2\pi N = \varphi(T) = \varphi(H) + \varphi(S_1) + \varphi(S_2) = \gamma - 2\pi,$$

which gives (15). If $\mu = -1$, the only change is that $\varphi(S_1) = \varphi(S_2) = \pi$, and (15) again follows.

(Received December 21th, 1936.)

Annals of Mathematics
Vol. 38, No. 4, October, 1937

ANALYTIC COORDINATE SYSTEMS AND ARCS IN A MANIFOLD[1]

By Hassler Whitney

(Received March 4, 1937)

1. **Introduction.** In a differentiable, or analytic, manifold,[2] each point is in a "coordinate system". Two overlapping systems are related by a differentiable, or analytic, transformation, with non-vanishing Jacobian. An important question is: are there "large" coordinate systems, containing say two given points, or a given arc, etc. In the analytic case, does there even exist an analytic arc joining given points? We answer some questions of this sort here. *We always suppose the given manifold may be imbedded in a Euclidean space* E_μ. This is no restriction in the differentiable case; see DM Theorem 1. The question is unsolved in the analytic case; see the Problem stated below.

We shall speak of manifolds of class C^r as in DM, and shall use $r = \omega$ for the analytic case; write $\infty < \omega$. By a *coordinate system* (of class C^r) in a C^r-m-manifold M we shall mean any regular (1-1) C^r-map (see DM, §§2, 3) of a "fundamental domain" Q of Euclidean m-space into M. For Q we shall take a spherical region or m-cube, or the region (for some k)

$$(1) \qquad \sum_{i=1}^{k} x_i^2 < 1, \qquad -1 < x_j < 1 \qquad (j = k+1, \cdots, m).$$

We begin by showing that a finite set of points in M may be joined by a differentiable arc. Next it is proved that any submanifold N of M satisfying certain conditions may be imbedded in a family of submanifolds, which fill out an open set in M. If N is an arc, or an n-cube etc., the conditions are always satisfied; the family of submanifolds then defines a coordinate system containing N. Now if two or more points are given in the analytic manifold M (in E_μ), we first join them by a differentiable arc A, and then find an analytic coordinate system R containing A. In R we may then easily find an analytic arc (in fact, an analytic closed curve) joining them. We end the main part of the paper by showing that an analytic closed curve in an analytic manifold M which bounds a differentiable surface element, also bounds an analytic surface element.

In Appendix I we give a simple proof of the imbedding theorem for closed

---

[1] Presented to the Am. Math. Soc., March, 1937.

[2] See for instance DM. We shall refer to the following papers by the author.

AE: *Analytic extensions* $\cdots$, Trans. Am. Math. Soc., vol. 36 (1934), pp. 63–89.

DM: *Differentiable manifolds*, Annals of Math., vol. 37 (1936), pp. 645–680.

IM: *The imbedding of manifolds* $\cdots$, Annals of Math., vol. 37 (1936), pp. 865–878.

FD: *Functions differentiable on the boundaries of regions*, Annals of Math., vol. 35 (1934), pp. 482–485.

manifolds of class C^r, $r \geqq 2$; a theorem on the projecting of a manifold in E_μ into a subspace $E_{\mu-1}$ is used. In Appendix II we clarify the proof of DM Lemma 16.

The simplest possible non-trivial analytic manifold is one of dimension 1, with just two overlapping coordinate systems. For it, the imbedding problem is equivalent to the following problem in function theory:

PROBLEM. Let $\phi(x)$ be any (real) analytic function for $-1 \leqq x \leqq 1$, with

$$\phi(-1) = -1, \qquad \phi(1) = 1, \qquad \phi'(x) > 0 \quad (-1 \leqq x \leqq 1).$$

Does there exist a function $f(x)$, analytic for $-2 < x < 1$, with $f'(x) > 0$ there, such that $g(x) = f(\phi(x))$, analytic for $-1 < x < 1$, may be extended analytically through $1 \leqq x < 2$, so that $g'(x) > 0$ there?

Much use will be made of the following lemma. See DM §6 for notations.

LEMMA 1. *Let M and N be C^r-m- and C^r-n-manifolds respectively,*[3] $1 < r \leqq \omega$, *and let ϕ be a C^t-homeomorphism of N into (a subset of) M, t finite,* $1 \leqq t < r$. *Then for any positive continuous function $\eta(p)$ in N there is a C^r-homeomorphism of N into M approximating (ϕ, N, t, η).*

Generally N will be an open subset of Euclidean space. For $r = \omega$, the lemma is a consequence of DM, Lemma 22; for any sufficiently close approximation to a C^1-homeomorphism is another one. For general r, we may either apply the lemma to C^r-homeomorphic analytic manifolds, or note that DM Lemma 22 holds with "analytic" replaced by "C^r".

COROLLARY. *Any coördinate system of class C^s in the C^r-m-manifold M is contained in one of class C^r, if* $1 \leqq s < r$.

For a generalization of this theorem, see Theorem 2.

2. Differentiable arcs and simple closed curves in M.

We prove here

LEMMA 2. *Theorem 3 is true if $r = 1$.*

Suppose we have found such an arc A' from p_0 as far as p_i $(0 \leqq i < k)$. If $m = 2$, let K be a differentiable simple closed curve running along close beside A' on one side, and returning on the other. (This is easily constructed.) As M is connected, there is a sequence of coordinate systems $U_1, \cdots, U_l$, such that U_1 contains p_i, U_l contains p_{i+1}, $U_j \cdot U_{j+1} \neq 0$, but no other U_j and U_h have common points. Set $q_0 = p_i$, $q_l = p_{i+1}$, and let q_j be a point of $U_j \cdot U_{j+1}$. Join q_j to q_{j+1} in U_j; this gives a path from p_i to p_{i+1}, from which we may pick out an arc B, differentiable except at a finite number of corners. We may make it start from p_i in a given direction. If $m = 2$ and B cuts A', then there is a first point q' and a last point q'' at which it cuts K; we replace the interior part $q'q''$ of B by one of the arcs of K. If $m \geqq 3$ and B cuts A', we may replace small pieces of B cutting A' by differentiable pieces avoiding A'. We now have an arc A^* from p_0 to p_{i+1}. At each corner, we may replace a small piece of A^* by an arc of a circle (using a coordinate system containing the corner); we find thus a differentiable arc from p_0 to p_{i+1}, and finally to p_k. If we want a simple

[3] We recall the hypothesis that the manifolds may be imbedded in Euclidean space.

closed curve, we take another differentiable arc running along beside the first, and join their ends by arcs of circles.

3. **Manifolds in regular position in** M. Let N be a C^s-n-manifold in the C^r-m-manifold M. We say N is in *regular position*[4] in M if the following condition is satisfied. If M is imbedded in some E_μ (DM Theorem 1), then there exist $m - n$ continuous vector functions $v_i(p)$ in N which are independent for each p, and each of which is tangent to M and normal to N. We may clearly take them orthogonal to each other if we wish.

The definition is independent of the choice of the imbedding. For suppose M is also imbedded in E'_ν, and N is regular position in M for M in E_μ. Staying in E_μ, there is a positive continuous function $\xi(p)$ in M' such that the vectors $\xi(p)v_i(p)$ project, for each p, into differentiable curves $c_i(p)$ in M (DM Lemma 23). Going over to E_μ, these curves define independent vectors $v'_i(p)$, the $(m - n)$-plane of which is independent of the tangent plane to N, for each p; for the maps of M into E_μ and into E'_ν are regular. Projecting each such plane onto the plane normal to N and tangent to M in E'_ν gives the required vector functions in E'_ν.

THEOREM 1. *Let N be a C^r-n-manifold in regular position in the C^r-m-manifold*[3] *$M (1 \leqq r \leqq \omega)$. Then there is an $(m - n)$-parameter family of manifolds $N(c_1, \cdots, c_{m-n})$ $(|c_i| < 1)$ C^r-homeomorphic with N, which fill out a neighborhood of N in M in a* (1-1) *way, and such that* $N(0, \cdots, 0) = N$.

REMARKS. We shall define the function $Z(p; c_1, \cdots, c_{m-n})$ of class C^r in the product space of N with the open $(m - n)$-cube $|c_i| < 1$, so that as p varies, we run over $N(c_1, \cdots, c_{m-n})$, and as $c_1, \cdots, c_{m-n}$ vary, we run over an "$(m - n)$-cube" $Q(p)$ in M. If M is of class C^s, $s > r$, we may make the $N(c_1, \cdots, c_{m-n})$ of class C^s for $(c_1, \cdots, c_{m-n}) \neq (0, \cdots, 0)$ (see IM, Theorem I); or we may make all the $N(c_1, \cdots, c_{m-n})$ of class C^s, and have N lie in the open set which they fill out (compare Theorem 2). If r is finite, we may make the $N(c_1, \cdots, c_{m-n})$ approximate to N as in IM, Theorem I, (3).

To prove the theorem, we suppose M is in E_μ, and replace the $v_i(p)$ defined above by independent vector functions $w_i(p)$ as in IM, §4.[5] The $w_i(p)$ are analytic in an open set in E_μ containing N, and hence are of class C^r in N (DM, Lemma 3). Then if $P^*(p)$ is the $(m - n)$-plane through p determined by the $w_i(p)$, the function $P^*(p)$ of p is of class C^r in N (see DM, §24). From the proof of DM Lemma 21 it is clear that for some positive continuous function $\xi(p)$ in N, the points

$$q = p + \sum \alpha_i w_i(p) \qquad (|\alpha_i| < \xi(p))$$

fill out a C^r-m-manifold N^* containing N. Moreover, if the $w_i(p)$ approximate to the $v_i(p)$ closely enough and $\xi(p)$ is small enough, then N^* is nearly tangent

[4] Compare IM, §3.

[5] Extend the $v_i(p)$ throughout a neighborhood of N in E_μ, and approximate to them with analytic functions.

to M at each p in N, and projects into M(DM Lemma 23), giving a C^r-m-manifold N' in M (i.e. an open subset of M). Extend $\xi(p)$ throughout a neighborhood of N in E_μ, and approximate to $\frac{1}{2}\xi(p)$ by an analytic function $\zeta(p)$ (DM Lemma 9); then $\zeta(p)$, considered in N alone, is of class C^r. Let $Z(p; c_1, \cdots, c_{m-n})$ be that point of N' corresponding to the point

$$p + \sum c_i \zeta(p) w_i(p)$$

in N^*. Z clearly has the required properties.

4. **Coordinate neighborhoods containing given arcs, etc.** Theorem 2 will follow from the last theorem and the following lemma. We shall call a manifold with boundary a manifold also.

Lemma 3. *Any manifold N in M is in regular position in M if it is C^1-homeomorphic with an open or closed fundamental domain.*

This will be proved in a paper on "sphere-spaces". Because of its importance, we shall give the proof for the case that N is an open or closed arc. Suppose $v_1(p), \cdots, v_{k-1}(p)$ (or no vector functions) have been defined properly on N; we shall define $v_k(p)$. Taking M in E_μ, let $P(p)$ be, for each p on N, the $(m-k)$-plane through p tangent to M and orthogonal to N and to $v_1(p), \cdots, v_{k-1}(p)$; $P(p)$ varies continuously with p. Cut N into subarcs $A_1, A_2, \cdots$ so that no two planes $P(p)$ and $P(q)$ are perpendicular for p and q on the same A_i, and so that each $A_i (i > 1)$ has exactly one point in common with the preceding arcs. Suppose $v_k(p)$ has been defined on $A' = A_1 + \cdots + A_{i-1}$; we shall define it on A_i. Say $A_i = p_0 q_0$, p_0 in A'. Corresponding to p_0 and $P_0 = P(p_0)$, define the transformation $T_{p,P}$ as in DM, §19. For each p in A_i, let $v_k(p)$ be the vector into which $v_k(p_0)$ is carried by $T_{p,P(p)}$. Proceed in this manner until all the $v_k(p)$ are defined throughout N.

Remark. It is not hard to show that a simple closed curve in M is in regular position if and only if it "preserves orientation" in M.

Theorem 2. *Any C^s-n-manifold N in the C^r-m-manifold M $(1 \leqq s \leqq r \leqq \omega)$ which is C^s-homeomorphic with an open or closed fundamental domain is contained in a coordinate system of class C^r. If $s = r$, we may make N correspond to the points $x_{n+1} = \cdots = x_m = 0$; in any case, N will approximate to this plane.*

We may suppose N is open; if it is closed, we may extend the map[6] of the closed fundamental domain Q into N to a map of a slightly larger region Q', and apply the proof to Q'. If the theorem with $s = r$ is proved, we may prove the first part of the theorem as follows. Applying the theorem with r replaced by s gives a coordinate system of class C^s containing N; Lemma 1 (applied to the fundamental domain defining this coordinate system) replaces it by one of class C^r.

Suppose then that Q is open and $s = r$. By Theorem 1, Remark, there is a C^r-homeomorphism $Z(p; x_{n+1}, \cdots, x_m)$ of the product S of Q and an open

[6] Supposing that M is in E, we apply the theorem of FD to each coordinate of the given map. The resulting map of Q' is then projected onto M.

$(m - n)$-cube into M. As Q is a fundamental domain, so is S; the map of S is the required coordinate system.

5. **Analytic arcs, etc.** We can now prove

THEOREM 3.[7] *Let $p_0, p_1, \cdots, p_k$ be points of the connected C^r-m-manifold*[3] *M ($1 \leqq r \leqq \omega$, $m \geqq 2$). Then there is an arc, in fact a simple closed curve, of class C^r, containing them in the given order. We may make the arc (or curve) have given directions at the given points.*

REMARK. If there is given a differentiable arc or simple closed curve through the points, it is clear from the proof that we may make the new arc or curve approximate to the given one as closely as we please, both in position and in direction.

We first find an arc A of class C^1 by Lemma 2. By Theorem 2, there is a coordinate system $(x_1, \cdots, x_m)$ of class C^r containing A[8]. In this system, A will approximate to the x_1-axis; in fact, by choosing a good enough approximation in the proof of Theorem 2, we may make the p_i lie as close to the x_1-axis as we please. Say $p_i = (x_1^i, \cdots, x_m^i)$. Let $f_j(x)$ be a polynomial (in one variable) such that

$$f_j(x_1^i) = x_j^i \qquad (i = 0, \cdots, k; j = 2, \cdots, m). \tag{2}$$

The set of points $(x_1, \cdots, x_m)$ satisfying

$$x_j = f_j(x_1) \qquad (j = 2, \cdots, m)$$

is an arc A^* through the points $(x_1^i, \cdots, x_m^i)$. If they are close enough to the x_1-axis, A^* will lie in the coordinate system. A^* is analytic in the coordinate system, and hence of class C^r in M. If we wish A^* to have given directions at the points, we assign the proper derivatives to the polynomials $f_j(x)$.

Suppose a closed curve is desired. We find an arc A^* as above. By Theorem 2, we may suppose it is the x_1-axis in a coordinate system. Join the ends of a corresponding arc $x_2 = \epsilon$, $x_3 = \cdots = x_m = 0$ to the ends of this one by semicircles, giving the closed curve C. By Lemma 1 there is an approximating closed curve C' of class C^r. The part A' of C' corresponding to A^* is given by functions $x_j = g_j(x_1)$ $(j = 2, \cdots, m)$; set $\bar{x}_j^i = g_j(x_j^i)$, the x_1^i determining the points p_i. Using the $\bar{x}_j^i$, find polynomials $\bar{f}_j(x)$ as before. The transformation of m-space

$$x_1' = x_1, \qquad x_j' = x_j - \bar{f}_j(x_1) \qquad (j = 2, \cdots, m) \tag{3}$$

carries C' into the required closed curve C^*.

6. **Analytic cells with analytic boundaries.** Our object is to prove Theorem 4. For the case $r = \omega$, we shall need a lemma from potential theory.

LEMMA 4. *Let $f(p)$ be an analytic function on the unit circle A in the plane.*

---

[7] This Theorem has recently been obtained under weaker hypotheses by T. Y. Thomas, Annals of Math., vol. 38 (1937), pp. 120.

[8] The existence of the required arc is now evident if $k = 1$.

Then the harmonic function $F(p)$ in the interior taking on the boundary values $f(p)$ is analytic in the closed region. There is a constant $K > 0$ such that if $f(p)$ and its first and second derivatives along A are bounded in absolute value by η, then $F(p)$ and its first partial derivatives are bounded in absolute value by $K\eta$.

The first part is well-known.[9] We shall prove the second part with K chosen as follows: Let $u(\theta)$, $|\theta| \leqq \pi$, denote any differentiable function defined on A, measuring θ from the point $(x, y) = (0, -1)$; let $u'(\theta)$, $u''(\theta)$ denote its first and second derivatives with respect to θ. Then there is a $K \geqq 4$ such that

$$(4) \qquad |u'(\theta)| \leqq \zeta \text{ (all } \theta) \text{ implies } \left|\frac{u(\theta) - u(0)}{y + 1}\right| \leqq \tfrac{1}{2}K\zeta \qquad (\theta \geqq \tfrac{1}{3}\pi).$$

Now take any $f(\theta)$, with $|f(\theta)|$, $|f'(\theta)|$, $|f''(\theta)| \leqq \eta$. We shall show first that if $F(x, y)$ is the harmonic function in A with the boundary values $f(\theta)$, then

$$(5) \qquad \left|\frac{\partial}{\partial y} F(0, -1)\right| \leqq K\eta.$$

Define $G(x, y)$ by

$$(6) \qquad G(x, y) = F(x, y) - F(0, -1) - x\frac{\partial}{\partial x}F(0, -1);$$

Then G is harmonic within A, and

$$G(0, -1) = \frac{\partial}{\partial x}G(0, -1) = 0.$$

Let $g(\theta)$ denote the boundary values of G; then

$$g(\theta) = f(\theta) - f(0) - f'(0)\sin\theta, \qquad g'(0) = 0.$$

Using a law of the mean twice (see for instance Osgood, *Advanced Calculus*, p. 209) with $0 < \theta < \pi$ gives, for some α and β,

$$\frac{g(\theta)}{y + 1} = -\frac{g(\theta) - g(0)}{\cos\theta - \cos 0} = -\frac{g'(\alpha)}{-\sin\alpha} \qquad (0 < \alpha < \theta)$$

$$= \frac{g'(\alpha) - g'(0)}{\sin\alpha - \sin 0} = \frac{g''(\beta)}{\cos\beta} \qquad (0 < \beta < \alpha).$$

A similar relation holds with $-\pi < \theta < 0$. Clearly $|g''(\theta)| \leqq 2\eta$; hence

$$(7) \qquad \left|\frac{g(\theta)}{y + 1}\right| \leqq \frac{2\eta}{\cos\frac{1}{3}\pi} = 4\eta \qquad (0 < |\theta| \leqq \tfrac{1}{3}\pi).$$

Using the definition of K with this inequality gives

$$(8) \qquad \left|\frac{g(\theta)}{y + 1}\right| \leqq K\eta \qquad (\theta \neq 0).$$

[9] See for instance W. F. Osgood, *Funktionentheorie*, 5th. ed. (1928), p. 703, 3. Satz. It must be pointed out that $F(p)$ may be defined so as to be analytic throughout a region containing A, not merely through a neighborhood of any arc of A.

Set

(9) $$H_{\pm}(x, y) = G(x, y) \pm K\eta(y + 1);$$

Then $H_{\pm}$ is harmonic within A. If $h_{\pm}(\theta)$ denotes its boundary values, then (8) gives

$$h_{+}(\theta) \geqq 0, \qquad h_{-}(\theta) \leqq 0 \qquad \text{(all } \theta).$$

Therefore the same inequalities hold for H_{+} and H_{-} within A. As they vanish at $(0, -1)$, we have

(10) $$\frac{\partial}{\partial y} H_{+}(0, -1) \geqq 0, \qquad \frac{\partial}{\partial y} H_{-}(0, -1) \leqq 0.$$

Hence (5) holds for G, and therefore also for F. By symmetry, the normal derivative of F at each point of A is bounded by $K\eta$; hence the same is true of any partial derivative of F on A.

Now suppose there were a point p within A at which some partial derivative, say $\partial F/\partial y$, were not bounded by $K\eta$. As $\partial F/\partial y$ is harmonic within A and continuous on the boundary, we have a contradiction with the above inequality and the maximum-minimum theorem.

THEOREM 4. *Let C be a simple closed curve of class C^r in the C^r-m-manifold*[3] *$M(1 \leqq r \leqq \omega)$. If C C^1-bounds*[10] *the 2-cell D, then it C^r-bounds a 2-cell D^*.*

Let A be the unit circle in the plane, with interior B, and let ϕ be the given C^1-homeomorphism of $A + B$ into D. Suppose first that r is finite or $r = \infty$. We may suppose ϕ defined over a slightly larger region B' (compare footnote 6). Let ψ be a C^r-map of B' into M approximating $(\phi, B', 1, \eta)$ for some $\eta > 0$ (Lemma 1). M being in E_μ, ϕ and ψ each have μ coordinates ϕ_i and ψ_i ; consider the functions

(11) $$f_i(p) = \phi_i(p) - \psi_i(p) \qquad (i = 1, \cdots, \mu)$$

on A. Extend the $f_i(p)$ through a neighborhood R of A by letting them be constant on the radii; then they are clearly of class C^r in R (see the proof of DM, Lemma 4). This defines the $f_i(p)$ and their partial derivatives of order $\leqq r$ on A; the $f_i(p)$ are now of class C^r on A, as a subset of the plane (see AE §3). Extend them so as to be of class C^r throughout B', by AE Lemma 2. It is clear from the proof of the lemma referred to that there is a constant K such that if the $f_i(p)$ and their first derivatives are bounded by $\eta > 0$, then the same is true of the extensions $F_i(p)$ of the $f_i(p)$, with η replaced by $K\eta$. Set

$$\Phi_i(p) = \psi_i(p) + F_i(p) \text{ in } A + B.$$

Then $\Phi_i(p)$ is of class C^r in B', and equals $\phi_i(p)$ in A. This set of functions maps B' into E_μ, A going into C; if η and hence $K\eta$ is small enough, this projects (see DM, Lemma 23) into the required cell in M.

[10] See DM, §13. Except for the case that $r = \omega$, similar theorems are easily proved in n dimensions. Note that, by the definition in DM, the function ϕ below coincides in A with the function defining C; hence ϕ, considered in A alone, is of class C^r.

Suppose now that $r = \omega$. We first replace $\phi(p)$ by a map of class C^2, by the above proof; call this ϕ again. Define ψ and the $f_i(p)$ again; the $f_i(p)$ are now analytic on A. Extend them through some B' containing $A + B$ by Lemma 4. Making the $f_i(p)$ and their first and second partial derivatives $< \eta$ in absolute value makes the $F_i(p)$ and their first partial derivatives $< K\eta$ in absolute value. Again map B' into E_μ, using the functions $\Phi_i(p)$, and project onto M.[11]

APPENDIX I

THE IMBEDDING THEOREM FOR CLOSED MANIFOLDS

We shall give a simple proof that any closed C^r-m-manifold M $(1 \leqq r \leqq \infty)$ may be C^r-imbedded in a Euclidean space E_μ. We shall then show that if M is either open or closed, $\mu > 2m + 1$, and $2 \leqq r \leqq \omega$, then M may be projected in some direction into a subspace $E_{\mu-1}$, and hence into E_{2m+1}. The latter fact is useful also in other connections. The two facts together prove the imbedding theorem for closed manifolds with $2 \leqq r \leqq \infty$.[12]

To prove the first statement, let $U_1, U_2, \cdots$ be coordinate systems of class C^r defining M, using the closed m-cube Q: $-1 \leqq x_i \leqq 1$ $(i = 1, \cdots, m)$. Say $U_i = \theta_i(Q)$. Let Q' be the cube $-\frac{1}{2} \leqq x_i \leqq \frac{1}{2}$ and set $U_i' = \theta_i(Q')$. As M is closed, there is an s such that $U_1', \cdots, U_s'$ cover M. Let $h(x)$ be a function of class C^r of the real variable x such that

$$h(x) = \begin{cases} 0 \text{ if } x \leqq -1 \text{ or } x \geqq 1, \\ 1 \text{ if } -\frac{1}{2} \leqq x \leqq \frac{1}{2}, \end{cases}$$

$$0 < h(x) < 1 \qquad \text{for all other } x.^{13}$$

Set

$$H(x_1, \cdots, x_m) = h(x_1) \cdots h(x_m),$$

$$H_i(x_1, \cdots, x_m) = x_i H(x_1, \cdots, x_m) \qquad (i = 1, \cdots, m).$$

Then $H \equiv 1$, $H_i \equiv x_i$ in Q'. For each $i = 1, \cdots, m$ and $j = 1, \cdots, s$, set

---

[11] The use of the Dirichlet problem in this connection was suggested to me by L. Ahlfors. The n-dimensional case, with $r = \omega$, might be more easily handled as follows. Suppose N is a compact analytic manifold in E_μ, and $f(p)$ is analytic in N; we wish to extend f throughout E_μ so as to be analytic (i.e. A and B above are replaced by N and $E_\mu - N$). Perhaps one can show that for K sufficiently large, some integral equation such as

$$f(x) = T'K^n \int_N \phi(y) e^{-K^2 r_{xy}^2} dy$$

(compare AE §15) may be solved for $\phi(y)$; $f(x)$ is then analytic in E_μ.

[12] This is the proof originally discovered by the author. The method of proof can also be extended to the case of open manifolds.

[13] We may use AE, Lemma 2, or use

$$g(y) = \frac{e^{1/y}}{e^{1/y} + e^{1/(1-y)}}, \qquad h(x) = g(2x - 1) \qquad (1/2 < x < 1).$$

$$f_{0j}(p) = \begin{cases} H(\theta_j^{-1}(p)) \text{ in } U_j, \\ 0 \text{ in } M - U_j, \end{cases} \qquad f_{ij}(p) = \begin{cases} H_i(\theta_j^{-1}(p)) \text{ in } U_j, \\ 0 \text{ in } M - U_j. \end{cases}$$

This is a set of $\mu = (m + 1)s$ functions of class C^r in M. Arrange them in a definite order.

Let ϕ map each point of M into that point of E_μ with these functions as coordinates; ϕ is of class C^r. To show that ϕ is regular, take any p in M; say p is in U_j'. Project $\phi(U_j')$ onto the m-plane P of the functions $f_{1j}, \cdots, f_{mj}$, in E_μ, and let ψ be the map of U_j' into P thus defined. As $H_i \equiv x_i$ in Q', ψ maps each point $(x_1, \cdots, x_m)$ of U_j' into the point of P with the coordinates $f_{ij} = x_i$. Hence ψ is regular, and consequently ϕ is regular at p. To show that ϕ is (1-1), let p and q be distinct points of M. If they are both in some U_j', the above proof shows that $\phi(p) \neq \phi(q)$. If not, say p is in U_j' while q is not. Then $f_{0j}(p) = 1$ and $f_{0j}(q) < 1$, and again $\phi(p) \neq \phi(q)$.

In the proof of Theorem 5 we shall use the following extension of DM Lemma 13.

LEMMA 5. *If $A = A_1 + A_2 + \cdots$ is a subset of the C^1-k-manifold S, and each A_i is of zero k-extent, then $S - A$ is dense in S.*

Each A_i is nowhere dense in S (apply DM Lemma 13 to $\bar{A}_i$); the statement is a consequence of the Baire Theorem.[14]

We shall now prove the second statement.

THEOREM 5. *If M is a C^r-m-manifold* ($2 \leqq r \leqq \omega$) *in E_μ, $\mu > 2m + 1$, we may imbed it by projection in a subspace $E_{\mu-1}$. If $\mu = 2m + 1$, we may project it into a local manifold in E_{2m}. The direction along which we project may in either case be taken arbitrarily near a given direction.*[15]

Let $S_{\mu-1}$ be the unit $(\mu - 1)$-sphere in E_μ. For p and q in M, $p \neq q$, there is a point $\phi(p, q)$ of $S_{\mu-1}$ whose direction from the origin is the same as that of q from p. For p in M, there is an $(m - 1)$-sphere $S(p)$ in $S_{\mu-1}$, the intersection of $S_{\mu-1}$ with the m-plane through the origin and parallel to the tangent plane to M at p. We shall find a point x arbitrarily near any x_0 in $S_{\mu-1}$ which is on no $\phi(p, q)$ or $S(p)$; this gives a direction along which we may project. Let U_1, $U_2, \cdots$ be coordinate neighborhoods in M such that each p is in arbitrarily small U_i. For each i, j such that $\bar{U}_i \cdot \bar{U}_j = 0$, let A_{ij} be the set of all points $\phi(p, q)$ for p in U_i and q in U_j; then $\sum A_{ij}$ covers all the $\phi(p, q)$. Each A_{ij} is of finite $2m$-extent and hence of zero $(\mu - 1)$-extent (DM, Lemmas 15 and 14). Let B_i be all points on all $S(p)$ for p in U_i. As M is of class C^2, B_i is a C^1-map of the product of the m-cell Q and an $(m - 1)$-sphere, and hence is of zero

[14] We may take open spheres G_i in S whose diameters approach 0, such that $\bar{G}_i$ is in G_{i-1} and has no points in A_i. Then there is a point p in each G_i; p is not in A.

[15] The theorem is not true for $r = 1$. For let $\phi(t)$ map the unit interval $0 \leqq t \leqq 1$ continuously into the whole of the unit sphere $S_{\mu-1}$ in E_μ. There is a curve of class C^1 in E_μ, $x = \psi(t)$, whose direction for each t is given by $\phi(t)$, with t = arc length approximately; it may be made (1-1). Joining the ends gives a C^1-1-manifold for which there is no projection as in the theorem.

$2m$-extent. There is a point x near x_0 and not on any A_{ij} or B_i, by Lemma 5. To prove the second statement in the theorem, we consider merely the B_i.

Appendix II

It seems worth while to give more details in the *proof of* DM *Lemma* 16, in particular, of the statements at the top of p. 662.

(a) If no ξ existed, there would be for each integer n an $\epsilon_n > 0$ and points x_n and y_n of A within ϵ_n of each other, whose transforms under T_{α^0} are distant apart at least $n\epsilon_n$. As A is closed and bounded, $T_{\alpha^0}(A)$ is bounded; hence $\epsilon_n \to 0$. We may therefore suppose $x_n \to x'$, $y_n \to x'$, x' in A. But this clearly contradicts the differentiability of $T_{\alpha^0}(x)$ at $x = x'$. The same proof (using α^n in place of α^0) shows that the statement holds for all α within η of α^0.

(b) A directional derivative $\partial x'/\partial\bar{\alpha}_i$ is a partial derivative after a rotation of axes: $\alpha_j = \sum a_{ji}\bar{\alpha}_i$, $\| a_{ij} \|$ an orthogonal matrix. Hence

$$\frac{\partial x'}{\partial\bar{\alpha}_i} = \sum c_j \frac{\partial x'}{\partial \alpha_j}, \qquad \sum c_j^2 = \sum_j a_{ji}^2 = 1.$$

As the $\partial x'/\partial\alpha_j$ are independent and the sets of c_j run over the unit $(h-1)$-sphere, which is self-compact, the possible $\| \partial x'/\partial\bar{\alpha}_i \|$ have a lower bound $\beta > 0$. As in (a), we may choose β so that this holds for all α within η of α^0.

(c) To make one of $T_\alpha(A^*)$, $T_{\alpha'}(A^*)$ fail to intersect B^*, it is clearly sufficient to take α and α' such that for each x in A,

$$\Delta = \| T_{\alpha'}(x) - T_\alpha(x) \| \geqq \xi\epsilon + \epsilon.$$

If α' approaches α along a straight line lying within η of α^0, then $\lim \Delta/\|\alpha' - \alpha\|$ is a directional derivative, and hence is $\geqq \beta$. Hence, for α and α' within some ζ of $\alpha^0 (\zeta \leqq \eta)$, $\Delta \geqq \frac{1}{2}\|\alpha' - \alpha\|\beta$. Now if $\|\alpha' - \alpha\| \geqq 4\xi\epsilon/\beta$, then $\Delta \geqq 2\xi\epsilon \geqq \xi\epsilon + \epsilon$.

Harvard University.

ANNALS OF MATHEMATICS
Vol. 45, No. 2, April, 1944

THE SELF-INTERSECTIONS OF A SMOOTH n-MANIFOLD IN $2n$-SPACE*

BY HASSLER WHITNEY

(Received August 19, 1943)

1. Introduction

Let f be a regular mapping (see the definitions below) of the simple closed curve M^1 into the plane E^2. The resulting curve $C = f(M)$ may cut itself a number of times; if this number is finite, and the "positive" and "negative" self-intersections are distinguished, the algebraic number I_f of them is invariant under "regular deformations," which keep the mapping always regular.[1] We may determine I_f by considering the space $\mathfrak{T}$ of ordered pairs of points of M^1, mapping it into E^2 in a manner determined by f (see below), and counting the coverings of the origin. The object of this paper is to study the situation in n dimensions, for regular mappings of a manifold M^n into E^{2n}. The main theorem (which is trivial for $n = 1$ and well known for $n = 2$) is that M^n may be imbedded in E^{2n}.

An outline of the paper follows. Take $E^n \subset E^{2n}$. It is shown how E^n may be distorted near the origin, giving a mapping $f(E^n) \subset E^{2n}$ such that there is just one self-intersection; specific equations (2.2) for f are given. By introducing such self-intersections into the mapping f of a manifold M^n into E^{2n}, we may alter I_f as we please. As in the case $n = 1$, we express I_f in terms of a mapping F of $\mathfrak{T}$ into V^{2n} = all vectors in E^{2n}, provided that n is even and M is orientable. In any other case, I_f is only determined mod 2; we must identify (p, q) with (q, p) in $\mathfrak{T}$ $(p, q \in M)$, forming a space $\mathfrak{T}^*$, and let F^* map it into V^*, formed from V by identifying each vector with its negative. The determination of the "divided degree" needed is accomplished through a study of "locally derivable" mappings F^* of spaces into V^*; these are obtainable, locally, as the projection into V^* of mappings into V. We carry this theory somewhat further (in §§5, 6) than is needed in the present paper; the notions are capable of great generalization.

After the proof of the invariance of I_f under regular deformations comes the key theorem (Theorem 4) in the proof of the imbedding theorem; it says essentially that the number of self-intersections may be reduced by two, if the number present is greater than $| I_f |$ (or is $\geqq 2$, if I_f is taken mod 2). It is a curious fact that this theorem is proved only in the case $n \geqq 3$. The proof of the imbedding theorem is now immediate if M is closed, and is easily carried out if M is open. In an appendix, we prove that the complex projective plane (a manifold of di-

---

* Presented to the American Mathematical Society, Sept. 9, 1942. The numbers in brackets refer to the bibliography at the end of the following paper.

[1] See [2]. It is shown further there that if $I_{f_0} = I_{f_1}$, then there is a regular deformation of f_0 into f_1. It would be interesting to know if there is such a theorem in n dimensions.

mension 4) may be imbedded in E^7; since[2] its characteristic is odd, it cannot be imbedded in E^6.

We give a brief description of $\mathfrak{T}$ and F. The set of all pairs (p, u), $p \in M$, u a unit vector tangent to M at p, is the "tangent space" $\mathfrak{S}$ of M. The space $\mathfrak{T}$ consists of $\mathfrak{S}$, and all pairs (p, q), $p \neq q$ $(p, q \in M)$; if $q \to p$ in the direction of a unit vector u at p, we let $(p, q) \to (p, u)$. Thus $\mathfrak{T}$ is bounded by $\mathfrak{S}$. F maps (p, q) into a multiple of $f(q) - f(p)$, so chosen that $\mathfrak{S}$ is mapped into vectors $\neq 0$.

NOTATIONS. A mapping of a manifold M^n is *smooth* if it is of class C^1 (has continuous first partial derivatives). It is *regular* if the Jacobian matrix is everywhere of rank n, or, n independent vectors at any point are carried into n independent vectors, or, each non-zero vector is carried into a non-zero vector. It is an *imbedding* if it is proper (see §24 of the following paper; we may omit this if M is closed) and regular, and is one-one; it is an *immersion* if it is proper and regular. Let I denote the unit interval $0 \leqq x \leqq 1$. We use ∂ for boundary in both the point set and combinatorial sense. If f maps chains in the complex K into chains in K', its dual f' is defined by

$$f'(\sum a'_i \sigma'_i) = \sum_j a'_i \alpha_{ji} \sigma_j, \quad \text{where} \quad f(\sigma_i) = \sum_j \alpha_{ij} \sigma'_j .$$

Scalar products are defined by $(\sum \alpha_i \sigma_i^r) \cdot (\sum \beta_i \sigma_i^r) = \sum \alpha_i \beta_i$. A subscript 2 commonly denotes reducing mod 2.

I. COMBINATORIAL STUDY OF SELF-INTERSECTIONS

2. A particular self-intersection

We shall define a regular mapping f of class C^∞ of the interior Q^n of the unit sphere in E^n into E^{2n} such that there is just one self-intersection, and so that f equals the identity, with all derivatives, on the boundary of Q^n.

DEFINITIONS. Let f be a regular mapping of M^n into E^{2n}. Then for each $p \in M$, there is a plane $T_f(p)$ in E^{2n} tangent to $f(M)$ at $f(p)$, of dimension n. Suppose $f(p_1) = f(p_2)$. We say this point is a *regular self-intersection* if $T_f(p_1)$ and $T_f(p_2)$ have only $f(p_1)$ in common. Then if $u_1, \cdots, u_n$ and $v_1, \cdots, v_n$ are sets of independent vectors in M at p_1 and at p_2 respectively, f carries these into a set of independent vectors in E^{2n} (for definitions, see (4.1) and (4.2)). Conversely, with two such sets of vectors, the self-intersection is regular. If f has only regular self-intersections, and there are no triple points $f(p) = f(q) = f(r)$, we say f is *completely regular*. See also the end of §4.

For $n = 1$, a curve with one loop, like the written letter e, gives the required self-intersection. Such a mapping of x-space into (y, z)-space may be expressed in the form

$$y = x - \frac{2x}{1 + x^2}, \qquad z = \frac{1}{1 + x^2}. \tag{2.1}$$

[2] H. Seifert, *Algebraische Approximationen von Mannigfaltigkeiten*, Math. Zeitschrift, vol. 41 (1936), pp. 1–17, Satz 2; compare E. Stiefel, *Richtungsfelder und Fernparallelismus in n-dimensionalen Mannigfaltigkeiten*, Comp. Math. Helvetici, vol. 8 (1936), Anhang II; also [3].

For $x = \pm 10$ say, these are approximately $y = x$, $z = 0$; hence we may flatten out $f(M)$ near $x = \pm 10$, to obtain exactly these equations there. For general n, the example will again be of a mapping near the identity for large values of the x's, and we shall not trouble to mention the final alterations necessary. (The detailed proof may be given with the help of Lemma 11 of the following paper.)

Consider E^4 as given by moving an E^3 parallel to itself in the y_2-direction. Take $E^2 \subset E^3 \subset E^4$, and take the above $f(M^1) \subset E^2$. As E^3 moves, say with increasing y_2, let the two parts of $f(M^1)$ near the self-intersection be pulled apart, into opposite sides of E^2 in E^3. As y_2 continues to increase, the curve may be flattened out into the line parallel to the y_1-axis. If we let y_2 decrease from 0, we pull the two parts of $f(M^1)$ away in the opposite directions, etc. This clearly defines a self-intersection of the required type. The case $n = 2$ in (2.2) is easily seen to define this mapping.

For general n, the equations are[3]

$$u = (1 + x_1^2) \cdots (1 + x_n^2);$$

$$y_1 = x_1 - \frac{2x_1}{u}, \qquad y_i = x_i \qquad (i = 2, \cdots, n), \tag{2.2}$$

$$y_{n+1} = \frac{1}{u}, \qquad y_{n+i} = \frac{x_1 x_i}{u} \qquad (i = 2, \cdots, n).$$

The matrix of partial derivatives, for example for $n = 3$, is, transposed,

$$\left\| \begin{matrix} 1 - \dfrac{2(1 - x_1^2)}{u(1 + x_1^2)} & 0 & 0 & \dfrac{-2x_1}{u(1 + x_1^2)} & \dfrac{x_2(1 - x_1^2)}{u(1 + x_1^2)} & \dfrac{x_3(1 - x_1^2)}{u(1 + x_1^2)} \\ \dfrac{4x_1 x_2}{u(1 + x_2^2)} & 1 & 0 & \dfrac{-2x_2}{u(1 + x_2^2)} & \dfrac{x_1(1 - x_2^2)}{u(1 + x_2^2)} & \dfrac{-2x_1 x_2 x_3}{u(1 + x_2^2)} \\ \dfrac{4x_1 x_3}{u(1 + x_3^2)} & 0 & 1 & \dfrac{-2x_3}{u(1 + x_3^2)} & \dfrac{-2x_1 x_2 x_3}{u(1 + x_3^2)} & \dfrac{x_1(1 - x_3^2)}{u(1 + x_3^2)} \end{matrix} \right\|$$

To show that f is regular, we must find, for each $p = (x_1, \cdots, x_n)$, a determinant of order n which is $\neq 0$. It is clearly sufficient to find an element $\neq 0$ in the first row. Suppose the element $\partial y_{n+1}/\partial x_1$ $(= \partial y_4/\partial x_1)$ is 0. Then $x_1 = 0$, and hence $\partial y_1/\partial x_1 = 1 - 2/u$. If this is 0, then $u = 2$; hence not all the x_i are 0, and (since $x_1 = 0$) some $\partial y_{n+i}/\partial x_1$ is $\neq 0$.

Next we find the self-intersections. Suppose

$$f(x_1, \cdots, x_n) = f(x_1', \cdots, x_n'), \qquad (x_1, \cdots, x_n) \neq (x_1', \cdots, x_n').$$

Since $y_i' = y_i$ $(i = 2, \cdots, n)$, $x_i' = x_i$ for $i > 1$. Since $y_{n+1}' = y_{n+1}$, $u' = u$; it follows that $x_1'^2 = x_1^2$; hence $x_1' = -x_1 \neq 0$. Since $y_{n+i}' = y_{n+i}$ $(i > 1)$, $x_i =$

[3] The equations are seen to be closely related to those defining the singularity (3.3) of the following paper. It would be possible to define the required self-intersections and singularities in turn, mapping E^1 into E^2, E^2 into E^3, E^2 into E^4, E^3 into E^5, etc., using each time the preceding mapping.

$0\ (i > 1)$. Since $y_1' = y_1$,

$$x_1 - \frac{2x_1}{u} = -x_1 + \frac{2x_1}{u}, \qquad u = 1 + x_1^2 = 2,$$

and $x_1 = \pm 1$. Thus the only self-intersection is

$$f(1, 0, \cdots, 0) = f(-1, 0, \cdots, 0).$$

To show that this is a regular intersection, we note that the matrix $\| \partial y_i / \partial x_j \|$ transposed, at $(\pm 1, 0, \cdots, 0)$, is (for instance for $n = 3$)

$$\begin{Vmatrix} 1 & 0 & 0 & \mp\frac{1}{2} & 0 & 0 \\ 0 & 1 & 0 & 0 & \pm\frac{1}{2} & 0 \\ 0 & 0 & 1 & 0 & 0 & \pm\frac{1}{2} \end{Vmatrix}.$$

The elements of the ith row are the components of the vector $\partial f / \partial x_i$ at $(\pm 1, 0, \cdots 0)$. The two matrices with the two sets of signs give a $2n$-rowed square matrix whose determinant we must prove $\neq 0$. If we subtract the ith row from the $(n + i)^{\text{th}}$ row and expand in terms of the first n columns, we obtain a determinant with diagonal terms equal to 1 or -1, and zeros below the diagonal; hence it is $\neq 0$.

Finally, in the boundary of the region $\sum x_i^2 \leqq 100n$ for example, the equations are practically

$$y_i = x_i\ (i = 1, \cdots, n), \qquad y_i = 0\ (i > n);$$

hence we may flatten out the mapping on the boundary.

3. The spaces $\mathfrak{T}$ and $\mathfrak{T}^*$

Consider M^n as lying in E^{2n+1} (see [1], Theorem 1). The points of $\mathfrak{S}$ are the pairs (p, v), $p \in M$, v a unit vector tangent to M at p. The points of $\mathfrak{T} - \mathfrak{S}$ are the pairs (p, q), $q \neq p$. We define neighborhoods in $\mathfrak{T} - \mathfrak{S}$ in the obvious fashion. To define neighborhoods in $\mathfrak{T}$ at points of $\mathfrak{S}$, we first define a subsidiary space $\mathfrak{T}'$ as follows. Choose a positive continuous function $\eta(p)$ (which may be made constant if M is compact) such that for each $p \in M$ and sphere $S(p)$ of radius $\eta(p)$ in the tangent plane $T(p)$ to M at p, $S(p)$ plus interior projects (along the planes perpendicular to $T(p)$) onto M in a one-one way. Let π_p be this projection. The points of $\mathfrak{T}'$ are the triples (p, v, λ), $p \in M$, v is a unit vector tangent to M at p, and $0 \leqq \lambda < \eta(p)$. The definition of neighborhoods in $\mathfrak{T}'$ is clear. We map $\mathfrak{T}'$ onto a portion of $\mathfrak{T}$ by setting

$$\Phi(p, v, 0) = (p, v) \in \mathfrak{S},$$

$$\Phi(p, v, \lambda) = (p, \pi_p(p + \lambda v)) \in \mathfrak{T} - \mathfrak{S} \qquad (\lambda > 0).$$

Neighborhoods in $\mathfrak{T}'$ map into neighborhoods in $\mathfrak{T}$.

If M is a manifold of class C^r, $\mathfrak{T}$ is a manifold with boundary, of class C^{r-1}. Clearly the choice of the imbedding of M^n in E^{2n+1} does not affect the definition of $\mathfrak{T}$, but only the particular neighborhoods chosen.

The space $\mathfrak{S}^*$ is formed from $\mathfrak{S}$ by identifying "opposite points" (p, v) and $(p, -v)$. The space $\mathfrak{T}^*$ is formed from $\mathfrak{T}$ by the above identification, and also the identification $(p, q) = (q, p)$.

LEMMA 1. *$\mathfrak{T}$ is orientable if and only if M is.*

Let C be a closed curve in $\mathfrak{T} - \mathfrak{S}$ running through (p, q). Let $u_1, \cdots, u_n$ and $v_1, \cdots, v_n$ be independent sets of vectors in M at p and at q respectively. Then keeping q fixed and letting p move in the direction of u_i defines a vector $\bar{u}_i$ in $\mathfrak{T}$; similarly $\bar{v}_i$ is defined, with p fixed. Now the ordered set

$$(3.1) \qquad (\bar{u}_1, \cdots, \bar{u}_n, \bar{v}_1, \cdots, \bar{v}_n)$$

defines an orientation of $\mathfrak{T}$ near (p, q). As we follow around C, the points p and q move along closed paths C_1 and C_2 in M.

Suppose first that M is orientable. Then we may carry the vectors of each set around so that they come back to the same vectors, showing that $\mathfrak{T} - \mathfrak{S}$ is orientable.[4] Now take any closed path in $\mathfrak{T}$. Clearly any points on $\mathfrak{S}$ may be pushed away from $\mathfrak{S}$ (replace $\lambda = 0$ by $\lambda > 0$ in $\mathfrak{T}'$). Applying the above proof shows that $\mathfrak{T}$ is orientable.

If M is non-orientable, let C_1 be a closed curve in M reversing the orientation. Then $n > 1$, so that we may choose a point q not on C_1. As p runs around C_1, (p, q) runs arouhd a curve C in $\mathfrak{T}$, which clearly reverses the orientation in $\mathfrak{T}$.

LEMMA 2. *If n is even and M is orientable, $\mathfrak{T}^*$ is orientable; otherwise, $\mathfrak{T}^*$ is non-orientable.*

As before, we need only consider $\mathfrak{T}^* - \mathfrak{S}^*$.

If M is not orientable, the proof above shows that $\mathfrak{T}^*$ is not. Now take M orientable.

Suppose $n > 1$ is odd. Let σ^n be a cell in M, represented with coordinates $x_1, \cdots, x_n$. Let e_i be the unit vector parallel to the x_i-axis in E^n; this maps into a vector $u_i(p)$ at p for each $p \in \sigma$. Let C be a simple closed curve in σ, and let p and q be distinct points of C. Then $(u_1(p), \cdots, u_1(q), \cdots)$ defines an orientation of $\mathfrak{T}^*$ at (p, q). Now move p and q along the two arcs of C into q and p respectively; these vectors are carried into $u_1(q), \cdots, u_1(p), \cdots$ respectively. Thus a permutation of the vectors is defined, formed by n^2 transpositions; n^2 is odd. Since $(q, p) = (p, q)$, we have defined a closed curve in $\mathfrak{T}^*$ reversing the orientation. Note that if n is even, then C preserves the orientation. If $n = 1$, essentially the same proof applies, with $C = M$.

Suppose finally that M is orientable and n is even. Let C^* be any closed curve in $\mathfrak{T}^* - \mathfrak{S}^*$. Take $(p, q) = (q, p)$ in C^*. As we run around C^*, the points p, q separately may be followed in M, and thus (p, q) may be followed in $\mathfrak{T}$. If (p, q) comes back into (p, q) in $\mathfrak{T}$, the proof in the last lemma shows that the orientation is unchanged. If (p, q) comes back into (q, p) in $\mathfrak{T}$, let C be a simple closed curve in M containing p and q, and as in the proof just above, carry p and q into q and p respectively in σ^n, forming a curve D^* in $\mathfrak{T}^*$. We just saw

[4] This follows also at once from the fact that $\mathfrak{T} - \mathfrak{S}$ is part of the cartesian product $M \times M$. We shall need later the vectors defined here.

that C^* followed by D^* does not reverse the orientation, and we noted above that D^* does not either; hence C^* does not.

DEFINITION. If $\mathfrak{T}$ or $\mathfrak{T}^*$ is orientable and M is oriented, the *corresponding orientation* of $\mathfrak{T}$ or $\mathfrak{T}^*$ is defined by the vectors (3.1).

4. The mappings of $\mathfrak{T}$ and $\mathfrak{T}^*$ corresponding to a mapping f of M into an E^m

Take $M \subset E^{2n+1}$ again. Then a vector u tangent to M at p may be defined as follows by a parametrized curve $P(\lambda)$ starting from p:

$$u = \lim_{\lambda \to 0} \frac{P(\lambda) - p}{\lambda}. \tag{4.1}$$

If ∇f is the mapping of vectors induced by f, than

$$\nabla f(u, p) = \lim_{\lambda \to 0} \frac{f(P(\lambda)) - f(p)}{\lambda}. \tag{4.2}$$

Note that ∇f is a generalization of the gradient of a real-valued function.

Let $|u|$ denote the length of u; $|q - p|$ is the distance from p to q in E^{2n+1}. Define

$$F(p, q) = \frac{f(q) - f(p)}{|q - p|}, \qquad p \neq q, \tag{4.3}$$

$$F(p, u) = \nabla f(u, p) \qquad \text{if } |u| = 1; \tag{4.4}$$

then, taking $P(\lambda)$ as before, if $|u| = 1$ in (4.1),

$$\lim_{\lambda \to 0} F(p, P(\lambda)) = \lim_{\lambda \to 0} \frac{f(P(\lambda)) - f(p)}{|P(\lambda) - p|} = \lim_{\lambda \to 0} \frac{f(P(\lambda)) - f(p)}{\lambda}$$
$$= \nabla f(u, p) = F(p, u).$$

Thus it is clear that F is a continuous mapping of $\mathfrak{T}$ into the space V^m of vectors in E^m.

Clearly F defines a mapping F^* of $\mathfrak{T}^*$ into V^{m*}.

LEMMA 3. *If f is regular, then F maps no point of $\mathfrak{S}$ into O, and maps a point (p, q) of $\mathfrak{T} - \mathfrak{S}$ into O if and only if $f(p) = f(q)$.*

This is obvious.

Suppose $f(p) = f(q)$, $p \neq q$, and u is a vector in M at p. Then if $\bar{u}$ in $\mathfrak{T}$ is defined as in §3,

$$\nabla F(\bar{u}, (p, q)) = \lim_{\lambda \to 0} \frac{F(P(\lambda), q) - F(p, q)}{\lambda}$$
$$= \lim_{\lambda \to 0} \frac{1}{\lambda}\left[\frac{f(q) - f(P(\lambda))}{|q - P(\lambda)|} - \frac{f(q) - f(p)}{|q - p|}\right];$$

clearly

$$\nabla F(\bar{u}_i, (p, q)) = -\frac{1}{|q - p|} \nabla f(u_i, p). \tag{4.5}$$

Similarly

$$\nabla F(\bar{v}_i, (p, q)) = \frac{1}{|q - p|} \nabla f(v_i, q). \tag{4.6}$$

LEMMA 4. *Let f be a completely regular mapping of M^n into E^{2n}. Then if $f(p) = f(q)$ $(p \neq q)$, F is regular in a neighborhood of (p, q), and maps it over O in V^{2n} with the degree ± 1.*

Since f is completely regular, the vectors

$$\nabla f(u_1, p), \cdots, \nabla f(u_n, p), \qquad \nabla f(v_1, q), \cdots, \nabla f(v_n, q) \tag{4.7}$$

are independent; hence so are the $\nabla F(\bar{u}_i, (p, q))$ and $\nabla F(\bar{v}_i, (p, q))$. The last part of the lemma is a consequence of this.

We consider briefly manifolds with boundary. If f is regular in M and on the boundary ∂M of M (with the obvious definitions, assuming for example that ∂M is a manifold), and $f(p) = f(q)$, $p \neq q$, implies that neither p nor q is in ∂M, we call f *completely regular*. If this is so, F maps no part of $\mathfrak{T}$ coming from points on or near ∂M into O; we need not be explicit as to the definitions of $\mathfrak{T}$ and $\mathfrak{S}$ for such points. Similarly for F^*.

5. Mappings into a vector space with opposite vectors identified

Let $V = V^m$ be the space of vectors in E^m. Let ω be the mapping $\omega(v) = -v$ of V into itself. Form $V^* = V^{m*}$ from V by identifying v with $-v$ for all v; let $\pi(v) = \pi(-v)$ be the corresponding point of V^*. Let O be the zero vector of V, and set $O^* = \pi(O)$.

An arc in $V - O$ from $v \neq O$ to $-v$ maps under π into a closed curve, defining the single element γ not the identity of the fundamental group of $V^* - O^*$.

Let T be an $(m - 1)$-plane in V through O; set $\pi(T) = T^*$. For any subdivision K^* of $V^* - O^*$ with no vertex in T^*, let W^{*1} be the cocycle mod 2 containing those 1-cells cutting T^*; let $\mathbf{W}^{*1}$ be its cohomology class (which is invariantly determined).

DEFINITION. Let f^* map the space R into V^*. If there is a mapping f of R into V such that $f^* = \pi f$, we say f^* is *derivable from a mapping into* V, or is *derivable* simply.

The following lemmas are easily proved.

LEMMA 5. *f^* is derivable if and only if it is derivable over each component of $f^{*-1}(V^* - O^*)$.*

LEMMA 6. *A mapping f^* of a connected complex K into $V^* - O^*$ is derivable if and only if any of the following equivalent conditions are satisfied:*

(α) *Each closed curve in K maps into a curve in $V^* - O^*$ defining the identity element of the fundamental group.*

(β) *Each closed curve in K maps into a curve cutting T^* an even number of times (if this number is finite).*

(γ) *Each 1-cycle of K maps into a bounding 1-cycle of $V^* - O^*$.*

(δ) $f^{*\prime}\mathbf{W}^{*1} = 0$.

DEFINITION. A mapping f^* of a space R into V^* is *locally derivable* (in terms of the f_i below) if there exist open sets $U_1, U_2, \cdots$, covering R and mappings f_i of U_i into V such that:

(a) For each i, $f^* = \pi f_i$ in U_i.

(b) For each i and j, either $f_i = f_j$ or $f_i = \omega f_j$ in $U_i \cap U_j$.

REMARKS. If $\{U_i\}$ and $\{f_i\}$ are replaced by $\{U'_i\}$ and $\{f'_i\}$, the definition is considered as equivalent if (b) holds for the old and new sets together. (See Example 3 below.) Say a set $A \subset R$ is *r-thin* if any singular r-cell in R may be moved away from A by an arbitrarily slight deformation. If $f^{*-1}(O^*)$ is 1-thin, and each non-void $U_i \cap U_j$ is connected, then clearly (b) is a consequence of (a), and any sets of f_i are equivalent. Any mapping of a space into $V^* - O^*$ is clearly locally derivable.

EXAMPLES. The first four illustrate the meaning of local derivability; the fifth illustrates its use in this paper. S and S^* are defined in §6.

(1) Let f_1^* be the identity mapping of $Q^* = S^*$ plus interior into itself. Let f_t map each radius from O^* into itself, shrinking it by the factor t; then $f_0(p) = O^*$ (all p). Then f_0 is locally derivable, while (as proved in Example 9) f_t is not if $t > 0$.

(2) Take f_1 as before, and let f_t map all points of any radius within a distance $1 - t$ of O^* into O^*, extending f_t in an obvious manner over Q^*; again $f_0(p) = O^*$. Then f_1^* is not locally derivable, while f_t is if $t < 1$. However, if a slight change is made for $t > 0$, forming g_t^*, such that each $g_t^{*-1}(O^*)$ is 1-thin in Q^*, then no g_t^* $(t > 0)$ is locally derivable; for the corollary to Lemma 7 below would be contradicted. Hence, by a Remark above, we cannot make (a) hold.

(3) Let $R_1, R_2, \cdots$ be a sequence of disjoint spheres plus interiors in some E^r, approaching a point p_0. Let f map all of E^r but these interiors into O, and map the interiors into $V - O$; set $f^* = \pi f$. Then if $p_0 \in U_i$, there are an infinite number of choices for f_i, making f^* locally derivable in distinct ways.

(4) Let $C_1, C_2, \cdots$ be a sequence of simple closed curves in E^r, with disjoint neighborhoods $U'_1, U'_2, \cdots$. Map U'_i into V^* so that $f^*(C_i)$ defines γ and so that if $f^*(p) = O^*$ $(p \in E^r - \sum U'_i)$, then f^* is continuous. If the C_i converge smoothly to a simple closed curve C, then f^* is locally derivable (but see a remark in Example 2); if the C_i converge to a line segment, it is easily seen that f^* is not locally derivable, even though we can make (a) hold.

(5) Let Q^2 be the unit circle S^1 plus interior in E^2, and let g be the identity mapping of Q^2 into E^2. (We could equally well map Q^r into E^m, $2 \leqq r \leqq m$.) If we identify opposite points of S^1, Q^2 becomes a projective plane P^2; πg becomes a mapping f^* of P^2 into V^*. f^* is locally derivable but not derivable. A diameter of Q^2 becomes a "projective line" L of P^2; set $f^*(L) = L^*$. Subdivide P^2 so that L lies on no vertex a_i, and set $U_i = St(a_i)$ (St = star). If f_i in U_1 is chosen so that $f^* = \pi f_1$ there, then there is a unique choice for each f_i such that if $a_i a_j$ is a simplex in P^2 not cutting L, then $f_i = f_j$ in $St(a_i a_j)$. Now the W^1 of (5.1) below contains just those $a_i a_j$ cutting L; $\mathbf{W}^1 \neq 0$. If we omit a small concentric disc from Q^2, and hence from P^2, a Möbius strip M^2 illustrating Lemma 6 is formed, with $f'\mathbf{W}^{*1}$ being the cohomology class of the part of the above W^1 lying in M^2.

(6) Take Q^2, S^1, as in (5). Map S^1 into a curve in $V^* - O^*$ so as to define γ; map radii of Q^2 into segments from O^*. Then f^* is not locally derivable; in the notations of Lemma 8 below,

$$\delta f^{*\prime} W^1 \cdot Q^2 = f^{*\prime} W^1 \cdot \partial Q^2 = W^1 \cdot f^* S^1 = 1_2 .$$

If f^* is any mapping of Q_2 into $V^* - O^*$ with an inner point p going into O^*, then a slight alteration of f^* may be made which will give a mapping g^* for which a neighborhood of p is mapped as above; then g^* is not locally derivable.

DEFINITION. Let f^* be a locally derivable mapping of the complex K into V^* such that $f^{*-1}(O^*)$ is 0-thin, i.e. nowhere dense in K. Then the cohomology class W^1 of f^* is defined as follows. Take the U_i and f_i as before. Let K_1 be a subdivision of K so fine[5] that each star $St_1(a_i)$ of a vertex a_i of K_1 is in some U_{λ_i}. Set

$$W^1(a_i a_j) = \begin{cases} 0_2 & \text{if } f_{\lambda_i} = f_{\lambda_j} \text{ in } St_1(a_i a_j), \\ 1_2 & \text{if } f_{\lambda_i} = w f_{\lambda_j} \text{ in } St(a_i a_j). \end{cases} \tag{5.1}$$

Then $\mathbf{W}^1$ is the cohomology class of $W^1 = \sum W^1(a_i a_j) a_i a_j$.

LEMMA 7. *W^1 is well defined, and is a cocycle (coefficients* mod 2). *Altering the f_i alters it by a coboundary. Its class $\mathbf{W}^1$ is independent of the choice of K, the U_i and the f_i, provided that the new f_i are equivalent to the old. If $f^{*-1}(O^*)$ is 1-thin, then f^* is derivable if and only if $\mathbf{W}^1 = 0$.*

REMARKS. Any such W^1 is called a *characteristic cocycle* of f^*. By standard theory, that $\mathbf{W}^1$ is independent of K is a consequence of the following statement: If K_1 and K_2 are such that each $St_2(a_{2i})$ lies in some $St_1(a_{1\nu_i})$, so that a simplicial mapping $\phi(a_{2i}) = a_{1\nu_i}$ of K_2 into K_1 is defined, and W_1^1 is a characteristic cocyle of K_1, then $W_2^1 = \phi' W_1^1$ (ϕ' = dual of ϕ) is a characteristic cocycle of K_2; by definition, $W_2^1(a_{2i}a_{2j}) = W_1^1(\phi(a_{2i}a_{2j})) = W_1^1(a_{1\nu_i}a_{1\nu_j})$.

First, since $f^{*-1}(O^*)$ is nowhere dense in K_1, the definition of W^1 is obviously unique. Next, suppose a definite f_k is changed to ωf_k. Then if $i_1, \cdots, i_r$ are the i such that $\lambda_{i_s} = k$, $W^1(a_i a_j)$ is changed if and only if just one of i, j lies among the i_s; thus, if δ denotes coboundary,

$$\text{new } W^1 = \text{old } W^1 + \delta \textstyle\sum (a_{i_s})_2 .$$

Now for any sets $\{U_i\}$, $\{U_i'\}$ and equivalent $\{f_i\}$, $\{f_i'\}$, choose K_1 so that $St_1(a_i) \subset U_{\lambda_i} \cap U'_{\mu_i}$. Let f_{ij} and f'_{ij} denote f_{λ_i} and f'_{μ_i} respectively, considered in $U_{\lambda_i} \cap U'_{\mu_j}$ (if this is not void); all the f are equivalent. Clearly the f_{λ_i} and the f_{ij}, also the f'_{μ_i} and f'_{ij}, define the same W^1. The proof above shows that changing the f_{ij} to the f'_{ij} alters W^1 by a coboundary.

Now suppose K_1 and K_2 are as described above. In using (5.1) to define W_2^1, since $St_2(a_{2i}) \subset St_1(a_{1\nu_i})$, we may take for the f_k defined about a_{2i} the same as the f_k defined about $a_{1\nu_i}$; then clearly $W_2^1(a_{2i}a_{2j}) = W_1^1(a_{1\nu_i}a_{1\nu_j})$.

[5] From the corollary below it is clear that the f_i may be defined in the stars in K, so that no subdivision is necessary in reality.

Suppose $f^* = \pi f$. Since $f^{*-1}(O^*)$ is 1-thin, the given set $\{f_i\}$ is equivalent to the single mapping f; hence W^1 may be defined using f. Since $W^1 = 0$, $\mathbf{W}^1 = 0$.

Suppose finally that $\mathbf{W}^1 = 0$. For a given subdivision K_1, and sets $\{U_i\}$, $\{f_i\}$, we have $W^1 = \delta X^0$ for some $X^0 = \sum \alpha_i a_i$ ($\alpha_i = 0_2$ or 1_2). If $St_1(a_i) \subset U_{\lambda_i}$, set $f'_i = f_{\lambda_i}$ in $St_1(a_i)$. For each i with $\alpha_i = 1_2$, alter f'_i to $\omega f'_i$; then (see a proof above) the new $W^1 = 0$. Now for any p in K, say $p \in St(a_i)$; set $f(p) = f'_i(p)$. This makes f continuous in K, and $f^* = \pi f$.

EXAMPLE (7). Let f map a circle C into V so that just one point p_0 goes into O, and set $f^* = \pi f$. Let U_1 and U_2 cover C, p not being in U_2. Then f_1 and f_2 may be chosen so that W^1 is not $\backsim 0$, though f^* is derivable. Note that $f^{*-1}(O^*) = p_0$, which is not 1-thin in C.

COROLLARY. *Any locally derivable mapping f^* of a cell (in fact, of a complex with vanishing* 1*-cohomology group* mod 2) *is derivable, if $f^{*-1}(O^*)$ is* 1*-thin in K.*

A direct proof for a cell may be given simply if we build it up as in (b) of Lemma 12.

We shall give a lemma which shows immediately whether or not a mapping is locally derivable or derivable, provided that $f^{*-1}(O^*)$ is 1-thin. Since it involves notions[6] not yet generally known, and the lemma is not needed here, we shall be rather brief.

A geometric r-G-cochain $X = X^r$ in a point set R is a pair $\bar{X} \supset \bar{X}'$ (nucleus and nuclear boundary) of closed sets, $(r - 1)$-thin and r-thin respectively, and a group of homomorphisms $X \cdot A^r$ of those integral singular chains A^r for which $A \cap \bar{X}' = \partial A \cap \bar{X} = 0$ into G, with the properties $X \cdot A = 0$ if $A \cap \bar{X} = 0$, $X \cdot \partial B^{r+1} = 0$ if $B \cap \bar{X}' = 0$. If G and H are paired to J, set $X \cdot \sum h_i \sigma_i^r = \sum (X \cdot \sigma_i^r) \cdot h_i$. The coboundary δX of X is the pair of sets $\bar{X}'$, 0, with the definition $(\delta X) \cdot A^{r+1} = X \cdot \partial A^{r+1}$. To any cochain X^r in a complex corresponds in a fairly obvious fashion a geometric cochain with nucleus and nuclear boundary formed from duals of r- and $(r + 1)$-cells.

Since T^* and O^* are 0-thin and 1-thin in V^* respectively, they give a geometric cochain W^1 of V^*, such that $W^1 \cdot \sigma^1$ counts the number of times mod 2 that σ^1 crosses T^*. This is the geometric analogue of the W^1 first defined. If f^* is as in the lemma below, then

$$W^1_{f^*} \cdot A^1 = W^1 \cdot f^* A^1$$

defines a geometric cocycle $W^1_{f^*} = f^{*\prime} W^1$ in K. Note that $\delta f^{*\prime} W^1 = f^{*\prime} \delta W^1$.

LEMMA 8. *Let f^* be any mapping of a complex K into V^* such that $f^{*-1}(T^*)$ and $f^{*-1}(O^*)$ are* 0*-thin and* 1*-thin respectively. Then f^* is locally derivable if and only if $\delta f^{*\prime} W^1 = 0$, and is derivable if and only if $f^{*\prime} W^1 \backsim 0$.*

REMARK. Using ordinary cocyles, we may replace the condition $\delta f^{*\prime} W^1 = 0$ by the following: For any arbitrarily fine subdivision K_1 of K there is an arbitrarily small deformation of f^* into f_1^* such that $f_1^{*\prime} W^1$ is defined and is a cocycle.

Suppose f^* is locally derivable, using $\{U_i\}$ and $\{f_i\}$. Take any singular chain A^2, with $\partial A^2 \cap f^{*-1}(O^*) = 0$, so that $\delta W^1_{f^*} \cdot A$ is defined. Using a fine

[6] See H. Whitney, *Geometric methods in combinatorial topology*, not yet published.

subdivision of A, we may write $A = \sum \alpha_i \sigma_i^r$, each σ_i^r in some U_{λ_i}; since $f^{*-1}(O^*)$ is 1-thin, we may deform A slightly so that $\sum \alpha_i(\delta W_{f^*}^1 \cdot \sigma_i^r)$ is defined; it equals $\delta W_{f^*}^1 \cdot A$. Using f_{λ_i}, it is clear that $\delta W_{f^*}^1 \cdot \sigma_i^r = W^1 \cdot \partial f^* \sigma_i^r = O_2$; hence $\delta W_{f^*}^1 = 0$.

Suppose $\delta W_{f^*}^1 = 0$. Each point $p \in K$ is in a connected neighborhood U such that any 1-cycle in U bounds. Choose $p_0 \in U$ so that $f^*(p_0) \neq O^*$; choose $f(p_0)$ so that $\pi f(p_0) = f^*(p_0)$. For any $p' \in U$ with $f^*(p') \neq O^*$, choose an arc C in U from p_0 to p' with $f^*(C) \subset V^* - O^*$ (which is possible, since $f^{*-1}(O^*)$ is 1-thin); $f(p')$ is uniquely definable so that $f(p'')$ exists and is continuous in C, and $f^* = \pi f$ there. We must show that going around a closed curve A^1 in U and through p_0 ($f^*(A^1) \subset V^* - O^*$) brings $f(p_0)$ back to its original value. Say $A^1 = \partial B^2$. Then $W_{f^*}^1 \cdot A^1 = \delta W_{f^*}^1 \cdot B^2 = 0$, from which the statement clearly follows. Thus we prove (a); since $f^{*-1}(O^*)$ is 1-thin, (b) follows also for suitable $\{U_i\}$.

Suppose f^* is derivable; say $f^* = \pi f$. Then $\mathbf{W}_{f^*}^1 = f'\pi'\mathbf{W}^1 = f'0 = 0$, since the cohomology groups of V vanish.

Suppose finally $W_{f^*}^1 = \delta W^0$. Then for any closed curve A^1 in K with $f^*(A^1) \subset V^* - O^*$, $W_{f^*}^1 \cdot A^1 = W^0 \cdot \partial A^1 = 0$; it follows by Lemma 5 and Lemma 6 (β) that f is definable throughout K.

LEMMA 9. *Let f_0^* be a locally derivable mapping of the complex K into V^*, with $f^{*-1}(O^*)$ nowhere dense in K. Then there is an arbitrarily slight deformation f_t^* of f_0^* $(0 \leqq t \leqq 1)$ such that $F^*(t, p) = f_t(p)$ is locally derivable, with mappings F_i equivalent in $0 \times K$ to the given $f_{0,i}$ there, and such that $f_1^*(K^{2n-1}) \subset V^* - O^*$ If A is a closed subset of K with $f^*(A) \subset V^* - O^*$, we may make $f_t^* = f_0^*$ in A.*

REMARK. As a consequence, it is easily seen (using the various derived of K) that $f_1^{*-1}(O^*)$ may be made $(2n - 1)$-thin in K.

We may suppose $f_{0,i}$ is defined in $St(a_i)$; we shall define F_i in $I \times St(a_i)$. Let $St'(a_i)$ be the barycentric star of a_i, containing all points on cells of the first derived of K which have a_i as a vertex. Then $\overline{St}'(a_i) \subset St(a_i)$ and $\sum \overline{St}'(a_i) = K$. Take a fixed i. We may define $f_{t,i}$ in $\overline{St}(a_i)$ for $0 < t \leqq t_1$ so that

$$f_{t_1,i}(K^{2n-1} \cap \overline{St}'(a_i)) \subset V^* - O^*,$$

$$f_{t,i}(p) = f_{0,i}(p) \qquad (p \in \partial St(a_i) \cup (St(a_i) \cap A)),$$

with $f_{t,i}(p)$ arbitrarily close to $f_{0,i}(p)$. For each j such that $a_i a_j$ is a 1-cell of K, set

$$f_{t,j}(p) = f_{t,i}(p) \text{ or } \omega f_{t,i}(p) \qquad (p \in St(a_i a_j)),$$

according as which holds for $t = 0$, and set $f_{t,j}(p) = f_{0,j}(p)$ elsewhere. This clearly defines $F^*(t, p)$ for $p \in K$ and $0 < t \leqq t_1$. Do this for each a_i in turn, using successive subintervals of (0, 1). (K may be infinite, if it is "locally finite.") If the successive alterations are small enough, we will have automatically $F^*(t, p) \subset V^* - O^*$ ($p \in K^{2n-1} \cap \overline{St}'(a_i)$) for previous i. Thus all of F^* is defined.

6. Locally derivable mappings into V^{2n*}

DEFINITIONS. Let $S = S^{2n-1}$ be the unit sphere in V^{2n}, and set $S^* = S^{2n-1*} = \pi S$. Then S^* is a projective space, which is orientable, since $2n - 1$ is odd. We may suppose the spaces oriented so that, combinatorially, $\pi V = 2V^*$ and $\pi S = 2S^*$; thus π has the degree 2. Let P and P^* be the projections of $V - O$ and $V^* - O^*$ along rays from O and O^* onto S and S^* respectively.

Let L^* be a ray from O^* in V^*. Given any subdivision K^* of $V^* - O^*$ with $(K^*)^{2n-2} \subset V^* - L^*$, we may define a cocycle $W = W^{2n-1}$ in K^* as follows. For any $\sigma = \sigma^{2n-2}$, $W^{2n-1}(\sigma) = KI(\sigma, L^*)$. This is the algebraic number of times that $P^*\sigma$ covers $L^* \cap S^*$ in S^* or, that L^* cuts through σ.

Since $L^* - O^*$ is $(2n - 2)$-thin in $V^* - O^*$, we can consider W as a geometric cocycle in $V^* - O^*$. Since O^* is not $(2n - 1)$-thin in V^* for $n \geqq 2$, (in fact, not 2-thin; compare Example (6) and the corollary to Lemma 7), we cannot properly consider it as a cochain in V^*, whose coboundary has the nucleus O^*. In fact, δW^* may be considered in a general or in a restricted way:

(A) If a singular chain $A^{2n} \subset V^*$ has $\partial A^{2n} \subset V^* - O^*$, we may set $\delta W \cdot A^{2n} = W \cdot \partial A^{2n}$. For example, if $A^{2n} = S^*$ plus interior, then $\partial A = S^*$, and $\delta W \cdot A = 1$.

(B) If we define $\delta W \cdot A^{2n}$ only for singular chains A which may be expressed as $A = \sum \alpha_i \sigma_i^{2n}$, with $\partial \sigma_i^{2n} \subset V^* - O^*$, then we have always $\delta W \cdot A = W \cdot \partial A \equiv 0 \pmod 2$; see the next lemma.

DEFINITIONS. For any singular cycle $A^{2n-1} \subset V^* - O^*$, the *degree* $d(A)$ of A^{2n-1} is $W^{2n-1} \cdot A^{2n-1}$. If f^* maps a cycle A^{2n-1} into $V^* - O^*$, the *degree* $d_{f^*}(A)$ *of* f^* *in* A *about* O^* is $d(f^*A)$. This is clearly invariant under deformations; furthermore, combinatorially,

$$P^*f^*A = \alpha S^*, \qquad \alpha = d_{f^*}(A). \tag{6.1}$$

For any singular chain A^{2n} with $\partial A^{2n} \subset V^* - O^*$, the *degree* $d(A^{2n})$ *of* A^{2n} *over* O^* is $W^{2n-1} \cdot \partial A^{2n}$ ($= \delta W \cdot A$); define $d_{f^*}(A^{2n})$ similarly. These are obvious generalizations of the degree about O or over O in V.

LEMMA 10. *Suppose* $n \geqq 2$. *If* f^* *maps a complex* K *into* V^* *so that* $f^*K^{2n-1} \subset V^* - O^*$, *then* $\delta f^{*\prime} W^{2n-1} \equiv 0 \pmod 2$, *so that*

$$\tilde{W}_{f^*}^{2n} = \tfrac{1}{2}\delta f^{*\prime} W^{2n-1} \quad \textit{exists.} \tag{6.2}$$

The strict meaning of $\tilde{W}_{f^*}^{2n}$ is given by its action on any $2n$-G-chain $A^{2n} = \sum g_i \sigma_i^{2n}$, which is as follows:

$$\tilde{W}_{f^*}^{2n} \cdot \sum g_i \sigma_i^{2n} = \sum g_i (\tilde{W}_{f^*}^{2n} \cdot \sigma_i^{2n}), \tag{6.3}$$

$$\tilde{W}_{f^*}^{2n} \cdot \sigma = \tfrac{1}{2}\delta f^{*\prime} W^{2n-1} \cdot \sigma = \tfrac{1}{2}(W^{2n-1} \cdot f^* \partial\sigma). \tag{6.4}$$

Since $2n - 1 > 1$, $\partial\sigma$ is simply connected; hence there is a mapping f of $\partial\sigma$ into $V - O$ such that $f^* = \pi f$. Say $Pf\partial\sigma = \alpha S$. Then

$$P^*f^*\partial\sigma = P^*\pi f\partial\sigma = \pi P f\partial\sigma = 2\alpha S^*;$$

hence

$$\delta f^{*\prime} W^{2n-1}\cdot\sigma = W^{2n-1}\cdot f^*\partial\sigma = W^{2n-1}\cdot P^*f^*\partial\sigma = 2\alpha,$$

from which the lemma follows.

REMARKS. By Lemma 9 and the remark following it, a slight deformation of a locally derivable f^* will give an f_1^* such that $f_1^{*\prime}W^{2n-1}$ and $\tilde{W}_{f_1^*}^{2n}$ are geometric cochains.

It is not true that $\partial A^{2n} = 0$ implies $\tilde{W}_{f^*}^{2n}\cdot A = \frac{1}{2}(\delta f^{*\prime}W^{2n-1}\cdot A) = \frac{1}{2}(W^{2n-1}\cdot f^*\partial A) = 0$; see Example (8) below. The second expression has in general no meaning, say if the coefficients are integers mod 2. But it holds for integral A, as is apparent from (6.4).

DEFINITION. Given a locally derivable mapping f^* of K into V^*, and a $2n$-G-chain A in K such that $f^*\partial A \subset V^* - O^*$, make a slight deformation of f^* into f_1^* such that $f_1^*(K^{2n-1}) \subset V^* - O^*$, as in Lemma 9, keeping ∂A^* in $V^* - O^*$; let the *divided degree* of A under f be

$$\tilde{d}_{f^*}(A) = \tilde{W}_{f_1^*}^{2n}\cdot A \qquad \text{(which is an element of } G). \tag{6.5}$$

Note that this defines $\tilde{d}_{f^*}(A)$ for locally derivable singular chains A with $\partial A \subset V^* - O^*$. It need not vanish for cycles (see Example (8) below).

LEMMA 11. *The above degree is independent of the choice of f_1^*, vanishes for chains mapped into $V^* - O^*$, and is invariant under deformations f_t^* such that $F^*(t, p) = f_t^*(p)$ is locally derivable in $I \times K$ and $f_t^*(\partial A) \subset V^* - O^*$. It vanishes for all boundaries ∂B^{2n+1}, in fact, for all "cycles of the first kind," expressible as $\sum g_i A_i^{2n}$ ($g_i \subset G$, A_i^{2n} integral cycles).*

We prove the last part first. Supposing $f_1^*(K^{2n-1}) \subset V^* - O^*$, we have

$$\tilde{W}_{f_1^*}^{2n}\cdot\sum g_i A_i^{2n} = \sum g_i(\tilde{W}_{f_1^*}^{2n}\cdot A_i^{2n}) = \sum g_i[\tfrac{1}{2}(W^{2n-1}\cdot f^*\partial A_i^{2n})] = 0;$$

also, if $A^{2n} = \partial B^{2n+1}$, say $B^{2n+1} = \sum g_i\sigma_i^{2n+1}$, then $A^{2n} = \sum g_i\partial\sigma_i^{2n+1}$ is a cycle of the first kind, so that $\tilde{W}_{f_1^*}^{2n}\cdot A^{2n} = 0$.

To prove the rest, it is sufficient to show that if

$$f_t^*\partial A \cup f_0^*(K^{2n-1}) \cup f_1^*(K^{2n-1}) \subset V^* - O^*,$$

and $F^*(t, p) = f_t^*(p)$ is locally derivable, then

$$\tilde{W}_{f_1^*}^{2n}\cdot A = \tilde{W}_{f_0^*}^{2n}\cdot A.$$

By a slight deformation of F^*, keeping $F^*(0, p)$ and $F^*(1, p)$ fixed, we may suppose that F^* maps $(I \times K)^{2n-1}$ into $V^* - O^*$, and is locally derivable (see Lemma 9); suppose this is done. The chain A determines chains $0 \times A$, $1 \times A$, $I \times A$ in $I \times K$ (if we use the usual subdivision of $I \times K$). Using the fact just proved, we have $\tilde{W}_{F^*}^{2n}\cdot[\partial(I \times A)] = 0$; hence, since $\partial(I \times A) = 1 \times A - 0 \times A - I \times \partial A$ and $F^*(I \times \partial A) \subset V^* - O^*$, $\tilde{W}_{F^*}^{2n}\cdot(1 \times A) = W_{F^*}^{2n}\cdot(0 \times A)$, which give the result.

MORE EXAMPLES. (8) (Compare Examples (1) and (5).) Let Q^{2n} be S^{2n-1} plus interior R^{2n}, and form P^{2n} by identifying opposite points of S^{2n-1}. Thus

$P^{2n} = S^*$ plus R^{2n}. Set $\bar{f}$ = identity in Q^{2n}, and $\bar{f}^* = \pi\bar{f}$; this gives a mapping f^* of P^{2n} into V^*. Now P^{2n} is a projective space, of even dimension $2n$, and hence is non-orientable; it has a fundamental $2n$-cycle A^{2n} mod 2. If we subdivide it so that O is interior to a $2n$-cell σ_0^{2n} of Q^{2n}, it is clear that

$$\tilde{W}_{f^*}^{2n} \cdot A^{2n} = [\tfrac{1}{2}(W^{2n-1} \cdot \pi\partial\sigma_0^{2n})]_2 = 1_2 .$$

Of course A^{2n} is not a cycle of the first kind. In the case $n = 1$, this type of mapping illustrates Theorem 1; see the discussion there for $n = 1$.

We may deform f^* by contracting $f^*(K)$ to O^*, and then pulling along a single ray away from O^*. This shows that the assumption in Lemma 11 that F^* is locally derivable cannot be omitted.

(9) Let f^* be the identity mapping of S^* into itself. We may consider S^* as an integral cycle; $W^{2n-1} \cdot S^* = 1$. Of course S^* bounds its interior (combinatorially); but by Lemma 11, there exists no complex K containing S^* in which S^* bounds, such that there is a locally derivable mapping of K into V^* which equals the identity in S^*.

7. The intersection number of a mapping

Definitions. Let f be a completely regular mapping of M^n (with or without boundary) into E^{2n}; we suppose E^{2n} oriented. Suppose first that M is orientable and n is even. Choose an orientation of M. Suppose $f(p) = f(q)$; let $u_1, \cdots, u_n$ and $v_1, \cdots, v_n$ be sets of vectors in M as in §3, each set determining the positive orientation of M. Then we say this self-intersection is of *positive* or *negative* type according as the vectors (4.7) determine the positive or negative orientation of E^{2n}. Also, by (4.5) and (4.6) (see the end of §3), the type is positive or negative according as the degree in Lemma 4 is 1 or -1. Note that this is independent of the orientation chosen for M. The *intersection number* I_f of f is the algebraic number of self-intersections in this case, and the number of self-intersections mod 2 if M is non-orientable or n is odd. In the case $n = 1$, it is possible to determine I_f in a certain sense as an integer; see the discussion following Theorem 1.

Define F and F^* as in §4. Clearly F^* is locally derivable. Considering $\mathfrak{T}^*$ as a singular chain, with integers or integers mod 2 as coefficients according as it is orientable or not, define the degree $d_F(\mathfrak{T})$ and divided degree $\tilde{d}_{F^*}(\mathfrak{T}^*)$ as in §6. Let D denote the distant intersection of M (see [3], §18).

Theorem 1. *Let f be a completely regular mapping of M^n into E^{2n} (M may have a boundary). Let the number of self-intersections be finite. Then*

(a) *If M is orientable and n is even,*

$$I_f = \tilde{d}_{F^*}(\mathfrak{T}^*) = \tfrac{1}{2} d_F(\mathfrak{T}) = \tfrac{1}{2}(D \cdot M^n) \qquad \textit{(an integer)}. \tag{7.1}$$

(b) *If M is non-orientable or n is odd,*

$$I_f = \tilde{d}_{F^*}(\mathfrak{T}^*) \qquad \textit{(an integer mod 2)}. \tag{7.2}$$

(c) *If M is orientable and n is odd,*

$$d_F(\mathfrak{T}) = D \cdot M^n = 0. \tag{7.3}$$

The theorem shows that F^* has more interest than F in this connection. We discuss the case $n = 1$ further later.

Suppose first that M is orientable and n is even. Take a self-intersection $f(p) = f(q)$, $p \neq q$. The points of $\mathfrak{T}$ near (p, q) are mapped by F over O with the degree 1 or -1, according as the intersection is of positive or negative type. The same is true for the points of $\mathfrak{T}$ near (q, p). Thus each positive [negative] self-intersection contributes 2 [-2] to d_F, and $I_f = \frac{1}{2}d_F$, which is part of (7.1). That $I_f = \frac{1}{2}(D \cdot M)$ follows easily from the definition of D.

Next we prove (c). Take (p, q) again, and define the vectors u_i and v_i as before. Neighborhoods of (p, q) and (q, p) are oriented by $\bar{u}_1, \cdots, \bar{v}_n$ and $\bar{v}_1, \cdots, \bar{u}_n$ respectively; since n is odd, they are mapped over O in V^{2n} with opposite degrees; hence they contribute 0 to d_F. Thus $d_F = 0$. Similarly $D \cdot M = 0$.

Now consider F^* in $\mathfrak{T}^*$, $\mathfrak{T}^*$ orientable; clearly F^* is locally derivable. The only points mapped into O^* by F^* are points $(p, q) = (q, p), f(p) = f(q), p \neq q$. Let U be a neighborhood of (p, q) in $\mathfrak{T}$; it corresponds to a neighborhood U^* in $\mathfrak{T}^*$, say under the mapping ϕ. Then $F^* = \pi F \phi^{-1}$ in U^*. Let $\sigma^* = \sigma^{*2n}$ be a cell in U^* about (p, q), oriented like $\mathfrak{T}^*$. If $\sigma = \phi^{-1}\sigma^*$, then as noted above, F maps σ over O with the degree 1 or -1, according as the self-intersection is positive or negative. Since π is of degree 2, πF maps σ over O^* with the degree 2 or -2, i.e.

$$W^{2n-1} \cdot \pi F \partial\sigma = W^{2n-1} \cdot F^* \partial\sigma^* = 2 \text{ or } -2.$$

Hence, by (6.4),

$$\tilde{W}^{2n}_{F^*} \cdot \sigma^* = \tfrac{1}{2}(W^{2n-1} \cdot F^* \partial\sigma^*) = 1 \text{ or } -1;$$

using (6.5), we see that σ^* contributes 1 or -1 to $\tilde{d}_{F^*}$. Thus $I_f = \tilde{d}_{F^*}$ (M orientable and n even), proving the rest of (7.1).

Finally we prove (b). In this case, $\mathfrak{T}^*$ is non-orientable, so that it is considered as a chain mod 2 in forming $\tilde{d}_{F^*}(\mathfrak{T}^*)$. As before, $\tilde{W}^{2n}_{F^*} \cdot \sigma^* = 1$ or -1, so that (using (6.3)) $\tilde{W}^{2n}_{F^*} \cdot (1_2\sigma^*) = 1_2$; clearly (7.2) follows.

THE CASE $n = 1$. Though $\mathfrak{T}^*$ is non-orientable, we shall use integer coefficients. For M we may take the unit circle in the plane E^2, described by an angle θ ($0 \leq \theta < 2\pi$). $\mathfrak{T}^*$ is a Möbius strip with boundary curve $\mathfrak{S}^*$. Any point of $\mathfrak{T}^* - \mathfrak{S}^*$ is a pair (θ, θ'); we may take $\theta < \theta'$. We may let (θ, θ) denote the points of $\mathfrak{S}^*$. If we express $\mathfrak{T}^*$ as the image of the triangle (x, y), $0 \leq x \leq y \leq 1$, the vertical and horizontal sides will each map into the set $\mathfrak{S}'$ of points $(0, \theta)$. Let $\mathfrak{T}_1^*$ be an integral chain covering $\mathfrak{T}^*$; then, combinatorially, $\partial\mathfrak{T}_1^* = \mathfrak{S}^* + 2\mathfrak{S}'$, so that by (6.3) and (6.4), with $W^1 = W^{2n-1}$ as in §6, using (7.1) to define I_f as an integer,

$$I_f = \tilde{W}^2_{F^*} \cdot \mathfrak{T}_1^* = \tfrac{1}{2}(W^1 \cdot F^* \partial\mathfrak{T}_1^*) = \tfrac{1}{2}(W^1 \cdot F^* \mathfrak{S}^*) + W^1 \cdot F^* \mathfrak{S}'.$$

$W^1 \cdot A^1$ is the degree of A^1 about O^* in V^{2*}; clearly $\frac{1}{2}(W^1 \cdot F^*\mathfrak{S}^*)$ is the number of times $F\mathfrak{S}^*$ (or more properly, ∇F^v, $v \in \mathfrak{S}$) winds about O in V^2, i.e. is the "rotation number" of f in M. Now if the point $\theta = 0$ is chosen so that $f(0)$ is on a line L of support to $f(M)$ in E^2 (i.e. $f(M)$ lies on one side of L), it is easy to see that $F\mathfrak{S}'$ makes half a turn about O, so that $F^*\mathfrak{S}'$ makes one turn about O^*, and $W^1 \cdot F^*\mathfrak{S}' = \pm 1$. Hence I_f always differs by unity from the rotation number of $f(M)$, again illustrating the warning in §6.

THEOREM 2. *Let M^n be compact (closed, or compact and with boundary). Then the intersection number I_f is invariant under regular deformations; if M has a boundary, we assume that during the deformation, the boundary is not carried across M:*

$$F_t(\partial M) \cap f_t(M - \partial M) = 0 \qquad \text{(all } t). \tag{7.4}$$

This means $I_{f_1} = I_{f_0}$, f_0 and f_1 being completely regular; other f_t need only be regular.

REMARKS. If we omit (7.4), it would be simple to give a formula expressing the change in I_f in terms of the crossings of ∂M through M. Thus if M^1 is a closed arc, mapped into the written letter e, $I_f = \pm 1$; if we pull it out into a line segment, $I_f = 0$, and ∂M^1 has crossed M^1 once.

If M is not compact (and no further conditions are assumed), the theorem fails. For example, let M^1 be the open segment $0 < \lambda < 4$. For $0 \leqq t \leqq 1$, let f_t be the mapping given by $(x, y) = f_t(\lambda)$:

$$x = \lambda, \qquad y = t/\lambda \qquad (0 < \lambda < 2),$$

and let the rest of M^1 be carried counter-clockwise around onto the line $y = 1$, and back across the y-axis along $y = 1$. Then $I_{f_t} = -1$ if $t > 0$, but $I_{f_0} = 0$.

Of course I_f is not invariant under deformations not assumed regular.

Suppose first that M is a manifold (without boundary). Since each F_t^* maps $\mathfrak{S}^* = \partial\mathfrak{T}^*$ into $V^* - O^*$, and is continuous, $\tilde{d}_{F_1^*} = \tilde{d}_{F_0^*}$, by Lemma 11. By the last theorem, $I_{f_0} = I_{f_1}$.

Now suppose M has a boundary. Then $\partial\mathfrak{T}^*$ consists of points of $\mathfrak{S}^*$ and points (p, q), $p \in \partial M$ or $q \in \partial M$. The hypothesis shows again that F_t leaves $\partial\mathfrak{T}^*$ in $V^* - O^*$, and again $I_{f_0} = I_{f_1}$.

II. THE IMBEDDING THEOREM

8. Discussion of the theorem

We first give theorems on the possibility of altering I_f, and of deforming f (keeping I_f fixed) so as to reduce the number of self-intersections. The imbedding theorem follows from these.

THEOREM 3. *Given any compact M (with or without boundary), there is a proper completely regular mapping f of M^n into E^{2n} with any desired I_f.*

Let f_0 be a completely regular mapping of M^n into E^{2n} (see [1], Theorem 3; it is not hard to extend this theorem to the case that M has a boundary, for example if ∂M is a manifold). Now take a small piece of M, flatten it, cut it out, and replace it by a piece with a single self-intersection, defined in §2, or by this piece altered by a reflection in E^{2n}. Then I_{f_0} is changed by 1 or -1 at will. Thus I_f may be increased or decreased by 1, and hence may be brought to any desired value.

THEOREM 4. *Let n be $\geqq 3$, and let f_0 be a proper regular mapping of a closed M^n into E^{2n}, with at most a finite number of self-intersections. Then there is a regular deformation f_t of f into a proper completely regular mapping f_1, such that the number of self-intersections under f_1 is increased by 2. If the number is $> |I_f|$, or is >0 if n is odd or M is non-orientable, we may decrease the number by 2. If M has a boundary, we may keep ∂M fixed, and keep M from cutting through ∂M.*

PROBLEM. Does the theorem hold for $n = 2$?

For the moment we shall prove merely that the number of self-intersections may be increased by 2. The rest of the proof will occupy §§9 through 12.

First, f may be made completely regular, as in [1]. Suppose this is done. Take a small nearly flat portion σ^n of $f(M)$. Take $p \in \sigma^n$, and pull p out away from σ^n, around and towards another point q of σ^n, pulling a small portion of σ^n along with it so we always have a regular mapping. We may make the moving part of M avoid any other points of M or ∂M. Say σ^n was approximately in the $(x_1, \cdots, x_n)$-plane in E^{2n}, and say q is the origin O, and we are pulling p around in the (x_1, x_{n+1})-plane, and down along the x_{n+1}-axis. We may tip the moving portion of σ^n near p to be parallel to the $(x_1, x_{n+2}, \cdots, x_{2n})$-axis, as we move p. Now pull p through O. Then if the mapping f_1 is obtained, the new self-intersections under f_1 are clearly on the x_1-axis T. The part of $f_1(\sigma^n)$ near p intersects T at points on each side of O. These latter points are new self-intersections, as required.

The rest of the proof of Theorem 4 runs as follows. We need merely show that two self-intersections of opposite types (or any two, if M is non-orientable or n is odd) can be gotten rid of by a regular deformation. Say

$$f(p_1) = f(p_2) = q, \qquad f(p_1') = f(p_2') = q',$$

these being of opposite types. Let C_1 and C_2 be non-intersecting curves in M, C_i joining p_i and p_i', neither passing through any other point where f has a self-intersection. Then $B_i = f(C_i)$ joins q to q', and $B = B_1 \cup B_2$ is a simple closed curve in $f(M)$. We let B bound a smooth 2-cell σ, which touches M only at B, and show that the part of M near C_2 may be deformed along near σ so that it passes beyond B_1, thus removing the two self-intersections. Until §12 we assume that M is orientable and n is even.

THEOREM 5. *Any smooth n-manifold may be imbedded in E^{2n}.*

REMARK. This theorem may be generalized to the case of manifolds with boundary, but we shall not discuss in what generality. For example, if ∂M is a manifold, the extension is not difficult. For the proof, see the proof of Case IV of Theorem 6 in the following paper.

For $n = 1$, the theorem is trivial. For $n = 2$, we imbed the sphere, projective plane, or Klein bottle, in E^4, and add the necessary number of handles to obtain the given manifold. (An imbedding of the projective plane is easily found from [3] p. 108.)

Now we suppose that $n \geqq 3$. First, by [1], Theorem 3, there is a proper completely regular mapping f_0 of M^n into E^{2n}. Suppose M is closed. Then, by Theorem 3, we alter f_0 to f_1 so that $I_{f_1} = 0$, and by Theorem 4, we may get rid of all of the intersections, giving the required imbedding.

Suppose now that M is open. We may suppose that M is connected. Let $(p_i, q_i)(i = 1, 2, \cdots)$ be the points with $f_0(p_i) = f_0(q_i)$. For each i, let C_i be a smooth curve in M from p_i to infinity, and let U_i be a neighborhood of C_i, so that U_i touches no other U_j or q_j. Let

$$(p, r, s)(p \in S_0^{n-2}, 0 \leqq r \leqq 1, 0 \leqq s < 1)$$

represent the points of U_i (supposing $n \geqq 2$), with $r = 0$ and $2\delta \leqq s < 1$ representing C_i; the points with $r = 1$ or $s = 0$ give the boundary of U_i. Deform U_i into itself by setting

$$\phi_t(p, r, s) = [p, r, \{1 - (1 - \delta)(1 - r)t\}s];$$

this leaves the boundary of U_i fixed. It may easily be replaced by a smooth deformation. Doing this for each i gives a mapping f with $f(M) \subset f_0(M) - \sum C_i$. Clearly f has no self-intersections, and is still proper.

PROBLEM. Does there exist an imbedding, for M open, with no limit set?

9. A lemma on bundles of vector spaces

A bundle $\mathfrak{B}$ *of vector spaces* is defined as follows. Let V_0^m be a fixed vector space of dimension n. Let A be a topological space, the *base space*, and let $\{F_i\}$ be a covering of A by closed sets (we could use open sets). To each $p \in A$ there corresponds a set of points $V^m(p)$; $V^m(p) \cap V^m(q) = 0$ if $p \neq q$. For each F_i, and each $p \in F_i$, $\xi_{F_i}(p, q)$ $(q \in V_0^m)$ is a one-one mapping of V_0^m onto $V^m(p)$. Let $q = \xi_{F_i}^{-1}(p, r)$ mean $r = \xi_{F_i}(p, q)$. For $p \in F_i \cap F_j$, we assume $\xi_{F_i}^{-1}(p, \xi_{F_j}(p, q))$ is a linear mapping of V_0^m onto itself, which varies continuously with p. Because of this, each $V^m(p)$ may be considered as a vector space. The ξ_{F_i} are the *coordinate systems*; the set of all points on all $V^m(p)$ is the *total space*. We write $\mathfrak{B} = (\mathfrak{B}; V_0^m, A)$. If there exists a coordinate system $\zeta(p, q)$, defined for p in all of A, then $\mathfrak{B}$ is *simple*; ζ expresses the total space as the topological product of A and V_0^m.

If we replace V_0^m by a sphere S_0^{m-1}, and linear mappings by orthogonal mappings, we have a *sphere bundle* (see [3]).

LEMMA 12. *Let* $(\mathfrak{B}; V_0^m, K)$ *be a bundle with a complex* K *for base space. Let* K' *be a (possibly void) subcomplex of* K. *Let* $e_1, \cdots, e_m$ *be the unit vectors in* V_0^m. *Let* $\zeta(p, e_1), \cdots, \zeta(p, e_{i-1})$ *be defined for* $p \in K$, *and let* $\zeta(p, e_i)$ *be defined for* p *in* K', *so that these are continuous and independent where defined. Then*

$\zeta(p, e_i)$ may be extended continuously over K so as to be independent of the other $\zeta(p, e_i)$ if either of the following two conditions hold:

(a) dim $(K) \leqq m - i$.

(b) *K can be built up from K' by adding cells, each having a cell, or nothing, in common with the subcomplex of K' plus the cells already chosen.*

REMARKS. In (b), it follows that $\mathfrak{B}$ is simple. If dim $(K) = m - i + 1$, the existence of $\zeta(p, e_i)$ is expressible simply in terms of one of the "characteristic classes" of $\mathfrak{B}$ (see [3]).

The proof in Case (b) is due essentially to Wazewski.[7] In Case (a), we extend $\zeta(p, e_i)$ over cells of K of higher and higher dimensions. That this can be done at each step is a simple consequence of the dimensions under consideration (see a similar proof in §10), and is fundamental in the existence of the characteristic classes (which exist equally well for bundles of spheres or vector spaces).

10. The 2-cell σ

Choose the C_i as described in §8; this is possible, since $n \geqq 3$. Let M_1 and M_2 be neighborhoods of C_1 and C_2 in M. Let p_{it} $(0 \leqq t \leqq 1)$ move from p_i along C_i to p'_i as t goes from 0 to 1. We can suppose p_{it} is an imbedding of the interval (0, 1) in M. Then $q_{it} = f(p_{it})$ is an imbedding.

Let E^2 have coordinates x, y. Let A_1 be the interval $0 \leqq x \leqq 1, y = 0$. Let A_2 be the arc of a circle of radius 1 and with center at $(1/2, -\sqrt{3}/2)$, which joins the ends r and r' of A_1 and lies in the half plane $y \geqq 0$. Let $A = A_1 \cup A_2$, let τ' be a small neighborhood of A in E^2, and let τ be τ' plus the interior region τ''. We shall find an imbedding ψ of τ in E^{2n} such that

$$\psi(r) = q, \qquad \psi(r') = q', \qquad \psi(A_i) = B_i \qquad (i = 1, 2),$$

and such that $\psi(\tau) \cap f(M) = B$, and no tangent plane to $\psi(\tau)$ at a point q^* of B lies in a tangent plane to $f(M)$ at q^*. We do this first with τ replaced by τ'.

Let T_1^n and T_2^n be the planes tangent to $f(M_1)$ and $f(M_2)$ at q. Let T^2 be the plane tangent to B_1 and B_2 at q; it cuts T_1^n and T_2^n in lines T_1^1 and T_2^1. By a slight deformation of f, we may suppose that f maps a neighborhood of p_i in M_i onto a neighborhood of q in T_i^n, and maps a neighborhood of C_i (extended) in M_i onto a neighborhood of T_i^1 $(i = 1, 2)$. We may map a closed neighborhood $\bar{V}$ of r in τ linearly onto a closed neighborhood of q in T^2 so that a closed neighborhood of r in A_i goes into a closed neighborhood of q in T_i^1 (and hence in B_i). To this end, we suppose the two ends of A_2 have been straightened. Call this mapping ϕ, and define ϕ similarly near r'.

Let r_{it} run from r to r' along A_i, choosing it so that $\phi(r_{it}) = q_{it}$ where ϕ is defined. Let u_{it} be a smooth vector function defined in A_i and with values in τ, such that u_{i0} and u_{i1} point in the forward and backward directions along A_j at r and r' $(j \neq i)$, and letting u_{1t} and u_{2t} turn in the positive and negative senses respectively as t runs from 0 to 1. Then u_{it} is not tangent to A_i at r_{it}. Let

[7] W. Wazewski, Compositio Math. vol. 2 (1935), pp. 63–68.

R_{it} be a segment of length ρ and center r_{it}, in the direction of u_{it}. We may suppose $\bar{V}$ and $\bar{V}'$ chosen so that part of their boundaries are formed by segments R_{it} extended, and may choose ρ so that no R_{1t_1} touches an R_{2t_2} if r_{1t_1} and r_{2t_2} are outside $\bar{V} \cup \bar{V}'$.

Set $v_{it} = \nabla\phi(u_{it}, r_{it})$ at points r_{it} of $\bar{V} \cup \bar{V}'$. Extend v_{it} over the rest of B_i so that it is smooth and not tangent to $f(M_i)$ at q_{it}. This may be done as follows (or we could use Lemma 12, (a)). Extend v_{it} so as to be smooth over all of B_i except a small piece D_i. With a curvilinear coordinate system about D_i, we may suppose D_i lies on the y_1-axis, and $f(M_i)$ lies in the $(y_1, \cdots, y_n)$-plane E^n. Now $v_{it} = v'_{it} + v''_{it}$, v'_{it} and v''_{it} being tangent and orthogonal to E^n respectively; v''_{it} is $\neq 0$ at the ends of D_i. Extend v'_{it} to be smooth over D_i. Since $2n - n > 1 = \dim (D_i)$, we may extend v''_{it} over D_i so as to be smooth and $\neq 0$ there. Now set $v_{it} = v'_{it} + v''_{it}$ in D_i.

Since ϕ is linear so far,

$$\phi(r_{it} + \alpha u_{it}) = q_{it} + \alpha v_{it} \quad (r_{it} \in \bar{V} \cup \bar{V}', \mid \alpha \mid \leqq \rho).$$

We may use this formula to extend ϕ over a closed neighborhood $\bar{\tau}'$ of A.

Let ψ' be a continuous extension of ϕ over τ. Since $2n \geqq 5$, we may[8] replace ψ' by an imbedding ψ such that $\psi(r)$ is arbitrarily close to $\psi'(r)$, and as we approach points of A, ψ approaches ψ' closer and closer, together with first derivatives. It follows that ψ is an imbedding, and no tangent plane to $\psi(\tau)$ at a point $\psi(r^*)$ $(r^* \in A)$ lies in a tangent plane to $f(M)$ at $\psi(r^*)$. Further, since $n + 2 < 2n$, we may suppose that $\psi(\tau) \cap f(M) = B$. Now $\sigma = \psi(\tau)$ is the required 2-cell.

11. The neighborhood of σ

It is here we shall use the fact that the two self-intersections are of opposite types. Our present purpose is to define smooth vector functions $w_1(q^*), \cdots, w_{2n}(q^*)$, for $q^* \in \sigma$, so that these are independent, and if $q^* = \psi(r^*)$,

(a) $w_1(q^*) = \nabla\psi(e_1, r^*)$, $w_2(q^*) = \nabla\psi(e_2, r^*)$,

(b) $w_3(q^*), \cdots, w_{n+1}(q^*)$ are tangent to $f(M_1)$ at q^* for $q^* \in B_1$,

(c) $w_{n+2}(q^*), \cdots, w_{2n}(q^*)$ are tangent to $f(M_2)$ at q^* for $q^* \in B_2$.

These show how σ, $f(M_1)$ and $f(M_2)$ lie together in E^{2n}.

Since ψ is regular, $w_1(q^*)$ and $w_2(q^*)$ are independent. Let V_0^{2n} be the space of vectors in E^{2n}. Let $e_3, \cdots, e_{n+1}$ and $e_{n+2}, \cdots, e_{2n}$ determine the subspaces V_1^{n-1} and V_2^{n-1} respectively. Consider, at each point $q^* \in B_1$, the vector space $V_1^{n-1}(q^*)$ of vectors tangent to $f(M_1)$ but orthogonal to B_1. Considering $V_1^{n-1}(q^{**})$ as disjoint from $V_1^{n-1}(q^*)$ if $q^{**} \neq q^*$, we have a bundle of vector spaces $\mathfrak{B}_1$ with the cell B_1 as base space; use $\mathfrak{B}_1 = (\mathfrak{B}_1; V_1^{n-1}, B_1)$. By Lemma 12, (b), we may find a coordinate system ζ_1 in it; choose it so that $w_1(q^*), w_3(q^*), \cdots, w_{n+1}(q^*)$ determine the positive orientation of $f(M)$ at $q^* \in B_1$, if we set

$$w_i(q^*) = \zeta_1(q^*, e_i) \quad (q^* \in B_1; i = 3, \cdots, n + 1). \tag{11.1}$$

[8] H. Whitney, Trans. Am. Math. Soc. vol. 36 (1934), pp. 63–89, Lemma 6; compare [1], Theorem 5, or Theorem 2, with the remarks following it.

Since the only vectors at $q^* \in B_1$ which are tangent to both τ and $f(M)$ are tangent to B_1, the w_i are independent so far, and all properties hold.

Similarly, taking $V_2^{n-1}(q^*)$ tangent to $f(M)$ but orthogonal to B_2 at $q^* \in B_2$ gives a bundle $(\mathfrak{B}_2 ; V_2^{n-1}, B_2)$ with a coordinate system ζ_2. Define $w_i(q^*)$ by (11.1), with ζ_2, for $q^* \in B_2$, $i \geqq n + 2$. If $w_2'(q^*)$ points positively along B_2 at q^*, we may suppose $w_2'(q^*)$, $w_{n+2}(q^*), \cdots, w_{2n}(q^*)$ determine the positive orientation of $f(M_2)$ at q^*. The properties still hold. (At q and q', clearly all the vectors are independent.)

Let $(\mathfrak{B}; V_0^{2n}, \sigma)$ be the bundle with each $V(q^*) = V_0^{2n}$. (The total space is $V_0^{2n} \times \sigma$.) In the subbundle $(\mathfrak{B}; V_0^{2n}, B_2)$ over B_2, the vectors $w_1(q^*)$, $w_2(q^*)$, $w_{n+2}(q^*), \cdots, w_{2n}(q^*)$ are independent; $w_1(q^*), \cdots, w_{2n}(q^*)$ are independent for $q^* = q$ and for $q^* = q'$. By Lemma 12, using (a), with $K = B_2$, $K' = q \cup q'$, we may extend $w_3(q^*), \cdots, w_n(q^*)$ over B_2 so that the $2n - 1$ vectors are independent. Now recall that the two self-intersections are of opposite types (we are supposing that M is orientable and n is even). By the choice of ζ_1 and ζ_2, this means that

$$w_1(q^*), \qquad w_3(q^*), \cdots, w_{n+1}(q^*), \qquad w_2'(q^*), \qquad w_{n+2}(q^*), \cdots, w_{2n}(q^*),$$

define opposite orientations of E^{2n} for $q^* = q$ and for $q^* = q'$. Now $w_2(q)$ and $w_2(q')$ may be deformed into $w_2'(q)$ and $-w_2'(q')$ respectively, keeping them tangent to σ and independent of B_1; they therefore remain independent of the other vectors listed above. Hence

$$w_1(q^*), w_2(q^*), \cdots, w_{2n}(q^*)$$

define the same orientation of E^{2n} at $q^* = q$ as at $q^* = q'$. It follows that $w_{n+1}(q^*)$ may be defined over B_2 so as to remain independent of the other vectors. (If we deform $w_{n+1}(q^*)$ for q^* near q and q' to become a unit vector orthogonal to the other vectors, then $w_{n+1}(q^*)$ is uniquely determined so as to be continuous and have this property in the rest of B_2.)

By Lemma 12, using (a) again, with $K = \sigma$ and $K' = B$, since $\dim(\sigma) = 2 \leqq 2n - (n + 1)$, we may extend $w_3(q^*), \cdots, w_{n+1}(q^*)$ over σ so that $w_1(q^*), \cdots, w_{n+1}(q^*)$ are independent there. Finally, by Lemma 12, using (b), since σ and B_2 clearly have the property of K and K', we may extend $w_{n+2}(q^*), \cdots, w_{2n}(q^*)$ over σ. The $w_i(q^*)$ determine a neighborhood of σ in E^{2n}, as will be apparent from (12.1).

12. Completion of the proof of Theorem 4

Consider $\tau \subset E^2$ as lying in E^{2n}. For each point $(a_1, a_2, a_3, \cdots, a_{2n})$, set $r^* = (a_1, a_2, 0, \cdots, 0)$, and

$$\psi\left(r^* + \sum_{i=3}^{n} a_i e_i\right) = \psi(r^*) + \sum_{i=3}^{n} a_i w_i(\psi(r^*)). \tag{12.1}$$

Since $w_1(q^*), \cdots, w_{2n}(q^*)$ are independent and smooth, this is a mapping with non-vanishing Jacobian; hence, in a neighborhood of the interior of σ, we may invert, and define ψ^{-1}. (The $w_i(q^*)$ are taken of class C^2 for this to hold.) If we set

$$N_1 = \psi^{-1}(f(M_1)), \qquad N_2 = \psi^{-1}(f(M_2)),$$

then N_1 and N_2 will lie in a neighborhood U of τ. If we define a deformation of N_2 in U which brings it into a position not intersecting N_1, applying ψ defines a corresponding deformation of f; there will then be two self-intersections fewer, as required.

Set $\pi(x_1, \cdots, x_{2n}) = (x_1, 0, x_3, \cdots, x_{2n})$. By (a), (b) and (c) of §11, and (12.1), for any $r^* \in A_1$, the tangent plane to N_1 and hence to $\pi(N_1)$ at r^* lies in the $(x_1, x_3, \cdots, x_{n+1})$-plane, and the tangent plane to $\pi(N_2)$ at r^* lies in the $(x_1, x_{n+2}, \cdots, x_{2n})$-plane, it follows that

(12.2) $$\pi(N_1) \cap \pi(N_2) \quad \text{is on the } x_1\text{-axis.}$$

Let $\mu(x_1)$ be a smooth real function whose graph $x_2 = \mu(x_1)$ lies just above A_2, and lies on the x_1-axis just outside A_2. Take $\epsilon > 0$ so that the points of N_2 within ϵ of the (x_1, x_2)-plane E^2 form a set interior to N_2. Let $\nu(\lambda)$ be a smooth real function with

$$|\nu(\lambda)| \leqq 1, \qquad \nu(0) = 1, \qquad \nu(\lambda) = 0 \quad \text{for} \quad |\lambda| \geqq \epsilon^2.$$

For $r^* = (x_1, \cdots, x_{2n})$, set

(12.3) $$\theta_t(r^*) = r^* - t\nu(x_3^2 + \cdots + x_{2n}^2)\mu(x_1)e_2,$$

and consider this as applying only to points of N_2. This is clearly a regular deformation. For $t = 1$, it moves the part of N_2 lying on A_2 into a position with $x_1 < 0$. Since $\pi(\theta_t(N_2)) = \pi(N_2)$, (12.2) gives

$$N_1 \cap \theta_1(N_2) \quad \text{is on the } x_1\text{-axis.}$$

But $\theta_1(N_2)$ does not touch the x_1-axis; hence this set is void. Furthermore, since $\psi(\tau) \cap f(M) = B$, taking ϵ small enough will insure that no new self-intersections are introduced. This completes the proof in the case that M is orientable and n is even.

In the other cases, we can get rid of any pair of self-intersections by a regular deformation. To show this, it is only necessary to show that C_1 and C_2 may be chosen in such a fashion that M_1 and M_2, when each is oriented, intersect in points of opposite types.

Suppose M is not orientable. Try a pair C_1, C_2. If the points are of the same type, choose a curve C_2' such that $C_2 \cup C_2'$ reverses the orientation in M; then C_1 and C_2' will do.

Suppose n is odd, and M is orientable. If C_1 and C_2 will not do, choose C_1' and C_2' so that (with the notation of §9) C_1' joins p_1 and p_2', and C_2' joins p_2 and p_1'. We may suppose C_i' coincides with C_i near one end and with C_j $(j \neq i)$ at the other, and $M_i = M_i'$ near p_i, and $M_i = M_j'$ $(j \neq i)$ near p_i'. Orient the M_i, and the M_i' accordingly (near p_i). Then the intersection of the pair (M_1', M_2') is like that of (M_2, M_1) at q'; hence (compare the proof in Lemma 2) the intersection is of opposite type from that of (M_1, M_2) at q', as required. The theorem is now proved.

Appendix

IMBEDDING THE COMPLEX PROJECTIVE PLANE M^4 IN E^7

Let M^6 be the space of all triples (z_1, z_2, z_3) of complex numbers, not all 0. If we identify any two triples (z_1, z_2, z_3), $(\lambda z_1, \lambda z_2, \lambda z_3)$ ($\lambda \neq 0$ complex), M^4 is obtained. Let $[z_1, z_2, z_3]$ denote points of M^4.

We first give briefly a topological description of M^4, then discuss how an imbedding may be found from the topological point of view, and finally give equations which are proved to be an imbedding. In terms of the real variables, the imbedding will be analytic; it of course cannot be so in terms of the complex variables, since 7 is odd.

Set $z_k = x_k + iy_k$, and

$$N(z_1, z_2, z_3) = |z_1|^2 + |z_2|^2 + |z_3|^2 = \sum (x_k^2 + y_k^2). \tag{1}$$

We remark that, since $[z_1, z_2, z_3]$ may be normalized to make $N = 1$, uniquely except for a factor λ with $|\lambda| = 1$, the 5-sphere $N = 1$ is expressed as the total space of a bundle of 1-spheres with M^4 as base space. A similar fact holds for the $(2n + 1)$-sphere.

First we divide M^4 into three subsets; the points with $z_3 \neq 0$, those with $z_3 = 0$ but $z_2 \neq 0$, and those with $z_3 = z_2 = 0$. These sets, with normal forms for their points, are:

$$\begin{aligned} &Q^4\text{: all } [z_1, z_2, 1],\\ &\left.\begin{aligned} &Q^2\text{: all } [z_1, 1, 0],\\ &Q^0\text{: all } [1, 0, 0];\end{aligned}\right\} S^2.\end{aligned}$$

Since these are cells, the Euler-Poincaré characteristic of M^4 is 3. S^2 is complex projective 1-space, equivalent to the extended complex plane, and is topologically a 2-sphere. For any $[z_1, z_2, 0]$ in S^2, set

$$P^*[z_1, z_2, 0] = \text{all} \quad [z_1, z_2, z_3].$$

Set

$$\rho[z_1, z_2, z_3] = |z_3|^2/(|z_1|^2 + |z_2|^2), \tag{2}$$

(ρ may be ∞), and define the subsets of M^4

$$A_-\colon \rho < 1, \qquad A\colon \rho = 1, \qquad A_+\colon \rho > 1.$$

Set $P_-[z_1, z_2, 0] = P^*[z_1, z_2, 0] \cap A_-$, and define P, P_+ similarly.

Clearly A is a 3-sphere (see below), and A_+ is a 4-cell bounded by it; each is in Q^4. Since the sets $P_-(p) \cup P(p)$ are disjoint, these form a bundle of closed 2-cells, with S^2 as base space. The $P(p)$ form a bundle of 1-spheres with S^2

as base space; the total space is the 3-sphere A. We remark that this proves the 2-dimensional invariant of the bundle to have the value ± 1; also, that mapping each point q of A into $p \in S^2$ such that $q \in P(p)$ is exactly Hopf's essential mapping of a 3-sphere into a 2-sphere.[9]

We now discuss the imbedding problem. If we set $F^*[z_1, z_2, 1] = (z_1, z_2)$, this imbeds Q^4 in 4-space. Of course F^* cannot be extended over S^2. To make this possible, let us pull back towards the origin those points with small ρ; we may set

$$F'[z_1, z_2, 1] = \frac{2}{N}(z_1, z_2) \tag{3}$$

or equivalently (the last expression allowable even for $z_3 = 0$)

$$F'[z_1, z_2, z_3] = \frac{2\,|z_3|^2}{N}\left(\frac{z_1}{z_3}, \frac{z_2}{z_3}\right) = \frac{2}{N}(z_1\bar{z}_3, z_2\bar{z}_3). \tag{4}$$

This mapping fails to be regular at points of A, where the vectors pointing away from $[0, 0, 1]$ (using (3)) are mapped into the origin.

Let us introduce three new coordinates into the image space, map S^2 into the unit sphere in this space, and extend the mapping over M^4. Using stereographic projection, and one complex and one real coordinate, we find

$$F''[z_1, z_2, 0] = \left(2\frac{z_1\bar{z}_2}{|z_1|^2 + |z_2|^2}, \frac{|z_1|^2 - |z_2|^2}{|z_1|^2 + |z_2|^2}\right),$$

or more generally,

$$F''[z_1, z_2, z_3] = \frac{1}{N}(2z_1\bar{z}_2, |z_1|^2 - |z_2|^2).$$

Combining this with (4), with a slight change, we define our mapping by

$$F[z_1, z_2, z_3] = \frac{1}{N}(z_2\bar{z}_3, z_3\bar{z}_1, z_1\bar{z}_2, |z_1|^2 - |z_2|^2), \tag{5}$$

using three complex and one real coordinate, N being defined by (1). In terms of the seven real variables,

$$\begin{aligned} F = \frac{1}{N}(&x_2x_3 + y_2y_3, y_2x_3 - x_2y_3, x_1x_3 + y_1y_3, x_1y_3 - y_1x_3, \\ &x_1x_2 + y_1y_2, y_1x_2 - x_1y_2, x_1^2 + y_1^2 - x_2^2 - y_2^2). \end{aligned} \tag{6}$$

From (5) it is clear that $F[z_1, z_2, z_3] = F[\lambda z_1, \lambda z_2, \lambda z_3]$, λ complex, i.e. it is independent of the manner of writing a point in M^4.

We now prove that F is regular. Consider first the F' of (3). Since the set

[9] H. Hopf, *Über die Abbildungen der dreidimensionalen Sphäre auf die Kugelfläche*, Math. Annalen, vol. 104 (1931), pp. 637–665. The mapping also is that determined in the Remark following (1).

of points (z_1, z_2) with $|z_1|^2 + |z_2|^2 = r^2$, which forms a 3-sphere of radius r if $r > 0$, is mapped into a concentric sphere, F' is regular wherever the vector $v_{z_1 z_2}$ at $[z_1, z_2, 1]$ pointing away from the origin is mapped into a non-zero multiple of itself. Set

$$\alpha = x_1^2 + y_1^2 + x_2^2 + y_2^2, \qquad N_\lambda = 1 + \lambda^2\alpha^2 \quad (\lambda \text{ real}).$$

Then the image of $\frac{1}{2}v_{z_1 z_2}$ is the value when $\lambda = 1$ of

$$\frac{\partial}{\partial\lambda}\left[\frac{1}{N_\lambda}(\lambda x_1, \lambda y_1, \cdots)\right] = -\frac{2\lambda\alpha^2}{N_\lambda^2}(\lambda x_1, \cdots) + \frac{1}{N_\lambda}(x_1, \cdots)$$
$$= -\frac{1}{N_\lambda^2}(2\lambda^2\alpha^2 - N_\lambda)(x_1, y_1, x_2, y_2).$$

Taking $\lambda = 1$, we see that F' is regular except possibly on a set B_3, where $2\alpha^2 - N = 0$, i.e. $2\alpha^2 = 1 + \alpha^2$, $\alpha = 1$, or else $z_3 = 0$. With $[z_1, z_2, z_3]$, we see that B_3 is given by the condition:

(7) $\qquad$ Either $z_3 = 0$ or $|z_3|^2 = |z_1|^2 + |z_2|^2$,

i.e. $B_3 = S^2 \cup A$.

Define B_1, B_2 similarly, and set $B = B_1 \cap B_2 \cap B_3$. Then replacing F' by F (where we need not consider the last variable $|z_1|^2 - |z_2|^2$), we see that F is regular except possibly at points of B. Now B contains only the points

(8) $\qquad p_{1z} = [0, 1, z], \qquad p_{2z} = [z, 0, 1], \qquad p_{3z} = [1, z, 0], \qquad |z| = 1.$

For $[0, 0, 1]$ is not in B_3; similarly for $[0, 1, 0]$ and $[1, 0, 0]$. In case at most one $z_i = 0$, say $|z_1| \leqq |z_2| \leqq |z_3|$; then $z_2 \neq 0$. If $z_1 \neq 0$, or if $z_1 = 0$ and $|z_3| > |z_2|$, then $|z_2|^2 < |z_1|^2 + |z_3|^2$, and $[z_1, z_2, z_3]$ is not in B_2, while if $z_1 = 0$ and $|z_3| = |z_2|$, then $[z_1, z_2, z_3] = [0, 1, z_3/z_2] = p_{1z}$, $z = z_3/z_2$.

We must prove still that F is regular at the points p_{iz}. To do this, it is sufficient to show that the $7 \times 6 = 42$ partial derivatives of the components of F, considered as mapping M^6 into E^7, form a matrix of rank 4 at each such point. For this proves that the tangent space to M^6 at any such point is mapped into a space of dimension 4 in E^7 by F; but the tangent space to M^4 is clearly mapped into the same space in E^7. It is of course sufficient to find a submatrix of rank 4.

For points of M^6 representing points p_{iz}, we choose points with $N = 1$, obtained from (8) by multiplying by a real positive factor. Now

$$p_{1z} = (0, 0, x_2, 0, x_3, y_3), \qquad x_2^2 = x_3^2 + y_3^2 = \tfrac{1}{2},$$
$$p_{2z} = (x_1, y_1, 0, 0, x_3, 0), \qquad x_3^2 = x_1^2 + y_1^2 = \tfrac{1}{2},$$
$$p_{3z} = (x_1, 0, x_2, y_2, 0, 0), \qquad x_1^2 = x_2^2 + y_2^2 = \tfrac{1}{2}.$$

The corresponding values of F are:

$$\begin{array}{llllllll} \text{at } p_{1z}: & (x_2x_3, & -x_2y_3, & 0, & 0, & 0, & 0, & -x_2^2 \quad), \\ \text{at } p_{2z}: & (\ 0, & 0, & x_1x_3, & -y_1x_3, & 0, & 0, & x_1^2 + y_1^2 \quad), \\ \text{at } p_{3z}: & (\ 0, & 0, & 0, & 0, & x_1x_2, & -x_1y_2, & x_1^2 - x_2^2 - y_2^2). \end{array}$$

Consider first a point p_{1z}. Form F_1 from F by omitting the third and fourth of the seven variables. Then since $N = 1$ and

$$-2x_2^2x_3 + x_3 = 2x_2^2y_3 - y_3 = 0$$

at the point under consideration, differentiating (6) gives the following matrix for $\partial F_1/\partial x_1, \cdots, \partial F_1/\partial y_2$:

$$\left\| \begin{matrix} 0 & 0 & x_2 & 0 & 0 \\ 0 & 0 & 0 & x_2 & 0 \\ 0 & 0 & 0 & 0 & 2x_2(x_2^2 - 1) \\ y_3 & x_3 & 0 & 0 & -2y_2 \end{matrix} \right\|.$$

This is obviously of rank 4. Similarly, at p_{2z}, let F_2 omit the fifth and sixth variables from F. Then the matrix of $\partial F_2/\partial x_2, \cdots, \partial F_2/\partial y_3$ is

$$\left\| \begin{matrix} x_3 & 0 & 0 & 0 & 0 \\ 0 & x_3 & 0 & 0 & 0 \\ 0 & 0 & 0 & 0 & -2x_3(x_1^2 + x_1^2) \\ 0 & 0 & y_1 & x_1 & 0 \end{matrix} \right\|,$$

again of rank 4. At p_{3z}, letting F_3 omit the first and second variables, the matrix of $\partial F_3/\partial x_1, \partial F_3/\partial y_1, \partial F_3/\partial x_3, \partial F_3/\partial y_3$ is

$$\left\| \begin{matrix} 0 & 0 & 0 & 0 & 2x_1 \\ 0 & 0 & y_2 & x_2 & 0 \\ x_1 & 0 & 0 & 0 & 0 \\ 0 & x_1 & 0 & 0 & 0 \end{matrix} \right\|,$$

again of rank 4. This completes the proof that F is regular.

To show that F is one-one, take any point

$$q = (w_1, w_2, w_3, \gamma) = (u_1, v_1, \cdots, v_3, \gamma) = F(p)$$

for some $p \in M^4$; we shall show that $p = [z_1, z_2, z_3]$ is uniquely determined, i.e. the z_i are determined up to a complex factor.

Suppose first that $w_3 = 0$, $\gamma = 0$. Then one of z_1, z_2 is 0, and since $|z_1|^2 - |z_2|^2 = 0$, both are 0; hence p is unique.

Suppose next that $w_3 = 0$, $\gamma \neq 0$. If $\gamma > 0$, then $z_2 = 0$, $|z_1| > 0$ (since $N > 0$). Since $z_1 \neq 0$, we can normalize p to $[1, 0, z_3]$. Now

$$\gamma = \frac{|z_1|^2}{N} = \frac{1}{N}, \qquad w_2 = \frac{z_3\bar{z}_1}{N} = \gamma z_3;$$

hence $z_3 = w_2/\gamma$ is determined. If $\gamma < 0$, then $z_1 = 0$, $z_2 \neq 0$, etc.

Suppose now that $w_3 \neq 0$; then $z_1 \neq 0$, $z_2 \neq 0$. In this case we normalize so that

$$N = 1, \qquad z_2 = x_2 > 0.$$

If first $w_2 \neq 0$, then

$$w_1w_3 = x_2^2z_1\bar{z}_3 = x_2^2\bar{w}_2, \qquad x_2^2 = \frac{w_1w_3}{\bar{w}_2},$$

determining x_2. Now

$$z_3 = \frac{\overline{w}_1}{x_2}, \qquad z_1 = \frac{w_3}{x_2},$$

so that p is determined.

Suppose finally that $w_3 \neq 0$, $w_2 = 0$. Normalize as before. Now $z_3 = 0$; we find, from $w_3 = x_2 z_1$, $\gamma = |z_1|^2 - x_2^2$,

$$|z_1|^2 = \frac{|w_3|^2}{x_2^2} = \gamma + x_2^2,$$

$$x_2^2 = \tfrac{1}{2}[-\gamma \pm (\gamma^2 + 4|w_3|^2)^{1/2}].$$

Since the square root is $> |\gamma|$, the minus sign is impossible, and x_2^2 and hence $z_2 = x_2$ is uniquely determined, as is $z_1 = w_3/x_2$. This completes the proof.

HARVARD UNIVERSITY.

ANNALS OF MATHEMATICS
Vol. 45, No. 2, April, 1944

THE SINGULARITIES OF A SMOOTH n-MANIFOLD IN $(2n - 1)$-SPACE*

BY HASSLER WHITNEY

(Received August 19, 1943)

1. Introduction

We showed in the preceding paper that any smooth n-manifold M^n may be imbedded in $2n$-space E^{2n}. Our primary purpose here is to show that it may be immersed in E^{2n-1}, provided that $n \geqq 2$. Then near any point of M, the mapping f into E^{2n-1} is one-one, but there may be self-intersections (which may be required to lie along curves). Equally important perhaps is the combinatorial study of singularities (points where the mapping is not regular). Along with true manifolds we study also manifolds with boundary. By a *partial manifold*, we mean a manifold with or without boundary. In simple cases, the boundary ∂M of the partial manifold M will be a manifold. (∂M means the point set boundary; it need not coincide with the boundary of the chain M if M is non-orientable.) Since the question of how general ∂M may be allowed to be (we insist at any rate that it be a complex) is a rather difficult one, which we expect to study further in another paper, we will use the term somewhat loosely here. Any special assumptions on ∂M which may be needed will be made at the time.

It is a highly difficult problem to see if the imbedding and immersion theorems of the preceding paper and the present one can be improved upon. Practically the only knowledge we have of this is found in the author's Michigan lecture, [3]. The most important result there for the present problem is the existence of a closed M^4 which cannot be imbedded in E^7. (We have not studied the possibility of immersing it in E^6.) This M^4 is non-orientable; it seems possible that any orientable, or any open or partial, M^4 may be imbedded in E^7, and immersed in E^6. Possibly also any M^3 may be imbedded in E^5!

We touch briefly on the case $n = 1$. Here M is a circle, or a closed, open, or half-open arc. Locally, the mapping f into the line E^1 is expressible as a differentiable real-valued function $x' = f(x)$; the singularities of f are the points where $df/dx = 0$. If the mapping is "semi-regular," the only singularities are maxima and minima of f. It is obvious that a slight alteration of any f will give a mapping g in which this holds. The combinatorial conditions as stated in this paper apply only to the case $n \geqq 2$, but analogues could easily be given for the case $n = 1$.

Suppose now that $n \geqq 2$. If we take a general smooth mapping f of M^2 into E^3, the singularities may be quite wild. But again, a slight alteration of f will reduce them to a single type; see §3 and Fig. 1. The new mapping is semi-regular; these are the mappings which concern us here.

The combinatorial part of the paper consists essentially in counting the

---

* Presented to the American Mathematical Society, Sept. 9, 1942.

algebraic number $\mathfrak{L}_f(M)$ of singular points (mod 2 if n is even) by means of the mapping f in the boundary of M. (As an immediate corollary, $\mathfrak{L}_f(M) = 0$ or $\equiv 0$ mod 2 for all closed manifolds, orientable or non-orientable.) The manner of counting may be seen from Fig. 1. If we follow around the boundary ∂M, it cuts through M at one point, p^*; hence $\mathfrak{L}_f(M) \equiv 1$ (mod 2). Or, again, let us cut off a strip around the edge of $f(M)$, and spread it out in the form of a circle; we will find that it has a single twist. If we cut this strip into two strips, the twist will show up in the linking of the two new strips; combinatorially, in the fact that (each being considered as a curve) their looping coefficient is ± 1. These two facts are expressible in the form:

$$\mathfrak{L}_f(M) = KI(fM^*, f\partial M) = LC(f\partial M^*, f\partial M), \tag{1.1}$$

M^* being M with a narrow neighborhood of ∂M removed. (1.1) holds in fact for any chain A, using A^*, which is A with a neighborhood of all $(n - 1)$-cells removed. The fundamental theorem states that this is the algebraic number of singular points of M, taken mod 2 if n is even. The proof in the case that M is non-orientable and n is odd is difficult to handle; the intersection theory in Part III is needed to unravel the situation.

The reader may wonder why we are willing to lose preciseness in the results by reducing mod 2 whenever n is even. The answer is, the formulas are only correct after reducing mod 2 (and in fact, this is all we need in the proof of the immersion theorem). This is well illustrated in Fig. 2. There are no singularities; yet $f(\partial M)$ cuts through $f(M)$ twice, each time in the same sense, so that $KI(fM^*, f\partial M) = \pm 2$. In the proof of the immersion theorem, we cut out pieces of $f(M)$ and alter them; it may be necessary to insert twisted pieces such as in Fig. 2 to gain the desired end in case n is even.

Fundamental definitions are as in the preceding paper (including its §4). We note also the following (see also §15 and elsewhere). A vector is *tangent* to M at $p \in M$ if it points into M or along ∂M (if $p \in \partial M$) at p. It is *independent* of M if it is not in the tangent plane to M at p. A vector field $f(p)$ is *independent* of M if each $v(p)$ is independent of M at p. We may use a complex K in place of M; then v is *independent* of K at p if it is independent of each cell σ of K with $p \in \bar{\sigma}$, etc.

2. Outline of the paper

A typical singularity is presented in (3.3), and a mapping of a sphere or plane with just two of them is given in §4. Though there is only one kind of singular point under a semi-regular mapping, we may differentiate between positive and negative ones in case n is odd; see the definition in §5. If a single cell is mapped by f so that it has just one singularity, as illustrated in Fig. 1, the relation to intersections is fairly simple, as noted above. For a partial manifold, or more generally, a complex, the relation is worked out in §7. When we express M and ∂M as chains, and sum, both pairs $\partial\sigma_i^n$, σ_j^n and $\partial\sigma_j^n$, σ_i^n will appear if $i \neq j$. By the commutation rule for Kronecker indices, such terms will cancel out for

n odd, giving the exact value of $\mathfrak{L}_f(M)$ in (8.2); for n even, we get this result only mod 2. If f is deformed, $f(\partial M)$ may cut through itself; yet if n is odd, this does not affect $\mathfrak{L}_f(M)$, as noted in Theorem 4.

To prove the immersion theorem, we need some detailed results on the type of looping coefficients which we mentioned above in cutting a strip into two strips. Lemmas are given which state that certain alterations of f are possible which map ∂M into a given position and let M have given directions at points of ∂M. Next we give a mapping f of an n-cube M_0 (n even) without singularities and with $\mathfrak{L}_f(M_0) = \pm 2$; compare Fig. 2. To prove the immersion theorem in case M is closed and n is odd, we use $\mathfrak{L}_f(M) = 0$ to show that the singular points may be paired, p_i and p_i', the two in a pair being of opposite type. If A is an arc from p_i to p_i', a neighborhood M_1 of A then has the property that $\mathfrak{L}_f(M_1) = 0$. We may therefore alter f in M_1 to remove these two singularities. The other cases do not require much further treatment. Two theorems are then given which discuss the position of ∂M under an immersion of a partial manifold.

Suppose M is non-orientable. Let M_1 be a chain formed by adding together the n-cells of M. Then $\partial M_1 = A + 2B$, where A is the sum of the $(n - 1)$-cells of ∂M, and B is a sum of cells interior to M. Thus, if M^2 is a Möbius strip, ∂M_1^2 is the boundary curve plus twice an arc cutting across the strip. Suppose n is odd. It was proved in Part I that (1.1) counts the singularities, provided that M_1, M_1^* and ∂M_1 are used. But we do not wish to use any interior cells of M; it is necessary to show that these always cancel out. It is clear that in this case we cannot use $KI(fM^*, f\,\partial\,M)$, since by an alteration of f we might move $f\,\partial\,M$ across itself, which would alter the Kronecker index. Moreover, $LC(f\,\partial\,M^*, f\,\partial\,M)$ is not defined, since ∂M^* and ∂M cannot be made into cycles. But if we choose that n-chain $\rho^+ f\,\partial\,M$ formed by deforming $f\,\partial\,M$ in the y_{2n-1}-direction to infinity, we may study $KI(f\,\partial\,M^*, \rho^+ f\,\partial\,M)$. This leads finally to the required result. The definition of $\mathfrak{L}_f(M)$ required, in (20.5), is more complicated than before; its necessity is shown by an example in §21.

In an appendix we take up some topics which are less fundamental in the paper.

I. SINGULARITIES AND INTERSECTIONS

3. The general type of singularity

DEFINITION. The mapping f of the n-manifold M^n (without boundary) into E^{2n-1} is *semi-regular* if it is of class C^{12} (so that we may apply Lemma 2) and is proper, and for each $p \in M$, either f is regular at p or the following holds: With a suitable coordinate system about p,

$$\left.\frac{\partial f}{\partial x_1}\right|_p = 0, \tag{3.1}$$

and the $2n - 1$ vectors

$$\left.\frac{\partial^2 f}{\partial x_1^2}\right|_p, \left.\frac{\partial f}{\partial x_2}\right|_p, \cdots, \left.\frac{\partial f}{\partial x_n}\right|_p, \left.\frac{\partial^2 f}{\partial x_1 \partial x_2}\right|_p, \cdots, \left.\frac{\partial^2 f}{\partial x_1 \partial x_n}\right|_p \tag{3.2}$$

are independent. This condition holds then in any coordinate system for which (3.1) holds; see [4]. The x_1-direction is uniquely determined except as to sense. If M is a partial manifold, we assume also that f is one-one in a neighborhood of the boundary. (See the appendix, Lemma 25.)

DEFINITION. The semi-regular mapping f is *completely semi-regular* if: (a) For any double point $f(p) = f(q)$ (p or q may be in ∂M) the two tangent planes to $f(M)$ there have only a line in common. (b) $f(\partial M)$ does not contain the image of any singular point. (c) If $n \geqq 3$, there are no triple points $f(p) = f(q) = f(r)$; if $n = 2$, there is no such triple point with $p \in \partial M$, and there are no quadruple points. The self-intersections are then along smooth curves; see §22.

LEMMA 1. *Arbitrarily close to any f there is a completely semi-regular f'; we may make f' be one-one in a neighborhood of ∂M, and may make f' of class C^∞.*

This is proved without the "completely," for manifolds, in [4]. For partial manifolds, we first imbed a neighborhood of the boundary (using the methods in [1]), then extend the mapping over the interior of M, and apply the proof mentioned to the interior. It is now easy to make the mapping completely semi-regular (see [1], especially §9, (D)).

LEMMA 2. *Let f be semi-regular. Then for any singular point p there exist (curvilinear) coordinate systems $(x_1, \cdots, x_n)$ about p and $(y_1, \cdots, y_{2n-1})$ about $f(p)$ such that f is given near p by*

$$(3.3) \qquad y_1 = x_1^2, \quad \left.\begin{aligned} y_i &= x_i, \\ y_{n+i-1} &= x_1 x_i, \end{aligned}\right\} \quad (i = 2, \cdots, n).$$

This also is proved in [4]. If f is of class C^{4r+8}, of class C^∞, or analytic, the new coordinate systems will be of class C^r, of class C^∞, or analytic respectively.

In case $n = 2$, the mapping is

$$(3.4) \qquad x = u^2, \qquad y = v, \qquad z = uv;$$

eliminating u and v gives $z = \pm y\sqrt{x}$. For each y, the cross-section is a parabola; as y passes through 0, the parabola degenerates to a half-ray, and opens out again (with sense reversed); see Figure 1. The only self-intersection is at $v = 0$, mapping into the positive x-axis; in the general case, at $x_2 = \cdots = x_n = 0$, mapping into the positive y_1-axis.

4. A mapping of a sphere or plane with just two singularities

The examples we give here not only are interesting as illustrating mappings of whole manifolds with definite singularities, but are useful in the proof of the fundamental Theorem 6, in the case of an open manifold.

Let S_0^n be the n-sphere $x_1^2 + \cdots + x_{n+1}^2 = 1$ in E^{n+1}. We define a smooth mapping f of E^{n+1}, and hence of S_0^n, into E^{2n-1} by the equations

$$(4.1) \qquad y_1 = x_{n+1}, \quad \left.\begin{aligned} y_i &= x_i, \\ y_{n+i-1} &= x_1 x_i, \end{aligned}\right\} \quad (i = 2, \cdots, n).$$

For $n = 2$, transposing terms gives

$$f(x, y, z) = (x, xy, z).$$

The effect of f is to turn the part $x < 0$ of the sphere inside out. More explicitly, f squeezes the cross-sections S_x^1 for each x so that for $x = 0$, the circle S_x^1 turns into a line segment, and for $x < 0$, into an ellipse with sense reversed. There are obviously two singularities, at $(0, 0, \pm 1)$.

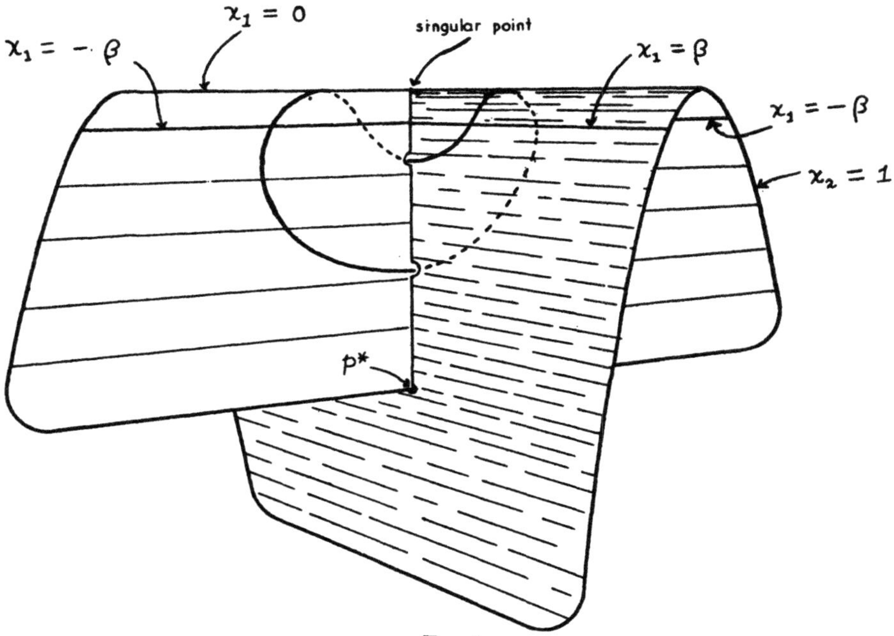

FIG. 1

The matrix of first partial derivatives in (4.1), transposed, is

$$\begin{Vmatrix} 0 & 0 & \cdots & 0 & x_2 & \cdots & x_n \\ 0 & 1 & \cdots & 0 & x_1 & \cdots & 0 \\ \cdots & \cdots & \cdots & \cdots & \cdots & \cdots & \cdots \\ 0 & 0 & \cdots & 1 & 0 & \cdots & x_1 \\ 1 & 0 & \cdots & 0 & 0 & \cdots & 0 \end{Vmatrix}.$$

To find the singularities, take any $p = (x_1, \cdots, x_{n+1}) \in S_0^n$, and any vector $v = (v_1, \cdots, v_{n+1})$ tangent to S_0^n at p; then v is orthogonal to $p - O$ (O = origin), so that $\sum v_i x_i = 0$. The vector v is mapped by f into a vector we shall call $\nabla f(v, p)$ (which may be considered as the derivative of f along v; see the preceding paper §4); it is

$$\begin{aligned} \nabla f(v, p) &= \sum v_i \frac{\partial f}{\partial x_i} \\ &= (v_{n+1}, v_2, \cdots, v_n, v_1 x_2 + v_2 x_1, \cdots, v_1 x_n + v_n x_1). \end{aligned}$$

Suppose this vanishes, with $v \neq 0$. Then $v_2 = \cdots = v_n = v_{n+1} = 0$, hence $v_1 \neq 0$, and since $v_1x_i + v_ix_1 = v_1x_i = 0$ $(i > 1)$, we have $x_2 = \cdots = x_n = 0$. Also $\sum v_ix_i = v_1x_1 = 0$, and hence $x_1 = 0$, and $x_{n+1} = \pm 1$. Thus the only singular points are

$$p_1 = (0, \cdots, 0, 1), \qquad p_2 = (0, \cdots, 0, -1).$$

Near each p_k we may determine x_{n+1} in terms of $x_1, \cdots, x_n$, and thus write

$$f(x_1, \cdots, x_n, x_{n+1}) = F(x_1, \cdots, x_n) \quad \text{in} \quad S_0^n;$$

we find $\partial F/\partial x_1 |_{p_k} = 0$. Computing $\partial F/\partial x_i$ and $\partial^2 F/\partial x_1 \partial x_i$ at p_1 and at p_2 shows at once that these singularities are of the required type.

From the mapping f in (4.1) we obtain a mapping ϕ of E^n into E^{2n-1} as follows. First, interchange x_1 and x_{n+1}:

$$f_1(x_1, \cdots, x_n, x_{n+1}) = (x_1, \cdots, x_n, x_{n+1}x_2, \cdots, x_{n+1}x_n).$$

Near the point $p_0 = (0, 1, 0, \cdots, 0)$, this is very close to the identity mapping of S_0^n into $S_0^n \subset E^{n+1} \subset E^{2n-1}$; a slight deformation of f_1 into f_2 will bring it to the identity in a neighborhood U of p_0 in S_0^n. By stretching $U - p_0$ into the part of E^n outside some $(n - 1)$-sphere, f_2 transforms into the required mapping.

A mapping f of E^n into E^{2n-1} with two singularities may also be defined as follows:

$$\begin{aligned} &u = (1 + x_1^2) \cdots (1 + x_n^2), \\ (4.2) \qquad &y_1 = x_1 - \frac{2x_1}{u}, \qquad y_i = x_i \qquad (i = 2, \cdots, n), \\ &y_{n+1} = \frac{1}{u}, \qquad y_{n+i} = \frac{x_1x_i}{u} \qquad (i = 2, \cdots, n-1). \end{aligned}$$

Note that far from the origin, f is very near the identity. Hence a slight alteration of f will make it the identity outside some sphere. (Compare the proof of Lemma 11.) Comparing with the preceding paper, §2, we see easily that f is regular except at the points $p_\pm = (0, \cdots, 0, \pm 1)$; at these points, $\partial f/\partial x_1 = 0$. At p_+ for example, the vectors (3.2) form a diagonal determinant, whose elements d_i are $d_1 = 2$, $d_i = 1$ $(i = 2, \cdots, n)$, $d_{n+1} = -1$, $d_{n+i} = \frac{1}{2}$ $(i = 2, \cdots, n - 1)$; hence f is semi-regular.

5. The orientation of singular points

We shall discuss the following problem. Given a singular point p and a neighborhood U of p, are there any orientation properties of E or of U determined by the set of points $f(U)$? Let A be the arc of self-intersection through p (i.e. part of the x_1-axis, in the coordinate system of Lemma 2). It turns out that *for n odd, an orientation of E is determined, while for n even, an orientation of E is determined by one of A near p.* We shall show in fact that the following definitions are permissible.

DEFINITIONS. We use the above notations. If n is odd, the singular point is *positive* or *negative* according as the vectors (3.2) determine the negative or positive orientation of E. If n is even, the *positive side* of M at p is the direction along A such that, if the x_1-axis points in that direction, then the vectors (3.2) determine the negative orientation of E. Note that *M need not be oriented or even orientable.* The reason for the choice will appear in Lemma 6.

LEMMA 3. *The above definitions are independent of the coordinate systems employed.*

Take two systems $\{x_i\}$ and $\{x'_i\}$, with $\partial f/\partial x_1|_p = \partial f/\partial x'_1|_p = 0$. We may rotate the x'_i-axes ($i > 1$), obtaining $\{x''_i\}$, so that

$$\left.\frac{\partial f}{\partial x''_i}\right|_p = \alpha_i \left.\frac{\partial f}{\partial x_i}\right|_p, \qquad \alpha_i > 0 \text{ for } i > 2.$$

The definitions with the $\{x''_i\}$ are the same as with the $\{x'_i\}$.

If $n \geqq 2$ and $\alpha_2 < 0$, let us replace x''_2 by $-x''_2$. This does not affect the orientation of A, and since both $\partial f/\partial x''_2|_p$ and $\partial^2 f/\partial x''_1 \partial x''_2|_p$ are reversed in direction, the vectors (3.2) with the new $\{x''_i\}$ determine the same orientation of E as with the old $\{x''_i\}$.

Now if $\alpha_1 < 0$, replace x''_1 by $-x''_1$. Suppose first that n is odd. Then the $n - 1$ vectors $\partial^2 f/\partial x''_1 \partial x''_2|_p, \cdots, \partial^2 f/\partial x''_1 \partial x''_n|_p$ are reversed, but no others of (3.2) are changed. Thus the same orientation of E is determined. Suppose next that n is even. Then the orientation of E is reversed; but the new x''_i-axis now points in the other direction along A.

With all the $\alpha_i > 0$, we may deform the $\{x''_i\}$ system into the $\{x_i\}$ system. The vectors (3.2) remain always independent, so the same orientation of E and of A are determined, completing the proof.

LEMMA 4. *If f is given by* (3.3), *and the coordinate systems determine the positive orientations of M and of E, then for n odd, the singular point is negative, while for n even, the positive direction in M at p is along the negative x_1-axis.*

It is sufficient to show that the matrix formed from the vectors (3.2) has a positive determinant. Differentiating (3.3), we see that the determinant is diagonal, with one 2 and the rest 1's on the diagonal; hence the determinant is $2 > 0$.

6. The intersection $\mathfrak{L}_f(M)$ of M with ∂M under f

We shall count the number of times that $f(\partial M)$ cuts through $f(M)$ in E^{2n-1}. The definition given here suffices in the orientable case; an interpretation in terms of the manner in which M attaches to ∂M will be studied in §9. The latter discussion will apply also to non-orientable partial manifolds such that ∂M can be made into a cycle; see §14.

If A^r and B^s are singular chains in an oriented E^{r+s}, such that $A \cap \partial B = \partial A \cap B = 0$, then their Kronecker index $KI(A, B)$ is defined. In particular, let σ^r and σ^s be oriented cells with just one common interior point p, their tangent planes at p having only p in common. Let $u_1, \cdots, u_r$ be independent vectors tangent to σ^r at p, determining the positive orientation of σ^r; choose $v_1, \cdots, v_s$ similarly for σ^s. Then the intersection is positive or negative ($KI = 1$ or -1) according as $u_1, \cdots, u_r, v_1, \cdots, v_s$ determine the positive or negative orientation of E^{r+s}.

DEFINITION. Let f be a semi-regular mapping of the orientable partial manifold M^n into E^{2n-1}. Choose an orientation of M. With an infinite subdivision of $M - \partial M$, we obtain an infinite singular chain $M^{\odot}$. The boundary of M is oriented, and becomes a chain ∂M. We define

$$\mathfrak{L}_f(M) = KI(f\partial M, fM^{\odot}) = KI(fM^{\odot}, f\partial M), \tag{6.1}$$

if this is finite. If ∂M is compact, it will be finite, since f is proper.

We now give the definition without the help of the infinite chain $M^{\odot}$.

LEMMA 5. *Let U be a neighborhood of ∂M in which f is one-one. Let M^* be a singular chain such that*

$$M - M^* \subset U. \tag{6.2}$$

Then (LC = looping coefficient)

$$\mathfrak{L}_f(M) = KI(fM^*, f\,\partial\,M) = LC(f\,\partial\,M^*, f\,\partial\,M). \tag{6.3}$$

For we can write $M^{\odot} = M^* + M'$ where $M' \subset U$; since f is one-one in U, $KI(fM', f\,\partial\,M) = 0$.

LEMMA 6. *Let the situation be as in Lemma 2. Let σ be an oriented n-cell, lying in the coordinate system, and obtained from the sphere $\sum x_i^2 = \beta^2$ plus interior by cutting off the part with $x_1 > \alpha$, where $0 < \alpha < \beta$. Then for n odd, $\mathfrak{L}(\sigma) = 1$ or -1 according as the singular point is positive or negative, while for n even, $\mathfrak{L}(\sigma) = 1$ or -1 according as the x_1-axis extends in the positive or negative direction in M.*

REMARKS. It is easily seen that for n odd, $\mathfrak{L}(\sigma_1)$ is the same as $\mathfrak{L}(\sigma)$ if σ_1 is obtained from the sphere by cutting off the part $x_1 < -\alpha$; this follows also directly from Theorem 4. In the proof of Theorem 6 for n even, we need only the obvious fact that $\mathfrak{L}(\sigma) = \pm 1$.

Let $p = (-\alpha, 0, \cdots, 0)$, $q = (\alpha, 0, \cdots, 0)$. The only intersection of $f(\partial M)$ with $f(M^{\odot})$ is $f(p) = f(q)$. It is clear from Lemma 3 that we may suppose that the coordinate systems determine the positive orientations of M (or σ) and of E. Now the result of Lemma 4 holds, so that it is sufficient to show that $\mathfrak{L}(\sigma) = -1$.

Let $e_1, \cdots, e_n$ be the unit vectors in E^n; these determine the positive orientation of σ, while at q, the vectors $e_2, \cdots, e_n$ determine the positive orientation of $\partial\sigma$. We must show that the vectors

$$\left.\frac{\partial f}{\partial x_1}\right|_p, \cdots, \left.\frac{\partial f}{\partial x_n}\right|_p, \left.\frac{\partial f}{\partial x_2}\right|_q, \cdots, \left.\frac{\partial f}{\partial x_n}\right|_q \tag{6.4}$$

determine the negative orientation of E^{2n-1}. The two sets of vectors give the matrices

$$\left\|\begin{matrix} -2\alpha & 0 & \cdots & 0 & 0 & \cdots & 0 \\ 0 & 1 & \cdots & 0 & -\alpha & \cdots & 0 \\ \cdots & \cdots & \cdots & \cdots & \cdots & \cdots & \cdots \\ 0 & 0 & \cdots & 1 & 0 & \cdots & -\alpha \end{matrix}\right\|, \qquad \left\|\begin{matrix} 0 & 1 & \cdots & 0 & \alpha & \cdots & 0 \\ \cdots & \cdots & \cdots & \cdots & \cdots & \cdots & \cdots \\ 0 & 0 & \cdots & 1 & 0 & \cdots & \alpha \end{matrix}\right\|.$$

Putting the second below the first forms a determinant D which we must prove negative. Subtracting the ith row from the $(n + i - 1)^{\text{th}}$ row $(i = 2, \cdots, n)$ gives a determinant with zeros below the diagonal, whose value is

$$D = (-2\alpha)(2\alpha)^{n-1} = -(2\alpha)^n < 0,$$

as required.

7. The self-intersection of an n-complex mapped into E^{2n-1}

We shall consider mappings of a finite n-complex K into E^{2n-1} *which are one-one in U* for some neighborhood U of K^{n-1}. Of course K might be a subcomplex of a complex of higher dimension. We note that any mapping may be approximated to by one of the required type, even if the cells of K are replaced by more general bounded smooth manifolds; see §16. The considerations will be considerably generalized in Part III.

Let K' be a subdivision of K such that any cell of K' with a vertex in K^{n-1} lies in U. For each oriented σ_i^n of K, let σ_i^* be the sum of the similarly oriented n-cells of K' in σ_i^n which do not touch $\partial\sigma_i^n$. For any chain $A^n = \sum a_i\sigma_i^n$ set $A^* = \sum a_i\sigma_i^*$. The coefficients a_i are integers.

Definition. Generalizing the definition in §6, we set

$$\mathfrak{I}_f(A^n) = KI(fA^*, f\partial A^n) = LC(f\partial A^*, f\partial A^n). \tag{7.1}$$

Note that, if $M = \sum \sigma_i^n$, then $M^* = \sum \sigma_i^*$, which is not the M^* previously used; but the two definitions of $\mathfrak{I}_f(M)$ agree, as is apparent from (8.1) below.

Lemma 7. *Under the above conditions, we have the point set relations*

$$f(\partial\sigma_i^n) \cap f(\partial\sigma_i^*) = 0, \tag{7.2}$$

$$f(\sigma_i^n - \sigma_i^*) \cap f(\partial\sigma_j^*) = 0 \quad \textit{for} \quad i \neq j. \tag{7.3}$$

For $\partial\sigma_i^n$ and $\partial\sigma_i^*$, also $\sigma_i^n - \sigma_i^*$ and $\partial\sigma_j^*$, are disjoint point sets in U.

Theorem 1. *Let the mapping f of K into E^{2n-1} be one-one in $K^n \cap U$, U a neighborhood of K^{n-1} in K. Then for any n-chain $\sum a_i\sigma_i^n$,*

$$\mathfrak{I}_f(\textstyle\sum a_i\sigma_i^n) = \sum a_i^2\,\mathfrak{I}_f(\sigma_i^n) \qquad \textit{for } n \textit{ odd}, \tag{7.4}$$

$$\mathfrak{I}_f(\textstyle\sum a_i\sigma_i^n) \equiv \sum a_i\,\mathfrak{I}_f(\sigma_i^n) \pmod 2 \qquad \textit{for } n \textit{ even}. \tag{7.5}$$

Remark. We could allow double points $f(p) = f(q)$ with both p and q in

K^{n-1}, for this would not destroy the relations (7.2) and (7.3). But $\mathfrak{L}_f$ would not then be invariant under slight deformations.

We may suppose $K = K^n$ in the proof. Set $\tau_i^n = f(\sigma_i^n)$, $\tau_i^* = f(\sigma_i^*)$. First, by (7.2) and (7.3), $KI(\tau_i^*, \partial\tau_j^n)$ has meaning for all i and j. Hence

$$\mathfrak{L}_f(\sum_i a_i\sigma_i^n) = KI(\sum_i a_i\tau_i^*, \partial\sum_j a_j\tau_j^n) = \sum_{i,j} a_i a_j KI(\tau_i^*, \partial\tau_j^n)$$

$$= \sum_i a_i^2 KI(\tau_i^*, \partial\tau_i^n) + \sum_{i<j} a_i a_j [KI(\tau_i^*, \partial\tau_j^n) + KI(\tau_j^*, \partial\tau_i^n)].$$

Now by (7.3), $\partial\tau_j^*$ does not intersect $\tau_i^n - \tau_i^*$ if $i \neq j$, and $\partial\tau_j^n - \partial\tau_j^*$ bounds a chain $\tau_j^n - \tau_j^*$ which does not intersect the boundary of τ_i^*. Hence, by elementary properties of the Kronecker index,

$$KI(\tau_j^*, \partial\tau_i^n) = (-1)^n KI(\partial\tau_j^*, \tau_i^n) = (-1)^n KI(\tau_i^n, \partial\tau_j^*)$$

$$= (-1)^n KI(\tau_i^*, \partial\tau_j^*) = (-1)^n KI(\tau_i^*, \partial\tau_j^n).$$

Consequently

$$(7.6) \qquad \mathfrak{L}_f(\sum_i a_i\sigma_i^n) = \sum_i a_i^2 \mathfrak{L}_f(\sigma_i^n) + [1 + (-1)^n] \sum_{i<j} a_i a_j KI(\tau_i^*, \partial\tau_j^n),$$

from which the theorem follows.

8. Relation between singularities and self-intersections in M

We are now ready to prove the fundamental combinatorial theorem in the orientable case. It will be the primary object of Part III to prove the theorem in the non-orientable case; if n is odd, we need a new definition, (20.5), of $\mathfrak{L}_f(M)$, and a slight further restriction on f.

THEOREM 2. *Let f be a semi-regular mapping of the compact partial manifold M^n into E^{n-1}. Then for n odd, $\mathfrak{L}_f(M)$ is the algebraic number of singular points, while for n even, it is congruent to this number* mod 2.

REMARK. We could find an exact expression for $\mathfrak{L}_f(M)$ when n is even with the help of the classification of self-intersections in §22.

First replace f by a completely semi-regular mapping (Lemma 1), which we again call f. Let $p_1, \cdots, p_\mu$ be the singular points. About each p_i choose a cell σ_i^n as in Lemma 6. We may let these be cells of a subdivision of M (which need not be simplicial) into smooth cells.[1] Moreover, by first deforming the cells of the subdivision slightly so that $(n-1)$-cells touch the curves of self-intersection in interior isolated points only, and then deforming slightly again, we may obtain a subdivision K such that f is one-one over K^{n-1}. By Lemma 25, it is one-one in a neighborhood of K^{n-1}. Also, since $M - \sum_{i=1}^{\mu} \sigma_i^n$ contains no singular points, we may suppose that f is one-one in each σ_i^n, $i > \mu$.

Define K' and the σ_i^* as in §7, and M^* as in §6. Now

$$f(\partial M) \cap f[\sum(\sigma_i^n - \sigma_i^*) \cap M^*] = 0,$$

[1] Rather than prove this fact, we could easily construct a subdivision containing cells $\sigma_i'^n$ approximately like the cells σ_i^n, and note that $\mathfrak{L}(\sigma_i'^n) = \mathfrak{L}(\sigma_i^n)$.

and hence

$$KI(fM^*, f\partial M) = KI(\sum\sigma_i^*, f\partial M). \tag{8.1}$$

Applying theorem 1 gives

$$\begin{aligned} \mathfrak{L}_f(M) &= \sum_{i=1}^{\mu} \mathfrak{L}_f(\sigma_i^n) && \text{for } n \text{ odd,} \\ \mathfrak{L}_f(M) &\equiv \sum_{i=1}^{\mu} \mathfrak{L}_f(\sigma_i^n) (\text{mod } 2) && \text{for } n \text{ even,} \end{aligned} \tag{8.2}$$

since $\mathfrak{L}(\sigma_i^n) = 0$ for $i > \mu$, as f is one-one there. The theorem now follows from Lemma 6.

THEOREM 3. *Let f be a semi-regular mapping of the closed manifold M^n into E^{2n-1}. Then for n odd, the algebraic number of singular points vanishes, while for n even, it vanishes* mod 2.

This is a corollary of the last theorem.

EXAMPLE. For $n = 1$, the theorem says that a (semi-regular) real-valued function defined on a circle has the same number of maxima as minima.

We give finally an invariance theorem whose proof requires Lemma 9 below.

THEOREM 4. *Let f_t be a deformation of the compact partial manifold M such that f_0 and f_1 are semi-regular, and for some neighborhood U of ∂M, each f_t is regular in U. Then if n is odd, f_0 and f_1 have the same algebraic number of singular points, while for n even, they have the same number* mod 2.

The hypothesis on f_t shows that each $\mathfrak{L}_{f_t}(M)$ may be defined as in §9. By Lemma 9, it is constant for n odd, and is constant (mod 2) for n even. The theorem now follows from Theorem 2.

9. Looping coefficients of vector fields in manifolds in space

DEFINITIONS. Let K^r be a finite smooth complex in E^{2r+1}, and let $v(p)$ be a continuous vector field in K^r, independent of K^r. Then there is an $\epsilon_0 > 0$ with the following property. For every ϵ, $0 < \epsilon < \epsilon_0$, if

$$\phi_{v,\epsilon}(K) = \text{all } p + \epsilon v(p), \qquad p \in K, \tag{9.1}$$

then $K \cap \phi_{v,\epsilon}(K) = 0$. Now for any chain A^r in K, $\phi_{v,\epsilon}A$ is a chain in $\phi_{v,\epsilon}(K)$. For cycles A^r, we define

$$\mathfrak{L}(A, v) = LC(\phi_{v,\epsilon}A, A). \tag{9.2}$$

Because of (9.1) it is clear that the result is independent of the choice of ϵ. The definition will be extended in §14.

The following lemma gives the relation to the previous $\mathfrak{L}$.

LEMMA 8. *Let M^n be a partial orientable manifold, and let f be a semi-regular mapping of M^n into E^{2n-1}. Let $v(p)$ be defined in $f(\partial M)$, and independent of $f(\partial M)$, and point into $f(M)$ at each p. Then*

$$\mathfrak{L}_f(M) = \mathfrak{L}(f\partial M, v). \tag{9.3}$$

To prove the lemma, we note that for ϵ sufficiently small, the set of all $\phi_{v,\epsilon'}(\partial M)$ for $0 \leqq \epsilon' \leqq \epsilon$ projects in a one-one manner into a subset M' of M, and M' contains all of M within some neighborhood of ∂M. Setting $M^* = M - M'$ defines a chain M^* to which Lemma 5 applies. Clearly

$$\mathfrak{L}_f(M) = LC(f\partial M^*, f\partial M) = LC(\phi_{v,\epsilon}\partial M, f\partial M) = \mathfrak{L}(f\partial M, v).$$

Let f be an imbedding of the r-manifold M^r in E^{2r+1}, and let $v(p)$ be defined so that $v(p)$ is independent of $f(M)$ at $f(p)$. If deformations f_t and v_t are given so that each f_t is an imbedding and each v_t is independent of $f_t(M)$, clearly $\mathfrak{L}(fM, v)$ is constant. The next lemma (used in the proof of Theorem 4) generalizes this.

Lemma 9. *Let f_t be a regular deformation of M^r in E^{2r+1} such that f_0 and f_1 are imbeddings. Let v_t be a continuous vector field independent of $f_t(M)$ for each t. Then*

$$\mathfrak{L}(f_1M, v_1) = \mathfrak{L}(f_0M, v_0) \qquad \textit{for } r \textit{ even},$$

$$\mathfrak{L}(f_1M, v_1) \equiv \mathfrak{L}(f_0M, v_0) \pmod 2 \qquad \textit{for } r \textit{ odd}.$$

Remark. The lemma clearly holds if M^r is a cycle in a complex K^n, f_t being a regular deformation in each closed $\bar{\sigma}_i^n$. The crossings (see below) may be taken interior to n-cells $\sigma_i^n = U_i$ of K.

It is easily seen by the methods in [1] that a slight alteration of f_t for $0 < t < 1$ will give a new f_t with the following property. If, for a certain t_1, f_{t_1} is not one-one, say $q_0 = f_{t_1}(p_1) = f_{t_1}(p_2)$, then this is the only double point, and the portions U_1 and U_2 of M near p_1 and p_2 are crossing each other as t moves through t_1. That is, if $u_{i1}, \cdots, u_{ir}$ are independent vectors tangent to $f_{t_1}(U_i)$ at q $(i = 1, 2)$, and

$$u' = \left.\frac{\partial f_t(p_1)}{\partial t}\right|_{t=t_1} - \left.\frac{\partial f_t(p_2)}{\partial t}\right|_{t=t_1},$$

then these $2r + 1$ vectors are independent.

Take t' and t'' very close to t_1, with $t' < t_1 < t''$. Since $\mathfrak{L}$ is constant over intervals containing no such t_1, it is sufficient to prove that the relations hold with $f_{t'}$ and $f_{t''}$ replacing f_0 and f_1. We may clearly accomplish the deformation f_t in two steps: first, push $f_{t'}(U_1)$ in the direction of u' so that it crosses $f_{t'}(U_2)$ (pushing $f_{t'}(p_1)$ a distance $(t'' - t')u'$); second, deform the result into $f_{t''}(M)$. We may replace v_t by $v_{t'} = v$ in all this if t' and t'' are close enough together. We now need merely prove the relations for the mappings before and after the first deformation, which we call g_t, using $0 \leqq t \leqq 1$.

Set

$$M_t = g_t(M), \qquad N_t = \phi_{v,\epsilon}(M_t) \qquad (0 \leqq t \leqq 1),$$

with a small $\epsilon > 0$. Except in U_1, these are independent of t. They define singular chains A and B such that (with M oriented)

$$M_1 = M_0 + \partial A, \qquad N_1 = N_0 + \partial B.$$

We find

$$\begin{aligned}\Delta &= \mathfrak{L}(M_1, v) - \mathfrak{L}(M_0, v) = LC(N_1, M_1) - LC(N_0, M_0)\\ &= LC(\partial B, M_1) + LC(N_0, M_1 - M_0)\\ &= LC(\partial B, M_1) + (-1)^{r+1}LC(\partial A, N_0)\\ &= KI(B, M_1) + (-1)^{r+1}KI(A, N_0).\end{aligned}$$

Since $u_{11}, \cdots, u_{1r}, u'$ are independent, we may suppose $A \cap B = 0$. Also g_0 is one-one in $M - U_2$. Hence we may clearly suppose

$$KI(B, g_1(M - U_2)) = KI(A, \phi_{v,\epsilon}\, g_0(M - U_2)) = 0.$$

Therefore

$$\Delta = KI(B, g_1U_2) + (-1)^{r+1}KI(A, \phi_{v,\epsilon}\, g_0U_2) = [1 + (-1)^{r+1}]KI(A, g_0U_2),$$

which proves the lemma.

II. The immersion theorem

10. Some deformations related to certain vector fields

We first show when one vector field in $M^r \subset E^{2r+1}$ may be deformed into another one. In Lemma 11 we show how the boundary of a partial manifold may be moved over to a desired position, and in Lemma 12, we show how the boundary may be twisted to point in given directions.

LEMMA 10. *Let $M^r \subset E^{2r+1}$ $(r \geqq 1)$ be a connected closed orientable manifold, and let v_0 and v_1 be continuous vector fields in M, each independent of M. Then there is a deformation $v_t (0 \leqq t \leqq 1)$ of v_0 into v_1 so that each v_t is independent of M, if and only if $\mathfrak{L}(M, v_0) = \mathfrak{L}(M, v_1)$.*

REMARK. If the normal bundle of M^r in E^{2r+1} is simple,[2] the proof is easy to give.

The necessity of the condition in the theorem is clear; we shall prove the sufficiency. We first deform v_0 and v_1 into fields of unit normal vectors. Next, let K be a simplicial complex forming a fine enough subdivision of M so that each cell of K is nearly flat, and so that v_0 is nearly constant in each cell. Now as considerations of dimensionality show at once, we may deform v_1 so that $v_1 = v_0$ in K^{r-1}. We now suppose v_0 and v_1 are of this nature.

Let $\{\sigma_i^r\}$ denote the similarly oriented r-cells of K. For each $p \in M$, let $S(p)$ denote the unit r-sphere about p whose plane is normal to M at p, and let $S'(p) \subset S(p)$ denote the subsphere orthogonal to $v_0(p)$. Let S_0^r denote a fixed r-sphere, and q_0, a fixed point of it. For each σ_i^r we introduce a "coordinate system"

[2] See [3]. The present lemma belongs properly in the subject considered there. Some of the details omitted in the present proof may be found there. In the present paper we need only the case that M is a sphere. By cutting it into two cells and using the theorem that any sphere-bundle over a cell is simple, the proof could be materially simplified in this case.

into the $S(p)$ as follows. For each $p \in \bar{\sigma}_i^r$ and each $q \in S_0^r$, $\xi_i(p, q)$ is a point of $S(p)$; for each p, it is an orthogonal (distance preserving) mapping, and this mapping is continuous in p; furthermore, $\xi_i(p, q_0) = v_0(p)$. (Since the σ_i^r are nearly flat, it is easy to construct ξ_i first over the part S_0^{r-1} of S_0^r orthogonal to q_0, so as to map into the $S'(p)$; it is then uniquely extendable over S_0^r.) Let

$$\xi_i^{-1}(p, q') = q \quad \text{if} \quad q' = \xi_i(p, q).$$

If we orient M and S_0^r, orient the $S(p)$ so that the orientations of M and $S(p)$ at p determine the positive orientation of E, and choose the ξ_i so that they are rotations (i.e. sense-preserving), then each $\xi_i^{-1}(p, \xi_j(p, q))$ for each p will be a rotation.

Set

$$\psi_i(p) = \xi_i^{-1}(p, v_1(p)). \tag{10.1}$$

Since $v_1(p) = v_0(p)$ in $\partial\sigma_i^r$, $\psi_i(p) = q_0$ there. Hence ψ_i maps σ_i^r into S_0^r so that $\partial\sigma_i^r$ goes into q_0, and thus ψ_i has a degree d_i over σ_i^r. Set

$$X(v_1) = \sum d_i\sigma_i^r. \tag{10.2}$$

Since $\dim(K) = r$, this is a cocycle.

Suppose v_1 is deformed as follows. Take any σ_j^{r-1}; $\psi_i(p) = q_0$ here. As t runs from 0 to 1, let $\psi_{it}(\sigma_j^{r-1})$ sweep over S_0^r with the degree α_j, keeping $\psi_{it}(\partial\sigma_j^{r-1}) = q_0$ and $\psi_{i1}(\sigma_j^{r-1}) = q_0$. (Thus if I is the unit interval $0 \leqq t \leqq 1$, and $\Psi_i(t, p) = \psi_{it}(p)$, Ψ_i maps $I \times \sigma_j^{r-1}$ into S_0^r with the degree α_j.) We may extend[3] ψ_{it} over the rest of M, requiring that it be independent of t except in the cells of $St(\sigma_j^{r-1})$. Set

$$v_t'(p) = \xi_i(p, \psi_{it}(p)), \qquad p \in \bar{\sigma}_i^r, \text{ each } i;$$

then $v_0'(p) = v_1(p)$, and each v_t' is a field of unit normal vectors. Let d_i' denote the degree defined with the help of v_1'. Then clearly for any σ_i^r,

$$d_i' = d_i + [\sigma_j^{r-1}:\sigma_i^r]\alpha_j,$$

and hence

$$X(v_1') = X(v_1) + \alpha_j\delta\sigma_j^{r-1}.$$

Since we may carry out this process for each σ_j^{r-1} in turn, we may alter X by any coboundary.

For a small $\epsilon > 0$, if $\omega_k(\sigma_i^r)$ denotes the cell σ_i^r displaced in the direction of v_k a distance ϵ (this mapping need not be one-one), then

$$\mathfrak{L}(M, v_k) = LC(\sum \omega_k\sigma_i^r, M) \qquad (k = 1, 2). \tag{10.3}$$

We shall show that

$$LC(\omega_1\sigma_i^r - \omega_0\sigma_i^r, M) = d_i. \tag{10.4}$$

---

[3] See Alexandroff-Hopf, *Topologie I*, Berlin, 1936, p. 501 Hilfsatz Ia.

This quantity is defined, since $v_0 = v_1$ in $\partial\sigma_i^r$. Let us flatten σ_i^r into σ', lying in a space E^r. A slight alteration of v_0 will make it constant in $\bar{\sigma}_i^r$. Let E^{r+1} be a plane through a point $p_0 \in \sigma_i^r$, orthogonal to E^r, and consider S_0^r as the unit sphere in E^{r+1} about p_0. If we project the chain $\omega_1\sigma_i^r - \omega_0\sigma_i^r$ parallel to E^r into E^{r+1}, and then in E^{r+1} away from p_0 into S_0^r, $\omega_0\sigma_i^r$ will go into a point, say q_0, and $\omega_1\sigma_i^r$ will go into a chain $\omega'\sigma_i^r$. Now

$$LC(\omega_1\sigma_i^r - \omega_0\sigma_i^r, E^r) = LC(\omega'\sigma_i^r, E^r).$$

This also equals $LC(\omega'\sigma_i^r, p_0)$, considering this as defined in E^{r+1} (which is oriented like $S(p_0)$). We may suppose the ξ_i chosen so that after the above alterations and projection, $v_1(p) + p$ becomes the point $\psi_i(p)$. Hence

$$LC(\omega'\sigma_i^r, E^r) = d_i.$$

Interpreting the looping coefficients as Kronecker indices shows that the looping coefficients with E^r are the same as with M. Thus (10.4) is proved.

Adding the equations (10.4) gives

$$\begin{aligned}\sum d_i &= LC(\sum \omega_1\sigma_j^r - \sum \omega_0\sigma_i^r, M)\\ &= \mathfrak{L}(M, v_1) - \mathfrak{L}(M, v_0) = 0.\end{aligned}$$

Since M is closed, connected and oriented, there is a one-one correspondence h between the cohomology classes of dimension r of M and the integers, given by

$$h(\sum \alpha_i\sigma_i^r) = M\cdot\sum \alpha_i\sigma_i^r = \sum \alpha_i.$$

It follows that

$$h(X(v_1)) = \sum d_i = 0, \qquad X \backsim 0.$$

Consequently we may deform v_1 into v_1' so that $X(v_1') = 0$, i.e. $d_i' = 0$ for each i. Now by a theorem of Hopf,[4] we may deform ψ_i' into q_0 in each $\bar{\sigma}_i^r$, keeping it at q_0 in $\partial\bar{\sigma}_i^r$. This defines a corresponding deformation of v_1' into v_0 in $\bar{\sigma}_i^r$, keeping it fixed in $\partial\sigma_i^r$. Thus v_1 is deformed into v_0 in M, and the lemma is proved.

LEMMA 11. *Let f_0 and f_1 be imbeddings of the manifold M^r in E^ν. Let $L(p)$ be the segment $f_0(p)f_1(p)$. Let no two of these have common points. For each $p \in M$ let there be a plane $T(p) = T^{\nu-r}(p)$ in E^ν such that*

(a) *$f_0(p)$ and $f_1(p)$ are in $T(p)$,*

(b) *$T(p)$ has only $f_i(p)$ in common with the tangent plane to $f_i(M)$ at $f_i(p)$ ($i = 0, 1$),*

(c) *The function $T(p)$ is smooth (Compare* [1], §24).

Then there is a smooth deformation ϕ_t of E^ν ($0 \leqq t \leqq 1$) such that

(d) *each ϕ_t is an imbedding, and ϕ_0 is the identity,*

(e) *$\phi_1(f_0(p)) = f_1(p)$ ($p \in M$),*

(f) *for a given neighborhood U of the set of all segments $L(p)$, $\phi_t(p) = p$ for $p \in E^\nu - U$ and $0 \leqq t \leqq 1$.*

[4] See ALEXANDROFF-HOPF, loc. cit., p. 504, Satz III$_n$, or H. WHITNEY, Duke Math. J., vol. 3 (1937), pp. 46–50, Appendix.

REMARKS. Any segment $L(p)$ may reduce to a single point $f_0(p) = f_1(p)$. If M is a partial manifold, and $f_0 = f_1$, together with first partial derivatives, in ∂M, the proof below holds, and each ϕ_t is the identity, together with first partial derivatives, at all points of ∂M. The most important application of the lemma is to the case $M^r = \partial M^n$; a given mapping f_0 of M^n is then altered to f_1, so that f_1 is a given mapping in ∂M. If the mappings are of class C^γ, we may make each ϕ_t of class C^γ.

Take any p_0 in M. By (a), (b) and (c), it is easy to see that for some neighborhood U_0 of p_0 and some $\alpha_0 > 0$, the set of points q in planes $T(p)$ with $p \in U_0$ which are within a distance α_0 of $L(p)$ fills out a neighborhood of $L(p_0)$ in E^ν in a smooth one-one way (compare the proof of [1], Lemma 21). Hence, since the $L(p)$ are distinct, there is a positive continuous function $\alpha(p)$ (or a constant $\alpha > 0$ if M is compact) such that if $R(p)$ is the set of points of $T(p)$ within a distance $\alpha(p)$ of $L(p)$, then the $R(p)$ fill out a neighborhood of $\sum L(p)$ in a smooth one-one-way, and $\sum R(p) \subset U$. We may choose a smooth function $\eta(p) > 0$ such that if $L'(p)$ is the segment $L(p)$ extended in each direction by the amount $\eta(p)$, and $C(p)$ is the cylinder (of dimension $\nu - r$) in $T(p)$ with axis $L'(p)$ and of radius $\eta(p)$, then $C(p) \subset R(p)$ for $p \in M$ (see [1], Lemma 25). It is easy to set up an expression depending smoothly on $\eta(p)$ and the (locally oriented) length of $L(p)$, which defines a smooth deformation of $R(p)$ into itself with the properties that it is constant in $R(p) - C(p)$, carries $f_0(p)$ into $f_1(p)$, and is an imbedding for each t. Letting this define ϕ_t in $\sum R(p)$ and setting $\phi_t(p) = p$ in $E^\nu - \sum R(p)$ proves the lemma.

LEMMA 12. *Let $M^n (n \geqq 2)$ be a partial manifold, let ∂M be a closed manifold, and let f be a mapping of class C^2 of M into E^{2n-1} such that in some neighborhood U of ∂M, f is an imbedding. Let $u(p)$ be a smooth vector field in ∂M, pointing into M at p in ∂M; set*

$$v(p) = \nabla f(u, p), \qquad p \in \partial M.$$

Let $v'(p)$ be a smooth vector field in $f(\partial M)$, independent of $f(\partial M)$, such that

$$\mathfrak{L}(f\partial M, v') = \mathfrak{L}(f\partial M, v).$$

Then there is a smooth mapping f' of M into E such that $f' = f$ in $M - U$, f' is an imbedding in U, $f' = f$ in ∂M, f' is arbitrarily close to f (but not together with first derivatives) in M, and

$$\nabla f'(u, p) = v'(p), \qquad p \in \partial M.$$

REMARKS. The assumption that ∂M is closed could be easily removed. A more accurate statement about the class of f and of f' could be given, but we shall not need it.

Since we need define f' in U only, we may consider U as lying in E^{2n-1}, and let f be the identity; then $v(p) = u(p)$. Set

$$p_t^* = p + tv(p), \qquad 0 \leqq t \leqq 1.$$

This is a smooth mapping of $I \times \partial M$ into E^{2n-1}. (Since f is of class C^2, v and p_t^* are of class C^1 in terms of the original coordinate systems in M.) For some $t_0 >$ 0, this is an imbedding for the values $0 \leqq t \leqq t_0$, since u is smooth. For t_0 small enough, we may project p_t^* into U. Say p_t^* projects into p_t. Now the points of U near ∂M are uniquely expressible in the form p_t ($p \in \partial M$, $0 \leqq t \leqq t_0$), and

$$\left.\frac{\partial p_t}{\partial t}\right|_{t=0} = v(p).$$

By Lemma 10, there is a deformation $v_t'(p)$ $(0 \leqq t \leqq 1)$ of $v'(p) = v_0'(p)$ into $v(p) = v_1'(p)$, such that each v_t' is independent of ∂M. We may replace $v_t'(p)$ by a smooth function $v_t^*(p)$ as follows. First set

$$v_t''(p) = v_0'(p) \qquad (t \leqq \tfrac{1}{3}), \qquad\qquad v_t''(p) = v_1'(p) \qquad (t \geqq \tfrac{2}{3}),$$

$$v_t''(p) = v_{t'}'(p) \qquad (t' = 3(t - \tfrac{1}{3}),\ \tfrac{1}{3} \leqq t \leqq \tfrac{2}{3}).$$

Then $v_t''(p)$ is smooth except for $\frac{1}{3} \leqq t \leqq \frac{2}{3}$. Now approximate to $v_t''(p)$ by a smooth vector function $v_t^*(p)$ for $\frac{1}{6} \leqq t \leqq \frac{5}{6}$, the approximation being closer and closer, together with first partial derivatives, as $t \to \frac{1}{6}$ or $t \to \frac{5}{6}$. (See [1], Theorem 2, (a) and (d). We could either make use of Theorem III of the author's paper in Trans. Am. Math. Soc., vol. 36 (1934), pp. 63–89, using first derivatives for $t < \frac{1}{3}$ and $t > \frac{2}{3}$, or note simply that in the approximation in Lemma 6, loc. cit., with $m = 0$, the first partial derivatives have automatically the desired approximation property.) Setting $v_t^*(p) = v_t''(p)$ for $t \leqq \frac{1}{6}$ and $t \geqq \frac{5}{6}$ makes v_t^* smooth for $0 \leqq t \leqq 1$ (see [1], Lemma 10). Moreover, with a close enough approximation, $v_t^*(p)$ is independent of ∂M for $0 \leqq t \leqq 1$.

Set

$$v_t(p) = \frac{\partial p_t}{\partial t}.$$

For certain numbers α and β to be determined later, with $0 < \alpha < \beta < t_0$, set

$$p_t' = p + \int_0^t v_{s/\alpha}^*(p)\, ds \qquad (0 \leqq t \leqq \alpha), \tag{10.5}$$

$$p_t' = p + \int_0^\alpha v_{s/\alpha}^*(p)\, ds + \int_0^{t-\alpha} v_s(p)\, ds \qquad (\alpha \leqq t \leqq \beta). \tag{10.6}$$

Cover ∂M with a finite set of coordinate systems $\{x_i\}$. Let V be the maximum of $|v_t^*(p)|$, $|v_t(p)|$, $|\partial v_t^*(p)/\partial x_i|$, $|\partial v_t(p)/\partial x_i|$. Now

$$\frac{\partial p_t'}{\partial t} = v_{t/\alpha}^*(p) \qquad (t \leqq \alpha), \qquad\qquad \frac{\partial p_t'}{\partial t} = v_{t-\alpha}(p) \qquad (t \geqq \alpha);$$

since $v_1^*(p) = v_1'(p) = v(p) = v_0(p)$, the mapping thus defined is smooth. Also, since

$$\frac{\partial p_t'}{\partial x_i} = \frac{\partial p}{\partial x_i} + \int_0^t \frac{\partial v_{s/\alpha}^*(p)}{\partial x_i}\, ds \qquad (t \leqq \alpha),$$

and similarly for $t \geqq \alpha$, we find

$$\left| \frac{\partial p_t'}{\partial x_i} - \frac{\partial p}{\partial x_i} \right| \leqq V\beta \qquad (0 \leqq t \leqq \beta).$$

Since the $\partial p/\partial x_i$ are independent for each p, and $\partial p_t'/\partial t$ is independent of them, by choosing β small enough we may insure that the $\partial p_t'/\partial x_i$ and $\partial p_t'/\partial t$ are independent for each p and t; hence the mapping is regular. Moreover, since

$$p_t = p + \int_0^t v_s(p)\, ds,$$

we have

$$p_\beta' - p_\beta = \int_0^\alpha v_{s/\alpha}^*(p)\, ds - \int_{\beta-\alpha}^\beta v_s(p)\, ds,$$

and

$$| p_\beta' - p_\beta | \leqq 2V\alpha, \qquad \left| \frac{\partial p_\beta'}{\partial x_i} - \frac{\partial p_\beta}{\partial x_i} \right| \leqq 2V\alpha,$$

$$\left| \frac{\partial p_t'}{\partial t} - \frac{\partial p_t}{\partial t} \right|_\beta = | v_{\beta-\alpha}(p) - v_\beta(p) | .$$

Hence, keeping β fixed, we may choose α so small that the mapping p_t' at $t = \beta$ is arbitrarily close to that of p_t at β, together with first derivatives.

Set

$$\phi(t, p) = p_t' - p_t \qquad (t \leqq \beta), \qquad \phi(t, p) = 0 \qquad (t \geqq t_0).$$

This is a mapping of the part of U outside $\beta < t < t_0$, which we have just seen may be taken arbitrarily small, with first derivatives. Hence, by [5], there is a smooth extension of ϕ through $\beta \leqq t \leqq t_0$, which may be taken arbitrarily small, together with first derivatives. Setting

$$p_t' = p_t + \phi(t, p) \qquad (\beta < t < t_0)$$

completes the definition of f' in U. By making ϕ and its first derivatives small enough, we insure that we have a close approximation to f, and that the new mapping is regular. Since

$$\left. \frac{\partial p_t'}{\partial t} \right|_{t=0} = v_0^*(p) = v_0'(p) = v'(p),$$

we have $\nabla f'(u(p), p) = v'(p)$, completing the proof.

11. A twisted cube

We wish to show how, for n even, an n-cube in E^{2n-1} can be slightly altered in position so that, on one face, there will be a "double twist".

THEOREM 5. *Let M_0 be an n-cube in $E^n \subset E^{2n-1}$, n even and $\geqq 2$, and let N_0 be one of its faces. Then there is an immersion f of M_0 in E^{2n-1} with the following properties:*

(1) *f is arbitrarily near the identity Θ.*

(2) *$f = \Theta$ in ∂M_0..*

(3) *$f = \Theta$, together with first derivatives, in $\partial M_0 - N_0$.*

(4) *$\mathfrak{I}_f(M_0) = 2$ or -2 at will.*

It would be easy to make f of class C^∞.

That it is possible to have $\mathfrak{I}_f(M_0) \neq 0$ can be seen at once as follows. Take the mapping of an $(n - 1)$-cube into E^{2n-2} with just one self-intersection as defined in the preceding paper; translating E^{2n-2} in E^{2n-1} gives a mapping f of M_0 into E^{2n-1} with a line of self-intersections, and with ∂M_0 intersecting itself in two points. Make slight deformations so as to remove the self-intersections of ∂M_0. Since n is even, it is easily seen that pulling ∂M_0 away from itself at one of these points in opposite directions has the opposite effect on $\mathfrak{I}_f(M_0)$; hence we may obtain $\mathfrak{I}_f(M_0) \neq 0$.

We must show how a mapping may be obtained to have also the remaining properties. We shall first describe geometrically the case $n = 2$. Take a long rectangle of paper, carry the right hand end up, towards the left, down (cutting through itself), and to the right again; it will then be approximately in its original position, except for the presence of a somewhat cylinder shaped portion near the middle. This mapping may be defined by

$$y_1 = x_1 - \frac{2x_1}{1 + x_1^2}, \qquad y_2 = \frac{1}{1 + x_1^2}, \qquad y_3 = x_2, \tag{11.1}$$

the y_1-axis pointing East (to the right), the y_2-axis up, and the y_3-axis South. By pulling the right half fairly taut, and a little to one side, the cylindrical piece is made very narrow, and is pulled to a sharp angle, say to about 12° from the direction from left to right. This renders the two long edges nearly straight again. (If a thin strip is cut off one of the long edges, it is found to have no self-intersections, and may be formed from a straight strip by simply twisting one end.)

The two long edges are now made into straight lines by a slight distortion of 3-space. A contraction in one direction turns the edge (now a rectangle) into a square. The resulting mapping has all the required properties except that there is a twist along two edges instead of along only one. Let us round off the corners. We could now either curl over all the right hand edge (see the proof below), or greatly contract the lower and right hand portions, pulling one twisted part of the edge all the way around the right hand end to a position near the other twisted part (see the figure). The figure shows all these operations except for the straightening of the wavy edge.

We turn now to the general case. By analogy with the above, we shall take the self-intersection defined in the preceding paper, for E^{n-1} mapped into E^{2n-2}, and translate E^{2n-2} in E^{2n-1} but moving it at an angle $\theta = \tan^{-1}(1/10)$ instead of $\pi/2$.

Set

$$u' = (1 + x_1^2) \cdots (1 + x_{n-1}^2),$$

(11.2)
$$y_1 = x_1 - \frac{2x_1}{u'}, \qquad y_i = x_i \qquad (i = 2, \cdots, n-1),$$
$$y_n = \frac{1}{u'}, \qquad y_{n+i-1} = \frac{x_1 x_i}{u'} \qquad (i = 2, \cdots, n-1),$$
$$y_{2n-1} = x_n + 10x_1 .$$

For x_n fixed, we obtain the mapping referred to. Since that mapping is regular, and x_n appears in y_{2n-1} only, the present mapping is regular. Let us call it f_0. The self-intersections are:

(11.3) $$f_0(1, 0, \cdots, 0, \alpha - 10) = f_0(-1, 0, \cdots, 0, \alpha + 10).$$

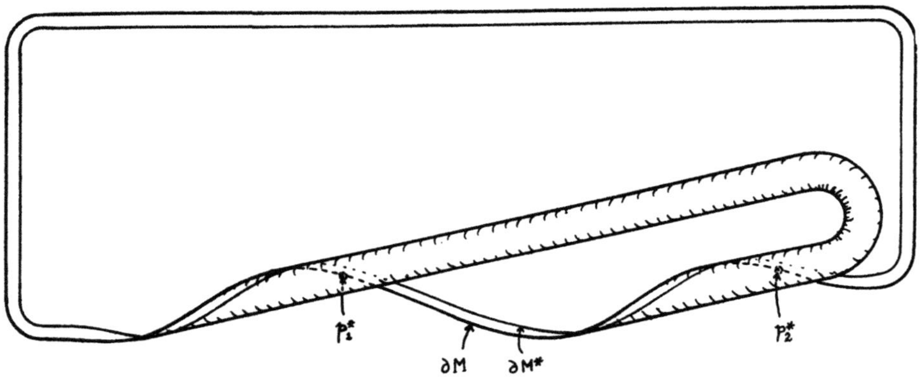

FIG. 2

Let M_1 be the part of E^n defined by $| x_n | \leqq 100$. Consider the affine mapping f' of M_1 into E^{2n-1} defined by omitting the terms in (11.2) containing u':

(11.4) $$y_i = x_i, \qquad y_{n+i-1} = 0 \qquad (i = 1, \cdots, n-1),$$
$$y_{2n-1} = x_n + 10\, x_1 .$$

Then f_0 is close to f', together with first partial derivatives, except near the x_n-axis. (Taking $n = 2$, the reader is advised to plot the parallelogram $y_1 = x_1$, $y_3 = x_2 + 3x_1$, for $| x_1 | \leqq 3$ and $| x_2 | \leqq 9$.)

Our next object is to replace the mapping f_0 by a mapping f_2 with the same kind of self-intersections, and such that for any $p = (x_1, \cdots, x_n)$, and some a,

(11.5) $$f_2(p) = f'(p) \quad \text{if} \quad | x_n | = 100 \quad \text{or} \quad \sum_{i=1}^{n-1} x_i^2 \geqq a^2.$$

We shall do this by defining a deformation ϕ_t of E^{2n-1}, and setting

(11.6) $$f_t(p) = \phi_t(f_0(p)) \qquad (0 \leqq t \leqq 2).$$

Let N_1^+ and N_1^- be the parts of ∂M_1 with $x_n = 100$ and $x_n = -100$ respectively. For each $q \in E^{2n-1}$, let $T(q)$ be the n-plane containing q which is parallel to the axes of $y_1, y_n, y_{n+1}, \cdots, y_{2n-2}$. We shall define ϕ_t in two parts. First, ϕ_t $(0 \leqq t \leqq 1)$ will carry each point $q = f_0(p)$ near $f'(\partial M_1)$ in the plane $T(q)$ into $f'(p)$, and will be the identity in E^{2n-1} outside a neighborhood of $f'(\partial M_1)$. Then $\phi_1(q)$ is near q if q is far from the y_{2n-1}-axis. Consequently it is easy to define ϕ_t $(1 \leqq t \leqq 2)$ so as to make (11.5) hold.

First, note that

$$\lambda(u) = \lambda(u; a, b) = 1 - 3\left(\frac{u-a}{b-a}\right)^2 + 2\left(\frac{u-a}{b-a}\right)^3 \qquad (a \leqq u \leqq b)$$

has the properties

$$\lambda(a) = 1, \qquad \lambda'(a) = \lambda(b) = \lambda'(b) = 0;$$

hence, if $\lambda = 1$ for $x < a$ and $\lambda = 0$ for $x > b$, λ is smooth. The maximum derivative of $-\lambda$ is at $x = \frac{1}{2}(a + b)$, and has the value $3/[2(b - a)]$.

For each $q = (y_1, \cdots, y_{2n-1})$ in E^{2-1} there is a unique point

(11.7) $$\pi^+(q) = (x_1, y_2, \cdots, y_{n-1}, 0, \cdots, 0, y_{2n-1}) \qquad \left(x_1 = \frac{y_{2n-1} - 100}{10}\right)$$
$$= f'(x_1, y_2, \cdots, y_{n-1}, 100)$$

in $f'(N_1^+) \cap T(q)$. Also,

(11.8) $$\sigma^+(q) = f_0(x_1, y_2, \cdots, y_{n-1}, 100)$$

is in $T(q)$, and is clearly the only point of $f_0(N_1^+)$ in $T(q)$. Set

$$\rho^+(q) = | \sigma^+(q) - q |,$$

and

(11.9) $$\phi_t(q) = q + t\lambda(p^+(q); 1, 9)[\pi^+(q) - \sigma^+(q)], \qquad \rho^+(q) \leqq 9.$$

Thus for any $q_0 = f_0(p_0) \in f_0(N_1^+)$ and any q in $T(q_0)$ within a distance 1 of q_0, ϕ_1 moves q by that vector which carries $f_0(p_0)$ into $f'(p_0)$; if q is at a distance 9 from q_0, then $\phi_t(q) = q$.

Set $\omega = y_{2n-1} - 10y_1$, and define the half-spaces

$$E^+ : \omega > 0, \qquad E^- : \omega < 0.$$

Since $\omega = 100$ at points of $f'(N_1^+)$, and max $[2x_1/u'] = 1$, $\omega \geqq 99$ at points of $f_0(N_1^+)$. Hence if $\rho^+(q) \leqq 9$, $\omega(q) \geqq 9$, and $q \in E^+$. We may therefore, using N_1^- in place of N_1^+, define $\pi^-(q)$ etc., and define $\phi_t(q)$ for $\rho^-(q) \leqq 9$ so as to have corresponding properties. Setting $\phi_t(q) = q$ in the rest of E^{2n-1} completes the definition of ϕ_t for $t \leqq 1$.

By direct substitution, we find that for any $p = (x_1, \cdots, x_n)$, taking $x_n > 0$,

$$f_1(p) - f_0(p) = \lambda[\rho^+(p); 1, 9]v(p),$$

$$v(p) = \left(\frac{2x_1'}{u''}, 0, \cdots, 0, -\frac{1}{u''}, -\frac{x_1' x_2}{u''}, \cdots, -\frac{x_1' x_{n-1}}{u''}, 0\right),$$

$$x_1' = x_1 + \frac{x_n - 100}{10}, \qquad u'' = (1 + x_1'^2)(1 + x_2^2) \cdots (1 + x_{n-1}^2).$$

Hence, for any $p \in M_1$ with sufficiently large $x_1^2 + \cdots + x_{n-1}^2$, $f_1(p)$ is close to $f_0(p)$ and hence to $f'(p)$, together with first partial derivatives. Consequently, for some a, if $q = f_1(p)$ is at least a distance a from the y_{2n-1}-axis, and hence $y_1^2 + \cdots + y_{n-1}^2$ is large, then $x_1^2 + \cdots + x_{n-1}^2$ is large, and the above statement holds. For such values of q, set

$$\text{(11.10)} \qquad w(q) = f'(q) - f_1(q);$$

Then w and its first derivatives are small if q is at least a distance a from the y_{2n-1}-axis; it vanishes in $f'(\partial M_1) = f_1(\partial M_1)$. It follows that w may be extended over E^{2n-1} so that it is small everywhere, together with first partial derivatives, and vanishes in $f'(\partial M_1)$. (This fact may be proved as follows. By a contraction in each $(2n-1)$-plane $y_{2n-1} - 10y_1 = \alpha$, we may bring the set A_a of points distant at least a from the y_{2n-1}-axis into A_1; then w is carried into w_a say, defined in $[f'(M_1) \cap A_1] \cup f'(\partial M_1)$. By taking a large enough, we may make w_a and its first partial derivatives arbitrarily small. We now apply the theorem of [5]—the fact that A_1 is not bounded is clearly inconsequential,—and reverse the above contraction.) We now set

$$\text{(11.11)} \qquad \phi_t(q) = q + (t-1)w(q) \qquad (1 \leqq t \leqq 2).$$

Then each ϕ_t is an imbedding, and (11.5) holds.

Consider E^n as a subspace of E^{2n-1}. If we define the affine mapping of E^{2n-1}

$$\psi\colon y_i' = y_i \ (i = 1, \cdots, 2n-2), \qquad y_{2n-1}' = y_{2n-1} - 10y_1,$$

this carries f_2 into a mapping f_3, where

$$f_3(p) = \psi(f_2(p)) \qquad (p \in M_1)$$

such that, by (11.5),

$$\text{(11.12)} \qquad f_3(p) = p \quad \text{if} \quad |x_n| = 10 \quad \text{or} \quad \sum_{i=1}^{n-1} x_i^2 \geqq a^2.$$

Next we shall change f_3 to f_4 so that $f_4 =$ identity, together with first derivatives, at all points of N_1^-. Let M_2 be a partial manifold contained in M_1 and containing all points of M_1 with $x_1^2 + \cdots + x_{n-1}^2 \leqq a^2$. For instance, let M_2 be the set of all points of E^n whose distance from the $(n-1)$-cell $x_n = 0, x_1^2 + \cdots + x_{n-1}^2 \leqq a^2$, is at most 100. Then $f_3(p) = p$ in ∂M_2. Let $u(p)$ be the inward normal at $p \in \partial M_2$. It is carried into a vector field $v(p)$ in E^{2n-1} by f_3. Let

$v'(p)$ be defined in ∂M_2 and lie in E^{2n-1}, let it equal $v(p)$ in $\partial M_2 - N_1^+ \cup N_1^-$, and let it be the inward normal in $\partial M_2 \cap N_1^-$; define it in $\partial M_2 \cap N_1^+$ so that

$$\mathfrak{K}(\partial M_2, v') = \mathfrak{K}(\partial M_2, v). \tag{11.13}$$

Let f_4 be the mapping given by Lemma 12. Then $f_4(p) = p$ in ∂M_2, and setting $f_4(p) = p$ in $M_1 - M_2$ gives the required mapping f_4. We may let $f_4(p) = p$ if $\sum_{i=1}^{n-1} x_i^2 \geqq a^2$.

Now contract E^{2n-1} with a factor 100 in the y_{2n-1}-direction, and with a factor $b \geqq a$ in the other directions. This carries f_4 into a mapping

$$f(p) = \theta(f_4(\theta^{-1}(p))).$$

A certain rectangular parallelopiped M_b' lying in M_1 and containing M_2 is carried into the cube

$$M_0\colon\ |x_i| \leqq 1 \qquad (i = 1, \cdots, n).$$

Since $f_4(p) = p$ in $M_1 - M_a'$ and f_4 leaves y_{2n-1} unchanged, choosing b large enough makes f arbitrarily near the identity Θ in M_0. Clearly $f(p) = p$ in ∂M_0, and $\partial f(p)/\partial x_i = \partial\Theta(p)/\partial x_i$ in $\partial M_0 - N_0$, where $N_0 = \partial M_0 \cap \theta(N_1^+)$. There remains to prove (4) of the theorem.

Since a reflection in E^{2n-1} will cause a change in sign in $\mathfrak{K}$, it is sufficient to show that $\mathfrak{K}_f(M_0) = \pm 2$. Clearly

$$\mathfrak{K}_f(M_0) = \mathfrak{K}_{f_4}(M_b') = \mathfrak{K}_{f_4}(M_2).$$

Since $u(p)$ points into M_2, and f_4 carries $u(p)$ into $v'(p)$ while f_3 carries $u(p)$ into $v(p)$, Lemma 8 and (11.13) give

$$\mathfrak{K}_{f_4}(M_2) = \mathfrak{K}(\partial M_2, v') = \mathfrak{K}(\partial M_2, v) = \mathfrak{K}_{f_3}(M_2).$$

Since the changes from f_0 to f_2 and to f_3 in M_1, and hence in M_2, are obtainable by deformations of E^{2n-1}, which leave Kronecker indices unchanged,

$$\mathfrak{K}_{f_3}(M_2) = \mathfrak{K}_{f_0}(M_2).$$

Hence there remains to prove

$$\mathfrak{K}_{f_0}(M_2) = KI(f_0 M_2^*, f_0 \partial M_2) = \pm 2. \tag{11.14}$$

The intersections of $f_0(\partial M_2)$ with $f_0(M_2)$ are:

$$p_1 = (-1, 0, \cdots, 0, 100), \qquad p_1^* = (1, 0, \cdots, 0, 80),$$

$$p_2 = (\ 1, 0, \cdots, 0, -100), \qquad p_2 = (-1, 0, \cdots, 0, -80),$$

$$p_i \in \partial M_2, \qquad p_i^* \in M_2^*, \qquad f_0(p_i) = f_0(p_i^*).$$

Let A and B be the parts of M_2 satisfying:

$$A\colon -2 \leqq x_1 \leqq -\tfrac{1}{2}, \qquad B_2\colon \tfrac{1}{2} \leqq x_1 \leqq 2,$$

respectively; we let these be chains, oriented like M_2. Then A and B contain neighborhoods of p_2^* and p_1^* in M_2^* respectively, and ∂A and ∂B contain neighborhoods of p_1 and p_2 in ∂M_2 respectively. Since $f_0(\partial A)$ and $f_0(\partial B)$ do not intersect B and A respectively at other points, and n is even, we have

$$\begin{aligned} KI(f_0M_2^*, f_0\partial M_2) &= KI(f_0B, f_0\partial A) + KI(f_0A, f_0\partial B) \\ &= KI(f_0B, f_0\partial A) + (-1)^n KI(f_0\partial A, f_0B) \\ &= [1 + (-1)^n]KI(f_0B, f_0\partial A) = \pm 2, \end{aligned}$$

completing the proof.

12. Proof of the immersion theorem

We can now prove the following theorem, with the help of Theorem 2; recall that that theorem is proved with the exception of the case M non-orientable, n odd.

THEOREM 6. *Given any n-manifold or partial manifold M ($n \geqq 2$) of class C^ν ($\nu \geqq 1$ finite or infinite), and a continuous mapping f of M into E^{2n-1}, there is an immersion g of M arbitrarily close to f, of class C^ν.*

REMARK. By Lemma 1, we may make the mapping completely semi-regular.

We suppose M is connected; otherwise, we would apply the theorem to each component of M separately. It is sufficient to find an immersion g of class C^1; for a sufficiently close approximation to g by a mapping g' of class C^ν is automatically an immersion also. We may suppose that M is of class C^{16}; if it were not, we could take a manifold or partial manifold M^* of class C^{16}, and an imbedding ϕ of M onto M^*, set $f^*(q) = f(\phi^{-1}(q))$ ($q \in M^*$), find an immersion g^* of M^*, and set $g(p) = g^*(\phi(p))$ ($p \in M$). Finally, by Lemma 1, we may suppose that f is of class C^{16} and semi-regular. We must now divide the proof into four cases.

CASE I. M is closed, and n is odd. By Theorem 3, we may name the singular points $p_1, p_1', \cdots, p_s, p_s'$ in such a manner that p_i and p_i' are of opposite types. If we show how to approximate to f by a function g' in which p_1 and p_1' are no longer singular points, and with no new singular points, a repetition of this process gives a function g approximating to f and without singular points; g is then an immersion.

The method of proof is as follows. Join p_1 to p_1' by an arc A in M. (This may be taken as an arc along which f has a self-intersection.) A neighborhood of A in M may be expressed as the image of a sphere plus interior Q_0^n; f then gives a smooth mapping f^* of Q_0^n into E^{2n-1}, with two singular points q_1 and q_1', of opposite types. If we approximate to f^* by an immersion g^*, which agrees with f^*, together with first partial derivatives, in ∂Q_0^n, this gives the required g' in M.

We shall describe the construction of g^* in the case $n = 2$, in spite of the fact that 2 is not odd. By Lemma 2, we may suppose that f^* near q_1 and q_1' is as shown in Fig. 1. We may choose the mapping of Q_0^n into M so that near $f^*(q_1)$

(and similarly, near $f^*(q_1')$)$f^*(Q_0^n)$ goes up to the top line $x_1 = 0$ in the figure, and down again a short distance, to $x_1 = -\beta$. If we cut off small pieces R and R' of Q_0^n containing q_1 and q_1' respectively, so that the new figure ends at $x_1 = \beta$ near these points, then it is pretty clear that by pushing part of $x_1 = \beta$ into part of $x_1 = -\beta$, we can change f^* to an immersion h', so that $h'(\partial(Q_0^n - R \cup R')) = f^*(\partial Q_0^n)$. Since Q_0^n is easily mapped into $Q_0^n - R \cup R'$, we obtain h with $h(q) = f^*(q)$, $q \in \partial Q_0^n$.

Noting that the part $\partial' R$ of ∂R inside Q_0^n must, under f^*, curl over the top edge in Fig. 1, we see that watching the image of the vector normal to ∂Q_0^n, under f^* and under h', as we go along $\partial R - \partial' R$ and along $\partial' R$, along the latter we obtain one complete twist more than along the former. The same is true near q_1'. As a result, since the singular points are of opposite types, $\mathfrak{L}_h(Q_0^n) = \mathfrak{L}_{f^*}(Q_0^n)$. Therefore we may apply Lemma 12, and obtain from h a mapping g^* which, like h, is an immersion, but agrees with f^*, together with first derivatives, in ∂Q_0^n.

We turn now to the proof for Case I. Turning to §22, (c), we see that given p_1, p_1' may be chosen so that these two points are the ends of arcs A and A', forming a smooth curve in M, and both mapping into a single arc in E^{2n-1}. (We could take, for A, any smooth arc in M which ends like the above arc at p_1 and at p_1'.) Choose coordinate systems about p_1 and p_1' as in Lemma 2; they are of class C^2. With the proper choice of the x_1-direction in each case, the ends of A are given by

$$(12.1) \qquad 0 \leqq x_1 \leqq c, \qquad x_2 = \cdots = x_n = 0.$$

It is easy to define independent smooth vector functions $v_i(p)$ $(i = 2, \cdots, n)$ along A so that $v_i(p_1)$ and $\pm v_i(p_1')$ are in the x_i-directions at p_1 and p_1' respectively. By considering M as lying in E^{2n+1}, and projecting the points $p + \sum \alpha_i v_i(p)$ into M, we may define a coordinate system throughout a neighborhood of A. We may extend the system beyond p_1 and p_1'. We have now a neighborhood of A expressed as the imbedding of a cylinder in M, and the ends of the cylinder are mapped approximately into the sets defined by

$$(12.2) \qquad x_1 = -\beta, \qquad x_2^2 + \cdots + x_n^2 \leqq \alpha^2.$$

(Note that x_1 may be replaced by $-x_1$ in Lemma 2.) If we alter the two ends of the cylinder slightly so that they coincide with $x_1 = -\beta$, and then round off the two edges (which are near $x_1 = -\beta$, $x_2^2 + \cdots + x_n^2 = \alpha^2$), we obtain a partial manifold Q_0^n, expressible as a sphere plus interior.

Let ψ be the imbedding of Q_0^n into M; set

$$f^*(q) = f(\psi(q)) \qquad (q \in Q_0^n),$$
$$q_1 = \psi^{-1}(p_1), \qquad q_1' = \psi^{-1}(p_1').$$

We may clearly suppose that f^* is an imbedding in some neighborhood of ∂Q_0^n. We now consider $(x_1, \cdots, x_n)$ as coordinates in Q_0^n near q_1 and near q_1'; then the sets defined by (12.2) with α replaced by $\alpha' = \alpha/2$ are parts N_0 and N_0' of ∂Q_0^n. Let λ be a real-valued function of class C^2 such that

$$\lambda(t) = 1 \quad \text{if} \quad |t| \leqq \alpha'/2, \qquad \lambda(t) = 0 \quad \text{if} \quad |t| \geqq \alpha'.$$

Let R and R' be the parts of Q_0^n about q_1 and q_1' respectively, defined by

$$(12.3) \qquad -\beta \leqq x_1 \leqq -\beta + 2\beta\lambda[(x_2^2 + \cdots + x_n^2)^{1/2}].$$

Then if $R_0 = \overline{Q_0^n - R \cup R'}$, ∂R_0 is a manifold of class C^1, and f^* immerses R_0 in E^{2n-1}. For each $q \in E^{2n-1}$, let $T_1(q)$ be the n-plane through q parallel to the axes of $y_1, y_{n+1}, y_{n+2}, \cdots, y_{2n-1}$. If $T(p) = T_1(f^*(p))$, $p \in N_0$, these planes cut $f^*(N_0)$ in the manner prescribed in Lemma 11. Moreover, if

$$N_1 = \partial R - N_0, \qquad N_1' = \partial R' - N_0',$$

then each $T(p)$ $(p \in N_0)$ cuts $f^*(N_1)$ in exactly one point $f_1^*(p)$. Pushing N_0 and N_0' onto N_1 and N_1' clearly defines an imbedding θ of Q_0^n onto R_0 such that $f_1^*(p) = f^*(\theta(p))$ agrees with the above f_1^* in N_0 and N_0'. Now f_1^* and f^*, defined in $N_0 - \partial N_0$, satisfy the conditions for f_0 and f_1 in Lemma 11; define ϕ_t by that lemma, and set

$$h_1(p) = \phi_1(f_1^*(p)), \qquad p \in Q_0^n.$$

Carry out a similar deformation of E^{2n-1} about $f_1^*(R')$, forming the mapping h of Q_0^n in E^{2n-1}. Then (see the remarks following the lemma) h is of class C^2, and $h(p) = f^*(p)$ $(p \in \partial Q_0^n)$. Since f^* is an immersion of R_0, f_1^* and h are immersions of Q_0^n. By taking β small enough, we may make h an arbitrarily good approximation to f^* (not of course with first derivatives).

Let $u(p)$ be the inward normal vector at p in ∂Q_0^n. If we replace h by g^*, using Lemma 12, so that

$$g^*(p) = f^*(p), \qquad \nabla g^*(u(p), p) = \nabla f^*(u(p), p) \qquad (p \in \partial Q_0^n),$$

then $g^* = f^*$, together with first partial derivatives, in ∂Q_0^n; hence, if

$$g'(p) = g^*(\psi^{-1}(p)) \quad (p \in \psi(Q_0^n)), \qquad g'(p) = f(p) \quad \text{otherwise},$$

g' will be smooth in M. Since h and hence g^* (see Lemma 12) is an immersion in Q_0^n, g' has no singular points in $\psi(Q_0^n)$, and the proof for Case I will be complete.

To apply Lemma 12, we need merely prove

$$(12.4) \qquad \mathfrak{L}(h\partial Q_0^n, \nabla h(u)) = \mathfrak{L}(h\partial Q_0^n, \nabla f^*(u)).$$

By Theorem 2 and Lemma 8, these numbers are the algebraic number of singular points of h and f^* in Q_0^n. It is 0 for h, since h is regular. It is also 0 for f^*, since f^* has just two singular points, q_1 and q_1', and these are of opposite types.

CASE II. *M is closed and n is even.* By Theorem 3, the number of singular points is even; call them $p_1, p_1', \cdots, p_s, p_s'$. If we proceed as under Case I, the only difficulty is at the last step; the two numbers in (12.4) may differ from each other by any even integer $2k$. Set $M^* = \psi(Q_0^n)$. Choose k points $r_1, \cdots, r_k$ in ∂M^*. For each i, we may express a small piece M_i' of $f(\overline{M - M^*})$ about r_i as the imbedding ϕ_i of an n-cube M_i, one face N_i going into part of $f(\partial M^*)$. If the M_i' are small, and we take $M_i \subset E^{2n-1}$, we may in an obvious fashion extend ϕ_i to be an imbedding of a neighborhood U_i of M_i in E^{2n-1} into E^{2n-1}. Let θ_i be the mapping of M_i into E^{2n-1} given by Theorem 5, with $\theta_i =$ identity in ∂M_i, together with first partial derivatives in $\partial M_i - N_i$. Then

$$F(p) = \phi_i\theta_i\phi_i^{-1}f(p) \quad (p \in f^{-1}(M_i')), \quad F(p) = f(p) \quad \text{otherwise},$$

defines a smooth mapping of $M - M^*$, agreeing with f except in the $f^{-1}(M_i')$. Choosing the correct sign in (4) of Theorem 5 in each case, we obtain

$$\mathfrak{L}(h\partial Q_0^n, \nabla h(u)) = \mathfrak{L}(h\partial Q_0^n, \nabla F^*(u)), \tag{12.5}$$

where F^* and its first partial derivatives are defined at points of ∂Q_0^n in Q_0^n in terms of F with the help of the imbedding ψ. We may now apply Lemma 12 as before.

Case III. M is open. Choose compact partial manifolds M_i in M by Lemma 20, Appendix. We shall define mappings $f_0 = f, f_1, f_2, \cdots$ with f_i arbitrarily close to f_{i-1}, such that f_i is regular in M_i, and $f_i = f_{i-1}$ in M_{i-1}. Then $g = \lim f_i$ exists and is an immersion.

Suppose f_{i-1} is properly defined. The number of singular points of f_{i-1} in M_i is finite; none are in M_{i-1}. It is sufficient to show how an arbitrarily slight alteration of f_{i-1} will get rid of one of these, say p_1. By (c) of Lemma 20, we may join p_1 to a point p_2 in $M - M_i$ by an arc $A \subset M - M_{i-1}$; we may clearly keep A away from ∂M. Take a small neighborhood U of p_2, and express a neighborhood of $f_{i-1}(U)$ in E^{2n-1} as the image ϕ of part of E^{2n-1}, so that for $E^n \subset E^{2n-1}$, $\phi(E^n)$ contains $f_{i-1}(U)$. Using the last mapping of §4, we may then alter f_{i-1} in U so that it has two singular points, say p_1' and p_1''. If n is odd, then by Theorem 2, these are of opposite types; hence one of these, say p_1', is of opposite type to that of p_1. Applying the proof in Case I, we alter f_{i-1} in a neighborhood of A, together with U, getting rid of the singularities at p_1 and p_1'. If n is even, we apply the proof in Case II, using p_1 and either of p_1', p_1''.

Case IV. M is compact but not closed; then $\partial M \neq 0$. Add a small piece onto M along part of ∂M, and remove a closed n-cell from the new portion, obtaining a new open manifold M', with $M \subset M'$, and extend f through M'. By Case III, we may alter f to give an immersion g of M'; this defines an immersion g of M, and completes the proof of the Theorem.

13. Further immersion theorems

We consider here what may be done with ∂M in an immersion of M.

Theorem 7. *Let f be a smooth mapping of the connected partial manifold M^n into E^{2n-1} ($n \geqq 2$) which is an imbedding in a neighborhood of ∂M. If M is not compact, there exists an immersion g of M in E^{2n-1} which is arbitrarily close to f*

and equals f in a neighborhood of ∂M. If M is compact and n is odd [is even], g exists if and only if $\mathfrak{L}_f(M)$ *is* $= 0$ [*is* $\equiv 0 \bmod 2$].

Using Lemma 1, we first replace f by a semi-regular mapping, which we again call f. If M is not compact, we may clearly apply the proof in Case III of the last theorem. If M is compact, the proof in Case I or in Case II applies. That the condition is necessary is a consequence of Theorem 2.

THEOREM 8. *Any compact partial manifold M may be immersed in* E^{2n-1} *so that the mapping f is an imbedding in a neighborhood of* ∂M, *and*

$$f(\partial M) \cap f(M - \partial M) = 0.$$

We shall not discuss the case of open manifolds. The theorem being clear if $n = 2$ (all M^2 being known), we assume $n \geqq 3$; also we may suppose M is connected. Let f_0 be the immersion given by Theorem 6; by Lemma 1, we may suppose f_0 is completely semi-regular. By §22, Appendix, the intersections of $f_0(\partial M)$ with $f_0(M - \partial M)$ are on ends of arcs as described in (e_3) and (e_4). Since the number of such arcs is finite, it will be sufficient to show how to get rid of an intersection of either kind.

Consider first an intersection as in (e_4). Let A and B be the arcs of M with $f_0(A) = f_0(B)$, and let U be a neighborhood of A in M. It is easy to define a smooth imbedding g of M in itself which is the identity outside U, and squeezes the whole arc A up into a part of U beyond the end point of A which is in $M - \partial M$, so that

$$g(M) \cap A = 0.$$

(Use a coordinate system about A, as in the proof for Case I of Theorem 6.) If we choose U so small that

$$f_0(U - A) \cap f_0(M - U) = 0,$$

and set $f(p) = f_0(g(p))$, we will clearly have removed the arc of self-intersection without introducing any further intersections.

Now take the case (e_3). Say $f_0(A) = f_0(B)$, all of A being in $M - \partial M$. Since $n \geqq 3$, we may extend A in one direction, forming a smooth arc A' with one end in ∂M, so that

$$f_0(A' - A) \cap f_0(M - A) = 0.$$

We now define g and f as above, with A' in place of A.

III. FURTHER INTERSECTION THEORY

14. Looping coefficients of vector fields in complexes in space

Using the definition of §9, we shall discuss "looping coefficients" of vector fields and pairs of vector fields with pairs of cycles in a smooth (not necessarily finite) complex $K^r \subset E^{2r+1}$. We derive two formulas which are useful, in particular, in studying $\mathfrak{L}_f(M)$ for a partial manifold $M^n, f(M) \subset E^{2n-1}, n = r + 1$.

In the rest of Part III, we study the situation when the above cycles are replaced by chains.

For any r-chain A of K, define the r-chain $\phi_{v,\epsilon}A$ as in §9. *Assuming v independent of K*, we shall take ϵ (or a positive continuous function $\epsilon = \epsilon(p)$ if K is infinite) so small that

$$(14.1) \qquad \phi_{v,\alpha}(K) \cap \phi_{v,\beta}(K) = 0 \quad \text{if} \quad -\epsilon < \alpha < \beta < \epsilon.$$

Generalizing (9.2), set (for finite r-cycles A, B)

$$(14.2) \qquad \mathfrak{L}(A, B, v) = LC(\phi_{v,\epsilon}A, B).$$

If we cut a closed manifold M^n into two parts M_1 and M_2, and $v(p)$ points into $f(M_1)$ at $f(p)$ in $f(\partial M_1)$ (where $f(M) \subset E^{2n-1}$), then the following lemma, with $r = n - 1$, $A = B = f\partial M_1$, relates $\mathfrak{L}_f(M_1)$ to $\mathfrak{L}_f(M_2)$. The lemma will be generalized in (19.10).

LEMMA 13. *Take K and v as above. Then for finite r-cycles A and B,*

$$(14.3) \qquad \mathfrak{L}(A, B, v) = (-1)^{r+1}\mathfrak{L}(B, A, -v).$$

By (14.1), $B = \phi_{v,0}B$ may be deformed into $\phi_{v,-\epsilon}B$ without touching $\phi_{v,\epsilon}A$, $\phi_{v,\epsilon}A$ may be deformed into A without touching $\phi_{v,-\epsilon}B$. Hence

$$LC(\phi_{v,\epsilon}A, B) = (-1)^{r^2+1}LC(B, \phi_{v,\epsilon}A)$$
$$= (-1)^{r+1}LC(\phi_{v,-\epsilon}B, \phi_{v,\epsilon}A) = (-1)^{r+1}LC(\phi_{-v,\epsilon}B, A),$$

which proves the lemma.

REMARK. Let α, β, γ be oriented arcs in E^3 joining the points p and q. Set $A = \alpha - \beta$, $B = \beta - \gamma$. Then it is easy to define v so that

$$\mathfrak{L}(A, B, v) = 1, \qquad \mathfrak{L}(B, A, v) = 0.$$

Suppose there is a small "fin" stretching out from K, in the direction of a vector field $u(p)$ (independent of K). Then $\mathfrak{L}(A, B, v)$, or $\mathfrak{L}_f(M)$ etc., may be determined by studying the intersection of $\phi_{v,\epsilon}A$ etc. with this fin. To show this, note first that by deforming $\phi_{v,\alpha}A$ into $\phi_{v,\beta}A$, we define a chain $\psi_{v,\alpha,\beta}A$ such that

$$(14.4) \qquad \partial\psi_{v,\alpha,\beta}A = \phi_{v,\beta}A - \phi_{v,\alpha}A - \psi_{v,\alpha,\beta}\partial A.$$

Now let $u(p)$ be a continuous vector function in K, independent of K, such that K, u, v are independent in K^{r-1}. That is, for each σ^r, face σ^{r-1}, and $p \in \bar{\sigma}^{r-1}$, the 2-plane through $u(p)$ and $v(p)$ has only p in common with the tangent plane to $\bar{\sigma}^r$ at p. We may suppose u, v are unit vector functions. Then we may take any η, $0 < \eta < \epsilon$, and define

$$(14.5) \qquad \mathfrak{L}(A, B, u, v) = KI(\phi_{u,\eta}A, \psi_{v,0,\epsilon}B).$$

Note that if $\partial A = \partial B = 0$, then this is defined without the restriction that K, u and v be independent in K^{r-1}.

This quantity is reducible to the former:

LEMMA 14. *If* $\partial A = \partial B = 0$, *then*

$$\mathfrak{L}(A, B, u, v) = \mathfrak{L}(B, A, v) + (-1)^r \mathfrak{L}(A, B, u). \tag{14.6}$$

In particular,

$$\begin{aligned} \mathfrak{L}(A, A, v, v) &= 2\mathfrak{L}(A, A, v) && (r \text{ even}), \\ &= 0 && (r \text{ odd}). \end{aligned} \tag{14.7}$$

By the method of proof of the last lemma, we find

$$\begin{aligned} \mathfrak{L}(A, B, u, v) &= KI(\psi_{v,0,\epsilon}B, \phi_{u,\eta}A) \\ &= LC(\phi_{v,\epsilon}B - B, \phi_{u,\eta}A) = LC(\phi_{v,\epsilon}B, A) - LC(B, \phi_{u,\eta}A) \\ &= LC(\phi_{v,\epsilon}B, A) + (-1)^r LC(\phi_{u,\eta}A, B), \end{aligned}$$

which gives (14.6).

15. The type of complex we shall use[5]

In the rest of Part III we shall use only complexes that are simplicial, or at least have certain properties of simplicial complexes. In particular:

(a_1) Each closed cell $\bar\sigma^n$ may be represented as the (smooth) one-one image $\theta(\bar\sigma_0^n)$ of a convex closed cell in E^n.

(a_2) Each closed r-face $\bar\sigma^r$ of σ^n is the intersection of $n - r$ $(n - 1)$-faces $\bar\sigma_1^{n-1}, \cdots, \bar\sigma_{n-r}^{n-1}$ of $\bar\sigma^n$.

(a_3) For $p \in \sigma^r$, the tangent planes to the $\bar\sigma_i^{n-1}$ there have only the tangent plane to $\bar\sigma^r$ in common.

It will be convenient to call the *tangent cone* $\Gamma(\bar\sigma^r, p)$ of $\bar\sigma^r$ at $p \in \bar\sigma^r$ the set of all vectors v tangent to $\bar\sigma^r$ at p; i.e. the set of all possible limits $\lim\, [\phi(t) - p]/t$, $\phi(0) = p$, $\phi(t) \in \bar\sigma^r$. The *tangent space* $\bar\Gamma(\bar\sigma^r, p)$ is the set of all linear combinations of vectors of $\Gamma(\bar\sigma^r, p)$. We say a vector of $\bar\Gamma$ is *parallel* to $\bar\sigma^r$ at p. In terms of this, (a_3) is equivalent to

$$\bar\Gamma(\bar\sigma^r, p) = \bar\Gamma(\bar\sigma_1^{n-1}, p) \cap \cdots \cap \bar\Gamma(\bar\sigma_{n-r}^{n-1}, p) \qquad (p \in \bar\sigma^r). \tag{15.1}$$

A cell may be in the form of a cube for example. Note that we may not subdivide a proper face of a cell without subdividing the cell itself, for then (a_3) would be contradicted.

Given $p \in K$, let $\Gamma(K, p)$ denote the set of all tangent cones of closed cells containing p, at p. These form parts of linear spaces which are unrelated except for those corresponding to incident cells. It is not so easy to give meaning to $\bar\Gamma(K, p)$.

Suppose f is a smooth mapping of K into E^ν. If f is regular in the part of a closed cell $\bar\sigma$ near $p \in \bar\sigma$, then

$$\Gamma(f(\bar\sigma), f(p)) = \nabla f \Gamma(\bar\sigma, p).$$

This is true with K in place of $\bar\sigma$; now $\Gamma(f(K), q)$ is formed of cones lying in E^ν.

[5] The author expects to give a more general theory of this subject in a paper on "Complexes of manifolds."

We shall say f is an *immersion* of K if it is regular (in each closed cell) and proper, and is an *imbedding* if, in addition, it is one-one, and further, ∇f is one-one in each $\Gamma(K, p)$. In the latter case, we say K is a complex *in* the second space.

EXAMPLE. No complex may be imbedded onto a pair of tangent circles.

LEMMA 15. *Any complex K^n may be imbedded in E^{2n+1} so that it has no limit set.*

It is not difficult to define an imbedding cell by cell, by the methods in [1]; of course it is easy and standard if K is simplicial. For the last statement (trivial if K is finite), compare [1], p. 665, footnote 32.

We shall let σ_i^r denote the cells of K, and set $\tau_i^r = f(\sigma_i^r)$. If K lies in E^ν, we may let τ_i^r denote its cells, thinking of f as the identity mapping.

LEMMA 16. *Let K lie in E^ν. Then each $\bar\tau_i^r$ may be enclosed in a larger cell $\tau_i'^r$, of the same class C^γ as K, such that if τ_j^s is a face of τ_i^r, then $\tau_j'^s$ lies in $\tau_i'^r$.*

REMARK. The lemma extends in an obvious way to the case, if $K = K^n$, that f is an imbedding in a neighborhood of each $\partial\sigma_i^n$.

Of course we take $\tau_i'^0 = \tau_i^0$. Suppose the $\tau_i'^1, \cdots, \tau_i'^{s-1}$ have been constructed. Take any $\bar\tau_i^s$, the image $\theta(\bar\tau_0^s)$ of a convex cell $\bar\tau_0^s$ in E^s. It is easy to define extensions of the boundary cells of τ_0^s, and define θ over these, mapping into the extended faces $\tau_j'^t$ of τ_i^s. Because of (a_3), we may now extend θ over a neighborhood of $\bar\tau_i^s$. In a sufficiently small neighborhood, which may be taken as a convex cell, θ is an imbedding.

16. On general position of a complex and vector field in space

Let f be a smooth mapping of a complex K (see §15) into E^ν. We say f is in *general position* if it is proper, and:

(b_1) For each s, f is regular in K^s at all points of $K^{\nu-s}$.

(b_2) For each s, each $p \in K^s$ and each $q \in K^{\nu-s-1}$, if $p \neq q$ then $f(p) \neq f(q)$.

(b_3) If $u_1 \in \Gamma(K^s, p)$, $u_2 \in \Gamma(K^{\nu-s}, p)$, $u_1 \neq u_2$, then $\nabla f(u_1, p) \neq \nabla f(u_2, p)$.

One could combine (b_1) and (b_3) in a more complicated statement. Note that (b_3) uses Γ, not $\bar\Gamma$. Any imbedding of K^n into E^ν, $\nu \geqq 2n$, is clearly a mapping in general position. From the above we deduce:

(b_1') f is regular in K^s if $\nu \geqq 2s$.

(b_2') f is one-one in K^s if $\nu \geqq 2s + 1$.

(b_3') ∇f is one-one in $\Gamma(K^s, p)$ if $\nu \geqq 2s$.

Let $(y_1, \cdots, y_\nu)$ be a coordinate system in E^ν, and let π be the projection: $\pi(y_1, \cdots, y_{\nu-1}, y_\nu) = (y_1, \cdots, y_{\nu-1})$ of E^ν into $E^{\nu-1}$.

Say f is in *general position with reference to the y_ν-direction* if the above holds, and in addition, πf is in general position in $E^{\nu-1}$. We could obviously generalize this, using a set of independent directions.

Let v be a continuous vector field, defined in a closed subcomplex of K, and with values in E^ν. Then $\nabla\pi v$ has values in $E^{\nu-1}$. We say f and v are in *general position* if f is, and also:

(b_4) For each s, $v(p)$ is independent of any $f(\bar\sigma^s)$ at $f(p)$ for all points $p \in K^{\nu-s-1}$.

(b_5) For each s, $v(p)$ is independent of $\nabla f[\bar\Gamma(\bar\tau_i^s, p) + \bar\Gamma(\bar\tau_j^{\nu-s-1}, p)]$, where defined.

Letting $\sigma^{\nu-s-1}$ be a face of σ^s shows that (b_4) is a consequence of (b_5). We shall often omit (b_5).

Finally, f and v are in *general position with reference to the y_ν-direction* if the above holds also for πf and $\nabla \pi v$.

LEMMA 17. *Let K be of class C^2, let f map K into E^ν with no limit set, and let $v_1, v_2, \cdots$ be smooth vector functions, each defined in a subcomplex of K and with values in E^ν. Let f and each v_k be in general position. Then by an arbitrarily small rotation of the axes we may make this hold with reference to the y_ν-direction. The lemma holds if we omit* (b_5) *in the hypothesis and conclusion.*

REMARKS. If K is not of class C^2, or the v_k are not smooth, we may use a C^1-homeomorphic K' which is of class C^2, and smooth v_k' approximating to the v_k. The application will be to the case that K_1 is a subdivision of a smooth manifold M^n, $K = K_1^{n-1}$, and the v_k are independent of the $(n-1)$-cells of K and tangent to the n-cells of K. We could use several independent directions in place of the single y_ν-direction. If $K = K^r$, we could allow the limit set L_f of f to exist, provided it is of zero $(\nu - r - 1)$-extent.

EXAMPLE. Let K be a subdivision of an open arc, and let f map K into E^3 so that it winds like a ball of string, having a 2-sphere as limit set. Then f is proper, but no projection into E^2 is proper.

To prove the lemma, let $S^{\nu-1}$ be the unit sphere in E^ν; its points may be thought of as directions in E^ν. Let R_{s1} be the set of all those directions parallel to $f(K^s)$ at a point of $f(K^{\nu-s-1})$. Let R_{s2} be all those through points $f(p)$ and $f(q)$ with $p \in K^s$, $q \in K^{\nu-s-2}$, $p \neq q$, (which implies $f(p) \neq f(q)$). Let R_{s3} be all directions of vectors $u' = \nabla f(u, p)$, $u = u_1 - u_2$, $u_1 \in \Gamma(K^s, p)$, $u_2 \in \Gamma(K^{\nu-s-1}, p)$, $u_1 \neq u_2$ (which implies $u' \neq 0$). Let R_{s4} be all those defined by $v_k(p)$ plus a vector parallel to some $f(\bar\sigma^s)$ at $p \in K^{\nu-s-2}$. If we are using (b_5), let R_{s5} be all those defined by $v_k(p) + u_1' + u_2'$, $u_i' = \nabla f(u_i)$, $u_1 \in \bar\Gamma(\bar\tau_i^s, p)$, $u_2 \in \bar\Gamma(\bar\tau_j^{\nu-s-2}, p)$.

Since the directions parallel to $f(K^s)$ at a point p form a set of finite $(s-1)$-extent in $S^{\nu-1}$, and this set varies smoothly with p, R_{s1} is a finite or denumerable sum of sets of finite $(\nu-2)$-extent.[6] In R_{s2}, since p and q range over sets of finite s- and $(\nu - s - 2)$-extent respectively, R_{s2} is a (denumerable) sum of sets of finite $(\nu-2)$-extent. In R_{s3}, take any p in any σ^t, and suppose σ^t is a face of σ^s and of $\sigma^{\nu-s-1}$. If $u_1 \in \Gamma(\bar\sigma^s, p)$ and $u_2 \in \Gamma(\bar\sigma^{\nu-s-1}, p)$, these range over sets of finite s- and $(\nu-s-1)$-extent respectively. But adding a vector of $\Gamma(\bar\sigma^t, p)$ to each leaves their difference unchanged; hence $u_1 - u_2$ ranges over sets of finite $(\nu - 1 - t)$-extent. Letting p vary shows that the directions vary over a sum of sets of finite $(\nu-2)$-extent. Since the directions defined by $v_k(p) + \nabla f(u, p)$, $u \in \bar\Gamma(\bar\sigma^s, p)$, form a sum of sets of finite s-extent, R_{s4} is as required. In R_{s5}, $v_k(p) + u_1' + u_2'$, p fixed in σ^t, defines directions of finite $(\nu - 2 - t)$-extent; hence R_{s5} is as required.

Consequently by [1], Lemmas 13 and 14, $R = \sum R_{si}$ has no inner points in

[6] See [1], in particular, Lemma 15. It would be possible to use dimension instead of extent.

$S^{\nu-1}$, so that arbitrarily near the y_ν-direction there is a direction not in R. Rotate axes so as to make this direction the new y_ν-direction. Then since any vector not in this direction projects into a non-zero vector, it is easily seen that f and each v_k are in general position with reference to the new y_ν-direction, completing the proof.

LEMMA 18. *Any smooth complex K^n may be mapped into any E^ν so that it is in general position with reference to a given direction.*

REMARK. This is a generalization of the imbedding and immersion theorems of [1] for manifolds.

By the remark to the last lemma, we may suppose that K is of class C^2. Imbed K^n in E^μ, $\mu = \max[\nu, 2n + 1]$, so that it has no limit set (Lemma 15). It is then in general position. By Lemma 17, applying a small rotation makes it in general position with reference to any chosen direction. If $\mu = \nu$, we let this be the given direction. If $\mu > \nu$, we choose any direction; projecting in this direction into $E^{\mu-1}$ gives a mapping in general position in $E^{\mu-1}$. Repeat the process till we reach E^ν.

17. The fins and corresponding projections

Take a smooth complex $K \underset{\sim}{=} K^r \subset E^{2r+1}$, and let K and v be in general position with reference to the y_{2r+1}-direction (using the identity mapping), omitting (b_5). We shall suppose K and v are of class C^2; if this is not so, and we do not wish to change K or v, we could replace the normal planes $T(q)$ below by a smooth function $T(q)$; the properties given will then hold.

Let $\{\tau_i^s\}$ denote the cells of K. Each τ_i^r may be enclosed in a larger cell τ_i' of class C^2 (Lemma 16). Suppose v is defined over $\bar{\tau}_i^r$; extend it to be of class C^2 over τ_i'. With small enough τ_i', we will still have general position. Set

$$\phi_{v,t}(p) = p + tv(p) \qquad (p \in \tau_i'). \tag{17.1}$$

For any point set $R \subset \tau_i'$, let $\Phi_v(R, t)$ denote all points $\phi_{v,t'}(p)$ with $p \in R$ and with $0 \leqq t' \leqq t$; let $\Phi_v^*(R, t)$ denote the same, with $0 < t' \leqq t$. We call these the *fin* and *deleted fin* respectively of τ_i', v and t. Since K and v are of class C^2, t_i' may be chosen so that the *double fin* $\Phi_v(\tau_i', t_i') \cup \Phi_{-v}(\tau_i', t_i')$ is expressed by ϕ as an imbedding of the product of τ_i' and the interval $-t_i' \leqq t \leqq t_i'$. Moreover, for each q in the double fin, if $T_0(q)$ is the set of all points in the normal plane $T(q)$ to the fin at q which are at a distance $\leqq t_i'$ from q, then these fill out a closed neighborhood $W_v(\tau_i', t_i')$ of the interior of the double fin in a one-one way, being again the imbedding of a product. If we set

$$P_v(q') = q \quad \text{if} \quad q' \in T_0(q), \tag{17.2}$$

this is a smooth projection of $W_v(\tau_i', t_i')$ onto the double fin. Finally, let $\Lambda_v(R, t)$ denote all points $q' \in T(q)$, with $q \in \Phi_v^*(p, t)$, $p \in R$, such that $|q' - q| \leqq t|q - p|$.

Since K is in general position with reference to the y_{2r+1}-direction, omitting (b_5), the (b_k) give

(b_1^*) π is regular in K.
(b_2^*) If $p \in K^{r-1}$, $q \in K$, $p \neq q$, then $\pi p \neq \pi q$.
(b_3^*) Distinct vectors in K at any point map under π into distinct vectors.
(b_4^*) $\nabla\pi v(p)$ is independent of any $\pi\bar{\sigma}^s (s \leqq r)$ at p for $p \in K^{r-1}$.
(b_6^*) There is a neighborhood V^* of K^{r-1} in K such that if $p \in V^*$, $q \in K$, $p \neq q$, then $\pi p \neq \pi q$.

To prove (b_6^*), we use (b_1^*) and (b_3^*) to show that π is one-one in a neighborhood of any point of K^{r-1}, use (b_2^*) to show that π is one-one in K^{r-1}, and apply Lemma 25, §24.

Let $v_1, v_2, \cdots$ be vector functions in K with the same properties as v, such that at most a finite number are defined in any τ_i^r, and for each v_k, some v_l equals $-v_k$. For any R, let $N_t R$ denote all points whose distance from R is $\leqq t$. We shall choose numbers $t_i > 0$ such that the following properties hold, for any τ_i^r and v_k defined in τ_i^r.

(c_1) $\Lambda_{v_k}(\tau_i' \cap N_{t_i}\tau_i^r, t_i) \subset W_{v_k}(\tau_i', t_i')$.
(c_2) $\Phi_{v_k}^*(\tau_i^r, t_i) \cap \Phi_{-v_k}(\tau_i^r, t_i) = 0$.
(c_3) $\pi\Phi_{v_k}^*(\tau_i^r, t_i) \cap \pi\Phi_{-v_k}(\partial\tau_i^r, t_i) = 0$.
(c_4) $\Lambda_{v_k}(\tau_i' \cap N_{t_i}\tau_i^r, t_i) \cap \tau_i^r = 0$.
(c_5) $\pi\Lambda_{v_k}(\tau_i' \cap N_{t_i}\tau_i^r, t_i) \cap \pi\partial\tau_i^r = 0$.
(c_6) $\pi\Lambda_{v_k}(\tau_i' \cap N_{t_i}\partial\tau_i^r, t_i) \cap \pi\tau_i^r = 0$.

It is sufficient to prove each property (c_i) separately. We shall write Λ in place of Λ_{v_k}. We may use a single k, since but a finite number of v_k are defined in τ_i^r. (c_1), (c_2) and (c_4) are clear. The proof of (c_3) is essentially contained in that of (c_5), so we turn to (c_5) and (c_6).

Take $p \in \partial\tau_i^r$. For a small enough t_1, if $q' \in \Lambda(\tau_i' \cap N_{t_1}p, t_1)$, say $q' \in T_0(q)$, $q \in \Phi_{v_k}^*(p')$, $p' \in \tau_i' \cap N_{t_1}p$, then $q - p' = \alpha v_k(p')$, and $|q' - q| \leqq t_1 |q - p'|$, so that the angle between $p'q'$ and $v_k(p)$ is small; also $p' - p$ is approximately a vector $u_1 \in \Gamma(\bar{\tau}_i^r, p)$; furthermore, for any $p'' \in \tau_i^r \cap N_{t_1}p$, $p'' - p$ is approximately a vector $u_2 \in \Gamma(\bar{\tau}_i^r, p)$; therefore, by ($b_4^*$), we may suppose that

$$\nabla\pi[(q' - p') + (p' - p)] \neq \nabla\pi(p'' - p), \qquad \pi q' \neq \pi p''.$$

This gives

$$\pi\Lambda(\tau_i' \cap N_{t_1}p, t_1) \cap \pi(\tau_i^r \cap N_{t_1}p) = 0.$$

Since $\tau_i^r - N_{t_1}p$ is compact, because of (b_2^*) we may clearly take $t_2 \leqq t_1$ so that

$$\pi\Lambda(\tau_i' \cap N_{t_2}p, t_2) \cap \pi(\tau_i^r - N_{t_1}p) = 0.$$

The last two relations give

$$\pi\Lambda(\tau_i' \cap N_{t_2}p, t_2) \cap \pi(\tau_i^r) = 0,$$

and hence (with $t_3 \leqq t_2$)

$$\pi\Lambda(\tau_i' \cap N_{t_3}\partial\tau_i^r, t_3) \cap \pi(\tau_i^r) = 0,$$

proving (c_6). Next, since $(\tau_i' - N_{t_3}\partial\tau_i^r) \cap K^{r-1} = 0$, ($b_2^*$) gives, for some $t_4 \leqq t_3$,

$$\pi\Lambda(\tau_i' - N_{t_3}\partial\tau_i^r, t_4) \cap \pi(\partial\tau_i^r) = 0.$$

Combining this with the last relation gives

$$\pi\Lambda(\tau_i', t_4) \cap \pi(\partial\tau_i^r) = 0,$$

proving (c_5).

REMARK. With the help of (b_5) we could prove (recalling that K is proper)

$$\pi\Lambda_{v_k}(\tau_i', t_i) \cap \pi K^{r-1} = 0.$$

18. The numbers $\xi_{i,v}^{\pm}$ and $\zeta_{ij}^{\pm}$

We give here, and with the $\mu_{ij,v}^{\pm}$ of §19, the promised generalization of the looping coefficient of §14. Let K^r and each v_i be in general position in E^{2r+1} with reference to the y_{2r+1}-direction, as in §17. For any (finite) singular chain A in E, let ρ^+A and ρ^-A be the (infinite) singular chains formed by deforming A to infinity in the y_{2r+1}-direction and the negative y_{2r+1}-direction respectively, oriented so that

$$\partial\rho^{\pm}A = -A - \rho^{\pm}\partial A. \tag{18.1}$$

We shall show that the following definitions are permissible.

DEFINITIONS. Choose the t_i so that (c_1) through (c_6) hold. Let $\phi_{v,t}A$ denote the singular chain A, pushed a distance $t \mid v \mid$ in the direction of v, and oriented like A; see (17.1). Set

$$\xi_{i,v_k}^{\pm} = KI(\phi_{v_k,t_i}\tau_i^r, \rho^{\pm}\tau_i^r) \quad \text{(if } v_k \text{ is defined in } \tau_i^r), \tag{18.2}$$

$$\zeta_{ij}^{\pm} = KI(\tau_i^*, \rho^{\pm}\tau_j^*) \qquad (i \neq j), \tag{18.3}$$

where τ_i^* is any cell lying in τ_i^r and slightly smaller than τ_i^r, oriented like τ_i^r. Note that $\xi_{i,v_k}^{\pm}$ does not depend on the orientation of τ_i^r.

We shall prove commutation properties of these:

$$\xi_{i,v_k}^{+} = (-1)^{r+1}\xi_{i,-v_k}^{-}, \tag{18.4}$$

$$\zeta_{ij}^{+} = (-1)^{r+1}\zeta_{ji}^{-}. \tag{18.5}$$

In the proofs of these and other relations, the following properties of Kronecker indices are useful:

(d_1) $KI(A, \rho^{\pm}B)$ is defined (for finite A and B) whenever $\dim A + \dim B = 2r$, and

$$A \cap B = 0, \qquad \pi A \cap \pi\partial B = 0, \qquad \pi\partial A \cap \pi B = 0.$$

We must show that

$$\partial A \cap \rho^{\pm}B = 0, \qquad A \cap \partial\rho^{\pm}B \subset (A \cap B) \cup (A \cap \rho^{\pm}\partial B) = 0;$$

these follow from the above relations.

(d_2) If A_λ and B are singular chains, A_λ being continuous in λ ($\lambda_0 \leqq \lambda \leqq \lambda_1$), and each $KI(A_\lambda, B)$ is defined, then $KI(A_{\lambda_0}, B) = KI(A_{\lambda_1}, B)$.

This follows either from continuity, or by making use of chains formed by deforming A_{λ_0} and ∂A_{λ_0}, with standard properties of the index.

(d_3). If $KI(A, B)$ is defined, and C_λ ($\lambda_0 \leqq \lambda \leqq \lambda_1$) satisfies $C_{\lambda_0} = \partial A$, $C_\lambda \cap B = 0$, then this defines a deformation A_λ of A such that $\partial A_\lambda = C_\lambda$ and $KI(A_\lambda, B)$ is defined.

This is clear if we let A_λ equal A plus the "path" of $C_{\lambda'}$, $\lambda_0 \leqq \lambda' \leqq \lambda$.

To prove (18.4), and show incidentally that (18.2) is permissible, choose l so that $-v_k = v_l$. By (d_1), we see that each of

$$KI(\phi_{v_k, t_i}\tau_i^r, \rho^{\pm}\phi_{-v_k, t}\tau_i^r), \qquad KI(\phi_{v_k, t}\tau_i^r, \rho^{\pm}\phi_{-v_k, t_i}\tau_i^r)$$

is defined for $0 \leqq t \leqq t_i$; for the three relations in (d_1) follow from (c_2), (c_3) and (c_5), using both v_k and v_l. Consequently, by (d_2),

$$\xi_{i,v_k}^+ = KI(\tau_i^r, \rho^+\phi_{-v_k, t_i}\tau_i^r) = KI(\rho^+\phi_{-v_k, t_i}\tau_i^r, \tau_i^r).$$

In the above proof, if $A = \phi_{-v_k, t_i}\tau_i^r$, we used $\pi A \cap \pi\partial\tau_i^r = \pi\partial A \cap \pi\tau_i^r = 0$. These give

$$KI(\rho^+A, \rho^-\partial\tau_i^r) = KI(\rho^+\partial A, \rho^-\tau_i^r) = 0,$$

and therefore, since dim $\rho^+A = r + 1$,

$$\begin{aligned}\xi_{i,v_k}^+ &= -KI(\rho^+A, -\tau_i^r - \rho^-\partial\tau_i^r) = -KI(\rho^+A, \partial\rho^-\tau_i^r)\\ &= (-1)^r KI(\partial\rho^+A, \rho^-\tau_i^r) = (-1)^{r+1}KI(A + \rho^+\partial A, \rho^-\tau_i^r)\\ &= (-1)^{r+1}KI(\phi_{-v_k, t_i}\tau_i^r, \rho^-\tau_i^r) = (-1)^{r+1}\xi_{i,-v_k}^- .\end{aligned}$$

Next we discuss (18.3) and (18.5). Because of (b_6^*), ζ_{ij}^+ is defined and independent of the choice of any τ_k^*, so long as $\tau_k^r - \tau_k^* \subset V^*$. Since

$$\pi\tau_i^* \cap \pi\partial\tau_j^* = \pi\partial\tau_i^* \cap \pi\tau_j^* = 0,$$

we find, as in the proof of (18.4),

$$\begin{aligned}\zeta_{ij}^+ &= KI(\rho^+\tau_j^*, \tau_i^*) = -KI(\rho^+\tau_j^*, -\tau_i^* - \rho^-\partial\tau_i^*)\\ &= -KI(\rho^+\tau_j^*, \partial\rho^-\tau_i^*) = (-1)^r KI(\partial\rho^+\tau_j^*, \rho^-\tau_i^*)\\ &= (-1)^{r+1}KI(\tau_j^* + \rho^+\partial\tau_j^*, \rho^-\tau_i^*) = (-1)^{r+1}\zeta_{ji}^- .\end{aligned}$$

We shall prove still a lemma regarding the $\xi_{i,v_k}^\pm$. Since $\bar\tau_i^r = \theta(\bar\sigma_0^r)$, $\bar\sigma_0^r$ convex in E^r, the parts near $\partial\sigma_0^r$ of radii from an inner point of σ_0^r map under θ into arcs which we may suppose fill out $\tau_i' \cap N_{t_i}\partial\tau_i^r$. With these arcs, we may define a deformation g_λ such that

(e_1) $g_\lambda(p)$ carries p along the arc in $\tau_i' \cap N_{t_i}\partial\tau_i^r$ on which it lies into $\partial\tau_i^r$.

LEMMA 19. *Let A be any singular chain such that (writing N for N_{t_i} and $\Lambda_v(R)$ for $\Lambda_v(R, t_i)$)*

(18.6) $$A \subset \Lambda_{v_k}(\tau_i' \cap N\tau_i^r), \qquad \partial A \subset \Lambda_{v_k}(\tau_i' \cap N\partial\tau_i^r).$$

Then there is a number a and a chain B such that

(18.7) $$KI(A, \rho^{\pm}\tau_i^r) = a\xi_{i,v_k}^{\pm},$$

(18.8) $$\partial A - a\partial\tau_i^r = \partial B, \qquad B \subset \Lambda_{v_k}(\tau_i' \cap N\partial\tau_i^r) \cup (\tau_i' \cap N\partial\tau_i^r).$$

The last relation determines a uniquely, even if we assume merely $B \subset W_{v_k}(\tau_i' \cap N\partial\tau_i^r, t_i)$.

First, by (c_4), (c_5), (c_6) and (d_1), $KI(A, \rho^{\pm}\tau_i^r)$ is defined. We shall define a deformation h_λ of A ($0 \leqq \lambda \leqq 3$) such that

(18.9) $$h_3 A = a\phi_{v_k,t_i}\tau_i^r.$$

Also $KI(h_\lambda A, \rho^{\pm}\tau_i^r)$ is defined for each λ; (18.7) follows from this, together with (18.2) and (d_2).

For any $q \in \Phi_{v_k}^*(\tau_i' \cap N\tau_i^r)$ and $q' \in T_0(q)$, set

$$h_\lambda(q') = (1 - \lambda)q' + \lambda q \qquad (0 \leqq \lambda \leqq 1);$$

this is defined in $\Lambda_{v_k}(\tau_i' \cap N\tau_i^r)$, by ($c_1$), and $h_1(q') = P_{v_k}(q)$, by (17.2). From the definition of Λ_{v_k} we see that $h_\lambda A$ obeys (18.6) so far, simply because A does. Moreover, $h_1 A$ is in the deleted fin $\Phi_{v_k}^*(\tau_i' \cap N\tau_i^r)$, and $h_1\partial A \subset \Phi_{v_k}^*(\tau_i' \cap N\partial\tau_i^r)$.

Next, for each $q = \phi_{v_k,t}(p)$ ($p \in \tau_i'$) in the deleted fin, set

$$h_{1+\lambda}(q) = (1 - \lambda)q + \lambda\phi_{v_k,t_i}(p) \qquad (0 \leqq \lambda \leqq 1);$$

this is a deformation of $\Phi_{v_k}^*(\tau_i')$ in itself; applying it to $h_1 A$ defines $h_\lambda A$ ($0 \leqq \lambda \leqq 2$) so that (18.6) continues to hold; now

$$h_2 A \subset \phi_{v_k,t_i}(\tau_i' \cap N\tau_i^r), \qquad h_2\partial A \subset \phi_{v_k,t_i}(\tau_i' \cap N\partial\tau_i^r).$$

Next, applying ϕ_{v_k,t_i} to the deformation g_λ defines h_λ, $2 \leqq \lambda \leqq 3$; by (e_1), it keeps $\phi_{v_k,t_i}(\tau_i' \cap N\partial\tau_i^r)$ in itself. This defines a deformation of $\partial h_2 A$, and hence of $h_2 A$, so that (18.6) continues to hold; see (d_3). Now

$$h_3 A \subset \phi_{v_k,t_i}(\tau_i^r), \qquad h_3\partial A = \partial h_3 A \subset \phi_{v_k,t_i}(\partial\tau_i^r).$$

Since ϕ_{v_k,t_i} is one-one in τ_i^r, the only $(r - 1)$-cycles in $\phi_{v_k,t_i}(\partial\tau_i^r)$ are multiples of the cycle $\phi_{v_k,t_i}\partial\tau_i^r$; this proves (18.9) and hence (18.7).

The deformation h_λ of ∂A ($0 \leqq \lambda \leqq 3$) defines a chain B_1 such that

$$\partial B_1 = \partial A - a\phi_{v_k,t_i}\partial\tau_i^r;$$

since $\Phi_{v_k}(\partial\tau_i^r)$, properly oriented, is a chain B_2 bounded by $\phi_{v_k,t_i}\partial\tau_i^r - \partial\tau_i^r$, $B = B_1 + aB_2$ satisfies (18.8).

To prove the uniqueness of a in (18.8), suppose it held with a' and B' also; then

$$(a' - a)\partial\tau_i^r = \partial C, \qquad C = B - B' \subset W_{v_k}(\tau_i' \cap N\partial\tau_i^r, t_i).$$

If we contract the fin $\Phi_{v_k}(\tau_i')$ onto τ_i', we may carry C into $C' \subset \tau_i'$, and the above holds with C' in place of C. Since $N_{t_i}\partial\tau_i^r$ does not cover all of τ_i^r, this implies $a' - a = 0$.

19. Application to complexes K^n mapped into E^{2n-1}

In this section we suppose that f, of class C^3, maps K^n in regular position in E^{2n-1} with reference to the y_{2n-1}-direction. We may suppose K imbedded in E^{2n+1}. Extend each cell $\bar{\sigma}_i^s$ to a cell $\sigma_i'^s$ as in Lemma 16; we may extend f over these in turn so that the properties in Lemma 16 and the remark following it hold. Set $\tau_i'^s = f(\sigma_i'^s)$. Taking these cells small enough, we may suppose the properties (b_1), (b_2), (b_3) still hold.

We shall determine open sets U_i and U' in K (which now consists of the $\sigma_i'^s$) such that:

(f_1) $\bar{\sigma}_i^{n-1} \subset U_i$, $K^{n-2} \subset U'$.

(f_2) $U_i \cap U_j \subset U'$ if $i \neq j$.

(f_3) f is an imbedding in $U = \sum U_i$.

(f_4) If $p \in U$, $p' \in U'$, $p \neq p'$, then $\pi f(p) \neq \pi f(p')$.

We shall restrict the U_i and U' further later. That f is locally an imbedding at points of K^{n-1} is an immediate consequence of (b_1) and (b_3), using $s = n$, $\nu - s = n - 1$. The proof of (f_3) and (f_4) is now the same as the proof of (b_6^*). It is easy to choose sets U_i satisfying (f_1) and (f_2) (and thus redefining U) if we make use of the sets $\sigma_j^{n-1} - U'$.

Set $r = n - 1$. For each pair of incident cells σ_i^r and σ_k^n, let $\sigma_{ik}'^n$ be the part of $\sigma_k'^n$ on the side of $\sigma_i'^r$ which includes σ_k^n, and let $u_{ik}(p)$ ($p \in \sigma_i'^r$) be the unit vector tangent to $\sigma_{ik}'^n$ and normal to $\sigma_i'^r$ at $p \in \sigma_i'^r$. Set $v_{ik}(p) = \nabla f(u_{ik}(p), p)$. The following property shows that we may use the results of §§17 and 18.

(f_5) f and each v_{ik}, also f and each $-v_{ik}$, are in general position with reference to the y_{2n-1}-direction, (b_5) being omitted. (Here, f is considered in K^{n-1} only.)

We need merely prove (b_4) for f and πf; this follows at once from the properties of general position; compare (f_3) and (f_4).

Set $V_i = f(U_i)$, $V = f(U)$, $V' = f(U')$.

Next, by further restricting U' and then the U_i, we may obtain:

(f_6) $\tau_{ik}'^n \cap V_i \subset \Lambda_{v_{ik}}(\tau_i'^r \cap N_{t_i}\tau_i^r, t_i)$,

(f_7) $\tau_{ik}'^n \cap V' \subset \Lambda_{v_{ik}}(\tau_i'^r \cap N_{t_i}\partial\tau_i^r, t_i)$.

These are obvious consequences of the definitions of $\tau_{ik}'^n$, v_{ik} and Λ_v.

Now let K' be a subdivision of K so fine that the following holds:

(f_8) Any cell of K' with a vertex in $\bar{\sigma}_i^r$ lies in U_i; any cell with a vertex in K^{r-1} lies in U'.

Each oriented cell σ of K becomes now a chain $Sd\sigma$ of K'. Let $[\sigma_i^{s-1}: \sigma_k^s]$ denote incidence numbers. Define the following chains of K', and use $\tau = f(\sigma)$ as before:

σ_i^{*s} = sum of s-cells of K' in σ_i^s with no vertex in $\partial\sigma_i^s$, each oriented like σ_i^s.

σ_{ik}^* = sum of n-cells of σ_{ik}^n which have a vertex in σ_i^r but none in $\partial\sigma_i^r$, each oriented like $[\sigma_i^r: \sigma_k^n]\sigma_k^n$.

σ_{ik}^{**} = that part of $-\partial\sigma_{ik}^*$ whose cells have no vertex in $\bar{\sigma}_i^r$.

We prove the following relations:

$$\partial \tau_{ik}^{*} = \tau_i^{*r} - \tau_{ik}^{**} + A_{ik}, \qquad A_{ik} \subset V'. \tag{19.1}$$

$$\partial \tau_k^{*n} = \sum_i [\sigma_i^r : \sigma_k^n] \tau_{ik}^{**} + B_k, \qquad B_k \subset V'. \tag{19.2}$$

First, supposing σ_i^r a face of σ_k^n, note that each cell of the chain

$$Sd\tau_k^n - (\tau_k^{*n} + [\sigma_i^r : \sigma_k^n]\tau_{ik}^{*})$$

has a vertex in $\partial \tau_k^n - \tau_i^r$; hence the chain lies in V_i', where $U_i' = \sum_{j \neq i} U_j$, $V_i' = f(U_i')$. Also, clearly

$$\partial Sd\tau_k^n = Sd\partial\tau_k^n = [\sigma_i^r : \sigma_k^n]\tau_i^{*} + C_{ik}, \qquad C_{ik} \subset V_i'.$$

It follows that

$$\partial(\tau_k^{*n} + [\sigma_i^r : \sigma_k^n]\tau_{ik}^{*}) = [\sigma_i^r : \sigma_k^n]\tau_i^{*r} + D_{ik}, \qquad D_{ik} \subset V_i'.$$

This last relation, with (19.1), gives

$$\partial \tau_k^{*n} = [\sigma_i^r : \sigma_k^n]\tau_{ik}^{**} + D_{ik} \pm A_{ik}.$$

Now neither τ_k^{*n} nor τ_{ik}^{**} have any cells with vertices in $\bar{\tau}_i^r$. Hence neither has $D_{ik} \pm A_{ik}$, and it follows that each cell of A_{ik} is a cell of D_{ik}. Therefore $A_{ik} \subset V_i'$. Since τ_{ik}^{*}, τ_i^{*r} and τ_{ik}^{**} are in V_i, each point of A_{ik} is in some $V_i \cap V_j = f(U_i \cap U_j)$ (see (f_3)) with $j \neq i$, and therefore in V', by (f_2). This proves (19.1).

Next, since

$$\sum_{j \neq i} [\sigma_j^r : \sigma_k^n]\tau_{jk}^{**}, \quad D_{ik}, \quad A_{ik} \text{ are in } V_i',$$

the last relation for $\partial\tau_k^{*n}$ gives

$$B_k = \partial\tau_k^{*n} - \sum_l [\sigma_l^r : \sigma_k^n]\tau_{lk}^{**} \subset V_i'.$$

Since B_k is independent of i, this holds for each i. Now if B_k had a cell with a point p in just one V_{i_0} (clearly $B_k \subset \sum V_j$), using i_0 in the last relation would give a contradiction. Hence each p is in at least two, and hence in V'. This proves (19.2).

With the help of the new chains, we shall find new expressions for the numbers in §18:

$$\xi_{i,v_{ik}}^{\pm} = KI(\tau_{ik}^{**}, \rho^{\pm}\tau_i^r) \qquad ([\sigma_i^r : \sigma_k^n] \neq 0), \tag{19.3}$$

$$\zeta_{ij}^{\pm} = KI(\tau_{ik}^{**}, \rho^{\pm}\tau_j^r) \qquad ([\sigma_i^r : \sigma_k^n] \neq 0, i \neq j). \tag{19.4}$$

First, since each cell of σ_{ik}^{**} is a face of a cell of σ_{ik}^{*}, which has a vertex in σ_i^r, (f_8) gives:

$$\tau_{ik}^{**} \subset \tau_{ik}'^{n} \cap V_i. \tag{19.5}$$

Taking the boundary of (19.1) and using (f_8) gives

$$\partial\tau_{ik}^{**} = \partial\tau_i^{*r} + \partial A_{ik} \subset \tau_{ik}^{\prime n} \cap V'. \tag{19.6}$$

These relations, with (f_6) and (f_7), show that we may apply Lemma 19, which gives

$$KI(\tau_{ik}^{**}, \rho^{\pm}\tau_i^r) = a\xi_{i,v_{ik}}^{\pm}.$$

We must show that $a = 1$. By (19.6),

$$\partial\tau_{ik}^{**} - \partial\tau_i^r = \partial[A_{ik} - (\tau_i^r - \tau_i^{*r})].$$

By (f_8) and (19.1), $A_{ik} - (\tau_i^r - \tau_i^{*r}) \subset V'$. Since this chain is also in $\tau_i^{\prime r} \cup \tau_{ik}^{\prime n}$, applying ($f_7$) gives (18.8) with $a = 1$. The uniqueness of a in (18.8) completes the proof of (19.3).

Next, taking $i \neq j$, since

$$\tau_j^r - \tau_j^{*r} \subset V', \qquad \tau_{ik}^{**} \subset V, \qquad (\tau_j^r - \tau_j^{*r}) \cap \tau_{ik}^{**} = 0,$$

(f_4) gives $\pi(\tau_j^r - \tau_j^{*r}) \cap \pi(\tau_{ik}^{**}) = 0$. Hence

$$KI(\tau_{ik}^{**}, \rho^{\pm}\tau_j^r - \rho^{\pm}\tau_j^{*r}) = 0.$$

Similarly, using (19.1), we find the two relations

$$KI(A_{ik}, \rho^{\pm}\tau_j^{*r}) = KI(\tau_{ik}^{*}, \rho^{\pm}\partial\tau_j^{*r}) = 0.$$

By (f_3), $KI(\tau_{ik}^{*}, \tau_j^{*r}) = 0$. This, with the last relations, gives

$$0 = KI(\tau_{ik}^{*}, \partial\rho^{\pm}\tau_j^{*r}) = \pm KI(\tau_i^{*r} - \tau_{ik}^{**}, \rho^{\pm}\tau_j^{*r}).$$

Hence

$$KI(\tau_{ik}^{**}, \rho^{\pm}\tau_j^r) = KI(\tau_{ik}^{**}, \rho^{\pm}\tau_j^{*r}) = KI(\tau_i^{*r}, \rho^{\pm}\tau_j^{*r}),$$

which gives (19.4).

We now give an extension of the results in §14, in particular, of Lemma 13.

DEFINITION. Let K^r and v be in general position in E^{2r+1}, with reference to the y_{2r+1}-direction. With the notations of §18, and a sufficiently small positive continuous $\eta(p)$ in K, set

$$\mu_{ij,v}^{\pm} = KI(\phi_{v,\eta}\tau_i^r, \rho^{\pm}\tau_j^r). \tag{19.7}$$

Save possibly for sign, this is a direct generalization of (14.2). The commutation rule is:

$$\mu_{ij,v}^{+} = (-1)^{r+1}\mu_{ji,-v}^{-}. \tag{19.8}$$

The proof is easily carried out, with the help particularly of (b_5), as follows:

$$\begin{aligned}\mu_{ij,v}^{+} &= KI(\rho^{+}\tau_j^r, \phi_{v,\eta}\tau_i^r) = KI(\rho^{+}\phi_{-v,\eta}\tau_j^r, \tau_i^r)\\ &= -KI(\rho^{+}\phi_{-v,\eta}\tau_j^r, \partial\rho^{-}\tau_i^r) = (-1)^r KI(\partial\rho^{+}\phi_{-v,\eta}\tau_j^r, \rho^{-}\tau_i^r)\\ &= (-1)^{r+1}KI(\phi_{-v,\eta}\tau_j^r, \rho^{-}\tau_i^r) = (-1)^{r+1}\mu_{ji,-v}^{-}.\end{aligned}$$

Definition. With the above notations, set

$$\mathfrak{K}^{\pm}(A, B, v) = KI(\phi_{v,\eta}A, \rho^{\pm}B). \tag{19.9}$$

The commutation rule, generalizing (14.3), is

$$\mathfrak{K}^{+}(A, B, v) = (-1)^{r+1}\mathfrak{K}^{-}(B, A, -v). \tag{19.10}$$

To prove this, suppose $A = \sum a_i\tau_i^r$, $B = \sum b_j\tau_j^r$. Then, writing ϕ for $\phi_{v,\eta}$,

$$\begin{aligned}\mathfrak{K}^{+}(A, B, v) &= \sum_{i,j} a_i b_j KI(\phi\tau_i^r, \rho^{+}\tau_j^r)\\ &= \sum_i a_i b_i \xi_{i,v}^{+} + \sum_{i<j} (a_i b_j \mu_{ij,v}^{+} + a_j b_i \mu_{ji,v}^{+})\\ &= (-1)^{r+1}[\sum_i b_i a_i \xi_{i,-v}^{-} + \sum_{i<j} (b_i a_j \mu_{ij,-v}^{-} + b_j a_i \mu_{ji,-v}^{-})]\end{aligned}$$

which, by the same proof, equals $(-1)^{r+1}\mathfrak{K}^{-}(B, A, -v)$.

20. Application to partial manifolds M^n mapped into E^{2n-1}

The following theorem contains essentially Theorem 2 if n is odd. Let $\mathfrak{K}_f'(M)$ denote the algebraic number of singular points if n is odd.

Theorem 9. *Let M^n be a partial manifold, let K be a subdivision of M (with the properties of §15), and let f be a semi-regular mapping of M into E^{2n-1} so that f maps K in general position. For each σ_i^{n-1} in ∂M, let $v_i(p)$ point into M at $f(p)$, $p \in \bar{\sigma}_i^{n-1}$. Then letting $\sum'$ denote the sum taken over just those cells in ∂M, we have:*

(a) *If n is even,*

$$\sum{}' \xi_{i,v_i}^{+} = \sum{}' \xi_{i,v_i}^{-} = \sum{}' \xi_{i,-v_i}^{+}. \tag{20.1}$$

(b) *If n is odd,*

$$2\mathfrak{K}_f'(M) = \sum{}' (\xi_{i,v_i}^{+} + \xi_{i,v_i}^{-}) = \sum{}' (\xi_{i,v_i}^{+} - \xi_{i,-v_i}^{+}). \tag{20.2}$$

We consider first the case that M^n is a single cell σ_k^n. Using the previous notations, setting $\partial_{ik} = [\sigma_i^{n-1}:\sigma_k^n]$ and noting that $\partial_{ik}^2 = 0$ or 1, we find, with the help of (19.2), (19.3) and (19.4),

$$\begin{aligned}\mathfrak{K}_f(\sigma_k^n) &= KI(\tau_k^{*n}, \partial\tau_k^n) = -KI(\tau_k^{*n}, \partial\rho^{\pm}\partial\tau_k^n)\\ &= (-1)^{n-1}KI(\partial\tau_k^{*n}, \rho^{\pm}\partial\tau_k^n) = (-1)^{n-1}\sum_{i,j} \partial_{ik}\partial_{jk}KI(\tau_{ik}^{**}, \rho^{\pm}\tau_j^r)\\ &= (-1)^{n-1}[\sum_i \xi_{i,v_{ik}}^{\pm} + \sum_{i<j} \partial_{ik}\partial_{jk}(\zeta_{ij}^{\pm} + \zeta_{ji}^{\pm})].\end{aligned}$$

Using first $+$ and then $-$, we find, with the help of (18.5),

$$(-1)^{n-1}\mathfrak{K}_f(\sigma_k^n) = \sum_i \xi_{i,v_{ik}}^{+} + \sum_{i<j} \partial_{ik}\partial_{jk}(\zeta_{ij}^{+} + \zeta_{ji}^{+}),$$

$$\mathfrak{K}_f(\sigma_k^n) = (-1)^{n-1}\sum_i \xi_{i,v_{ik}}^{-} - \sum_{i<j} \partial_{ik}\partial_{jk}(\zeta_{ji}^{+} + \zeta_{ij}^{+}).$$

Adding these and using Theorem 2, gives

$$[1 + (-1)^{n-1}]\mathfrak{L}'_f(\sigma_k^n) = \sum_i [\xi^+_{i,v_{ik}} + (-1)^{n-1}\xi^-_{i,v_{ik}}], \tag{20.3}$$

from which the theorem for this case follows, with the help of (18.4).

By (20.3) and (18.4), we have, in the general case,

$$[1 + (-1)^{n-1}] \sum_k \mathfrak{L}'_f(\sigma_k^n) = \sum_i \sum_k (\xi^+_{i,v_{ik}} - \xi^+_{i,-v_{ik}}),$$

where for each i we sum over all k such that σ_i^{n-1} is a face of σ_k^n. If σ_i^{n-1} is interior to M, then σ_i^{n-1} is a face of two cells σ_k^n and σ_l^n, and $v_{il} = -v_{ik}$; hence

$$(\xi^+_{i,v_{ik}} - \xi^+_{i,-v_{ik}}) + (\xi^+_{i,v_{il}} - \xi^+_{i,-v_{il}}) = 0.$$

Thus all these values of i drop out. For each i with $\sigma_i^{n-1} \subset \partial M$, there is just one k, and $v_i = v_{ik}$. Hence the theorem follows.

REMARK. Clearly both sides of (20.2) are independent of the subdivision K of M employed. All the terms are independent of the chosen orientations of cells.

DEFINITION. Let M be a partial n-manifold, n odd, with the property that there exists a continuous vector field $u(p)$ defined in ∂M, $u(p)$ being independent of each closed cell of ∂M containing p, and pointing into M. Let K be a subdivision of M and let f be a semi-regular mapping of M into E^{2n-1} such that f maps K in general position with reference to the y_{2n-1}-direction. Set $v(p) = \nabla f(u, p)$. Set

$$A^{n-1} = \sum{}' \sigma_i^{n-1}, \text{ summed over the cells in } \partial M, \tag{20.4}$$

these cells being oriented arbitrarily. Let e be the unit vector in the y_{2n-1}-direction. Then for $\alpha(p) > 0$ sufficiently small in ∂M, and $\eta(p) > 0$ sufficiently much smaller than $\alpha(p)$, define (using the definition of ψ in §14 and recalling that n is odd)

$$\begin{aligned}\mathfrak{L}_f(M) &= \tfrac{1}{2}[KI(\phi_{v,\eta}fA,\ \psi_{e,0,\alpha}fA) + KI(\phi_{v,\eta}fA,\ \psi_{-e,0,\alpha}fA)] \\ &= \tfrac{1}{2}[KI(\phi_{v,\eta}fA,\ \psi_{e,0,\alpha}fA) - KI(\phi_{-v,\eta}fA,\ \psi_{e,0,\alpha}fA)].\end{aligned} \tag{20.5}$$

REMARK. If the hypothesis of general position in Theorem 9 does not hold, we may apply Lemma 17 to make it hold; that $\mathfrak{L}_f(M)$ is independent of the rotation chosen follows from the proof below, in which general position is assumed.

COMPLETION OF THE PROOF OF THEOREM 2. We must prove $\mathfrak{L}'_f(M) = \mathfrak{L}_f(M)$. Because of (20.2), it is only necessary to show that (for n odd)

$$KI(\phi_{v,\eta}fA,\ \psi_{e,0,\alpha}fA) = \sum{}' \xi^+_{i,v_i}. \tag{20.6}$$

For each σ_i^{n-1} in ∂M, we may deform $v_i(p)$ into $v(p)$, keeping it tangent to M and independent of $\bar\sigma_i^{n-1}$; hence ξ^+_{i,v_i} may be replaced by $\xi^+_{i,v}$. Since both sides of (20.6) are independent of the subdivision employed, and πf is regular in K^{n-1} and hence in ∂M, we may suppose that K is so fine that for each pair σ_i^{n-1},

σ_j^{n-1} of cells of ∂M with a common vertex, πf is an imbedding in $\bar{\sigma}_i^{n-1} \cup \bar{\sigma}_j^{n-1}$. Since also πf is regular in M at points of K^{n-2}, it follows at once that for small enough $\alpha(p)$ and $\eta(p)$,

$$KI(\phi_{v,\eta}\tau_i^r, \psi_{e,0,\alpha}\tau_j^r) = 0 \qquad (i \neq j).$$

Exactly as in corresponding proofs in §19, we see that

$$KI(\phi_{v,\eta}\tau_i^r, \psi_{e,0,\alpha}\tau_i^r) = KI(\phi_{v,\eta}\tau_i^r, \rho^+\tau_i^r) = \xi_{i,v}^+ .$$

We find therefore

$$KI(\phi_{v,\eta}fA, \psi_{e,0,\alpha}fA) = \sum_{i,j}{}' KI(\phi_{v,\eta}\tau_i^r, \psi_{e,0,\alpha}\tau_j^r)$$

$$= \sum_i{}' KI(\phi_{v,\eta}\tau_i^r, \psi_{e,0,\alpha}\tau_i^r) = \sum_i{}' \xi_{i,v}^+ ,$$

taken over the cells of ∂M, which completes the proof.

21. The necessity for the type of formula in (20.5) (n odd)

One might expect that a single term on the right in (20.5) would suffice' without the factor $\frac{1}{2}$. This is the case if the cells σ_i^r of ∂M can be so oriented that $\sum{}'\sigma_i^r$ is a cycle, but not in the general case, as we shall show.

To compare the two terms, note that

$$\psi_{-e,0,\alpha} = -\psi_{-e,\alpha,0} = -\psi_{e,-\alpha,0};$$

hence, subtracting one of the terms from the other,

$$(21.1)\qquad \begin{aligned} \Delta &= KI(\phi_{v,\eta}fA, \psi_{e,0,\alpha}fA - \psi_{-e,0,\alpha}fA) \\ &= KI(\phi_{v,\eta}fA, \psi_{e,-\alpha,\alpha}fA). \end{aligned}$$

Let e' be the unit vector in the y_{2n-2}-direction. By deforming $\phi_{v,\eta}\, C$ into $\phi_{e',\eta}C$, we define θC for all $C \subset \partial M$, with the property

$$\partial\theta C = \phi_{e',\eta}C - \phi_{v,\eta}C - \theta\partial C;$$

we may take this in general position together with the chains ψ considered, with reference to the y_{2n-1}-direction. From the obvious relations (for small $\eta(p)$)

$$\phi_{e',\eta}fA \cap \psi_{e,-\alpha,\alpha}fA = 0,$$

$$\theta fA \cap (\phi_{e,\alpha}fA \cup \phi_{e,-\alpha}fA) = 0,$$

we obtain

$$\begin{aligned} \Delta &= KI(\phi_{v,\eta}fA - \phi_{e',\eta}fA, \psi_{e,-\alpha,\alpha}fA) \\ &= -KI(\partial\theta fA + \theta\partial fA, \psi_{e,-\alpha,\alpha}fA) \\ &= -KI(\theta\partial fA, \psi_{e,-\alpha,\alpha}fA) - (-1)^n KI(\theta fA, \partial\psi_{e,-\alpha,\alpha}fA), \end{aligned}$$

$$(21.2)\qquad \Delta = -KI(\theta f\partial A, \psi_{e,-\alpha,\alpha}fA) + (-1)^n KI(\theta fA, \psi_{e,-\alpha,\alpha}f\partial A).$$

As a corollary, $\Delta = 0$ if the cells σ_i^r of ∂M can be so oriented that their sum is a cycle.

REMARK. The last fact follows also on applying Lemma 14 to $\mathfrak{L}(fA, fA, v, \pm e)$.

EXAMPLE. We shall show that Δ may be $\neq 0$. Take a cylinder plus interior:

$$x^2 + y^2 \leqq 1, \ -1 \leqq z \leqq 1;$$

identify the two ends, after reflecting one across a diameter: set

$$(x, y, -1) = (x, -y, 1).$$

A "Klein bottle B^2 plus interior" is formed; this is a partial manifold M^3, with $\partial M = B^2$. Let K be a subdivision of M^3. We may imbed B^2 in E^4 in such a fashion that for any continuous vector field $v(p)$ in B^2, independent of B^2 at points of $K^1 \cap B^2$,

$$KI(\phi_{v,\eta}B_0^2, B_0^2) = -4 \text{ in } E^4,$$

B_0^2 being $\sum' \sigma_i^2$, σ_i^2 in B^2, and η being small; see [3], p. 107, Fig. 4, and the top of p. 108. Now take $E^4 \subset E^5$. Then clearly, from the above,

$$\Delta = KI(\phi_{v,\eta}B_0^2, \psi_{e,-\alpha,\alpha}B_0^2) = -4 \text{ in } E^5.$$

If, for instance, we take $v = (e + e')/2^{1/2}$, then (20.5) reads

$$\mathfrak{L}_f(M^3) = \tfrac{1}{2}[-4 + 0] = -2.$$

Mapping M^3 into E^5 so that v points into $f(M)$ gives the result stated.

APPENDIX

22. The self-intersections under a completely semi-regular mapping

Let A be the set of points p of M such that for some $q \neq p$, $f(p) = f(q)$. We shall discuss A and $f(A)$.

Let $p_1, p_2, \cdots$ be the singular points. By Lemma 2 and the definition of completely semi-regularity, there is a neighborhood U_i^* of p_i such that $A \cap U_i^*$ is given by $x_2 = \cdots = x_n = 0$; thus this is an open arc, mapped by f doubly into a half-open arc. We may suppose $\bar{U}_i^* \cap \bar{U}_j^* = 0$ for $i \neq j$.

Choose U_i^{**}, $p_i \in U_i^{**}$, $\bar{U}_i^{**} \subset U_i^*$, and set $M' = M - \sum \bar{U}_i^{**}$. Take p, q in M', $p \neq q$, $f(p) = f(q)$. If $p, q \in M' - \partial M'$, there are neighborhoods U and V of p and q in M' such that $f(U) \cap f(V)$ is a smooth open arc in E, the image of smooth open arcs in M'; if one of p, q is in $\partial M'$, these are half-open arcs. We may cover M' by neighborhoods U_i such that each $f(U_i) \cap f(U_j)$ $(U_i \cap U_j = 0)$ is void or a smooth open or half-open arc. These arcs in M' together with those in $\sum \bar{U}_i^{**}$ fit together to form simple closed curves or arcs; in each direction on each arc, we end either at a singular point or at a point whose image is also the image of a point of ∂M, or we reach no limit point of M (this can occur only if M is open). Moreover, it is not hard to see that there is a grouping of these curves into pairs (the two in a pair need not be distinct), in one of the following ways:

(a) There may be two distinct closed curves, with one closed curve as image.

(b) A single closed curve may be mapped doubly over a closed curve.

(c) If the closed curve contains any singular points, it contains two, and is mapped into an arc.

(d) An arc may have one singular point interior to it; then (d_1) it is open at both ends, or (d_2) one end stops at ∂M and the other stops interior to M.

(e) A pair of arcs, each containing no singular points, may map into an arc. Then (e_1) both are open, or (e_2) each is half open, one ends in ∂M, and the other ends interior to M, or (e_3) one has both ends in ∂M and the other has both ends interior to M, or (e_4) each has one end in ∂M and one end interior to M.

If $n = 2$, we have the same subdivision into cases, but the curves may cut through each other. The case $n = 1$ is trivial. We could further subdivide the cases by taking into account orientation properties.

REMARK. If f is not proper, the set A need not be closed; for example, we may map a strip M^2 around and around in the interior of a torus so that A is dense in M^2. Still A is expressible as a union of curves.

23. On the covering of an open partial manifold by a sequence of compact partial manifolds

Let M be a partial manifold (in particular, a manifold). Since it may be covered by a denumerable number of coordinate systems, and each is compact, M may be covered by a sequence of compact partial manifolds. We wish to show how partial manifolds with certain properties may be chosen. Let a *proper half-open arc* A in a point set R mean the one-one continuous and proper image $\phi(A_0)$ of the half-line $0 \leqq x$. We shall say A runs from $\phi(0)$ to infinity if, for any compact subset B of R, there is an x_0 such that $\phi(x) \in R - B$ for $x \geqq x_0$.

The following lemma is used in the proof of the immersion theorem for open manifolds.

LEMMA 20. *Let M be an open connected partial manifold. Then there is a sequence $M_1, M_2, \cdots$ of compact partial manifolds in M such that*

(a) $M_i \subset M_{i+1} - \partial M_{i+1}$ $(i = 1, 2, \cdots)$,

(b) $M = \sum M_i$,

(c) *any point of $M - M_i$ may be joined to infinity by a half open arc which does not touch M_i.*

REMARKS. It may be shown that each M_i may be taken as connected, and such that ∂M_i is a closed manifold (not necessarily connected); but we do not need these facts here. For the present purpose, by a "partial manifold" we shall mean merely the closure of an open subset of some open manifold.

To start with, let $M_1'', M_2'', \cdots$ be the sets in M covered by a fixed set of coordinate systems, so that $M = \sum M_i''$. Set $M_1' = M_1''$. Since M_1' is compact, we may choose μ_2 so that $M_1' \subset \sum_{k=1}^{\mu_2}(M_k'' - \partial M_k'')$. Set $M_2' = \sum_{k=1}^{\mu_2} M_k''$. In general, choose μ_{i+1} so that $M_i' \subset \sum_{k=1}^{\mu_{i+1}} (M_k'' - \partial M_k'')$, and set $M_{i+1}' = \sum_{k=1}^{\mu_{i+1}} M_k''$. Then $M_i' \subset M_{i+1}' - \partial M_{i+1}'$, and $M = \sum M_i'$.

Let M_i be the set of all points p of M for which the following is not true: For each j there is an arc A joining p to a point $q \in M - M'_j$, such that $A \subset M - M'_i$. We show first that M_i is compact. If not, then there is a sequence p_1, $p_2, \cdots$ of points of M_i with no subsequence which converges in M. Since each M'_j is compact, it contains at most a finite number of the p_k; hence we may suppose the p_k chosen so that $p_j \in M - M'_j$, all j. Since M is connected, there is an arc A_j in M joining p_1 to p_j. Let q_j be the last point of A_j in M'_{i+1} for $j > i$, and let $A(p_j)$ be the arc $q_j p_j$. Since M'_{i+1} is compact and $\partial M'_{i+1}$ is closed in M'_{i+1}, there is a subsequence $q'_1, q'_2, \cdots$ of $q_1, q_2, \cdots$ converging to a point q of $\partial M'_{i+1}$. For some connected neighborhood U of q, $U \subset M - M'_i$. Choose s so that $q'_k \in U$ for $k \geqq s$. Let p'_k correspond to q'_k. For each $j > s$, $p'_j \in M - M'_j$, and from $A(p'_s)$, $A(p'_j)$, and an arc in U, we find an arc in $M - M'_i$ joining p'_s to p'_j. Hence p'_s is not in M'_i, a contradiction, proving that M_i is compact.

Clearly all boundary points of M_i are in M'_i; hence (a) holds. Since $M'_i \subset M_i$, (b) is true. To prove (c), take $p \in M - M_i$. By definition of the M_j, there is an arc A_j joining p to a point p_{i+1} in $M - M'_{i+1}$, such that $A_i \subset M - M'_i$, there is an arc A_{i+1} joining p_{i+1} to a point p_{i+2} in $M - M'_{i+2}$, such that $A_{i+1} \subset M'_{i+1}$, etc. Since $\sum M'_j = M$, these arcs give a proper half open arc A joining p to infinity, such that $A \subset M - M'_i$. Moreover, $A \subset M - M_i$; for if $q \in A \cap M_i$, then part of A gives an arc in $M - M'_i$ joining q to a point in an arbitrary $M - M_j$, contradicting the definition of M_i. This completes the proof.

24. On proper mappings

We recall the definition from [1]: The *limit set* L_f of a mapping f of a space R into a space R' is the set of points $q \in R'$ such that for some sequence $\{p_k\}$ in R, $f(p_k) \to q$, while $\{p_k\}$ has no limiting point in R. f is *proper* if $f(R) \cap L_f = 0$. Note that if R is compact, the limit set under any mapping is void. If f is proper in R, it is proper in any closed subset of R. If f maps R into a single point, f is proper if and only if R is compact.

REMARK. If f is one-one and continuous, then clearly f is proper if and only if f^{-1} is continuous.

We give first a characterization of proper mappings without use of sequences.

LEMMA 21. *A mapping f of a locally compact separable metric space R into a similar space R' is proper if and only if for each point $p \in R$ (or equally well, each self-compact subset $A \subset R$) there is a neighborhood U of $f(p)$ (or of $f(A)$) in R' and a self-compact subset B of R such that*

$$f(R - B) \cap U = 0. \tag{24.1}$$

We may suppose R is not compact, the lemma being trivial otherwise. Let $R_1, R_2, \cdots$ be self-compact subsets of R with $R_i \subset R_{i+1}$ and $R = \sum R_i$ (see the proof of Lemma 20). Suppose first that the condition in its strong form does not hold; then a self-compact subset A is given; say $A \subset R_k$. Let $\{U_i\}$ be a sequence of neighborhoods of $f(R_k)$ such that $\prod U_i = f(R_k)$. In each $R - R_{k+i}$ choose a point p_i so that $f(p_i) \in U_i$; we may suppose that the p_i are distinct. A subsequence may be chosen so that $f(p_{\lambda_i}) \to q \in f(R_k)$; thus f is not proper.

If, conversely, f is not proper, say $f(p_i) \to q = f(p)$, $\{p_i\}$ having no limit in R. For some subsequence $\{p_{\lambda_i}\}$, $p_{\lambda_i} \in R - R_i$. Since each compact subset of R is in some R_j, the condition does not hold, using p.

We state without proof:

LEMMA 22. *A mapping f of R into R' is proper if and only if antecedents of sets compact in $f(R)$ are sets compact in R.*

LEMMA 23. *A continuous proper mapping of R into R' maps sets closed in R into sets closed in $f(R)$.*

Suppose A is closed, while $f(A)$ is not closed in $f(R)$; then $q_1, q_2, \cdots$ exist in $f(A)$, $q_i \to q$, $q \in f(R) - f(A)$. Choose p_i so that $f(p_i) = q_i$. If the sequence $\{p_i\}$ had a limiting point p, then say $p_{\lambda_i} \to p$; since f is continuous, $q_{\lambda_i} = f(p_{\lambda_i}) \to f(p)$. But $q_{\lambda_i} \to q$; hence $f(p) = q$, and $q \in f(A)$, a contradiction. This shows that f is not proper.

We state without proof:

LEMMA 24. *Let f be continuous and map closed sets into closed sets, and let the antecedents of single points be finite sets of points. Then f is proper.*

The following lemma is needed in one or two places in the present paper.

LEMMA 25. *Let R and R' be locally compact separable metric spaces. Let f, mapping R into R', be continuous, proper, and locally one-one. Let A and B be closed subsets of R such that*

$$\text{(24.2)} \qquad \text{if } p \in A, \quad q \in B, \quad p \neq q, \quad \text{then } f(p) \neq f(q).$$

Then there are neighborhoods U of A and V of B such that

$$\text{(24.3)} \qquad \text{if } p \in U, \quad q \in V, \quad p \neq q, \quad \text{then } f(p) \neq f(q).$$

The proof is not very difficult; we expect to give it as an application of much more general ideas in another paper.

HARVARD UNIVERSITY

BIBLIOGRAPHY

1. H. WHITNEY, *Differentiable manifolds*, Annals of Math. vol. 37 (1936), pp. 645–680.
2. ———, *On regular closed curves in the plane*, Compositio Math., Vol. 4 (1937), pp. 276–284.
3. ———, *On the topology of differentiable manifolds*, in Lectures in Topology, University of Michigan Press, 1941.
4. ———, *The general type of singularity of a set of 2n − 1 smooth functions of n variables*, Duke Math. J., vol. 10 (1943), pp. 161–172.
5. ———, *On the extension of differentiable functions*, Bulletin of the Am. Math. Soc., 1943.

THE WORK OF JOHN W. MILNOR

By HASSLER WHITNEY

To aid in understanding the significance of Milnor's work, I will first say a few words about the recent history of algebraic topology. In the early thirties, the subject seemed to have reached a certain level of completeness in its basic methods and results; the famous textbooks of Lefschetz, of Seifert and Threlfall, and of Alexandroff and Hopf, gave a very good picture of the field. The term "algebraic" had not yet been applied. Then in 1935, the sudden explosion of "cohomology theory", with its various kinds of applications, provided a great impetus to research. In developing these applications to new heights, various algebraic problems arose; one could often write out a formula which described a situation, but could not understand the geometric meaning of the formula. In the forties, powerful and general algebraic machinery came into being; this enabled problems, which had before seemed hopelessly complex, to be answered in relatively simple algebraic terms. The subject of "algebraic topology" was in full swing.

The new algebraic methods now began to grow by themselves; new concepts, such as exact sequences, sheaves, homological algebra, spread out not only through topology but through neighboring domains of algebra and analysis, where they have had extraordinary success. As pointed out by H. Hopf, in presenting the work of the Fields Medalist R. Thom at the last Congress at Edinburgh, the geometric point of view tended to be swamped in the algebraic machinery. Papers in algebraic topology had commonly the appearance of being pure algebra. Then in the mid fifties, some great discoveries of a more geometric nature took place, which brought on a resurgence of the geometric point of view. The definition of cobordism and its basic properties and applications by Thom in 1954 was one such discovery; for this he was awarded a Fields Medal in 1958, as mentioned above. In 1956, the mathematical world was astounded by Milnor's proof that the 7-dimensional sphere S^7 was capable of several differential structures. This at first might have seemed like an isolated fact; but through Milnor and others it has led to a vast field of work, which has now acquired a name: *differential topology*.

To explain this work, I must give some definitions. A *differential manifold* is a topological manifold with given local coordinate systems, such that the transformation from one coordinate system to an overlapping one is differentiable. I (and some others) have purposely altered the standard term, "differentiable manifold", precisely because Milnor's result shows that one must tell, not merely whether a manifold is "differentiable", but which differential structure one has in mind. (Recently M. Kervaire has shown that there are topological manifolds which admit no differential structure at all.) A homeomorphism between two topological manifolds is a one–one correspondence between them which is continuous in both directions; a *diffeomorphism* is a homeomorphism that is differentiable in both directions.

We may now state Milnor's result: There are differential manifolds which are homeomorphic to the 7-sphere S^7, but not diffeomorphic to it.

A first reaction to such a theorem might be: What is the use in showing that S^7 can be distorted to such a degree that the original differential structure is lost, but a new one is taken on, distinct from the first? The fact is, the new manifolds are constructed not by pathological means, but by simple and well known methods; the actual differential manifolds had certainly been considered, but one had not suspected that they were homeomorphic to the sphere, and yet had different differential structures.

The manifolds are defined as follows. Start with the Cartesian product $S^4 \times S^3$. Cut S^4 apart at its equator S^3, thus cutting the product manifold apart. Now glue the two resulting bounded manifolds together again at their boundaries, identifying (u,v) with (u,v'), where $v' = u^h v u^j$; the multiplication here is quaternion multiplication in $S^3 \subset R^4$. Taking $h+j=1$, $h-j=k$, for any odd integer k, gives a manifold M_k^7; it is the total space of a sphere bundle over S^4, and has an obvious differential structure. It is easy to show that there is a differentiable real valued function on M_k^7 which admits just two critical points, these being nondegenerate; from this it follows that M_k^7 is homeomorphic to S^7.

An invariant λ for closed oriented differential manifolds M^7 is now defined, as follows: As part of Thom's cobordism theory, M^7 is the boundary of a differential manifold B^8. (For the M_k^7, one may take B_k^8 to be the total space of the associated disk bundle.) Now $\lambda(M^7)$ is defined as a certain function of the "index" and the first Pontrjagin class of B^8; it is an integer mod 7, independent of the choice of B^8. Moreover, $\lambda(M_k^7) \equiv k^2 - 1$ mod 7, and $\lambda(S^7) \equiv 0$; hence M_3^7 for instance is homeomorphic but not diffeomorphic to S^7, and it thus may be considered as defining a new differential structure on the topological manifold S^7.

Milnor later extended this work to a study of spheres of dimension $4m-1$, with $m>2$. The 31-sphere, for example, was shown to admit more than sixteen million distinct differential structures.

The next job in building up a theory of these structures was to form them into some sort of mathematical system. Milnor showed that an abelian group θ^n could be constructed from them, for each dimension n. We give a brief description of θ^n. The *connected sum* $M_1 \# M_2$ of two differential n-manifolds M_1 and M_2 is formed by removing a small n-cell from each, and then gluing together the resulting boundaries, with proper regard to orientation. Two closed differential manifolds M and M' are *h-cobordant* (homotopy cobordant) if the disjoint sum $M + (-M')$ is the boundary of some differential manifold W, such that both M and M' are deformation retracts of W; this is an equivalence relation. Now the group θ^n is the abelian group of h-cobordism classes of differential manifolds which are homotopy n-spheres (that is, have the same homotopy type as S^n), under the connected sum operation.

Through some deep work of S. Smale and others, it is known that the elements of θ^n correspond precisely to the distinct diffeomorphism classes of S^n. Hence, for instance, the order of θ^n is the number of distinct differential structures on S^n. These groups have wide importance; for instance, they serve as coefficient groups in the study of obstructions to the construction of differential structures on a manifold. They play a role, relative to differentiability, similar to the role of the homotopy groups in more purely topological questions.

Kervaire and Milnor, in a joint study, have shown that the groups θ^n

are always finite, except that the structure of θ^3 is unknown. (If the Poincaré conjecture were proved, it would follow that θ^3 is trivial.) They have found the orders of many of these groups; for instance:

n	1	2	3	4	5	6	7	8	9	10	11	12	13	14	15
order of θ^n	1	1	?	1	1	1	28	2	8	6	992	1	3	2	16256

The order of θ^{4m-1} is given in terms of the order of the stable homotopy group $\pi_{k+4m-1}(S^k)$ (k large) and the Bernoulli number B_m.

We must mention the recent solution by Milnor of two outstanding problems of topology; in each case the solution was made possible by deep work of someone else. First, when R. Bott proved a theorem about the divisibility of the highest Pontrjagin class of the $4m$-sphere, Milnor saw that, combining this with results of Wu Wen-Tsün, it followed that the n-sphere is parallelizable only if $n=1$, 3 or 7. (Kervaire and Hirzebruch also discovered this independently.) A further consequence of this is that there exists a division algebra of rank n over the real numbers if and only if $n=1$, 2, 4 or 8. (In these cases, examples are: the real numbers, the complex numbers, the quarternions, and the Cayley numbers.)

Second, the so-called "Hauptvermutung" for complexes, that homeomorphic polyhedra have isomorphic cell subdivisions, was formulated in a paper by H. Tietze in 1908 (it was conjectured earlier). In this paper, he studied the "lens spaces" $L(p,q)$. These are manifolds, defined as follows. Set $\omega=e^{2\pi i/p}$. Taking the unit sphere S^3 in the space of pairs (z_1,z_2) of complex numbers and identifying any two points (z_1,z_2), $(\omega z_1,\omega^q z_2)$ of S^3 defines $L(p,q)$. Tietze showed that p is a topological invariant. With the help of a very recent theorem of B. Mazur on the existence of a diffeomorphism between certain product spaces, Milnor shows that $L(p,q)\times R^n$ is diffeomorphic to $L(p,q')\times R^n$ provided that $\pm qq'$ is a quadratic residue mod p, and $n>3$. Now construct the finite complex X_q from the product $L(7,q)\times\sigma^n$, σ^n being an n-simplex, by adjoining a cone over the boundary $L(7,q)\times\partial\sigma^n$. Then for $n\geqslant 3$, X_1 is homeomorphic with X_2, as can be proved from the result stated above. Now the Hauptvermutung is *disproved* by showing that X_1 and X_2 have no isomorphic cell subdvisions. This is accomplished by slightly generalizing the torsion invariant of Reidemeister, Franz and de Rham, and following some methods developed by J. H. C. Whitehead.

One possible peculiarity of differential structures appears as a corollary of the work just mentioned: The bounded manifolds $L(7,1)\times D^5$ and $L(7,2)\times D^5$ (where D^5 is the unit 5-disk) are not diffeomorphic, but their interiors are.

To look ahead, Milnor is now developing a theory of "microbundles", which injects a new and basic tool into the subjects discussed above. You will hear about this from him in his address at this Congress.

In conclusion, it is apparent that differential topology is a strong young field, with a bright present and a great outlook, and that its vitality is largely due to the fine achievements of Milnor. It gives me especial pleasure to report on this; in the thirties I was much concerned with the relations between differential and topological properties, and I felt that my own contributions were merely some beginnings of what might come at some time in the future. The future has arrived!

Chapter 2

Bundles and Characteristic Classes

TOPOLOGICAL PROPERTIES OF DIFFERENTIABLE MANIFOLDS*

BY HASSLER WHITNEY

I. INTRODUCTION

1. *The Problems*. In many fields of work one is led to the consideration of n-dimensional spaces. A given dynamical system has a certain number of "degrees of freedom"; thus a rigid body, with one point fixed, has three. A line in euclidean space is determined by four "parameters." We therefore consider the positions of the rigid body, or the straight lines, as forming a space of three, or four, dimensions.† But when we try to determine the points of the space by assigning to each a set of three, or four, numbers, we are doomed to failure. This is possible for a small region of either space, but not for the whole space at once. The best we can do is to cover the space with such regions, define a coordinate system in each, and state how the coordinate systems are related in any two overlapping regions. They will be related in general by means of differentiable,‡ or analytic, transformations, with non-vanishing Jacobian. Any such space we shall call a differentiable, or analytic, manifold.

For a complete study of such spaces, we must know not only properties of euclidean n-space E^n, which we may apply in each coordinate system separately, but also properties which arise from the manifold being pieced together from a number of such systems. It is these latter properties, essentially topological in character, which form the subject of the present address.

Suppose we wish to study differential geometry in the n-dimensional manifold M^n. At each point p of M^n, the possible differentials (or "tangent" vectors) form an n-dimensional vec-

---

* An address delivered before the Society on September 9, 1937, in State College, by invitation of the Program Committee.

† The first space forms the group G^3 of rotations in 3-space; it is homeomorphic with projective 3-space P^3. The second space is homeomorphic with the total space (see §2) of the tangent vector space of the projective plane P^2, or of the 2-sphere S^2 if we use oriented lines. This is easily seen by considering together all parallel lines.

‡ Differentiable will always mean continuously differentiable.

tor space $V(p)$, the so-called tangent space at p. For topological considerations, it is sufficient to consider vectors of unit length,* or directions, which form a sphere $S(p)$ of dimension $n-1$. This set of spheres forms the *tangent sphere-space* of M^n. Most of our work will be on the problem, how do the spheres fit together over the whole manifold? Suppose M^n is imbedded in a higher dimensional manifold M^m (for instance, euclidean E^m). Then we may consider the normal unit vectors at each point, forming an $(m-n-1)$-sphere, and thus the *normal sphere-space*. The methods we use in Part II were discovered independently by E. Stiefel [11]† and myself [15]. Here we shall not attempt always to describe the most elegant methods, but lean rather to the intuitive side.

In the last part we give some fundamental results, due to de Rham,‡ on the theory of multiple integration in a manifold. If we wish to integrate over an r-dimensional subset, and no measure function is given, we integrate "differential forms," in other words, "alternating covariant tensors" of order r; we shall call these simply r-functions. It is necessary for various purposes to find what "exact" r-functions exist. This problem, in any small region, is rather trivial; again, we must consider the whole manifold to answer it. The methods we describe are somewhat different from those of de Rham.

2. *Sphere-Spaces.* A *sphere-space* $S(K)$ is defined as follows. Let S_0^ν be the unit ν-sphere§ in $E^{\nu+1}$. Let K be a complex. To each point p of K let there correspond a ν-sphere $S(p)$; if $p\neq q$, we assume that $S(p)$ and $S(q)$ have no common points. For each closed cell σ of K and each point p of σ, let $\xi_\sigma(p)$ be a (1-1)map of S_0^ν into $S(p)$; let $\xi_\sigma(p, q)$ be the point into which the point q of S_0^ν goes. If the cells σ and σ' have the common point p, then

---

* As every differentiable manifold may be imbedded in a euclidean space (see the bibliography, Whitney [14], Theorem 1), we may define in it a Riemannian metric and thus lengths of vectors, and so on.

† Numbers in square brackets refer to the bibliography at the end of this paper.

‡ See [6], [7], and [8]. The main theorem was suggested by E. Cartan, Comptes Rendus de l'Académie des Sciences, Paris, vol. 187 (1928), pp. 196–198.

§ That is, the set of points $\sum x_i^2=1$ in $E^{\nu-1}$. A 0-sphere is a pair of points, a 1-sphere is a circle, and a 2-sphere is a spherical surface.

$\xi_\sigma(p)$ and $\xi_{\sigma'}(p)$ are both defined. We assume that

$$(1) \qquad \xi_{\sigma'}^{-1}(p)\xi_\sigma(p), \quad \text{that is,} \quad \xi_{\sigma'}^{-1}(p, \xi_\sigma(p, q)),$$

which is a (1-1) map of S_0^ν into itself, is orthogonal,* and varies continuously with p. We call K the *base space*, and the set $\mathfrak{S}(K)$ of all points on all $S(p)$, the *total space*.† If K were a general point set, we would replace the cells σ by open sets covering K. We might replace the $S(p)$ by vector spaces $V(p)$ (in the obvious manner for tangent and normal spaces); however, the topological properties of these are reducible to those of sphere-spaces. For any sphere-space $S(K)$ and subset L of K, there is a corresponding sphere-space $S(L)$, the *part of* $S(K)$ *over* L. We shall call the map $\xi_\sigma(p)$, (p in σ), the *σ-coordinate system* in $\mathfrak{S}(\sigma)$. It determines (for a fixed orientation of S_0^ν) one of two possible orientations of each $S(p)$, (p in σ).

We find invariants which serve to distinguish between sphere-spaces with the same base space by studying the following questions. First, is it possible to choose, for each p in K, a point $\phi(p)$ of $S(p)$, so that $\phi(p)$ is continuous in K? Any such map we shall call a *projection* of K into $\mathfrak{S}(K)$. More generally, is it possible to find k projections $\phi_1(p), \cdots, \phi_k(p)$ of K into $\mathfrak{S}(K)$, so that these are orthogonal in each $S(p)$, $(1 \leqq k \leqq \nu+1)$? If so, we call $S(K)$ *k-simple*. If $S(K)$ is $(\nu+1)$-simple, we call it *simple*.

For tangent sphere-spaces, the existence of a projection is obviously equivalent to the existence of a continuous field of non-vanishing (tangent) vectors. The only closed orientable surface on which there is such a field is the torus.‡ (If we try, on the sphere, making all vectors point north or east for instance, this fails at the two poles; see §6.) Note that if we can find one such field on an orientable surface, we can at once find a second field of orthogonal vectors; at each point, looking along the first vector, we let the second vector point to the left. It follows that the tangent space is simple in this case.

* It is sufficient to assume that the map is linear. However, this general case is easily reducible to that given. Because of the assumption, we may define orthogonality on any $S(p)$.

† For a study of some total spaces, see Hotelling [4], Seifert [9] and [10], Stiefel [11], and Threlfall [12].

‡ For a study of this problem in manifolds, see Hopf [3]; also Alexandroff-Hopf, *Topologie* I, pp. 548–552, and Stiefel [11].

Suppose $S(K)$ is simple. Let $q_1=(1, 0, \cdots, 0), \cdots, q_{\nu+1}=(0, 0, \cdots, 1)$ be the unit points of S_0^ν. Set $\zeta(p, q_i)=\phi_i(p)$, $(i=1, \cdots, \nu+1)$. Then for each p in K, we can define $\zeta(p, q)$ for all other q in S_0^ν uniquely so that it is an orthogonal map of S_0^ν into $S(p)$. Clearly ζ is (1-1) and continuous. Hence *if $S(K)$ is simple, then $\mathfrak{S}(K)$ is the cartesian product of K and S_0^ν.*

If the tangent space to M^n is simple, then to each direction $\zeta(p, q)$ at p corresponds a direction $\zeta(p', q)$, (same q), at any other point p'. Thus *a parallelism may be defined throughout M^n if and only if the tangent sphere-space is simple.*

Suppose the normal space to M^n in M^m is simple. Then corresponding to each normal vector function $\phi_i(p)$ we may define a function $f_i(p)$ in M^m with $\phi_i(p)$ as its gradient at p, and vanishing in M^n. We may then define M^n as the set of points in M^m at which all the $f_i(p)$ vanish. Conversely, if it is possible to define M^m in this manner, with the gradients of the $f_i(p)$ linearly independent, then the normal space is simple.*

II. Invariants of sphere-spaces

3. *Elementary Properties of Complexes.*† Let ${}^r\partial_i^j$ be the incidence number of the cells σ_i^r and σ_j^{r-1}. Define the *boundary* and *coboundary* of the chain $A^r=\sum\alpha_i\sigma_i^r$ by

$$\text{(2)} \qquad \partial A^r = \sum {}^r\partial_i^j \alpha_i \sigma_j^{r-1}, \qquad \delta A^r = \sum {}^{r+1}\partial_j^i \alpha_i \sigma_j^{r+1}.$$

Then A^r is a *cycle* or *cocycle* if $\partial A^r=0$ or $\delta A^r=0$; A^r is *homologous* to B^r, $A^r \sim B^r$, if A^r-B^r is a boundary; similarly, for *cohomologous*, for which we use the symbol $\backsim$. If we identify homologous r-cycles, we obtain the rth *homology group*; similarly for the rth *cohomology group*. Define *scalar products* of r-chains by

$$\text{(3)} \qquad (\sum \alpha_i\sigma_i^r)\cdot(\sum \beta_i\sigma_i^r) = \sum \alpha_i\beta_i.$$

Note that $A^r\cdot\sigma_i^r$ is the coefficient of σ_i^r in the chain A^r. As $\partial\sigma_i^r\cdot\sigma_j^{r-1}$ and $\sigma_i^r\cdot\delta\sigma_j^{r-1}$ both equal the incidence number ${}^r\partial_i^j$,

* See Seifert [10], and Whitney, Annals of Mathematcis, vol. 37 (1936), pp. 865–878.

† For further details, see for instance Whitney, *Matrices of integers and combinatorial topology*, Duke Mathematical Journal, vol. 3 (1937), pp. 35–45, and *On products in a complex*, Annals of Mathematics, vol. 39 (1938). We shall refer to these papers as MI and PC respectively.

we have, on equating, multiplying by coefficients, and summing,

(4) $$\partial A^r \cdot B^{r-1} = A^r \cdot \delta B^{r-1}.$$

We shall use σ also for the set of points in the closed cell σ, and $\partial\sigma$ for its point set boundary. Let K^r be the r-dimensional part of K, the sum of all cells of dimension $\leqq r$.

4. *Orientability of Sphere-Spaces*. First note that the part of $S(K)$ over K^0, $S(K^0)$, is simple. We can define $\mathfrak{S}(K^0)$ as a product by setting $\zeta(a) = \xi_a(a)$ for each vertex a. We now ask, under what conditions is $S(K^1)$ simple? We attempt to extend $\zeta(p)$ through the 1-cells of K. Consider any 1-cell $\sigma = ab$. $S(a)$ and $S(b)$ are given orientations both by ζ and by ξ_σ. If the orientations given by ζ are both the same as, or both opposite to, those

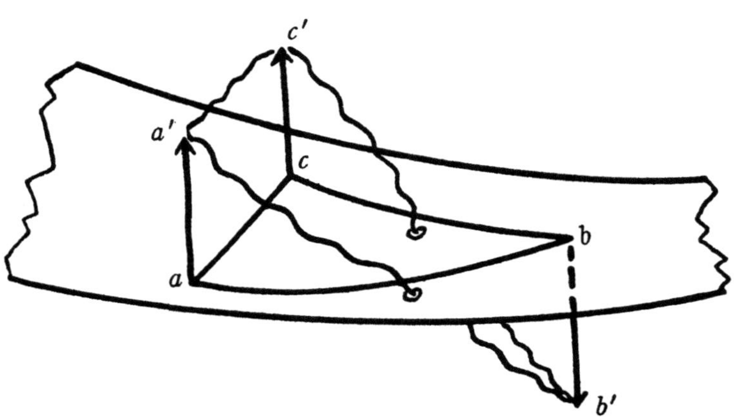

FIG. 1

given by ξ_σ, set $F^1(\sigma) = 0$; otherwise, $F^1(\sigma) = 1$. In Fig. 1,* if the end point of each vector, say at a, is $a' = \zeta(a, q_1)$, we have $F^1(ab) = F^1(bc) = 1$, $F^1(ca) = 0$. We may say, $F^1(\sigma)$ is 0 or 1 according as the orientations given to $S(a)$ and $S(b)$ by ζ are similar or opposite, when viewed through the σ-coordinate system. We define a *characteristic* 1-*cocycle* of K by letting $F^1(\sigma)$ be the coefficient of σ in a chain:

(5) $$F^1 = \sum_i F^1(\sigma_i^1)\sigma_i^1, \quad \text{coefficients integers mod 2.}$$

* For p in ab, all vectors $\xi_{ab}(p)$ point up or all point down.

To show that F^1 is a cocycle, we shall find the coefficient $\delta F^1\cdot\sigma^2=F^1\cdot\partial\sigma^2$ of δF^1 in any 2-cell σ^2 of K. For each 1-cell ab of $\partial\sigma^2$, the similarity or dissimilarity of the orientations given to $S(a)$ and $S(b)$ when viewed through the ab-coordinate system is clearly the same as when viewed through the σ^2-coordinate system. Say $\partial\sigma^2=ab+bc+\cdots+ea$. If we compare the orientation of $S(a)$ with that of $S(b)$, then of $S(b)$ with that of $S(c),\ldots,$ the number of changes of orientation is even, as we come back to the original orientation of $S(a)$. Hence $F^1\cdot\partial\sigma^2\equiv 0 \pmod 2$, as required.

We might have chosen different coordinate systems over K^1, and have thus found a different F^1. To find the change made in F^1, suppose the orientation of $S(a)$ is changed,* while that of each other $S(x)$ is unaltered. Then for each 1-cell σ with a as a vertex, $F^1(\sigma)$ is changed from 0 to 1 or from 1 to 0, that is, it is increased by 1 (mod 2). Thus the new F^1 is the old plus δa (mod 2). Thus we may change F^1 only by coboundaries, and *the cohomology class $h^1(F^1)$ is an invariant of $S(K)$*. Clearly if F^1 is a characteristic cocycle, we may obtain any other $F'^1\smile F^1$ by reorienting some of the $S(a)$ for vertices a.

We call the cohomology class $h^1=h^1(F^1)$ the *characteristic 1-class* of $S(K)$. If it vanishes, we call $S(K)$ *orientable*, otherwise *non-orientable*. Suppose $S(K)$ is orientable. Then we may reorient some of the $S(a)$ so that $F^1=0$. For each 1-cell ab, $S(a)$ and $S(b)$ now have either both the same or both the opposite orientations with the a- and b-coordinate systems as with the ab-coordinate system. In the latter case, change the orientation in the ab-coordinate system so that the orientations agree. Carrying out the same process in the 2-cells, and so on, we have finally chosen all coordinate systems so that for any point p of K in two cells σ and σ', their two coordinate systems give $S(p)$ the same orientation. This justifies the term "orientable."

To answer our original question, we shall show that *if $S(K)$ is orientable, and only then, it is simple over K^1*. If $h^1=0$, choose the ξ_a so that $F^1=0$ and set $\zeta(a)=\xi_a(a)$. Take any 1-cell ab. By hypothesis,

$$(6)\qquad \overset{-1}{\xi}_{ab}(a)\xi_a(a) \quad\text{and}\quad \overset{-1}{\xi}_{ab}(b)\xi_b(b)$$

* That is, let Ω denote a reflection in $S_0{}^r$, and set $\xi_a'(a)=\xi_a(a)\Omega$. Clearly F^1 is independent of the $\sigma_i{}^1$-coordinate systems.

are orthogonal transformations of S_0^ν into itself, that is, points $\psi(a)$ and $\psi(b)$ of the group $\bar{G}^{\nu+1}$ of orthogonal transformations in $(\nu+1)$-space. As $F^1(ab) \equiv 0 \pmod 2$, $\psi(a)$ and $\psi(b)$ are in the same component of $\bar{G}^{\nu+1}$ (of which there are two), and hence we may join them by an arc. Let $\psi(p)$ run along this arc as p runs along ab, and set

$$\zeta(p) = \xi_{ab}(p)\psi(p) \tag{7}$$

in ab. Thus ζ is defined in ab, and in the same manner, throughout K^1. Conversely, if ζ is defined in K^1, set $\xi_a(a) = \zeta(a)$ for each vertex a; this makes $F^1 = 0$.

Let A be any 1-cycle (mod 2) of K; then $A \cdot F^1$ *is a numerical invariant* (mod 2) of $S(K)$. For if we had used F'^1 instead of F^1, then $F'^1 - F^1 = \delta G^0$ for some G^0, and

$$A \cdot F'^1 - A \cdot F^1 = A \cdot \delta G^0 = \partial A \cdot G^0 \equiv 0 \pmod 2. \tag{8}$$

Any closed path in K defines a 1-cycle A; the orientation of $S(K)$ is "preserved" or "reversed" on going around the path according as $A \cdot F^1 \equiv 0$ or $1 \pmod 2$.

It is easily seen that the tangent space to a manifold is orientable if and only if the manifold is orientable; the orientation of $S(K)$ is reversed on going around a path if and only if the orientation of the manifold is reversed. For an M^n in an M^m, the 1-class of the tangent space of M^n plus that of the normal space gives that of the part of the tangent space of M^m over M^n (mod 2).* For an example, consider a Möbius strip in E^3.

In the future, unless the contrary is specifically stated, we consider only orientable sphere-spaces.

5. *Higher Dimensional Invariants.* One can find, for oriented sphere-spaces, a characteristic cohomology class h^r for $2 \leq r \leq \nu+1$, in the same manner as we defined orientability. We shall illustrate it with $\nu = 1$, $r = 2$; for instance, for the tangent space of S^2. As $S(K^1)$ is simple, we may define $\zeta(p, q)$ for p in K^1. Set $\phi(p) = \zeta(p, q_1)$; this is a projection of K^1 into $\mathfrak{S}(K^1)$. (For a general ν, we would use ν orthogonal projections in K^1.)

Now take any 2-cell σ of K, and consider the coordinate system ξ_σ in $\mathfrak{S}(\sigma)$. As p runs around $\partial\sigma$, $\psi(p) = \xi_\sigma(p, q_1)$ runs around a curve $C = \psi(\partial\sigma)$ on the torus $\mathfrak{S}(\partial\sigma)$ (see Fig. 2). Now consider

* Compare Seifert-Threlfall, *Topologie*, pp. 272–276.

$\phi(p)$. As p runs around $\partial\sigma$, it runs around a curve $D=\phi(\partial\sigma)$, cutting C in the positive sense (given by the orientation of $S(p)$) say α times; it is equivalent to running around C once, and α times around an $S(p)$. Set $F^2(\sigma)=\alpha$, and

$$F^2 = \sum_i F^2(\sigma_i^2)\sigma_i^2, \quad \text{coefficients integers.} \tag{9}$$

This chain is easily seen to be independent of the ξ_{σ^2} (the orientation of the $S(p)$ is preserved). Just as before, we show that

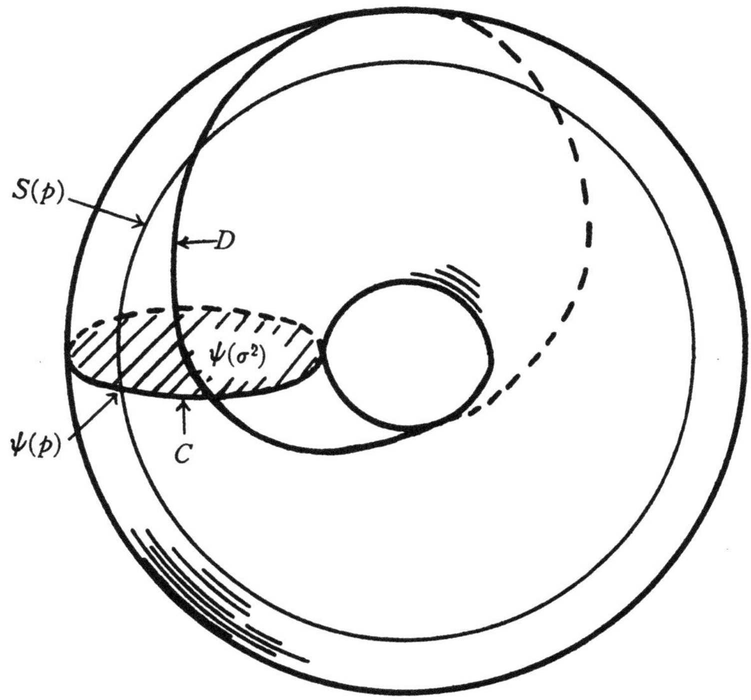

Fig. 2

it is a cocycle, and its cohomology class is an invariant of $S(K)$; also, we may obtain any $F'^2 \backsim F^2$.

For a general ν and r, we may always choose $\nu-r+2$ orthogonal projections of K^{r-1} into $\mathfrak{S}(K^{r-1})$; we then consider these maps in $\mathfrak{S}(\sigma^r)$, in relation to ξ_{σ^r}, to determine $F^r(\sigma^r)$ and thus F^r.

Suppose (see Stiefel [11], §6) we replace the first projection $\phi_1(p)$ used in defining F^r, for each p, by the diametrically opposite point of $S(p)$. It turns out that if r is odd, the new F'^r is $-F^r$. But $F'^r \backsim F^r$, hence $2F^r \backsim 0$, and $2h^r = 0$. We find the general theorem:

For each r, $2 \leqq r \leqq \nu+1$, there is a characteristic r-class h^r, which is an invariant of $S(K)$; the coefficients used are integers unless $r < \nu+1$ and r is even, in which case we reduce mod 2. *If r is odd, $2h^r = 0$; $S(K)$ is $(\nu - r + 2)$-simple over K^r if and only if $h^r = 0$.*

In regard to h^3, see §7.

For a non-orientable sphere-space, h^r is an invariant if we reduce mod 2 in all cases.*

As in §4, we can define a numerical invariant $A^r \cdot F^r$ corresponding to any cycle A^r. Further, if we do not reduce mod 2, then to any cycle A^r (mod μ), that is, such that $\partial A^r = \mu B^{r-1}$, corresponds the invariant (mod μ), $A^r \cdot F^r$; for if F'^r is another characteristic cocycle,

$$(10) \qquad A^r \cdot F'^r - A^r \cdot F^r = A^r \cdot \delta G^{r-1} = \mu B^{r-1} \cdot G^{r-1} \equiv 0 \pmod{\mu}.$$

We discuss the possibility of further invariants in Part II.

6. *Tangent and Normal Sphere-Spaces.* We shall consider (a) the highest dimensional invariant $F^{\nu+1}$ for tangent spaces, (b) the same for normal spaces, (c) relations between the spaces, and (d) other questions.

(a) In a closed orientable M^n, there is a "fundamental n-cycle" Z^n, whose multiples give all n-cycles. A single n-cell σ^n forms a cocycle, whose multiples $k\sigma^n$ determine all cohomology classes.† Any n-chain A^n is $\backsim k\sigma^n$ with $k = A^n \cdot Z^n$ = sum of coefficients of A^n. Hence, to find the characteristic n-class of $S(M^n)$, we need merely find $Z^n \cdot F^n$.

Let us do this for the tangent space of the 2-sphere S^2. The equator E cuts S^2 into the hemispheres σ_N and σ_S. Orient these so that $\partial\sigma_N = -\partial\sigma_S = E$; then $Z^2 = \sigma_N + \sigma_S$. To define F^2, we must choose a projection in $K^1 = E$, that is, a point $\phi(p)$ of $S(p)$ for each p in E. Let $\phi(p)$ be the direction at p pointing south. To find $F^2(\sigma_N)$, project σ_N stereographically onto the tangent plane

* Better, the theorem still holds if we replace chains in K by chains in an abstract complex K^* obtained as follows. The cells of K^* are those of K. Set ${}^r\partial_i'^j = \pm {}^r\partial_i^j$ (if the latter is $\neq 0$) according as the σ_i^r- and σ_j^{r-1}-coordinate systems determine the same or opposite orientations of $S(p)$ for p in σ_j^{r-1}.

† Compare MI, Appendix I.

P_N at the north pole N;* then the $\phi(p)$ go into directions $\phi'(p)$ pointing away from N. As we run around the projection of E, these directions make one complete turn in the same direction, that is, in the positive sense as given by σ_N. Hence $F^2(\sigma_N)=1$. Similarly, projecting onto the tangent plane at S, we find $F^2(\sigma_S)=1$. Hence

$$(11) \qquad Z^2 \cdot F^2 = F^2(\sigma_N) + F^2(\sigma_S) = 2 \neq 0.$$

It follows that a continuous field of directions on S^2 is impossible.

Similarly, for the n-sphere S^n, we find $Z^n \cdot F^n = 2$ if n is even and $=0$ if n is odd.† For any closed M^n (orientable or not), $Z^n \cdot F^n$ is its Euler-Poincaré characteristic χ.‡ By definition of the n-class, this equals the "index sum of singularities" of a vector field in M^n. As $2h^n=0$ for n odd, it follows that $2\chi=0$, $\chi=0$, for n odd, as is well known.

If $M^n=M^3$ is closed and orientable, then h^2 vanishes also, and it follows easily that the tangent sphere-space to any closed orientable 3-manifold is simple.§

(b) Consider first the normal space to a surface (say a Möbius strip) in E^3. Let K be a subdivision of the surface. To find F^1, we choose at each vertex a one of the two normals aa'. Suppose that for each 1-cell ab, we join the end points a' and b' of the respective normals by an arc which lies close to ab (see Fig. 1). Then clearly $F^1(ab)$ is the number of intersections (mod 2) of $a'b'$ with the surface. Thus F^1 may be found as follows. Deform the 1-dimensional part K^1 of K slightly so that the new positions a' of the vertices a are off the surface; then $F^1(ab)$ is the "Kronecker index" (mod 2) of the deformed $a'b'$ with the surface.

More generally, consider an M^n in an M^m with $m \leqq 2n$; if $m>n+1$, we assume that both are orientable, so that the nor-

* Taking $S_0{}^1$ in P_N, we may define $\xi_{\sigma N}$ as follows. Any direction d at a point p of σ_N projects into a direction d' in P_N, and determines thus a point d^* of $S_0{}^1$; we set $\xi_{\sigma N}(p, d^*)=d$.

† It is not known whether or not the tangent space of S^n is simple for all odd n.

‡ See (for the orientable case) Stiefel [11], §5, or Alexandroff-Hopf, *Topologie* I, p. 549. The associated complex K^* of M^n has a fundamental n-cycle even if M^n is non-orientable.

§ See Stiefel [11], §5.

mal space is orientable. To find a characteristic cocycle F^{m-n}, take a subdivision K of M^n, and deform the $(m-n)$-dimensional part into a new position K'^{m-n} in M^m so that K'^{m-n-1} does not touch M^n. Then $F^{m-n}(\sigma^{m-n})$ is the Kronecker index $(\sigma^{m-n}\cdot M^n)$ of σ^{m-n} with M^n in M^m; the coefficients are integers unless $m=n+1$. For any chain A^{m-n}, $A^{m-n}\cdot F^{m-n}$ is the Kronecker index $(A^{m-n}\cdot M^n)$, as this is true if A^{m-n} is an $(m-n)$-cell.

As an application, suppose M^n is a closed orientable manifold in E^m. By joining each point of M^n to a fixed point in E^m, we construct a "singular chain" Y^{n+1} bounded by M^n (or more properly, by the fundamental n-cycle Z^n of M^n). Set $s=m-n$, and

$$G^{s-1} = (-1)^s\sum_j (\sigma_j^{s-1}\cdot Y^{n+1})\sigma_j^{s-1}. \tag{12}$$

Then, by a simple property of Kronecker indices (see for instance Lefschetz, *Topology*, page 169, (20)),

$$\begin{aligned}\delta G^{s-1}\cdot\sigma_i^s &= G^{s-1}\cdot\partial\sigma_i^s = (-1)^s\sum_j (\sigma_j^{s-1}\cdot Y^{n+1})^s\partial_i{}^j \\ &= (-1)^s(\partial\sigma_i^s\cdot Y^{n+1}) = (\sigma_i^s\cdot M^n) = F^s\cdot\sigma_i{}^s,\end{aligned}$$

and it follows that $F^{m-n}=\delta G^{m-n-1}$, and $h^{m-n}=0$.

Consider two particular cases. In $m=2n$, the normal space is 1-simple, and it follows that there is a continuous field of normals to any closed orientable M^n in E^{2n}.* If $m=n+2$, then the 2-class vanishes; as the $S(p)$ are 1-dimensional and the normal space is orientable, it follows easily that $S(M^n)$ is simple.†

We remark that any h^r may be studied for M^n in E^m, $(m>n+r)$, by projecting into E^{n+r}.‡ Further, using methods in [**14**], we may construct a closed orientable M^4 in E^9 containing an S^2 with $S^2\cdot F^2\equiv 1 \pmod 2$; thus lower dimensional invariants may be non-vanishing.

(c) For an S^2 in E^3 or in E^4, the tangent space has the invariant 2, while the normal space is simple. With this in mind, we shall look only for relations between the invariants of the two spaces after reducing mod 2. A complete relation for orien-

* The theorem is probably not true in the generality stated in [**14**], footnote 25.

† See Seifert [**10**], Satz 1, and Whitney [**15**], §8, (*d*).

‡ See Whitney [**15**], §9.

tability was given at the end of §4. For an orientable M^n in an orientable M^m, it may be shown that the same statement holds for the characteristic 2-classes, after reducing mod 2. This is probably not true for the non-orientable case. (Consider a projective plane in E^4.) For still higher dimensional invariants, no complete results are known. But if the normal or tangent space is simple, the duality theorem holds for any dimension.* It follows immediately, by results in (a) and (b), that a closed orientable manifold M^n with odd characteristic χ cannot be imbedded in E^{n+2}.† This is the case, for instance, for the complex projective plane, for which $\chi=3$.

(d) Consider two sphere-spaces $S^\mu(K)$ and $S^\nu(K)$ with the same base space K. There is a unique sphere-space $S^{\mu+\nu+1}(K)$, their *product*, in which the first two spaces may be imbedded so that for each p in K, $S^\mu(p)$ and $S^\nu(p)$ are orthogonal great spheres in $S^{\mu+\nu+1}(p)$. As an example, for M^n in M^m, the product of the tangent and normal spaces gives the part $S^{m-1}(M^n)$ of the tangent space of M^m over M^n. A sphere-space which may be expressed as a product is *reducible*. A simple sphere-space is *completely reducible*, that is, it is a product of $(\nu+1)$ 0-sphere-spaces; any even number of these may be taken as non-orientable, the rest being simple. A k-simple sphere-space is the product of k simple 0-sphere-spaces and another space. See also §9.

The normal space to a closed orientable M^n in E^{n+1} is of course simple. But, one may ask, in what manner does this space lie in the tangent space of E^{n+1}? The outward normal at each point p of M^n determines a point $\psi(p)$ of the unit sphere S_0^n in E^{n+1}; thus M^n is mapped by ψ into S_0^n. The degree of ψ is the *generalized curvatura integra* of M^n in E^{n+1}.‡ If n is even, it is independent of the imbedding of M^n in E^{n+1}, and equals one half the characteristic of M^n.

For another problem on the imbedding of one sphere-space in another, consider a manifold M^n with a boundary B in a manifold M^m. Suppose a vector distribution is given over B; can it be extended over M^n?§ We may also consider sets of vec-

* See Stiefel [11], §6, No. 2.

† See Seifert [10], Satz 2.

‡ See H. Hopf [2], for various results in the direction noted.

§ See Morse [5].

tors, and so on. Of course, in defining the invariants, we considered exactly this problem, with M^n replaced by a single cell σ. The general problem of discussing how one sphere-space lies in another is almost untouched.

III. General properties of sphere-spaces

7. *On the Classification of Sphere-Spaces.* We shall now study in how far the invariants characterize a sphere-space with a given base space. But first we must clear up a matter of definition. We have allowed changes of coordinate systems in a sphere-space, considering the space to be unchanged thereby. We shall say that two sphere-spaces $S(K)$ and $S'(K)$ with the same base space K are *equivalent* if there is a continuous function $f(p)$ in K, such that, for each p, $f(p)$ is an orthogonal map of $S(p)$ into $S'(p)$. In other words, there is a homeomorphism between $\mathfrak{S}(K)$ and $\mathfrak{S}'(K)$ in which $S(p)$ and $S'(p)$ correspond for each p, the homeomorphism being orthogonal there. Then if we identify corresponding points of corresponding spheres, one sphere-space is obtained from the other merely by changing coordinate systems. If $S(K)$ and $S'(K)$ are both oriented, and there is an $f(p)$ which preserves orientations, we call them *positively equivalent.*

If K is an open or closed cell, then its cohomology groups vanish, so there can be no invariants as above described. But also any $S(K)$ is necessarily simple. It will turn out that the invariants characterize sphere-spaces also whenever $\nu=0$ or 1, and whenever dim $(K)\leqq 3$.

If $\nu=0$, there is just one type of $S(K)$ for each 1-cohomology class of K with coefficients mod 2. Suppose now that $\nu=1$ and $S(K)$ is oriented; if not, we use the methods in the footnote to §5. To show that $S(K)$ is characterized by its 2-class, suppose that $S'(K)$ has the same 2-class; we must prove $S(K)$ and $S'(K)$ to be positively equivalent, that is, we must find the map f of $S(K)$ into $S'(K)$, preserving the orientation of the $S(p)$. We first choose projections $\phi_1(p)$ and $\phi_1'(p)$ of K^1 into $\mathfrak{S}(K^1)$ and $\mathfrak{S}'(K^1)$ so as to determine the same $F^2=F'^2$. As $S(K)$ is oriented, we may choose a projection $\phi_2(p)$ of K^1 into $\mathfrak{S}(K^1)$ orthogonal to the first, so that, for each p, the ordered pair $\phi_1(p)$, $\phi_2(p)$ determines the positive orientation of $S(p)$. Do the same for $S'(p)$. These projections determine coordinate systems $\zeta(p)$ and $\zeta'(p)$

throughout K^1 with $\zeta(p, q_i) = \phi_i(p)$, $(i=1, 2)$, and similarly for ζ'; set

$$f(p) = \zeta'(p)\zeta^{-1}(p) \quad \text{in} \quad K^1. \tag{13}$$

Now take any 2-cell σ. For each p in $\partial\sigma$, set

$$\psi(p) = \xi_\sigma'^{-1}(p)f(p)\xi_\sigma(p); \tag{14}$$

this maps S_0^1 into $S(p)$, then into $S'(p)$, and then back into S_0^1, and is a rotation. Thus $\psi(p)$ is a point of the group G^2 of rotations of S_0^1 into itself; topologically, G^2 is a circle. It is easily seen that $F'^2(\sigma) = F^2(\sigma)$ implies that the degree of ψ is 0.* Hence we may extend ψ throughout σ so that it is continuous, and define f there by inverting (14):

$$f(p) = \xi_\sigma'(p)\psi(p)\xi_\sigma^{-1}(p). \tag{15}$$

Thus f is extended throughout K^2. The boundary of a 3-cell is a 2-sphere, and any map of a 2-sphere into a circle G^2 may be shrunk to a point;† hence the same process extends f throughout K^3, and so on.

Next take any ν, and suppose dim $(K) \leqq 3$. The above proof applies again, with G^2 replaced by $G^{\nu+1}$. That f may be extended through each σ^3 follows from the fact that any map of a 2-sphere into any $G^{\nu+1}$ may be shrunk to a point.‡ What has happened to the 3-class, which should differentiate further between sphere-spaces? The above proof shows that it can play no role. In fact, given the 2-class, the sphere-space (over K^3) is completely determined, and hence so is the 3-class.§

In the next simplest case, $\nu=2$ and dim $(K)=4$, the invariants are insufficient. To show this, note that for $K=S^4$, the cohomol-

---

* Define C and D in $\mathfrak{S}(\partial\sigma)$, using ξ_σ and ζ, as in §5, and define C' and D' similarly. Under f, D goes into D', while C goes into C^* say. As $F^1(\sigma) = F'^1(\sigma)$, C cuts D, and hence C^* cuts D', the same algebraic number of times that C' cuts D'. Hence C^* cuts C' zero times algebraically, $\psi(p, q_1)$ cuts q_1 in S_0^1 zero times algebraically, and the degree of ψ is 0.

† See for instance Alexandroff-Hopf, *Topologie* I, p. 516, Satz II.

‡ This does not seem to have been published yet. It has been proved independently by E. Cartan, Ch. Ehresmann, and myself.

§ The exact relation between h^2 and h^3 is as follows. As F^2 is a cocycle (mod 2), if we choose G^2 with integer coefficients so that reducing mod 2 gives F^2, then $\partial G^2 = 2F^3$ for some F^3; then F^3 is a characteristic 3-cocycle.

ogy groups of dimensions 1, 2, and 3 all vanish, so that there can be no invariants of the above sort. Now let σ_1 and σ_2 be the 4-cells into which S^4 is divided by a great 3-sphere τ. To define abstractly an $S(S^4)$, we suppose that coordinate systems ξ_{σ_1} and ξ_{σ_2} are given, and we shall define the relation between them, that is, the orthogonal maps $\phi(p)=\xi_{\sigma_2}^{-1}(p)\xi_{\sigma_1}(p)$ of S_0^2 into itself (p in τ). We must thus choose a map ϕ of the 3-sphere τ into the group G^3. As the latter is homeomorphic with projective 3-space, whose covering space is the 3-sphere, these maps may be chosen in an infinity of distinct ways, thus defining an infinity of non-equivalent sphere-spaces. To define a 4-dimensional invariant distinguishing between them, we must use, for instance, the "homotopy groups" of maps of a 3-sphere into a 2-sphere as coefficient group in the 4-chains considered.

To classify sphere-spaces with $\nu=2$ and dim $(K)=4$, more advanced methods must be used.*

8. *An Imbedding Theorem*. We shall show that all possible sphere-spaces can be generated from certain simple ones. If $S(K)$ is a sphere-space and K' is any complex (or point set), then each continuous map ϕ of K' into K generates a sphere-space $S'(K')$ as follows. For each p' in K', let $S'(p')$ be the sphere $S(\phi(p'))$.

The spaces $S[\nu, \mu]$ *of great* ν*-spheres in the* μ*-sphere* $S_0{}^\mu$† *form a set of universal sphere-spaces*; any $S^\nu(K)$ may be generated by mapping K suitably into the base space of $S[\nu, \mu]$, with $\mu=\nu+\dim(K)$.‡ We shall illustrate the proof for $\nu=1$. For each vertex a of K, let $\phi(a)$ be an arbitrary point of the base space of $S[1, \mu]$. Let $f(a)$ be an orthogonal map of $S(a)$ into the corre-

* Such methods have been found, but the classification has not been carried through. Similar methods were used in classifying the maps of a 3-complex into a 2-sphere; see Whitney, this Bulletin, vol. 42 (1936), p. 338.

† Note that the space $S[1, 2]$ of (non-oriented) great circles on $S_0{}^2$ is equivalent to the tangent space of P^2; the points on each circle give the directions from one of the corresponding poles. The base spaces of the $S[\nu, \mu]$ have been studied recently by C. Ehresmann, Journal de Mathématiques Pures et Appliquées, vol. 16 (1937), pp. 69–100. If we consider only oriented sphere-spaces, we may use the space of oriented great ν-spheres in S_0.

‡ If K is a compact metric space of finite dimension, we may easily prove the theorem by imbedding K in some E^2, defining a sphere-space with an open set in E^2 containing the image of K for base space, and applying the theorem to this space.

sponding sphere $S(\phi(a))$. We next extend ϕ and f over any 1-cell ab of K. If we define $f(p, q_1)$ and $f(p, q_2)$ for p in ab so that they are orthogonal points of S_0^μ for each p, and attach to the given values at a and b, then there is a unique orthogonal map $f(p)$ of $S(p)$ into a great circle of S_0^μ which takes on the values already given; this defines $\phi(p)$ also.

We may let $f(p, q_1)$ run along any curve on S_0^μ from $f(a, q_1)$ to $f(b, q_1)$ as p runs along ab. Now for each p in ab, there is a great $(\mu-1)$-sphere $S^{\mu-1}(p)$ of points on S_0^μ orthogonal to $f(p, q_1)$; $f(p, q_2)$ must lie on this sphere. These spheres form a sphere-space $S^{\mu-1}(ab)$, which is simple, as ab is a 1-cell (see §7). With the help of a coordinate system $\zeta(p, q)$ in it, it is easy to join $f(a, q_2)$ to $f(p, q_2)$ by the required arc, as $\mu-1\geqq 1$. The extension of f through cells of higher dimension is carried out in the same manner.

9. *On Simple Sphere-Spaces.* If $S(K)$ is simple, it can be expressed as a product $K\times S_0^\nu$, that is, a coordinate system $\zeta(p, q)$ may be defined for all p in K. In how many ways may we do this, preserving the orientation of the $S(p)$? To answer this, let ζ_0 be a fixed coordinate system. Then any other one, ζ, determines a rotation $\phi(p)=\zeta^{-1}(p)\zeta_0(p)$ of S_0^ν; there is thus defined a continuous map ϕ of K into $G^{\nu+1}$. Conversely, any ϕ determines a new coordinate system ζ from the given one ζ_0. Thus *the expressions of $S(K)$ as a product, preserving orientation, correspond to the maps of K into the group $G^{\nu+1}$ of rotations in $(\nu+1)$-space.*

For an example, take $\nu=1$, K a circle; $\mathfrak{S}(K)$ is a torus. G^2 is a circle, and there are an infinity of non-homotopic maps and corresponding coordinate systems. For $\nu=1$ and any K, the essentially distinct expressions of $S(K)$ as a product correspond to the elements of the 1-dimensional cohomology group of K with integer coefficients.

IV. r-FUNCTIONS IN A DIFFERENTIABLE MANIFOLD

10. *An Imbedding Problem; Vector Functions.* We begin by considering the problem of imbedding a differentiable manifold M^n in euclidean space. Suppose f is a (continuously) differentiable map of M^n into E^m. Then $f(p)$ has m coordinates $f_1(p), \cdots, f_m(p)$, and if we use a coordinate system $(x_1, \cdots, x_n)$

in a region of M^n, each $f_j(p)=f_j(x_1, \cdots, x_n)$ is differentiable. The ith contravariant coordinate vector at p is mapped into a contravariant vector in E^m with components $\partial f_1/\partial x_i, \cdots, \partial f_m/\partial x_i$. If these n vectors, $(i=1, \cdots, n)$, are independent, then clearly a neighborhood of p goes into E^m in a (1-1) way. If this holds for each p in M^n, we call the map *regular*. It is regular at p if the matrix

$$\left\| \begin{matrix} \dfrac{\partial f_1}{\partial x_1} \cdots \dfrac{\partial f_m}{\partial x_1} \\ \cdots\cdots \\ \dfrac{\partial f_1}{\partial x_n} \cdots \dfrac{\partial f_m}{\partial x_n} \end{matrix} \right\|, \qquad \begin{matrix} i\text{th row} = \text{map of } i\text{th coordinate vector}, \\ \\ j\text{th column} = \text{gradient } \nabla f_j \text{ of } f_j, \end{matrix}$$

is of rank n. The jth gradient ∇f_j, a covariant vector in M^n, may be considered as the "dual map" into M^n of the jth covariant coordinate vector of E^m. *A regular map is always possible if* $m \geqq 2n$ (see [**14**], Theorem 3). To study the question for $m<2n$, perhaps the best method is to look for m covariant vector functions in M^n such that (a) they are gradients, and (b) at each point of M^n, n of them are independent.*

We are thus led to the problem, when is a given (covariant) vector function v a gradient? As is well known, v is a gradient if and only if its integral $\int_C v = \int_C \sum v_i dx_i$ around every closed curve C in M^n vanishes. Now let us ask, when is a vector function such that integrating around any curve gives the same result as integrating around any sufficiently nearby curve? The answer is, if and only if the integral around any sufficiently small curve is 0; or, the function is *exact*, that is, all numbers $w_{ij}=\partial v_i/\partial x_j - \partial v_j/\partial x_i$ in any coordinate system in M^n vanish. In any simply connected region, a vector function is a gradient if and only if it is exact; but this does not hold in general over the whole manifold. For the relation of these functions to chains in a subdivision of the manifold, and generalization to higher dimensions, see the next section.

For a thorough study of gradients ∇f and their singularities

* If we use [**14**], Theorem 4, we need not distinguish between contravariant and covariant vectors.

(critical points of f), we refer to the work of Morse and others.* Almost no study has been made yet of sets of gradients.†

11. *r-Functions in a Manifold.*‡ By an *r-function* in a differentiable manifold M^n we shall mean a differentiable§ alternating covariant tensor α of order r. We may find the integral $\int_\sigma \alpha$ of α over an r-cell σ; then we define the integral over an r-chain by

$$(16) \qquad \int_{\sum a_i\sigma_i^r} \alpha = \sum a_i \int_{\sigma_i^r} \alpha.$$

Corresponding to any r-function α there is a *derived* $(r+1)$-function α; α is *exact* if $\alpha' = 0$. A 0-function is a scalar function; as its derived is its gradient, it is exact if and only if it is a constant (at least if M^n is connected). For a vector function, that is, a 1-function, with components v_i, the derived has the components w_{ij} as given above. The function α is *derived* if it is the derived of an $(r-1)$-function β. As $(\alpha')' = 0$, any derived function is exact. By Stokes's theorem, for any r-function α and any $(r+1)$-chain A,

$$(17) \qquad \int_{\partial A} \alpha = \int_A \alpha';$$

for this holds for any σ^r. This corresponds to (4).

Take any subdivision K of M^n. To each r-function α corresponds an r-chain, with *real numbers* for coefficients, defined by

$$(18) \qquad \phi(\alpha; \phi_i)^r = \int_{\sigma_i^r} \alpha, \qquad \phi(\alpha) = \sum_i \phi(\alpha; \sigma_i^r)\sigma_i^r;$$

then $\int_A \alpha = A \cdot \phi(\alpha)$. By (7) and (4), for any $\sigma = \sigma^{r+1}$,

$$\phi(\alpha') \cdot \sigma = \int_\sigma \alpha' = \int_{\partial\sigma} \alpha = \phi(\alpha) \cdot \partial\sigma = \delta\phi(\alpha) \cdot \sigma,$$

and hence, as an equivalent of Stokes's theorem,

$$(19) \qquad \phi(\alpha') = \delta\phi(\alpha).$$

* See for instance M. Morse, *The Calculus of Variations*, Ch. VI.

† See A. E. Pitcher, this Bulletin, vol. 43 (1937), p. 167.

‡ Compare de Rham [6], [7], and [8], and, for non-continuous functions with general coefficient groups, J. W. Alexander, Proceedings of the National Academy of Sciences, vol. 21 (1935), pp. 511–512.

§ The theory holds if we use tensors with continuous partial derivatives of order k, that is, of class C^k, the derived tensors being of class C^{k-1}.

From this formula it is apparent that *if α is exact, then $\phi(\alpha)$ is a cocycle, and if $\beta=\alpha'$ is derived, then $\phi(\beta)$ is a coboundary.*

We shall need an existence lemma: *If α is an exact r-function, and B is an $(r-1)$-chain such that $\delta B=\phi(\alpha)$, then there is an $(r-1)$-function β such that $\phi(\beta)=B$ and $\beta'=\alpha$.* We define β over K cell by cell, using an extended form of a lemma of de Rham* and a theorem on the extension of differentiable functions.† From this we deduce two facts: (a) If α is exact and $\phi(\alpha)$ is a coboundary, then α is derived. For B exists. (b) For any cocycle B there exists an exact function β with $\phi(\beta)=B$. For we may use $\alpha\equiv 0$.

We now prove the fundamental theorem: *If we identify any two exact r-functions whose difference is derived, we obtain the rth cohomology group of K with real coefficients.* We map each class of exact functions α into the cohomology class of $\phi(\alpha)$; this is easily seen to be a well defined homeomorphism. That it is actually an isomorphism follows at once from (a) and (b) above.

Each cocycle A may be considered as a linear function L defined on cycles, with $L(\partial B)=0$, if we set $L(B)=A\cdot B$. The group described is isomorphic with the set of all such functions. Hence, if $C_1, \cdots, C_p$ form a maximal set of independent cycles (p is the rth Betti number of the manifold), the integrals $\int_{C_i}\alpha$ may be chosen arbitrarily, α being determined up to derived functions. These integrals are called the *periods* of α.

12. *On Products of Functions.*‡ Given an r-function α and an s-function β, we define as usual their (alternating) product, giving an $(r+s)$-function $\alpha\beta$. We recall that

$$(\alpha\beta)' = \alpha'\beta + (-1)^r\alpha\beta'. \tag{20}$$

Also, chains of dimensions r and s may be multiplied, giving a chain of dimension $r+s$; if certain simple conditions are satisfied, including

$$\delta(A \cup B) = \delta A \cup B + (-1)^r A \cup \delta B, \tag{21}$$

* See de Rham [6], §21, Lemma II; we do not need his Lemma III in our method of proof.

† Whitney, Transactions of this Society, vol. 36 (1934), pp. 63–89, Lemma 2.

‡ Compare de Rham [6], §27, and [7]. The proof below clearly holds for both orientable and non-orientable manifolds.

then there is a uniquely defined corresponding product in the cohomology groups.* We wish to show that these products correspond.

First, it is possible to construct† an r-function $\omega(\sigma^r)$ corresponding to each r-cell σ^r (for all r), such that $\omega(\sigma^r)$ vanishes outside the star of σ^r, and setting $\omega(\sum a_i\sigma_i^r)=\sum a_i\omega(\sigma_i^r)$, we have

$$(22) \qquad \omega'(A) = \omega(\delta A), \qquad \phi(\omega(A)) = A, \qquad (\text{all } A).$$

Also, if I is the sum of the vertices of K, then $\omega(I)\equiv 1$.

Now to each r-chain A and s-chain B we construct a corresponding $(r+s)$-chain

$$(23) \qquad A \cup B = \phi[\omega(A)\omega(B)].$$

Applying (19) and (20), we derive (21). Also $I \cup B=B$. Noting also a certain local condition, we see that it follows that the product thus defined has the required properties. We now prove easily in turn: (a) If α is exact, then so is $\omega(\phi(\alpha))$, and $\phi[\alpha-\omega(\phi(\alpha))]=0$, so that $\alpha-\omega(\phi(\alpha))$ is derived. (Use (a) of §11.) (b) If β is exact also, then $\alpha\beta-\omega(\phi(\alpha))\beta$, and also $\alpha\beta-\omega(\phi(\alpha))\omega(\phi(\beta))$, are derived. (Use (20).) Finally,

$$(24) \qquad \phi(\alpha\beta) \backsim \phi[\omega(\phi(\alpha))\omega(\phi(\beta))] = \phi(\alpha) \cup \phi(\beta).$$

Hence *the products of functions and the products of cocycles determine the same products in the cohomology groups.*

BIBLIOGRAPHY

We give here only those papers most closely connected with the present paper.

1. Hadamard, J., *Sur quelques applications de l'indice de Kronecker*, Appendix to J. Tannery, *Théorie des Fonctions*, Paris, 1910.

2. Hopf, H., *Über die Curvatura Integra geschlossener Hyperflächen*, Mathematische Annalen, vol. 95 (1925), pp. 340–367.

3. Hopf, H., *Vektorfelder in n-dimensionalen Mannigfaltigkeiten*, Mathematische Annalen, vol. 96 (1927), pp. 225–250.

4. Hotelling, H., *Three dimensional manifolds of states of motions*, Transactions of this Society, vol. 27 (1925), pp. 329–344.

* See Whitney, PC, §11; also Proceedings of the National Academy of Sciences, vol. 23 (1937), pp. 285–291, Theorem 4.

† See de Rham [6], §24.

5. Morse, M., *Singular points of vector fields under general boundary conditions*, American Journal of Mathematics, vol. 51 (1929), pp. 165–178.

6. de Rham, G., *Sur l'analysis situs des variétés à n dimensions*, Journal de Mathématiques Pures et Appliquées, (10), vol. 9 (1931), pp. 115–200.

7. de Rham, G., *Sur la théorie des intersections et les intégrales multiples*, Commentarii Mathematici Helvetici, vol. 4 (1932), pp. 151–157.

8. de Rham, G., *Relations entre la topologie et la théorie des intégrales multiples*, l'Enseignement Mathématique, vol. 35 (1935), pp. 213–228.

9. Seifert, H., *Topologie dreidimensionaler gefaserter Räume*, Acta Mathematica, vol. 60 (1932), pp. 147–238.

10. Seifert, H., *Algebraische Approximation von Mannigfaltigkeiten*, Mathematische Zeitschrift, vol. 41 (1936), pp. 1–17.

11. Stiefel, E., *Richtungsfelder und Fernparallelismus in n-dimensionalen Mannigfaltigkeiten*, Commentarii Mathematici Helvetici, vol. 8 (1936), pp. 3–51.

12. Threlfall, W., *Räume aus Linienelementen*, Jahresbericht der Deutschen Mathematiker Vereinigung, vol. 42 (1932), pp. 88–110.

13. Wazewski, W., *Sur les matrices dont les éléments sont des fonctions continues*, Compositio Mathematica, vol. 2 (1935), pp. 63–68.

14. Whitney, H., *Differentiable manifolds*, Annals of Mathematics, vol. 37 (1936), pp. 645–680.

15. Whitney, H., *Sphere-spaces*, Proceedings of the National Academy of Sciences, vol. 21 (1935), pp. 64–68.

The work of Hopf, Rinow, Myers, and Cohn-Vossen on the relations between the topology and differential geometry of analytic Riemannian manifolds should also be noted.

HARVARD UNIVERSITY

ON THE THEORY OF SPHERE-BUNDLES

By Hassler Whitney

Department of Mathematics, Harvard University

Communicated January 13, 1940

1. *Introduction.*—We give here a brief sketch of some new results in the theory of sphere-bundles;[1] in particular, further properties of the characteristic classes, a duality theorem, theorems on tangent and normal bundles to a manifold, and some examples. The results will be published later in book form.

2. *Fibre-Bundles.*—Let S_0 be a space, and G, a group of homeomorphisms of S_0 into itself. Then over any space K, the *base space*, with neighborhoods U_i, we may define a *fibre-bundle* $\mathfrak{B}(K)$ as follows. For $p \in U_i$, let $\xi_i(p)$ be a homeomorphism of S_0 into a set of points $S(p)$. Let $S(p)\cdot S(q) = 0$ if $p \neq q$. Let $\mathfrak{S}(K)$ be the space of all points on all $S(p)$, the *total space* (gefaserte Raum). Let $\xi_i(p, q)$ be the image of q in S_0 under $\xi_i(p)$. For $p \in U_i\cdot U_j$, set $\xi_{ij}(p, q) = \xi_i^{-1}(p, \xi_j(p, q))$; assume $\xi_{ij}(p) \in G$ for each p, and that it varies continuously with p. Then a topology is easily defined in $\mathfrak{S}$. The part $\mathfrak{S}(U_i)$ of $\mathfrak{S}$ over U_i is a product $U_i \times S_0$.

If S_0 is a set of μ points, and G is the group of permutations, we obtain the covering spaces of K with μ sheets. If S_0 is a subgroup of a continuous group R_0, and $G = S_0$, then the left (or right) cosets of S_0 form a space K (factor group if S_0 is normal); the total space is R_0. If $S_0 = S_0^\nu$ is a ν-sphere, and $G = G^{\nu+1}$, the orthogonal group, we have a *sphere-bundle*. If S_0 is a vector space, and G, the affine or orthogonal group, an equivalent theory is obtained.

3. *Particular Coördinate Systems.*—We use this section in the proof of the duality theorem. Let K be a complex with ordered vertices. We may use ξ_σ, defined over closed cells σ. (See TP.) We may choose them so $\xi_\sigma = \xi_{\sigma'}$ if σ and σ' have the same first vertex. Let P^r be a small closed

region in K surrounding all $D_a(\sigma^r, \sigma')$, σ' of any dimension (see I, §2); set $Q^r = \overline{K - P^r}$. Then if K^r is the r-dimensional part of K, and $e_1, \ldots, e_{\nu+1}$ are the unit points of S_0^ν,

$$\xi_{\sigma,\sigma'}(p,e_i) = e_i \qquad (p \in Q^k\cdot\sigma\cdot\sigma', i = 1, \ldots, \nu - k).$$

Note that $\phi_i(p) = \xi_{\sigma^{r-1}}(p, e_i)$ ($p \in$ any σ^{r-1}, $i = 1, \ldots, \nu - r + 2$) defines orthogonal projections of K^{r-1} into $\mathfrak{S}$.

4. *Characteristic Classes.*—Choose $\phi_1, \ldots, \phi_{\nu-r+2}$ as above over K^{r-1}; then for each σ^r, studying these on $\partial\sigma^r$ gives $W^r\cdot\sigma^r$, which is an integer mod 2 if $r = 1$ or $r \leqq \nu$ is even, and an integer otherwise. W^r is a cocycle whose class $\mathbf{W}^r$ is an invariant of $\mathfrak{B}$; the $\mathbf{W}^r$ characterize $\mathfrak{B}$ if $\nu \leqq 1$ or dim $(K) \leqq 3$ (see TP). We may use a general type of subdivision of the polyhedron K in defining the $\mathbf{W}^r$.

If $\mathfrak{B}$ is not orientable (TP, §4), and K' is obtained from K by replacing $[\sigma^r: \sigma^{r+1}]$ by $-[\sigma^r:\sigma^{r+1}]$ when ξ_{σ^r} and $\xi_{\sigma^{r+1}}$ give opposite orientations to the $S(p)$ ($p \in \sigma^r$) (see TP, p. 793, footnote), then K' is locally isomorphic with K, and $\mathbf{W}' = \mathbf{W}^1$. We call K' the *complex associated with* $\mathfrak{B}$. The characteristic classes are taken in K'; the theorems above hold still.

If f maps K_1 into K_2, and $\mathfrak{B}_2(K_2)$ is defined, then a bundle $\mathfrak{B}_1(K_1)$ is defined (TP, §8), and $\mathbf{W}_1^r = f'\mathbf{W}_2^r$ (f' = dual of f).

If $\nu = 2$, $\dim(K) = 4$, and $\mathbf{W}^1 = 0$, $\mathbf{W}^2 = 0$, then an invariant characterizing $\mathfrak{B}$ is obtained as follows. A triple $\phi = (\phi_1, \phi_2, \phi_3)$ of orthogonal projections of K^3 into $\mathfrak{S}(K^3)$ exists. Let $\Phi(p)$ ($p \in K^3$) map S_0 into $S(p)$ so that $\Phi(p, e_i) = e_i$ ($i = 1, 2, 3$). For each σ^4, set

$$\Psi_{\sigma^4}(p) = \xi_{\sigma^4}^{-1}(p)\Phi(p) \ (p \in \partial\sigma^4).$$

This maps $\partial\sigma^4$ into the orthogonal group G^3; as G^3 is homeomorphic with projective 3-space P^3, this defines an integer $D^4\cdot\sigma^4$, the degree of Ψ_{σ^4}. D^4 is a cocycle. If we identify two cocycles if they are cohomologous, or differ by a cocycle of the form $X^1 \smile X^1 \smile X^1 \smile X^1$ (X^1 a 1-cocycle), the class determined is the invariant.

The classes W^{2r+1} are determined from the others as follows (see I, §11):

$$W^{2r+1} = \tfrac{1}{2}\partial\omega W^{2r} \qquad (\text{if } \nu \geqq 2r).$$

5. *On Mappings into* $G^{\nu+1}$.—In the theorem just stated, and in the duality theorem, we need the following (and other more complicated) theorems. (*a*) Let f map σ^ν into $G^{\nu+1}$ so that if $\phi(p) = f(p,e_1)$, then $\phi(p) = e_1$ in $\partial\sigma^\nu$; let ϕ be of degree α. Let ψ map σ^ν into S_0^ν with the degree β, and let $\psi(p) = e_1(p \in \partial\sigma^\nu)$. Then $\theta(p) = f(p, \psi(p))$ is of degree $\alpha + \beta$. (*b*) Take f as before; then $\phi'(p) = f^{-1}(p, e_1)$ (the point of S_0 mapped into e_1 by $f(p)$) is of degree $-\alpha$. (Use (*a*).) (*c*) Let f map σ^ν into $G^{\nu+1}$, let ϕ map $\partial\sigma^\nu$ into $S_0^{\nu-1}$ with the degree α, and suppose $f(p, \phi(p)) = e_1$ ($p \in$

$\partial\sigma^r$). Then $\psi(p) = f(p, e^{r+1})$ maps $\partial\sigma^r$ into the S_1^{r-1} orthogonal to e_1 with a degree $\equiv \alpha$ (mod 2).

6. *The Duality Theorem.*—Given bundles $\mathfrak{B}_1^\lambda(K)$ and $\mathfrak{B}_2^\mu(K)$, there is a uniquely determined bundle $\mathfrak{B}_3^\nu(K)$, their *product* ($\nu = \lambda + \mu + 1$; see TP, p. 796); thus if $M^m \subset M^n$, the tangent times the normal bundle gives the part of the tangent bundle of M^n over M^m. The formula for the characteristic classes of $\mathfrak{B}_3^\nu$ is

$$\mathbf{W}_3^r = \sum_i \mathbf{W}_1^i \smile \mathbf{W}_2^{r-i}, \text{ reducing mod 2 if necessary.}$$

(See §4 and I, §12; we use W^0 = sum of vertices.) The proof is very difficult if $r \geqq 4$. We use the special $\xi_{1,\sigma}$ of §3 in $\mathfrak{B}_1$, and $\xi_{2,\sigma}$ in $\mathfrak{B}_2$, with a replaced by $a' < a$, so the P_1 and P_2 will be in "general position" (see I, §5). The projections into $\mathfrak{S}_3(K^{r-1})$ are defined successively over $Q_1^0, Q_1^1, \ldots$. For each σ^r, they are now deformed in $\partial\sigma^r$ into a simpler position, except in each $A^i = P_1^i \cdot P_2^{r-1-i} \cdot \sigma^{r-1}$ (σ^{r-1} = face of σ^r opposite first vertex of σ^r). The terms shown come from the A^i, two coming from A^{r-1}. The results of §5 and the products of I, §6, are needed.

REMARK. We do not know whether or not the individual terms $\mathbf{W}_1^i \smile \mathbf{W}_2^{r-i}$ have topologicial significance.

Reducing everything mod 2, write, for any $\mathfrak{B}$, the formal power series

$$\mathbf{W} = \sum_i \mathbf{W}^i t^i, \quad \overline{\mathbf{W}} = 1/\mathbf{W} = \sum_i \overline{\mathbf{W}}^i t^i;$$

then

$$\overline{\mathbf{W}}^0 = I|_2, \overline{\mathbf{W}}^1 = \mathbf{W}^1, \overline{\mathbf{W}}^2 = \mathbf{W}^2 + \mathbf{W}^1 \smile \mathbf{W}^1,$$

etc. The duality theorem gives then, as $\mathbf{W}_N = \mathbf{W}/\mathbf{W}_T$, etc.,

$$\mathbf{W}_N^r = \sum \mathbf{W}^i \smile \overline{\mathbf{W}}_T^{r-i}, \quad \overline{\mathbf{W}}^r = \sum \overline{\mathbf{W}}_T^i \smile \overline{\mathbf{W}}_N^{r-i}, \text{ etc.} \quad (\text{mod } 2).$$

7. *Tangent Bundles.*—Let K be a simplicial subdivision of the manifold M^n, with ordered vertices. Each p in K may be written uniquely as $p = \sum \eta_{\lambda_i}(p) x_{\lambda_i}$, if $p \in x_{\lambda_0} \ldots x_{\lambda_r}$. Define

$$v_k(p) = \sum_{\lambda_0 < \ldots < \lambda_k} \eta_{\lambda_0}(p) \ldots \eta_{\lambda_k}(p)(x_{\lambda_k} - x_{\lambda_{k-1}}) \quad (k = 1, 2, \ldots).$$

These are continuous in K, and the first r are independent except in K^{r-1} (any r). If K^* is the usual complex dual to K, these may be used to define W^r, a cocycle in K^*. Its dual is a *characteristic cycle* C^{n-r} in K', the complex associated with $\mathfrak{B}$ (which was studied in I, §13). Note that $C^{n-(2r+1)} = \frac{1}{2}\partial\omega C^{n-2r}$. The value of $C^s \cdot \sigma^s$ ($s = n - r$, $\sigma^s = x_{\lambda_0} \ldots x_{\lambda_s}$) is as follows. Let K_1 be the subcomplex of the closed star of σ^s containing all vertices x_i with

$$\lambda_s > i > \lambda_{s-1} \text{ or } \lambda_{s-2} > i > \lambda_{s-3} \text{ or } \ldots.$$

(This includes all vertices below λ_0 if s is even.) Then $C^s \cdot \sigma^s = 1 - \chi(K_1)$ (χ = Euler-Poincaré characteristic), or this mod 2.

From this we prove: If K is the first derived of a simplicial subdivision of M, then C^s is the sum of all s-simplexes of K (properly oriented if integer coefficients are used).[2]

In the proofs of the following theorems, we study the classes over sub-manifolds of the given manifold, and use the duality theorem and results from §8. For $M^m \subset M^n$, let $\mathbf{W}^r$ mean the part of $\mathbf{W}_T^r(M^n)$ in M^m.

If a closed M^m can be imbedded in E^n (with or without singularities), then

$$0 = \mathbf{W}_N^{n-m}|_2 = \sum \mathbf{W}^i \smile \overline{\mathbf{W}}_T^{n-m-i} = \overline{W}_T^{n-m}.$$

Hence $\overline{W}_T^m = 0$ always. This gives, if $(X)^2 = X \smile X$, etc.,

closed M^2: $\mathbf{W}^2|_2 = (\mathbf{W}^1)^2$; closed M^3: $(\mathbf{W}^1)^3 = 0$;

closed M^4: $\mathbf{W}^4|_2 + (\mathbf{W}^2)^2 + \mathbf{W}^2 \smile (\mathbf{W}^1)^2 + (\mathbf{W}^1)^4 = 0$; etc.

In any M^2, for any 1-I_2-cocycle X^1, $X^1 \smile X^1 \frown X^1 \smile W^1$. In any M^3, $\mathbf{W}^2 = \mathbf{W}^1 \smile \mathbf{W}^1$; hence (Stiefel) for orientable M^3 (closed or not), the tangent bundle is simple. In any orientable M^4, $\mathbf{W}^3 = 0$. (The proof uses facts from §10.) For any 2-I_2-cocycle X^2 in any M^4, $X^2 \smile X^2 \frown X^2 \smile \overline{W}^2$. For any orientable M^4 in an orientable M^r, $\overline{W}_T^4 = W_N^4$.

8. *Normal Bundles.*—For any M^m in any M^n, $\mathbf{W}_N^{n-m}$ is the intersection of M^m with itself in M^n, which is a cohomology class of the complex associated with the normal bundle; if M^m is closed and $M^n = E^n$, then $\mathbf{W}_N^{n-m}|_2 = 0$, and if also M^m is orientable, then $\mathbf{W}_N^{n-m} = 0$. Compare PC, §20, and TP, p. 795.

If M^m is mapped regularly into M^n, but with singularities, we may deform M^m slightly into M'^m, and consider the intersections of the σ' with a neighborhood of σ' in M; then W^{n-m} is the *local intersection* thus defined. For a closed orientable $M^m \subset E^n$, the *intersection* vanishes, so that the *distant intersection* equals the local. This holds mod 2 in the non-orientable case. If $n = 2m$, and the singularities are isolated points, the distant intersection is of course 0 (mod 2); if m is odd, it vanishes, because $\{\sigma^m, \sigma'^m\} = -\{\sigma'^m, \sigma^m\}$.

Take an orientable $M^m \subset E^n$. Then the normal bundle is simple if $m = 1$ or 2, or $n = m + 1$ or $m + 2$, or $m = 3$ and M is closed, or M is a cell. This holds if M^m is merely mapped regularly, provided that if $n = m + 2$, then m is odd, and we omit $m = 2$ if M is closed.

9. *Examples.*—Consider a cylinder, the product $T^3 = T^1 \times T^2$ of a segment and a disc. Let P^2 be one end, let S_2^1 be a segment crossing P^2,

let Q^2 be a rectangle cutting through T^3 and ending on S_2^1, and let S_1^1 be the center of Q^2, ending at p_0, the center of S_2; rather, let these be the sets after the identifications below. (1) Join the ends of T^3, and shrink ∂T^2 to a point; this forms $M_1^3 = S^1 \times S^2$, with a simple tangent bundle. (2) Join the ends, and identify opposite points of ∂T^2; then characteristic cycles are (mod 2) $C^2 = Q^2$, $C^1 = S_1^1$; $\therefore \mathbf{W}^2 \neq 0$. (3) Join the ends, reflecting one so that P^2 is joined to itself with orientation reversed, and shrink ∂T^2 to a point. We obtain M_3^3, with $\mathbf{W}^2 = 0$. (4) Join the ends as in (3), and identify opposite points of ∂T^2, forming M_4^3. Now P^2 is a projective plane, and Q^2 is a Klein bottle. Intersections are (mod 2)

$$\{P^2, P^2\} \sim 0,\ \{Q^2, P^2\} \sim S_2^1,\ \{Q^2, Q^2\} \sim S_1^1 + S_2^1;$$

$$\{P^2, S_2^1\} \sim 0,\ \{P^2, S_1^1\} \sim p_0,\ \{Q^2, S_2^1\} \sim p_0,\ \{Q^2, S_1^1\} \sim 0;$$

characteristic classes are

$$C^2 \sim (P^2 + Q^2)_2,\ C^1 \sim (S_1^1 + S_2^1)_2.$$

Define $\mathfrak{B}_1^1(M_4^3)$, with $C_1^2 \sim P^2$, $C_1^1 \sim 0$, and $\mathfrak{B}_2^1(M_4^3)$, with $C_2^2 \sim Q^2$, $C_2^1 \sim 0$; let the total spaces be M_1^4, M_2^4. We may pretend M_4^3 is in either (because $C_i^1 = 0$). Then

$$C_1^3 \sim \mathfrak{S}(Q^2),\ C_1^2 \sim \mathfrak{S}(S_1^1),\ C_1^1 \sim \mathfrak{S}(p_0);\ \therefore \overline{C}_1^1 \sim 0;$$

$$C_2^3 \sim \mathfrak{S}(P^2),\ C_2^2 \sim \mathfrak{S}(S_2^1),\ C_2^1 \sim \mathfrak{S}(p_0);\ \therefore \overline{C}_2^1 \sim \mathfrak{S}(p_0);$$

hence in M_2^4, $\mathbf{W}^3 \neq 0$, $\overline{\mathbf{W}}^3 \neq 0$. Hence (see §7) M_2^4 cannot be imbedded in E^7.

Define $\mathfrak{B}^2(M_3^3)$, with $C^2 \sim P^2$, and $C^1 \sim S_1^1$. Then we may consider $M_3^3 \subset M^5 = \mathfrak{S}(M_3^3)$, and prove (*a*) M^5 is closed and orientable, (*b*) $C^2(M^5) \sim \mathfrak{S}(p)$; hence $\overline{\mathbf{W}}^3 = \mathbf{W}^3 \neq 0$, and M^5 cannot be imbedded in E^8.

We may define $M^8 = \mathfrak{S}(S_0^4)$, with $\overline{\mathbf{W}}^4 = \mathbf{W}^4 \neq 0$; hence M^8 cannot be imbedded in E^{12}.

The complex projective plane P^{*4} cannot be imbedded in E^6, as $\mathbf{W}^2 \neq 0$. ($\mathbf{W}^2 \smile \mathbf{W}^2 = \mathbf{W}^4|_2$; $W^4 \cdot P^{*4} = \chi(P^{*4}) = 3$.) But it can be in E^7.

For any closed orientable M^4, and any cocycles X^2, X^4 (mod 2) in M^4, we may imbed M^4 in an $M^8 \subset E^{17}$, so that the part over M^4 of the characteristic classes of the normal bundle of $M^8 \subset E^{17}$ are X^2 and X^4. Hence we may make $\mathbf{W}_N^3 \neq 0$ and $\mathbf{W}_N^4 \neq 0$ also.

If we put a Klein bottle Q^2 in E^3, then the distant and local intersections are equal (mod 2); these are $W_N^1 = W_T^1$; $C_T^1 =$ a closed curve in Q^2. Hence the distant intersection, as a cycle, is a certain curve in Q^2, as is clear in the usual immersion of Q^2 in E^3. For P^2 in E^3, we get the "projective line" similarly.

A direct study shows: If $M^2 \subset E^4$ is closed (with or without singularities), and C^2 is the fundamental cycle of the associated complex (integer coefficients), then $W^2 \cdot C^2 = 2[\chi(M^2) + 2k]$ for some k; any k may be obtained. Hence if $\chi \not\equiv 0 \pmod 2$, then a field of normal vectors never exists. Also one imbedding cannot be deformed into another with a different $W^2 \cdot C^2$.

10. *Homology Groups of Total Spaces, Etc.*—Let K be connected. Given $\mathfrak{B}(K)$, the homology groups satisfy $\mathbf{H}^r(\mathfrak{S}) \approx \mathbf{H}^r(K)$, $r < \nu$. $kS(p) \sim 0$ if and only if for some $A^{\nu+1}$, $A^{\nu+1} \cdot W^{\nu+1} = k$. (We may use $k_\lambda = k \bmod \lambda$, and $A^{\nu+1}|_\lambda$.) For $\mathfrak{B}$ oriented, a $(\nu + 1)$-I_μ-cycle A in K is the projection of such a cycle in $\mathfrak{S}$ if and only if $A \cdot W^{\nu+1} = 0_\lambda$. Now $\mathbf{H}^\nu(\mathfrak{S})$ may be described in terms of $\mathbf{W}^{\nu+1}$ and properties of $kS(p)$. If μ is the smallest integer such that $\mu S(p) \sim 0$, then $\mathbf{H}^\nu(\mathfrak{S}) \approx I_\mu \oplus \mathbf{H}^\nu(K)$ if and only if for each λ and each $(\nu + 1)$-I_λ-cycle A, $A \cdot W^{\nu+1} \equiv 0 \bmod (\lambda, \mu)$. A mapping f of a complex K' of dimension $\leqq \nu + 1$ into K is the projection of a mapping into $\mathfrak{S}$ if and only if $f'\mathbf{W}^{\nu+1} = 0$.

[1] We refer the reader to papers in these PROCEEDINGS, **21**, 464–468 (1935), and in *Bull. Am. Math. Soc.*, **43**, 785–805. We denote the latter by TP, and the preceding note, by I. Sphere-bundles were formerly called "sphere-spaces."

[2] This was a conjecture of Stiefel, *Comm. Math. Helv.*, **8**, 40 (1936).

ON THE TOPOLOGY OF DIFFERENTIABLE MANIFOLDS

By HASSLER WHITNEY

1. Introduction. This paper is concerned with problems about vector fields in manifolds and sets of vector fields; we may demand that the fields be tangent to the manifold or, if the manifold is in euclidean space, normal to the manifold.

In Part I we study the question of the existence of a field of vectors everywhere normal to a surface in 4-space E^4. It turns out that, although such a field always exists if the surface is orientable, this need not be the case for a non-orientable surface, for example, the projective plane. The proof requires but a minimum knowledge of topology on the part of the reader. In Part II the general theory of sphere bundles (formerly called "sphere spaces") is discussed. In particular, properties of the characteristic classes are studied. For further information see [3].[1] Part III gives applications of the theory to the tangent and normal bundles of a manifold—and hence to problems on vector fields. A formula which may be thought of as a duality theorem plays a fundamental rôle here. The characteristic classes of n-dimensional manifolds M^n with $n = 2, 3, 4$ are studied in particular. The classes of a certain M^3 and of a certain M^4 are determined; it is shown that the latter cannot be imbedded (see § 16 for definitions) in E^7.

PART I. NORMAL VECTOR FIELDS TO SURFACES IN 4-SPACE

2. The 2-dimensional invariant $W^2 \cdot M$. Consider first a surface[2] M in 3-space E^3. At each point $p \in M$ there are two unit

[1] Some further theorems will be found in [5]. The author expects to give a detailed account of the theory in book form.

[2] Only differentiable, or "smooth," manifolds will be considered. Thus M is supposed to have a continuously turning tangent plane.

normal vectors. At a point p_0 choose one of them, $v(p_0)$. We may choose $v(p)$ for all p near p_0 so that $v(p)$ is continuous. But can we so choose $v(p)$ over all of M? Clearly if and only if M is orientable (2-sided in E^3).[3] Thus, for the Möbius strip, if we choose $v(p_0)$ and carry p from p_0 along the strip back to p_0, a continuous $v(p)$, starting as $v(p_0)$, will come back into $-v(p_0)$; hence $v(p)$ cannot be made continuous over the strip.

Now suppose we have $M \subset E^4$. Then at each p there are a tangent plane $P_T(p)$ and a normal plane $P_N(p)$; a normal vector $v(p)$ may lie anywhere in $P_N(p)$. The end points of the unit normals at p form a 1-sphere (i.e. a circle) $S(p)$. Our problem may now be stated as follows: For each $p \varepsilon M$ is it possible to choose a point $\phi(p) \varepsilon S(p)$ so that ϕ is continuous in M? There is greater likelihood of a positive solution in this case than for $M \subset E^3$, since there is greater freedom of choice for the $v(p)$.

We shall attempt to construct v, or ϕ, over M as follows. Suppose M is triangulated into very small 2-cells, 1-cells, and vertices. Choose $\phi(x_i) \varepsilon S(x_i)$ arbitrarily at each vertex x_i. Now take any 1-cell $\sigma^1 = x_i x_j$. This cell is a small slightly curving line segment; the set of $S(p)$ for $p \varepsilon \sigma^1$ forms a cylinder $\mathfrak{S}(\sigma^1)$, slightly curved. If we let $\phi(p)$ run along the cylinder from $\phi(x_i)$ to $\phi(x_j)$ as p runs from x_i to x_j, so that $\phi(p) \varepsilon S(p)$, then ϕ will be properly defined over σ^1. Thus we define ϕ over all 1-cells.

Now take any 2-cell σ^2. Choose coördinates so that σ^2 is approximately in the (x_1, x_2)-plane; then each $S(p)$ is nearly parallel to the (x_3, x_4)-plane. Hence for each p we may determine the points of $S(p)$ by an angle θ, measuring from the x_3-axis toward the x_4-axis. We now have a $(1-1)$ correspondence between the pairs (p, θ) and the points $\xi(p, \theta)$ of $\mathfrak{S}(\sigma^2)$, where $\mathfrak{S}(\sigma^2)$ is the set[4] of all points on all $S(p)$, $p \varepsilon \sigma^2$.

[3] Compare Seifert-Threlfall, *Topologie* (Leipzig, 1934), pp. 272–276. See also [3], pp. 789–791, and formula (20. 4) of the present paper.

[4] Thus this set of points is expressed as the cartesian product of σ^2 and the values θ. Such a "coördinate system" in the $S(p)$ is studied more closely in Part II; see the $\xi_\sigma(p)$.

The values θ we may think of as being points of the unit circle S_0^1 in the (x_3, x_4)-plane. For each $p' \varepsilon S(p)$ let $\xi^{-1}(p, p')$ denote the point θ of S_0^1 such that $\xi(p, \theta) = p'$.

An orientation of E^4 is determined by choosing an ordered set of four independent vectors, w_1, w_2, w_3, w_4, for instance, the unit vectors along a fixed set of axes. We keep this orientation fixed from now on.

Select a definite orientation of σ^2 by choosing an ordered pair of independent vectors u_1, u_2 parallel to σ^2. Then there is a corresponding orientation of S_0^1, by vectors u_3, u_4, such that $u_1, \cdots, u_4$ determine the same orientation of E^4 as $w_1, \cdots, w_4$.

Since ϕ is defined over the boundary $\partial\sigma^2$ of σ^2 we may set

$$\psi(p) = \xi^{-1}(p, \phi(p)), \qquad p \varepsilon \partial\sigma^2. \tag{2.1}$$

This is a continuous mapping of $\partial\sigma^2$ into S_0^1, giving the "angle" of $\phi(p)$ in $S(p)$. Now $\partial\sigma^2$ is oriented by going from the direction of u_1 toward the direction of u_2. As p runs around $\partial\sigma^2$ once positively, $\psi(p)$ runs around S_0^1 a certain number of times,[5] which number we call $W(\sigma^2)$.

Suppose $W(\sigma^2) = 0$. Then it is not hard to see that it is possible to define $\psi(p)$ through σ^2 so that it is continuous. We may then set

$$\phi(p) = \xi(p, \psi(p)), \qquad p \varepsilon \sigma^2, \tag{2.2}$$

and thus extend $\phi(p)$ over σ^2. If, however, $W(\sigma^2) \neq 0$, then this is impossible. Of course, altering ϕ on 1-cells of M will change some of the $W(\sigma^2)$. Our problem is, then, *Can the* $\phi(p)$ *be chosen on the* 1-*cells so that all* $W(\sigma^2) = 0$? If so, and only if so, $\phi(p)$, and hence the normal vector field $v(p)$, can be defined all over M.

If the surface is open, or has a boundary, then it can be shown that $v(p)$ always exists.[6] Henceforth we consider only closed sur-

[5] This is the "degree" of ψ.

[6] In terms of the general theory the characteristic 2-class, determined by $W^2 \cdot M$ (see below), is a 2-cocycle; but in this case every cocycle is a coboundary. See Part II.

faces. It will turn out that $\sum W(\sigma_i^2)$ is independent of ϕ and is the only invariant; hence v exists if and only if $\sum W(\sigma_i^2) = 0$.

If we let W^2 be the chain whose coefficient on σ_i^2 is $W(\sigma_i^2)$, and let M be the chain $\sum \sigma_i^2$, then[7] (see §§ 9 and 15)

$$\sum W(\sigma_i^2) = W^2 \cdot M. \tag{2. 3}$$

3. The invariance of $W^2 \cdot M$. We shall show first that the integer $W(\sigma^2)$ is independent[8] of the orientation chosen for σ^2. For suppose the orientation of σ^2 had been chosen in the opposite manner, say by the ordered pair of vectors u_2, u_1. Then the orientation of S_0^1 would have been chosen in the opposite manner, for instance, by u_4, u_3, to obtain a set u_2, u_1, u_4, u_3 equivalent to $u_1, \cdots, u_4$. In defining $W(\sigma^2)$ we now let p run around $\partial\sigma^2$ in the opposite sense; but as the orientation of S_0^1 is reversed, the same number $W(\sigma^2)$ is determined.

We next consider the effect of altering ϕ to ϕ' on a 1-cell $\sigma^1 = x_i x_j$; let $\phi' = \phi$ elsewhere. There are two 2-cells with σ^1 as face; suppose these are $\sigma_1^2 = x_i x_j x_k$ and $\sigma_2^2 = x_i x_j x_l$; let them be oriented as shown. As p runs along $x_i x_j$, say $\phi'(p)$ runs around the cylinder $\mathfrak{S}(\sigma^1)$ from $\phi'(x_i) = \phi(x_i)$ to $\phi'(x_j) = \phi(x_j)$ a number α of times more than $\phi(p)$. Then, if a fixed circle S_0^1 is chosen as in § 2 and oriented like a circle of $\mathfrak{S}(\sigma^1)$, and ψ_1 and ψ_2 are defined in $\partial\sigma_1^2$ and $\partial\sigma_2^2$ as ψ was defined in § 2, and ψ_1' and ψ_2' are defined with ϕ replaced by ϕ', it is clear that $\psi_h'(p)$ runs around S_0^1 α times more than $\psi_h(p)$ $(h = 1, 2)$. But the chosen orientations of σ_1^2 and σ_2^2 give opposite orientations of the (x_1, x_2)-plane (see § 2); hence, in defining $W(\sigma_1^2)$ and $W(\sigma_2^2)$, S_0^1 is to be given opposite orientations. Therefore one of these $W(\sigma_h^2)$ is increased by α, and the other is decreased by α, so that $\sum W(\sigma_h^2)$ is unchanged.

[7] The chains must be considered chains in the complex associated with M^2 if M^2 is not orientable; see § 15.

[8] In a chain of an ordinary complex $W(-\sigma^2) = -W(\sigma^2)$. The present chain is in a complex K^* whose cells are the pairs (σ, θ), θ being an orientation of the $S(p)$, $p \in \sigma$. We have defined in reality $W(\sigma^2, \theta)$ rather than $W(\sigma^2)$; since $(-\sigma^2, -\theta) = (\sigma^2, \theta)$ (see § 15), $W(-\sigma^2, -\theta) = W(\sigma^2, \theta)$.

Thus altering ϕ on 1-cells has no effect on $W^2 \cdot M$. It is easily seen that the definition of ϕ on the vertices has no effect.[9] It can also be shown that the number $W^2 \cdot M$ comes out the same, no matter how M is subdivided into a complex. Hence it is a topological invariant of the imbedding of M in E^4; it changes sign with reversal of orientation of E^4. Finally, deforming M in E^4 does not alter $W^2 \cdot M$.

If $W^2 \cdot M = 0$, it is fairly simple to show that ϕ may be altered over the 1-cells, one after another, until a ϕ' is obtained, with $W_{\phi'}(\sigma_i^2) = 0$ (all i). Hence *for closed surfaces there exists a normal vector field to* M *in* E^4 *if and only if* $W^2 \cdot M = 0$.

4. Types of critical points. Take $M \subset E^4$, and let $p_1, \cdots, p_s$ be the points where the tangent plane is perpendicular to the x_4-direction; we may assume these are finite in number. (If not, a slight deformation of M will give this and the following properties.) Let $f_1(p), \cdots, f_4(p)$ be the coördinates of $p \in M$; set $c_i = f_4(p_i)$. We may assume that the c_i are distinct. Also, for a suitable parametrization (u, v) of the part of M near p_i, f_4 has the form

$$f_4(u, v) = \pm u^2 \pm v^2. \tag{4. 1}$$

If both signs are $+$ (or $-$), p_i is a relative minimum (or maximum) of f_4. If one is $+$ and one $-$, we have a saddle point.

Let $M(a, b)$ be the set of all p with $a \leqq f_4(p) \leqq b$. Let $M(a) = M(a, a)$. We may suppose, finally, that for $a \neq$ any c_i, $M(a)$ is a set of smooth closed curves, lying in $E^3(a)$ (the 3-space for which $f_4(p) = a$), and that $M(c_i)$ is a set of closed curves, together with a point p_i (in case of a minimum or a maximum), or a figure 8 through p_i (Fig. 1), or a pair of curves intersecting at right angles at p_i (Fig. 2).

Let c_i' and c_i'' be numbers close to c_i, $c_i' < c_i < c_i''$.

In case of a figure 8, $M(c_i')$ has one (or two) curves near the 8,

[9] If ϕ and ϕ' are defined, we may deform ϕ' into ϕ'', where $\phi'' = \phi$ on the vertices; clearly $W_{\phi''}^2 \cdot M = W_{\phi'}^2 \cdot M$. By the proof above $W_{\phi''}^2 \cdot M = W_{\phi}^2 \cdot M$.

while $M(c_i'')$ has two (or one); we then call p of type $1 \rightarrow 2$ (or $2 \rightarrow 1$). In the case of the two loops, $M(c_i')$ and $M(c_i'')$ have each a single curve near the loops; then p_i is of type $1 \rightarrow 1$. A minimum (maximum) is of type $0 \rightarrow 1$ $(1 \rightarrow 0)$. If p_i is of type $1 \rightarrow 1$ (and only then), $M(c_i', c_i'')$ consists of a pair of Möbius strips joining at p_i (Fig. 2).

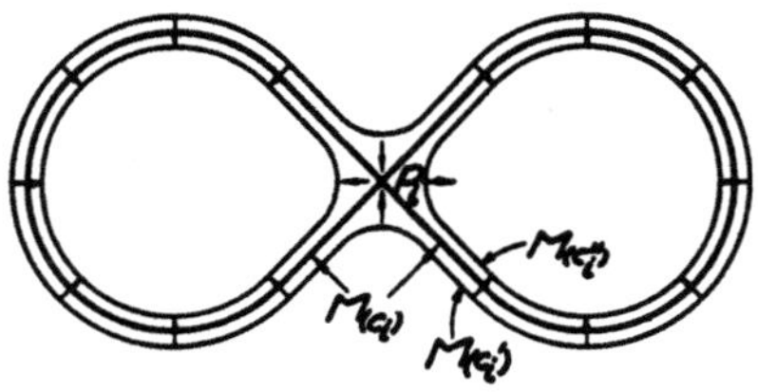

FIG. 1. Point p_i of type $1 \rightarrow 2$. Small arrows show direction of increase of f_4

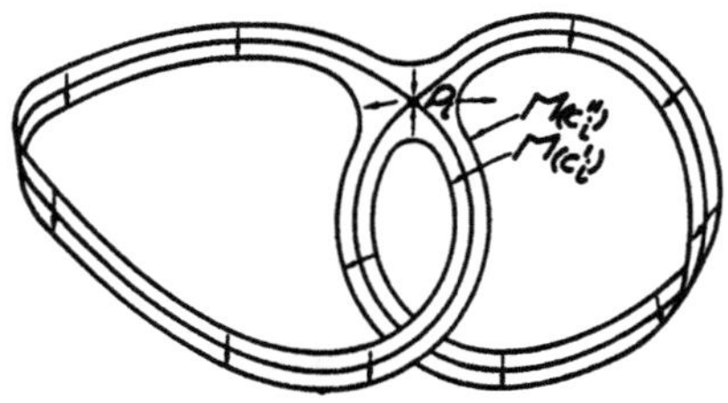

FIG. 2. Point p_i of type $1 \rightarrow 1$

5. A method of imbedding M in E^4. If we choose a manner of cutting a surface M into sets $M(a)$ as above, and choose a manner of placing each $M(a)$ in $E^3(a)$ so that the position varies continuously with a, we may thus define an imbedding of M in E^4.

For example, a 2-sphere S^2 may be cut into the north and south poles p_2 and p_1, and circles parallel to the equator. As a increases from c_1 to c_2, $M(a)$ is a curve springing out from p_1, and finally shrinking down rapidly to $p_2 = M(c_2)$.

Note that the part of $M(c_i')$ or $M(c_i'')$ (in E^4) near p_i is approximately an ellipse, or approximately a portion of a hyperbola. (See formula (4. 1).)

If a torus is "stood up on end" in E^4, so that $f_3(p) \equiv 0$, the $M(a)$ are as follows: $M(c_1) = p_1$; $M(c_1'')$ is approximately an ellipse; $M(c_2')$ is a closed curve, two points of which are near together (and near p_2): $M(c_2)$ is a figure 8; $M(c_2'')$ and $M(c_3')$ are pairs of curves; $M(c_3)$ is a figure 8; $M(c_3'')$ and $M(c_4')$ are curves; $M(c_4) = p_4$. The p_i are of types $0 \rightarrow 1$, $1 \rightarrow 2$, $2 \rightarrow 1$, $1 \rightarrow 0$.

For the Klein bottle B^2 we may take points p_i of the same types; but we turn over one of the curves of $M(a)(c_2 < a < c_3)$

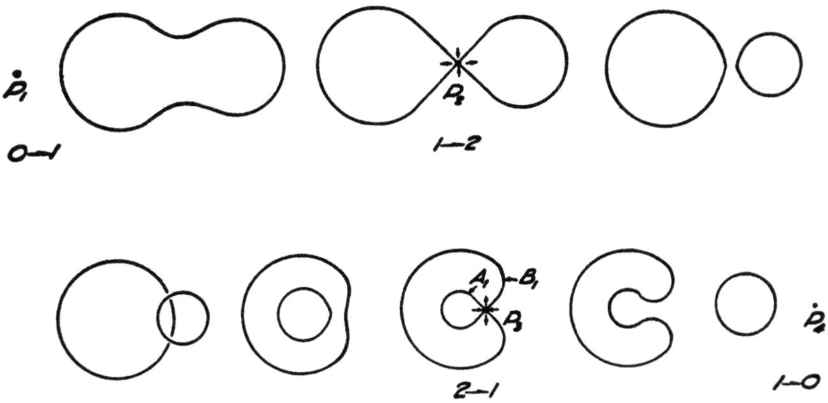

FIG. 3

before bringing it back to the other at p_3, or (see Fig. 3) we carry the curve into an "inside" position before joining it to the other.

Note that each $M(a)$ may be flattened into a plane $E^2(a)$ (the plane of the paper); we thus obtain a smooth mapping of B^2 into 3-space, though of course with singularities.[10] Also, if the $E^2(a)$

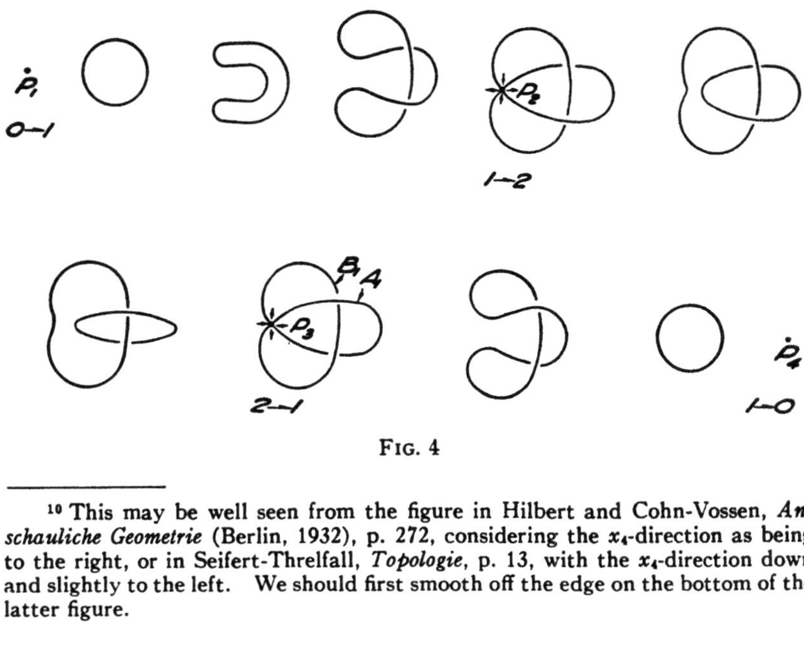

FIG. 4

[10] This may be well seen from the figure in Hilbert and Cohn-Vossen, *Anschauliche Geometrie* (Berlin, 1932), p. 272, considering the x_4-direction as being to the right, or in Seifert-Threlfall, *Topologie*, p. 13, with the x_4-direction down and slightly to the left. We should first smooth off the edge on the bottom of the latter figure.

are parallel to the (x_1, x_2)-plane, then, if $v(p)$ is in the positive x_3-direction at each p, $v(p)$ is never tangent to M, and hence may be deformed into a normal vector field in E^4. Therefore also $W^2 \cdot M = 0$.

In Figure 4 we show an imbedding of B^2 in E^4, for which (see formulas (7. 1) and (6. 5))

$$L_3'' - L_3' = -2[L(A_1', B_1) + L(A_1, B_1')] = -2(1+1) = -4;$$

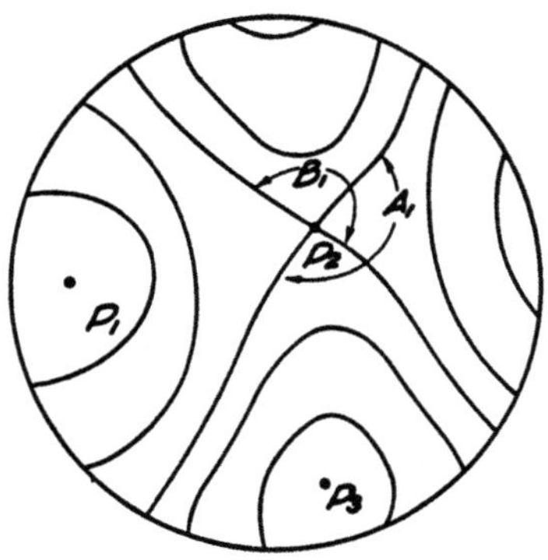

FIG. 5

hence $W^2 \cdot M = -4$. Therefore, in this case, no normal vector field exists.

A projective plane P^2 is formed by identifying opposite points of the circumference of a circular disc. In Figure 5 we show how P^2 may be cut into sets $M(a)$; the p_i are of types $0 \to 1$, $1 \to 1$, $1 \to 0$. Figure 6 shows an "immersion" of P^2 in E^4 with just one singular point, at p^*. We could avoid this singular point by untwisting the curve $M(a)$ for $a < f_3(p^*)$ and shrinking it to a point p_3.

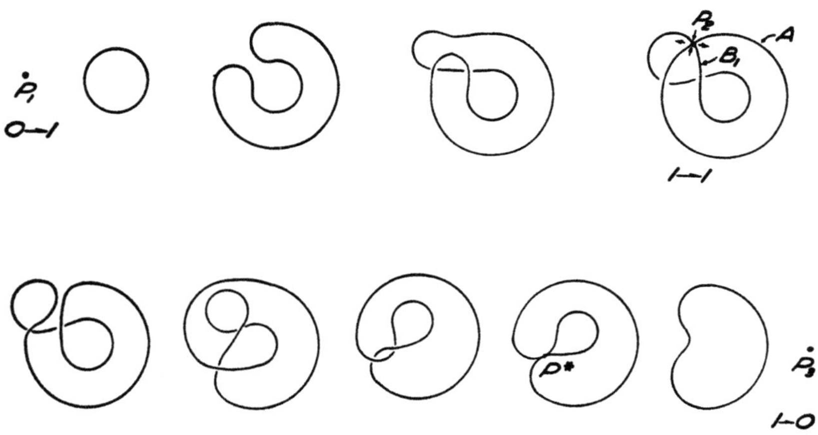

FIG. 6

In the immersion shown we may flatten the $M(a)$ into planes $E^2(a)$, and thus find a normal vector field, as in the case of the Klein bottle.

6. $W^2 \cdot M$ in terms of looping coefficients.[11] Take any $W \subset E^4$. In attempting to define a normal field $v(p)$, or in studying $W^2 \cdot M$, it is sufficient to consider non-tangent fields (see a remark in § 5). It is easy to choose a non-tangent vector field $v(p)$ over each $M(c_i', c_i'')$; moreover, we may let each $v(p)$ lie in $E^3(f_4(p))$. (If $v(p)$ is not tangent to $M(a)$ in $E^3(f_4(p))$ at p, it will not be tangent to M in E^4 at p.) Cut up each $M(c_i', c_i'')$ and each $M(c_i'', c_{i+1}')$ into cells. Considering $v(p)$ as defining $\phi(p)$ on the 1-cells of $M(c_i', c_i'')$, each $W(\sigma_h^2) = 0$ there, as ϕ can be (in fact, is) extended over the interior of σ_h^2. Hence, if we determine

$$D_i = \sum W(\sigma_h^2) \quad \text{over all } \sigma_h^2 \text{ in } M(c_i', c_{i+1}'), \tag{6. 1}$$

then we can find $W^2 \cdot M$ by the formula

$$W^2 \cdot M = \sum_{i=1}^{s-1} D_i. \tag{6. 2}$$

Whenever $v(p)$ is defined over an $M(a)$, and is in $E^3(f_4(p))$, we set $M'(a) =$ all points $p + v(p)$; this is in $E^3(f_4(p))$. Then $M'(a)$ does not intersect $M(a)$, and hence the looping coefficient, $L(M'(a), M(a))$ is defined.

Let $v^*(p) = v(p)$ on $M(c_i'')$. It is easy to extend the definition of $v^*(p)$ continuously through $M(c_i'', c_{i+1}')$ so that it is in $E^3(f_4(p))$, and is not tangent to M. Then clearly

$$L(M^*(c_{i+1}'), M(c_{i+1}')) = L(M'(c_i''), M(c_i'')), \tag{6. 3}$$

where $M^*(c_{i+1}') =$ all $p + v^*(p)$, $p \in M(c_{i+1}')$. Call the last term

[11] If C_1 and C_2 are two oriented non-intersecting closed curves in oriented 3-space, their looping coefficient $L(C_1, C_2)$ is the number of algebraic times that one "loops" about the other. If we let C_1 bound a piece of surface A^2, then L is the number of times C_2 cuts through A^2 in the positive sense, minus the number of times in the negative sense. It can be shown that $L(-C_1, C_2) = -L(C_1, C_2) = -L(C_2, C_1)$. We define $L(\sum C_i, \sum C_j) = \sum L(C_i, C_j)$.

L_i'', and define L_i' similarly. Also, if it should happen that $v^*(p) = v(p)$ in $M(c_{i+1}')$, we would have $D_i = 0$, as is shown by letting $v(p) = v^*(p)$ on the 1-cells of $M(c_i'', c_{i+1}')$. At least we may let $v(p) = v^*(p)$ except interior to 1-cells σ_h^1 of $M(c_{i+1}')$. Then changing $v^*(p)$ to $v(p)$ on σ_h^1 changes L_{i+1}^* by some integer α (which is the number of times $p + v(p)$ runs around σ_h^1 more than $p + v^*(p)$ does), and clearly changes $W(\sigma^2)$ and hence D_i by the same amount, if σ^2 is the 2-cell of $M(c_i'', c_{i+1}')$ with σ_h^1 as a face. Thus we find, if we pay due regard to sign, and use (6. 3),

$$(6.\,4)\qquad -D_i = L_{i+1}' - L(M^*(c_{i+1}'), M(c_{i+1}')) = L_{i+1}' - L_i'.$$

Summing over i, and noting that there are no curves in $M(c_1')$ or in $M(c_s'')$, we find

$$(6.\,5)\qquad W^2\cdot M = \sum_{i=1}^{s-1}(L_i'' - L_{i+1}') = \sum_{i=1}^{s}(L_i'' - L_i').$$

Observe that changing the orientation of $M(c_i')$ (or $M(c_i'')$) does not alter L_i' (or L_i''); however, we may not reorient only a part of $M(c_i')$ or $M(c_i'')$. Further, corresponding curves of $M(c_i'')$ and $M(c_{i+1}')$ must be similarly oriented for (6. 4) to hold. Of course, $M(a)$ and $M^1(a)$ are to be similarly oriented.

7. Study of the looping coefficients. It is possible to choose certain of the p_i, say $p_{\lambda_1}, p_{\lambda_2}, \cdots$, each of type $1 \to 1$ or $2 \to 1$, and one of the curves, say A_k, of each $M(c_{\lambda_k})$, so that A_k contains p_{λ_k} and so that the A_k form a minimal set of cuts rendering M orientable. That is, there is a curve in M cutting any single A_k which reverses orientation in M, but none cutting no A_k. (All p_i of type $1 \to 1$ must be used.)

For example, for the Klein bottle (see § 5), A_1 is part of the figure 8, $M(c_3)$; for the projective plane, A_1 is one of the loops of $M(c_2)$.

Let B_k be the part of $M(c_{\lambda_k})$ not in A_k. Then A_k and the curves of B_k may be oriented so that the following holds: If the curves of $M(c_{\lambda_k}')$ are oriented like $A_k + B_k$, the curves of $M(c_{\lambda_k}'')$

are oriented like $-A_k + B_k$, and the curves of other $M(c_i')$ and $M(c_i'')$ ($i \neq$ any λ_k) are properly, and similarly, oriented, then for each i the curves of $M(c_i'')$ and $M(c_{i+1}')$ are similarly oriented,[12] so that these orientations may be used in the definition of the L_i' and L_i''. Moreover, if $p_i \neq$ any p_{λ_k}, then $M(c_i')$ and $M(c_i'')$ are similarly oriented.

We show now that, if $M(c_i)$ contains no A_k, then $L_i'' = L_i'$. Take a p_i of type $1 \rightarrow 2$ or $2 \rightarrow 1$. We can pull $M'(c_i')$ into $M'(c_i)$ so that we never intersect $M(c_i')$; hence $L_i' = L(M'(c_i),$

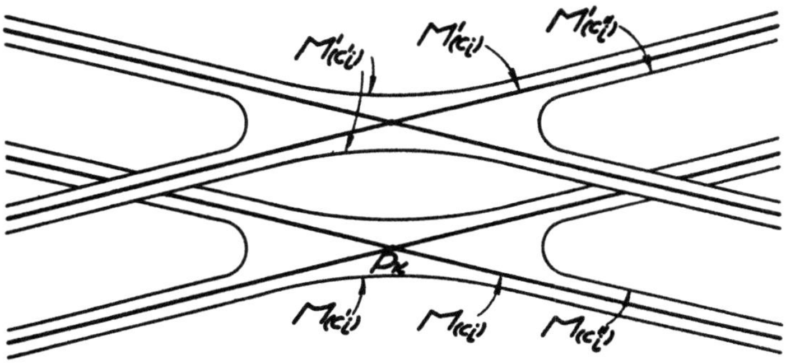

FIG. 7

$M(c_i'))$. Similarly, pull $M'(c_i'')$ into $M'(c_i)$, and $M(c_i')$ and $M(c_i'')$ into $M(c_i)$. (See Fig. 7.) We find thus

$$L_i' = L_i'' = L(M'(c_i), M(c_i)).$$

Clearly $L_i' = L_i''$ if p_i is of type $1 \rightarrow 0$ or $0 \rightarrow 1$.

Consider next a $p_i = p_{\lambda_k}$. Then the proof above holds, except that we must give the curves corresponding to A_k opposite orientations in $M(c_i')$ and $M(c_i'')$. Hence, by using the bilinearity and commutativity of L,

$$\begin{aligned} L_i' - L_i' &= L(B_k' - A_k', B_k - A_k) - L(B_k' + A_k', B_k + A_k) \\ (7.\ 1) \qquad &= -2L(B_k', A_k) - 2L(A_k', B_k) \\ &= -2[L(A_k', B_k) + L(A_k, B_k')]. \end{aligned}$$

[12] This may be expressed by the boundary relation $\partial M(c_i'', c_{i+1}') = M(c_{i+1}') - M(c_i'')$. Also $\partial M(c_{\lambda_k}', c_{\lambda_k}'') = M(c_{\lambda_k}) - M(c_{\lambda_k}') + 2A_k$.

Suppose first that p is of type $2 \to 1$. Then $L(A_k', B_k) = L(A_k, B_k')$, as is seen by pulling B_k away from A_k at p_i and deforming it up to B_k', at the same time pulling A_k' away from B_k' and down to A_k. (For instance, in Figure 7 we might have A_k and A_k' the parts of $M(c_i)$ and $M'(c_i)$ on the right, and B_k and B_k' the parts on the left.) Hence, by (7. 1),

$$\text{(7. 2)} \quad L_i' - L_i' \equiv 0 \pmod 4 \quad \text{if } p_i \text{ is of type } 2 \to 1.$$

If p_i is of type $1 \to 1$, then the parts of A_k and B_k near p_i are smooth curves intersecting at an angle; when we deform B_k into B_k' and A_k' into A_k, the two moving curves cut each other; hence $L(A_k, B_k') = L(A_k', B_k) \pm 1$, and

$$L_i'' - L_i' = -2[2L(A_k', B_k) \pm 1],$$

$$\text{(7. 3)} \quad L_i' - L_i' \equiv 2 \pmod 4 \quad \text{if } p_i \text{ is of type } 1 \to 1.$$

Let $N_{\alpha\beta}$ be the number of points p_k of type $\alpha \to \beta$. Then the results above give

$$\text{(7. 4)} \qquad W^2 \cdot M = \sum_{i=1}^{s} (L_i' - L_i') \equiv 2N_{11} \pmod 4.$$

8. $W^2 \cdot M$ in terms of the characteristic $\chi(M)$ of M. For any complex K, with α^i i-cells ($i = 0, 1, 2, \cdots$), $\chi(M)$ is defined as $\alpha^0 - \alpha^1 + \alpha^2 - \cdots$. This is a topological invariant. If K is cut into disjoint pieces K_i, then $\chi(K) = \sum \chi(K_i)$.

Closed curves and cylinders clearly have the characteristic 0. Hence

$$\text{(8. 1)} \qquad \chi[M(c_i'', c_{i+1}') - M(c_i'')] = 0.$$

It is easily seen that

$$\text{(8. 2)} \quad \chi[M(c_i', c_i') - M(c_i')] = \begin{cases} 1 \text{ if } p_i \text{ of type } 0 \to 1 \text{ or } 1 \to 0, \\ -1 \text{ if } p_i \text{ of any other type.} \end{cases}$$

For instance, if p_i is of type $1 \to 2$, $M(c_i', c_i'')$ is a set of cylinders, together with the part R of $M(c_i', c_i'')$ near p_k; R may be cut into two long strips and a small middle piece; it then has three 2-cells and four inner 1-cells, together with closed curves.

As each point of type $0 \to 1$ or $1 \to 2$ increases the number of curves in $M(a)$ by one while each point of type $2 \to 1$ or $1 \to 0$ decreases the number by one, we have $N_{01} + N_{12} - N_{21} - N_{10} = 0$. Therefore, by adding all relations (8. 1) and (8. 2),

$$\begin{aligned} \chi(M) &= \sum_i \chi[M(c_i', c_i') - M(c_i')] \\ &= N_{01} + N_{10} - (N_{12} + N_{21} + N_{11}) \\ &= 2(N_{01} - N_{21}) - N_{11} \equiv N_{11} \pmod 2. \end{aligned} \tag{8. 3}$$

Multiplying by 2 and comparing with (7. 4) we find

$$W^2 \cdot M \equiv 2\chi(M) \pmod 4. \tag{8. 4}$$

For a non-orientable M with even characteristic there exist imbeddings with $W^2 \cdot M = 0$, and imbeddings with $W^2 \cdot M \neq 0$ (see Figs. 3 and 4). From the proof above we see that, if M is orientable, then $W^2 \cdot M = 0$.[13] Summing up, and using the fundamental theorem on the classification of surfaces, we have:

THEOREM. *To any open surface, or surface with boundary, or any closed orientable surface, in E^4, there exists a normal vector field. Any other surface is homeomorphic to a projective plane P^2, or P^2 with handles, in which case there exists no normal field, or to a Klein bottle B^2, or to B^2 with handles, in which case there may or may not exist a normal field.*[14]

REMARK. For the immersion of P^2 in E^4 with one singularity shown in Figure 6 we noted that a normal vector field existed; hence $W^2 \cdot M = 0$. But for any imbedding without singularity $W^2 \cdot M \neq 0$; hence this singularity cannot be removed by a smooth deformation.

For more details on singularities see § 18.

It seems a reasonable conjecture that for any non-orientable closed surface of characteristic χ the possible values of $W^2 \cdot M$ are

$$2\chi - 4, 2\chi, 2\chi + 4, \cdots, 4 - 2\chi.$$

[13] Another proof of this will be given in § 18.

[14] This theorem may be proved from general theory also.

See W. S. Massey, Proof of a conjecture of Whitney, Pacific Jour. Math. 31 (1969), 143–156.

Part II. Characteristic Classes of Sphere Bundles

9. Homology and cohomology groups. Let K be a fixed complex, with oriented cells σ_i^r. If ${}^r\partial_i^j = [\sigma_j^{r-1} : \sigma_i^r]$ is the incidence number of σ_i^{r-1} and σ_j^r, then the *boundary* and *coboundary* of the chain $A^r = \sum \alpha_i \sigma_i^r$ are defined by

$$(9.1) \qquad \partial A^r = \sum {}^r\partial_i^j \alpha_i \sigma_j^{r-1}, \qquad \delta A^r = \sum {}^{r+1}\partial_j^i \alpha_i \sigma_j^{r+1},$$

where the α_i are elements of an Abelian group G. A^r is a *cycle* or *cocycle* if $\partial A^r = 0$ or $\delta A^r = 0$; A^r is *homologous* to B^r, $A^r \sim B^r$, or *cohomologous* to B^r, $A^r \backsim B^r$, if $A^r - B^r$ is a boundary or a coboundary. The sets of all chains, all cycles, all boundaries, etc., form groups. If we identify homologous r-cycles we obtain the rth *homology group* ${}^G H^r(K)$; similarly for the rth *cohomology group* ${}^G H_r(K)$.

Define *scalar products* of r-chains by

$$(9.2) \qquad (\sum \alpha_i \sigma_i^r) \cdot (\sum \beta_i \sigma_i^r) = \sum \alpha_i \beta_i,$$

supposing the multiplication $\alpha\beta$ has meaning. Note that

$$(9.3) \qquad \partial \sigma_i^r \cdot \sigma_j^{r-1} = [\sigma_j^{r-1} : \sigma_i^r] = \sigma_i^r \cdot \delta \sigma_j^{r-1};$$

$$(9.4) \qquad \partial A^r \cdot B^{r-1} = A^r \cdot \delta B^{r-1}; \qquad A^r \cdot B^r = B^r \cdot A^r.$$

The groups G we shall use mostly are the integers, I_0, and the integers mod m, I_m. Let $(a)_m = a_m$ be the integer a reduced mod m; thus 0_2 and 1_2 are the elements of I_2. For an r-I_0-chain $A^r = \sum a^i \sigma_i^r$, define A^r reduced mod m as $(A^r)_m = \sum (a^i)_m \sigma_i^r$.

To each vertex x_i corresponds the chain x_i; we may say x_i has a natural orientation. Set

$$(9.5) \qquad I = \sum x_i; \quad \text{then} \quad \delta I = 0,$$

at least in an ordinary geometric complex. For take any $\sigma^1 = x_i x_j$; then

$$\delta I \cdot \sigma^1 = I \cdot \partial \sigma^1 = I \cdot (x_j - x_i) = 1 - 1 = 0.$$

This need not be true in a general abstract complex; see § 15. Note that, if K is connected, then any 0-I_0-cocycle is $\backsim aI$ for some integer a; hence ${}^{I_0}\mathbf{H}_0(K) \approx I_0$ (i.e. is isomorphic to I_0) if K is connected.

Define A^r and B^r to be *orthogonal* if $A^r \cdot B^r = 0$. *Any cocycle*[15] X^r *is orthogonal to any boundary* $A^r = \partial B^{r+1}$; for $X^r \cdot A^r = \partial X^r \cdot B^{r+1} = 0$. Similarly, *any cycle is orthogonal to any coboundary.*

The following theorem may be proved.[16] *An* r-I_0-*chain* X^r *is a coboundary if and only if it is orthogonal (mod* m*) to every* r-I_m-*cycle* A^r *for every* m (i.e. $X^r \cdot A^r = 0_m$). *An* r-I_m-*chain is a coboundary if and only if it is orthogonal to every* r-I_m-*cycle.* Similarly for boundaries.

There is an important operation converting r-I_m-cocycles X^r into $(r+1)$-I_0-cocycles (or, similarly, r-I_m-cycles into $(r-1)$-I_0-cycles[17]), as follows: Let $\omega_m X^r$ be any r-I_0-cocycle such that $(\omega_m X^r)_m = X^r$. Then $\delta\omega_m X^r \equiv 0 \pmod{m}$, and hence $Y^{r+1} = (1/m)\delta\omega_m X^r$ exists. The cohomology class of Y^{r+1} is uniquely determined by that of X^r. Note that $mY^{r+1} \backsim 0$.

10. Chains and intersections in manifolds. Let K be a subdivision of the closed manifold M^n. If M^n is orientable,[18] we may orient its n-cells σ_i^n so that $\sum\sigma_i^n$ is a cycle, a *fundamental cycle* of M^n (which we also call M^n); every n-cycle is a multiple of M^n, and hence ${}^{I_0}\mathbf{H}^n(M^n) \approx I_0$. Clearly $(M^n)_2$ is a cycle even if M^n is not orientable.

If σ_1^n and σ_2^n have the common face σ^{n-1}, then $\delta\sigma^{n-1} = \pm\sigma_1^n \pm \sigma_2^n$; hence $\sigma_2^n \backsim \pm\sigma_1^n$, and, similarly, each σ_i^n is $\backsim \pm\sigma_1^n$. M^n is non-orientable if and only if $2\sigma_i^n \backsim 0$; clearly ${}^{I_0}\mathbf{H}_n(M^n) \approx I_0$ or I_2 ac-

[15] We generally use $X, Y, \cdots$ for cocycles, and $A, B, \cdots$ for cycles.

[16] See, for instance H. Whitney, *On matrices of integers and combinatorial topology*, Duke Math. Journ., 3 (1937), 35–45, Theorem 6. The theorem is due essentially to H. Hopf, *Abbildung der dreidimensionalen Sphäre auf die Kugelfläche*, Math. Ann., 104, 658.

[17] See Alexandroff-Hopf, *Topologie* (Berlin, 1935), pp. 222–223.

[18] For the non-orientable case see § 15.

cording as M^n is orientable or not. If M^n is oriented, then $\sigma_i^n \backsim \sigma_1^n$, and

(10. 1) $\sum \alpha_i\sigma_i^n \backsim (\sum \alpha_i)\sigma_1^n, \quad (\sum \alpha_i\sigma_i^n)\cdot M^n = \sum \alpha_i,$

and $X^n \backsim Y^n$ if and only if $X^n \cdot M^n = Y^n \cdot M^n$. (This corresponds to the fact that $A^0 \sim B^0$ in a connected complex if and only if $A^0 \cdot I = B^0 \cdot I$.)

Let K be a subdivision of M^n, which we assume closed and oriented, and let K' be the regular subdivision of K. Then the simplexes of K' may be fitted together into cells τ so that to each σ_i^r corresponds a $\tau_i^{n-r} = D(\sigma_i^n)$ cutting through it. Set $D(\tau_i^{n-r}) = \sigma_i^r$. Then

(10. 2) $\partial D(X^r) = (-1)^{r+1}D(\delta X^r), \quad \delta D(A^r) = (-1)^r D(\partial A^r).$

Hence D, the "dual" operator, maps cycles into cocycles, and conversely, and establishes an isomorphism between ${}^G\boldsymbol{H}^r(M^n)$ and ${}^G\boldsymbol{H}_{n-r}(M^n)$ (Poincaré duality theorem).

Let A^r and B^{n-r} be two singular chains, neither of which meets the boundary of the other. Then their *Kronecker index* $\{A^r, B^{n-r}\}^0$, an integer, is defined and gives their algebraic number of intersections. Thus $\{\sigma_i^r, \tau_i^{n-r}\}^0 = 1$. Let $\{\ \}_m^0$ mean $\{\ \}^0$ reduced mod m.

More generally, if $r + s \geqq n$, then $\{A^r, B^s\}$ may be defined as an $(r + s - n)$-chain which is linear in the two variables and satisfies the relation

(10. 3) $\partial\{A^r, B^s\} = (-1)^{n-s}\{\partial A^r, B^s\} + \{A^r, \partial B^s\}.$

Since $\{A^r, B^{n-r}\}\cdot I = \{A^r, B^{n-r}\}^0$, and $\delta I = 0$, this gives

(10. 4) $\{\partial A^r, B^{n-r+1}\}^0 = (-1)^r\{A^r, \partial B^{n-r+1}\}^0.$

In any geometric complex K, bilinear products $X^r \smile Y^s = Z^{r+s}$ may be defined with certain simple properties;[19] in particular,

(10. 5) $\delta(X^r \smile Y^s) = \delta X^r \smile Y^s + (-1)^r X^r \smile \delta Y^s,$

(10. 6) $I \smile X^r = X^r \smile I = X^r.$

[19] See H. Whitney, *On products in a complex*, Ann. of Math., 39 (1938), 397–432. See also [4].

If K is simplicial and its vertices are ordered, $x_1, x_2, \cdots$, we may use

$$(10.7) \qquad (x_{\lambda_0} \cdots x_{\lambda_r}) \smile (x_{\lambda_r} \cdots x_{\lambda_{r+s}}) = x_{\lambda_0} \cdots x_{\lambda_r} \cdots x_{\lambda_{r+s}},$$

for $\lambda_0 < \cdots < \lambda_r < \cdots < \lambda_{r+s}$, and $\sigma^r \smile \sigma^s = 0$ for other pairs of simplexes.

In a manifold, D turns r-cocycles into $(n-r)$-cycles; the product then goes into the Kronecker index in that the same products are defined in the homology groups by the two methods.

Suppose M^{n-r} and M^{n-s} are oriented smooth submanifolds of M^n such that at any point p of intersection, except on an $(n-r-s-1)$-dimensional set, the tangent planes to M^{n-r} and M^{n-s} at p have only a plane P^{n-r-s} in common;[20] then the geometric intersection is a manifold or set of manifolds N^{n-r-s}. Moreover, interpreting these manifolds as chains and orienting N^{n-r-s} properly, we obtain

$$(10.8) \qquad \{M^{n-r}, M^{n-s}\} = N^{n-r-s}.$$

If M^{n-r} and M^{n-s} do not satisfy the given condition, a small deformation of one of them will bring this about; of course the homology class of N is uniquely determined. In particular, $\{M^r, M^r\}$ has meaning for $2r \geqq n$. Note that $\{M^n, M^r\} = \{M^r, M^n\} = M^r$. corresponding to (10. 6).

11. Sphere bundles. For any smooth M^2 in E^4 we saw in § 2 how normal 1-spheres $S_N(p)$ were defined; these form the *normal sphere bundle of M^2 in E^4*. At each $p \in M$ there is also a tangent plane, and in it a unit circle $S_T(p)$; these 1-spheres form the *tangent sphere bundle* of M^2. Similarly, for $M^n \subset E^m$ the tangent and normal sphere bundles may be defined; the spheres are of dimension $n-1$ and $m-n-1$ respectively. See also § 19.

We now define a more general concept. A *bundle of spheres* or *sphere bundle* $\mathfrak{B}$ consists of a certain topological space T, the *base*

[20] This will have meaning if we imbed M^n in E^{2n+1}. See §16. The exceptional set may be removed by a slight deformation; see the paper in footnote 33, Theorem 2 (p. 654) and (D) (p. 655).

space (M in the examples above), a set of points $S(p)$ corresponding to each p in M, and certain relations between these sets, as follows: For $p \neq p'$, $S(p)$ and $S(p')$ have no common points. We denote by $\{U_i\}$ a certain set of open sets covering T. For each i and each $p \varepsilon U_i$ there is a $(1-1)$ correspondence

$$q' = \xi_{U_i}(p, q) \varepsilon S(p), \qquad q \varepsilon S_0^\nu, \tag{11. 1}$$

between the unit sphere S_0^ν in $E^{\nu+1}$ and the set of points $S(p)$. Let (11. 1) be written also $q = \xi_{U_i}^{-1}(p, q')$. Set

$$\xi_{U_i, U_j}(p, q) = \xi_{U_i}^{-1}(p, \xi_{U_j}(p, q)), \qquad p \varepsilon U_i \cdot U_j. \tag{11. 2}$$

We assume that $\xi_{U_i, U_j}(p)$ is an orthogonal mapping of S_0^ν into itself, which varies continuously with p. The set of all points of all $S(p)$ forms the *total space* $\mathfrak{S}$. It may be made into a topological space in an obvious manner.

Define orthogonality on any $S(p)$ by means of $\xi_{U_i}(p)$ for $p \varepsilon U_i$. Since each $\xi_{U_i, U_j}(p)$ is orthogonal, this definition is independent of i.

One may alter the U_i and ξ_{U_i}, obtaining a new sphere bundle. But as long as we obtain a homeomorphic total space, split into spheres in the same fashion, and with the same definition of orthogonality on them we call the bundles *equivalent*, and shall not distinguish between them. Each ξ_{U_i} may be thought of as a coordinate system in that part, $\mathfrak{S}(U_i)$, of $\mathfrak{S}$ for which $p \varepsilon U_i$.

We shall deal exclusively with the case that the base space is a complex K. If each closed cell[21] $\bar{\sigma}$ is in some U_i, we can use $\xi_\sigma(p) = \xi_{U_i}(p)$ $(p \varepsilon \bar{\sigma})$.

REMARK. If we order the vertices of K, say $x_1, x_2, \cdots$, and for each x_i choose a $U(x_i)$ containing all cells of K with x_i as a vertex, and let $U(\sigma) = U(x_i)$ for any σ whose first vertex is x_i, and use $\xi_\sigma = \xi_{U(\sigma)}$, then, if σ_1 and σ_2 have the same first vertex,

$$\xi_{\sigma_1}(p) = \xi_{\sigma_2}(p), \qquad p \varepsilon \sigma_1 \cdot \sigma_2. \tag{11. 3}$$

[21] As a point set, we let σ mean the open cell and $\bar{\sigma}$ the closed cell; $\partial\sigma$ is the point-set boundary of σ.

Similarly we may define a *bundle of fibers* by replacing S_0^ν by any topological space S_0, and the group $G^{\nu+1}$ of orthogonal mappings of S_0^ν into itself by any group of homeomorphisms of S_0 into itself. (See [5].)

The bundle $\mathfrak{B}$ is *simple* if we can choose the ξ_σ so that $\xi_{\sigma,\sigma'}(p)$ = identity always. (This is equivalent to replacing $G^{\nu+1}$ by the group containing the identity alone.) We can then set $\zeta(p) = \xi_\sigma(p)$ ($p \varepsilon \sigma$), and thus express $\mathfrak{S}$ as the Cartesian product $K \times S_0^\nu$. Let $e_1, \cdots, e_{\nu+1}$ be the unit points of the axes of S_0^ν. Then if $\mathfrak{B}$ is simple, $\phi_i(p) = \zeta(p, e_i)$ defines $\nu + 1$ orthogonal *projections* of K into $\mathfrak{S}$. Conversely, if such (continuous) projections exist, we may define $\zeta(p, e_i)$, and hence $\zeta(p, q)$, since orthogonality has meaning in the $S(p)$, and thus prove that $\mathfrak{B}$ is simple. The bundle $\mathfrak{B}$ *is always simple if the base space is a cell.*

12. The characteristic classes, $\nu = 1$. Let $\mathfrak{B}$ be a bundle of 1-spheres (as in Part I). For each vertex x_i of K we may choose a pair of orthogonal points $\phi_1(x_i)$, $\phi_2(x_i)$ in $S(x_i)$. This defines a *characteristic* 1-I_2-*cocycle* $W^1 = W_\phi^1$, as follows. For each 1-cell $\sigma^1 = x_ix_j$ let $W(\sigma^1) = 0_2$ or 1_2 according as ϕ_1 and ϕ_2 define the same or opposite orientations of the ends $S(x_i)$, $S(x_j)$ of the cylinder $\mathfrak{S}(\sigma^1)$; that is, according as the ordered pair of points

$$\psi_{l,\sigma^1}(p) = \overset{-1}{\xi_{\sigma^1}}(p, \phi_l(p)) \qquad (l = 1, 2) \tag{12.1}$$

of S_0^1 define the same or opposite orientations of S_0^1 for $p = x_i$ as for $p = x_j$. Set

$$\overset{1}{W} = \sum W(\overset{1}{\sigma_h})\overset{1}{\sigma_h} \quad \text{(coefficients in } I_2); \tag{12.2}$$

this is the required cocycle. For any ϕ', $W_{\phi'}^1 \backsim W_\phi^1$; hence the cohomology class $\boldsymbol{W}^1$ of W^1 is invariant; this is the *characteristic* 1-*class* of $\mathfrak{B}$. The bundle is *orientable* if we may make all $\xi_{\sigma,\sigma'}(p)$ rotations; this is so if and only if $\boldsymbol{W}^1 = 0$. (For further details see [3]. The W^r are there called F^r.)

We may not be able to extend the definition of both ϕ_1 and ϕ_2

continuously over the 1-cells of K so that they are orthogonal; certainly not if $W^1 \neq 0$. However, the definition of ϕ_1 can be extended, and this will define a *characteristic* 2-I_0-*cocycle* $W^2 = W^2_\phi$ as follows.[22] We suppose that $\mathfrak{B}$ is orientable until § 14, and that the $\xi_{\sigma,\sigma'}(p)$ are rotations.

Take any oriented 2-cell $\sigma^2 = x_i x_j x_k$ of K. Set

$$\psi_{\sigma^2}(p) = \overset{-1}{\xi_{\sigma_2}}(p, \phi(p)), \qquad p \,\varepsilon\, \partial\sigma^2; \tag{12. 3}$$

this is a mapping of $\partial\sigma^2$ into S^1_0. Let $W(\sigma^2)$ be the degree of $\psi\sigma_2$ (compare § 2). Then $W^2 = \sum W(\sigma^2_h)\sigma^2_h$; the cohomology class $\mathbf{W}^2$ of W^2 is the characteristic 2-class of $\mathfrak{B}$, and is uniquely determined.

The proof runs as follows: First, it is clear that $W(-\sigma^2) = -W(\sigma^2)$. Now let us find the effect of changing ϕ to ϕ' in $\sigma^1 = x_i x_j$; let $\phi' = \phi$ elsewhere. Then $\phi'(p)$ runs around the cylinder $\mathfrak{S}(\sigma^1)$ some number α of times more than $\phi(p)$ does as p runs from x_i to x_j. That is, if

$$\psi_{\sigma^1}(p) = \overset{-1}{\xi_{\sigma^1}}(p, \phi(p)), \qquad \psi'_{\sigma^1}(p) = \overset{-1}{\xi_{\sigma^1}}(p, \phi'(p)), \tag{12. 4}$$

then $\psi'_{\sigma^1}(p)$ runs around S^1_0 α times more than $\psi_{\sigma^1}(p)$.[23]

Now deform ξ_{σ^1} into $\xi^*_{\sigma^1}$, where $\xi^*_{\sigma^1}(p) = \xi_{\sigma^2}(p)$ ($p \,\varepsilon\, \sigma^1$), by the lemma below. The integer α is always defined, and is continuous, and hence constant. Therefore $\psi'_{\sigma^2}(p)$ runs around S^1_0 α times more than $\psi_{\sigma^2}(p)$; thus $W_{\phi'}(\sigma^2) - W_\phi(\sigma^2) = \alpha$. If $[\sigma^1 : \sigma^2]$ had been -1 instead of $+1$, we would have found the same relation, with $-\alpha$. Thus in any case, as $[\sigma^1 : \sigma^2] = \delta\sigma^1 \cdot \sigma^2$,

$$W_{\phi'}(\sigma^2) - W_\phi(\sigma^2) = (W^2_{\phi'} - W^2_\phi)\cdot\sigma^2 = \alpha\delta\sigma^1\cdot\sigma^2,$$

and hence

$$W^2_{\phi'} - W^2_\phi = \delta(\alpha\sigma^1), \qquad W^2_{\phi'} \backsim W^2_\phi. \tag{12. 5}$$

---

[22] For an example where W^2 is determined see [3], § 6, (a).

[23] For instance, if $\psi_{\sigma^1}(p)$ runs from e_1 to e_2 and $\psi'_{\sigma^1}(p)$ from e_1 through $-e_2$ and $-e_1$ to e_2, then $\alpha = -1$.

As in § 3, we see that W^2_ϕ is independent of ϕ; it is also independent of the subdivision into a complex K. Because of the following lemma it is independent of the ξ_σ chosen.

LEMMA. *If $u(p) = \xi^{-1}(p)\xi^*(p)$ is a rotation for $p \varepsilon \bar{\sigma}$, then ξ may be deformed into ξ^*.*

The rotation $u(p)$ is an element of the group $\overline{G}^{\nu+1}$ of rotations, and is defined in the closed cell $\bar{\sigma}$; thus it maps $\bar{\sigma}$ into $G^{\nu+1}$. Now $\bar{\sigma}$ may be deformed in itself to a point p_0; if u is applied, $u(p)$ is deformed into $u'(p) = u(p_0)$. Deform $u'(p)$ into $u^*(p) =$ identity. If $u_t(p)$ $(0 \leqq t \leqq 1)$ is the whole deformation, set $\xi_t(p) = \xi^*(p)u_t^{-1}(p)$.

We must still show that W^2_ϕ is a cocycle. Consider any 3-cell σ^3. Set $\phi'(p) = \xi_{\sigma^3}(p, e_1)$ on the 1-cells of σ^3. It may be extended over the rest of the 1-cells of K, and it determines $W^2_{\phi'}$. We showed above that $W^2_{\phi'} - W^2_\phi = \delta X^1$ for some X^1. Clearly $W^2_{\phi'}(\sigma^2_i) = 0$ for all faces σ^2_i of σ^3. Hence $W^2_{\phi'} \cdot \partial\sigma^3 = 0$, and

$$\delta \overset{2}{W}_\phi \cdot \overset{3}{\sigma} = (\delta \overset{2}{W}_{\phi'} - \delta\delta \overset{1}{X}) \cdot \overset{3}{\sigma} = \overset{2}{W}_{\phi'} \cdot \partial \overset{3}{\sigma} = 0.$$

Since σ^3 was arbitrary, $\delta W^2_\phi = 0$.

13. The characteristic classes, general ν. For any integer r, $1 \leqq r \leqq \nu + 1$, a characteristic cohomology class W^r may be defined as follows: Construct a set $\phi = (\phi_1, \cdots, \phi_\alpha)$ $(\alpha = \nu - r + 2)$ of orthogonal projections of K^{r-1} into $\mathfrak{S}$.[24] We construct them over the vertices, 1-cells, etc., in turn. Suppose $\phi_1, \cdots, \phi_{i-1}$ $(i \leqq \alpha)$ have been constructed over K^{r-1}, and ϕ_i over K^{s-1} $(s < r)$. Then ϕ_i is constructed over any σ^s, as follows: For each $p \varepsilon \bar{\sigma}^s$ let $S'(p)$ be the $(\nu - i + 1)$-sphere consisting of all points of $S(p)$ orthogonal to $\phi_1(p), \cdots, \phi_{i-1}(p)$; we must choose $\phi_i(p)$ in $S'(p)$. The $S'(p)$ form a sphere bundle $\mathfrak{B}'$ with $\bar{\sigma}^s$ as base space. Since $\bar{\sigma}^s$ is a cell, $\mathfrak{B}'$ is simple; hence we may choose a coördinate system ζ in it. Set

[24] The symbol K^s denotes the subcomplex of K containing all cells of dimension $\leqq s$.

$$\psi(p) = \zeta^{-1}(p, \phi_i(p)), \qquad p \in \partial\sigma^s;$$

this is a mapping of $\partial\sigma^s$ into S_0'. Now $\partial\sigma^s$ and S_0' are sacks[25] of dimensions $s - 1$ and $\nu - i + 1$ respectively. Since $s - 1 < \nu - i + 1$, it is possible to extend the definition of ψ over σ^s.[26] The projection ϕ_i is extended over σ^s by setting

$$\phi_i(p) = \zeta(p, \psi(p)), \qquad p \in \sigma^s.$$

To define $W^r = W^r_\phi$, take any σ^r and set

$$\text{(13. 1)} \qquad \psi_{i,\sigma^r}(p) = \overset{-1}{\xi_{\sigma^r}}(p, \phi_i(p)), \quad p \in \partial\sigma^r \qquad (i = 1, \cdots, \alpha).$$

The mapping $\psi_1 = \psi_{1,\sigma^r}$ may be deformed into the mapping $\psi_1^{(1)}(p) = e_1$. It is possible to deform the other $\psi_i(p)$ simultaneously into $\psi_i^{(1)}(p)$, so that the mappings are always orthogonal. Now $\psi_2^{(1)}(p), \cdots, \psi_\alpha^{(1)}(p)$ are in the $(\nu - 1)$-sphere $[e_2, \cdots, e_{\nu+1}]$ of S_0^ν orthogonal to e_1. Deform $\psi_2^{(1)}(p)$ in this sphere into $\psi_2^{(2)}(p) = e_2$, and deform the other $\psi_i^{(1)}(p)$ simultaneously, leaving $\psi_1^{(1)}(p)$ fixed, etc. Finally, $\psi_i^{(i)}(p) = e_i$ $(i < \alpha)$, and $\psi'(p) = \psi_\alpha^{(\alpha-1)}(p)$ is in $[e_\alpha, \cdots, e_{\nu+1}]$, a sphere of dimension $\nu + 1 - \alpha = r - 1$. Since $\partial\sigma^r$ is an $(r - 1)$-sack, ψ' has a degree[27] d. Then $W^r_\phi(\sigma^r) = d$ or d_2, according to the following table:

integers mod 2	if $r = 1$	or $1 < r < \nu + 1$ and r even,
integers	if $r = \nu + 1$	or $1 < r < \nu + 1$ and r odd.

We illustrate why coefficients mod 2 must be used in the case $\nu = 2$, $r = 2$, $\alpha = 2$. The mapping $\psi_1(p)$ may be deformed into $\psi_1^{(1)}(p) = e_1$ in various ways; when $\psi_2(p)$ is deformed simultaneously, the final $\psi'(p)$, mapping $\partial\sigma^2$ into $[e_2, e_3]$, may have different resulting degrees, differing from each other by any multiple of 2. For example, let us identity $\partial\sigma^2$ with $[e_2, e_3]$; suppose

$$\psi_1(p) = p, \qquad \psi_2(p) = -e_1, \qquad p \in \partial\sigma^2.$$

[25] Since we wish to reserve the term *sphere* for a set in which orthogonality is defined, we shall call any set homeomorphic to a sphere a *sack*.

[26] See Alexandroff-Hopf, *Topologie*, pp. 502 and 509.

[27] If $r = 1$, $\sigma^r = x_i x_j$, then $d = 0_2$ or 1_2, according as $\psi'(x_i)$ and $\psi'(x_j)$ are the same or opposite points of $[e_{\nu+1}] = \pm e_{\nu+1}$.

If we deform $\psi_1(p) = p$ directly into e_1, along the shortest arc of the circle $[p, e_1]$ through p and e_1, we may deform $\psi_2(p)$ into $\psi'(p) = p$, along $[p, e_1]$ simultaneously; then $d = 1$. But suppose we first deform $\psi_1(p)$ along $[p, e_1]$ into $\psi_1^*(p) = -e_1$, and deform this along a semicircle of $[e_1, e_2]$ into $\psi_1^{(1)}(p) = e_1$. Then we find $\psi_2^*(p) = -p$, and $\psi'(p) =$ the point of $[e_2, e_3]$ obtained by reflecting p across the x_3-axis; then $d = -1$.

14. Properties of the W^r. We recall that if $\nu = 1$, or if dim $(K) \leqq 3$, then $\mathfrak{B}$ is characterized by the classes W^1 and W^2 (by W^1 alone if $\nu = 0$).[28] This holds even if $\mathfrak{B}$ is not orientable; but see § 15. For $\nu = 2$ and dim $(K) = 4$ this is not true; but $\mathfrak{B}$ can be characterized in case $W^1 = 0$ and $W^2 = 0$, by means of a new invariant.[29]

THEOREM. *If W^{2r} is a characteristic cocycle, then a characteristic cocycle W^{2r+1} may be determined by the formula* (see § 9)

$$W^{2r+1} = \tfrac{1}{2}\delta\omega_2 W^{2r}. \tag{14. 1}$$

As a consequence, for any odd r, $2W^r = 0$.

Again $\mathfrak{B}$ need not be orientable. The formula is trivial if $2r = \nu + 1$, for we may then interpret $\omega_2 W^{2r}$ as W^{2r} itself. It holds for $r = 0$ if we interpret W^0 as I in the associated complex, and reduce mod 2 (see § 15).

15. Non-orientable sphere bundles. We first discuss abstract complexes. They have cells and incidence numbers; but the vertices need have no natural orientation. Let us call two complexes *locally isomorphic* if their cells are in $(1-1)$ correspondence, and for each σ and corresponding σ^* the complexes σ plus faces and σ^* plus faces are isomorphic. For instance, if K has the cells and boundary relations

$$\partial\sigma_1 = b - a, \qquad \partial\sigma_2 = c - b, \qquad \partial\sigma_3 = a - c,$$

and K^* is similar, except that $\partial\sigma_3^* = -a^* - c^*$, then K and K^*

[28] For this and further theorems see [3], § 7, etc. [29] See [5], § 4.

are locally isomorphic. Thus the corresponding sets σ_3, a, c, and σ_3^*, $-a^*$, c^* of oriented cells form isomorphic complexes. If we demand that any abstract complex be locally isomorphic to a geometric complex, then $\partial\partial A = 0$ and $\delta\delta A = 0$ always.

In the abstract complex K^* let I^* be a "unit 0-chain," i.e. the sum of all arbitrarily oriented vertices. Now δI^* can be $\neq 0$; but it is $\equiv 0 \pmod 2$, so that $W^* = (\frac{1}{2}\delta I^*)_2$ exists. Set

$$(15.1) \qquad W^* = \text{cohomology class of } (\tfrac{1}{2}\delta I^*)_2.$$

This is the *characteristic class* of K^*. Since $\delta(\frac{1}{2}\delta I^*)_2 = (\frac{1}{2}\delta\delta I^*)_2 = 0$, it is a cocycle. Since changing the orientation of a vertex x_i in I^* changes W^* by $(\frac{1}{2}\delta(2x_i))_2 = (\delta x_i)_2$, W^* is an invariant of K^*. *If K_1 and K_2 are locally isomorphic, then they are isomorphic if and only if $W_1 = W_2$.* The complex K is *normal*, i.e. like a geometric complex, if and only if $W = 0$. Given any K and any 1-I_2-cohomology class X in K,[30] there is a locally isomorphic K^* with $W^* = X$.[31] In the example above $(\frac{1}{2}\delta I^*)_2 = (-\sigma_3^*)_2 = (\sigma_3^*)_2$, and $W^* \neq 0$.

Set $\rho(0_2) = 1$, $\rho(1_2) = -1$. If K is simplicial, with ordered vertices, we may define incidences in K^* by letting $[\sigma^{*r-1} : \sigma^{*r}]^* = [\sigma^{r-1} : \sigma^r]$ if σ^{r-1} and σ^r have the same first vertex, while if their first vertices x_i, x_j are distinct, we set

$$(15.2) \qquad [\sigma^{*r-1} : \sigma^{*r}]^* = \rho[W^*(x_i x_j)][\sigma^{r-1} : \sigma^r].$$

Now take any sphere bundle $\mathfrak{B}$, with base space K. We define the *complex K^* associated with $\mathfrak{B}$* as follows: For each cell σ of K choose an orientation θ of the spheres $S(p)$, $p \in \bar{\sigma}$; let (σ, θ) be a cell of K^*, where, if $-\theta$ is the orientation opposite to θ, then

$$(15.3) \qquad (\sigma, \theta) = -(-\sigma, \theta) = -(\sigma, -\theta) = (-\sigma, -\theta).$$

Suppose θ and θ' are orientations of a sphere. Then set $(\theta, \theta') = 1$

[30] Clearly any I_2-chain in a complex may be considered a chain in any locally isomorphic complex.

[31] For further details see [4], § 7. The application to manifolds below is due to G. de Rham, *Sur la théorie des intersections et les intégrales multiples*, Comm. Math. Helv., 4 (1933), 151–157.

or -1 according as these orientations agree or not. Now if $\sigma = \sigma^r$ and $\sigma' = \sigma'^{r-1}$ are incident in K, let (σ, θ) and (σ', θ') be incident in K^*, and set

$$(15.4) \qquad [(\sigma', \theta') : (\sigma, \theta)] = (\theta, \theta')[\sigma' : \sigma].$$

The complexes K and K^* are locally isomorphic. For given any σ, we choose a single orientation of all $S(p)$, $p \in \bar{\sigma}$, and use the oriented cells (σ', θ) for faces σ' of σ. Then the incidences in this part of K^* are the same as in K.

If K is the complex defined in the example above, and $\mathfrak{B}$ is the non-orientable 1-sphere bundle over K (whose total space is a Klein bottle), then the associated complex is the K^* defined above.

The complex K^* may also be defined as follows: We may subdivide $\mathfrak{S}$ so that for each σ of K, $\mathfrak{S}(\sigma)$ is a subcomplex of $\mathfrak{S}$, and orient its cells so that $\partial\mathfrak{S}(\sigma) \subset \mathfrak{S}(\partial\sigma)$. Then we may consider each $\mathfrak{S}(\sigma_i^r)$ (which is of dimension $r + \nu$) a cell σ_i^{*r}, as incidence relations are defined for them. These form K^*.

If we choose a fixed set of ξ_σ in $\mathfrak{B}$, then these define orientations of the $S(p)$ $(p \in \bar{\sigma})$. If we identify the cells of K and of K^*, the definition above gives: $[\sigma^{r-1} : \sigma^r]^* = [\sigma^{r-1} : \sigma^r]$ or $-[\sigma^{r-1} : \sigma^r]$ according as $\xi_{\sigma^{r-1},\sigma^r}(p)$ is a rotation or not.

Consider now the case of the tangent bundle of a manifold M^n. Orienting an $S(p)$ is equivalent to orienting a neighborhood U of the manifold about p. Each n-cell of the associated complex K^* has a natural orientation, namely (σ^n, θ), where θ is the same orientation of U that σ^n gives. The sum $\sum(\sigma_i^n, \theta_i)$ of these cells is a cycle, the *fundamental n-cycle* of K^*. It corresponds to I in a geometric complex. The Poincaré duality theorem holds:

$$(15.5) \qquad {}^G\boldsymbol{H}_p(K) \approx {}^G\boldsymbol{H}^{n-p}(K^*), \quad {}^G\boldsymbol{H}_p(K^*) \approx {}^G\boldsymbol{H}^{n-p}(K).$$

We now show how the characteristic classes of $\mathfrak{B}$ may be determined as cohomology classes in the associated complex K^*. The present definition is preferable even to the one of § 13, since the $\boldsymbol{W}^r$ will not depend on any previously chosen orientation of the $S(p)$. Choose $\phi = (\phi_1, \cdots, \phi_{\nu-r+2})$ over K^{r-1} as before. Given any σ^r, choose an orientation θ of the $S(p)$, $p \in \bar{\sigma}^r$. Write all cells incident with (σ^r, θ) in the form (σ', θ); then the definition of $W(\sigma^r)$ and the derivation of the properties stated go exactly as before. We must note that the definition gives $W(\sigma^r, -\theta) = -W(\sigma^r, \theta)$, as it should.[32]

We now prove

(15. 6) $\qquad W^*(K^*) = W^1(\mathfrak{B})$ if K^* is associated with $\mathfrak{B}$.

Choose orthogonal points $\phi_1(x_i), \cdots, \phi_{\nu+1}(x_i)$ at each vertex x_i. These determine an orientation $\theta_i(x_i)$ of $S(x_i)$. Let $I^* = \sum(x_i, \theta_i)$; then $\delta I^* \cdot (x_i x_j) = 0$ or $\neq 0$ according as θ_i and θ_j agree or disagree in the cylinder $\mathfrak{S}(x_i x_j)$. Thus $(\frac{1}{2}\delta I^*)_2 \cdot (x_i x_j) = W^*(x_i x_j) = 0_2$ or 1_2 depending on which holds. But also $W^1(x_i x_j) = 0_2$ or 1_2 in the same circumstances.

Let K_1, K_2, K_3 be locally isomorphic complexes, with

(15. 7) $\qquad W_1 + W_2 + W_3 = 0.$

Then we may define products $X_1^r \smile Y_2^s = Z_3^{r+s}$, as follows: We suppose the locally isomorphic normal complex K is simplicial, with ordered vertices; then incidences in the K_i may be defined as in (15. 2). We use $W_i = (\frac{1}{2}\delta_i I_i)_2$, and

(15. 8) $$\begin{aligned}(x_{\lambda_0} \cdots x_{\lambda_r})_1 \smile (x_{\lambda_r} \cdots x_{\lambda_{r+s}})_2 \\ = \rho[W_2(x_{\lambda_0} x_{\lambda_r})](x_{\lambda_0} \cdots x_{\lambda_{r+s}})_3,\end{aligned}$$

the subscripts here referring to the complexes used.

Part III. Tangent and Normal Bundles of a Manifold

16. Differentiable manifolds.[33] A *differentiable n-manifold* M^n is a topological space, with a certain system of neighborhoods U_i, each in $(1-1)$ correspondence with the interior Q of the unit

[32] In Part I we used essentially the cells (σ, θ), θ being an orientation of tangent spheres instead of normal spheres. But since $W_T^1 = W_N^1$ (see § 20), the two associated complexes are isomorphic (see (15. 6)).

[33] For a detailed treatment see H. Whitney, *Differentiable manifolds*, Ann. of Math., 37 (1936), 645–680.

sphere in n-space, in such a manner that, if $U_i \cdot U_j \neq 0$, then in the common part the resulting mapping of a portion of Q into another portion is continuously differentiable, with non-vanishing jacobian. By a manifold we shall always mean a differentiable one. It may be shown that any M^n may be *imbedded* in E^{2n+1}, that is, mapped in a $(1-1)$ differentiable manner, so that the matrix of partial derivatives is everywhere of maximum rank. Also, it may be *immersed* in E^{2n}; this is the same as imbedding, except that we allow self-intersections. If M^n is immersed in E^m, we may, by a slight deformation, make the singularities the simplest possible (compare § 18). A differentiable M^n in E^m may be approximated arbitrarily closely by an analytic M^n in E^m. Any M^n may be triangulated into a simplicial complex, with differentiable cells.[34]

17. The characteristic classes of tangent bundles. Let K be a simplicial subdivision of M^n, with simplexes σ_i^r, and let L be the regular subdivision of K. Each vertex y_i^r of L is the center of a simplex σ_i^r of K, and $y_{\lambda_0}^{\alpha_0} y_{\lambda_1}^{\alpha_1} \cdots y_{\lambda_r}^{\alpha_r}$ is a "normally oriented" simplex if $\sigma_{\lambda_{i-1}}^{\alpha_{i-1}}$ is a face of $\sigma_{\lambda_i}^{\alpha_i}$ $(i = 1, \cdots, r)$.

The points p of any simplicial complex K may be given barycentric coördinates $\eta_1(p), \eta_2(p), \cdots$, which are continuous functions of p, and so that if p is interior to the simplex $x_{\lambda_0} \cdots x_{\lambda_r}$, then $\eta_{\lambda_0}, \cdots, \eta_{\lambda_r}$, and only these, are $\neq 0$. Moreover, if we think of each simplex as a flat space, p can be written in the form

$$p = \sum \eta_k(p) x_k, \qquad \eta_k(p) \geqq 0, \qquad \sum \eta_k(p) = 1.$$

In any such complex we may define a continuous vector field by

$$v(p) = \sum_{\lambda < \mu} \eta_\lambda(p) \eta_\mu(p)(x_\mu - x_\lambda). \tag{17.1}$$

This is tangent to K at each p, and vanishes only at the vertices of p. The vectors are shown in Figure 8.

---

[34] See S. S. Cairns, *Triangulation of the manifold of class one*, Bull. Amer. Math. Soc., 41 (1935), 549–552.

Define this field in M^n, using L. For example, if $n = 2$, then in a 1-cell of the form $y_i^r y_j^s$ $(r < s)$, $v(p)$ is a positive multiple of $y_j^s - y_i^r$. For $p \varepsilon y_i^0 y_j^1 y_k^2$, $v(p)$ is of the form

$$a(\overset{1}{y_j} - \overset{0}{y_i}) + b(\overset{2}{y_k} - \overset{0}{y_i}) + c(\overset{2}{y_k} - \overset{1}{y_j}),$$

a, b, c being positive. (Compare Stiefel [2].)

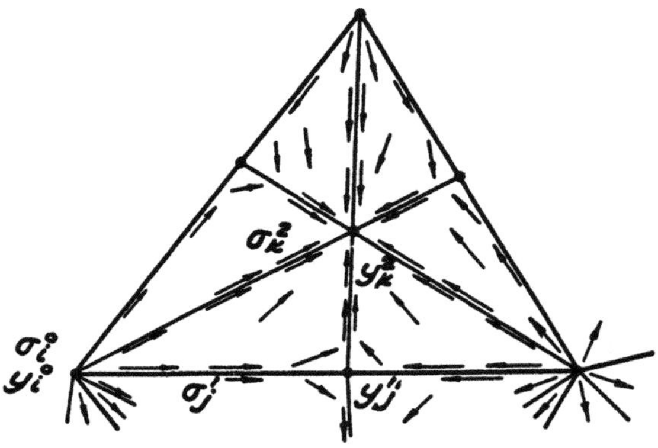

FIG. 8

Let L' be the dual complex of L; its $(n - 1)$-cells cut the 1-cells of L. Since $v(p)$ is defined and $\neq 0$ on L'^{n-1}, we may deform it into a field of unit vectors, and hence define W^n as a cocycle in L' (or in a locally isomorphic complex; see § 15). The dual $D(W^n) = C^0$ of W^n (see § 10) is a cycle in L, a *characteristic* 0-*cycle*.

We shall determine $W^n \cdot M^n = C^0 \cdot I$ in the case $n = 2$. First take a τ_i^2 of L' dual to y_i^0 in L; $W(\tau_i^2) = C(y_i^0)$ is easily seen to be 1 (see Fig. 8); for if we let p run around y_i^0 in a definite direction, $v(p)$ rotates once in the same direction. If p runs around a vertex y_j^1, then $v(p)$ rotates in the opposite direction, so that $C(y_j^1) = -1$. Similarly $C(y_k^2) = 1$. By adding, we obtain

$$\overset{0}{C} \cdot I = \sum C(\overset{r}{y_i}) = \chi(L),$$

χ being the characteristic of L and hence of M^n (see § 8). The same fact may be proved for general n; hence[35]

[35] This theorem is due to H. Hopf. Compare Alexandroff-Hopf, *op. cit.*, p. 549, and Stiefel, [2], p. 37.

(17.2) $$W^n \cdot M^n = C^0 \cdot I = \chi(M^n).$$

REMARK. If n is odd, then $2W^n$ is 0 (see § 14); hence $2\chi(M^n) = 0$, and $\chi(M^n) = 0$.

REMARK. If we push the points p in the direction of the vectors $v(p)$, we obtain a continuous mapping of M^n into itself, with the fixed points $y_i^\prime$; $C^0 \cdot I$ is the sum of the indices of these fixed points.

To determine W^r for $r < n$ we may define a set of $n - r + 1$ vector fields on M^n which are independent except on L^{n-r}; it may be shown that C^{n-r} is the sum of all $(n - r)$-cells of L properly oriented if integer coefficients are used (see [5]).

18. The characteristic classes of normal bundles. Given $M^n \subset M^m$ (compare § 19), we shall derive a formula for W_N^{m-n} of the normal bundle; we may suppose $m \leqq 2n$.

We may take $M^n \subset M^m \subset E^{2m+1}$, and suppose both manifolds are twice differentiable; then the normal spheres $S(p) = S^{m-n-1}(p)$, at present spheres in E^{2m+1}, may be taken so small that after projecting them into M^m they form a "tube" $\mathfrak{S}$ surrounding M^n in M^m.

Let K be a subdivision of M^n, and let ϕ be a projection of K^{m-n-1} into $\mathfrak{S}$. For each σ^{m-n} let $\phi(p)$ be chosen on or inside $S(p)$ in M^m, so that ϕ is continuous; we may make $\phi(p) \neq p$ except at most at a single point of σ^{m-n}. Let $K^{\prime m-n}$ denote the set of all $\phi(p)$, $p \in K^{m-n}$; this is a deformed position of K^{m-n}. Now the Kronecker index $\{\sigma^{\prime m-n}, M^n\}^0$ is defined, and it is easily seen that it equals $W_{N,\phi}(\sigma^{m-n})$, reduced mod 2 if necessary; thus $W_{N,\phi}^{m-n}$ is the "intersection chain" of $K^{\prime m-n}$ with M^n.[36] Clearly, for any chain A^{m-n},

(18.1) $$W_N^{m-n} \cdot A^{m-n} = \{A^{\prime m-n}, M^n\}^0.$$

Suppose that $m = 2n$. Then this gives

[36] Compare H. Whitney, *On products in a complex, loc. cit.*, § 20.

(18. 2) $$W_N^n \cdot M^n = \{M'^n, M^n\}^0.$$

For example, let M^2 be a Möbius strip, and let M^1 be a curve running around it. We may push M^1 into a position M'^1 in M^2 so that it crosses M^1 at just one point. Hence $W_N^1 \cdot M^1 = 1_2$. (As $m - n = 1$, coefficients mod 2 must be used.)

Suppose $M^n \subset E^m$ is orientable. Then M^n may be thought of as a singular cycle bounding a singular chain A^{n+1} in E^m. Now for any $(m-n)$-I_μ-cycle B'^{m-n} of the deformed K'^{m-n} (we use I_2 only if coefficients are mod 2),

$$W_N^{m-n} \cdot B'^{m-n} = \{B'^{m-n}, \partial A^{n+1}\}_\mu^0 = \pm \{\partial B'^{m-n}, A^{n+1}\}_\mu^0 = 0_\mu;$$

hence $W_N^{m-n} \sim 0$, and $W_N^{m-n} = 0$ for the normal bundle of an orientable $M^n \subset E^m$.[37] If M^n is not orientable, then this holds mod 2, but not necessarily otherwise; see Part I.

It is also possible for us to define W_N^r for $r < m - n$ by intersection properties, using the cartesian product of M^n and an $(m - n - r)$-cell.

For $M^n \subset M^m$ we may determine a characteristic cycle $C_N^{n-(m-n)} = C_N^{2n-m}$ directly, as follows: Deform M^n slightly into a position M'^n in M^m as free from intersections with M^n as possible. Then any σ'^{m-n} intersects M^n in at most a single point, as before. The intersections will be along a set of submanifolds M^{2n-m}, which may be taken as formed of cells of the dual subdivision of M^n. If $\tau = \tau_i^{2n-m}$ is dual to $\sigma = \sigma_i^{m-n}$, then τ is in M^{2n-m} if and only if $W_N(\sigma) \neq 0$; thus M^{2n-m}, as a chain, is a characteristic cycle:

(18. 3) $$C_N^{2n-m} = \{M'^n, M^n\}.$$

For example, let M^2 be a Möbius strip in E^3. Then we may deform it away from itself, except along an arc crossing the strip. Then C_N^1 is this arc (taken mod 2).

Now suppose M^n is immersed in M^m; then we may define the

[37] This is proved slightly differently in [3], p. 795. The proof may be made still more direct with the help of intersection chains; see footnote 36.

(18.4) *distant intersection* $D\{M^n\}$ of M^n with itself,

as being the set of submanifolds M^{2n-m} along which M^n cuts itself; each part of M^{2n-m} must be considered a chain in each of the two pieces of M^n intersecting there. For instance, for an M^1 in the form of a figure 8 in the plane E^2, $D\{M^1\}$ is a point p, counted once in each piece of the curve; but, as is easily seen, with opposite signs. In fact, *if* $M^n \subset E^{2n}$ *is orientable, and* n *is odd, then* $D\{M^n\} \sim 0$; for each point counts twice, with opposite signs. For M^n orientable or not, and any n, if $M^n \subset E^{2n}$, then $(D\{M^n\})_2 \sim 0$.

For another example, consider the usual immersion of the Klein bottle B^2 in E^3. Here $D\{B^2\}$ (reduced mod 2) is a circle, counted once as a circle on one side of part of the tube, and once as a circle going around another part of the tube. Note that $D\{B^2\}$ is not ~ 0.

If M^n is immersed in M^m, and we wish to determine C_N^{2n-m} as before, we must use only the local intersection, and hence remove the distant intersection. We find, therefore,

$$(18.5) \quad C_N^{2n-m} = \{M'^n, M^n\} - D\{M^n\} \quad (M^n \text{ immersed in } M^m).$$

This may be checked for the $B^2 \subset E^3$ above.

19. Product bundles. Take $M^n \subset M^m \subset E^s$. Then at each $p \in M^n$ there are tangent planes $P^m(p)$ and $P^n(p) \subset P^m(p)$, and a normal plane $P^{m-n}(p)$ to $P^n(p)$ in $P^m(p)$. The sphere $S(p) \subset P^m(p)$ we shall call the *join* (or product) of the spheres $S_T(p) = S^{n-1}(p)$ and $S_N(p) = S^{m-n-1}(p)$. Thus the part of the tangent bundle of M^m over M^n is the product of the tangent bundle of M^n and the normal bundle of M^n in M^m. We shall write $\mathfrak{B} = \mathfrak{B}_T \times \mathfrak{B}_N$, or

$$(19.1) \quad \mathfrak{B}(M^m; M^n) = \mathfrak{B}(M^n) \times \mathfrak{B}_N(M^n, M^m).$$

In general, given $\mathfrak{B}_1$ and $\mathfrak{B}_2$, each with the base space K, we may define $S(p)$ as the join of $S_1(p)$ and $S_2(p)$, and thus define the product $\mathfrak{B}$, with the base space K.

20. The characteristic classes of a product bundle. Recall that $(a)_2$ is the integer a reduced mod 2; set $(0_2)_2 = 0_2$, $(1_2)_2 = 1_2$. Also set $\omega(0_2) = 0$, $\omega(1_2) = 1$ (we could use any odd number here), and $\omega(a) = a$ for integers a. Set $\omega(\sum a_i\sigma_i^r) = \sum \omega(a_i)\sigma_i^r$, etc. Define

$$X^r \underline{\smile} Y^s = \omega X^r \smile \omega Y^s. \tag{20.1}$$

This has topological significance only in certain cases (see [4], § 12); it need not even be a cocycle. Finally, set $W^0 = I = \sum$ vertices. We must be careful about meanings if non-normal complexes are used.

THEOREM. *The characteristic classes of $\mathfrak{B} = \mathfrak{B}_1 \times \mathfrak{B}_2$ are given by the following formulas, according as the coefficient group I_2 or I_0 is used in W^r; in the latter case the chains are chains of the complexes associated with the bundles.*

$$W^r = \begin{cases} \sum (W_1^i \smile W_2^{r-i})_2 & \text{if } I_2 \text{ is used,} \\ \sum W_1^i \underline{\smile} W_2^{r-i} & \text{if } I_0 \text{ is used.} \end{cases} \tag{20.2}$$

We shall call this the *duality theorem*; see (21. 9).

If K is a manifold, then we may define characteristic cycles; in the properly chosen complexes (20. 2) gives

$$C^{n-r} = \sum \{C_1^{n-i}, C_2^{n-r+i}\}_2 \quad \text{or} \quad \sum \{\omega C_1^{n-i}, \omega C_2^{n-r+i}\}. \tag{20.3}$$

We give a few special cases of the theorem:

$$W^1 = W_1^1 \smile W_2^0 + W_1^0 \smile W_2^1 = W_1^1 + W_2^1. \tag{20.4}$$

For example, if $M \subset E^n$, then $W^1 = 0$, and $W_N^1 = W_T^1$, as is well known.

(20. 5) $W^2 = (W_1^2)_2 + W_1^1 \smile W_2^1 + (W_2^2)_2$ if $\nu \geqq 2$;

(20. 6) $W^2 = W_1^1 \underline{\smile} W_2^1$, $2W^2 = 0$, if $\nu_1 = \nu_2 = 0$, $\nu = 1$.

The W^r for r odd are best found with the help of (14. 1).

21. Formal power series expressions. In this section we shall always reduce mod 2, and use $\equiv$. Write XY for $X \smile Y$, $(X)^2$ for $X \smile X$, etc. Given any $\mathfrak{B}$, define

$$(21.1) \qquad W = \sum W^i t^i, \qquad \overline{W} = \frac{1}{W} = \sum \overline{W}^i t^i;$$

W^0 takes the place of 1 here. Equating coefficients in $W\overline{W} = 1$ gives

$$(21.2) \qquad \overline{W}^0 W^0 \equiv 1, \qquad \overline{W}^0 \equiv W^0 (\equiv 1);$$

$$(21.3) \qquad \overline{W}^1 W^0 + \overline{W}^0 W^1 \equiv \overline{W}^1 + W^1 \equiv 0, \qquad \overline{W}^1 \equiv W^1;$$

$$(21.4) \qquad \overline{W}^2 + \overline{W}^1 W^1 + W^2 \equiv 0, \qquad \overline{W}^2 \equiv W^2 + (W^1)^2;$$

$$(21.5) \qquad \overline{W}^3 + \overline{W}^2 W^1 + \overline{W}^1 W^2 + W^3 \equiv 0, \qquad \overline{W}^3 \equiv W^3 + (W^1)^3;$$

$$(21.6) \qquad \overline{W}^4 \equiv W^4 + (W^2)^2 + W^2(W^1)^2 + (W^1)^4, \text{ etc.}$$

The duality theorem (mod 2) gives:

$$(21.7) \qquad W \equiv W_1 W_2, \ \overline{W} \equiv \overline{W}_1 \overline{W}_2, \ W_1 \equiv W\overline{W}_2, \ \overline{W}_1 \equiv \overline{W} W_2,$$

etc. Hence, for example,

$$(21.8) \qquad W_1^r \equiv \sum W^i \overline{W}_2^{r-i}, \qquad \overline{W}_1^r \equiv \sum \overline{W}^i W_2^{r-i}.$$

For $M^n \subset E^m$ we have tangent and normal bundles, $\mathfrak{B}_T$ and $\mathfrak{B}_N$. The duality theorem gives, as $W \equiv 1$ for E^m,

$$(21.9) \qquad W_T^r \equiv \overline{W}_N^r, \qquad W_N^r \equiv \overline{W}_T^r \qquad (\text{for } M^n \subset E^m).$$

22. Proof of (20. 6). This case of the duality theorem is relatively simple to prove. We choose it also since it shows why products must be used, and it illustrates the use of non-normal complexes.

Choose coördinate systems $\xi_{\sigma,1}$ for $\mathfrak{B}_1$ so that (11. 3) holds. The system $\xi_{\sigma,1}(p)$ maps e_1 and $-e_1$ (the points of S_0^0) into the two points of $S_1(p)$. Choose $\xi_{\sigma,2}$ for $\mathfrak{B}_2$ so that (11. 3) holds, and so that $\xi_{\sigma,2}(p)$ maps e_2 and $-e_2$ into $S_2(p)$. Then if we map S_0^1 into $S(p)$, $\xi_\sigma(p)$ may be defined as follows:

$$\xi_\sigma(p, ae_1 + be_2) = a\xi_{\sigma,1}(p, e_1) + b\xi_{\sigma,2}(p, e_2) \qquad (a^2 + b^2 = 1);$$

then (11. 3) holds for the ξ_σ.

Set, for each vertex x_i of K,

$$\phi(x_i) = \xi_{x_i,1}(x_i, e_1) = \xi_{x_i}(x_i, e_1), \tag{22. 1}$$

$$\phi'(x_i) = \xi_{x_i,2}(x_i, e_2) = \xi_{x_i}(x_i, e_2) \tag{22. 2}$$

then (compare (12. 1), etc.)

$$\psi_{x_i}(x_i) = e_1, \qquad \psi'_{x_i}(x_i) = e_2. \tag{22. 3}$$

Let these define W_1^1 and W_2^1 respectively. Let S^+ be the upper semicircle of S_0^1, from e_1 through e_2 to $-e_1$. Take any $\sigma^1 = x_ix_j$ $(i < j)$. By (11. 3) and (22. 3) $\psi_{\sigma^1}(x_i) = e_1$. Hence, by definition of $W_1^1(\sigma^1)$, $\psi_{\sigma^1}(x_j) = e_1$, or $-e_1$, according as $W_1^1(\sigma^1) = 0_2$ or 1_2. In the former case set $\phi(p) = \xi_{\sigma^1}(p, e_1)$ $(p \in \sigma^1)$; in the latter case define $\phi(p)$ so that $\psi_{\sigma^1}(p)$ runs across S^+ as p runs from x_i to x_j.

Let ϕ define W^2. Take any $\sigma^2 = x_ix_jx_k$ $(i < j < k)$. We shall show that, if $W_2^1(x_ix_j) = 0_2$ or $W_1^1(x_jx_k) = 0_2$, then $W^2(\sigma^2) = 0$, while if $W_2^1(x_ix_j) = W_1^1(x_jx_k) = 1_2$, then

$$\begin{aligned} W^2(\sigma^2) &= -1 \quad \text{if} \quad W_1^1(x_ix_j) = 0_2, \\ W^2(\sigma^2) &= +1 \quad \text{if} \quad W_1^1(x_ix_j) = 1_2. \end{aligned} \tag{22. 4}$$

This will prove

$$\omega W_2^1(x_ix_j)\omega W_1^1(x_jx_k) = -\rho[W_1^1(x_ix_j)]W^2(\sigma^2),$$

and hence, using (15. 8), we find the required formula,

$$W^2 = -(W_2^1 \smile W_1^1) \smile W_2^1 \smile W_1^1 \smile W_1^1 \smile W_2^1.$$

Now $W(\sigma^2)$ is the degree d of the mapping $\psi_{\sigma^2}(p)$ of $\partial\sigma^2$ into S_0^1. We shall determine d by counting the algebraic number of times that $\psi_{\sigma^2}(p)$ crosses the semicircle S^- opposite S^+; the positive direction of S^- is from $-e_1$ through $-e_2$ toward e_1.

On account of (11. 3) $\psi_{\sigma^2}(p) = \psi_{x_ix_j}(p)$ in x_ix_j, and $= \psi_{x_ix_k}(p)$ in x_ix_k; hence ψ_{σ^2} maps $x_ix_j + x_ix_k$ into S^+. We must still consider x_jx_k.

Suppose $W_2^1(x_ix_j) = 0_2$. Then $\psi'_{x_ix_j}(x_j) = e_2$, i.e. $\xi_{x_ix_j,x_j,2}(x_j, e_2)$

$= e_2$. Since $\xi_{x_ix_j,2} = \xi_{\sigma^2,2}$ and $\xi_{x_j,2} = \xi_{x_jx_k,2}$, this gives $\xi_{\sigma^2,x_jx_k,2}(x_j, e_2) = e_2$, and hence

$$\xi_{\sigma^2,x_jx_k}(p, e_2) = e_2, \qquad p \in x_jx_k.$$

Since $\psi_{x_jx_k}$ maps x_jx_k into S^+, the same is true of

$$\psi_{\sigma^2}(p) = \xi_{\sigma^2,x_jx_k}(p, \psi_{x_jx_k}(p));$$

hence $d = 0$. We suppose now that $W_2^1(x_ix_j) = 1_2$, in which case $\psi_{\sigma^2}(p)$ maps x_jx_k into S^- (or into an end of it).

Suppose that $W_1^1(x_jx_k) = 0_2$. Then as $W_1^1 \cdot \partial\sigma^2 = \delta W_1^1 \cdot \sigma^2 = 0$, $W_1^1(x_ix_k) = W_1^1(x_ix_j)$, and $\psi_{x_ix_k}(x_k) = \psi_{x_ix_j}(x_j)$; hence, by (11. 3), $\psi_{\sigma^2}(x_j) = \psi_{\sigma^2}(x_k)$, so that $\psi_{\sigma^2}(p) =$ constant for $p \in x_jx_k$, and again $d = 0$.

Suppose, finally, that we have $W_2^1(x_ix_j) = W_1^1(x_jx_k) = 1_2$. If $W_1^1(x_ix_j) = 0_2$, then $\psi_{\sigma^2}(x_j) = \psi_{x_ix_j}(x_j) = e_1$, $\psi_{\sigma^2}(x_k) = -e_1$, and $\psi_{\sigma^2}(p)$ runs along S^- from e_1 to $-e_1$; hence $d = -1$. If $W_1^1(x_ix_j) = 1_2$, then $\psi_{\sigma^2}(x_j) = -e_1$, $\psi_{\sigma^2}(x_k) = e_1$, and $\psi_{\sigma^2}(p)$ runs in the other direction, and $d = 1$. This proves (22. 4) and hence (20. 6).

23. The characteristic classes in low-dimensional manifolds. We shall study the characteristic classes of the tangent bundles of manifolds of dimensions 2, 3, and 4. Certain results follow from the fact that M^n may be immersed in E^{2n}. More detailed results follow from a study of the parts of the classes in submanifolds of M^n.

Suppose M^n is imbedded in E^m. Then $(W_N^{m-n})_2 = 0$, by § 18. Hence, by use of (21. 9),

$$(23.\ 1) \qquad \overline{W}_T^{m-n} \equiv 0 \quad \text{if } M^n \text{ can be imbedded in } E^m.$$

Recall that M^n can always be immersed in E^{2n}, so that the distant intersection $D\{M^n\}$ is ~ 0 (mod 2) (see § 18); hence $(W_N^n)_2 = 0$, and $\overline{W}_T^n \equiv 0$. Using a remark in § 17, and (21. 4), etc., we have, therefore,[38]

[38] These relations are trivial if M^n is open or bounded; for then the highest cohomology group vanishes.

(23. 2) in any M^2, $\overset{2}{W_T} + (\overset{1}{W_T})^2 \equiv 0$;

(23. 3) in any M^3, $(\overset{1}{W_T})^3 \equiv 0$;

(23. 4) in any M^4, $\overset{4}{W_T} + (\overset{2}{W_T})^2 + \overset{2}{W_T}(\overset{1}{W_T})^2 + (\overset{1}{W_T})^4 \equiv 0$, etc.

Formulas (23. 2) and (17. 2) show that in any closed M^2, $(W^1 \smile W^1)\cdot M^2 = 0_2$ or 1_2 according as $\chi(M^2)$ is even or odd; hence, if M^2 is orientable, $\chi(M^2)$ is even. Formula (23. 3) gives no information in case M^3 is orientable.

LEMMA. *For any smooth closed curve M^1 in M^2,*

$$\{M'^1, M^1\}_2^0 = \overset{1}{W}\cdot \overset{1}{M}. \tag{23. 5}$$

For the tangent bundle of M^1 is simple; hence, by (18. 2) and (20. 4),

$$\{M'^1, M^1\}_2^0 = \overset{1}{W_N}\cdot \overset{1}{M} = (\overset{1}{W} + \overset{1}{W_T})\cdot \overset{1}{M} = \overset{1}{W}\cdot \overset{1}{M}.$$

LEMMA. *For any 1-I_2-cocycle X^1 of any M^2,*

$$(X^1 \smile X^1)\cdot M^2 = (X^1 \smile W^1)\cdot M^2. \tag{23. 6}$$

This follows from the last lemma, on using a cycle dual to X, which is homologous to a sum of smooth curves in M^2.

LEMMA. *Any 2 - I_0-cycle (2 - I_2-cycle) in any M^n is homologous to the sum of a set of disjoint closed oriented surfaces (closed surfaces) imbedded in M^n.*

This lemma, due essentially to Stiefel,[39] is proved by considering the cycle a singular cycle, and making small deformations until surfaces are obtained.

THEOREM. *In any M^3,*

$$W^2 = W^1 \smile W^1. \tag{23. 7}$$

---

[39] Stiefel, [2], p. 48. It would be very useful to obtain such a lemma for r-cycles, $r > 2$. Possibly the use of the lemma can be avoided below by the following method: Define the "characteristic tangent classes" of a cycle by the formulas of § 17; define "characteristic normal classes" of a cycle by formulas of § 18; prove that the duality theorem holds with these definitions.

REMARK. Thus $\overline{W}^2 = 0$, a seemingly stronger condition than (23. 3).

If we use a theorem in § 9 and the last lemma it is sufficient to show that for any closed M^2 in M^3, $W^2 \cdot M^2 = (W^1 \smile W^1) \cdot M^2$. Now $W_N^2 = 0$, as the normal bundle of M^2 in M^3 consists of 0-spheres. Hence, using (20. 5), (20. 4), (23. 2), and (23. 6) with $X^1 = W^1$, we find

$$\begin{aligned} W^2 \cdot M^2 &= [W_T^2 + W_T^1 \smile (W^1 + W_T^1) + W_N^2]_2 \cdot M^2 \\ &= 2(W_T^1 \smile W_T^1) \cdot M^2 + (W^1 \smile W^1) \cdot M^2 = (W^1 \smile W^1) \cdot M^2. \end{aligned}$$

COROLLARY. *In any orientable* M^3, $\mathbf{W}^2 = 0$.

A different proof, not using (20. 3), is given by Stiefel [2, p. 39]. It follows that the tangent bundle of any orientable M^3 is simple (see [2, p. 7] or [3, p. 788]; hence a "parallelism" may be defined in M^3.

THEOREM. *In any orientable* M^4, $\mathbf{W}^3 = 0$.

We shall sketch the proof. By § 9 it is sufficient to prove that for any 3-I_μ-cycle A^3,

$$A^3 \cdot W^3 = A^3 \cdot (\tfrac{1}{2}\delta\omega W^2) = 0_\mu.$$

Choose a 3-I_0-chain B^3 with $(B^3)_\mu = A^3$; say $\partial B^3 = \mu C^2$. Then $A^3 \cdot W^3 = [\frac{1}{2}\mu(C^2 \cdot \omega W^2)]_\mu$; hence it is sufficient to show that $(C^2 \cdot \omega W^2)_2 = 0_2$, i.e. that $C^2 \cdot W^2 = 0_2$.

Let M_1^2, M_2^2, $\cdots$ be disjoint oriented surfaces in M^4 with $C^2 \sim \sum M_i^2$. Since $\mu \sum M_i^2 \sim \partial B^3 \sim 0$, we can choose a singular chain D^3 in M^4 with $\partial D^3 = \sum \mu M_i^2$. Let D_i^3 be the part of D^3 near M_i^2; say $\partial D_i^3 = \mu M_i^2 + E_i^2$. The chain E_i^2 may be considered a cycle in the total space $\mathfrak{S}_N$ of the normal bundle of M_i^2 in M^4.

Let PE_i^2 be the projection of E_i^2 into M_i^2; then $\mu M_i^2 \sim PE_i^2$ in M_i^2. But it may be shown that any $(\nu + 1)$-cycle in a complex K which may be obtained as the projection of a cycle in $\mathfrak{S}$, for a bundle $\mathfrak{B}$ of ν-spheres over K, is orthogonal to $W^{\nu+1}(K)$. Hence $\mu M_i^2 \cdot W_{N,i}^2 = 0$. Since I_0 is used in $W_{N,i}^2$, $M_i^2 \cdot W_{N,i}^2 = 0$. Since M_i^2 is orientable,

$W^1_{T,i} = 0$. Also, by (23. 2), $W^2_{T,i} \cdot M^2_i \equiv 0 \pmod 2$. Therefore, by (20. 5),

$$W^2 \cdot M^2_i = (W^2_{T,i} + W^1_{T,i} \cup W^1_{N,i} + W^2_{N,i})_2 \cdot M^2_i = 0_2.$$

Hence $C^2 \cdot W^2 = 0_2$, as required.

24. Example of an M^3 with $W^2 \neq 0$. Let P^2 be the projective plane. Let M^3 be the cartesian product of P^2 and a 1-sphere. We may regard P^2 as imbedded in M^3; then clearly $W^1_N = 0$. Hence $W^1_T = W^1$ over P^2, and since $\chi(P^2) = 1$, (23. 7), (20. 4), (23. 2), and (17. 2) give

$$W^2 \cdot P^2 = (W^1 \cup W^1) \cdot P^2 = (W^1_T \cup W^1_T) \cdot P^2 = W^2_T \cdot P^2 = 1_2;$$

hence $W^2 \neq 0$. We could also use (20. 5) to prove this.

We shall study more fully an example to be used in § 25. Consider the cylinder plus interior L^3:

$$x^2 + y^2 \leqq 1, \qquad -1 \leqq z \leqq 1.$$

Now make the following identifications:

$$(x, y, z) = (-x, -y, z) \ (x^2 + y^2 = 1), \ (x, y, -1) = (x, -y, 1).$$

We obtain M^3, which may easily be expressed as a differentiable manifold. The identifications define a continuous mapping f of L^3 into M^3. Define, in L^3, the subsets

$$\text{(24. 1)} \qquad \bar{P}^2 : z = 0; \qquad \bar{Q}^2 : x = 0;$$
$$\bar{S}^1_1 : x = y = 0; \qquad \bar{S}^1_2 : x = z = 0; \qquad \bar{p}_0 : x = y = z = 0.$$

Set $P^2 = f(\bar{P}^2)$, etc. Then P^2 is a projective plane, Q^2 is a Klein bottle, and S^1_1 and S^1_2 are simple closed curves.

We may show that

$$\text{(24. 2)} \qquad \begin{gathered} \{P^2, P^2\}_2 \sim 0, \qquad \{P^2, Q^2\}_2 \sim (S^1_2)_2, \\ \{Q^2, Q^2\}_2 \sim (S^1_1 + S^1_2)_2, \\ \{P^2, S^1_1\}_2 \sim (p_0)_2, \qquad \{P^2, S^1_2\}_2 \sim \{Q^2, S^1_1\}_2 \sim 0, \\ \{Q^2, S^1_2\}_2 \sim (p_0)_2. \end{gathered}$$

For instance, by increasing z we deform $\overline{P}^2$ away from itself; applying f deforms P^2 away from itself; hence $\{P^2, P^2\} \sim 0$. If for $z \geqq \frac{1}{2}$ we twist $\overline{Q}^2$ in one sense through an angle α, and for $z \leqq -\frac{1}{2}$ we twist it in the opposite sense by the same amount, and for $-\frac{1}{2} \leqq z \leqq \frac{1}{2}$ we twist it by an amount varying continuously, then applying f deforms Q^2 into Q'^2; we clearly have $\{Q'^2, Q^2\}_2 = (S_1^1 + S_2^1)_2$.

We shall determine the characteristic cycles $\mathbf{C}^r$ of the tangent bundle of M^3. Of course, $C^3 = M^{*3}$, in the associated complex. Also $\mathbf{C}^0 = 0$, since 3 is odd. Further,

$$(24.3) \qquad \overset{2}{C} = (\overset{2}{P} + \overset{2}{Q})_2, \qquad \overset{1}{C} = (\overset{1}{S_1} + \overset{1}{S_2})_2.$$

To prove the first relation we note that S_1^1 and S_2^1 both reverse the orientation in M^3; this checks with

$$\{\overset{1}{S_1}, \overset{2}{P} + \overset{2}{Q}\}_2^0 = \{\overset{1}{S_2}, \overset{2}{P} + \overset{2}{Q}\}_2^0 = 1_2.$$

The second relation follows from this, (23. 7), and (24. 2). (In the dual form $\mathbf{C}^1 = \{C^2, C^2\}$.)

Note that, if the cells of M^3 are properly oriented, we find[40]

$$(24.4) \qquad \partial M^3 = 2(P^2 + Q^2).$$

25. Example of an M^4 with $W^3 = \overline{W}^3 \neq 0$. By (23. 1) this M^4 *cannot be imbedded in E^7.*

First define M_0^3 as in § 24. Determine $\mathfrak{B}_1$ as the 0-sphere bundle with M_0^3 as base space for which

$$\overset{2}{C_1} = (\overset{2}{Q_0})_2.$$

Let $\mathfrak{B}$ be the product of $\mathfrak{B}_1$ and a simple 0-sphere bundle $\mathfrak{B}_2$ over M_0^3. We set $M^4 = \mathfrak{S}(\mathfrak{B})$.

Since $\mathfrak{B}_2$ is simple, $f(p) = \xi_{\sigma,2}(p, e_2)$ defines a projection of M_0^3 into M^4; let M^3 be the set $f(M_0^3)$. Use P_0^2 in M_0^3, etc.

Consider the normal bundle $\mathfrak{B}_N(M^3, M^4)$ of M^3 in M^4. For

[40] It may be shown in general that, if $\partial M^n = 2C^{n-1}$, then C^{n-1} is a characteristic cycle, dual to W^1.

any $q \varepsilon M^3$, say $q \varepsilon S(p)$, $p \varepsilon M_0^3$, we may regard the normal 0-sphere $S_N(q)$ of $\mathfrak{B}_N$ as being realized by the pair of points of $S(p)$ orthogonal to q. But these are just the points of $S_1(p)$ of the bundle $\mathfrak{B}_1$; hence we may consider $\mathfrak{B}_N = \mathfrak{B}_1$. In particular,

$$C_N^2(M^3, M^4) = (Q^2)_2. \tag{25. 1}$$

If $C^3(M^4)$ is a characteristic cycle of M^4, dual to W^1, then

$$C^2(M^4; M^3) = \{C^3(M^4), M^3\}$$

is dual, in M^3, to the part of W^1 over M^3. By the duality theorem, (24. 3), and (25. 1),

$$C^2(M^4; M^3) \sim C^2(M^3) + C_N^2(M^3, M^4) = (P^2)_2 \quad \text{in} \quad M^3.$$

From this it follows that

$$C^3(M^4) \sim \mathfrak{S}(P_0^2)_2 \qquad \text{(total space over } P_0^2 \text{ in } \mathfrak{B}). \tag{25. 2}$$

As P_0^2 can be deformed away from itself in M_0^3, so can $\mathfrak{S}(P_0^2)$ in M^4; hence

$$\{C^3(M^4), C^3(M^4)\} \sim 0, \quad W^1 \smile W^1 = 0, \quad (W^1)^3 = 0. \tag{25. 3}$$

Next, since $C_N^1(M^3, M^4) = 0$ (the spheres being 0-spheres), the duality theorem and (25. 1) give

$$\begin{aligned} C^1(M^4; M^3) &\sim C^1(M^3) + \{C^2(M^3), C_N^2(M^3, M^4)\} \\ &\sim (S_1^1 + S_2^1)_2 + \{P^2 + Q^2, Q^2\}_2 \sim (S_2^1)_2. \end{aligned}$$

Hence

$$\{C^2(M^4), M^3\} \sim (S_2^1)_2, \qquad C^2(M^4) \sim \mathfrak{S}(S_{20}^1)_2. \tag{25. 4}$$

We must determine $C^1(M^4)$. Since $W^3(M^4)$ is a cocycle of the complex associated with the tangent bundle of M^4, its dual, $C^1(M^4)$, is a cycle of the geometric complex of M^4 (compare (15. 5)). We determine it by (14. 1) and (25. 4):

$$C^1(M^4) = \tfrac{1}{2}\partial\omega C^2(M^4) = \tfrac{1}{2}\partial\mathfrak{S}(S_{20}^1).$$

Now in M_0^3,

$$\{\overset{1}{S}_{20}, \overset{2}{C}(\mathfrak{B})\}_2^0 = \{\overset{1}{S}_{20}, \overset{2}{C}(\mathfrak{B}_1) + \overset{2}{C}(\mathfrak{B}_2)\}_2^0 = \{\overset{1}{S}_{20}, \overset{2}{Q}_0\}_2^0 = 1_2;$$

hence the orientation of $\mathfrak{B}$ is reversed over $\overset{1}{S}_{20}$. Therefore $\mathfrak{S}(\overset{1}{S}_{20})$ is a Klein bottle, and we may orient its cells so that $\partial\mathfrak{S}(\overset{1}{S}_{20}) = 2\mathfrak{S}(p_{00})$. (This is equivalent to a choice of ω above.) Hence

(25. 5) $$\overset{1}{C}(\overset{4}{M}) = \tfrac{1}{2}\partial\mathfrak{S}(\overset{1}{S}_{20}) = \mathfrak{S}(p_{00}).$$

Using (21. 5) and (25. 3) we have

(25. 6) $$\bar{C}^1(M^4) = C^1(M^4) = \mathfrak{S}(p_{00}).$$

Since this is not ~ 0 in M^4 (although twice it is), $\bar{\mathbf{C}}^1(M^4) \not\approx 0$, and $\overline{W}^3(M^4) \not\approx 0$.

REMARK. If M^4 is immersed in E^7, then we may find the distant intersection (mod 2) from (18. 5), (25. 6), and (21. 9); since $\{M'^4, M^4\} \sim 0$ (mod 2),

(25. 7) $$\begin{aligned} D\{\overset{4}{M}\} &\equiv \{\overset{\prime 4}{M}, \overset{4}{M}\} + \overset{1}{C}_N(\overset{4}{M}, \overset{7}{E}) \\ &\equiv \overset{1}{\bar{C}}_T(\overset{4}{M}) \equiv \mathfrak{S}(p_{00}). \end{aligned}$$

For further examples see [5].

HARVARD UNIVERSITY

LITERATURE

Only the papers most directly connected with the present problem are cited here. Further references may be found in [3].

1. SEIFERT, H., Algebraische Approximation von Mannigfaltigkeiten, Math. Zeitschr., 41 (1936), 1–17.
2. STIEFEL, E., Richtungsfelder und Fernparallelismus in n-dimensional Mannigfaltigkeiten, Comm. Math. Helv., 8 (1936), 3–51.
3. WHITNEY, H., Topological properties of differentiable manifolds, Bull. Amer. Math. Soc., 43 (1937), 785–805.
4. —— Some combinatorial properties of complexes, Proc. Nat. Acad. Sci., 26 (1940), 143–148.
5. —— On the theory of sphere-bundles, *ibid.*, 26 (1940), 148–153.
6. WHITEHEAD, J. H. C., On the homotopy type of manifolds, Ann. Math., 41 (1940), 825–832.

ANNALS OF MATHEMATICS
Vol. 69, No. 3, May, 1959
Printed in Japan

CLASSIFICATION OF ORIENTED SPHERE BUNDLES OVER A 4-COMPLEX

BY A. DOLD AND H. WHITNEY

(Received April 4, 1958)

1. Introduction

The classification of $(n-1)$-sphere bundles with structure group $SO(n)$ (special orthogonal group) over a complex K of dimension at most 4 has been carried out in several special cases. If $n = 2$ or if the dimension of K does not exceed 3 then the characteristic class W_2 is a complete invariant (see [17]; note that W_2 is an integer class for $n = 2$, and is a class mod 2 if $n \geqq 3$). 2-sphere bundles with vanishing class W_2 were classified by the second author in an unpublished manuscript* (1938; announced in [19], 4); these bundles are not determined by their characteristic classes (see our example in Section 3). Pontrjagin in 1945 (see [9]) gave a solution for arbitrary n provided that $H^4(K; Z)$ has no 2-torsion; in this case $(n-1)$-sphere bundles are characterized by W_2, W_4 (for $n \geqq 4$) and P_4 (see [9] or the Corollary in Section 3). In this paper we give the classification for the general case. Throughout the paper we assume $n \geqq 3$.

The $(n-1)$-sphere bundles over a complex K are generated by mappings of K into some Grassmann manifold G_n (see [17]); in fact, they are in natural one-to-one correspondence with the homotopy classes (see [14], [8], and [15], 19). The manifold G_n has vanishing 1st and 3rd homotopy groups (see (3)); since we assume $\dim(K) \leqq 4$, our problem reduces to the standard problem of classifying mappings of a complex into a space (G_n) with only two non-vanishing homotopy groups. This problem has been studied in many papers, beginning with special cases in [7], [11], and in general in [13], [4]. In particular there are explicit solutions if the homotopy groups of the image space are in dimensions m and $m+2$ (see [13]), but unfortunately not for $m = 2$, the case in which we are interested. A partial solution in this case was given in [2].

---

* This classification involves the cohomology operation Ψ: $H^1(K; Z_2) \to H^4(K; Z)$ which is obtained by applying the Bockstein homomorphism first and then taking the cup square; see (23). In the manuscript Ψ was defined by a cochain formula (in a simplicial complex with ordered vertices) as follows. For any cochain y mod 2, let ωy be the corresponding integral cochain (compare [18, 11]) whose coefficients are 0 or 1. Then $\omega(y \smile y') = \omega y \smile \omega y'$, and for 1-cocycles c mod 2, we see that $\frac{1}{2}\delta\omega c = \omega' c \smile c$). Therefore the function $\phi(c) = \omega(c \smile c \smile c \smile c)$ coincides with $\frac{1}{2}\delta\omega c \smile \frac{1}{2}\delta\omega c$ and induces the operation Ψ. Unfortunately the operation ω was omitted in [19], giving a wrong statement, as noted by Pontrjagin [9].

The general result in [4, Theorem 14.2], leaves one with the task of expressing the k-invariant $k(G_n)$ of G_n by familiar cohomology operations. To be more precise, it is only the suspension of $k(G_n)$ and the deviation of $k(G_n)$ from additivity which is needed in [4, Theorem 14.2]. In the following Section 2, we make these computations and formulate the corresponding classification theorem. In Section 3 we relate the resulting invariants to the characteristic classes of the bundles involved. As an application we give a short proof of a recent theorem of Massey (see [6, Theorem V]). The last Section contains a correction of Pontrjagin's Theorem 2 in [9].

2. The classification theorem ($n \geqq 3$)

Let G_n be the Grassmann manifold of oriented n-planes through the origin in a euclidean space of dimension at least $n+5$. Then G_n is the base of a 5-universal $SO(n)$-bundle (see [15, 19.6–19.7]), and we have natural isomorphisms (see [15, 19.9])

$$\pi_i(G_n) \approx \pi_{i-1}(SO(n)) , \qquad i \leqq 4 . \tag{1}$$

According to [15, 24.11],

$$\begin{cases} \pi_0(SO(n)) \approx \pi_2(SO(n)) = 0 , \\ \pi_1(SO(n)) \approx Z_2 = \text{cyclic of order } 2 , \\ \pi_3(SO(n)) \approx \begin{cases} Z = \text{free cyclic for } n \neq 4 , \\ Z + Z \text{ for } n = 4 ; \end{cases} \end{cases} \tag{2}$$

hence

$$\begin{cases} \pi_1(G_n) \approx \pi_3(G_n) = 0 , \\ \pi_2(G_n) \approx Z_2 , \\ \pi_4(G_n) \approx \begin{cases} Z \text{ for } n \neq 4 , \\ Z + Z \text{ for } n = 4 . \end{cases} \end{cases} \tag{3}$$

We choose generators for $\pi_3(SO(n))$ as in [15, 22.3, 22.7 and 23.6] and thereby identify this group with Z (for $n \neq 4$) or $Z+Z$ (for $n=4$); similarly for the groups $\pi_i(G_n)$, using the isomorphisms (1).

The k-invariant of G_n is an element $k(G_n) \in H^5(Z_2, 2; \pi_4(G_n))$ (see [4, 11]). We want to compute the *suspension* and the *deviation from additivity* of $k(G_n)$.

The suspension of $k(G_n)$ is an element of $H^4(Z_2, 1; \pi_4(G_n)) = H^4(Z_2, 1; \pi_3(SO(n)))$. Up to sign it coincides with $k(SO(n))$, the k-invariant of $SO(n)$ (see [16]).

Consider the case $n = 3$ first. Now $SO(3)$ is homeomorphic with real projective 3-space P_3. A space $K(Z_2, 1)$ is given by real projective space P_∞ of infinite dimension. The k-invariant of P_3 is the first obstruction for retracting P_∞ onto P_3, i.e., the obstruction for retracting P_4 onto P_3. Since such a retraction is not possible the invariant $k(SO(3)) = k(P_3)$ is not zero; hence it is the only non-vanishing element in $H^4(Z_2, 1; Z) = H^4(P_\infty; Z)$. If v is the non-trivial element of $H^1(Z_2, 1; Z_2)$, then since $\beta v \smile \beta v \neq 0$, we have

$$k(SO(3)) = \beta v \smile \beta v \; ; \tag{4}$$

β is the Bockstein homomorphism associated with the coefficient sequence

$$0 \longrightarrow Z \xrightarrow{2} Z \longrightarrow Z_2 \longrightarrow 0 \; . \tag{5}$$

Since v is the basic cohomology class of $P_\infty = K(Z_2, 1)$, the formula (4) also describes the cohomology operation defined by $k(SO(3))$ (see [4, 7] or [12, 4]): If X is any space and $x \in H^1(X, Z_2)$, then

$$k(SO(3)) \vdash x = \beta x \smile \beta x \; . \tag{6}$$

For $n = 4$ the situation is similar since $SO(4)$ is homeomorphic with $SO(3) \times S^3 = P_3 \times S^3$ (see [15, 22.6]). We have

$$\begin{aligned} k(SO(4)) \in H^4(Z_2, 1; \pi_3(SO(4))) &= H^4(Z_2, 1; Z + Z) \\ &= H^4(Z_2, 1; Z) + H^4(Z_2, 1; Z) \; . \end{aligned}$$

If we write $k(SO(4)) = k_1(SO(4)) + k_2(SO(4))$ according to this decomposition, then

$$k_1(SO(4)) \vdash x = \beta x \smile \beta x \; , \qquad k_2(SO(4)) \vdash x = 0 \; . \tag{7}$$

For $n \geqq 5$ the invariant $k(SO(n)) \in H^4(Z_2, 1; Z)$ is zero since there exists a mapping $r : P_{n-1} \to SO(n)$ with $r_* : \pi_1(P_{n-1}) \approx \pi_1(SO(n))$ (see [15, 23.3–23.4]);

$$k(SO(n)) = 0 \qquad \text{for } n \geqq 5 \; . \tag{8}$$

We have now to compute the deviation of $k(G_n)$ from additivity. We claim that

$$k(G_3) \vdash (y, z) = \beta(y \smile z) \; , \tag{9}$$

$$k_1(G_4) \vdash (y, z) = \beta(y \smile z) \; , \qquad k_2(G_4) \vdash (y, z) = 0 \; , \tag{10}$$

$$k(G_n) \vdash (y, z) = 0 \qquad \text{for } n \geqq 5 \; . \tag{11}$$

Here y and z denote 2-dimensional cohomology classes mod 2 of an arbitrary space Y, and $k_1(G_4)$, $k_2(G_4)$ are the components of $k(G_4)$ with respect to the decomposition

$$H^5(Z_2, 2; \pi_4(G_4)) = H^5(Z_2, 2; Z + Z) = H^5(Z_2, 2; Z) + H^5(Z_2, 2; Z) \; .$$

The formulas (9)-(11) follow from our computation (6)-(8) of the suspension of $k(G_n)$, together with the

LEMMA. *Take* $a \in H^5(Z_2, 2; Z)$ *and let* $\sigma a \in H^4(Z_2, 1; Z)$ *be the suspension of* a. *Then*

$$a \vdash (y, z) = \beta(y \smile z) \text{ or } a \vdash (y, z) = 0 \qquad (\textit{for all } y, z \textit{ as above})$$

according to whether

$$\sigma a \vdash x = \beta x \smile \beta x \text{ or } \sigma a \vdash x = 0 \qquad (\textit{for all } x \textit{ as in } (6)).$$

We remark first that $\beta(y \smile z)$ is a non-trivial cohomology operation, even when reduced mod 2: If we denote reduction mod 2 by a horizontal bar, then

$$\overline{\beta(y \smile z)} = Sq^1(y \smile z) = Sq^1 y \smile z + y \smile Sq^1 z \,, \tag{12}$$

and this is different from zero for instance in $S^2 \times X$, where S^2 is a 2-sphere and X is a space consisting of a 2-sphere with a 3-cell attached by a mapping of its boundary of degree 2 (i.e., X is the suspension of the real projective plane).

To prove the lemma we look at the universal example

$$Y = K(Z_2, 2) \times K(Z_2, 2)$$

(see [4]). Let u be the basic class of $H^2(Z_2, 2; Z_2)$. Then in Y we have, by definition of $a \vdash (y, z)$,

$$a \vdash (u \otimes 1 + 1 \otimes u) = a \otimes 1 + 1 \otimes a + a \vdash (u \otimes 1, 1 \otimes u) \,. \tag{13}$$

If we compute $H^5(Y; Z)$ by applying the Künneth formula, using $H^2(Z_2, 2; Z) = 0$, $H^3(Z_2, 2; Z) = Z_2$ (see [3, Ch. IV]), we find

$$\begin{aligned} H^5(Y; Z) = {} & H^5(Z_2, 2; Z) \otimes H^0(Z_2, 2; Z) \\ & + H^0(Z_2, 2; Z) \otimes H^5(Z_2, 2; Z) + Z_2 \,; \end{aligned} \tag{14}$$

the cross term Z_2 has only one non-zero element. This element must be

$$\beta(u \otimes 1 \smile 1 \otimes u) = \beta(u \otimes u) \,, \tag{15}$$

since (15) is a cross term element of $H^5(Y; Z)$ which is different from zero, as we have seen. (Cross term elements can be characterized as lying in the intersection of the kernels of the two homomorphisms $H^*(Y) \to H^*(Z_2, 2)$ which are induced by the two natural inclusions $K(Z_2, 2) \to Y$.) Since $a \vdash (u \otimes 1, 1 \otimes u)$ is also a cross term element, we have

$$a \vdash (u \otimes 1, 1 \otimes u) = \lambda \beta(u \otimes u) \qquad \text{with } \lambda = 0 \text{ or } 1 \,. \tag{16}$$

Reducing this equation modulo 2 gives

$$\bar{a} \vdash (u \otimes 1, 1 \otimes u) = \lambda Sq^1(u \otimes u) \,. \tag{17}$$

The element $\bar{a}$ lies in $H^5(Z_2, 2; Z_2)$. This module (over Z_2) has a base consisting of Sq^2Sq^1u and $u \smile Sq^1u$ (see [12, 9]). Therefore

(18) $$\bar{a} = \mu Sq^2Sq^1u + \nu(u \smile Sq^1u) \quad \text{with } \mu, \nu = 0 \text{ or } 1 .$$

If we apply Sq^1 to (18) the left side gives zero because $\bar{a}$ comes from an integer class. Sq^1 applied to either of Sq^2Sq^1u or $u \smile Sq^1u$ gives $Sq^3Sq^1u = Sq^1u \smile Sq^1u \neq 0$. Therefore the coefficients μ and ν must be equal:

(19) $$\bar{a} = \mu(Sq^2Sq^1u + u \smile Sq^1u) .$$

If we compute $\bar{a} \vdash (u \otimes 1, 1 \otimes u)$ from (19) we find $\mu Sq^1(u \otimes u)$, hence (by (17)) $\mu = \lambda$, and

(20) $$\bar{a} = \lambda(Sq^2Sq^1u + u \smile Sq^1u) .$$

Now apply the suspension homomorphism σ to (20). It commutes with Steenrod squares and kills cup products; therefore

(21) $$\sigma\bar{a} = \lambda Sq^2Sq^1v = \lambda(Sq^1v \smile Sq^1v) ,$$

where v is the basic class in $H^1(Z_2, 1; Z_2)$. Comparing (16) and (21) now proves the lemma; we have only to remark that $\sigma\bar{a} = \overline{\sigma a}$ and $\overline{\beta x \smile \beta x} = Sq^1x \smile Sq^1x$.

REMARK. With some more effort the invariant $k(G_n)$ itself can be computed. It turns out that

(22) $$\begin{aligned} k(G_3) &= \pm\beta_4\mathfrak{p}(u) , \\ k_1(G_4) &= \pm\beta_4\mathfrak{p}(u) , \quad k_2(G_4) = 0 , \\ k(G_n) &= 2\beta_4\mathfrak{p}(u) \qquad \text{for } n \geqq 5 , \end{aligned}$$

where $u \in H^2(Z_2, 2; Z_2)$ is the basic class, $\mathfrak{p} : H^2(Z_2, 2; Z_2) \to H^4(Z_2, 2; Z_4)$ is the Pontrjagin square, and β_4 is the Bockstein homomorphism associated with the coefficient sequence $0 \longrightarrow Z \xrightarrow{4} Z \longrightarrow Z_4 \longrightarrow 0$. The formula for $k(G_4)$ was given to the authors by F. Peterson.

Inserting our computations into [4, 14.2] gives the

THEOREM 1 (CLASSIFICATION THEOREM). *Let $\mathfrak{B}_1$, $\mathfrak{B}_2$ be principal $SO(n)$-bundles, $n \geqq 3$ (or equivalently $(n-1)$-sphere bundles with structure group $SO(n)$*; see [15, 8.2]) *over a complex K whose dimension is at most 4, and let $h_i : K \to G_n$ be a classifying mapping for $\mathfrak{B}_i$* (see [15, 19]). *Assume $W_2(\mathfrak{B}_1) = W_2(\mathfrak{B}_2) = w_2$. This implies that the parts of the $\mathfrak{B}_i$ over the 3-skeleton K^3 of K are equivalent* (see [17]); *we can therefore assume* (see [15, 19]) *that h_1 and h_2 agree on K^3. Then the difference cocycle of (h_1, h_2) is defined; its cohomology class $d(h_1, h_2)$ is an element of $H^4(K; \pi_4(G_n))$.*

The bundles $\mathfrak{B}_1$ and $\mathfrak{B}_2$ are equivalent if and only if:

(a) *for* $n = 3$ *there exists a cohomology class* $x \in H^1(K; Z_2)$ *such that*

$$d(h_1, h_2) = \beta x \smile \beta x + \beta(x \smile w_2) \tag{23}$$

where β *is the Bockstein homomorphism associated with the coefficient sequence* (5), *and* $\pi_4(G_3)$ *has been identified with* Z *by* (3);

(b) *for* $n = 4$ *there is an* $x \in H^1(K; Z_2)$ *such that*

$$d_1(h_1, h_2) = \beta x \smile \beta x + \beta(x \smile w_2)\ , \qquad d_2(h_1, h_2) = 0\ , \tag{24}$$

where $d(h_1, h_2) = d_1(h_1, h_2) + d_2(h_1, h_2)$ *is the decomposition corresponding to* $\pi_4(G_4) = \pi_3(SO(4)) = Z + Z$, *using the chosen generators*;

(c) *for* $n \geqq 5$ *we have*

$$d(h_1, h_2) = 0\ . \tag{25}$$

Since for given $w_2 \in H^2(K; Z_2)$ and $d \in H^4(K; \pi_4(G_n))$ one can always find bundles $\mathfrak{B}_1$, $\mathfrak{B}_2$ and classifying mappings h_1, h_2 with $W_2(\mathfrak{B}_i) = w_2$ and $d(h_1, h_2) = d$, this theorem gives a complete classification.

REMARK. Since $\pi_5(G_n) = \pi_4(SO(n)) = 0$ for $n \geqq 6$, $\pi_6(G_n) = 0$ for $n \geqq 7$ (see [15, 24.11]), and $\pi_7(G_n) = 0$ for $n \geqq 6$ (see [1, Proposition 19.3]), we see that the classification theorem extends to 5-complexes K for $n \geqq 6$, and to 7-complexes for $n \geqq 7$ (for G_n we have now to take oriented n-planes in a euclidean space of dimension at least $n+8$).

3. Characteristic classes

The following theorem is essentially contained in Pontrjagin [9].

THEOREM 2. *Let* $\mathfrak{B}_1$, $\mathfrak{B}_2$ *be bundles with classifying mappings* h_1, h_2 *as in Theorem* 1. *Let* $P(\mathfrak{B}_i) \in H^4(K; Z)$ *denote the Pontrjagin class of* $\mathfrak{B}_i$, *and* (*for* $n = 4$) $W_4(\mathfrak{B}_i) \in H^4(K; Z)$ *respectively* (*for* $n > 4$) $W_4(\mathfrak{B}_i) \in H^4(K; Z_2)$, *its* 4th *Stiefel-Whitney class.*

(a) *If* $n = 3$ *then*

$$P(\mathfrak{B}_1) - P(\mathfrak{B}_2) = -4d(h_1, h_2).$$

(b) *If* $n = 4$ *then*

$$P(\mathfrak{B}_1) - P(\mathfrak{B}_2) = -4d_1(h_1, h_2) - 2d_2(h_1, h_2),$$
$$W_4(\mathfrak{B}_1) - W_4(\mathfrak{B}_2) = d_2(h_1, h_2).$$

(c) *If* $n \geqq 5$ *then*

$$P(\mathfrak{B}_1) - P(\mathfrak{B}_2) = -2d(h_1, h_2),$$
$$W_4(\mathfrak{B}_1) - W_4(\mathfrak{B}_2) \equiv d(h_1, h_2) \bmod 2.$$

PROOF. Let $\rho_n \in \pi_4(G_n)$ for $n \neq 4$ respectively ρ_4^1, $\rho_4^2 \in \pi_4(G_4)$ for $n = 4$

be the generators as in Section 2 (see (3)). We show that

(26) $\langle P, \rho_3\rangle = -4$; $\langle P, \rho_4^1\rangle = -4$, $\langle P, \rho_4^2\rangle = -2$; $\langle P, \rho_n\rangle = -2$ for $n \geqq 5$;

(27) $\langle W_4, \rho_4^1\rangle = 0$, $\langle W_4, \rho_4^2\rangle = 1$; $\langle W_4, \rho_n\rangle \equiv 1 \bmod 2$ for $n \geqq 5$,

where P, W_4 are the characteristic classes of the universal $SO(n)$-bundle over G_n, and $\langle\ \rangle$ denotes the evaluation of a cohomology class on a homotopy class (considered as a homology class).

Consider W_4 first. The generators ρ_4^1, ρ_4^2 correspond to the elements α_3, $\beta_3 \in \pi_3(SO(4))$ (see [15, 22.7]) under the isomorphisms (1); therefore the equations (27) follow from the definition of W_4 as an obstruction (see [15, 38.2]).

The class P is invariant under the inclusion $SO(n) \subset SO(n+1)$, i.e., the class $P(\mathfrak{B})$ of an $SO(n)$-bundle $\mathfrak{B}$ remains the same if we consider $\mathfrak{B}$ as an $SO(n+1)$-bundle (see [5, § 4]). Since the element $\alpha_3 - 2\beta_3$ goes into zero under the inclusion $SO(4) \to SO(5)$ (see [15, 23.5 – 23.6]), this shows that $\langle P, \rho_4^1 - 2\rho_4^2\rangle = 0$, or $\langle P, \rho_4^1\rangle = 2\langle P, \rho_4^2\rangle$. Next we show that $\langle P, \rho_4^2\rangle = -2$. Then the remaining equations (26) follow since ρ_3 goes into ρ_4^1 and ρ_4^2 goes into ρ_n ($n \geqq 5$) under the appropriate inclusions $SO(m) \to SO(m+k)$ (see [15, 22.7 and 23.6]).

To prove $\langle P, \rho_4^2\rangle = -2$ we consider the inclusion $U(2) \to SO(4)$, where $U(2)$ is the unitary group in 2 variables. Thereby every $U(2)$-bundle defines an $SO(4)$-bundle; if $\mathfrak{B}$ is an $SO(4)$-bundle over the 4-sphere S^4 which is obtained in this way then $P(\mathfrak{B}) = -2W_4(\mathfrak{B})$ (see [20, Theorem 9]). In particular this holds for the bundle which is defined by ρ_4^2 (see [15, 25.1]), hence $\langle P, \rho_4^2\rangle = -2\langle W_4, \rho_4^2\rangle = -2$.

Using (26) we now prove the case (a) of the theorem; (b) and (c) follow similarly. Assume that h_1, h_2 are cellular mappings (i.e., map the i-skeleton of K into the i-skeleton of G_3 for some cellular decomposition of K and G_3), which agree on the 3-skeleton of K. Let $\Delta(h_1, h_2)$ be their difference cocycle (its cohomology class is $d(h_1, h_2)$), and let p be a cellular cocycle in the cohomology class P. We show that

$$(28) \qquad h_1^*(p) - h_2^*(p) = -4\Delta(h_1, h_2)\,,$$

where h_i^* is the homomorphism on cellular cochains which is induced by h_i. This equation implies (a) since $h_i^*(P) = P(\mathfrak{B}_i)$.

It is sufficient to verify (28) on every 4-cell of K; we can therefore assume that K consists of a single (closed) 4-cell e, and h_1, h_2 are mappings of e into the 4-skeleton $G_3^{(4)}$ of G_3 which agree on the boundary ∂e of e. Then h_1, h_2 determine an element $\alpha \in \pi_4(G_3^{(4)})$, the "difference" of h_1 and

h_2, which under the injection $G_3^{(4)} \to G_3$ goes into $\Delta(h_1, h_2)\cdot e$.

Let $e \vee S^4$ be the wedge (= union with a single common point) of e with a 4-sphere. Let $f: e \to e \vee S^4$ be a mapping which is the identity on ∂e, and which covers both e and S^4 with degree one. Finally let $h: e \vee S^4 \to G_3^{(4)}$ be a mapping whose restriction to e is h_2 and whose restriction to S^4 represents the class α. Then the composite mapping

$$e \xrightarrow{f} e \vee S^4 \xrightarrow{h} G_3^{(4)}$$

is homotopic rel ∂e to h_1 (i.e., there is a homotopy which does not move the image of ∂e); hence

$$f^*h^*(p) = (hf)^*(p) = h_1^*(p) \ . \tag{29}$$

But $h^*(p)$ restricted to e is $h_2^*(p)$, and the value of $h^*(p)$ on S^4 is $-4\Delta(h_1, h_2)\cdot e$, by (26). Since f covers both e and S^4 with degree one this shows that $f^*h^*(p) = h_2^*(p) - 4\Delta(h_1, h_2)$, as asserted in (28).

COROLLARY (Pontrjagin [9]). *If $H^4(K; Z)$ has no 2-torsion, then oriented $(n-1)$-sphere bundles $\mathfrak{B}$ over a 4-complex K are determined by their characteristic classes $W_2(\mathfrak{B})$, $W_4(\mathfrak{B})$ (if $n \geqq 4$) and $P(\mathfrak{B})$.*

Indeed, if $\mathfrak{B}_1, \mathfrak{B}_2$ have the same characteristic classes then the left sides of the equations in Theorem 2 are zero, and hence $4d(h_1, h_2) = 0$, where h_1, h_2 are classifying mappings. If there is no 2-torsion in $H^4(K; Z)$ this implies that $d(h_1, h_2) = 0$, and hence $\mathfrak{B}_1$ and $\mathfrak{B}_2$ are equivalent.

APPLICATION. We re-prove the following theorem ([6, Theorem V]) of Massey.

THEOREM 3 (Massey). *Let $\mathfrak{B}$ be a 2-sphere bundle with structure group $SO(3)$ over a 4-complex K, and assume $H^4(K; Z)$ has no 2-torsion. Then $\mathfrak{B}$ admits a cross-section if and only if there exists an integer class $\gamma \in H^2(K; Z)$ such that*

(a) $\gamma \equiv W_2(\mathfrak{B}) \bmod 2$,

(b) $\gamma \smile \gamma = P(\mathfrak{B})$.

PROOF. $\mathfrak{B}$ admits a cross-section if and only if its structure group can be reduced to $SO(2)$ (see [15, 9.5]); For an $SO(2)$-bundle $\mathfrak{B}$ the class $W_2(\mathfrak{B})$ is an integer class whose cup-square $W_2(\mathfrak{B}) \smile W_2(\mathfrak{B})$ is the Pontrjagin class $P(\mathfrak{B})$ (see [20, Theorem 9] or [5, Satz 4.5.1 with $q = 1$]). This proves the "only if" part of the theorem. Assume now that a class γ with the properties (a) and (b) exists, and let $\mathfrak{B}'$ be a $SO(2)$-bundle over K with $W_2(\mathfrak{B}') = \gamma$. Then, as $SO(3)$-bundles, $\mathfrak{B}$ and $\mathfrak{B}'$ have the same characteristic classes, and hence are equivalent by the Corollary above.

EXAMPLE of a non-trivial $SO(3)$-bundle over a 4-complex with vanishing characteristic classes: Let K be a complex consisting of a 3-sphere S^3 with a 4-cell e^4 attached by a mapping $\varphi : \partial e^4 \to S^3$ of its boundary of degree 2. Then $H^i(K; Z) = 0$ for $i \neq 0$, 4, and $H^4(K; Z) = Z_2$. Let $h : K \to G_3$ be a mapping which sends S^3 into a point and is such that $d(h, c)$, the difference cohomology class with a constant mapping $c : K \to G_3$, is the non-zero element of $H^4(K; Z)$. The $SO(3)$-bundle $\mathfrak{B}$ over K which is induced by h is non-trivial, by Theorem 1, but its characteristic classes are zero (see Theorem 2). A similar example can be constructed with the product $P_2 \times P_2$ of two real projective planes as base.

4. Remark on Pontrjagin's Theorem 2 in [9]

This theorem can be formulated as follows. Let $\mathfrak{B} = \{E, p, B, \Gamma, \Gamma\}$ be a principal bundle, with B a complex, Γ a connected topological group. Assume the part of $\mathfrak{B}$ over the m-skeleton B^m of B is trivial, i.e., there exists a cross-section f over B^m (see [15, 9]). The obstruction for extending f over the $(m + 1)$-skeleton B^{m+1} is an $(m + 1)$-cocycle $z(f)$ of B with coefficients in $\pi_m(\Gamma)$. Let Σ_m be the image of the Hurewicz homomorphism $\Phi : \pi_m(\Gamma) \to H_m(\Gamma)$; we can factor Φ as follows:

$$\Phi : \pi_m(\Gamma) \xrightarrow{\varphi} \Sigma_m \subset H_m(\Gamma) .$$

Now $\varphi \circ z(f)$ is a cocycle with coefficients in Σ_m, and the theorem states that the cohomology class w_f^{m+1} of $\varphi \circ z(f)$ is an invariant of $\mathfrak{B}$, i.e., does not depend on f. We show by a counter-example that this is not correct. However, following the indication for a proof as given by Pontrjagin one finds that *the cohomology class of* $\Phi \circ z(f)$ *is an invariant* (see below).

For an example, let Γ be the group $SO(3)$, take $m = 3$, and let B be a 4-complex. Under the Hurewicz homomorphism the group $\pi_3(SO(3))$ maps isomorphically onto the subgroup $\Sigma_3 \subset H_3(\Gamma)$ of spherical homology classes. Therefore we may consider w_f^4 as the cohomology class of $z(f)$ which in turn can be identified with the class $d(h, c)$ of Theorem 1, where $h : B \to G_3$ is a classifying mapping for the given bundle $\mathfrak{B}$, and $c : B \to G_3$ is a constant mapping. But if B is real projective 4-space P_4, and if $d(h, c)$ is the non-zero element of $H^4(P_4; Z)$, then $\mathfrak{B}$ is trivial by Theorem 1 (a); hence $d(h, c)$ is not an invariant.

We now show that the cohomology class of $\Phi \circ z(f)$ (in the notation above) does not depend on the choice of the cross-section f. Let $g : B^m \to E$ be a second cross-section. Then for every $b \in B^m$ there is a unique $\gamma(b) \in \Gamma$ which (under right translation) transforms $g(b)$ into $f(b)$: $g(b)\gamma(b) = f(b)$. This defines a continuous mapping $\gamma : B^m \to \Gamma$; let $z(\gamma)$ be its obstruction cocycle. We claim that

(30) $$z(g) + z(\gamma) = z(f)\,.$$

It is sufficient to verify (30) on every $(m + 1)$-cell σ of B. But over σ the bundle $\mathfrak{B}$ is trivial and (30) expresses the fact that the addition in $\pi_m(\Gamma)$ is induced by the multiplication in Γ (see [15, 16.7]).

Next we show that $\Phi \circ z(\gamma)$ is a coboundary; then the invariance of the class of $\Phi \circ z(f)$ follows if we apply Φ to (30). The group $C_m(B)$ of cellular m-chains of B can be written as a direct sum $C_m(B) = Z_m(B) + D_m$, where $Z_m(B)$ denotes the cycles and D_m is a complementary summand. Now define an m-cochain y of B with coefficients in $H_m(\Gamma)$ as follows: y maps each cycle $\mathfrak{Z}_m \in Z_m$ into the homology class of $\gamma(\mathfrak{Z}_m)$, and y maps D_m into zero. Then for every $(m + 1)$-cell σ, $(\delta y)\cdot\sigma = y\cdot(\partial\sigma)$ is the homology class of Γ which is represented by $\gamma|\partial\sigma$, and this is nothing but $(\Phi \circ z(\gamma))\cdot\sigma$; hence $\delta y = \Phi \circ z(\gamma)$, q.e.d. (This is virtually the same proof as in [10, proof of Satz 3, § 3]).

In our example ($\Gamma = SO(3)$, $m = 3$) the inclusion $Z = \Sigma_3 \to H_3(SO(3)) = Z$ is multiplication by 2; therefore $2d(h, c)$ (notation from above) is an invariant of $\mathfrak{B}$. This can also be deduced from Theorem 1.

INSTITUTE FOR ADVANCED STUDY

BIBLIOGRAPHY

1. A. BOREL and J.-P. SERRE, *Groupes de Lie et puissances réduites*, Amer. J. Math. 75 (1953), 409–448.
2. R. E. CHAMBERLIN, *On the mappings of a 4-complex into certain simply connected spaces*, in Proc. Intern. Math. Congress 1950, Vol. I, 522–523.
3. S EILENBERG and S. MACLANE, *On the groups* $H(\pi, n)$, II, Ann. of Math. 60 (1954), 49–139.
4. ———, *On the groups* $H(\pi, n)$, III, Ann. of Math. 60 (1954), 513-557.
5. F. HIRZEBRUCH, Neue topologische Methoden in der algebraischen Geometrie, Ergebnisse der Mathematik und ihrer Grenzgebiete, Neue Folge Heft 9, 1956.
6. W. S. MASSEY, *On the cohomology ring of a sphere bundle*, J. Math. and Mech. 7 (1958), 265-289.
7. L. PONTRJAGIN, *A classification of mappings of the 3-dimensional complex into the 2-dimensional sphere*, Rec. Math. (Mat. Sbornik) N.S. 9 (51) (1941), 331-363.
8. ———, *Characteristic cycles*, Dokl. Acad. Nauk SSSR., NS 47 (1945), 242–245.
9. ———, *Classification of some skew products*, Dokl. Acad. Nauk SSSR., NS 47 (1945), 322-325.
10. D. PUPPE, *Zur Homotopie von Abbildungen eines Polyeders*, Math. Zeits. 61 (1954), 303–323.
11. H. ROBBINS, *On the classification of the mappings of a 2-complex*, Trans. Amer. Math. Soc. 49 (1941), 308-324.
12. J.-P. SERRE, *Cohomologie* mod 2 *des complexes d'Eilenberg-MacLane*, Comment. Math. Helv., 27 (1953), 198–231.
13. N. SHIMADA and H. UEHARA, *Classification of mappings of an* $(n+2)$*-complex into*

an $(n-1)$-connected space with vanishing $(n+1)^{st}$ homotopy group, Nagoya Math. J., 4 (1952), 43-50.

14. N. STEENROD, *Classification of sphere bundles*, Ann. of Math. 45 (1944), 294-311.
15. ———, The Topology of Fibre Bundles, Princeton University Press, 1951.
16. H. SUZUKI, *On the Eilenberg-MacLane invariants of loop spaces*, J. Math. Soc. Japan, 8 (1956), 93-101.
17. H. WHITNEY, *Topological properties of differentiable manifolds*, Bull. Amer. Math. Soc., 43 (1937), 785-805.
18. ———, *Some combinatorial properties of complexes*, Proc. Nat. Acad. Sci. U.S.A., 26 (1940), 143-148.
19. ———, *On the theory of sphere bundles*, Proc. Nat. Acad. Sci. U.S.A., 26 (1940), 148-153.
20. WU WEN-TSUN and G. REEB, *Sur les espaces fibrés et les variétés feuilletées*, in Actualités Sci. Ind. 1183, Hermann, Paris, 1952.

Chapter 3

Topology and Algebraic Topology

A CHARACTERIZATION OF THE CLOSED 2-CELL*

BY

HASSLER WHITNEY†

1. Introduction. A number of characterizations have been given of the simple closed surface.‡ The proofs involve considerable point set difficulties. We give here a characterization of the closed 2-cell, that is, a point set homeomorphic with a circle and its interior. The fundamental theorem is partly of a combinatorial and partly of a continuity nature. It reads

THEOREM I. *Let R be a continuous curve* § *containing the simple closed curve J, such that*

(1) *J is irreducibly homologous to zero in R, and*

(2) *If γ is an arc with just its two end points a and b on J, then $R-\gamma$ is not connected.*

Let R' and J' be defined similarly. Then R and R' are homeomorphic, with J corresponding with J'.

That R is a closed 2-cell then follows immediately from the following theorem. We note that J corresponds with the circle, that is, J is the boundary of R.

THEOREM II. *If I is a circle in the plane and S is I with its interior, then S and I satisfy the conditions prescribed for R and J in the above theorem.*

The exact meaning of Condition (1) of Theorem I is given in §4; a stronger condition is the following: For every $\epsilon>0$ and any two points a and b on J, there is a set of points a_{ij} in R, $1\leqq i\leqq m$, $1\leqq j\leqq n$, such that all points a_{1j} coincide with a, all points a_{mj} coincide with b, all points a_{i1} lie on one arc ab of J, all points a_{in} lie on the other arc ab of J, and||

$$\rho(a_{ij}, a_{i+1,j}) < \epsilon, \quad \rho(a_{ij}, a_{i,j+1}) < \epsilon;$$

moreover, this does not hold in any proper subset of R containing J.

* Presented to the Society, October 31, 1931; received by the editors April 13, 1932.

† National Research Fellow.

‡ That is, a point set homeomorphic with the surface of a sphere. See L. Zippin, American Journal of Mathematics, vol. 52 (1931), pp. 331–350; these Transactions, vol. 31 (1929), pp. 744–770; C. Kuratowski, Fundamenta Mathematicae, vol. 13 (1929), pp. 307–318; also references in these papers.

§ See Lemma A.

|| $\rho(p, q)$ = distance from p to q, or in general, distance between two point sets; $\delta(S)$ = diameter of S; $V_\epsilon(S)$ = those points p for which $\rho(p, S)<\epsilon$; $W_\epsilon(S)$ = those points p for which $\rho(p, S)\leqq\epsilon$.

Notations and preliminary theorems are given in §§2, 3 and 4; an outline of the proof of Theorem I will be found in §5. The Jordan and related theorems follow of course from the above theorems.

2. **Point set background.** Elementary properties of point sets we shall need may be found in Hausdorff, *Mengenlehre*, chapter VI. A continuous curve is a metric space which can be expressed as the continuous image of a closed line segment. An arc is the topological image of a closed line segment; a simple closed curve, the topological image of a circle.

Two fundamental lemmas are the following:

LEMMA A.* *A compact, connected and locally connected metric space is a continuous curve, and conversely.*

LEMMA B.† *A continuous curve is arcwise connected.*

That is, any two points p and q in the set are end points of an arc pq in the set. Using the definition of a continuous curve, it is easily seen that two continuous curves which have common points form a continuous curve.

From these lemmas we deduce the following known theorems.

LEMMA C. *Any continuum C of diameter $<\epsilon$ in a continuous curve R is contained in a continuous curve C' in R of diameter $<\epsilon$.*

Say $\delta(C)=\epsilon-\epsilon'$. R being the continuous image of a closed line segment, we can divide this segment into segments so small that the diameter of the image of each is $<\epsilon'$. We let C' be the union of all of these images which have points in common with C.

LEMMA D. *A continuous curve R is locally arcwise connected.*

That is, given a point p and an $\epsilon>0$, there is a $\delta>0$ such that if $q\subset V_\delta(p)$, then there is an arc pq in R of diameter $<\epsilon$. As R is locally connected, we can take δ so that if $q\subset V_\delta(p)$, there is a continuum C in R of diameter $<\epsilon$ containing p and q. The continuum C is contained in a continuous curve C' of diameter $<\epsilon$, and C' is arcwise connected; hence there is an arc $pq\subset C'\subset R$, and $\delta(pq)<\epsilon$.

R is of course uniformly locally arcwise connected, by the Borel Theorem.

LEMMA E. *A connected open subset R' of a continuous curve R is arcwise connected.*

---

* See G. T. Whyburn, *Concerning continuous images of the interval*, American Journal of Mathematics, vol. 53 (1931), pp. 670–674.

† See references in R. L. Moore, *Report on continuous curves*, Bulletin of the American Mathematical Society, vol. 29 (1923), p. 293, footnote (†).

If there are two points p and q in R' which are joined by no arc in R', let A contain p and all points of R' joined to p by an arc in R', and put $B=R'-A$; then there is no arc joining a point of A to a point of B in R'. As R' is connected, there is a point p' in one of these sets, say B, which is a limit point of points of the other set, A. As R' is open in R, $\rho(p', R-R')=\epsilon>0$. We can take q' in A so close to p' that there is an arc $p'q'$ in R of diameter $<\epsilon$. But then $p'q' \subset R'$, a contradiction.

Suppose R is connected, and $p \subset R$ is such a point that $R-p$ is not connected. Then p is called a cut point of R.

LEMMA F. *Let R be a continuous curve without a cut point. Then for every $\epsilon>0$ there is a $\delta>0$ such that if $\rho(q, p) \geqq \epsilon$ and $\rho(q', p) \geqq \epsilon$, then there is an arc qq' with no points in $V_\delta(p)$.*

Suppose the contrary. Then there are three sequences of points $\{p_n\}$, $\{q_n\}$, $\{q_n'\}$, approaching points p, q, q', respectively, with $\rho(q_n, p_n) \geqq \epsilon$, $\rho(q_n', p_n) \geqq \epsilon$, and such that for each n, any arc q_nq_n' must contain points in $V_{\delta_n}(p_n)$, where $\lim_{n\to\infty}\delta_n=0$. By Lemma D it is seen that for any n greater than some N there are arcs q_nq, $q_n'q'$, with no points in $V_{\epsilon/2}(p)$. It follows that any arc qq' must pass through p, contradicting Lemma E (as p is not a cut point).

3. **Combinatorial background.*** A *k-simplex*, or *abstract k-simplex*, is a set of k elements (say points) $a_1a_2 \cdots a_k$. The order in which we write the points is immaterial. For $k=0$, 1 and 2 we use also the terms *vertex*, *segment* and *triangle* respectively. A *k-chain* is a set of k-simplexes, and is written as the sum of these simplexes. The *sum* (mod 2) of several k-chains is the k-chain containing those simplexes which occur in an odd number of the k-chains.

The *boundary* K of a *k-simplex* L, $k>0$, is the sum of all $(k-1)$-simplexes formed by dropping out one of the vertices of the simplex. We write $L\to K$. A 0-simplex has no boundary. Thus

$$a \to 0,\ ab \to a+b,\ abc \to ab+ac+bc.$$

The *boundary* of a k-chain is the sum (mod 2) of the boundaries of the simplexes of the chain. Thus

$$ab+bc+cd \to a+d,\ abc+bcd \to ab+ac+bd+cd.$$

Evidently *the boundary of a sum of several k-chains is the sum of the boundaries of the chains.* If a k-chain has no boundary, it is called a k-cycle. (Any 0-chain is a 0-cycle.) *The boundary of a k-chain* $(k>0)$ *is a* $(k-1)$*-cycle.* This is evi-

* Compare L. Vietoris, *Über den höheren Zusammenhang kompakter Räume*, Mathematische Annalen, vol. 97 (1927), pp. 454–472.

dent if the k-chain is a k-simplex. The general case then follows from the last theorem.

LEMMA G. *If $K \to a+b$ is a 1-chain, then there is a chain of segments aa_1, $a_1a_2, \cdots, a_nb$ in K.*

For otherwise we could divide the segments of K into two groups $K_1 \supset a$ and $K_2 \supset b$, no two simplexes from different groups having a common vertex. But then $K_1 \to a$, $K_2 \to b$, which cannot be, as the boundary of any 1-chain contains an even number of vertices.

A 1-circuit is a 1-cycle of the form $a_1a_2, a_2a_3, \cdots, a_{n-1}a_n, a_na_1$, the vertices being distinct except as shown.

LEMMA H. *Any 1-cycle K is a sum of 1-circuits.*

If a_1a_n is a segment of K, then $K+a_1a_n \to a_1+a_n$, as $K \to 0$ and $a_1a_n \to a_1+a_n$. We can thus find a set of distinct segments and vertices $a_1a_2, \cdots, a_{n-1}a_n$ in $K+a_1a_n$ not containing a_1a_n. This with a_1a_n is a 1-circuit K_1. As $K_1 \to 0$, $K+K_1$ is a 1-cycle containing no segments of K_1, and it contains a 1-circuit K_2. Continuing, we find $K=K_1+K_2+\cdots+K_m$.

4. A k-chain K is said to *lie in a point set R* if each vertex of K is in R. Any vertex now has both a name and a position. Two vertices are distinct if their names are distinct, irrespective of whether they coincide in position or not. ϵ being a positive number, a k-simplex $K \subset R$ is called an *(ϵ, k)-simplex* in R if $\delta(K) < \epsilon$, i.e. if any two vertices of K are within ϵ of each other. A k-chain is an *(ϵ, k)-chain* if each of its simplexes is an (ϵ, k)-simplex. A k-cycle K in S is said to be *ϵ-homologous to zero* ($K\epsilon\sim 0$) in R if there is an $(\epsilon, k+1)$-chain L in R of which K is the boundary. *If $K_1\epsilon\sim 0$ and $K_2\;\epsilon\sim 0$, then K_1+K_2* $\epsilon\sim 0$. We write also $K_1\epsilon\sim K_2$ for $K_1+K_2\epsilon\sim 0$. *If $K_1\epsilon\sim K_2$ and $K_2\epsilon\sim K_3$, then $K_1\epsilon\sim K_3$.*

Suppose the closed set R contains the simple closed curve J. If for every $\epsilon>0$ there is a $\delta>0$ such that any $(\delta, 1)$-cycle on J is $\epsilon\sim 0$ in R, then we say that *$J\sim 0$ in R*. If J is ~ 0 in R but is not ~ 0 in any proper closed subset of R containing J, then we say that J is *irreducibly* ~ 0 in R.

LEMMA I. *Given a simple closed curve J, let us divide it into the arcs* $\overline{a_1a_2}$, $\overline{a_2a_3}, \cdots, \overline{a_{n-1}a_n}, \overline{a_na_1}$, each of diameter $<\epsilon/2$. Let δ be smaller than the distance between any two of these arcs which have no common points. Then if $K'=a_1a_2+a_2a_3+\cdots+a_na_1$ and K is any $(\delta, 1)$-cycle on J, K is either $\epsilon\sim 0$ or $\epsilon\sim K'$ on J.*

By Lemma H, K is a sum of 1-circuits $K_1, \cdots, K_m$. If we show that each

* Here, $\overline{a_1a_2}$ denotes an arc, and a_1a_2, a segment.

K_i is $\epsilon\sim\alpha_i K'$, $\alpha_i=0$ or 1, it will follow that $K=\sum K_i\ \epsilon\sim\sum\alpha_i K'=0$ or K' (depending on whether $\sum\alpha_i$ is even or odd), and the lemma will be proved.

Consider any $K_i=b_1b_2+b_2b_3+\cdots+b_sb_1$, say. If a vertex b_j of K_i does not lie on any point a_k, say $b_j\subset\overline{a_ka_{k+1}}$; add to K_i the boundary of the ϵ-triangles $b_{j-1}b_ja_k'+b_jb_{j+1}a_k'$, where a_k' is a new vertex lying on a_k. The result is an $(\epsilon, 1)$-circuit $K_i^{(1)}\ \epsilon\sim K_i$, the vertex b_j having been replaced by the vertex a_k'. Repeat the process till we have an $(\epsilon, 1)$-circuit $K''=c_1c_2+c_2c_3+\cdots+c_sc_1\ \epsilon\sim K_i$.

Now any two consecutive vertices c_j, c_{j+1} lie on the same or consecutive vertices of K'. Suppose c_j is on a_k and c_{j+2} is on a_{k+p}, $p\neq 2$ or -2. Then add the boundary of $c_jc_{j+1}c_{j+2}$, replacing the segments $c_jc_{j+1}+c_{j+1}c_{j+2}$ by the single segment c_jc_{j+2}. Continue till we arrive at a (possibly void) $(\epsilon, 1)$-circuit $K^*=d_1d_2+\cdots+d_rd_1\epsilon\sim K_i$. If d_1 lies on a_k, then d_{j+1} lies on $a_{k\pm j}$, where we put $n+p=p$, etc.

If K^* contains no segments, $K_i\ \epsilon\sim 0$. Otherwise, following the vertices $d_1, d_2, \cdots, d_r, d_1$ of K^*, we have gone around J p times say. Add to K^* the boundaries of all the $2r$ ϵ-triangles of the following sort. If d_j lies on a_k, and d_{j+1} on $a_{k\pm 1}$, two of the triangles are $d_jd_{j+1}a_k$ and $d_{j+1}\,a_ka_{k\pm 1}$. The result is an $(\epsilon, 1)$-cycle $pK'=0$ or K'. Thus $K_i\epsilon\sim 0$ or K', and the proof is complete.

An immediate consequence of this lemma is

LEMMA J. *Let the simple closed curve J lie in the closed set R. If for every $\epsilon>0$ there is a 1-cycle K' in J as above described which is $\epsilon\sim 0$ in R, then $J\sim 0$ in R.*

LEMMA K. *If γ is an arc, then for every $\epsilon>0$ there is a $\delta>0$ such that any $(\delta, 1)$-cycle on γ is $\epsilon\sim 0$ on γ.*

The proof below holds in fact if γ is a closed k-cell, any k. It is sufficient to prove it for the case that γ is a closed line segment, in which case we can take $\delta=\epsilon/2$.†

Let K be a $(\delta, 1)$-cycle on γ, let a_0b_0 be a segment of K, and say $\delta(\gamma)=\alpha$. Choose a fixed point p in γ, and an integer $n>\alpha/\delta$. Let the vertices $a_1, a_2, \cdots, a_{n-1}$ divide the segment a_0p into n equal parts, and similarly for the vertices $b_1, b_2, \cdots, b_{n-1}$. Add to K the boundaries of all triangles of the form $a_ia_{i+1}b_i$, $a_{i+1}b_ib_{i+1}$, $a_{n-1}b_{n-1}p$, and of all similar triangles corresponding to the other segments of K. The result is 0. As all the triangles employed are ϵ-triangles, $K\epsilon\sim 0$ in γ.

5. **Outline of the proof of Theorem I.** The proof runs as follows.

---

† The essential point in the proof below is that γ is convex: any two points of γ are end points of a line segment in γ. The proof is then easily extended to the case of any set homeomorphic with γ.

(a) In §6 we show how an arc γ can be drawn in R crossing J,† avoiding two given closed sets. $R-\gamma$ is not connected.

(b) In §7 we prove some lemmas. These show (§8) that $R-\gamma$ contains exactly two components A' and B'. If $A=A'+\gamma$, then A and its boundary curve J_A (which is γ plus a part of J) satisfy condition (1) of the theorem; similarly for $B=B'+\gamma$ and J_B. Further, A and B are continuous curves.

(c) In §9 it is shown that any arc in A (or B) crossing J_A (J_B) divides A (B). Thus A and J_A (B and J_B) satisfy all the conditions of the theorem. Hence we can cut up each set just as we cut up R, and can continue indefinitely.

(d) The object of §10 is to prove that R may be cut into pieces of arbitrarily small diameter.

(e) The homeomorphism between R and R' is now easily established. We cut R up indefinitely, and cut R' in a corresponding fashion. Any point p of R lies in a descending sequence of pieces; the corresponding sequence in R' determines a point p', which we let correspond to p.

We turn now to the detailed proof.

6. **An arc crossing J.** We prove here

LEMMA L.‡ *Let the simple closed curve J be ~ 0 in the continuous curve R. Let c and d be two points of J, dividing J into the two arcs η_1 and η_2. If C and D are two closed sets in R containing c and d respectively, and $C\cdot D=0$, then there is an arc γ in R joining η_1 to η_2 which has no points in C or in D.*

Say $\rho(C, D)=3\epsilon$, and put $C'=W_\epsilon(C)$, $D'=W_\epsilon(D)$; then $\rho(C', D')=\epsilon$. Take σ so small that any two points in R within σ of each other are joined by an arc of diameter $<\epsilon$ (Lemma D). Take δ so small that any $(\delta, 1)$-cycle on J is $\sigma\sim 0$ in R. Construct the $(\delta, 1)$-cycle $K=cc_1+c_1c_2+\cdots+c_md+dd_1+d_1d_2+\cdots+d_nc$, $c_i\subset\eta_1$, $d_i\subset\eta_2$. There is a $(\sigma, 2)$-chain

$$L = L_C + L_D \rightarrow K$$

in R, where we let L_C contain all those triangles of L with vertices in C', and let L_D be the rest of L.

Say

$$L_C \rightarrow K_C = K_C' + K^*,$$

where we let K_C' contain all those segments of K_C which are also in K. As $L_C\subset V_\sigma(C')$, $K^*\cdot D'=0$. Define K_D' by the relation

† That is, γ lies in R, and has only its end points on J.

‡ Compare P. Urysohn, *Über Räume mit verschwindender erster Brouwerscher Zahl*, Proceedings, Amsterdam Akademie van Wetenschappen, vol. 31 (1928), pp. 808–810.

$$L_D \to K_D = K_D' + K^*.$$

Adding these relations gives L on the left, and hence K on the right:

$$K = K_C' + K_D'.$$

As all the segments of K_C' are in K, K_D' must contain just those segments of K not in K_C'; in particular, it contains no segments of K^*. Hence all the segments of K^* are present in $K_D' + K^*$, the boundary of L_D (i.e. none have canceled out with segments of K_D'). Hence, as $L_D \cdot C' = 0$,

$$K^* \cdot C' = K^* \cdot D' = 0.$$

As K_C is the boundary of L_C, it is a 1-cycle; hence

$$K_C + cc_1 = K_C' + K^* + cc_1 \to c + c_1.$$

By Lemma G, $K_C + cc_1$ contains a chain of segments joining c_1 to c. Following this chain, let p_s be the first vertex in η_2, and p_0, the last vertex before p_s in η_1, and say $p_0p_1, p_1p_2, \cdots, p_{s-1}p_s$ are the segments in between. We shall show that these segments are in K^*. If $s>1$ this is obvious, as then $p_1, \cdots p_{s-1}$ exist and are not on J. Suppose $s=1$ and p_0p_1 is not in K^*; then it is in $K_C' + cc_1$. It could only be the segment cc_1. But cc_1 lies in K and not in K_D, hence it is in K_C; it is not in K^*, hence it is in $K_C + K^* = K_C'$, and therefore not in $K_C' + cc_1$. This proves the statement.

Now let $\overline{p_ip_{i+1}}$ be an arc of diameter $<\epsilon$ in R, $i=0, \cdots, s-1$. These arcs form a continuous curve, from which we can pick out an arc γ (Lemma B) joining η_1 to η_2; we can take γ so only its end points are on J. As $p_ip_{i+1} \subset K^*$ and $\delta(\overline{p_ip_{i+1}}) < \epsilon$, γ has no points in C or in D, and the lemma is proved.

7. We prove three lemmas.

LEMMA M. *If $J \subset C$, $J \sim 0$ in $C+D$, and $C \cdot D =$ an arc γ, then $J \sim 0$ in C.*

Given an $\epsilon > 0$, choose first ϵ_1 so small that any $(3\epsilon_1, 1)$-cycle on γ is $\epsilon\sim 0$ in γ (Lemma K). Take next $\epsilon_2 < \epsilon_1$ so that if $p \subset D$ and $\rho(p, C) < \epsilon_2$, then $\rho(p, \gamma) < \epsilon_1$. (If $D_1 = D - D \cdot V_{\epsilon_1}(\gamma)$, take $\epsilon_2 < \rho(D_1, C)$.) Take finally $\delta < \epsilon_2$ so that any $(\delta, 1)$-cycle K on J is $\epsilon_2\sim 0$ in $C+D$; we shall show that $K \epsilon\sim 0$ in C.

Let $L \to K$ be an $(\epsilon_2, 2)$-chain in $C+D$. Take any vertex p of L in $D \cdot V_{\epsilon_2}(C) - \gamma$, and replace it by a vertex $p' \subset \gamma$, where $\rho(p, p') < \epsilon_1$. L is thus replaced by a $(3\epsilon_1, 2)$-chain L', in which each triangle lies wholly in either C or D. Moreover, $L' \to K$, as no vertices of K have been moved.

Put $L' = L_C + L_D$, where L_C contains those triangles of L' in C. Say

$$L_C \to K + K^*; \text{ then } L_D \to K^*.$$

K^* is a $(3\epsilon_1, 1)$-cycle lying in $C \cdot D = \gamma$; it bounds an $(\epsilon, 2)$-chain L^* in γ. Hence

$$L_C + L^* \to (K + K^*) + K^* = K.$$

L_C+L^* is an $(\epsilon, 2)$-chain in C, and the lemma is proved.

LEMMA N. *Let $A \cdot B=\gamma$, an arc whose end points are a and b. Let the arcs α and β join a and b in A and B respectively, neither having any points other than a and b in common with γ. If $\alpha+\beta\sim 0$ in $A+B$, then $\alpha+\gamma\sim 0$ in A.*

Given an $\epsilon>0$, choose ϵ_1, ϵ_2 and δ as in the last lemma. Take $(\delta, 1)$-chains K_α, K_β and K_γ in α, β and γ respectively, each bounded by $a+b$; by Lemma J, it is sufficient to show that $K_\alpha+K_\gamma\ \epsilon\!\sim 0$ in A.

$K_\alpha+K_\beta$ bounds an $(\epsilon_2, 2)$-chain L in $A+B$; we move each vertex of L in $B\cdot V_{\epsilon_2}(A)-\gamma$ onto γ, giving a $(3\epsilon_1, 2)$-chain $L'\to K_\alpha+K_\beta'$. Say $L'=L_A+L_B$, where $L_A\subset A$, $L_B\subset B$. If $L_A\to K_\alpha+K^*$, then $L_B\to K_\beta'+K^*$, and $K^*\subset\gamma$. K^*+K_γ is a $(3\epsilon_1, 1)$-cycle on γ bounding an $(\epsilon, 2)$-chain L^* in γ. Hence $L_A+L^*\to K_\alpha+K_\gamma$ in A, completing the proof.

LEMMA O. *Let α, β and γ be three arcs such that $\alpha\cdot\beta=\alpha\cdot\gamma=\beta\cdot\gamma=a+b$. Say $\alpha+\gamma\subset A$ and $\beta+\gamma\subset B$. If $\alpha+\gamma\sim 0$ in A and $\beta+\gamma\sim 0$ in B, then $\alpha+\beta\sim 0$ in $A+B$.*

Define K_α, K_β, K_γ as before; we need merely show that $K_\alpha+K_\beta\,\epsilon\!\sim 0$ in $A+B$. There are $(\epsilon, 2)$-chains L_A and L_B such that $L_A\to K_\alpha+K_\gamma$ in A and $L_B\to K_\beta+K_\gamma$ in B; hence $L_A+L_B\to K_\alpha+K_\beta$ in $A+B$.

8. **The set $R-\gamma$.** Let γ be any arc in R crossing J; say the end points of γ divide J into the two arcs α and β. By condition (2) of the theorem, $R-\gamma$ is not connected. Let A' and B' be those components of $R-\gamma$ containing $<\alpha>$† and $<\beta>$ respectively. These are not the same component. For if they were, putting $A = A'+\gamma, D = R-A'$, we have $J\subset A$, $J\sim 0$ in $R=A+D$, and $A\cdot D=\gamma$; hence, by Lemma M, $J\sim 0$ in A, a proper subset of R, contrary to condition (1) of the theorem.

The same reasoning shows that R has no cut point p; we need merely replace γ by p in Lemma M and above.

Put

$$A = A' + \gamma, B = B' + \gamma.$$

If $D=R-A'$, then $A\cdot D=\gamma$ and $J=\alpha+\beta\sim 0$ in $R=A+D$. Hence, by Lemma N, $\alpha+\gamma\sim 0$ in A. Similarly, $\beta+\gamma\sim 0$ in B. Consequently, by Lemma O, $J\sim 0$ in $A+B$, from which follows that $A+B=R$.

Moreover, $\alpha+\gamma$ is irreducibly ~ 0 in A. For if $\alpha+\gamma\sim 0$ in A^*, $\alpha+\gamma\subset A^*\subset A$, then, by Lemma O, $\alpha+\beta\sim 0$ in A^*+B; hence $A^*+B=R$, which is only possible if $A^*=A$. Similarly, $\beta+\gamma$ is irreducibly ~ 0 in B.

† $<\alpha>$ is α except for its end points, etc.

Let us show that A is a continuous curve. It is connected, as A' is; it is self-compact, being a closed subset of a compact space. A is locally connected. For if p and q are points of A close enough together, there is an arc pq in R of small diameter; if pq lies partly in B', we can replace that part of it by an arc of γ of small diameter. Lemma A now applies. Similarly, B is a continuous curve.

9. We shall now show that any arc δ crossing $J_A = \alpha + \gamma$ in A divides A. The following two lemmas will be useful.

LEMMA P. *If η_1 and η_2 are arcs contained within the arcs γ and β respectively, then there is an arc pq crossing $J_B = \beta + \gamma$ in B, with $p \subset \eta_1$, $q \subset \eta_2$.*

This is an immediate consequence of Lemma L, if we take, for the closed sets of that lemma, the closed intervals of J_B complementary to η_1 and η_2.

LEMMA Q. *There are no two arcs ab and cd in R without common points, each crossing J, whose end points are in the order $acbd$ on J.*

This follows directly from what we have seen above.

To show that δ divides A, we must consider four cases.

Case 1. Both end points of δ lie on α. Suppose $A - \delta$ is connected; then it is arcwise connected, by Lemma E. Hence there is an arc in $A - \delta$ joining a point p of α lying between the two end points of δ and a point q within γ. If η_1 is an arc within γ containing q, there is an arc rs in B joining η_1 to a point s within β, with only its end points r and s on J_B, by Lemma P. The arc $pqrs$ crosses J and does not touch δ. But the end points of this arc alternate with those of δ on J, contradicting Lemma Q.

Case 2. δ is an arc cd, where c lies within α, d lies within γ. If $A - \delta$ is connected, let pq be an arc in this set joining points of α on opposite sides of c. If η_1 is an arc of γ containing d but not touching pq, let the arc rs join η_1 to β in B; then the arcs pq and $cdrs$ contradict Lemma Q.

Case 3. The end points c and d of δ lie within $\gamma = ab$, say in the order $acdb$. If $A - \delta$ is connected, let pq be an arc in this set joining a point p within α to a point q in γ between c and d. If η_1 is an arc of γ containing q but not touching δ, let r_1s_1 be an arc in B joining η_1 to a point s_1 within β.

The arcs acr_1 of γ and r_1s_1 form an arc acr_1s_1 crossing J; hence

$$R - acr_1s_1 = C_1 + C_2,$$

where C_1 contains the open arc $<as_1>$ of β, and C_2 contains b and points connected with b. As r_1s_1 lies in B, $A' \subset C_1 + C_2$; the connected set $A' + b$ lies thus in C_2. If η_2 is an arc of γ containing c but not touching η_1, and r_2s_2 is an arc in C_1 joining η_2 to a point s_2 of β between a and s_1, then $\eta_2 + r_2s_2$ does not touch pqr_1s_1, and has only the point c in common with δ.

Similarly, if η_3 is an arc of γ containing d but not touching η_1, there is an arc r_3s_3 in $R-bdr_1s_1$ such that r_3 lies in η_3, s_3 lies in β between s_1 and b, and $\eta_3+r_3s_3$ does not touch pqr_1s_1 and has only the point d in common with δ. The arc r_3s_3 does not touch r_2s_2, as it lies in C_2. Thus the two arcs pqr_1s_1 and $s_2r_2cdr_3s_3$ $(cd=\delta)$ contradict Lemma Q.

Case 4. The same as Case 3, except that $c=a$ or $d=b$, say the latter. Then, in the notation of Case 3, the arcs pqr_1s_1 and s_2r_2cb $(cb=\delta)$ contradict Lemma Q.

This completes the proof that A and J_A (B and J_B) satisfy the conditions of Theorem I.

10. **The cutting up of R.** We are concerned with the following lemma.

LEMMA R. *R may be cut into a finite number of pieces of arbitrarily small diameter.*

Given an $\epsilon>0$, choose $\delta<\epsilon$ so as to satisfy the requirement in Lemma F. Suppose R is cut up so that the diameter of the boundary of each piece is $<\delta$. Then each piece is of diameter $<3\epsilon$. For otherwise there is a point q of some piece R_i at a distance $\geqq\epsilon$ from its boundary J_i. Let p be a point of J_i, and q', a point of $R-R_i$ at a distance $\geqq\epsilon$ from p. Every arc from q to q' must cut the boundary J_i of R_i and thus must pass within δ of p, contradicting Lemma F.

The lemma thus follows from

LEMMA S. *Given a $\delta>0$, R can be cut up so that the diameter of the boundary of each piece is $<\delta$.*

Express R as the union of a finite number of continua:

$$R = K_1 + K_2 + \cdots + K_m,\ \delta(K_i) < \delta/2.$$

We shall cut up R in such a manner that no two of these continua K_i and K_j have points on the boundary of the same piece of R, if $K_i\cdot K_j=0$; the lemma will then follow.

Suppose we have cut R up a certain amount (perhaps not yet at all), into the pieces $R_1, R_2, \cdots, R_n$, with boundaries $J_1, J_2, \cdots, J_n$ (we may have R and J alone). Of course *each boundary J_i separates R_i from the rest of R.* Take any two continua, say K_1 and K_2, with $K_1\cdot K_2=0$, each of which has points on one of these J_i, say J_1. We shall cut R up further so that in the new pieces there is no one (i.e. no *piece*, not merely no *boundary* of a piece) which has any points in common with both K_1 and K_2; then on any further cutting up of R, this will still be true.

Divide the points of J_1 into three sets, as follows. We put a point x into the first set if it lies in K_1, or if following J_1 in both directions we reach points of K_1 before reaching points of K_2; we put x into the second set if the same conditions hold with K_1 and K_2 interchanged; all other points we put into the third set. This set L_3' consists of open intervals of J_1, each being bounded by a point of K_1 on one end and a point of K_2 on the other. The points of the first set together with the points $K_1 \cdot R_1$ form a closed set L_1, and those of the second set together with $K_2 \cdot R_1$ form a closed set L_2. Then $\rho(L_1, L_2) > 0$ as $L_1 \cdot L_2 = 0$, from which follows that there are but a finite number of intervals in L_3'. As K_1 is connected, each component of L_1 has points on J_1, and thus on one of the intervals L_3 of J_1 complementary to the intervals of L_3'. Thus there are a finite number of components $L_{11}, L_{12}, \cdots, L_{1m_1}$ in L_1. Similarly there are a finite number of components $L_{21}, L_{22}, \cdots, L_{2m_2}$ in L_2.

We shall now cut R_1 into a number of pieces, in each of which either K_1 has no points or K_2 has no points. Suppose $L_{31}', \cdots, L_{3m_3}'$ and $L_{31}, \cdots, L_{3m_3}$ are the intervals of L_3' and L_3 respectively, and say they lie in the order L_{31}, L_{31}', L_{32}, L_{32}', $\cdots$, L_{3m_3}, L_{3m_3}' on J_1. If we go around J_1, the intervals of L_3 lie alternately in L_1 and L_2. Starting at L_{31}, which lies in L_{11} say, go around J_1 till we reach another interval L_{3k} in L_{11} (we may have gotten back to L_{31}). Put L_{32}, $L_{3,k-1}$ and all of J_1 between these into a set M_2' (which may be L_{32} alone), and put L_{3k}, L_{31}, and all of J_1 between these on the other side from L_{32} into a set M_1' (which may be L_{31} alone). L_{31}' and $L_{3,k-1}'$ are the two intervals of J_1 complementary to M_1' and M_2'.

No set L_{1i} or L_{2j} has points in both M_1' and M_2'. This follows for L_{11} by construction. If it were false for some other set, say L_{1s}, then L_{1s} would have points on two intervals L_{3p} and L_{3q} separated by L_{31} and L_{3k} on J_1. Now $L_{11} \cdot L_{1s} = 0$, hence $\rho(L_{11}, L_{1s}) > 0$. As R_1 is a continuous curve, there are continuous curves L_{11}^* and L_{1s}^* in R_1 containing L_{11} and L_{1s} and such that $L_{11}^* \cdot L_{1s}^* = 0$ (see Lemma C). These sets are arcwise connected, and we can draw arcs contradicting Lemma Q.

Let M_1 be M_1' plus all components L_{1i} and L_{2j} containing points of M_1', and define M_2 similarly. Then M_1 and M_2 are closed, $M_1 \cdot M_2 = 0$, and $M_1 + M_2 \supset L_1 + L_2$. By Lemma L we can draw an arc γ_1 from L_{31}' to $L_{3,k-1}'$ which has no points in M_1 or in M_2. R_1 is thus cut into two pieces, in each of which there is at least one component L_{1i} or L_{2j}; for one contains L_{11}, and the other contains that L_{2j} containing L_{32}. Thus in each piece there are less than $m_1 + m_2$ components, the number in R_1.

If one of the resulting pieces contains more than one component, we cut it up, etc. Finally each new piece of R_1 has points of only one component, and thus K_1 and K_2 are separated in R_1. We now separate K_1 and K_2 in each other piece R_i of R also. This is possible, for if K_i $(i=1, 2)$ has points in any R_k, it also has points on J_k.

If now there are any other two of the continua K_i and K_j, $K_i \cdot K_j = 0$, each of which has points on some new J_k, we cut R further till this is no longer true, etc. This completes the proof.

11. **The homeomorphism.** Cut R into pieces of diameter $<$ some σ. We make corresponding cuts in R' as follows. The first arc γ drawn in R cuts R into the two pieces R_1 and R_2 with boundaries J_1 and J_2 say. Draw any arc γ' crossing J' in R', cutting R' into the pieces R_1' and R_2' with boundaries J_1' and J_2'. We note that $J_1' + J_2'$ is homeomorphic with $J_1 + J_2$, with J_k' corresponding to J_k, $k = 1, 2$. Say γ_1 is an arc in R_1, cutting R_1 into pieces R_{11} and R_{12} with boundaries J_{11} and J_{12}. If a_1 and b_1 are the end points of γ_1, let $a_1{}^*$ and $a_2{}^*$ be the corresponding points of J_1' in the above homeomorphism. Draw an arc γ_1' crossing J_1' in R_1', with end points a_1' and b_1' close to $a_1{}^*$ and $b_1{}^*$ respectively (Lemma P); R_1' is divided thereby into the pieces R_{11}' and R_{12}' with boundaries J_{11}' and J_{12}'. Moreover, $J_{11}' + J_{12}' + J_2'$ is homeomorphic with $J_{11} + J_{12} + J_2$, with boundaries with the same subscripts corresponding.

In general, suppose $R_{i_1 i_2 \cdots i_m}$ is a piece that is present after R is cut a certain amount, and say the arc $\gamma_{i_1 \cdots i_m}$ divides this set into the pieces $R_{i_1 \cdots i_m 1}$ and $R_{i_1 \cdots i_m 2}$, with boundaries $J_{i_1 \cdots i_m 1}$ and $J_{i_1 \cdots i_m 2}$. If $a_{i_1 \cdots i_m}$ and $b_{i_1 \cdots i_m}$ are the end points of $\gamma_{i_1 \cdots i_m}$, let $a_{i_1}{}^*{}_{\cdots i_m}$ and $b_{i_1}{}^*{}_{\cdots i_m}$ be the corresponding points on $J'_{i_1 \cdots i_m}$ in the homeomorphism we have already. Draw an arc $\gamma_{i_1}'{}_{\cdots i_m}$ crossing $J_{i_1}'{}_{\cdots i_m}$, with end points $a_{i_1}'{}_{\cdots i_m}$ and $b_{i_1}'{}_{\cdots i_m}$ close to the above points, dividing $R_{i_1}'{}_{\cdots i_m}$ into the pieces $R_{i_1}'{}_{\cdots i_m 1}$ and $R_{i_1}'{}_{\cdots i_m 2}$, with boundaries $J_{i_1}'{}_{\cdots i_m 1}$ and $J_{i_1}'{}_{\cdots i_m 2}$. The set of boundaries with primes is now homeomorphic with the set of boundaries without primes, boundaries with the same subscripts corresponding. We note that *if* $R_{i_1 \cdots i_m}$ *and* $R_{j_1 \cdots j_m}$ *have common points, then* $R_{i_1}'{}_{\cdots i_m}$ *and* $R_{j_1}'{}_{\cdots j_m}$ *have common points, and conversely.*

Having cut R into pieces of diameter $<\sigma$ and having cut R' in a corresponding fashion, we now cut each piece of R' into pieces of diameter $<\sigma/2$ and cut each piece of R in a corresponding fashion. Next we cut each resulting piece of R into pieces of diameter $<\sigma/4$, etc. Now for any $\epsilon > 0$ there is an m such that

$$\delta(R_{i_1 \cdots i_m}) < \epsilon, \quad \delta(R'_{i_1 \cdots i_m}) < \epsilon,$$

for any m-fold subscript.

We now establish the homeomorphism between R and R'. Let p be any point of R. It lies in either R_1 or R_2 (perhaps in both), say in R_{i_1}. Then it lies in either $R_{i_1 1}$ or $R_{i_1 2}$ (perhaps in both), say in $R_{i_1 i_2}$, etc. Thus we have a sequence of pieces

$$R \supset R_{i_1} \supset R_{i_1 i_2} \supset \cdots \supset p.$$

The corresponding pieces in R' have a single limit point:

$$R' \supset R'_{i_1} \supset R'_{i_1 i_2} \supset \cdots \supset p'.$$

This point p' we let correspond to p.

If there are different sequences of pieces in R containing p, we have different sequences in R' defining points p'. However, all these points p' are the same. For if R, R_{i_1}, $R_{i_1 i_2}$, $\cdots$, and R, R_{j_1}, $R_{j_1 j_2}$, $\cdots$, are two sequences containing p, then each piece $R_{i_1 \cdots i_m}$ has points in common with $R_{j_1 \cdots j_m}$, namely, the point p; hence, as we saw above, $R'_{i_1 \cdots i_m}$ and $R'_{j_1 \cdots j_m}$ have common points. Thus the corresponding sequences in R' close down on a single point. Similarly, to each point p' in R' corresponds a single point p in R.

Finally, the correspondence is continuous. For take a point p in R and an $\epsilon > 0$. Let p' be the corresponding point in R', and choose an m so that $\delta(R'_{i_1 \cdots i_m}) < \epsilon$ for all m-fold subscripts. Consider all the $R_{i_1 \cdots i_m}$ with m-fold subscripts which contain p; these include all points of R in some $V_\delta(p)$. Then if $q \subset V_\delta(p)$, the corresponding point q' is in $V_\epsilon(p')$, and the continuity is established. This completes the proof of Theorem I.

12. **Proof of Theorem II.** Let I be a circle in the plane, and let S be I plus its interior. S is self-compact, connected and locally connected, and is thus a continuous curve. That $I \sim 0$ in S follows from Lemma K.†

To show that I is irreducibly ~ 0 in S, suppose that $I \sim 0$ in S', a proper closed subset of S; we can suppose that S' is a continuous curve. Let p be a point of S not in S', and let $V_\delta(p)$ have no points in S'. Let ab be a segment of a straight line passing through p with its ends on I. Let a_1b_1 and a_2b_2 be parallel segments enclosing ab, and lying at a distance δ from ab. Then in that portion of S' between a_1b_1 and a_2b_2, the (short) arcs a_1a_2 and b_1b_2 are not connected. But if C and D are those parts of S' outside a_1b_1 and a_2b_2, by Lemma L we can draw an arc joining a_1a_2 to b_1b_2 in $S' - (C + D)$, a contradiction.

Finally, that an arc crossing I in S divides S is a special (and easily proved) case of the Jordan theorem. This completes the proof.

13. **The Jordan theorem.** Let J be a simple closed curve in the plane. Let I be a circle containing J in its interior. Draw two non-intersecting line segments from I to J. $S = I$ plus its interior is thus cut into three closed 2-cells, one of which, say R, has the boundary J. Then $R - J$ is the inside of J. The points of J are obviously accessible from either side.

† For S is a closed 2-cell.

PRINCETON UNIVERSITY,
PRINCETON, N. J.

REGULAR FAMILIES OF CURVES.[1]

By Hassler Whitney.[2]

1. **Introduction.** The topological properties of families of curves, occurring as solutions of differential equations, were first studied extensively by Poincaré. Other authors have continued these investigations, building up a considerable theory of the behavior of such curves, as determined by the differential equations considered.[3] These equations usually are, or may be, taken in the form

$$\frac{dx_i}{dt} = X_i(x_1, \cdots, x_n) \qquad (i = 1, \cdots, n);$$

the solutions may then be considered as the paths of particles in n-space, t being the time.

In this paper we study families of curves, given abstractly. The curves are required to satisfy a certain condition, the "regularity condition" (§ 6); it is certainly satisfied by solutions of differential equations. The main object of the paper is first, to prove the existence of cross-sections of these curves (§ 17), and secondly, to show how a function $q = f(p, t)$ can be defined (under slight further restrictions), which may be interpreted as defining a motion of particles along the curves of the family (§ 27).

2. **Summary of results.** Part I is devoted to a function of point sets μ; it is this function which makes possible the proofs of the main theorems.

In Part II regular families of curves are defined, and necessary and sufficient conditions that a family be regular are given. A definition of the distance between two arcs due to Fréchet[4] is fundamental. The "motions" are defined, and the orientability of a regular family is discussed. A function $q = g(p, t)$ is introduced; corresponding to a point p on a curve C and a number t, there is a point q, which moves away from p as t increases. By means of this function, a set of arcs lying between two cross-sections can be mapped topologically into a family of straight

---

[1] Received October 26, 1932.—Presented to the American Mathematical Society, March 25, 1932. An outline of the paper will be found in the Proceedings of the National Academy of Sciences, vol. 18 (1932), pp. 275-278 and 340–342.

[2] National Research Fellow.

[3] See for instance L. Bieberbach, Differentialgleichungen, Berlin, 1930, Part I, Ch. IV.

[4] M. Fréchet, Sur quelques points du calcul fonctionnel; Rendiconti del Circolo Matematico di Palermo, vol. 22 (1906), pp. 1–74.

lines. A general definition of cross-sections is given, their existence is proved, and a few of their properties are discussed.

Suppose we have a steady flow of fluid through space. The path of each moving particle is a curve C; these curves are fixed in position. There may also be invariant points. Given a position p of a particle and a length of time t, we may represent by $q = f(p, t)$ the position of the particle after the time t has elapsed. The set of positions the particle moves over during this time defines the motion (pq). In Part III we show under what conditions such a flow may be defined over a given family of curves and invariant points.

A family of paths is first defined. Tubes are introduced, and it is shown that the ends of a tube are cross-sections, and conversely, a tube can be formed from a cross-section. A sequence of tubes can be found covering the curves of the family, and having certain properties. Using the function g of Part II, we define the function f over each tube in turn, till it is defined over the whole set of curves. Letting f be constant at the invariant points completes the definition of the flow. A function $\tau(pq)$ is also defined for each motion (pq); it may be interpreted as the time taken by a particle in travelling from p to q. We prove finally that any two flows defined over a family of paths are equivalent in a certain general sense.[5]

I. The function of sets μ.

3. Definition and elementary properties of μ. Let R be a separable metric space,[6] and let $a_1, a_2, \cdots$ be a sequence of points dense in R. Corresponding to these points we define functions $f_1(p), f_2(p), \cdots$ by the equations

$$f_i(p) = \frac{1}{1+\varrho(a_i, p)} \qquad (i = 1, 2, \cdots). \tag{3.1}$$

We note the following inequalities:

$$\begin{gathered} 0 < f_i(p) \leqq 1; \\ |f_i(p) - f_i(q)| \leqq |\varrho(a_i, p) - \varrho(a_i, q)| \leqq \varrho(p, q). \end{gathered} \tag{3.2}$$

[5] We make use of the following notation:

$\varrho(p, q)$ = distance between the points (or point sets) p, q.

$\delta(A)$ = diameter of the point set A.

$V_\varepsilon(A)$ = all points p such that $\varrho(p, A) < \varepsilon$.

$W_\varepsilon(A)$ = all points p such that $\varrho(p, A) \leqq \varepsilon$.

We say a point p (or point set P) lies within, or interior to, the set A if p is (each point of P is) an inner point of A. P lies within an arc if it lies in the arc minus its end points. $A \subset A'$ signifies that each point of A is a point of A'.

[6] See for instance Hausdorff, Mengenlehre, Berlin, 1927, Ch. VI.

16

Let A be any non-void point set in R. For a given i, consider the various values of $f_i(p)$ for $p \subset A$; these values have a least upper bound U. B. and a greatest lower bound L. B. We put

$$(3.3) \quad \mu_i(A) = \underset{p \subset A}{\text{U. B.}} f_i(p) - \underset{p \subset A}{\text{L. B.}} f_i(p) = \underset{p, q \subset A}{\text{U. B.}} |f_i(p) - f_i(q)|,$$

and

$$(3.4) \quad \mu(A) = \sum_{i=1}^{\infty} \frac{\mu_i(A)}{2^i}.$$

Using (3.2), we find

$$(3.5) \quad \mu_i(A) \leqq \delta(A), \quad \mu(A) \leqq \delta(A), \quad \mu(A) \leqq 1.$$

In the following theorems, we give some elementary properties of the function μ.

THEOREM 3A. *If A contains but a single point, then $\mu(A) = 0$.*

For each $\mu_i(A) = 0$.

THEOREM 3B. *If $A \subset A'$, then $\mu(A) \leqq \mu(A')$.*

THEOREM 3C. *If $\bar{A} = A$ plus its limit points, then $\mu(\bar{A}) = \mu(A)$.*

For this reason, the function μ is of interest mainly in dealing with closed sets.

THEOREM 3D. *If $A \subset W_\varepsilon(A')$, then*

$$\mu(A) < \mu(A') + 2\varepsilon.$$

Consider a fixed i, and take any point $p \subset A$. There is a corresponding point $p' \subset A'$ for which $\varrho(p, p') \leqq \varepsilon + \beta$, where $\beta > 0$ is so small that for some $\alpha > 0$, $|f_i(p) - f_i(p')| < \varepsilon - \alpha$. Therefore

$$\underset{p \subset A}{\text{U. B.}} f_i(p) \leqq \underset{p' \subset A'}{\text{U. B.}} f_i(p') + \varepsilon - \alpha, \qquad \underset{p \subset A}{\text{L. B.}} f_i(p) \geqq \underset{p' \subset A'}{\text{L. B.}} f_i(p') - \varepsilon + \alpha,$$

and thus $\mu_i(A) < \mu_i(A') + 2\varepsilon$. Dividing by 2^i and summing over i gives the theorem.

THEOREM 3E. *If $A \subset A'$ and A' contains a point p at a positive distance d from A, then*

$$\mu(A) < \mu(A').$$

This is true, in particular, if A is closed and $A \subset A'$, $A \neq A'$.

Of course $\mu_i(A) \leqq \mu_i(A')$ for each i; the theorem will follow when we have found an i such that $\mu_i(A) < \mu_i(A')$. As the sequence $a_1, a_2, \cdots$ is dense in R, there is a point a_i such that $\varrho(a_i, p) < d/3$; then $f_i(p) > 1/(1 + d/3)$. Now $\varrho(a_i, A) > 2d/3$, hence $f_i(q) < 1/(1 + 2d/3)$ for $q \subset A$. Hence $\underset{q \subset A}{\text{U. B.}} f_i(q) < \underset{p' \subset A'}{\text{U. B.}} f_i(p')$, and thus $\mu_i(A) < \mu_i(A')$.

We can state the last two theorems in the form: $\mu(A)$ *is a continuous and increasing function of* A.

THEOREM 3F. *If* $A \cdot A' \neq 0$, *then*

$$\mu(A + A') \leqq \mu(A) + \mu(A').$$

The proof is simple.[7]

4. Comparison of the function μ with other functions. Functions obeying both Theorems 3D and 3E do not seem to have been used before. We mention here some functions which obey these theorems partially.

(a) The function $\delta(A)$ obeys Theorem 3D but not Theorem 3E.

(b) Let R^n be Euclidean n-space, and let $m_n(A)$ be the n-dimensional Lebesgue measure of a set A. Theorem 3D, in a weakened form, is obeyed by this function. Theorem 3E will be obeyed if, for instance, we restrict ourselves to the closures of open sets.

(c) The capacity of a set, say in three dimensions, obeys Theorem 3D in a weakened form, but not Theorem 3E: The capacity of a sphere is the same as the capacity of its surface.

(d) Let R be a separable space, defined by the sequence of neighborhoods $U_1, U_2, \cdots$. (If R is metric and the points $a_1, a_2, \cdots$ are dense in R, we can take for these neighborhoods open spheres of rational diameters about these points.) For any set A, put $\alpha_i(A) = 1$ or 0 according as $U_i \cdot A \neq 0$ or $= 0$; put $\nu(A) = \sum_1^\infty \alpha_i/2^i$. This function satisfies Theorem 3E, as stated for closed sets, but does not satisfy Theorem 3D. It has the remarkable property that any two distinct closed sets have distinct numbers ν.

5. The function μ'. Let R be a Euclidean space, or more generally, a metric space in which every bounded set is totally bounded.[8] Take any

---

[7] As a simple application of the function μ, we prove *Brouwer's Reduction Theorem.* We consider only *closed* subsets of a separable metric space R. If a (closed) set A has the property P, we write: $P(A)$. Assume of P that if $A_0 \supset A_1 \supset \cdots \supset A$ (A is the limiting set of the sequence), and $P(A_i)$, each i, then $P(A)$. *If* $P(B_0)$, *then there is a set* $B \subset B_0$ *such that* (1) $P(B)$, *and* (2) *if* $B' \subset B$ *and* $P(B')$, *then* $B' = B$.

Say $\mu(B_0) = \mu_0$. Let μ_0' be the lower bound of numbers $\mu(B')$ for $B' \subset B_0$ and $P(B')$, and take $B_1 \subset B_0$ so that $P(B_1)$ and $\mu(B_1) = \mu_1 < \mu_0' + 1$. In general, let μ_i' be the lower bound of numbers $\mu(B')$ for $B' \subset B_i$ and $P(B')$, and take $B_{i+1} \subset B_i$ so that $P(B_{i+1})$ and $\mu(B_{i+1}) = \mu_{i+1} < \mu_i' + 1/i$. Then $B_0 \supset B_1 \supset \cdots \supset B$, and $P(B)$. As $B_{i+1} \subset B_i$, $\mu_{i+1} \leqq \mu_i$. As any set in B_{i+1} lies also in B_i, $\mu_{i+1}' \geqq \mu_i'$. As $\mu_{i+1} \leqq \mu_i' + 1/i$, $\lim \mu_i = \lim \mu_i' =$ a number μ. As $B \subset B_i$, $\mu(B) \leqq \mu$ (in fact, $\mu(B) = \mu$). Suppose $B' \subset B$, $B' \neq B$ and $P(B')$. By Theorem 3E, $\mu(B') < \mu(B)$; hence, for some k, $\mu(B') < \mu_k'$. But this contradicts the definition of μ_k'.

As we used only Theorem 3E, not Theorem 3D, the function ν (§ 4) could be used in place of the function μ in this proof.

[8] See Hausdorff, loc. cit., p. 108.

16

bounded set A in R, and a fixed i. If $p_1, p_2, \cdots, p_i$ are points of A, there is associated with this set of points P_i a number $d(P_i) = \min \varrho(p_k, p_l)$ (which is 0 if two of the points coincide). Let $\mu_i'(A)$ be the upper bound of the numbers $d(P_i)$ for such sets P_i in A, and put

$$\mu'(A) = \sum_2^\infty \frac{\mu_i'(A)}{2^{i-1}}. \tag{5.1}$$

This function (for bounded sets) is easily seen to obey both Theorems 3 D and 3 E.[9]

In a general separable space, Theorem 3 E will not be satisfied by this function. For set $A = p_1 + p_2 + \cdots$, each p_i being that point in Hilbert space which has the ith coördinate equal to 1, the other coördinates being 0. If A' is any infinite subset of these points, $\mu'(A') = \mu'(A) = \sqrt{2}$.

A compensating advantage in the function μ' is the following. If A and A' are any two congruent sets, then $\mu'(A') = \mu'(A)$. If A' is similar to A (in the common geometric sense), each distance in A' being k times the corresponding distance in A, then $\mu'(A') = k\mu'(A)$.

II. Regular families of curves.

6. **Definitions.** Consider two arcs pq, $p'q'$. Let H be a homeomorphism between these arcs, in which p and q correspond to p' and q' respectively. Let $d(H)$ be the upper bound of all numbers $\varrho(r, r')$ for corresponding points r, r' in this homeomorphism, and let

$$\sigma(pq, p'q') = \sigma(p'q', pq) = \sigma(qp, q'p') \tag{6.1}$$

be the lower bound of the numbers $d(H)$ for all such homeomorphisms between the two arcs. This number we call the *span* between the two (oriented) arcs (écart in the sense of Fréchet).[10] Some simple properties of the span are given in the following theorems.

THEOREM 6A. *If pqr and $p'q'r'$ are arcs and $\sigma(pq, p'q') < \varepsilon$, $\sigma(qr, q'r') < \varepsilon$, then $\sigma(pqr, p'q'r') < \varepsilon$.*

THEOREM 6B. *If $\sigma(pr, p'r') < \varepsilon$ and q lies within pr, then there is a point q' within $p'r'$ such that $\sigma(pq, p'q') < \varepsilon$ and $\sigma(qr, q'r') < \varepsilon$.*

THEOREM 6C. *For any three arcs pq, $p'q'$, $p''q''$,*

$$\sigma(pq, p''q'') \leqq \sigma(pq, p'q') + \sigma(p'q', p''q''). \tag{6.2}$$

We shall use the word *curve* in this part as the topological image of a line segment with or without either end point, or of a circle. Consider

[9] See the first Proceedings note referred to above.

[10] In Fréchet's definition, all homeomorphisms are allowed. In terms of σ, his écart is the lesser of the two numbers $\sigma(pq, p'q')$, $\sigma(pq, q'p')$.

a set of curves, the points of which form a separable metric space. We shall say these curves form a *regular family* F if no two of them intersect, and if the following condition is satisfied: Given any arc pq of a curve C of the family and any $\varepsilon > 0$, there is a $\delta > 0$ such that if p' lies on C' and $\varrho(p, p') < \delta$, then there is an arc $p'q'$ of C' such that

$$\sigma(pq, p'q') < \varepsilon.$$

It is easily shown that the condition that no two of the curves intersect may be replaced by the following: No two of the curves differ, as point sets, by at most two points.

Any arc lying on a curve of the regular family we shall call an *arc* of the regular family.

7. **Conditions under which a family of curves is regular.** We prove here

THEOREM 7A. *A necessary and sufficient condition that a family of non-intersecting curves be regular is that each point p of any curve C be on an arc pq of C (where q can be taken on either side of p if p is not an end point of C), with the following property. For every $\varepsilon > 0$ there is a $\delta > 0$ such that if p' lies on C' and $\varrho(p, p') < \delta$, then there is an arc $p'q'$ of C' such that*

(1) *$p'q' \subset V_\varepsilon(pq)$ and $q' \subset V_\varepsilon(q)$, and*

(2) *if r' and s' are on $p'q'$ and $\varrho(r', s') < \delta$, then the arc $r's'$ is of diameter $< \varepsilon$.*

We prove first the necessity of the condition. Given a point p of C, take any arc pq of C and any $\varepsilon > 0$. Divide pq into arcs of diameter $< \varepsilon/4$, and take η less than one third the distance between any two of these arcs without common points. Take $\delta < \eta$ so that if $p' \subset V_\delta(p)$, then there is an arc $p'q'$ with $\sigma(pq, p'q') < \eta$. (1) is certainly satisfied. To prove (2), take r' and s' on $p'q'$ with $\varrho(r', s') < \delta$; let r and s be corresponding points of pq so that $\sigma(pr, p'r')$, $\sigma(rs, r's')$ and $\sigma(sq, s'q')$ are $< \eta$. Then $\varrho(r', r) < \eta$ and $\varrho(s', s) < \eta$, hence $\varrho(r, s) < 3\eta$. Consequently r and s are either on the same or adjacent arcs of our subdivision of pq, and $\delta(rs) < \varepsilon/2$. But $\sigma(rs, r's') < \eta < \varepsilon/4$, and hence $\delta(r's') < \varepsilon$, as required.

We turn now to the sufficiency of the condition. The theorem gives us certain arcs pq with q on either side of p; we shall show that the regularity condition is satisfied, first, for these special arcs pq, then, for any arc pq. Divide $pq = p_0 p_m$ into arcs $p_0p_1, \cdots, p_{m-1}p_m$, each of diameter $< \varepsilon/6$, and let ε_1 be less than half the distance between any two of the points $p_0, \cdots, p_m$. Take a number δ_1 corresponding to ε_1, as given in the theorem. From each point p_i ($i \neq 0, m$) follow C towards

p_0 and towards p_m to the first points r_i and s_i respectively not in $V_{\delta_1/4}(p_i)$, and define the arcs $C_i^- = p_0 r_i$, $C_i = r_i s_i$, $C_i^+ = s_i p_m$. Take $\varepsilon_2 < \delta_1/4$ and $< \varrho(C_i^-, C_i^+)/2$ $(i = 1, \cdots, m-1)$. Take δ corresponding to ε_2 as given in the theorem.

If p_0' is any point in $V_\delta(p_0)$, there is an arc $p_0' p_m'$ satisfying (1) and (2), with ε replaced by ε_2; we shall show that $\sigma(p_0 p_m, p_0' p_m') < \varepsilon$. We note first that $p_0' p_m'$ has points in $V_{\delta_1/2}(p_i)$, $i = 1, \cdots, m-1$. For $p_0' \subset V_\delta(p_0) \subset V_{\varepsilon_2}(C_i^-)$, $p_m' \subset V_{\varepsilon_2}(p_m) \subset V_{\varepsilon_2}(C_i^+)$, $\varrho(C_i^-, C_i^+) > 2\varepsilon_2$, and $p_0' p_m'$ is connected; hence there is a point r' of $p_0' p_m'$ not in $V_{\varepsilon_2}(C_i^- + C_i^+)$. But $p_0' p_m' \subset V_{\varepsilon_2}(p_0 p_m)$, and thus $r' \subset V_{\varepsilon_2}(C_i) \subset V_{\delta_1/2}(p_i)$.

Let p_i' be a point of $p_0' p_m'$ in $V_{\delta_1/2}(p_i)$, $i = 1, \cdots, m-1$. These points are in the order $p_0', p_1', \cdots, p_{m-1}', p_m'$ (they are of course distinct). For if they are not, then there is a point p_j' lying between p_0' and p_i', $j > i$. We see then just as above that $p_0' p_j'$ contains a point $r' \subset V_{\delta_1/2}(p_i)$ (evidently $p_j' \subset V_{\varepsilon_2}(C_i^+)$); hence $\varrho(r', p_i') < \delta_1$. But also $\varrho(p_i', p_i) < \delta_1/2$, $\varrho(p_j', p_j) < \delta_1/2$, $\varrho(p_i, p_j) > 2\varepsilon_1$, and $\delta_1 < \varepsilon_1$; hence $\varrho(p_i', p_j') > \varepsilon_1$ and $\delta(r' p_j' p_i') > \varepsilon_1$, contradicting our choice of δ_1.

Establish a homeomorphism between each arc $p_i p_{i+1}$ and the corresponding arc $p_i' p_{i+1}'$; this gives a homeomorphism between $p_0 p_m$ and $p_0' p_m'$. If we show that $\delta(p_i p_{i+1} + p_i' p_{i+1}') < \varepsilon$, it will follow that $\sigma(p_i p_{i+1}, p_i' p_{i+1}') < \varepsilon$, and hence $\sigma(p_0 p_m, p_0' p_m') < \varepsilon$, as required. If we put $r_{-1} = r_0 = p_0$ and $s_m = s_{m+1} = p_m$ and note that $r_{i-1} s_{i+2} \subset V_{\varepsilon/6+\delta_1}(p_i p_{i+1})$, we see it is sufficient to show that $p_i' p_{i+1}' \subset V_{\varepsilon_2}(r_{i-1} s_{i+2})$; for then $p_i' p_{i+1}' \subset V_{\varepsilon/3}(p_i p_{i+1})$, and thus $\delta(p_i p_{i+1} + p_i' p_{i+1}') < \varepsilon$. If this inclusion is false, then there is a point s' of $p_i' p_{i+1}'$ in $V_{\varepsilon_2}(C_{i-1}^-)$ or $V_{\varepsilon_2}(C_{i+2}^+)$, say the former. Just as before, we see then that $s' p_m'$ contains a point r' in $V_{\delta_1/2}(p_{i-1})$. It follows that $\varrho(r', p_{i-1}') < \delta_1$ and $\delta(r' s' p_{i-1}') > \varepsilon_1$, which cannot be. This completes the proof for the special arcs pq.

Consider now any arc pq. We saw above that there is an arc pq_0 satisfying the required condition, with q_0 lying on the same side of p as q. If there is such an arc with q_0 lying beyond q, then it follows from Theorem 6B that pq has the required property. If not, then consider all such arcs pq_0, and let q_1 be the bound of these points q_0 on C; q_1 lies on pq. There is an arc $q_1 q_2$ with the required property, q_2 lying beyond q_1 from p.

Given an $\varepsilon > 0$, take $\varepsilon_1 < \varepsilon$, $< \varrho(p, q_1 q_2)/2$ and $< \varrho(q_2, pq_1)/2$, and take $\eta < \varepsilon_1/2$ so that if $q_1' \subset V_\eta(q_1)$, then there is an arc $q_1' q_2'$ with $\sigma(q_1 q_2, q_1' q_2') < \varepsilon_1$. There is a point $q_0 \neq q_1$ on pq_1 such that $\delta(q_0 q_1) < \eta/3$, and pq_0 has the required property. Take δ so that for any $p' \subset V_\delta(p)$, there is an arc $p' q_0'$ with $\sigma(pq_0, p' q_0') < \eta/3$. Take a point q_1' beyond q_0' from p' so that $\delta(q_0' q_1')$

$<\eta/3$; then $\sigma(pq_1, p'q_1')<\eta$. Take an arc $q_1'q_2'$ with $\sigma(q_1q_2, q_1'q_2')<\varepsilon_1$. On account of our choice of ε_1, q_2' lies beyond q_0' from p'. Hence for any $p'\subset V_\delta(p)$ there is an arc $p'q_2'$ with $\sigma(pq_2, p'q_2')<\varepsilon$, and pq_2 has the required property. But this contradicts the definition of q_1. The theorem is now completely proved.

8. Examples of regular families.

(a) In the xy-plane, the parabolas $y = a(1+x^2)$, one curve for each value of a, form a regular family. We could also restrict a to any set of values we like, say to the values $0, \pm 1, \pm\frac{1}{2}, \pm\frac{1}{3}, \cdots$.

(b) Using polar coördinates, the curves

$$r = 1, \quad \text{and} \quad r = 1-\frac{1}{\theta+\gamma} \qquad (\theta>1-\gamma),$$

one curve for each γ, $0\leqq\gamma<2\pi$, form a regular family. We might replace the curve $r=1$ by a finite number of curves, formed by dropping out points or arcs from the curve.

(c) One curve C is an open segment in the plane, and each other curve consists of those points in the plane at a fixed distance from C.

(d) Let F be a set of curves forming a separable metric space, and let each curve be an arc. Suppose (1) for each point p on a curve C of F and each $\varepsilon>0$ there is a $\delta>0$ such that if p' is on a curve C' of F and $\varrho(p, p')<\delta$, then $C\subset V_\varepsilon(C')$ and $C'\subset V_\varepsilon(C)$,[11] and (2) the curves of F are equicontinuous.[12] Then F is a regular family.

To show this, we show that the conditions of Theorem 7A are satisfied; it is evident that no two of the curves can intersect. Let pq be any curve of F, and take any point p_1 interior to pq. We shall show that p_1q has the required property. (The case $p_1 = p$ or q admits a simpler proof.) Given an $\varepsilon>0$, take $p_2\subset pp_1$ so that $\delta(p_2p_1)<\varepsilon/2$, and put $2\varepsilon_1 = \varrho(p_1, pp_2)$. Take ε_2 so that if p' and r' are on some C' and $\varrho(p', r')<3\varepsilon_2$, then $\delta(p'r')<\varepsilon_1$. Take $r_1\subset p_2p_1$ and r_2 in p_1q so that $r_1p_1r_2 \subset V_{\varepsilon_2}(p_1)$. Put $2\varepsilon_3 = \varrho(pr_1, r_2q)$, and take δ so that if $p'\subset C'\cdot V_\delta(p_1)$, then $C\subset V_{\varepsilon_3}(C')$ and $C'\subset V_{\varepsilon_3}(C)$. Take any such p'; then there is a point $q'\subset C'\cdot V_{\varepsilon_3}(q)$. We shall show that condition (1) of Theorem 7A is satisfied; (2) holds automatically.

---

[11] We may express this by saying that the curves of F form a continuous (i. e. both lower and upper semicontinuous) collection.

[12] That is, for every $\varepsilon>0$ there is a $\delta>0$ such that if p and q are end points of an arc pq of any curve C of F and $\varrho(p, q)<\delta$, then $\delta(pq)<\varepsilon$; in other words, the curves of F are uniformly locally connected.

It is of course sufficient to assume that the curves are equicontinuous in the neighborhood of each curve of the family.

We need merely show that $p'q' \subset V_\varepsilon(p_1 q)$, or, $p'q' \subset V_{\varepsilon/2}(p_2 q)$. As $C' \subset V_{\varepsilon_3}(C)$, it suffices to show that $p'q' \cdot V_{\varepsilon_3}(pp_2) = 0$. Suppose $s' \subset p'q' \cdot V_{\varepsilon_3}(pp_2)$. Then as $C' \subset V_{\varepsilon_3}(C)$ and $\varrho(pr_1, r_2 q) = 2\varepsilon_3$, there is a point $r' \subset s'q' \cdot V_{\varepsilon_3}(r_1 p_1 r_2) \subset V_{2\varepsilon_2}(p_1)$. But then $\varrho(p', r') < 3\varepsilon_2$, and also $\varrho(p', s') \geqq \varepsilon_1$ and thus $\delta(p'r') \geqq \varepsilon_1$, contradicting our choice of ε_2.

In regard to this example, see § 14.

(e) Let C be a simple closed curve in the plane, containing inside it the point a. Let every point inside C except for a lie on a simple closed curve containing a inside it. If no two of these curves intersect, and they are equicontinuous, then they form a regular family. The proof is not difficult. See also § 20.

9. **The neighborhood of an arc.** Take an arc pq of the regular family F and a $\lambda > 0$. The λ-neighborhood of pq, $N_\lambda(pq)$, consists of all arcs $p'q'$ of F such that $\sigma(pq, p'q') < \lambda$. $N_\lambda(pq)$ contains, in particular, arcs contained in and arcs containing pq.

THEOREM 9A. *The set of points on $N_\lambda(pq)$ is open.*

For suppose $r' \subset p'q'$, $\sigma(pq, p'q') = \lambda' < \lambda$. Take δ small enough so that for every $r^* \subset V_\delta(r')$ there are arcs r^*p^* (if $r' \neq p'$) and r^*q^* (if $r' \neq q'$) with $\sigma(r'p', r^*p^*) < \lambda - \lambda'$ and $\sigma(r'q', r^*q^*) < \lambda - \lambda'$, and so that p^* and q^* are on opposite sides of r^*. Then $\sigma(p'q', p^*q^*) < \lambda - \lambda'$ and hence $\sigma(pq, p^*q^*) < \lambda$; thus every r^* in $V_\delta(r')$ is on an arc of $N_\lambda(pq)$, and this set of points is open.

THEOREM 9B. *If pq is interior to a curve C of F, then for each $\varepsilon > 0$ there is a $\lambda > 0$ such that if the arcs $p'q'$ and $p''q''$ are both in $N_\lambda(pq)$ and they have common points, then they differ at most by a point set lying in $V_\varepsilon(p) + V_\varepsilon(q)$.*

Let pq be contained within an arc $p_0 q_0$ and contain within it an arc $p_1 q_1$ so that $p_0 p p_1 \subset V_{\varepsilon/2}(p)$ and $q_1 q q_0 \subset V_{\varepsilon/2}(q)$. Take $2\lambda_0 < \varepsilon$ and less than any of the distances $\varrho(p_0, pq)$, $\varrho(q_0, pq)$, $\varrho(p, p_1 q_0)$, $\varrho(q, p_0 q_1)$, and take $\lambda < \lambda_0$ so that all points of $N_\lambda(pq)$ are on arcs of $N_{\lambda_0}(p_0 q_0)$. Suppose now $p'q'$ and $p''q''$ are in $N_\lambda(pq)$ and have the common point r'. Let $p_0'q_0'$ be an arc of $N_{\lambda_0}(p_0 q_0)$ containing r', and take p_1' and q_1' on $p_0'q_0'$ so $\sigma(p_0 p_1, p_0' p_1')$, $\sigma(p_1 q_1, p_1' q_1')$ and $\sigma(q_1 q_0, q_1' q_0')$ are all $< \lambda_0$. p_0' and q_0' are not on $p'q'$, hence $p'q' \subset p_0'q_0'$. $p_1'q_0' \subset V_{\lambda_0}(p_1 q_0)$, hence $p_1'q_0' \cdot V_{\lambda_0}(p) = 0$, and thus $p' \subset p_0' p_1'$; similarly $q' \subset q_1' q_0'$. In the same way we see that $p_1'q_1' \subset p''q'' \subset p_0'q_0'$. Therefore $p'q'$ and $p''q''$ differ at most by subsets of $p_0'p_1'$ and $q_1'q_0'$, which lie in $V_\varepsilon(p) + V_\varepsilon(q)$.

10. **Orientability of a regular family.** Suppose a positive direction is assigned to each curve of a regular family F. An arc pq of F is now *positively oriented* if q lies in the positive direction from p along pq; otherwise it is *negatively oriented.* pq and qp have then the opposite orientation.

We shall say the curves of F are *similarly oriented* in the neighborhood of the arc pq of F if the following condition is satisfied. If $p_1q_1, p_2q_2, \cdots$ is a sequence of oriented arcs of F approaching pq: $\sigma(pq, p_iq_i) \to 0$, then for every i greater than some N, pq and p_iq_i are similarly oriented. If it is possible to assign a direction to each curve of F so that the curves are similarly oriented in the neighborhood of each arc of F, we shall say F is *orientable.* We always assume hereafter that the curves of a regular orientable family are properly oriented.

Each regular family described in § 8 is orientable except that in (c).

11. **The motions of a regular oriented family.** Let C be a curve of the regular oriented family F. Let K be a correspondence between the numbers x, $0 \leqq x \leqq 1$, and points p_x of C. If this correspondence maps $(0, 1)$ continuously onto C, and 0 goes into p_0 and 1 into p_1, we say K defines the *motion* (p_0p_1). We shall say two correspondences K_0 and K_1 define the same motion (p_0p_1) if one is deformable into the other; that is, there is a correspondence K_y for each y, $0 \leqq y \leqq 1$, such that if p_{xy} denotes the point corresponding to x in K_y, then $p_{0y} = p_0, p_{1y} = p_1$, and p_{xy} is continuous in both variables x and y.

A *null motion* (pp) is determined by letting all of $(0, 1)$ correspond to a single point p.

Suppose the curve C of F is the image of a line segment and p and q are distinct points of C. Then if we put the arc pq into $1-1$ continuous correspondence with $(0, 1)$, this defines a motion (pq); evidently there is no motion $(pq)'$ different from this one. In this case the motion (pq) is uniquely determined by its end points p and q. However, if C is a simple closed curve, and we follow the points p_x as x increases from 0 to 1 in the motion (pq), we may pass around C a number of times, and thus find different motions (pq). In any case, we may define any non-null motion (pq) as follows. Let $pp_1, p_1p_2, \cdots, p_{m-1}q$ be a set of arcs of C, all positively or all negatively oriented. Putting $p = p_0$ and $q = p_m$ and putting each p_ip_{i+1} in $1-1$ continuous correspondence with the interval $(i/m, (i+1)/m)$ gives a correspondence between $(0, 1)$ and points of C, defining the motion (pq).

A non-null motion (pq) is *positively* or *negatively oriented* according as the above arcs defining (pq) are positively or negatively oriented.

If F is a family of paths and the functions τ and f are defined as in § 27, then we may define the motion (pq) uniquely by means of the following correspondence:

$$p_x = f[p, x\tau(pq)] \qquad (0 \leqq x \leqq 1). \tag{11.1}$$

Let (pq) and (qr) be two motions on a curve C, the first ending, the second starting, at the point q. Define these in terms of the intervals

$(0, \frac{1}{2})$ and $(\frac{1}{2}, 1)$ respectively. This gives a correspondence between $(0, 1)$ and points of C, defining the motion $(pr) = (pq) + (qr)$, the *sum* of (pq) and (qr). (qp) is the *negative* of (pq) if $(pq) + (qp)$ is the null motion.

12. **The space of motions.** Given two motions (pq) and $(p'q')$ on C and C' respectively, we can define the *span* $\sigma[(pq), (p'q')]$ between them just as we defined the span between arcs: Let K be any correspondence defining (pq), and K', one defining $(p'q')$; this induces a correspondence K^* between points of C and points of C'. $d(K^*)$ is the upper bound of $\varrho(r, r')$ for corresponding points in K^*, and $\sigma[(pq), (p'q')]$ is the lower bound of $d(K^*)$ for such correspondences K^*.

We may let the motions of F be points of a new space $\mathfrak{M}$, letting the distance between two of these points be the span between the two motions. $\mathfrak{M}$ is metric and separable. A sequence of motions dense in $\mathfrak{M}$ may be determined as follows. Let $a_1, a_2, \cdots$ be a sequence of points dense in R. For each $i, j > 0$ choose a point a_{ij} on the curve C_i through a_i so that $\varrho(a_j, a_{ij}) < \varrho(a_j, C_i) + 1/j$; then the points a_{ij} are dense on C_i. For any two points a_{ij} and $a_{ij'}$ on C_i, there is either one or a denumerable number of motions $(a_{ij} a_{ij'})$; the set S of all such motions is dense in $\mathfrak{M}$. For given an $\varepsilon > 0$ and a motion $(p_0 p_m)$ of $\mathfrak{M}$, consider a set of arcs $p_0 p_1, \cdots, p_{m-1} p_m$ defining $p_0 p_m$ (see above). If a_{i_0} is close enough to p_0, there is an arc $a_{i_0} a_{i_1}$ close to $p_0 p_1$, next an arc $a_{i_1} a_{i_2}$ close to $p_1 p_2$, etc. This set of arcs determines a motion $(a_{i_0} a_{i_m})$ within ε of $p_0 p_m$, as required.

13. **The function** g. Let p be a point on a curve C of a regular family, and consider a variable point q on C. For each arc pq there is a number $\mu(pq)$ (see Part I); $\mu(pq)$ increases continuously from zero as q moves away from p. Given a number α, if there is a point q such that $\mu(pq) = \alpha$, we shall say $q = g(p, \alpha)$:

$$(13.1) \qquad q = g(p, \alpha) \text{ equivalent to } \mu(pq) = \alpha.$$

g is in general two-valued: if p is interior to C and α is small enough, there will be two corresponding points q; but if α is large enough, there will be one or no corresponding points.

The importance of the function g lies in its continuity:

THEOREM 13A. *In a regular family of curves, $g(p, \alpha)$ is a continuous function of both variables.*

By this we mean that if the arc pq is interior to the curve C and $q = g(p, \alpha)$, then for every $\varepsilon > 0$ there is a $\delta > 0$ such that if $p' \subset V_\delta(p)$ and $|\alpha' - \alpha| < \delta$, then there is a point $q' = g(p', \alpha') \subset V_\varepsilon(q)$.

To prove this, take first $\eta < \varepsilon/2$ so small that there are points

$$q_1 = g(p, \alpha - 3\eta), \qquad q_2 = g(p, \alpha + 3\eta)$$

on the same side of p as q, and so that the arc $q_1 q q_2 \subset V_{\varepsilon/2}(q)$. Take $\delta < \eta$ so small that any point $p' \subset V_\delta(p)$ is an end point of an arc $p' q_2'$ with $\sigma(p q_2, p' q_2') < \eta$. Consider a homeomorphism between $p q_2$ and $p' q_2'$ in which corresponding points are distant $< \eta$; let q_1' be the point corresponding to q_1.

We shall now show that

$$(13.2) \qquad \mu(p' q_1') < \alpha - \eta, \qquad \mu(p' q_2') > \alpha + \eta.$$

To prove the first inequality, every point of $p' q_1'$ being within η of the corresponding point of $p q_1$, $p' q_1' \subset V_\eta(p q_1)$; hence, by Theorem 3D, $\mu(p' q_1') < \mu(p q_1) + 2\eta = \alpha - \eta$. Also $p q_2 \subset V_\eta(p' q_2')$, and thus the second inequality follows.

Now take any such point $p' \subset V_\delta(p)$ and any α', $|\alpha' - \alpha| < \delta < \eta$. From the above inequalities we see that there is a point $q' = g(p', \alpha')$ in the arc $q_1' q_2'$. But the correspondence between $p q_2$ and $p' q_2'$ shows that $q_1' q_2' \subset V_\eta(q_1 q_2) \subset V_\varepsilon(q)$; hence $q' \subset V_\varepsilon(q)$.

THEOREM 13B. *g is uniformly continuous over any compact and closed set of points p and bounded closed range of values α.*

This is an immediate consequence of the Borel Theorem.

14. An application. Assume that the regular family in (d), § 8 is orientable. Then each curve of F has a first point p_0 and a last point p_1. Let S_0 be the set of all first points, and S_1, the set of all last points. It is easily seen that S_0 and S_1 are closed, and are homeomorphic. We prove now

THEOREM 14A. *F can be mapped in Euclidean or Hilbert space so as to become a family of straight line segments.*

First map S_0 into a subset of such a space in which the first coördinate $x = 0$. Let p be any point of F, on the arc $p_0 p_1$ of F, and say p_0 is mapped into the point p_0'; then map p into the point whose first coördinate $x = \mu(p_0 p)/\mu(p_0 p_1)$, and whose other coördinates agree with those of p_0'. By Theorem 3D, we see that $\mu(p_0 p_1)$ is a continuous function of p_0; hence, using Theorem 13A, we see that the points of F are mapped topologically into space as required.

A special case of this theorem has been proved by R. L. Moore.[13]

15. The function g'. Let F be a regular oriented family. Take any point p of a curve C of F, and any number $\alpha > 0$. If q is such a point of C that $\mu(p q) = \alpha$, put $q = g'(p, \alpha)$ or $q = g'(p, -\alpha)$ according as q is in the positive or negative direction from p along pq.

[13] R. L. Moore, Fundamenta Mathematicae, vol. 4 (1923), pp. 106–117, Theorem I. See also his Colloquium Lectures, Foundations of point set Theory, p. 397. Note that it is fairly easy to show directly that the above mapping is continuous, without showing that the family is regular.

g' is single-valued, and is continuous in the same sense that g was.

THEOREM 15A. *In a regular oriented family, if* $p_i \to p$, $\alpha_i \to \alpha \neq 0$, *and* $q_i = g'(p_i, \alpha_i)$, $q = g'(p, \alpha)$, *then* $p_i q_i \to pq$.

This is a strengthening of Theorem 13A. Given an $\varepsilon > 0$, choose $\delta > 0$ as in the proof of that theorem, and using the same notation, note that $\sigma(pq_1, p'q_1') < \eta$. Take any correspondence between $q_1 q$ and $q_1' q'$; as $q_1' q' \subset q_1' q_2' \subset V_\varepsilon(q)$, we see thus that $\sigma(q_1 q, q_1' q') < \varepsilon$, and hence $\sigma(pq, p'q') < \varepsilon$. From this we see that if $q_i^* = g(p_i, \alpha_i)$ (using the notation of the present theorem) is taken on the right side of p_i, then $p_i q_i^* \to pq$. But as the family is oriented, pq and $p_i q_i^*$ have the same orientation for $i >$ some N; hence $q_i = q_i^*$ for $i > N$, and $p_i q_i \to pq$.

16. Cross-sections. Let p be a point interior to a curve C of the regular family F. A set of points S forms a *cross-section through* p if the following conditions are satisfied:

(1) Each point p' of S lies within an arc $q_0' q_1'$ of F such that for some $\lambda' > 0$, each arc of $N_{\lambda'}(q_0' q_1')$ (see § 9) contains at most one point of S.

(2) p lies within an arbitrarily small arc $q_0 q_1$ of F such that for some $\lambda > 0$, each arc of $N_\lambda(q_0 q_1)$ contains exactly one point of S (hence $p \subset S$).

(3) S is closed.

S is a cross-section if it is a cross-section through some point. We might call those points of S satisfying (2) *inner points* of the cross-section.

It would be reasonable to require further that all points of S be inner points or limit points of inner points; also if R is locally connected, we might demand that S be connected.

Example. Let F be a regular family of curves filling a region of the plane (or of n-space); assume the curves have directional derivatives which are continuous in the region. At the point p within the arc $q_0 q_1$ erect a small line segment (or portion of an $(n-1)$-plane) perpendicular to the direction of $q_0 q_1$ at p; this forms a cross-section through p.

17. Existence of cross-sections. We prove here

THEOREM 17A. *Through any point* p *interior to a curve* C *of the regular family* F *there is a cross-section* S.

Let $q_0 q_1$ be an arc interior to C and containing p within it. Take $4\sigma < \varrho(q_0, pq_1)$ and $< \varrho(q_1, pq_0)$, and put $U = W_{2\sigma}(q_0)$. For any superscript $*$, if the curve C^* of F contains a point p^* near p, let q_ζ^* $(\zeta > 0)$ be the first point of C^* in $W_\zeta(q_0)$ (if it exists), following C^* from p^* towards U; q_ζ is that point on C. Take η so small that the arc $q_{2\eta} q_0 \subset V_\sigma(q_0)$. Say $\mu(U + pq_\eta) = \alpha$; we shall show that on each arc near $q_0 q_1$ there is a point p' for which $\mu(U + p'q_\eta') = \alpha$, and that these points form a cross-section S.

Take $\lambda < \eta$ small enough so that any two arcs of $N_\lambda(q_0 q_1)$ with common points differ at most by a point set in $V_\eta(q_0) + V_\sigma(q_1)$ (see Theorem 9B). Then we can assign a positive direction to each arc $q_0' q_1'$ of $N_\lambda(q_0 q_1)$ so that $q_0' (\subset V_\eta(q_0))$ lies in the positive direction from q_1'.

Take $\beta < \eta/2$ so small that if p_1 and p_2 are the points of $q_0 q_1$ for which

$$(17.1) \qquad \mu(U + p_1 q_\eta) = \alpha - 2\beta, \quad \mu(U + p_2 q_\eta) = \alpha + 2\beta,$$

then $p_1 p p_2 \subset V_{\varepsilon/2}(p)$, where ε is some number $< \sigma$. If r^* is any point near p on an arc of $N_\lambda(q_0 q_1)$, put

$$r^*(t) = g'(r^*, t) \qquad (t \geqq 0),$$

in the positive direction from r^*, as defined above. Put also

$$t_{k\zeta} = \mu(p_k q_\zeta) \qquad (k = 1, 2;\ \zeta \geqq 0).$$

Then in particular,

$$p_k(t_{k\zeta}) = g'[p_k, \mu(p_k q_\zeta)] = q_\zeta.$$

By the uniform continuity of the function g', we can take $\gamma_1 < \varepsilon/2$ so that if $p_1' \subset V_{\gamma_1}(p_1)$, then

$$(17.2) \qquad \varrho[p_1(t), p_1'(t)] < \beta \qquad (0 \leqq t \leqq t_{10}).$$

We shall show that $U + p_1' q_\eta' \subset V_\beta(U + p_1 q_\eta)$; it will follow that

$$(17.3) \qquad \mu(U + p_1' q_\eta') < \mu(U + p_1 q_\eta) + 2\beta = \alpha.$$

It is sufficient to show that $p_1' q_\eta' \subset V_\beta(U + p_1 q_\eta)$. Putting $t = t_{10}$ in (17.2), we see that $p_1'(t_{10}) \subset V_\beta(q_0)$; hence q_η' lies between p_1' and $p_1'(t_{10})$, and $p_1' q_\eta' \subset p_1' p_1'(t_{10})$. By (17.2), $p_1' p_1'(t_{1\eta}) \subset V_\beta(p_1 q_\eta)$, and by the choice of η, $p_1'(t_{1\eta})\, p_1'(t_{10}) \subset V_\beta(q_\eta q_0) \subset V_\beta(q_{2\eta} q_0) \subset U$; thus (17.3) holds.

Take $\gamma_2 < \varepsilon/2$ so that if $p_2' \subset V_{\gamma_2}(p_2)$, then (17.2) holds with the subscript 1 replaced by 2. We shall show that $U + p_2 q_\eta \subset V_\beta(U + p_2' q_\eta')$; then

$$(17.4) \qquad \mu(U + p_2' q_\eta') > \mu(U + p_2 q_\eta) - 2\beta = \alpha.$$

As $p_2 q_\eta \subset p_2 q_0$ and $q_{2\eta} q_0 \subset U$, it is sufficient to show that $p_2 q_{2\eta} \subset V_\beta(U + p_2' q_\eta')$. By the choice of γ_2, $p_2' p_2'(t_{2,2\eta}) \subset V_\beta(p_2 q_{2\eta})$; thus, as $p_2 q_{2\eta} \cdot V_{2\eta}(q_0) = 0$ and $\beta < \eta$, $p_2' p_2'(t_{2,2\eta}) \cdot W_\eta(q_0) = 0$. Therefore the point q_η' lies beyond $p_2'(t_{2,2\eta})$. Hence $p_2 q_{2\eta} \subset V_\beta[p_2' p_2'(t_{2,2\eta})] \subset V_\beta(p_2' q_\eta')$, and (17.4) is proved.

Take $\gamma < \gamma_1, \gamma_2$ and λ, and take δ so small that any point of $W_\delta(p)$ lies on an arc $q_0' q_1'$ of $N_\gamma(q_0 q_1)$; this arc contains points p_1' and p_2' so that $\sigma(q_0 p_1, q_0' p_1')$, $\sigma(p_1 p_2, p_1' p_2')$ and $\sigma(p_2 q_1, p_2' q_1')$ are $< \gamma$. Then $p_1' \subset V_{\gamma_1}(p_1)$ and $p_2' \subset V_{\gamma_2}(p_2)$, and on account of (17.3) and (17.4) there will be a point p' of $p_1' p_2'$ for which

$$(17.5) \qquad \mu(U + p' q_\eta') = \alpha.$$

We shall show that the set S of all these points forms a cross-section through p.

We show first that (1) holds. Take any arc A' of $N\lambda(q_0 q_1)$ containing a point p'' of S. p'' was determined as follows: A point r'' was taken in $W_\delta(p)$, on the arc A'' say of $N\lambda(q_0 q_1)$, this determined a point q''_η, and this determined p''. A' and A'' have p'' in common; hence they differ at most by a point set in $V_\eta(q_0)+V_\sigma(q_1)$, because of the choice of λ. Hence $A'\cdot W_\delta(p)\neq 0$, and no matter what point r' is taken in $A'\cdot W_\delta(p)$, determining q'_η and $p'\subset S$, $q'_\eta=q''_\eta$ and $p'=p''$. Thus p' is the only point of A' on S, from which (1) follows immediately.

For some $\lambda_0>0$, each arc of $N_{\lambda_0}(q_0 q_1)$ cuts $W_\delta(p)$ and therefore contains a point of S. This evidently holds also for any smaller arc $r_0 r_1$ containing p (as S is closed); hence (2) holds.

We must prove still that S is closed. Let $p^1, p^2, \cdots$, be a sequence of points in S, lying on arcs $A^1, A^2, \cdots$, of $N_\gamma(q_0 q_1)$. These arcs all cut the closed set $W_\delta(p)$, hence if the sequence has a limit point, it lies on an arc A^* of $N_\gamma(q_0 q_1)$ cutting $W_\delta(p)$, and A^* contains a point p^* of S. Assume now that p^* exists. We shall show that for an arbitrary $\varepsilon^*>0$ there is a $\delta^*>0$ such that the points of S on all arcs of $N_\gamma(q_0 q_1)$ cutting $V_{\delta^*}(p^*)$ lie in $V_{\varepsilon^*}(p^*)$; it follows then that $p^i\to p^*$. If $p^*=p$, this follows from what we saw above. For replacing the ε used above by an arbitrary ε and choosing γ_1, γ_2 and δ accordingly, we have $\sigma(p_2 p_1, p'_2 p'_1)<\gamma<\varepsilon/2$, hence $p'\subset V_{\varepsilon/2}(p_1 p_2)\subset V_\varepsilon(p)$.

Suppose now $p^*\neq p$. Given an $\varepsilon^*>0$, take $\beta^*\leqq\beta$ and $<\varepsilon^*/2$ so that if p_3^* and p_4^* are the points of A^* for which

$$(17.6)\qquad \mu(U+p_3^* q_\eta^*)=\alpha-2\beta^*,\qquad \mu(U+p_4^* q_\eta^*)=\alpha+2\beta^*,$$

then $p_3^* p^* p_4^*\subset V_{\varepsilon^*/2}(p^*)$. Put $t_{k\zeta}^*=\mu(p_k^* q_\zeta^*)$ $(k=3,4)$. Note that by (17.2), the point q_β^* on A^* exists.

Take γ^* so that if $p'_3\subset V_{\gamma^*}(p_3^*)$ and $p'_4\subset V_{\gamma^*}(p_4^*)$ are points of an arc A' of $N\lambda(q_0 q_1)$, then

$$(17.8)\qquad \varrho[p_i^*(t), p'_i(t)]<\beta^*\qquad (0\leqq t\leqq t_{i\beta}^*,\ i=3,4).$$

We shall show that $p'_3 q'_\eta\subset V_{\beta^*}(U+p_3^* q_\eta^*)$ and $p_4^* q_\eta^*\subset V_{\beta^*}(U+p'_4 q'_\eta)$; then

$$(17.9)\qquad \mu(U+p'_3 q'_\eta)<\alpha,\qquad \mu(U+p'_4 q'_\eta)>\alpha,$$

and hence the point p' of A' in S (if it exists) lies on the arc $p'_3 p'_4\subset V_{\beta^*}(p_3^* p^* p_4^*)\subset V_{\varepsilon^*}(p^*)$. Therefore if we choose δ^* corresponding to γ^* as we choose δ corresponding to γ, then any point p' of S on an arc A' cutting $W_{\delta^*}(p^*)$ lies in $V_{\varepsilon^*}(p^*)$, completing the proof.

We show first that $q^*_{\eta+\beta}q^*_\beta \subset U$. Say p^*_1 of A^* is in $V_{\gamma_1}(p_1)$. As $p^*_1 p^*_1(t_{1,2\eta}) \subset V_\beta(p_1 q_{2\eta})$, and $p_1 q_{2\eta} \cdot V_{2\eta}(q_0) = 0$, and $\beta < \eta/2$, we have $p^*_1 p^*_1(t_{1,2\eta}) \cdot W_{\eta+\beta}(q_0) = 0$; hence $q^*_{\eta+\beta}$ lies beyond $p^*_1(t_{1,2\eta})$. As $p^*_1(t_{10}) \subset V_\beta(q_0)$, $p^*_1(t_{10})$ lies beyond q^*_β. Thus $q^*_{\eta+\beta}q^*_\beta \subset p^*_1(t_{1,2\eta})\,p^*_1(t_{10}) \subset V_\beta(q_{2\eta}q_0) \subset U$.

By (17.8), $p'_3(t^*_{3\beta}) \subset V_{\beta^*}(q^*_\beta) \subset V_\eta(q_0)$; thus $p'_3 q'_\eta \subset p'_3 p'_3(t^*_{3\beta})$. As $p'_3 p'_3(t^*_{3\eta}) \subset V_{\beta^*}(p^*_3 q^*_\eta)$ and $p'_3(t^*_{3\eta})\,p'_3(t^*_{3\beta}) \subset V_{\beta^*}(q^*_\eta q^*_\beta) \subset V_{\beta^*}(U)$, $p'_3 q'_\eta \subset V_{\beta^*}(U + p^*_3 q^*_\eta)$. As $p'_4 p'_4(t^*_{4,\eta+\beta}) \subset V_{\beta^*}(p^*_4 q^*_{\eta+\beta})$ and $p^*_4 q^*_{\eta+\beta} \cdot V_{\eta+\beta}(q_0) = 0$, q'_η lies beyond $p'_4(t^*_{4,\eta+\beta})$; hence $p^*_4 q^*_{\eta+\beta} \subset V_{\beta^*}[p'_4 p'_4(t^*_{4,\eta+\beta})] \subset V_{\beta^*}(p'_4 q'_\eta)$. As also $q^*_{\eta+\beta}q^*_\eta \subset U$, $p^*_4 q^*_\eta \subset V_{\beta^*}(U + p'_4 q'_\eta)$. This completes the proof of the theorem.

18. **Equivalence of two cross-sections.** Let p_1 and p_2 be two points, or the same point, of a curve C, and let S_1 and S_2 be cross-sections through p_1 and p_2 respectively. We suppose that the arc $p_1 p_2$ (if $p_1 \neq p_2$) contains no points of S_1 or of S_2 within it. Take an arc $q_1 q_2$ interior to C and containing p_1 and p_2 within it, short enough so that it contains no points of S_1 or S_2 other than p_1 and p_2. Take $\delta > 0$ so that every arc in $N_\delta(q_1 q_2)$ contains just one point of S_1 and one point of S_2. If S'_1 and S'_2 are those points of S_1 and S_2 respectively which lie on arcs of $N_\delta(q_1 q_2)$, plus limit points, then S'_1 and S'_2 are cross-sections through p_1 and p_2, and *they are homeomorphic*. For, to each point of one set, let correspond that point of the other set lying on the same arc of $N_\delta(q_1 q_2)$. If $p_{11}, p_{12}, \cdots$ are in S_1 and approach $p_1 \subset S_1$, and p_{1i} and p_1 lie on arcs A_i and A respectively of $N_\delta(q_1 q_2)$, then the corresponding points $p_{21}, p_{22}, \cdots$ of S_2 approach the arc A; as S_2 is closed, they approach the point p_2 of S_2 corresponding to p_1, and this correspondence is continuous.

19. **The local connectivity of cross-sections.** We show here that *if R is locally connected, then any cross-section is locally connected at all inner points* (see § 16).

If p is an inner point of S, and $q_0 q_1$ and λ are taken so as to satisfy (2), § 16, take $\lambda' \leqq \lambda$ so that the points of S on all arcs of $N_{\lambda'}(q_0 q_1)$ lie in $V_\varepsilon(p)$ ($\varepsilon > 0$ arbitrary). Take γ so that $V_\gamma(p) \subset N_{\lambda'}(q_0 q_1)$, and take δ so that any point q in $S \cdot V_\delta(p)$ is joined to p by a connected set $A \subset V_\gamma(p)$. Let A' consist of those points of S which are on arcs of $N_{\lambda'}(q_0 q_1)$ cutting A. Then A' is a continuous image of A, hence it is connected; also $A' \subset V_\varepsilon(p)$, and $p \subset A$, $q \subset A$. Therefore S is locally connected at p.

20. **Regular families in n-space.** Let F be a regular family of curves filling a region in Euclidean n-space. A fundamental question is:

Is there a cross-section through any inner point of a curve which is a closed $(n-1)$-cell (the topological image of a closed $(n-1)$-simplex)? If $n = 2$, the answer is yes. For any connected cross-section S contains two points p and q such that dropping out any other point of S disconnects these points; hence S is an arc.[14] The answer is also in the affirmative if $n = 3$. For $n > 3$, nothing is known.

Suppose, in a given regular family, there is a cross-section S through the point p which is a closed $(n-1)$-cell. Map S onto a portion S' of the $(n-1)$-plane $x_n = 0$ in n-space. If we consider only short arcs containing points of S, we can orient them similarly. Let pq be such an arc, containing the point p_0 of S; say p_0' is the corresponding point of S'. If $r \subset pq$, map r into that point of the line through p_0' perpendicular to S' for which $x_n = \pm \mu(p_0 r)$, using the $+$ or $-$ sign according as r lies in the positive or negative direction from p_0. By Theorem 13 A it is seen that this mapping is continuous; hence *in the neighborhood of any such point p, the curves are equivalent to a family of straight lines filling a region of n-space.*

Example. As we noted above, a cross-section in example (e) of § 8 is an arc; we can fit a denumerable number of these arcs together to form an arc joining C to a. If we map this arc onto the radius of a circle C', we can map the curves into circles concentric with C', as in Theorem 14A. We can thus prove a theorem of Kerékjártó.[15]

III. Families of paths.

21. **Definition.** In this part we shall denote by the word *curve* the topological image of an open line segment or of a circle only. If the points of a regular oriented family of curves form an open subset R' of a locally compact[16] separable metric space R, then the curves of F together with the points of $R^* = R - R'$ form a *family of paths.* The points of R^* we call *invariant points.*

22. **Tubes.** We call a set of arcs *self-compact* if any sequence chosen from these arcs contains a subsequence $p_1 q_1, p_2 q_2, \cdots$ approaching an arc pq of the set: $\lim_{n\to\infty} \sigma(pq, p_n q_n) = 0$. A set of arcs of F form a *tube* T if the following conditions are satisfied:

(1) The arcs are self-compact.

(2) No two distinct arcs have common points.

(3) The set of points in T contains inner points.

[14] See Hausdorff, loc. cit. [6], p. 222, III.

[15] B. v. Kerékjártó, Topologie I, Berlin, 1923, p. 246. See also pp. 238–239.

[16] R is locally compact if each point of R is contained in a compact neighborhood of R. Euclidean space is locally compact; Hilbert space is not. R' is evidently locally compact also.

One might require also that all points of T be either inner points or limit points of inner points, and that T be connected if R' is locally connected, as in the definition of cross-sections.

The sets S_1 and S_2 of points on the negative and positive ends of the arcs of T we call the *negative* and *positive ends* of T respectively. The point sets T, S_1 and S_2 are evidently compact and closed. S_1 and S_2 are connected if T is.

LEMMA. *If T is a tube with ends S_1 and S_2 and pq is an arc of F containing at least two points of S_1+S_2, then $\delta(pq) \geqq \varrho(S_1, S_2) > 0$.*

This is certainly true if pq contains points of both S_1 and S_2. Suppose pq contains two points p' and p'' of S_1 (or of S_2), and say p'' lies in the positive direction from p'. If $p'q'$ is the arc of T containing p', then $p'q' \subset p'p''$ and $p'p''$ contains also a point of S_2, hence $\delta(pq) \geqq \varrho(S_1, S_2)$.

THEOREM 22A. *The ends S_1 and S_2 of a tube T are cross-sections.*

Take S_1 for example. If $p' \subset S_1$, take an arc $q_0'q_1'$ of diameter $< \lambda'$ containing p' within it, where $\lambda' < \varrho(S_1, S_2)/3$. Then any arc of $N_{\lambda'}(q_0'q_1')$ contains at most one point of S_1, by the lemma; hence (1), § 16 holds. Let p be a point of S_1 on an arc pq of T containing inner points of T. Then all points of pq but p and q are inner points of T. Let $q_0 q_1$ be an arc of diameter $< \varrho(S_1, S_2)/2$ and containing p in its interior, q_1 being interior to pq. Take λ so small that $V_\lambda(q_0) \cdot T = 0$ and $V_\lambda(q_1) \subset T$. Then any arc of $N_\lambda(q_0 q_1)$ contains exactly one point of S_1, and (2), § 16 is proved. We noted that S_1 was closed; hence it is a cross-section.

THEOREM 22B. *If S_0 is a compact cross-section through p_0, then for some $\alpha > 0$ the set of arcs qr for which $q = g'(p, -\alpha)$ and $r = g'(p, \alpha)$ with $p \subset S_0$ form a tube T, containing p_0 as an inner point.*

Containing each point $p' \subset S_0$ is an arc $q'r'$ such that for some λ', each arc in $N_{\lambda'}(q'r')$ contains at most one point of S_0. Take $\mu < \mu(p'+q')$ and $< \mu(p'+r')$, and take $\lambda < \lambda'$ and $< \mu/3$. Take $\delta < \lambda$ so that every point p^* of $S_0 \cdot V_\delta(p')$ is on an arc q^*r^* of $N_\lambda(q'r')$. Suppose now A is any arc of F containing such a point p^*, and $\mu(A) \leqq \lambda$. As $p'+q' \subset V_\lambda(p^*q^*)$, $\mu(p^*q^*) > \mu(p'+q') - 2\lambda > \lambda$, hence A does not contain q^*. Similarly A does not contain r^*; hence $A \subset q^*r^*$, and therefore it contains at most one point of S_0. A finite number of such sets $V_\delta(p')$ cover S_0; if 4α is the least of the corresponding numbers λ, then any arc A with $\mu(A) \leqq 4\alpha$ contains at most one point of S_0. Then α is our required number, and the arcs pq constructed as above form the required tube T.

To prove this, we see first that the arcs are self-compact (use Theorem 15A). Note now that for any arc qpr of T, $\mu(qpr) \leqq \mu(pq) + \mu(pr) = 2\alpha$, by Theorem 3F. No two arcs A', A'' of T have a common point.

17

For if they did, they would form an arc A with two points in S_0; hence $\mu(A) > 4\alpha$. But $\mu(A) \leqq \mu(A') + \mu(A'') \leqq 4\alpha$, a contradiction. Finally, p_0 is an inner point of T. For there is an arc $q^* r^*$ containing p_0 such that for some λ^*, each arc of $N_{\lambda^*}(q^* r^*)$ contains exactly one point of S_0, and so that $\mu[N_{\lambda^*}(q^* r^*)] < \alpha$. All points of some $V_\delta(p_0)$ are on arcs of $N_{\lambda^*}(q^* r^*)$ and are thus in T, completing the proof.

23. **First sequence of tubes.** By the last theorem, if S is a compact cross-section through p, there is an $\alpha > 0$ such that if $S_3 = g'(S, -2\alpha)$, $S_1 = g'(S, -\alpha)$, $S_2 = g'(S, \alpha)$, $S_4 = g'(S, 2\alpha)$, then there are tubes T' and T containing p, with ends S_3 and S_4, S_1 and S_2 respectively.

If R' is self-compact, put $W_1 = R'$. Otherwise, let $W_1, W_2, \cdots$ be a sequence of compact and closed sets such that each W_i is interior to W_{i+1} and $W_1 + W_2 + \cdots$ covers R'.[17] Each point of W_1 is interior to a tube T as above; a finite number of these, say $T_1, \cdots, T_{l_1}$, cover W_1. If $W_1 \neq R'$, a finite number of tubes cover W_2; let the new tubes required be $T_{l_1+1}, \cdots, T_{l_2}$, etc. We may choose the tubes so that $\delta(T_k) < 1/i$ if $k > l_i$. We define thus a (possibly finite) sequence of tubes $T_1, T_2, \cdots$ with the following properties:

(A) They cover R'.

(B) Any compact and closed set in R' has points in common with but a finite number of the tubes.

(C) $\lim_{n \to \infty} \delta(T_n) = 0$.

Corresponding to the tube T_i is the tube T_i' and cross-sections S_{i3}, S_{i1}, S_{i0} (= the above S), S_{i2}, and S_{i4}.

Note that it is possible for an arc of one tube to contain a number of points on both ends of another tube.

24. **First sequence of pseudo-tubes.** Any set of arcs satisfying (1) and (2) of § 22 we shall call a *pseudo-tube*. We shall replace the tubes $T_1, T_2, \cdots$ by pseudo-tubes $T_1^2, T_2^2, \cdots$ which have the above properties, and also the following:

(D) No two ends of pseudo-tubes have common points.

(E) Any pseudo-tube T_i^2 contains points of at most one of the ends of any preceding pseudo-tube T_k^2 $(k < i)$.

From this it follows also that any arc of any pseudo-tube T_i^2 contains at most a single point of any end of any previous pseudo-tube.

Put $T_1^2 = T_1$. Suppose we have constructed the pseudo-tubes $T_1^2, \cdots, T_{m_{i-1}}^2$, covering the tubes $T_1, \cdots, T_{i-1}$, and having properties (D) and (E) above;

[17] About each point of R' is a compact neighborhood; a denumerable number of these, $U_1, U_2, \cdots$ cover R'. Put $W_i' = \bar{U}_1 + \cdots + \bar{U}_i$. We can take a subsequence $W_1, W_2, \cdots$ of this sequence so that W_i is interior to W_{i+1}; for W_i' is contained in a finite number of the sets U_k, and is thus interior to some W_j'.

we shall construct pseudo-tubes $T^2_{m_{i-1}+1}, \cdots, T^2_{m_i}$ covering T_i and having these properties. We shall call the ends of any pseudo-tube T^2_k S^2_{k1} and S^2_{k2}.

Take a number λ less than any $\varrho(S^2_{k1}, S^2_{k2})$ $(1 \leqq k \leqq m_{i-1})$. Given any arc pq of T'_i, divide it into arcs $pq_1, q_1q_2, \cdots, q_nq$, each of diameter $< \lambda/2$, so that none of the points $q_1, \cdots, q_n$ lie on ends of previous pseudo-tubes, and q_1 lies between S_{i3} and S_{i1}, and q_n lies between S_{i2} and S_{i4}. If p_0 is the point of pq on S_{i0}, say $q_k = g'(p_0, \beta_k)$ $(k = 1, \cdots, n)$. We can take a number η, and a cross-section $S^* \subset S_{i0}$ containing all points of S_{i0} in some $V_\delta(p_0)$, (or rather, a *pseudo-cross-section*, i. e. a set of points satisfying (1) and (3) of § 16), small enough so that no cross-section $g'(S^*, \beta_k \pm \eta)$ contains points of any end $S^2_{11}, \cdots, S^2_{m_{i-1}2}$, and so that each pseudo-tube T^*_k, with ends $g'(S^*, \beta_k - \eta)$ and $g'(S^*, \beta_{k+1} + \eta)$ $(k = 1, \cdots, n-1)$, is of diameter $< \lambda$. (T^*_k is obviously a pseudo-tube.) Then these pseudo-tubes cover all points of T_i on arcs cutting S^*, and each pseudo-tube T^*_k contains points on at most one end of any pseudo-tube $T^2_1, \cdots, T^2_{m_{i-1}}$.

For each point p_0 of S_{i0} there is such an S^*; a finite number of these cross-sections cover S_{i0}. We define pseudo-tubes as above for each such S^*; the resulting pseudo-tubes cover T_i. We can choose the numbers η so that no two ends of these pseudo-tubes have common points. If one of these pseudo-tubes has ends $g'(S^*, \beta_k - \eta)$ and $g'(S^*, \beta_{k+1} + \eta)$, call its length $(\beta_{k+1} + \eta) - (\beta_k - \eta)$. Then if we arrange the new pseudo-tubes $T^2_{m_{i-1}+1}, \cdots, T^2_{m_i}$ in order of decreasing length, the pseudo-tubes $T^2_1, \cdots, T^2_{m_i}$ have properties (D) and (E). Continuing in this manner, we obtain the sequence of pseudo-tubes $T^2_1, T^2_2, \cdots$ with all the desired properties.

25. **Second sequence of pseudo-tubes.** We replace here the pseudo-tubes of the last sequence by a sequence of pseudo-tubes $T^3_1, T^3_2, \cdots$, which have the properties (A), (B) and (C) above, and also the property

(F) If any inner point p' of an arc pq of a pseudo-tube T^3_k lies in a preceding pseudo-tube T^3_l $(l < k)$, then all of pq lies in preceding pseudo-tubes.

Put $T^3_1 = T^2_1$. Having constructed the pseudo-tubes $T^3_1, \cdots, T^3_{n_{i-1}}$ with the above property, and so that the point sets $T^2_1 + \cdots + T^2_{i-1}$ and $T^3_1 + \cdots + T^3_{n_{i-1}}$ are identical, we shall construct the pseudo-tubes $T^3_{n_{i-1}+1}, \cdots, T^3_{n_i}$ with the property (F), lying in and filling out T^2_i.

The pseudo-tube T^2_i may contain points of one or the other end of each of the pseudo-tubes $T^2_1, \cdots, T^2_{i-1}$. If $T^2_{\alpha_1}, \cdots, T^2_{\alpha_M}$ are those pseudo-tubes whose negative ends contain points of T^2_i, put for simplicity $S_k = T^2_i \cdot S^2_{\alpha_k 1}$, $k = 1, \cdots, M$; if there are no such pseudo-tubes, put

$M = 0$. Similarly let $S'_1, \cdots, S'_N$ be those points of positive ends of pseudo-tubes $T^2_1, \cdots, T^2_{i-1}$ in T^2_i; if there are no such pseudo-tubes, put $N = 0$. There are 2^M subsets of the numbers $1, \cdots, M$, including the whole set and the null set. Arrange these in order, choosing first the whole set, next all sets of $M-1$ of the numbers, etc. Call these sets $C_1, \cdots, C_{2^M}$. Arrange the 2^N subsets of the numbers $1, \cdots, N$ in order similarly; call them $D_1, \cdots, D_{2^N}$. Now arrange all the 2^{M+N} pairs of these sets $(C_1, D_1), (C_1, D_2), \cdots, (C_1, D_{2^N}), (C_2, D_1), \cdots, (C_{2^M}, D_{2^N})$ in the order shown.

We shall consider each of these pairs in turn, and construct a corresponding pseudo-tube if a certain condition is satisfied. For the pair (C_s, D_t), this condition is:

There is an arc pq of T^2_i such that

(α) pq contains a point of every S_k with subscript k in the set of numbers C_s, and of every S'_l with subscript l in D_t, and

(β) if p_1 is the last point of pq on an S'_l, l in D_t, or $p_1 = p$ if D_t is the null set, and if q_1 is the first point of pq on an S_k, k in C_s, or $q_1 = q$ if C_s is the null set, then q_1 lies in the positive direction from p_1.

The condition is satisfied automatically for (C_{2^M}, D_{2^N}), there being no corresponding sets S_k, S'_l.

If the condition is satisfied for the pair (C_s, D_t), then the corresponding pseudo-tube consists of all arcs $p_1 q_1$ as named above. (The pseudo-tube corresponding to (C_{2^M}, D_{2^N}) is just T^2_i itself.) We order these pseudo-tubes $T^3_{n_{i-1}+1}, \cdots, T^3_{n_i}$ in the same manner that we ordered the pairs.

We show first that these really are pseudo-tubes. As the S_k and the S'_l are compact and closed, we can pick out from any sequence of arcs of T^3_ν $(n_{i-1}+1 \leqq \nu \leqq n_i)$ a subsequence $p_1 q_1, p_2 q_2, \cdots$, approaching an arc $p' q'$, where p_i and p' are in some fixed S'_l, l in D_t, and q_i and q' are in some fixed S_k, k in C_s (see Theorem 15 A); we must show that $p'q'$ is an arc of T^3_ν. Say $p^k q^k$ is the arc of T^2_i containing $p_k q_k$, and $p^* q^*$ is the arc of T^2_i containing $p' q'$; then also $p^k p_k \to p^* p'$, $q_k q^k \to q' q^*$. If T^3_ν corresponded to (C_s, D_t), then each $p^k p_k$ contains a point of each S'_l, l in D_t; hence the same is true of $p^* p'$. Similarly $q' q^*$ contains a point of each S_k, k in C_s. Therefore, by property (E), $p' q'$ contains within it no points of S'_l, l in D_t, or of S_k, k in C_s. Thus $p' q'$ is an arc of T^3_ν, and the arcs of T^3_ν are self-compact. As any two distinct arcs of T^3_ν lie on distinct arcs of T^2_i and no two arcs of T^2_i have common points, no two arcs of T^3_ν have common points; hence T^3_ν is a pseudo-tube.

We show now that the sequence $T^3_1, T^3_2, \cdots$ has the required properties. As each $T^3_{n_i}$ is identical with T^2_i, (A) is satisfied. (B) and (C) hold obviously. To prove (F), we must show that, if p' is a point interior to an arc $p_1 q_1$

of a pseudo-tube T^3_ν, and p' lies in a pseudo-tube T^3_λ, $\lambda<\nu$, then all of $p_1 q_1$ lies in pseudo-tubes preceding T^3_ν. Choose i so that $n_{i-1}<\nu\leqq n_i$, and say pq is the arc of T^2_i containing $p_1 q_1$.

Suppose first $\lambda\leqq n_{i-1}$; then p' lies in a pseudo-tube T^2_η, $\eta\leqq i-1$. If pq contains no points of the ends of T^2_η, then pq lies in T^2_η, and therefore $p_1 q_1$ lies in $T^3_{n_\eta}=T^2_\eta$. Suppose pq contains a point of $S^2_{\eta 1}$ or $S^2_{\eta 2}$. Define the sets S_k and S_l' for T^2_i as before. If some point of pq on an S_l' lies after some point on an S_k, then all of $p_1 q_1$ lies in pseudo-tubes $T^3_1, \cdots, T^3_{n_{i-1}}$. Suppose then p_2 is the last point of pq on any S_l', q_2 is the first point of pq on any S_k, and p_2 lies before q_2; in this case we shall show that $p_2 q_2$ lies in a pseudo-tube T^3_μ, $n_{i-1}<\mu<\nu$, from which the statement will follow again.

If (C_s, D_t) is the pair of sets of numbers to which the pseudo-tube T^3_ν corresponds, then pq contains points on all the S_l', l in D_t, and on all the S_k, k in C_s. As p' lies in T^2_η and is interior to $p_1 q_1$, either $S^2_{\eta 2}$ has a point q^* in $p' q_1$, or $S^2_{\eta 1}$ has a point in $p_1 p'$; thus pq contains a point on some $S'_{l'}$, l' not in D_t, or a point on some $S_{k'}$, k' not in C_s. Therefore if the set of all $S_k\,(S_l')$ with points on pq gives the set of numbers $C_{s'}\,(D_{t'})$, $C_{s'}$ and $D_{t'}$ contain C_s and D_t respectively, and one of them contains more numbers; hence the pair $(C_{s'}, D_{t'})$ precedes the pair (C_s, D_t). As the condition is satisfied for this pair, a corresponding pseudo-tube T^3_μ, $\mu<\nu$, was constructed, and it contains the arc $p_2 q_2$.

Suppose now $n_{i-1}<\lambda<\nu$; then the pseudo-tube T^3_λ contains a certain portion $p_2 q_2$ of the arc pq. Then pp_2 (if p_2 is not p) and $q_2 q$ (if q_2 is not q) are both contained in pseudo-tubes $T^2_1, \cdots, T^2_{i-1}$, and thus in pseudo-tubes $T^3_1, \cdots, T^3_{n_{i-1}}$. The property (F) is now shown to hold.

26. **The function τ'.** With each motion (pq) of F (see § 11) we shall associate a number $\tau'(pq)$ with the following properties:

(1) $\tau'(pq)$ is greater than, less than, or equal to zero according as (pq) is positively oriented, negatively oriented, or is the null motion.

(2) τ' is continuous, in that if $(p_i q_i)\to(pq)$ (see § 12), then $\tau'(p_i q_i)\to\tau'(pq)$.

(3)

(26.1) $$\tau'[(pq)+(qr)] = \tau'(pq)+\tau'(qr).$$

Then also if (qp) is the negative of pq,

(26.2) $$\tau'(qp) = -\tau'(pq).$$

We shall define τ' successively through the pseudo-tubes T^3_1, T^3_2, $\cdots$. Consider first T^3_1. If p is any point of an arc $p_0 q_0$ of T^3_1 ($p_0\subset S^3_{11}$, $q_0\subset S^3_{12}$), determining the motion $(p_0 p)$, put

(26.3) $$\tau'(p_0 p) = \mu(p_0 p).$$

If q is another point of $p_0 q_0$, put

$$\tau'(pq) = \tau'(p_0 q) - \tau'(p_0 p). \tag{26.4}$$

τ' has so far all the required properties.

Suppose now we have defined $\tau'(pq)$ for all motions lying wholly in $T_1^3 + \cdots + T_{i-1}^3$; we shall extend its definition through T_i^3. Consider first motions (pq) lying wholly in T_i^3. If no arcs of T_i^3 are contained in former pseudo-tubes, we define $\tau'(pq)$ just as before. If all of T_i^3 is contained in former pseudo-tubes, $\tau'(pq)$ is already defined. Otherwise, we define $\tau'(pq)$ by a process similar to a method of extending the domain of continuous functions.[18]

On account of the fundamental property of the sequence of pseudo-tubes, if τ' is already defined over any part of an arc of T_i^3, it is defined over the whole arc. Let A be the set of points of S_{i1}^3 which are end points of arcs of T_i^3 over which τ' is already defined; A is closed. If p is any point of T_i^3, on the arc $p_0 q_0$ of T_i^3, put

$$p = h(p_0, s), \tag{26.5}$$

where

$$s = \frac{\mu(p_0 p)}{\mu(p_0 q_0)}.$$

s is a continuous function of p_0 and p; hence h is continuous in p_0 and s (see § 13). We shall define $\tau'(p_0 p)$ in terms of the values $\tau'(p_0' p')$, $p_0' \subset A$, $p' = h(p_0', s)$.

Given the point $p_0 \subset S_{i1}^3 - A$, put

$$a = a(p_0) = \varrho(p_0, A), \tag{26.6}$$

and

$$X_r(p_0) = A \cdot W_r(p_0) \qquad (a \leqq r \leqq 2a). \tag{26.7}$$

Given the point $p = h(p_0, s)$, let $\psi(p, r)$ be the maximum of the values $\tau'(p_0' p')$, $p' \subset h[X_r(p_0), s]$ (and hence $p_0' \subset X_r(p_0)$). We define $\tau'(p_0 p)$ by the equation

$$\tau'(p_0 p) = \frac{1}{a(p_0)} \int_a^{2a} \psi(p, r)\, dr. \tag{26.8}$$

Note that $\tau'(p_0 p)$ increases as p moves away from p_0 (as s increases).

We show now that the new τ' is continuous. Let G be the maximum of $\tau'(p_0' p')$, $p_0' \subset A$, $p' \subset T_i^3$. Take any two points $p = h(p_0, s)$ and $q = h(q_0, t)$, $p_0, q_0 \subset S_{i1}^3 - A$. If $p^* = h(p_0, t)$ and $\varrho(p_0, q_0) \leqq \delta < a(p_0)$, then, by Carathéodory, loc. cit. (end of § 541), we see that

[18] See C. Carathéodory, Reelle Funktionen, first edition, Leipzig, 1918, p. 617.

$$|a(q_0)\tau'(q_0 q) - a(p_0)\tau'(p_0 p^*)| \leqq 3G\delta,$$

which is independent of t. Obviously $\tau'(p_0 p)$ is continuous in p (p_0 fixed); hence we can take q so near p and t so near s that $3G\delta < \epsilon/2$ and so that $a(p_0)\,|\tau'(p_0 p^*) - \tau'(p_0 p)| < \epsilon/2$; it follows that

$$|a(q_0)\tau'(q_0 q) - a(p_0)\tau'(p_0 p)| < \epsilon.$$

Therefore $a(p_0)\tau'(p_0 p)$ and thus $\tau'(p_0 p)$ is continuous for $p_0 \subset S^3_{i1} - A$.

We must consider next the case that p_0 is in A and is a limit point of $S^3_{i1} - A$. Take points $p_1 = h(p_0, s_1)$ and $p_2 = h(p_0, s_2)$ on opposite sides of p so that $\tau'(p_0 p_1)$ and $\tau'(p_0 p_2)$ differ from $\tau'(p_0 p)$ by less than $\epsilon/2$. By Carathéodory, loc. cit., we see that if η is small enough and $p'_i = h(p'_0, s_i) \subset V_\eta(p_i)$ $(p'_0 \subset S^3_{i1} - A)$, then

$$|\tau'(p'_0 p'_i) - \tau'(p_0 p_i)| < \frac{\epsilon}{2} \qquad (i = 1, 2).$$

As $\tau'(p'_0 p')$ is an increasing function of p', it follows immediately that if p' is near enough p, it lies on such an arc $p'_1 p'_2$, and hence $\tau'(p'_0 p')$ differs from $\tau'(p_0 p)$ by less than ϵ. Thus τ' is continuous over its present domain of definition.

If the motion (pq) is interior to T^3_i, define $\tau'(pq)$ as in (26.4). Consider now any motion (pq) in $T^3_1 + \cdots + T^3_i$. If it does not lie in $T^3_1 + \cdots + T^3_{i-1}$, let $pp_1, p_1 p_2, \cdots, p_m q$ be a set of arcs defining (pq) (see § 11), chosen so that each $p_k p_{k+1}$ lies in some tube T^3_l. Corresponding to each $p_k p_{k+1}$ is a motion $(p_k p_{k+1})$ for which $\tau'(p_k p_{k+1})$ is defined. Noting that $(pq) = (pp_1) + (p_1 p_2) + \cdots + (p_m q)$, we put

$$\tau'(pq) = \tau'(pp_1) + \tau'(p_1 p_2) + \cdots + \tau'(p_m q). \tag{26.9}$$

This definition determines $\tau'(pq)$ uniquely. For if $pp'_1, p'_1 p'_2, \cdots, p'_n q$ is another representation of (pq), let $q_1, q_2, \cdots, q_s$ be the set of all points p_k and p'_l; the arcs $pq_1, q_1 q_2, \cdots, q_s q$ gives another representation of (pq). Using these representations, we get corresponding numbers $\tau'_1(pq)$, $\tau'_2(pq)$, $\tau'_3(pq)$. Using (26.1), we see that $\tau'_1 = \tau'_3$, also $\tau'_2 = \tau'_3$; hence $\tau'_1 = \tau'_2$.

It is easily seen that the resulting τ' is continuous and that (1) and (3) hold. Hence defining τ' successively through all the pseudo-tubes $T^3_1, T^3_2, \cdots$, we have defined it throughout F as required.

27. The functions τ and f. We shall replace the function τ' by a function τ which has all the properties of τ', and has the additional property: If pq is an arc of a pseudo-tube T^3_i $(p \subset S^3_{i1},\ q \subset S^3_{i2})$ and (pq) is the corresponding motion, then

$$\tau(pq) \geqq 1. \tag{27.1}$$

To do this, let Δ_i be the minimum of $\tau'(pq)$ for arcs pq of T_i^3 $(i = 1, 2, \cdots)$. We can define a continuous function $\varphi(p) > 0$ for $p \subset R'$ with the following property: If $p \subset T_i^3$, then $\varphi(p) \leqq \Delta_i$. For if Γ_k is the smallest Δ_i for any T_i^3 having points in W_k (see § 23), let $\varphi(p) = \Gamma_{k+1}$ on the boundary of each W_k, and in $W_k - W_{k-1}$, let $\varphi(p)$ be a continuous extension of its values on the boundaries of W_k and W_{k-1}, so that $\Gamma_k \geqq \varphi(p) \geqq \Gamma_{k+1}$ there. Now if $p \subset T_i^3$, and $p \subset W_k - W_{k-1}$ say, then $\varphi(p) \leqq \Gamma_k \leqq \Delta_i$, as required. (If the number of pseudo-tubes T_i^3 is finite, we need merely take $\varphi(p)$ equal to the smallest Δ_i.)

Take now any motion (pq), and express it by means of arcs $pp_1, \cdots, p_m q$, as in § 11. If we let p' move along these arcs from p to q, we obtain sub-motions (pp') of (pq); the corresponding numbers $\tau'(pp')$ increase from 0 to $\tau'(pq)$. Thus for any value of τ', $0 \leqq \tau' \leqq \tau'(pq)$, there is a corresponding point $p_{\tau'}$ for which $\tau'(pp_{\tau'}) = \tau'$. We put

$$\tau(pq) = \int_0^{\tau'(pq)} \frac{d\tau'}{\varphi(p_{\tau'})}. \tag{27.2}$$

Now if pq is any arc of a tube T_i^3, then

$$\tau(pq) \geqq \int_0^{\Delta_i} \frac{d\tau'}{\Delta_i} = 1.$$

Moreover, τ has the properties given in § 26.

We now define the function f. Take any point p of R', and any number t. Find the motion (pq) of F for which

$$\tau(pq) = t;$$

put then

$$q = f(p, t).$$

For any point p of $R^* = R - R'$, put $f(p, t) = p$.

Given a point p and a number t, there is at most a single point $q = f(p, t)$; we shall show that there always is such a point. Starting at p, follow along the curve C through p in the positive or negative direction according as t is positive or negative, until we have passed through $n \geqq |t|$ pseudo-tubes $T_{i_1}^3, \cdots, T_{i_n}^3$, such that the last point in $T_{i_k}^3$ precedes the first point in $T_{i_{k+1}}^3$. Then for the resulting motion (pq'), $|\tau(pq')| > n > |t|$; hence there is a sub-motion (pq) of (pq') for which $\tau(pq) = t$, and thus $q = f(p, t)$.

f is continuous in R', as is seen by using the proof in § 13. Of course f is continuous at all inner points of R^*. Take now any boundary point p of R^*, any $\varepsilon > 0$, and any positive integer n. Then, on account of property (C) of § 23, we can take δ so small that if q is any point of

R' in $V_\delta(p)$, on the curve C of F, we must pass through at least n distinct tubes T_i^3 before we pass out of $V_\varepsilon(p)$. It follows that for any t, $|t| \leqq n$, $f(q,t) \subset V_\varepsilon(p)$; hence $\varrho[f(p,t'), f(q,t)] = \varrho[p, f(q,t)] < \varepsilon$ for $q \subset V_\delta(p)$ and $|t| < n$, and f is shown to be continuous throughout R.

Take now any point p of R' and any two numbers t_1, t_2. Take motions (pq) and (qr) so that

$$\tau(pq) = t_1, \; \tau(qr) = t_2.$$

Then if $(pr) = (pq) + (qr)$,

$$\tau(pr) = t_1 + t_2,$$

by (26.1). As then $q = f(p, t_1)$, $r = f(q, t_2)$, and $r = f(p, t_1 + t_2)$, we have $f[f(p, t_1), t_2] = f(p, t_1 + t_2)$. As $f(p, 0) = p$, we have now proved

THEOREM 27 A. *A function $f(p, t)$ can be defined over any family of paths, with the following properties:*

(1) *For each point p of R and any number t, $-\infty < t < +\infty$, there is a unique point*

$$q = f(p, t)$$

which lies on the curve C through p, or coincides with p if p is an invariant point. Further, for each point q of C there is a t such that $q = f(p, t)$.

(2) *$f(p, t)$ is continuous in both variables; that is, for each point p, any number t, and any $\varepsilon > 0$, there is a $\delta > 0$ such that*

$$\varrho[f(p,t), f(p',t')] < \varepsilon \text{ if } \varrho(p,p') < \delta \text{ and } |t - t'| < \delta.$$

(3) *For $p \subset R'$, $f(p, t)$ moves in the positive (negative) direction along the curve C through p as t increases (decreases).*

(4) *For each point p of R and any two numbers t_1 and t_2,*

$$f[f(p, t_1), t_2] = f(p, t_1 + t_2); \tag{27.3}$$

also

$$f(p, 0) = p. \tag{27.4}$$

Any such function f we shall say defines a *flow* in R.

28. The equivalence of two flows in R. Let f_0 and f_1 be two flows in R. We show that *one can be deformed into the other* in the following sense: For each number x, $0 \leqq x \leqq 1$, there is a flow $f_x(p, t)$, and this set of functions is continuous in x, p and t together. That is, for any $\varepsilon > 0$, p, t and x, there is a $\delta > 0$ such that $\varrho[f_{x'}(p', t'), f_x(p, t)] < \varepsilon$ if $|x' - x| < \delta$, $\varrho(p', p) < \delta$ and $|t' - t| < \delta$.

Consider any motion (pq) of F; having defined τ_0 and τ_1 corresponding to f_0 and f_1, we define the function $u(pq)$ by the equation

$$\tau_1(pq) = \tau_0(pq) + u(pq). \tag{28.1}$$

Now for any x, $0 \leqq x \leqq 1$, put

$$\tau_x(pq) = \tau_0(pq) + xu(pq), \tag{28.2}$$

and define f_x in terms of τ_x. By writing (28.1) successively for motions (pq), (qr) and (pr), we see that

$$u(pq) + u(qr) = u(pr). \tag{28.3}$$

Noting also that $u(pq)$ is continuous, and that $\tau_0(pq) + xu(pq)$ increases as q moves in the positive direction, we see that τ_x, and hence f_x, has all the properties it should have; hence f_x defines a flow. f_x is obviously continuous in x, p and t, and the theorem is proved.

29. **Remark on cross-sections.** If we know of the existence of a function f in a family of paths, then we can find a cross-section through a point p of R' quite simply as follows. For any point q, put $q_t = f(q, t)$. Define the continuous function

$$\theta(q) = \int_0^1 \varrho(q_t, p)\, dt. \tag{29.1}$$

Using (27.3) we see that $(q_\varepsilon)_t = q_{\varepsilon+t}$. Hence, given a point q_ε on the curve through q, we find by a change of variable

$$\theta(q_\varepsilon) - \theta(q) = \int_1^{1+\varepsilon} \varrho(q_t, p)\, dt - \int_0^{\varepsilon} \varrho(q_t, p)\, dt,$$

and thus θ has a partial derivative along each curve:

$$\theta'(q) = \partial\theta(q)/\partial t = \varrho(q_1, p) - \varrho(q, p).$$

θ' is continuous and positive near p: $\theta'(p) = \varrho(p_1, p)$. Hence on each curve passing near p we can find a point q for which $\theta(q) = \theta(p)$; these points are easily seen to form a cross-section.

PRINCETON UNIVERSITY, HARVARD UNIVERSITY.

Reprinted from DUKE MATHEMATICAL JOURNAL
Vol. 3, No. 1, March, 1937

ON MATRICES OF INTEGERS AND COMBINATORIAL TOPOLOGY

BY HASSLER WHITNEY

1. **Introduction.** Our object is to give an elementary account of some algebraic theorems, with some immediate applications in combinatorial topology, in particular, in the theory of homology and cohomology[1] groups. The theorems are to a certain extent known, if in somewhat different forms.

The main tool in the algebraic part is the theory of group pairs, and in particular the question of when one group "resolves" or "completely resolves" another. The main theorems are on the existence of extensions of a homomorphism, and the existence of solutions of linear equations, with a matrix of integers and elements of an abelian group as unknowns. In each theorem, two types of conditions are employed, one using mod m properties, the other using group pairs. We shall use only discrete groups (except in Theorem 1).

The applications to topology are concerned with the relations between the ordinary homology theory, and the newer "dual" theory. An illustration of the convenience of the newer theory is given in Appendix I. For other illustrations, we mention the duality theorems (Kolmogoroff, Alexander, Čech, etc.), and properties of maps (see a following paper).

I. Group pairs and homomorphisms

2. **Group-pairs.** All groups will be abelian. 0 will denote the identity in any group. Let

$$mG = \text{all elements } mg,\ g \text{ in } G \ (m \text{ an integer}),$$

$$_mG = \text{all } g \text{ in } G \text{ such that } mg = 0.$$

Then $0G$ contains 0 alone. $G - G'$ is the difference (factor) group of G over the subgroup G'.

Let G, H, Z be three groups. If to each g in G and h in H there corresponds a $z = gh$ in Z, and both distributive laws are satisfied, we say G and H form a *group*

Received December 1, 1936; presented to the American Mathematical Society December, 1936.

[1] The relation of "coboundary" is becoming of increasing importance. Other terms have been used: dual boundary, upper boundary, inverse boundary, boundary in the dual subdivision, derived (of a function). The present term (accepted by E. Čech) offers definite advantages. In differential geometry, the (exact alternating) covariant tensors and the contravariant differentials play the same rôle as cocycles and (contra)cycles in the combinatorial theory; in fact, they may be obtained directly by a passage to the limit. Moreover, the prefix co is very convenient to handle.

pair with respect to Z. If $gh = 0$, we say g and h are *orthogonal.* For any subgroup H' of H, set

$(G, H') =$ *nullifier*[2] of H' in $G =$ all elements g in G orthogonal to every h in H'.

Define (H, G') similarly.

If $(G, H') = 0$, we say H' *resolves* G.[3] If $(G, {}_mH') = mG$, we say H' *m-resolves* G.[4] Clearly "H' 0-resolves G" and "H' resolves G" are equivalent statements. If H' m-resolves G for all integers $m \geqq 0$, we say H' *resolves G completely.* Note that

$$\text{(2.1)} \qquad \text{if } H' \text{ resolves } G, \text{ then } (G, mH') = {}_mG \qquad \text{(all } m).$$

(For $m = 1$, this is the definition.) For if $mg \neq 0$, choose h in H' so that $(mg)h \neq 0$; then $g(mh) \neq 0$.

Let G and Z be groups. A *Z-character* of G is a homomorphism of G into (part of) Z.[5] Let $Ch_Z(G)$ be the group of Z-characters of G. Given g in G and ϕ in $Ch_Z(G)$, set $g \cdot \phi = \phi(g)$; then these groups form a pair. Clearly *any group resolves any of its character groups.* The converse is false.

G is *infinitely divisible* if $mG = G$ for each m.

3. **Examples.** Let I_0 be the group of integers, and $I_\mu = I_0 - \mu I_0$ the group of integers mod μ. Using ordinary multiplication into $I_{(\mu,\nu)}$, it is seen that I_μ resolves I_ν if and only if μ is a multiple of ν. I_0 m-resolves no group G unless $m = 0$ or 1 or $mG = G$. The general question of when I_μ m-resolves I_ν is rather complicated. However, *I_μ completely resolves I_μ* if $\mu > 0$ and ordinary multiplication mod μ is used. This is a consequence of Theorem 1 below.[6] If R_1 is the group of the real (or the rational) numbers mod 1, then I_0 and R_1 completely resolve each other, as is easily seen; this also follows from Theorem 1.

R_1 is infinitely divisible; no group with a finite number of generators is.

THEOREM 1. *A discrete or compact (or locally bicompact*[7]*) group completely resolves and is completely resolved by its R_1-character group.*

[2] The term annihilator has been used, but this word seems unnecessarily barbarous.

[3] If H' resolves G, and $g_1 \neq g_2$, then $(g_1 - g_2)h \neq 0$ for some h in H', and $g_1h \neq g_2h$. Thus distinct elements of G may be shown to be distinct by multiplying by elements of H'.

[4] Thus, by multiplying by elements of H', we can tell whether a given element of G is divisible by m or not. Any H' 1-resolves G.

[5] It is often important to introduce a topology into groups and character groups; we shall not do this here.

[6] Direct proof: Note that

$$mI_\mu = \text{all numbers in } I_\mu \equiv 0 \bmod (m, \mu),$$

$${}_mI_\mu = \text{all numbers in } I_\mu \equiv 0 \bmod \mu/(m, \mu).$$

Suppose $ab \equiv 0 \bmod \mu$ for all b in ${}_mI_\mu$. Taking $b = \mu/(m, \mu)$ shows that $a \equiv 0 \bmod (m, \mu)$, and a is in mI_μ.

[7] See L. Pontrjagin, Annals of Math., vol. 35 (1934), pp. 361–388; also van Kampen, Annals of Math., vol. 36 (1935), pp. 448–463.

That G and $H = Ch_{R_1}(G)$ resolve each other is shown by Pontrjagin, loc. cit., Theorem 5 and Remark 4. Now set $H' = mH$. As H resolves G, $(G, mH) = {}_mG$, by (2.1). By Pontrjagin, Theorem 4, $(H, (G, mH)) = mH$; hence $(H, {}_mG) = mH$, and G m-resolves H. As m is arbitrary, G completely resolves H. Similarly H completely resolves G.

Note that a group may resolve and be resolved by its character group, and not resolve it completely. For an example, take $G = H = Z = I_0$.

4. **Extensions of homomorphisms.** Suppose part of a group is mapped into another group. Can the map be extended through the first group so that it is a homomorphism? We shall give two answers to the question.

THEOREM 2.[8] *Let A be a free group with a finite number of generators, let A' be a subset of A, and let f map the elements of A' into the group G. The definition of f may be extended over A so that it is a homomorphism if and only if* (a) *holds:*

(a) *For any elements $a_1, \cdots, a_s$ of A' and integers $m \geqq 0, \alpha_1, \cdots, \alpha_s$,*

$$\sum \alpha_i a_i \textit{ in } mA \textit{ implies } \sum \alpha_i f(a_i) \textit{ in } mG. \tag{4.1}$$

Suppose that A and H, and G and H, form group pairs, and let H resolve G completely. If the following condition holds, the extension of f over A is possible:

(b) *For any elements $a_1, \cdots, a_s$ of A', integers $\alpha_1, \cdots, \alpha_s$, and element h of H,*

$$(\sum \alpha_i a_i)h = 0 \textit{ implies } (\sum \alpha_i f(a_i))h = 0. \tag{4.2}$$

The necessity of (a) is clear:

$$\sum \alpha_i f(a_i) = f(\sum \alpha_i a_i) = f(ma) = mf(a).$$

We turn to the sufficiency. Let A^* be the subgroup of A generated by the elements of A'. For each $a = \sum \alpha_i a_i$ in A^*, set

$$f(a) = \sum \alpha_i f(a_i).$$

To prove that f is uniquely determined in A^*, suppose

$$\sum \alpha_i a_i = \sum \beta_i a_i, \qquad \sum (\beta_i - \alpha_i)a_i = 0.$$

[8] The theorem will be strengthened in Appendix II. We note from the proof that (b) implies (a)

The theorems in Alexandroff-Hopf, *Topologie*, I. pp. 591-3, are consequences of the strengthened theorem (using footnote 6). Our proof has naturally much in common with theirs.

(We may carry out both sums over the same terms.) Suppose (a) holds. Then as 0 is in $0A$,

$$\sum (\beta_i - \alpha_i)f(a_i) = \sum \beta_i f(a_i) - \sum \alpha_i f(a_i)$$

is in $0B$, hence it is 0, and the last two terms are equal. Now suppose (b) holds. Then for any h in H,

$$[\sum (\beta_i - \alpha_i)a_i]h = 0;$$

hence

$$[\sum \beta_i f(a_i) - \sum \alpha_i f(a_i)]h =' 0,$$

and as H resolves G, the term in brackets equals 0.

Obviously f is a homomorphism in A^*. Moreover, as each linear combination of elements of A^* is a linear combination of elements of A', (a) or (b) is easily seen to hold in A^* if it holds in A'.

Let $b_1, \cdots, b_r, b_{r+1}, \cdots, b_s$ form a base for A, and let $m_1, \cdots, m_r$ be positive integers such that $a_1 = m_1 b_1, \cdots, a_r = m_r b_r$ form a base for A^*.[9] Suppose (a) holds. As a_i is in $m_i A$, $f(a_i)$ is in $m_i G$, and we may choose g_i so that $m_i g_i = f(a_i)$ $(i = 1, \cdots, r)$. If (b) holds, $a_i h = b_i(m_i h) = 0$ for all h in ${}_{m_i}H$; hence $f(a_i)h = 0$ for all h in ${}_{m_i}H$, and as H resolves G completely, $f(a_i)$ is in $m_i G$. Again g_i exists. Set

$$f(b_i) = g_i \ (i = 1, \cdots, r), \quad \text{and} \quad f(b_i) = 0 \ (i = r + 1, \cdots, s).$$

The resulting homomorphism of A into G obviously agrees with the one already found in A^*.

Remarks. If A' is a subgroup of A with division (i.e., ma in A', $m \neq 0$, implies a in A'), and f is a homomorphism of A' into G, we may always extend f over A. For we may choose a base $b_1, \cdots, b_r, b_{r+1}, \cdots, b_s$ for A such that $b_1, \cdots, b_r$ form a base for A', and set $f(b_i) = 0$ $(i > r)$. Further, any homomorphism of any subgroup A' of any group A into G may be extended over A, if G is infinitely divisible; see Pontrjagin, loc. cit., proof of Lemma 1.

II. Linear equations with integer coefficients

5. Cycles, etc., of a matrix. Let η be a matrix of integers:

$$(5.1) \qquad \eta = \begin{Vmatrix} \eta_1^1 & \cdots & \eta_n^1 \\ \cdots & \cdots & \cdots \\ \eta_1^p & \cdots & \eta_n^p \end{Vmatrix}.$$

Given a group X, let $X^s = X + \cdots + X$, a direct sum with s terms. The elements of X^s may be written $(x_1, \cdots, x_s)$ or $(x^1, \cdots, x^s)$. We define the *boundary* and *coboundary* (with respect to η) of elements of X^n and X^p by

$$(5.2) \qquad \partial(x^1, \cdots, x^n) = (\sum \eta_j^1 x^j, \cdots, \sum \eta_j^p x^j) \text{ in } X^p,$$

$$(5.3) \qquad \delta(x_1, \cdots, x_p) = (\sum \eta_1^j x_j, \cdots, \sum \eta_n^j x_j) \text{ in } X^n.$$

[9] See Alexandroff-Hopf, loc. cit., p. 568, no. 24.

Any element (5.2) we call an *X-boundary*; these form a group $\mathbf{B}^X(\eta)$. Define similarly *X-coboundaries* and $\mathbf{B}_X(\eta)$.

Any element of $X^n[X^p]$ whose boundary [coboundary] vanishes we shall call an *X-cycle* [*X-cocycle*]. We form with these the groups $\mathbf{C}^X(\eta)$ and $\mathbf{C}_X(\eta)$.

Given a group pair G, H, we form a group pair G^s, H^s by

$$(5.4) \qquad (g^1, \cdots, g^s)\cdot(h_1, \cdots, h_s) = \sum g^i h_i .$$

THEOREM 3. *If H resolves G, or G resolves H, then*

$$(5.5) \qquad \mathbf{C}^G(\eta) = (G^n, \mathbf{B}_H(\eta)), \textit{ or } \mathbf{C}_H(\eta) = (H^p, \mathbf{B}^G(\eta)).$$

We shall prove the first; the other is obtained by transposing η. For any G-cycle $(g^1, \cdots, g^n)$ and H-coboundary $\delta(h_1, \cdots, h_p)$,

$$(5.6) \qquad (g^1, \cdots, g^n)\cdot\delta(h_1, \cdots, h_p) = \sum_i g^i \sum_j \eta_i^j h_j = \partial(g^1, \cdots, g^n)\cdot(h_1, \cdots, h_p) = 0.$$

Now suppose $(g^1, \cdots, g^n)$ satisfies this relation for all $(h_1, \cdots, h_p)$. Taking all the $h_k = 0$ but the j-th gives $(\sum \eta_i^j g^i)h = 0$ (all h); as H resolves G, $\sum \eta_i^j g^i = 0$ (all j), and $(g^1, \cdots, g^n)$ is a G-cycle.

6. Linear equations in integers. We shall now prove

THEOREM 4.[10] *Let η and G be given; if we use (β), assume H resolves G completely. $(g^1, \cdots, g^p)$ is a G-boundary, i.e., the equations*

$$(6.1) \qquad \sum_{j=1}^{n} \eta_j^i \bar{g}^j = g^i \qquad (i = 1, \cdots, p)$$

are solvable for $\bar{g}^1, \cdots, \bar{g}^n$, if and only if one of the following is true:

(α) *For any integers $m \geqq 0, \alpha_1, \cdots, \alpha_p$,*

$$(6.2) \qquad \sum_{i=1}^{p} \alpha_i \eta_j^i \equiv 0 \bmod m \textit{ (all } j) \textit{ implies } \sum_{i=1}^{p} \alpha_i g^i \textit{ is in } mG;$$

in other words, every I_0-cocycle mod m[11] *(or, every I_m-cocycle) is orthogonal* mod m *to $(g^1, \cdots, g^p)$.*

(β) *For any $(h_1, \cdots, h_p)$ in H^p,*

$$(6.3) \qquad \sum_{i=1}^{p} \eta_j^i h_i = 0 \textit{ (all } j) \textit{ implies } \sum_{i=1}^{p} g^i h_i = 0;$$

in other words, every H-cocycle is orthogonal to $(g^1, \cdots, g^p)$.

Example. Let $G = I_\mu$, μ a prime; then η may be considered as a matrix of integers mod μ. Using footnote 7, both (α) and (β) (with $H = Z = I_\mu$) reduce to testing (α) for the single integer $m = \mu$.

[10] A proof of the first half, (α), of this theorem, using a standard theorem on linear equations (see Veblen, *Analysis Situs*, Second edition, Appendix 2) has been furnished to me by H. T. Engstrom. See H. Hopf, Math. Annalen, 104, p. 658.

[11] x is a cocycle mod m if its coboundary is divisible by m; it is orthogonal mod m to y if xy is divisible by m. We might state the condition as follows: $(g^1, \cdots, g^m)$ is in the "complete nullifier" of $\mathbf{C}_{I_0}(\eta)$.

The necessity of either condition is trivial; we shall prove the sufficiency. Set $A = I_0^n$ = all $(a_1, \cdots, a_n)$ (a's integers). Let A' be the rows $\eta^1, \cdots, \eta^p$ of η, and set

$$f(\eta^i) = f(\eta_1^i, \cdots, \eta_n^i) = g^i \qquad (i = 1, \cdots, p). \tag{6.4}$$

Suppose f has been extended over A so that it is a homomorphism of A into G. Then set

$$\bar{g}^1 = f(1, 0, \cdots), \qquad \bar{g}^2 = f(0, 1, \cdots), \qquad \cdots. \tag{6.5}$$

Then

$$\sum_{j=1}^{n} \eta_j^i \bar{g}^j = \sum_{j=1}^{n} \eta_j^i f(\overset{1}{0}, \cdots, \overset{j}{1}, \cdots, \overset{n}{0}) = f(\eta_1^i, \cdots, \eta_n^i) = g^i.$$

Thus we need merely show that (a) or (b) of Theorem 2 is satisfied.

Suppose first (α) holds, and $\sum \alpha_i \eta^i$ is in mA. Then $\sum \alpha_i \eta_j^i \equiv 0 \bmod m$ (all j), hence $\sum \alpha_i g^i = \sum \alpha_i f(\eta^i)$ is in mG, and (a) holds.

Suppose next (β) holds. Let A and H form a group pair by setting $(a_1, \cdots, a_n)h = (a_1 h, \cdots, a_n h)$. Suppose

$$(\sum \alpha_i \eta^i)h = \sum \alpha_i(\eta_1^i h, \cdots, \eta_n^i h) = 0.$$

Then

$$\sum \alpha_i \eta_j^i h = \sum \eta_j^i h_i = 0 \qquad \text{(all } j\text{)},$$

where $h_i = \alpha_i h$. Hence, by (β),

$$(\sum \alpha_i f(\eta^i))h = \sum f(\eta^i)h_i = \sum g^i h_i = 0,$$

and (b) holds. This completes the proof.

Remark. (β) of this theorem and a similar theorem with η transposed say that

$$\mathbf{B}^G(\eta) = (G^p, \mathbf{C}_H(\eta)) \text{ if } H \text{ resolves } G \text{ completely,} \tag{6.6}$$

$$\mathbf{B}_H(\eta) = (H^n, \mathbf{C}^G(\eta)) \text{ if } G \text{ resolves } H \text{ completely.} \tag{6.7}$$

III. Applications to topology

7. Cycles, boundaries, cocycles, and coboundaries. Let K be a (finite) complex, with cells σ_i^r ($i = 1, \cdots, \alpha^r$) of dimension r, and incidence matrices ${}^r\partial = \| {}^r\partial_j^i \|$. The *boundary* and *coboundary* of σ_i^r are

$$\partial\sigma_i^r = \sum_j {}^r\partial_i^j \sigma_j^{r-1}, \qquad \delta\sigma_i^r = \sum_j {}^{r+1}\partial_j^i \sigma_j^{r+1}. \tag{7.1}$$

In terms of the coefficients of chains, boundaries and coboundaries are given by (5.2) and (5.3) with η_j^i replaced by ${}^r\partial_j^i$ and ${}^{r+1}\partial_j^i$.

An *r-X-chain* is a linear form $\sum x_i \sigma_i^r$ (x_i in X); we may consider it as an element of X^{α^r}. Define the groups of

$$\begin{aligned} &\textit{r-X-cycles:} && {}^r\mathbf{C}^X = \mathbf{C}^X({}^r\partial), \\ &\textit{r-X-boundaries:} && {}^r\mathbf{B}^X = \mathbf{B}^X({}^{r+1}\partial), \\ &\textit{r-X-cocycles:} && {}^r\mathbf{C}_X = \mathbf{C}_X({}^{r+1}\partial), \\ &\textit{r-X-coboundaries:} && {}^r\mathbf{B}_X = \mathbf{B}_X({}^r\partial). \end{aligned}$$

Then the *r-X-homology groups* and *r-X-cohomology groups* are the difference groups

$$(7.2) \qquad {}^r\mathbf{H}^X = {}^r\mathbf{C}^X - {}^r\mathbf{B}^X, \qquad {}^r\mathbf{H}_X = {}^r\mathbf{C}_X - {}^r\mathbf{B}_X .$$

If G and H form a group pair, then r-G-chains and r-H-chains may be multiplied, using (5.4). By (5.6),

$$(7.3) \qquad \partial A \cdot B = A \cdot \delta B$$

for any $(r + 1)$-chain A and r-chain B.

Equations (5.5), (6.6) and (6.7) give

$$(7.4) \qquad {}^r\mathbf{C}^G = (G^{\alpha^r}, {}^r\mathbf{B}_H) \qquad \text{if } H \text{ resolves } G,$$

$$(7.5) \qquad {}^r\mathbf{C}_H = (H^{\alpha^r}, {}^r\mathbf{B}^G) \qquad \text{if } G \text{ resolves } H,$$

$$(7.6) \qquad {}^r\mathbf{B}^G = (G^{\alpha^r}, {}^r\mathbf{C}_H) \qquad \text{if } H \text{ resolves } G \text{ completely},$$

$$(7.7) \qquad {}^r\mathbf{B}_H = (H^{\alpha^r}, {}^r\mathbf{C}^G) \qquad \text{if } G \text{ resolves } H \text{ completely}.$$

Expressed in words, we have (using (7.3))

THEOREM 5. *With suitable G and H, an r-chain (with coefficients in G or H) is an r-G-cycle, or an r-G-boundary, or an r-H-cocycle, or an r-H-coboundary, if and only if it is orthogonal to every r-H-coboundary, or r-H-cocycle, or r-G-boundary, or r-G-cycle.*

THEOREM 6. *An r-G-chain is a cycle, or a cocycle, if and only if it is orthogonal to every I_0-coboundary, or I_0-boundary. It is a boundary, or a coboundary, if and only if it is orthogonal* mod m *to every I_m-cocycle, or I_m-cycle, for all m ($\neq 1$).*

The proof of the first half is like that of Theorem 3; the second half follows from Theorem 4, (α).

8. **Homology and cohomology groups.** We shall find a case in which the homology groups are determined by the cohomology groups or vice versa.

THEOREM 7. *For any G and H, the groups ${}^r\mathbf{H}^G$, ${}^r\mathbf{H}_H$ form a pair, if G and H form a pair, as follows. Given elements ξ of ${}^r\mathbf{H}^G$ and ζ of ${}^r\mathbf{H}_H$, choose a cycle C of ξ and a cocycle D of ζ, and set $\xi\zeta = CD$.*

Theorem 5 shows that the definition depends on ξ and ζ alone.

THEOREM 8. *Let Z be infinitely divisible. If $H = Ch_Z(G)$ and G resolves H completely, or if $G = Ch_Z(H)$ and H resolves G completely, then*

$$(8.1) \qquad {}^r\mathbf{H}_H = Ch_Z({}^r\mathbf{H}^G), \text{ or } {}^r\mathbf{H}^G = Ch_Z({}^r\mathbf{H}_H).$$

We shall prove the first equation; the proof of the second is the same. Each element ζ of ${}^r\mathbf{H}_H$ determines a Z-character ϕ_ζ of ${}^r\mathbf{H}^G$ by the last theorem. Suppose $\zeta \neq \zeta'$. Take cocycles D in ζ and D' in ζ'; then $D' - D$ is not a coboundary. By (7.7) there is a cycle C such that $C(D' - D) \neq 0$, $CD \neq CD'$. Let ξ be the homology class of C; then $\xi\zeta \neq \xi\zeta'$, and $\phi_\zeta \neq \phi_{\zeta'}$. Therefore ${}^r\mathbf{H}_H$ determines in a (1-1) way a subgroup of the Z-characters of ${}^r\mathbf{H}^G$.

To show that this is the whole group, take any Z-character ϕ of ${}^r\mathbf{H}^G$. For any cycle C with homology class ξ, set $\psi(C) = \phi(\xi)$; this is a Z-character of ${}^r\mathbf{C}^G$. By the remark following Theorem 2, we may extend it to a Z-character ψ of G^{α^r}, the group of all r-G-chains. Set

$$\psi_i(g) = \psi(0, \cdots, g, \cdots, 0) \qquad (g \text{ in the } i\text{-th place}).$$

As $H = Ch_Z(G)$, there is an h_i such that $gh_i = \psi_i(g)$ (all g). Set

$$D = (h_1, \cdots, h_{\alpha^r}) = \sum h_i \sigma_i^r;$$

then for any r-G-chain $C = (g^1, \cdots, g^{\alpha^r})$,

$$CD = \sum g^i h_i = \sum \psi_i(g^i) = \psi(g^1, \cdots, g^{\alpha^r}) = \psi(C).$$

As ψ maps all boundaries into 0, D is a cocycle, by (7.5); let ζ be its homology class. Then for any ξ in ${}^r\mathbf{H}^G$, $\xi\zeta = \phi(\xi)$. This completes the proof.

COROLLARY. *${}^r\mathbf{H}_{I_0}$ is isomorphic to the direct sum of the reduced r-th homology group and the $(r - 1)$-th torsion group, all with integer coefficients.*

This follows from the last theorem, Theorem 1, and Alexandroff-Hopf, *Topologie* I, p. 234, equation (17′). A direct proof is not hard to give, using the ordinary elementary divisor theory.

Remarks. The theorem does not hold for arbitrary Z. For, consider the projective plane, using $G = R_1$, $Z = I_2$; then $H = 0$, and hence ${}^2\mathbf{H}_H = 0$. But ${}^2\mathbf{H}^G$ = torsion group of dimension $1 = I_2$ (see Alexandroff-Hopf, loc. cit., p. 234), and hence $Ch_Z({}^2\mathbf{H}^G) = CH_{I_2} I_2 = I_2$.

Given H, set $G = Ch_{R_1}(H)$; then $H = Ch_{R_1}(G)$, and G completely resolves H, by Theorem 1. Hence ${}^r\mathbf{H}_H = Ch_{R_1}({}^r\mathbf{H}^G)$. As ${}^r\mathbf{H}^G$ is a topological invariant, ${}^r\mathbf{H}_H$ is a topological invariant of K.

APPENDIX I

On closed complexes and pseudomanifolds

For the general theory, see Alexandroff-Hopf, *Topologie* I, Ch. VII, §1. We refer to this work as AH. K will always denote a homogeneous[12] n-complex, and $\sigma_1, \sigma_2, \cdots$, its n-cells. The chain σ_i means the n-chain $\sum \delta_{ij}\sigma_j$ with coefficient 1 for σ_i and 0 for all other σ_j. Write $C \backsim D$ if C is cohomologous to D.

[12] That is, each k-cell ($k < n$) is on some n-cell.

9. **Closed complexes.** We say K is *closed* if no σ_i is $\backsim 0$ in K. That this definition agrees with the ordinary one is shown by Theorem 9 below.

LEMMA. *σ_i is not $\backsim 0$ if and only if it is contained in some cycle C* mod m *for some $m \neq 1$.*[13]

By Theorem 6, σ_i is not $\backsim 0$ if and only if for some $m \neq 1$ there is a cycle C mod m such that $C \cdot \sigma_i \not\equiv 0$ mod m. Write $C = a\sigma_i + \cdots$; then $C \cdot \sigma_i = a$, and the above holds if and only if $a \not\equiv 0$ (mod m), i.e., if and only if C contains σ_i.

THEOREM 9. *K is closed if and only if each σ_i is contained in a cycle* mod m *for an $m \neq 1$.*

This is a consequence of the lemma. Alexandroff-Hopf, p. 275, Satz Ia, shows that our definition is equivalent to the ordinary one.

10. **Irreducibly closed complexes.** We say K is *irreducibly closed* if it is closed but no proper subcomplex is.

THEOREM 10.[14] *If K is irreducibly closed, then*

(a) *the cohomology group ${}^n\mathbf{H}_{I_0}$ is cyclic (of order $\neq 1$), and*

(b) *no σ_i is $\backsim 0$, but each σ_i is $\backsim$ some multiple of each σ_j.*

To prove (b) we shall show first that $\sigma_i \backsim 0$ in $K - \sigma_j$. If not, then by the lemma σ_i is contained in a cycle C mod m, $m \neq 1$, C in $K - \sigma_j$. Let K' be the complex containing the cells of C; then each of its cells is contained in C, and hence K' is closed, by Theorem 9; but K' is in $K - \sigma_j$, a contradiction. Say $\sigma_i = \delta A$ in $K - \sigma_j$; then

$$\delta A = \sigma_i - p_{ij}\sigma_j \text{ in } K \qquad \text{(some } p_{ij}\text{)},$$

and $\sigma_i \backsim p_{ij}\sigma_j$, as required.

To prove (a), take a σ_i, and let m be the smallest positive integer such that $m\sigma_i \backsim 0$; if there is none, set $m = 0$. Each $\mu\sigma_i$ is a cocycle (as $\dim(K) = n$), and determines an element of ${}^n\mathbf{H}_{I_0}$. Further, given any element ξ of ${}^n\mathbf{H}_{I_0}$, determined by the cocycle C,

$$C = \sum_j \alpha_j\sigma_j \backsim \sum_j \alpha_j(p_{ji}\sigma_i) = \mu\sigma_i,$$

and ξ is determined by $\mu\sigma_i$. Hence ${}^n\mathbf{H}_{I_0}$ is cyclic, of order m.

Remark. There exist complexes in which some p_{ij} is $\neq \pm 1$; see AH, p. 280, Bemerkung.

THEOREM 11.[15] *For the homogeneous n-complex K to be irreducibly closed, either of the two following conditions is (necessary and) sufficient.*

(α) *This is* (b) *of Theorem* 10.

(β) ${}^n\mathbf{H}_{I_0}(K) \neq 0$, *but* ${}^n\mathbf{H}_{I_0}(K') = 0$ *for any proper subcomplex K' of K.*

[13] The condition $m \neq 1$ could be omitted; for a cycle mod 1 contains no cells, i.e., is $\equiv 0$ mod 1.

[14] Compare AH, p. 277, Satz IV and Satz V.

[15] See AH, Theorems on p. 280.

Suppose (α) holds; then K is closed. If K is not irreducibly closed, then there is a proper (homogeneous) subcomplex K' which is closed. Take σ_i in K', σ_j in $K - K'$, and say

$$\sigma_i - a\sigma_j = \delta A \text{ in } K.$$

Write

$$A = A_1 + A_2, \qquad A_1 \text{ in } K', \qquad A_2 \text{ in } K - K';$$

we shall show that $\sigma_i = \delta A_1$ in K'. As no σ_k^{n-1} of A_2 is on a cell of K', δA_2 has no cells in K'; hence the part of δA in K' is the part of δA_1 in K', and this is σ_i. But then $\sigma_i \backsim 0$ in K', contradicting the assumption that K' is closed.

Suppose (β) holds. Let D be a cocycle not $\backsim 0$. By Theorem 6, there is a cycle C mod m for some $m \neq 1$ with $C \cdot D \neq 0$; hence $C \neq 0$. The cells contained in C form a closed complex K', by Theorem 9. Now let K^* be any proper subcomplex of K. Then ${}^n\mathbf{H}_{I_0}(K^*) = 0$; consequently each σ_i of K^* is $\backsim 0$ in K^*, and K^* is not closed. Therefore $K' = K$, and K is irreducibly closed.

11. **Pseudomanifolds.** A *pseudomanifold* is a strongly connected[16] homogeneous complex in which each $(n - 1)$-cell is on either one or two n-cells; it is *closed* if each $(n - 1)$-cell is on two n-cells.

THEOREM 12.[17] *If the pseudomanifold K is not closed, then ${}^n\mathbf{H}_{I_0}$ vanishes; if it is closed, then it is an irreducibly closed complex, and ${}^n\mathbf{H}_{I_0}$ is cyclic of order* 0 *or* 2 *according as K is orientable or not.*

Take any succession of distinct n-cells $\sigma_0^n, \sigma_1^n, \cdots, \sigma_s^n$, σ_i^n and σ_{i+1}^n having the common $(n - 1)$-face σ_i^{n-1}. We may orient the cells so that σ_i^{n-1} is on σ_i^n negatively and on σ_{i+1}^n positively. Let A be the sum of these σ_i^{n-1}; then

$$\delta A = \sigma_s^n - \sigma_0^n.$$

It follows that any two n-cells of K, with the proper orientations, are $\backsim$.

If K is not closed (as a pseudomanifold), there is an $(n - 1)$-cell σ^{n-1} on just one n-cell σ^n; then $\delta\sigma^{n-1} = \pm\sigma^n$, and $\sigma^n \backsim 0$; hence each $\sigma_i^n \backsim 0$, and ${}^n\mathbf{H}_{I_0}$ vanishes. Suppose K is closed. Then each $\delta\sigma^{n-1}$ and hence each δA^{n-1} is a chain the sum of whose coefficients is even, and hence cannot be any σ_i^n; as each $\sigma_i \backsim \pm$ each σ_j, K is an irreducibly closed complex, by Theorem 11, (α).

If K is not orientable, a cell σ_0^n may be joined to itself by a succession of n- and $(n - 1)$-cells so as to reverse the orientation. Let A be the sum of the properly oriented $(n - 1)$-cells of the chain; then $\delta A = 2\sigma_0^n$, so that $2\sigma_0^n$ and hence any $2\sigma_i^n$ is $\backsim 0$; hence ${}^n\mathbf{H}_{I_0}$ is of order 2. Otherwise, we may orient the cells of K concurrently; then each $\delta\sigma_i^{n-1}$ and hence each δA^{n-1} is a chain, the sum of whose coefficients vanishes, and no $m\sigma_i^n$ is $\backsim 0$ for $m \neq 0$. It follows that ${}^n\mathbf{H}_{I_0}$ is the infinite cyclic group.

[16] That is, each two n-cells are joined by a succession of n- and $(n - 1)$-cells.

[17] Compare AH, p. 281, Satz VIII.

Appendix II

We shall show that Theorem 2 holds if A is any group with a finite number of generators. Extend f over A^* as before. Let $b^1, \cdots, b^r$ form a base for A, and $a^1, \cdots, a^s$, a base for A^*. Say

$$a^i = \sum_{j=1}^{r} \eta_j^i b^j \qquad (i = 1, \cdots, s).$$

Let μ_i be the order of b^i ($\mu_i = 0$ if b^i is of infinite order). Set

$$\eta_j^{s+i} = \mu_i \text{ if } j = i, \ = 0 \text{ if } j \neq i \qquad (i, j = 1, \cdots, r);$$

then

$$\sum_{j=1}^{r} \eta_j^{s+i} b^j = \mu_i b^i = 0 \qquad (i = 1, \cdots, r).$$

Set

$$g^i = f(a^i) \qquad (i = 1, \cdots, s), \qquad g^{s+i} = 0 \qquad (i = 1 \cdots, r).$$

Suppose $\sum_{i=1}^{s+r} \alpha_i \eta_j^i = mk_j$ (all j). Then

$$\sum_{i=1}^{s} \alpha_i a^i = \sum_{i=1}^{s} \sum_{j=1}^{r} \alpha_i \eta_j^i b^j = \sum_{i=1}^{s+r} \sum_{j=1}^{r} \alpha_i \eta_j^i b^j = m \sum_{j=1}^{r} k_j b^j$$

is in mA, and hence, using (a) or (b) (see footnote 8),

$$f\left(\sum_{i=1}^{s} \alpha_i a^i\right) = \sum_{i=1}^{s} \alpha_i f(a^i) = \sum_{i=1}^{s+r} \alpha_i g^i$$

is in mG. Therefore, by Theorem 4, there are elements $\bar{g}^1, \cdots, \bar{g}^r$ of G such that

$$\sum_{j=1}^{r} \eta_j^i \bar{g}^j = g^i \qquad (i = 1, \cdots, s + r).$$

Setting $f'(\sum \alpha_i b^i) = \sum \alpha_i \bar{g}^i$ defines uniquely a homomorphism of A into G; for

$$f'(\mu_i b^i) = \sum_{j=1}^{r} \eta_j^{s+i} \bar{g}^j = g^{s+i} = 0 \qquad (i = 1, \cdots, r).$$

Further,

$$f'(a^i) = f'(\sum \eta_j^i b^j) = \sum \eta_j^i \bar{g}^j = g^i = f(a^i) \quad (i = 1, \cdots, s),$$

so that f' is an extension of f.

Harvard University.

ON THE MAPS OF AN n-SPHERE INTO ANOTHER n-SPHERE

BY HASSLER WHITNEY

1. **Introduction.** It is well known that to each map[1] f of an n-sphere S^n into another one S_0^n ($n \geqq 1$ always) there corresponds a number d_f, the *degree* of f, and $d_f = d_g$ if f and g are homotopic (see §2). H. Hopf[2] has proved the converse theorem, that if $d_f = d_g$, then f and g are homotopic. The object of this note is to give an elementary proof of the latter theorem. The methods will be used and extended in later papers.

In an appendix we give somewhat briefly a proof of the theorem for the case that $d_f = 0$. This is the only case needed in the following paper; the general theorem then follows from that paper. The second proof is more intuitive geometrically than the first, but complete details would make it perhaps more lengthy.

2. **On deformations.** A *deformation* of one space S in another S_0 is a family $\phi_t(p)$ $(0 \leqq t \leqq 1,\ p$ in $S)$ of maps of S into S_0, continuous in both variables together. Given maps f and g of S into S_0, if there exists a deformation ϕ_t such that $\phi_0 \equiv f$ and $\phi_1 \equiv g$, we say f and g are *homotopic.* If f is homotopic to g, where $g(p) \equiv P_0$ (all p in S), we say f is *homotopic to zero,* and f may be *shrunk to the point* P_0.

Suppose S and S_0 are complexes, K_0 is a simplicial subdivision of S_0, and f maps S into S_0. Then, for a sufficiently fine simplicial subdivision K of S, the following is true. To each vertex V of K we may choose a vertex $g(V)$ of a cell of K_0 which contains $f(V)$, so that the vertices of any cell of K go into the vertices of a cell of K_0. This determines uniquely a "simplicial map" g of K into K_0, affine in each cell (see §5); moreover, f is homotopic to g.

3. **The degree of a map.** Let S_0^n be the unit n-sphere in $(n+1)$-space, let K_0^n be a simplicial triangulation of S_0^n, and let σ_0^n be an n-cell of K_0^n. We choose K_0^n so that if P_1 is a point of σ_0^n and P_0 is the antipodal point of S_0^n, each great semicircle from P_1 to P_0 intersects the boundary $\partial\sigma_0^n$ of σ_0^n in exactly one point. By pushing along these semicircles, we define a deformation Ω_t of the identity $\Omega_0(p) \equiv p$ into a map Ω_1, where $\Omega_1(p) \equiv P_0$ for p in $S_0^n - \sigma_0^n$.

Let σ^k be a k-cell $(k \leqq n)$, in fixed correspondence with a k-simplex, and let

Received December 1, 1936; presented to the American Mathematical Society, October, 1936.

[1] All maps will be assumed continuous.

[2] See Alexandroff-Hopf, *Topologie*, I, Berlin, 1935, pp. 501–505. See also the reference to Lefschetz in the following paper.

f map σ^k into S_0^n. We say f is *standard* if $f(p) \equiv P_0$, or, $k = n$ and for some affine map ϕ of $\sigma^k = \sigma^n$ into $\sigma_0^n, f(p) \equiv \Omega_1(\phi(p))$. In any case, $f(p) \equiv P_0$ in $\partial\sigma^k$.[3] The map f of an n-complex K^n into S_0^n is *standard* if it is standard over each k-cell ($k \leqq n$).

We may orient S_0^n by orienting σ_0^n. Let K^n be a simplicial triangulation of the oriented n-sphere S^n, and let f be a standard map of K^n into S_0^n. Let σ^n be an (oriented) n-cell of K^n. If $f(p) \equiv P_0$ in σ^n, we set $d_f(\sigma^n) = 0$. Otherwise, there is a simplicial map ϕ of σ^n into (the whole of) σ_0^n such that $f(p) \equiv \Omega_1(\phi(p))$ in σ^n; we set $d_f(\sigma^n) = 1$ or -1 according as ϕ is positive or negative. We define the *degree* of f by

$$d_f = \sum_{\sigma^n} d_f(\sigma^n). \tag{3.1}$$

4. **The theorem.** In homology theory it is shown how to attach to each map f of S^n into S_0^n (both spheres oriented) an integer d_f, the degree of the map. Moreover, if f is homotopic to g, then $d_f = d_g$, and if S^n and S_0^n are triangulated and f is standard, then d_f is given by (3.1).

Suppose f and g map S^n into S_0^n, and $d_f = d_g$. Then for a sufficiently fine subdivision K^n of S^n, both f and g can be deformed into simplicial maps and hence into standard maps ϕ and ψ. As f and ϕ, also g and ψ, are homotopic, $d_\phi = d_\psi$. By Theorem 1 below, ϕ is homotopic to ψ; hence f is homotopic to g. Therefore this theorem furnishes the converse of the statements above.

THEOREM 1. *If ϕ and ψ are standard maps of S^n into S_0^n, using the same subdivision K^n of S^n, and $d_\phi = d_\psi$, then ϕ is homotopic to ψ.*

From the proof below, the following corollary is apparent.

COROLLARY. *If $\phi(V) = \psi(V) = P_0$ for a fixed vertex V of K^n, we can make V remain at P_0 throughout the deformation.*

In fact, if $n \geqq 2$, all vertices of K^n remain at P_0. If $n = 1$, we may choose the chains of cells in §8 so that in no chain do we pass over V; then V is never moved.

THEOREM 2. *For any integer γ there is a map f of S^n into S_0^n with $d_f = \gamma$.*

To prove this, subdivide S^n into $\alpha \geqq |\gamma|$ n-cells. Let ϕ map $|\gamma|$ of these cells simplicially into σ_0^n, positively or negatively according as $\gamma > 0$ or $\gamma < 0$ (if $\gamma \neq 0$), and set $f(p) = \Omega_1(\phi(p))$ in these cells and $f(p) = P_0$ elsewhere. Clearly $d_f = \gamma$. Note that the degree of the identity map of S_0^n into itself is 1.

The remainder of the paper is devoted to the proof of Theorem 1.

5. **Coördinates p_t in a cell.** Any simplicial complex K^n is homeomorphic to a complex $\bar{K}^n$ in euclidean space whose cells are straight. Using $\bar{K}^n$, we define straightness in K^n, the center of a cell (i.e., center of mass of its vertices), etc. Hence an "affine map" of one cell into another has meaning. Let σ be a cell of K^n, and a, the center of σ. For each point p of the boundary $\partial\sigma$ of σ let p_t be the point of the segment ap such that $ap_t/ap = t$.

[3] There are $(n + 1)!$ standard maps ϕ of σ^n into S_0^n with $\phi(p) \not\equiv P_0$.

6. Certain deformations of simplexes. We prove first a combinatorial lemma, needed in Lemma 2.

LEMMA 1. *Any even permutation of the letters* $a_0a_1 \cdots a_n$ $(n \geqq 2)$ *may be made by means of a succession of cyclic permutations, each on three of the letters.*

This is clear if $n = 2$; then any even permutation is cyclic. Suppose $n > 2$, and let $B = a_{\alpha_0} \cdots a_{\alpha_n}$ be any even permutation. If $\alpha_n \neq n$, bring a_{α_n} to the right end by a cyclic permutation; bring $a_{\alpha_{n-1}}$ next to a_{α_n}. Suppose $\alpha_0 \neq 0$. We then perform the two cyclic permutations

$$a_0 \cdots a_{\alpha_0} \cdots a_{\alpha_{n-1}} a_{\alpha_n} \rightarrow a_{\alpha_0} \cdots a_{\alpha_{n-1}} \cdots a_0 a_{\alpha_n} \rightarrow a_{\alpha_0} \cdots a_0 \cdots a_{\alpha_n} a_{\alpha_{n-1}} .$$

If $n \geqq 4$ and a_{α_1} is not now in the second place, we perform two cyclic permutations to bring it there, again interchanging $a_{\alpha_{n-1}}$ and a_{α_n}, etc. When $a_{\alpha_0}, \cdots, a_{\alpha_{n-3}}$ are in their correct places, $a_{\alpha_{n-2}}$ is also; as B is even and the above permutations are even, $a_{\alpha_{n-1}}$ and a_{α_n} are also in their correct positions.

LEMMA 2. *Let* $\sigma^n = a_0 \cdots a_n$ *be a simplex, and let* $a_{\alpha_0} \cdots a_{\alpha_n}$ *be an even permutation of its vertices. Then there is a deformation* ϕ_t *of* σ^n *in itself, such that* $\phi_0(p) \equiv p$, $\phi_1(a_s) = a_{\alpha_s}$, ϕ_1 *is affine, and* ϕ_t *for each t is a homeomorphism both in* σ^n *and in its boundary.*

If $n = 0$ or 1, the lemma is trivial. Suppose that $n = 2$; say $a_{\alpha_0}a_{\alpha_1}a_{\alpha_2} = a_1a_2a_0$. Let $\phi_t(a_i)$ be the point p of a_ia_{i+1} (setting $2 + 1 = 0$) for which $a_ip/a_ia_{i+1} = t$. Let ϕ_t map the segment a_ia_{i+1} into the broken line $\phi_t(a_i)a_{i+1}\phi_t(a_{i+1})$ so that, if the line were straightened, the map would be linear. For any point p_u (see §5) interior to σ^2, set $\phi_t(p_u) = (\phi_t(p))_u$. As $\phi_t(a) = a =$ center of mass of σ^2, ϕ_1 is easily seen to be affine.

Now suppose $n > 2$; consider first a cyclic permutation, changing say $a_0a_1a_2$ into $a_1a_2a_0$. Set $\sigma = a_0a_1a_2$, $\sigma' = a_3 \cdots a_n$, and let $[p, q, u]$ for p in σ, q in σ', $0 \leqq u \leqq 1$, be the point r of the segment pq for which $pr/pq = u$. Define ϕ_t in σ as above. For any point $[p, q, u]$ not in σ, set $\phi_t[p, q, u] = [\phi_t(p), q, u]$. We show that ϕ_t is a homeomorphism. Suppose $\phi_t[p, q, u] = \phi_t[p', q', u']$; then $[\phi_t(p), q, u] = [\phi_t(p'), q', u']$, which implies $\phi_t(p) = \phi_t(p')$, $q = q'$, $u = u'$;[4] as ϕ_t is a homeomorphism in σ, $p = p'$ also. Further, given $[p, q, u]$ and t, we may find a p^* for which $\phi_t(p^*) = p$; then $\phi_t[p^*, q, u] = [p, q, u]$. The other properties of ϕ_t are clear, and the lemma for this case is proved. Now take any permutation. We may obtain it by cyclic permutations as in Lemma 2; the corresponding deformations together give the required deformation.

7. Two types of deformations of S^n in S_0^n. Let ϕ' be a standard map of S^n into S_0^n, and let σ and σ' be oriented n-cells of K^n with the common $(n-1)$-face τ:

$$\sigma = a_0a_1 \cdots a_n, \qquad \sigma' = -a_0'a_1 \cdots a_n, \qquad \tau = a_1 \cdots a_n .$$

(a) Suppose $d_{\phi'}(\sigma) = 1$, $d_{\phi'}(\sigma') = 0$; we shall deform ϕ' into ϕ'' so that $d_{\phi''}(\sigma) = 0$, $d_{\phi''}(\sigma') = 1$, leaving $K^n - (\sigma + \sigma')$ fixed.

(b) Suppose $d_{\phi'}(\sigma) = 1$, $d_{\phi'}(\sigma') = -1$; we shall obtain $d_{\phi''}(\sigma) = d_{\phi''}(\sigma') = 0$.

[4] This is so if $0 < u < 1$, as we may assume.

In each case ϕ'' will be a standard map.

(a) Set $\sigma_1 = a_0a_2 \cdots a_n$, $\sigma_1' = a_0'a_2 \cdots a_n$, or if $n = 1$, then $\sigma_1 = a_0$, $\sigma_1' = a_0'$. Let θ_1 and θ_2 be the affine maps of σ_1 into τ and σ_1' determined by sending a_0 into a_1 and a_0' respectively. For each p in σ_1, let $\alpha(p, u)$ run linearly along the segments $p\theta_1(p)$ and $\theta_1(p)\theta_2(p)$ as u runs from 0 to 1 and from 1 to 2. Set

$$\phi_t'[\alpha(p, u)] = \begin{cases} \phi'[\alpha(p, u - t)] & (t \leqq u), \\ \phi'[\alpha(p, 0)] & (t > u), \end{cases} \tag{7.1}$$

and $\phi_t'(p) = \phi'(p)$ in $K^n - (\sigma + \sigma')$. As $\phi'(p) \equiv P_0$ in $\partial\sigma_1 + \partial\sigma_2$, this is clearly a deformation of $\phi' = \phi_0$ into a map $\phi'' = \phi_1$. The map ϕ'' in σ' is obtained from the map ϕ' in σ by replacing $a_0, a_1, \cdots, a_n$ (which form $+\sigma$) by $a_1, a_0', \cdots, a_n$ (which form $+\sigma'$); hence $d_{\phi''}(\sigma') = d_{\phi'}(\sigma)$. Also $d_{\phi''}(\sigma) = 0$ as $\phi''(p) \equiv P_0$ in σ, and (a) is proved.

(b) Let λ and λ' be the affine maps of σ and σ' into σ_0^n such that $\phi'(p) = \Omega_1(\lambda(p))$ in σ and $= \Omega_1(\lambda'(p))$ in σ'. Say $\sigma_0^n = b_0 \cdots b_n$,

$$\lambda(a_i) = b_{k_i}, \qquad \text{and } \lambda'(a_0') = b_{l_0}, \qquad \lambda'(a_i) = b_{l_i} \qquad (i > 0).$$

As $d_{\phi'}(\sigma') = -d_{\phi'}(-\sigma')$, and hence

$$d_{\phi'}(\sigma) = d_{\phi'}(a_0a_1 \cdots a_n) = -d_{\phi'}(\sigma') = d_{\phi'}(a_0'a_1 \cdots a_n),$$

$b_{l_0} \cdots b_{l_n}$ is an even permutation of $b_{k_0} \cdots b_{k_n}$. Applying Lemma 2, we find a deformation λ_t' of σ' in σ_0^n such that $\lambda_0' \equiv \lambda'$, λ_1' is affine, and

$$\lambda_1'(a_0') = \lambda(a_0), \qquad \lambda_1'(a_i) = \lambda(a_i) \qquad (i > 0). \tag{7.2}$$

Set

$$\phi_t'(p) = \begin{cases} \Omega_1(\lambda_t'(p)) & p \text{ in } \sigma', \\ \phi'(p), & p \text{ in } K^n - \sigma'. \end{cases} \tag{7.3}$$

Then as $\Omega_1(\lambda_t'(p)) \equiv P_0$ in $\partial\sigma'$, ϕ_t' is a deformation of ϕ' into a map $\phi^* \equiv \phi_1'$.

For each p in τ, let $\beta(p, u)$ be the point q of the segment a_0p of σ such that $a_0q/a_0p = u$, and let $\beta'(p, u)$ be the corresponding point of the segment $a_0'p$ in σ'. As λ and λ_1' are affine, (7.2) and (7.3) give

$$\phi^*[\beta(p, u)] = \phi^*[\beta'(p, u)] \qquad (p \text{ in } \tau, 0 \leqq u \leqq 1). \tag{7.4}$$

We deform ϕ^* into ϕ'' by setting

$$\phi_t^*[\beta(p, u)] = \phi_t^*[\beta'(p, u)] = \phi^*[\beta(p, (1 - t)u)], \tag{7.5}$$

and $\phi_t^*(p) = \phi^*(p)$ in $K^n - (\sigma + \sigma')$. This is clearly a deformation; (7.4) shows that $\phi_0^* \equiv \phi^*$. As $\phi''(p) \equiv \phi_1^*(p) \equiv P_0$ in $\sigma + \sigma'$, $d_{\phi''}(\sigma) = d_{\phi''}(\sigma') = 0$.

8. **Proof of Theorem** 1. Suppose there are cells of K^n mapped positively over S_0^n by ϕ, and also cells mapped negatively. Then we can find a chain $\sigma_0, \cdots, \sigma_\nu$ of adjacent n-cells of K^n such that

$$d_\phi(\sigma_0) = 1, \quad d_\phi(\sigma_1) = 0, \quad \cdots, \quad d_\phi(\sigma_{\nu-1}) = 0, \quad d_\phi(\sigma_\nu) = -1.$$

Using (a), §7, we deform ϕ in $\sigma_0 + \sigma_1$, then in $\sigma_1 + \sigma_2$, etc.; then, using (b), §7, we deform the map in $\sigma_{\nu-1} + \sigma_\nu$. The new map ϕ' has $d_{\phi'}(\sigma_i) = 0$ ($i = 0, \cdots, \nu$). Continue in this manner till no cells are mapped positively or none are mapped negatively over S_0^n; for definiteness, say the latter holds. Do the same for ψ. The new maps ϕ^* and ψ^* each have exactly $d_\phi = d_\psi$ cells mapped positively over S_0^n.

Suppose $d_{\phi^*}(\sigma) \neq d_{\psi^*}(\sigma)$ for some σ. Then let $\sigma_0, \sigma_1, \cdots, \sigma_\nu$ be a chain of adjacent n-cells such that

$$d_{\phi^*}(\sigma_0) = d_{\psi^*}(\sigma_\nu) = 1, \qquad d_{\phi^*}(\sigma_\nu) = d_{\psi^*}(\sigma_0) = 0,$$
$$d_{\phi^*}(\sigma_i) = d_{\psi^*}(\sigma_i) \qquad (0 < i < \nu).$$

Let $\sigma_0, \sigma_{k_1}, \cdots, \sigma_{k_s}$ be the cells of the chain for which $d_{\phi^*} = 1$. Using (a), §7, we deform ϕ^* over $\sigma_{k_s} + \sigma_{k_s+1}$ etc. until we have $d_{\phi_1^*}(\sigma_{k_s}) = 0$, $d_{\phi_1^*}(\sigma_\nu) = 1$; another succession of deformations makes $d_{\phi_2^*}(\sigma_{k_{s-1}}) = 0$, $d_{\phi_2^*}(\sigma_{k_s}) = 1$, etc. Finally $d_{\phi^{**}}(\sigma_0) = 0$, $d_{\phi^{**}}(\sigma_{k_i}) = 1$ (all i), and $d_{\phi^{**}}(\sigma_\nu) = 1$. $d_{\phi^{**}}(\sigma)$ differs from $d_{\psi^*}(\sigma)$ over fewer cells than $d_{\phi^*}(\sigma)$. Continuing in this manner, we deform ϕ^* into a map ϕ' with $d_{\phi'}(\sigma) = d_{\psi^*}(\sigma)$, all σ. ϕ' and ψ^* are standard. Applying Lemma 2, we deform ϕ' over each n-cell where necessary, to obtain ψ^*. (Compare the first half of the proof of (b), §7.) This completes the proof.

Appendix[5]

Let f be a map of S^n into S_0^n with the degree 0. We first deform it into a simplicial map and then into a standard map ϕ (see §§ 2, 3). To shrink ϕ to a point is equivalent to extending ϕ through the interior R of S^n (see the following paper, § 4). Let $\sigma_1, \cdots, \sigma_s$ and $\sigma_1', \cdots, \sigma_s'$ be the simplexes of S^n mapped positively and negatively over S_0^n respectively. Let T_i be a tube joining σ_i to σ_i' inside R. We may choose these so no two intersect, and also (to prove the corollary) so no one cuts the radius of R to the vertex V. Let $a_0 \cdots a_n$ and $a_0' \cdots a_n'$ be positive and negative orientations of σ_i and σ_i' respectively, such that $\lambda(a_i) = b_i$ and $\lambda'(a_i') = b_i$ determine simplicial maps of σ_i and σ_i' into σ_0^n, which in turn determine ϕ in σ_i and σ_i'. Now carry σ_i through T_i to σ_i', turning it so that a_i goes into a_i'; let $g_t(\sigma_i)$ be the position of σ_i after the time t. We do this so that $g_t(\sigma_i)$ does not intersect $g_{t'}(\sigma_i)$ if $t \neq t'$. (We are using a deformation theorem on simplexes in euclidean space, similar to but simpler than Lemma 2.) The definition of ϕ in R is as follows. For p not in any $g_t(\sigma_i)$, set $\phi(p) = P_0$. For p in $g_t(\sigma_i)$, choose q in σ_i so that $p = g_t(q)$, and set $\phi(p) = \phi(q)$.

HARVARD UNIVERSITY.

[5] Added in proof.

THE MAPS OF AN n-COMPLEX INTO AN n-SPHERE

By Hassler Whitney

1. **Introduction.** The classes of maps of an n-complex into an n-sphere were classified by H. Hopf[1] in 1932. Recently, W. Hurewicz[2] has extended the theorem by replacing the n-sphere by much more general spaces. Freudenthal[3] and Steenrod[4] have noted that the theorem and proof are simplified by using real numbers reduced mod 1 in place of integers as coefficients in the chains considered. We shall give here a statement of the theorem which seems the most natural; the proof is quite simple. As in the original proof by Hopf, we shall base it on a more general extension theorem.

The fundamental tool of the paper is the relation of "coboundary";[5] it has come into prominence in the last few years.

In later papers we shall classify the maps of a 3-complex into a 2-sphere and of an n-complex into projective n-space.

I. Elementary facts

2. **Boundaries and coboundaries.** Let K be a complex, with oriented cells σ_i^r (not necessarily simplicial) of dimension r, $r = 0, \cdots, n$. Let $\partial_{ij}^r = 1, -1$, or 0 according as σ_i^{r-1} is positively, negatively, or not at all, on the boundary of σ_j^r. An r-chain C^r is a linear form $\Sigma\alpha_i\sigma_i^r$, the α_i being integers (or elements of an abelian group). The *boundary* (or *contraboundary*) and *coboundary* of C^r are defined by

$$(2.1)\qquad \partial\Big(\sum_i \alpha_i\sigma_i^r\Big) = \sum_{i,j} \alpha_i\partial_{ji}^r\sigma_j^{r-1}, \qquad \delta\Big(\sum_i \alpha_i\sigma_i^r\Big) = \sum_{i,j} \alpha_i\partial_{ij}^{r+1}\sigma_j^{r+1}.$$

As in the ordinary theory, we say C^r is a *cocycle* if its coboundary vanishes, and C^r is *cohomologous* to D^r, $C^r \backsim D^r$, if $C^r - D^r$ is a coboundary. The relation $\delta\delta C^r = 0$ (easily proved; equivalent to $\partial\partial C^r = 0$) says that every coboundary

Received December 1, 1936; presented to the International Topological Conference in Moscow, September 1935, and to the American Mathematical Society, October 1935.

[1] H. Hopf, Commentarii Mathematici Helvetici, vol. 5 (1932), pp. 39–54. See also Alexandroff-Hopf, *Topologie* I, Ch. XIII. A recent proof has been given by S. Lefschetz, Fund. Math., vol. 27 (1936), pp. 94–115. In Lemma 3 he gives a new proof of the theorem of the preceding paper; the author does not understand how the final map is made simplicial.

[2] W. Hurewicz, Proc. Kön. Akad. Wet. Amsterdam, vols. 38–39 (1935–36); in particular, vol. 39, pp. 117–126. The full paper will appear in the Annals of Math.

[3] H. Freudenthal, Compositio Math., vol. 2 (1935), footnote 8.

[4] Unpublished.

[5] This is discussed briefly in §2. For further details, see our paper *On matrices of integers*, pp. 35–45 of this volume of this Journal. We refer to this paper as I. The relation of Theorems 2, 3 and 4 to the theorems as stated by Hopf are made apparent by the theorems in I. *The present paper is independent of* I.

is a cocycle. Hence we may define the difference group of the group of r-cocycles over the group of r-boundaries, forming the r-th *cohomology group*.[6]

3. **Normal maps of cells into S_0^n.** Let S_0^n be the (oriented) unit n-sphere in $(n+1)$-space. Let f map the (oriented) n-cell σ^n into S_0^n. We say f is *normal* if $f(p) \equiv P_0$, a fixed point of S_0^n, for p in the boundary $\partial\sigma^n$ of σ^n. This is equivalent to identifying the points of $\partial\sigma^n$ in σ^n, forming an n-sphere S^n, and mapping this sphere into S_0^n. Hence we may define the degree[7] $d_f(\sigma^n)$. If f and g are normal in σ^n and $d_f(\sigma^n) = d_g(\sigma^n)$, then we may deform f into g, keeping $\partial\sigma^n$ at P_0, by II, corollary.

Any map f of σ^r into S_0^n, $r < n$, may be shrunk to P_0: we deform f into a simplicial map, and apply Ω_t (see II, §3). P_0 being assumed a vertex of K_0^n, if $\partial\sigma^r$ is at P_0 it remains there during the deformation.

If K is any complex, let K^r be the subcomplex of K containing all its cells of dimension $\leqq r$. The map f of K into S_0^n is *normal* if $f(p) \equiv P_0$ for p in K^{n-1}. Suppose σ^n or S^n is subdivided into cells σ_i^n, and f is a normal map of it into S_0^n. Then the $d_f(\sigma_i^n)$ are defined, and

$$d_f(\sigma^n) \text{ or } d_f(S^n) = \sum_i d_f(\sigma_i^n). \tag{3.1}$$

To show this, subdivide σ^n or S^n further, so that we can deform f into a simplicial map, and apply Ω_1 (see II, §3). The above quantities are unchanged, and (3.1) is now a consequence of II, (3.1).

4. **On deformations.** We shall need the following elementary results. Let $K \times I$ be the product of K and the unit interval I, consisting of all pairs (p, t), p in K, $0 \leqq t \leqq 1$. The deformation $\phi_t(p)$ of K in S_0^n is equivalent to the map $\Phi(p, t) = \phi_t(p)$ of $K \times I$ into S_0^n. Hence ϕ_0 is homotopic to ϕ_1 if and only if Φ, defined over $K \times 0 + K \times 1$, may be extended over $K \times I$.

Let f map the boundary $\partial\sigma^r$ of σ^r into S_0^n. Then f is homotopic to zero (in $\partial\sigma^r$) if and only if it may be extended through σ^r. For the deformation $f_t(p)$ (p in $\partial\sigma^r$) into $f_1(p) \equiv P$ is equivalent to the map $f(p_{1-t})$ (see II, §5) $= f_t(p)$ of σ^r into S_0^n.

LEMMA 1. *If $\phi \equiv \phi_0$ maps σ^n into S, and the deformation ϕ_t of ϕ is defined over $\partial\sigma^n$, then its definition may be extended over σ^n.*

We define ϕ_t in σ^n by

$$\phi_t(p_u) = \begin{cases} \phi(p_{(1+t)u}) & \left(0 \leqq t \leqq \dfrac{1}{u} - 1\right), \\ \phi_{t+1-\frac{1}{u}}(p) & \left(\dfrac{1}{u} - 1 \leqq t \leqq 1\right). \end{cases} \tag{4.1}$$

[6] This is the character group of the homology group with numbers mod 1 as coefficient group.

[7] See pp. 46–50 of this volume of this Journal; we refer to this paper as II.

LEMMA 2. *Any map ϕ of K into S_0^n may be deformed into a normal one; all cells already at P_0 we may keep fixed.*

We deform the map successively so that $K^0, K^1, \cdots, K^{n-1}$ are at P_0. Suppose K^{r-1} is at P_0 (if $0 < r < n$). As each $\partial\sigma^r$ is at P_0, we may deform each σ^r into P_0, keeping $\partial\sigma^r$ at P_0 (see §3). This deformation, defined over K^r, is extended over all $(r + 1)$-cells, $(r + 2)$-cells, etc., by Lemma 1. It is now defined over K, and K^r is at P_0.

5. **Parts of cocycles.** Let K' be a subcomplex of K. Any r-chain C of K may be written $C' + C''$, the coefficients of cells of $K - K'$[8] [of K'] being zero in C' [in C'']. We say C' is *part* of C. Clearly the chain C' in K' is part of a cocycle if and only if $\delta C'$ cobounds in $K - K'$, i.e., if and only if for some chain C'' in $K - K'$, $\delta C' = \delta C''$. The $(r + 1)$-chains are chains of K.

6. **The product** $K \times I$. We subdivide $K \times I$ (see §4) by means of all cells $\sigma_i^r \times I$ (σ_i^r in K). Orient the cells $\sigma_i^r \times 0$ and $\sigma_i^r \times 1$ like the σ_i^r, and orient each $(r + 1)$-cell $\sigma_i^r \times I$ so that $\sigma_i^r \times 1$ is on its boundary positively. Then

$$(6.1) \qquad \delta(\sigma_i^r \times 0) = -\sigma_i^r \times I + \cdots, \qquad \delta(\sigma_i^r \times 1) = \sigma_i^r \times I + \cdots,$$

$$(6.2) \qquad \delta(\sigma_i^r \times I) = -\sum_j \partial_{ij}^{r+1}(\sigma_j^{r+1} \times I).$$

To prove (6.2), say $\delta(\sigma_i^r \times I) = A_{ij}^{r+1}(\sigma_j^{r+1} \times I) + \cdots$. Then

$$\begin{aligned}\delta\delta(\sigma_i^r \times 1) &= \delta[(\sigma_i^r \times I) + \sum_j \partial_{ij}^{r+1}(\sigma_j^{r+1} \times 1)] \\ &= (A_{ij}^{r+1} + \partial_{ij}^{r+1})(\sigma_j^{r+1} \times I) + \cdots = 0,\end{aligned}$$

and $A_{ij}^{r+1} = -\partial_{ij}^{r+1}$. The first equation in (6.1) is clear for $r = 0$; it is proved in succession for $r = 1, 2, \cdots$ by considering the coefficient of $\sigma_j^r \times I$ in $\delta\delta(\sigma_i^{r-1} \times 0)$.

THEOREM 1. *Let C_0 and C_1 be n-chains in $K = K^n$, and let D_0 and D_1 be the corresponding chains in $K \times 0$ and $K \times 1$. Then $D_0 + D_1$ (as a chain in $K \times I$) is part of a cocycle if and only if $C_0 \backsim C_1$ in K.*

Say

$$C_0 = \sum a_i\sigma_i^n, \qquad C_1 = \sum b_i\sigma_i^n.$$

Consider any n-chain

$$(6.3) \qquad D = D_0 + D_1 + \sum h_j(\sigma_j^{n-1} \times I);$$

then, by (6.1) and (6.2),

$$\begin{aligned}\delta D &= -\sum a_i(\sigma_i^n \times I) + \sum b_i(\sigma_i^n \times I) - \sum h_j\partial_{ji}^n(\sigma_i^n \times I) \\ (6.4) \qquad &= \sum_i [b_i - a_i - \sum_j h_j\, \partial_{ji}^n](\sigma_i^n \times I).\end{aligned}$$

[8] $K - K'$ is in general not a subcomplex of K, i.e., is not closed in K.

Suppose $D_0 + D_1$ is part of a cocycle D; then (6.4) set $= 0$ gives

$$\delta\Big(\sum_j h_j\sigma_j^{n-1}\Big) = \sum_{i,j} h_j\partial_{ji}^n\sigma_i^n = \sum_i (b_i - a_i)\sigma_i^n = C_1 - C_0,$$

and $C_0 \backsim C_1$. Conversely, suppose $C_1 - C_0 = \delta(\sum h_j\sigma_j^{n-1})$; then the last set of equations shows that the bracket in (6.4) vanishes, and hence D, defined by (6.3), is a cocycle.

II. The theorems

7. The extension theorem. We shall prove

THEOREM 2. *Let f be a normal map of the subcomplex K' of $K = K^{n+1}$ into S_0^n. Then f can be extended over K if and only if the chain*

$$(7.1) \qquad D' = \sum_{\sigma_i^n \text{ in } K'} d_f(\sigma_i^n)\sigma_i^n$$

in K' is part of a cocycle.

First suppose D' is part of a cocycle $D = \sum a_i\sigma_i^n$:

$$(7.2) \qquad a_i = d_f(\sigma_i^n) \quad (\sigma_i^n \text{ in } K'), \qquad \sum_i a_i\partial_{ij}^{n+1} = 0 \ (\text{all } j).$$

f maps $(K')^{n-1}$ into P_0; set $f(p) \equiv P_0$ in K^{n-1}. Let f map each σ_i^n not in K' into S_0^n with the degree a_i (see II, Theorem 2); then (7.2) holds for all σ_i^n. Consider any $(n+1)$-cell σ_j^{n+1} of $K - K'$. Using (3.1), we find

$$(7.3) \qquad \begin{aligned} d_f(\partial\sigma_j^{n+1}) &= d_f\Big(\sum_i \partial_{ij}^{n+1}\sigma_i^n\Big) = \sum_i \partial_{ij}^{n+1}d_f(\sigma_i^n) \\ &= \sum_i \partial_{ij}^{n+1}a_i = 0. \end{aligned}$$

Hence f, considered only in $\partial\sigma_j^{n+1}$, is homotopic to zero (II, Theorem 1), and f may be extended over σ_j^{n+1} (see §4). Thus we extend f throughout K.

Now suppose f is extended throughout K. By Lemma 2, we deform f into a normal map, leaving $(K')^{n-1}$, and hence also K', fixed. Call the new map f again, and define the a_i and D by (7.2). Then D' is part of D. By §4, f, in each $\partial\sigma_j^{n+1}$, is homotopic to zero; hence (7.3) holds, and D is a cocycle.

Remark. If f is any map of K' into S_0^n, we may deform it into a normal map ϕ, by Lemma 2. From Lemma 1, it is apparent that f can be extended over K if and only if ϕ can be. Define D' by (7.1). By Theorem 2, $\delta D'$ has zero coefficients over cells of K', and is therefore a chain, which is clearly a cocycle, of $K'' = K - K'$. By Theorem 3, Remark, if f is also deformed into the normal map ψ, defining the chain C' of K', then $C' \backsim D'$ in K', and hence for some H in K',

$$C' - D' = (\delta H)' = \delta H - (\delta H)''.$$

Therefore $\delta C' - \delta D' = \delta[(\delta H)'']$, which lies in K''. Thus the cohomology class in K'' of $\delta D'$ is uniquely determined by f, and we have (using Theorem 2): *f may be extended over K if and only if its cohomology class thus defined in K'' is $\backsim 0$ in K''.*

8. **The classes of maps of K^n into S_0^n.** If we put two maps of K^n into S_0^n into the same class if they are homotopic, the maps fall into classes, the *homotopy classes.* To any normal map f of K^n into S_0^n we let correspond a chain C_f as in (7.1).

THEOREM 3. *The normal maps ϕ and ψ of $K = K^n$ into S_0^n are homotopic if and only if $C_\phi \backsim C_\psi$.*

Set $\Phi(p \times 0) = \phi(p), \Phi(p \times 1) = \psi(p)$; then ϕ is homotopic to ψ if and only if Φ may be extended through $K \times I$ (see §4). If D_0 and D_1 correspond to C_ϕ and C_ψ in $K \times 0$ and $K \times 1$, Theorem 2 shows that this is possible if and only if $D' = D_0 + D_1$ is part of a cocycle in $K \times I$. By Theorem 1, this is true if and only if $C_\phi \backsim C_\psi$.

Remark. If K is of any dimension and ϕ and ψ are homotopic, then C_ϕ and C_ψ are cocycles and $C_\phi \backsim C_\psi$. The first statement follows from Theorem 2; the second follows on considering ϕ and ψ in K^n alone.

THEOREM 4. *The classes of maps of K^n into S_0^n are in* (1 − 1) *correspondence with the elements of the n-th cohomology group of K with integer coefficients. The correspondence is given by deforming the map f into a normal one and taking the cohomology class of the resulting cocycle. In particular, f is homotopic to zero if and only if the corresponding cohomology class is zero.*

The deformation is possible, by Lemma 2. The cohomology class is uniquely determined by f, and non-homotopic maps determine different classes, by Theorem 3. Finally, to each cohomology class corresponds a map; we take a cocycle C of the class, and let f map each σ^n normally into S_0^n with the degree equal to its coefficient in C (see II, Theorem 2).

9. **The Theorem of Hurewicz.** Let Q_0 be a fixed point of a space S. Then the classes of maps of S_0^r into S for which P_0 goes into Q_0 form an abelian group, the *r-th homotopy group* of S.[9] If f maps σ^n [or S_0^n] into S, and $f(p) \equiv Q_0$ in $\partial\sigma^n[f(P_0) = Q_0]$, we may call the corresponding homotopy element the *degree* $d_f(\sigma^n)[d_f(S_0^n)]$ of f. (If $S = S_0^n$, the n-th homotopy group is the group of integers, as was seen in II, so that this is a natural generalization of the term degree.) The fundamental formula (3.1) holds still. The theorems of the preceding paper become matters of definition. The proofs in the present paper hold without change, and we have a new version of the Theorem of Hurewicz:

THEOREM 5. *Theorems* 2, 3 *and* 4 *hold if we replace S_0^n by any locally contractible space S_0 whose r-th homotopy groups vanish for $r < n$, and replace the integers by the n-th homotopy group of S as coefficient group in the chains and cohomology classes.*

Hurewicz also shows that in the above space S_0 the n-th homotopy group is the same as the n-th homology group with integer coefficients.

HARVARD UNIVERSITY.

[9] See Hurewicz, loc. cit. We assume a knowledge of the fundamental properties of homotopy groups.

Reprinted from Duke Mathematical Journal
Vol. 4, No. 1, March, 1938

CROSS-SECTIONS OF CURVES IN 3-SPACE

By Hassler Whitney

1. **Introduction.** We consider in this paper "regular families of curves", that is, families F of non-intersecting curves such that if two curves are sufficiently close at a point, then they remain close for a "finite time"; see [3] (i.e., the third reference below), Theorem 7A. It was shown in [3] that a cross-section may always be found through an arbitrary point of the family. If the family fills a region of Euclidean 3-space E, it is natural to suspect that *any cross-section contains a cross-section which is a 2-cell*; our object here is to prove this fact. It follows (see [3], §20) that *locally, a family of curves in E is equivalent to a family of straight lines.* The proof is arranged so that a minimum of preliminary material is assumed. We use rather fully the methods in [1], [2], and the first half of [3], and a method of proof in [4].

2. **Two types of homology.** We relate the two types of homology used in [1] and [2], and give some simple properties of the second type. For a curve (= simple closed curve) J in a closed set, $J \sim 0$ was defined in [2]. For J in a general set, say $J \sim 0$ if it is ~ 0 in some bounded closed subset. Call a chain of a subdivision from some fixed sequence of simplicial subdivisions (as in [1]) a *polygonal* chain. Say two curves are equivalent in a set G if, using fixed parametrizations $f_0(\theta)$ and $f_1(\theta)$ for them, one can be deformed into the other in G, i.e., $f_t(\theta)$ $(0 \leqq t \leqq 1)$ exists and is continuous.

Lemma 1. *If J and J' are equivalent in G, then $J \sim 0$ in G if and only if $J' \sim 0$ in G.*

This is easily seen, using [2], Lemma I, if we subdivide the θ-circle and the t-segment, and consider triangles of the forms

$$f_{t_i}(\theta_j)f_{t_{i+1}}(\theta_j)f_{t_i}(\theta_{j+1}), \qquad f_{t_{i+1}}(\theta_j)f_{t_i}(\theta_{j+1})f_{t_{i+1}}(\theta_{j+1}).$$

Lemma 2. *If $J \sim 0$ in A, and A is deformed into A', leaving J fixed, then $J \sim 0$ in A'.*

For if $L \to K$, K a 1-cycle in J, we need merely consider the deformed L.

Lemma 3. *Let J be equivalent to a polygonal J' in an open set G. Then $J \sim 0$ in G if and only if J' bounds a polygonal chain in G.*

By Lemma 1, $J \sim 0$ if and only if $J' \sim 0$. Suppose J' bounds a polygonal chain. Then, using a fine enough subdivision and [2], Lemma I, we see that $J' \sim 0$. Suppose $J' \sim 0$; say $J' \sim 0$ in the closed subset A of G. Set $\epsilon =$

Received December 3, 1937; presented to the American Mathematical Society, October 29, 1932. The theorem was first proved when the author was a National Research Fellow.

$\rho(A, E - G)$, and take a simplicial subdivision of E of norm $< \epsilon$. Take δ so that a δ-chain K as in [2], Lemma I, is $\epsilon \sim 0$ in A. A simplicial approximation of this (abstract) chain gives a polygonal chain in G bounded by the polygonal J'.

LEMMA 4. *Let B be a (finite) linear graph. Let each arc α_i of B be divided into arcs of diameter $< \epsilon$, defining an ϵ-chain K_i. Let δ be the minimum distance between any two such subarcs of B without common points. Then any $(\delta, 1)$-cycle on B is $\epsilon \sim$ some linear combination of the K_i.*

The proof of [2], Lemma I, applies almost without change.

LEMMA 5. *Let Z and Z' be closed sets with only the curve J in common. If $J \sim 0$ in $Z + Z'$, then $J \sim 0$ in one of Z, Z'.*

It is sufficient to show that for every $\epsilon > 0$ there is a 1-cycle K in J as described in [2], Lemma I, which is $\epsilon \sim 0$ in one of Z, Z'. Take a sufficiently fine K in J, bounding L in $Z + Z'$. Pushing vertices near J onto J gives an $L_1 \to K$, such that $L_1 = L_2 + L_2'$, L_2 in Z, L_2' in Z'. If $L_2 \to K_2$, $L_2' \to K_2'$, then one of K_2, K_2' is $\epsilon \sim K$ in J (see [2], Lemma I; hence $K \epsilon \sim 0$ in one of Z, Z'. (Compare the proof of [2], Lemma M.)

3. **A local separation theorem.** We shall prove two lemmas on how a portion of a cylinder separates a region. Let H denote a curved cylinder in E, given by (t, θ), $-2 \leqq t \leqq 2$, θ on a circle. Let H_{t_1,t_2} denote that part of H with $t_1 \leqq t \leqq t_2$; set $H_t = H_{t,t}$. Let H'_{t_1,t_2} and H''_{t_1,t_2} denote those parts of H_{t_1,t_2} corresponding to two complementary arcs of the θ-circle.

LEMMA 6. *Let R be a spherical region, and let H have its ends H_{-2} and H_2 outside of R. Then arbitrarily near any point q of $H \cdot R$ there are points r_1 and r_2 in different components of $R - H$.*

Choose coördinates in H so that q is on H_0. Let L be a 1-chain in $R - H_0$ linking H_0 (see [1], Theorem X^i). Say $L = L_- + L_+$, $L_- \cdot H_{-2,0} = 0$, $L_+ \cdot H_{0,2} = 0$; we may choose L_- and L_+ so that each is bounded by $a + b$, a pair of points. If a and b are in the same component of $R - H$, there is a polygonal chain M joining them. As R is spherical, we may choose $N_1 \to L_- + M$ in R; N_1 is in $E - H_{-2}$. As $H'_{-2,0}$ is a cell, we may choose $N_2 \to L_- + M$ in $E - H'_{-2,0}$ ([1], Theorem T^i). As $H_{-2} \cdot H'_{-2,0}$ is an arc, $N_1 + N_2$ cannot link it; hence there is a chain $N_3 \to L_- + M$ in $E - (H_0 + H'_{-2,0})$([1], Corollary W^i). Similarly, using $H''_{-2,0}$, we find a chain $N_4 \to L_- + M$ in $E - H_{-2,0}$. In the same manner,[1] find $N_5 \to L_+ + M$ in $E - H_{0,2}$. Adding gives $N_4 + N_5 \to L$ in $E - H_0$, a contradiction.

LEMMA 7. *Let R and R' be concentric spherical regions, $R' \subset R$. Let $H_{-1,1}$ lie in R, and let $H_{-2,-1} + H_{1,2}$ lie in $E - R'$. Then of any three points in $R' - H$, some two can be joined in $R - H$. If a cell is omitted from $H \cdot R'$, or H is replaced by $H'_{-2,2}$, then any two points can be joined.*

Take three points a, b, c in $R' - H$, and suppose no two can be joined in $R - H$.

[1] Replace H_{-2} and $H'_{-2,0}$ by $H_{-2} + H'_{-2,0}$ and $H''_{-2,0}$, respectively.

Say $L_1 \to b + c$ in R'; L_1 is in $E - (H_{-2,-1} + H_{1,2})$. Using $H'_{-1,0}$ and $H''_{-1,0}$ in R as above, etc., we find chains

$$L_{a-} \to b + c \text{ in } R - H_{-2,0}, \qquad L_{a+} \to b + c \text{ in } R - H_{0,2},$$
$$L_{b-} \to a + c \text{ in } R - H_{-2,0}, \qquad L_{b+} \to a + c \text{ in } R - H_{0,2}.$$

Let $\bar{E}$ be the sphere formed by making all of $E - R$ a single point. In $\bar{E}$, by [1], Corollary W^i, $L_{a-} + L_{a+}$ and $L_{b-} + L_{b+}$ both link H_0. Hence (Theorem X^i) their sum L_c does not. But

$$L_c = (L_{a-} + L_{b-}) + (L_{a+} + L_{b+}) = L_{c-} + L_{c+},$$
$$L_{c-} \subset R - H_{-2,0},\ L_{c+} \subset R - H_{0,2},\ L_{c-} \to a + b,$$

and hence, applying [1], Corollary W^i, in $\bar{E}$, $a + b$ bounds in $R - H$, a contradiction.

To prove the last two statements, we need merely note that the part of H remaining may be built up by starting with $H_{-2} + H_2$, and adding cells, each having one or two arcs in common with the preceding set; for a 1-cycle cannot link a set of arcs.

4. **The curve** J. Let S be a cross-section through p. Follow the curve of the family F through p in both directions to the first points p' and p'' on some sphere about p; joining p' to p'' by an arc outside the sphere gives a curve J^*. Take $\lambda > 0$ so that any arc[2] of $N_\lambda(p'p'')$ has just one point in S (see [3], §§9 and 16); assume $\lambda < \frac{1}{3}\rho(p, p')$ and $< \frac{1}{3}\rho(p, p'')$. The arcs of N_λ may be oriented; for any q in $S \cdot N_\lambda$, set $q_t = g'(q, t)$ ([4], §15). We may choose the function μ ([3], §3) so that $p_{-2} = p'$, $p_2 = p''$. For any q in the set of points on N_λ, let $\phi(q)$ be the point of S on the arc of N_λ through q. ϕ is continuous (compare [3], §18).

Let R_1 be a spherical region about p in N_λ. Choose a concentric R_2 so that for any q in R_2, the arc $q\phi(q)$ is in R_1. Choose a concentric R_3 so that (using a new μ if necessary) $p_{-1}pp_1 \subset R_2$ and $p_{-2}p_{-1} + p_1p_2 \subset E - \bar{R}_3$. Choose a concentric R_4 so that the last two relations hold with p and p_t replaced by any q and q_t in $S \cdot R_4$. Choose R_5 and R_6 so that $\phi(R_5) \subset R_4$, $\phi(R_6) \subset R_5$.

Take a 1-cycle A in R_6 linking J^*; we may suppose it a polygonal curve. $A' = \phi(A)$ is in R_5. We shall find a curve J in A', J not ~ 0 in $E - J^*$. Take $\eta > 0$ so small that any two points of A' within 2η of each other may be joined by an arc in A' of diameter $< \rho = \rho(A', J^*)$. Divide A into arcs $A_1, A_2, \cdots, A_n$ so that their images $\phi(A_1), \phi(A_2), \cdots$ are of diameter $< \eta$; let $x_i = \phi(A_{i-1} \cdot A_i)$. Let B_i be an arc in A' of diameter $< \rho$ joining x_i to x_{i+1}. We may replace B_i by arcs from $B_1 + \cdots + B_i$ so that $B = B'_1 + B'_2 + \cdots + B'_n$ is a linear graph (see [4], Appendix). We may define a deformation ψ

[2] In [3], §16, (2), we assume the existence of an arc $q_0 q_1$ with a certain property; but it is easily seen that any arc, in particular $p'p''$, may be used.

carrying A along arcs of F into A', and then along short line segments into B; all this is in $E - J^*$.

Suppose that every simple closed curve of B were ~ 0 in $E - J^*$. Then there is a bounded closed set C in $E - J^*$ in which every simple closed curve of B is ~ 0; we may include in C the set of points passed over by A in the deformation ψ. As A is not ~ 0 in $E - J^*$ (Lemma 3), there is an $\epsilon > 0$ such that for every $\delta > 0$ there is a $(\delta, 1)$-cycle on A which is not $\epsilon \sim 0$ in C. We may easily choose $\zeta_1, \zeta_2, \zeta_3$ in turn so that the following is true. Take any $(\zeta_3, 1)$-cycle K_3 on A. Using ψ (see the proof of Lemma 1), we find a $(\zeta_2, 1)$-cycle K_2 on B, with $K_2 \zeta_2 \sim K_1$ in C. By Lemma 4, K_2 is $\zeta_1 \sim K_1$, a $(\zeta_1, 1)$-cycle $=$ a linear combination of the K_i as described in the lemma. Express this cycle as a sum of circuits; each lies on a simple closed curve of B, and hence is $\epsilon \sim 0$ in C. Thus any such K_3 is ~ 0 in C, a contradiction. Therefore there is a curve J in B, J not ~ 0 in $E - J^*$.

5. **The 2-cell Q.** J is in $S \cdot R_5$ and is not ~ 0 in $E - J^*$. Joining the points of J to p by line segments gives a point set D in which $J \sim 0$; applying ϕ shows that $J \sim 0$ in $S \cdot R_4$ (Lemma 2). Let P be the component of $S \cdot R_1 - J$ which contains p. Set $Q = P + J$, $P' = S \cdot R_1 - Q$. By Lemma 5, $J \sim 0$ in one of Q, $P' + J$. As J is not ~ 0 in $S - p$, $J \sim 0$ in Q.

For any point set A in S, let $H_{t_1,t_2}(A)$ denote all points q_t, q in A, $t_1 \leqq t \leqq t_2$; set $H(A) = H_{-2,2}(A)$.

We show next that J is irreducibly ~ 0 in Q. If not, then there is a point q of P so that $J \sim 0$ in $Q - q$. Let α be an arc in P joining p to q ([2], Lemma E). $D = H(\alpha)$ is a closed 2-cell. Set

$$J_1^* = (J^* - p_{-2}p_2) + H(q) + H_{-2}(\alpha) + H_2(\alpha).$$

J_1^* is a curve containing only q in P; hence $J \sim 0$ in $E - J_1^*$. Take a polygonal J_1 equivalent to J in $E - (D + J^*)$; then J_1 bounds in $E - J_1^*$ (Lemma 3). As D is a cell, J_1 bounds in $E - D$ ([1], Theorem T^i). As $D \cdot J_1^*$ is an arc, no chain can link it; hence J_1 bounds in $E - (D + J_1^*)$ ([1], Corollary W^i). Hence $J \sim 0$ in $E - (D + J_1^*)$ (Lemma 3) and hence in $E - J^*$, a contradiction.

As Q is a continuous curve (compare [2], top of p. 269), and is a cross-section through p (for the proof that J is irreducibly ~ 0 in Q shows that Q contains all points of S near p), there remains to show only that any arc γ in Q with just its ends a and b on J divides Q.

If $Q - \gamma$ is connected, let γ' be an arc in $Q - \gamma$ joining the two arcs into which a and b divide J. Let α and β be arcs of J about a and b not touching γ'. Let A and B be cells of $H(J)$ about a and b, such that $\phi(A)$ and $\phi(B)$ are subsets of α and β. By Lemma 6, there are two points in $R_4 - H(J)$ which cannot be joined by an arc in $R_1 - H(J)$. By Lemma 7, any point of $R_3 - H(J)$ can be joined to one of these by an arc in $R_2 - H(J)$; we may now write

$$R_3 - H(J) = G + G',$$

G containing p. Then two points of the same set can be joined in $R_2 - H(J)$, and two points of different sets cannot be joined in $R_1 - H(J)$.

Also, if A is omitted, G and G' may be joined; hence there is a point a' of G' arbitrarily close to a, and we may find an arc α' joining a' to A with only an end in $H(J)$. Choose β' similarly, and let γ'' join the ends of α' and β' in $R_2 - H(J)$. Projecting $\alpha' + \beta' + \gamma''$ onto S, we find an arc γ^* in $S \cdot R_1$ with its two ends on α and β. As the points of γ^* may be joined to a' through arcs of F and of $\alpha' + \beta' + \gamma''$, and a' is in G' while p is in G, no point of γ^* may be joined to p in $R_1 - H(J)$; hence γ^* has no points in P. Now γ and γ^*, together with arcs α^* in α and β^* in β, give a curve I in $S \cdot R_1$, with only $\gamma_1 = \gamma + \alpha^* + \beta^*$ in Q.

Take q within γ. By Lemma 6, there are points r_1 and r_2 arbitrarily near q which cannot be joined in $R_1 - H(I)$; $r_1' = \phi(r_1)$ and $r_2' = \phi(r_2)$ are in P. By Lemma 7, r_1' and r_2' can be joined in $R_2 - H(\gamma_1)$; applying ϕ, we find an arc δ joining them in $S \cdot R_1 - \gamma_1$. By the choice of r_1 and r_2, r_1' and r_2' cannot be joined in $Q - \gamma_1$; hence δ has points not in Q. As P is a component of $S \cdot R_1 - J$, δ has points in J. Run along δ from r_1' and r_2' to the first points s_1 and s_2 in J. As s_1 and s_2 can be joined by an arc in $J + \gamma' - (\alpha + \beta)$, r_1' and r_2' can be joined in $Q - \gamma_1$. But this is a contradiction; hence $Q - \gamma$ is not connected. All the hypotheses of Theorem I of [2] are now satisfied; hence Q is a cross-section through p which is 2-cell.

References

1. J. W. Alexander, *A proof and extension of the Jordan-Brouwer separation theorem*, Transactions of the American Mathematical Society, vol. 23 (1922), pp. 333–349.
2. H. Whitney, *A characterization of the closed 2-cell*, Transactions of the American Mathematical Society, vol. 35 (1933), pp. 261–273.
3. H. Whitney, *Regular families of curves*, Annals of Mathematics, vol. 34 (1933), pp. 244–270.
4. R. L. Wilder, *On the linking of Jordan continua in E_n by $(n-2)$-cycles*, Annals of Mathematics, vol. 34 (1933), pp. 441–449. We note a typographical error. On p. 449, line 17, $\lambda \subset t_k \cdot t_i$ should read $\lambda \supset t_k \cdot t_i$.

Harvard University.

ANNALS OF MATHEMATICS
Vol. 39, No. 2, April, 1938

ON PRODUCTS IN A COMPLEX[1]

BY HASSLER WHITNEY

(Received June 10, 1937; Revised November 9, 1937)

1. Introduction

In classical homology theory, founded by Poincaré, the fundamental operation is that of forming the boundary ∂A^p of a chain A^p. This is found by multiplying the coefficients of A^p into a matrix of incidence. Algebraically, an equally obvious operation, using the same matrix of incidence, forms the "coboundary" δA^{p-1} from a given A^{p-1}. It has recently been discovered that the algebraic part of the theory of intersections of chains in a manifold, when interpreted with the other operation, could be generalized to arbitrary complexes. It is the object of this paper to give a complete treatment of the fundamentals of this theory. We use a general type of complex and general coefficient groups, and prove the required invariance theorems. Parts of the paper are new only in form. Various notions used here appear first in Tucker's thesis, [12].[2]

In Part I we define, after Tucker, the complexes to be used. The cohomology groups are defined, and then elementary properties of "dual homomorphisms" are given. The latter are used throughout the paper.

Suppose we ask for a product of p-chains A^p and q-chains B^q, giving $(p + q)$-chains $A^p \smile B^q$, which shall have topological significance. A p-cell times a q-cell far away from the p-cell should certainly give nothing; hence (P_1) of §5 is a natural assumption. Considering δ as the fundamental operation, if we wish the multiplication to give a result in the cohomology groups, we must have cocycle $\smile$ cocycle = cocycle. Hence $\delta(A \smile B)$ must be expressible in terms such as $\delta A \smile B$ and $A \smile \delta B$. (P_2) is the natural form.[3] Suppose we ask that a vertex times itself equal itself. (Hence the γ of §5 is 1.) Then (using Theorem 1), at least in an ordinary connected complex, the products exist, and when carried out in the cohomology groups, are uniquely determined (Theorem 5). This may be considered the fundamental theorem of the present paper.

---

[1] Presented to the American Mathematical Society, under a different title, March 27, 1937. An outline of the paper appeared in Proc. Nat. Ac. Sci., vol. 23 (1937), pp. 285–291.

[2] The numbers in square brackets refer to the bibliography at the end of the paper.

[3] Note that the dimension in each term is correct. It is easily seen that the signs must alternate in one of the terms. We might call $\smile$ and $\frown$ "cup" and "cap". Equivalent formulas (see also (Q_2) and (19.2)) occur in a fundamental manner in the classical theory; see for instance Lefschetz, [10], pp. 111, 169 and 226. The same formula appears in the theory of differential forms.

Another product, $A^p \frown B^q = C^{q-p}$, is considered; it is algebraically equivalent to the $\smile$ product. We also consider briefly a particular definition of the products which may be used in simplicial complexes. The rest of Part II is devoted to proving other elementary properties and to showing how general coefficient groups may be used.

In Part III we answer two questions. First, in a polyhedron, are homology and cohomology groups and products independent of the particular simplicial subdivision chosen? The proof §16 that they are (which may be considered well-known) is relatively simple; the considerations show that the groups and products may be associated with an abstract space. Secondly, to find the groups and products, it is often very inconvenient to have to use simplicial subdivisions; but we must then show that a general complex gives the same theory as a simplicial subdivision. This combinatorial theorem (Theorem 14) occupies the rest of Part III. It turns out that complexes which may be used for determining the groups often may not be used for determining the products.[4] However, most of the invariance proof may be carried out in the general type of complex. Similar (but slightly weaker) theorems for homology groups have been proved by Tucker, [12], and Alexandroff-Hopf, [4], Ch. VI.

The relation of the two products to intersection theory in a manifold is considered briefly in Part IV. In contrast with Čech, [5], we use the classical method of dual complexes. In Part V, the products are considered in product complexes and in Euclidean space. As an application of preceding results, some mapping theorems are proved, due in part to H. Hopf.

Some special topics are considered in the Appendix.

Historical note. The coboundary of a chain, when a passage to the limit is applied, becomes the derived of a covariant alternating tensor (compare Alexander, [1]). In this form, of course, it has long been known. From the algebraic standpoint, cocycles appear in a different form in the "pseudocycles" of S. Lefschetz, [10]. Cocycles may be interpreted as cycles in the "dual complex," considered in papers by W. Mayer[5] and A. W. Tucker, [12]. An application of cocycles in their direct form was given in our note on Sphere-spaces.[6]

The work of L. Pontrjagin on character groups led to the realization that not only the homology but also the cohomology groups might be important. At the International Topological Conference, Moscow, 1935, J. W. Alexander and A. Kolmogoroff presented papers giving not only the theory of cohomology groups (with different notations), but also defining a product (for simplicial

---

[4] For instance, in a torus, for determining the groups, we may subdivide into one vertex, two 1-cells, and one 2-cell; but for determining the products, we must have say four vertices, eight 1-cells, and four 2-cells. (Or we may use Part V.)

[5] Monatsh. f. Math. u. Phys., vol. 36 (1929), pp. 1–42 and 219–258.

[6] Proc. Nat. Ac. of Sci., vol. 21 (1935), pp. 464–468, §6.

complexes) in the groups.[7] It appears that D. van Danzig[8] and E. Čech also had a portion of these results. However, the Kolmogoroff-Alexander product was not wholly satisfactory, it being too large by a numerical factor. In studying their product at the end of 1935, the author discovered the $\smile$ product of §6. In an effort to generalize a theorem of H. Hopf (see footnote 17), the remaining results of §6 and §7 were found.[9] At the same time, Čech discovered the same products; see [5]. Alexander studied the revised $\smile$ product; see [3]. Finally, the present Part II may be considered an outgrowth of Čech's paper [5]. Our assumptions (P_1) and (P_2) are closely allied to those of Čech, §2. Our proof of Theorem 5 was obtained after a study of a corresponding proof in Čech's paper. In the mean time, H. Freudenthal, [6], also found the results of §7, and studied the relation of these products to other known products.

Recently S. Lefschetz[10] has shown that the products of §6, when properly translated into the residual space of a sphere containing the complex, become the classical intersections. As in Tucker, [12], he gives postulates which a "chain-product" should satisfy. Neither author proves a uniqueness theorem (such as our Theorem 5).

I. Preliminaries

2. The complexes used

Complexes, in topology, are certain algebraic structures which may be given a geometric significance. For a given algebraic structure to be geometrically realizable, certain conditions must be satisfied. For instance, we may demand that there be a simplicial subdivision (algebraically defined); any simplicial complex determines a geometric complex in Euclidean space. Complexes of this nature we shall say "admit a simplicial subdivision" (see Part III). To define a product theory in the complex, stronger conditions are necessary; a complex satisfying these conditions "admits a product theory." Most of the paper will be concerned with these complexes.

A *complex K admitting a product theory* is a system as follows. It has cells[11]

---

[7] See [2], [8] and [9]. The Kolmogoroff-Alexander product is not that given in [2].

[8] See Recueil Math., Moscow, vol. 1 (43), (1936), pp. 672–674. Cohomology groups with general coefficient groups are studied by Steenrod, Am. Journal of Math., vol. 58 (1936), pp. 661–701.

[9] These results were applied in classifying the maps of a 3-sphere into a 2-sphere; see Bull. Am. Math. Soc., vol. 42 (1936), p. 338. They were communicated in letters to L. Zippin early in 1936.

[10] Bull. Am. Math. Soc., vol. 43 (1937), pp. 345–359, §5, (d). It should be noted that without a postulate such as our (P_1), the product may give pratically any product in the cohomology groups.

[11] We assume the cells are finite in number; however, most of the results extend to the infinite case, at least with the proper definitions.

σ_i^p of dimension p. There are two types of relations. A cell may be a *face* of a cell of higher dimension, and is a face of itself. Two cells σ_i^{p-1} and σ_j^p of neighboring dimension have an *incidence number* ${}^p\partial_j^i$, which is an integer. In terms of these numbers we define boundaries ∂A^p etc. as usual (see §3). The *closure* $\bar{\sigma}_i^p$ of a cell σ_i^p is the set of all its faces. The *star* St (σ) of σ is the set of cells with σ as a face. The closure and star of any subcomplex are defined similarly. We say a cycle $A^p = \sum \alpha_i \sigma_i^p$ is *boundary-like* if $p > 0$ or if $p = 0$ and $\sum \alpha_i = 0$ (or $I \cdot A^1 = 0$; see (2.1) and (3.2)). The definition as here given is useful only in the case that δI contains no 1-cells which have two vertices as faces; see Lemma 5.

We make the following assumptions.

(K_1) If σ_1 is a face of σ_2 and σ_2 is a face of σ_3, then σ_1 is a face of σ_3.

(K_2) If ${}^p\partial_j^i \neq 0$, then σ_i^{p-1} is a face of σ_j^p.

(K_3) $\partial\partial A^p = 0$ always (or equivalently, $\delta\delta A^p = 0$).[12]

(K_4) Each boundary-like cycle (with integer coefficients) in any $\bar{\sigma}_i^p$ bounds a chain in $\bar{\sigma}_i^p$.

Note that, using (4.7) below, (K_3) is equivalent to the vanishing (for all A^p, B^{p-2}) of any of $\partial\partial A^p \cdot B^{p-2}$, $\partial A^p \cdot \delta B^{p-2}$, $A^p \cdot \delta\delta B^{p-2}$, or $\delta\delta B^{p-2}$.

Certain elementary properties of these complexes are the following:

LEMMA 1. *If $p > 0$, then $\partial \sigma_i^p \neq 0$.*

For if $\partial \sigma_i^p = 0$, then σ_i^p is a cycle in $\bar{\sigma}_i^p$. By (K_4), it must bound a $(p + 1)$-chain in $\bar{\sigma}_i^p$. But $\bar{\sigma}_i^p$ contains no $(p + 1)$-cells.

LEMMA 2. *Every cell of dimension > 0 has a lower dimensional face, and hence a vertex as a face.*

This follows from the last lemma and (K_2) and (K_1).

LEMMA 3. *Each 1-cell has just one or two vertices as faces.*

By the last lemma, we need merely prove that any σ^1 has not three vertices a, b, c as faces. If it had, then by (K_4), $b - a = k\partial\sigma^1$ and $c - b = l\partial\sigma^1$. Hence $l(b - a) = k(c - b)$, which is impossible (as $k \neq 0$).

The definitions of open and closed subcomplexes are as usual, in terms of "being a face of"; for σ_1 and σ_2 in the subcomplex K' of K, we say σ_2 is a face of σ_1 in K' if it is in K.

LEMMA 4. *If K admits a product theory, so does every closed subcomplex.*

The proof is simple.

Note that, if

$$I = \sum \sigma_i^0 = \text{sum of all the vertices of } K, \tag{2.1}$$

we may have $\delta I \neq 0$. But see Theorem 2. If K is connected, and $\delta I = 0$, the only 0-cocycles are the multiples of I.

---

[12] Complexes satisfying (K_1), (K_2) and (K_3) are exactly those considered by Tucker, [12]. If (K_4) is satisfied also, and $\delta I = 0$ (that is, K is "augmentable"), then Tucker shows that K admits a simplicial subdivision and hence is geometrically realizable.

LEMMA 5. *If each* 1-*cell is on two vertices, then* $\delta I = 0$ *and* K *admits a simplicial subdivision* (see §13).

For take any 1-cell σ; say $\partial\sigma = \alpha a + \beta b$. As $a - b$ is boundary-like, there is a k such that $a - b = k\partial\sigma$; hence $\beta = -\alpha$. Therefore $\delta I\cdot\sigma = I\cdot\partial\sigma = \alpha + (-\alpha) = 0$, and $\delta I = 0$. The last statement has been proved by Tucker (see footnote 12).

3. Homology and cohomology groups[13]

The *boundary* and *coboundary* of chains are defined by

$$(3.1)\quad \partial\left(\textstyle\sum_i \alpha_i\sigma_i^p\right) = \textstyle\sum_{i,j} \alpha_i{}^p\partial_i^j\sigma_j^{p-1}, \qquad \delta\left(\textstyle\sum_i \alpha_i\sigma_i^p\right) = \textstyle\sum_{i,j} \alpha_i{}^{p+1}\partial_j^i\sigma_j^{p+1}.$$

The α_i are integers or elements of an abelian group G.[14] The chain A^p is a *cycle* [*cocycle*] if $\partial A^p = 0$ [$\delta A^p = 0$]. We say

A is *homologous to* B, $A \sim B$, if $A - B$ is a boundary,
A is *cohomologous to* B, $A \backsim B$, if $A - B$ is a coboundary.

As $\partial\partial A^p = 0$, $\delta\delta A^p = 0$ (see (K_3)), we may define as usual the homology and cohomology groups. Using the coefficient group G, we denote these by ${}^p\mathbf{H}^G$, ${}^p\mathbf{H}_G$. The group of p-chains of a complex with integer coefficients will be denoted by L^p.

We define the *scalar product* of two chains of the same dimension by

$$(3.2)\qquad \left(\textstyle\sum \alpha_i\sigma_i^p\right)\cdot\left(\textstyle\sum \beta_i\sigma_i^p\right) = \textstyle\sum \alpha_i\beta_i.$$

Note that $A^p\cdot\sigma^p$ is the coefficient of σ^p in the chain A^p.

4. Dual homomorphisms[15]

Let G and G' be free groups with *fixed* sets of generators $a_1, \cdots, a_p$ and $a_1', \cdots, a_q'$. Then to any matrix of integers $\|\phi_{ij}\|$ $(i = 1, \cdots, p, j = 1, \cdots, q)$ correspond homomorphisms of G into G' and of G' into G, defined by

$$(4.1)\qquad \phi a_i = \textstyle\sum_j \phi_{ij}a_j', \qquad \phi' a_i' = \textstyle\sum_j \phi_{ji}a_j.$$

If ϕ is any homomorphism of G into G', then $\phi a_i = \sum \phi_{ij}a_j'$ for some integers ϕ_{ij}; thus the matrix $\|\phi_{ij}\|$ is defined, and hence the homomorphism ϕ'. Each of ϕ, ϕ' determines the other uniquely; the matrix of one is the transposed matrix of the other. We call these homomorphisms *dual*, and write

$$(4.2)\qquad \phi' = D(\phi), \qquad \phi = D(\phi') = D(D(\phi)).$$

---

[13] For further details, see for instance Whitney, [13].

[14] Except where otherwise stated, we use integers: $G = I_0$.

[15] Compare the Appendix, §27. Dual homomorphisms correspond to adjoint linear transformations in algebra. They have been used in topology by Tucker, [12], §25.

If ϕ maps G into G' and ψ maps G' into G'', then $\theta = \psi\phi$ maps G into G''. Its matrix is clearly

$$\| \theta \| = \| \psi \| \, \| \phi \|: \qquad \theta_{ij} = \textstyle\sum_k \psi_{ik}\phi_{kj} .$$

On transposing, we find

$$\theta' = D(\theta) = D(\psi\phi) = D(\phi)D(\psi) = \phi'\psi'. \tag{4.3}$$

Linear combinations of homomorphisms are defined as usual. We find

$$D(c\phi + d\psi) = cD(\phi) + dD(\psi). \tag{4.4}$$

We shall apply these homomorphisms to the groups L^p, the individual cells forming the generators. Clearly ∂ *and* δ *are dual.* If ϕ maps $L^p(K)$ into $L^{p+k}(K')$ for each p (k fixed), and $\phi' = D(\phi)$, then

$$\delta\phi'A' = \phi'\delta A' \text{ (all } A'\text{) if and only if } \partial\phi A = \phi\partial A \text{ (all } A\text{)}. \tag{4.5}$$

For one relation follows from the other on taking duals. These relations hold in particular for simplicial maps. Using (3.2), we have

$$A \cdot \phi' B' = \phi A \cdot B' \text{ if and only if } \phi' = D(\phi). \tag{4.6}$$

For

$$\phi\sigma_i \cdot \sigma'_j = \left(\textstyle\sum_k \phi_{ik}\sigma'_k\right) \cdot \sigma'_j = \phi_{ij}, \qquad \sigma_i \cdot \phi'\sigma'_j = \phi'_{ji} .$$

Hence

$$A^p \cdot \delta B^{p-1} = \partial A^p \cdot B^{p-1}. \tag{4.7}$$

Remark. The above definitions and results hold equally well if G and G' are vector spaces with fixed bases.

Finally, note that

$$A \cdot B = 0 \text{ if } \delta A = 0 \text{ and } B \sim 0 \text{ or if } A \backsim 0 \text{ and } \partial B = 0. \tag{4.8}$$

For if $\delta A = 0$ and $B = \partial C$, then $A \cdot B = A \cdot \partial C = \delta A \cdot C = 0$, etc. Hence, if $\delta A = 0$ and $B \sim B'$, then $A \cdot B = A \cdot B'$, etc.

II. The Products

5. Definition and properties of products

We shall use only integer coefficients until §11. Corresponding to each p-cell σ_i^p, q-cell σ_j^q, and $(p+q)$-cell σ_k^{p+q}, we wish to find an integer ${}^{pq}\Gamma_k^{ij}$, such that the following properties hold.

(Γ_1). If σ_i^p and σ_j^q are not both faces of σ_k^{p+q}, then ${}^{pq}\Gamma_k^{ij} = 0$.

(Γ_2). For all p, q, i, j, k,

$$\textstyle\sum_l {}^{p+q+1}\partial_k^l \, {}^{pq}\Gamma_l^{ij} = \sum_l {}^{p+1}\partial_l^i \, {}^{p+1,q}\Gamma_k^{lj} + (-1)^p \sum_l {}^{q+1}\partial_l^j \, {}^{p,q+1}\Gamma_k^{il}.$$

(Γ_3). For some integer γ, and all q and j,

$$\sum_i {}^{0q}\Gamma_j^{ij} = \gamma.$$

The important case is the case $\gamma = 1$; after Part II, we always take $\gamma = 1$.

In terms of the quantities ${}^{pq}\Gamma_k^{ij}$, we define two bilinear operations on chains, as follows.

$$\sigma_i^p \smile \sigma_j^q = \sum_k {}^{pq}\Gamma_k^{ij} \sigma_k^{p+q}, \tag{5.1}$$

$$\sigma_j^q \frown \sigma_k^{p+q} = \sum_l {}^{pq}\Gamma_k^{lj} \sigma_l^p. \tag{5.2}$$

Clearly

$$(\sigma_i^p \smile \sigma_j^q) \cdot \sigma_k^{p+q} = {}^{pq}\Gamma_k^{ij} = \sigma_i^p \cdot (\sigma_j^q \frown \sigma_k^{p+q}), \tag{5.3}$$

and hence

$$(A^p \smile B^q) \cdot C^{p+q} = A^p \cdot (B^q \frown C^{p+q}). \tag{5.4}$$

Two products $\smile$ and $\frown$ correspond if and only if (5.4) holds. Given one of $\smile$, $\frown$, we can find the corresponding Γ by (5.3), and hence the other.

The above properties translate into the following, for the $\smile$ product.

(P_1) $\sigma_i^p \smile \sigma_j^q$ is a $(p + q)$-chain in $\mathrm{St}\,(\sigma_i^p) \cdot \mathrm{St}\,(\sigma_j^q)$.[16]

(P_2) $\delta(A^p \smile B^q) = \delta A^q \smile B^q + (-1)^p A^p \smile \delta B^q$.

(P_3) For some integer γ, $I \smile \sigma_j^q = \gamma \sigma_j^q$ (all σ_j^q).

Also, for the $\frown$ product,

(Q_1) $\sigma_j^q \frown \sigma_k^{p+q}$ is a p-chain in $\overline{\mathrm{St}\,(\sigma_j^q) \cdot \sigma_k^{p+q}}$.

(Q_2) $\partial(A^q \frown B^{p+q}) = (-1)^p \delta A^q \frown B^{p+q} + A^q \frown \partial B^{p+q}$.

(Q_3) For some integer γ, $I \cdot (\sigma_j^q \frown \sigma_j^q) = \gamma$ (all σ_j^q).

We shall prove the equivalence of the three sets of properties. (P_1) is clearly equivalent to (Γ_1). If we write (P_2) with σ_i^p and σ_j^q, use (5.1), and consider the coefficient of σ_k^{p+q+1} on each side, we obtain (Γ_2); conversely, (Γ_2) implies (P_2). (Γ_3) and (P_3) are clearly equivalent. Suppose (Γ_1) holds; we shall prove (Q_1). If σ_j^q is not a face of σ_k^{p+q}, then ${}^{pq}\Gamma_k^{ij} = 0$ for every i, hence $\sigma_j^q \frown \sigma_k^{p+q} = 0$, and ($Q_1$) holds. If σ_j^q is a face of σ_k^{p+q}, then $\overline{\mathrm{St}\,(\sigma_j^q) \cdot \sigma_k^{p+q}} = \bar{\sigma}_k^{p+q}$; by ($\Gamma_1$), $\sigma_j^q \frown \sigma_k^{p+q}$ is a chain in $\bar{\sigma}_k^{p+q} = \overline{\mathrm{St}\,(\sigma_j^q) \cdot \sigma_k^{p+q}}$. Suppose ($Q_1$) holds. If σ_j^q is not a face of σ_k^{p+q}, then $\mathrm{St}\,(\sigma_j^q) \cdot \sigma_k^{p+q} = 0$, $\sigma_j^q \frown \sigma_k^{p+q} = 0$, and ${}^{pq}\Gamma_k^{ij} = 0$ (all i). If σ_i^p is not a face of σ_k^{p+q}, then (Q_1) shows that it does not occur in $\sigma_j^q \frown \sigma_k^{p+q}$; hence ${}^{pq}\Gamma_k^{ij} = 0$. (Q_2) is seen to be equivalent to (Γ_2), if we replace p by $p + 1$. Finally, (Q_3) is equivalent to (Γ_3).

---

[16] $S \cdot T$ is the subcomplex of K containing those cells in both S and T. Thus $\mathrm{St}(\sigma_1) \cdot \sigma_2 = \sigma_2$ if σ_1 is a face of σ_2, and $= 0$ otherwise. Čech, [5], assumes merely that the chain is in $\mathrm{St}(\sigma_i^p)$.

THEOREM 1. *If K is connected and $\delta I = 0$, then (Γ_3) is a consequence of (Γ_1) and (Γ_2).*

By Lemmas 1 and 2, it is sufficient to show (using the (P_i)) that if

$$\partial\sigma^q = \alpha\sigma^{q-1}\ (\alpha \neq 0), \qquad I \smile \sigma^{q-1} = \gamma\sigma^{q-1}, \qquad I \smile \sigma^q = \gamma'\sigma^q,$$

then $\gamma = \gamma'$. (That $I \smile \sigma^h = \theta\sigma^h$ for some θ follows from (P_1).) By (P_2),

$$\delta(I \smile \sigma^{q-1}) = I \smile \delta\sigma^{q-1} = I \smile \alpha\sigma^q + \cdots = \alpha\gamma'\sigma^q + \cdots;$$

also

$$\delta(I \smile \sigma^{q-1}) = \delta\gamma\sigma^{q-1} = \gamma\alpha\sigma^q + \cdots,$$

as $\alpha \neq 0, \gamma = \gamma'$.

THEOREM 2. For all σ_j^q, $\delta I \smile \sigma_j^q = 0$.

For, by (P_2) and (P_3),

$$\delta I \smile \sigma_j^q = \delta(I \smile \sigma_j^q) - I \smile \delta\sigma_j^q = \delta\gamma\sigma_j^q - \gamma\delta\sigma_j^q = 0.$$

THEOREM 3. *If $\delta I = 0$, then*

$$I \smile \sigma^p = \sigma^p \smile I = I \frown \sigma^p = \gamma\sigma^p \qquad (\text{all } \sigma^p). \tag{5.5}$$

First, $\sigma^0 \smile I = \sigma^0 \smile \sigma^0 = I \smile \sigma^0 = \gamma\sigma^0$. Suppose $\sigma^{p-1} \smile I = \gamma\sigma^{p-1}$ and $\delta\sigma^{p-1} = \alpha\sigma^p + \cdots\ (\alpha \neq 0)$. Clearly $\sigma^p \smile I = \theta\sigma^p$ for some θ. Then as $\delta I = 0$, (P_2) and (P_3) give

$$\delta(\sigma^{p-1} \smile I) = \delta\sigma^{p-1} \smile I = \alpha\theta\sigma^p + \cdots,$$

$$\delta(\sigma^{p-1} \smile I) = \delta(\gamma\sigma^{p-1}) = \gamma\alpha\sigma^p + \cdots,$$

and $\theta = \gamma$. That $\sigma^p \smile I = \gamma\sigma^p$ now follows, by Lemma 1. The last relation is proved similarly, considering $\partial(I \frown \sigma^p)$.

THEOREM 4. *The $\smile$ and $\frown$ products define products among the cohomology and homology groups, thus:*

cohomology class $\smile$ cohomology class = cohomology class,

cohomology class $\frown$ homology class = homology class.

Explicit formulas are given in (11.6) and (11.7). That cocycle $\smile$ cocycle = cocycle, cocycle $\frown$ cycle = cycle, follows from (P_2) and (Q_2). Also

$$A_1 \smile B \backsim A_2 \smile B \quad \text{if} \quad A_1 \backsim A_2 \quad \text{and} \quad \delta B = 0, \tag{5.6}$$

$$A \smile B_1 \backsim A \smile B_2 \quad \text{if} \quad B_1 \backsim B_2 \quad \text{and} \quad \delta A = 0, \tag{5.7}$$

$$A_1 \frown B \sim A_2 \frown B \quad \text{if} \quad A_1 \backsim A_2 \quad \text{and} \quad \partial B = 0, \tag{5.8}$$

$$A \frown B_1 \sim A \frown B_2 \quad \text{if} \quad B_1 \sim B_2 \quad \text{and} \quad \delta A = 0. \tag{5.9}$$

For instance, if $A_2 - A_1 = \delta C$ and $\partial B = 0$, then by (Q_2),

$$A_2 \frown B - A_1 \frown B = \delta C \frown B = \pm\partial(C \frown B).$$

Hence the definitions in the theorem are unique.

THEOREM 5. *For each integer γ, the products of Theorem 4 exist in any complex and are uniquely determined.*

This will be proved in §8 and §9. The relation between products with different γ's is given in Theorems 9 and 10.

In §10 we shall prove the two *associative* laws and the *commutative* law:

If K admits a simplicial subdivision (see §13), *then*

$$A \smile (B \smile C) \backsim (A \smile B) \smile C \quad \text{if} \quad \delta A = \delta B = \delta C = 0, \tag{5.10}$$

$$A \frown (B \frown C) \sim (A \smile B) \frown C \quad \text{if} \quad \delta A = \delta B = \partial C = 0, \tag{5.11}$$

it being understood in (5.11) that $\frown$ and $\smile$ have the same γ. *If $\delta I = 0$, then*

$$A^p \smile B^q \backsim (-1)^{pq} B^q \smile A^p \quad \text{if} \quad \delta A^p = \delta B^q = 0. \tag{5.12}$$

THEOREM 6. *Let K and K' be simplicial, let f be a simplicial map of K into K', and let f' be the dual of f. Then*

$$f'(A' \smile B') \backsim f'A' \smile f'B' \quad \textit{if} \quad \delta A' = \delta B' = 0, \tag{5.13}$$

$$f(f'A' \frown B) \sim A' \frown fB \quad \textit{if} \quad \delta A' = \partial B = 0. \tag{5.14}$$

These generalize a theorem of Hopf to arbitrary complexes.[17] For the proof, see §10. For the use of different coefficient groups, see §11.

6. The products in simplicial complexes

If K is simplicial, a very simple definition of Γ, $\smile$, $\frown$ is possible. Order the vertices of K in a fixed manner. Each simplex σ^p may now be written in the normal form $\sigma^p = x_{i_0} \cdots x_{i_p}$, $i_0 < \cdots < i_p$. We define, if $x_{i_0} \cdots x_{i_p} \cdots x_{i_{p+q}}$ is a simplex,

$${}^{pq}\Gamma^{i_0 \cdots i_p, i_p \cdots i_{p+q}}_{i_0 \cdots i_p \cdots i_{p+q}} = 1 \qquad (i_0 < \cdots < i_p < \cdots < i_{p+q}), \tag{6.1}$$

and $\Gamma = 0$ for any other triple of simplexes. (The meaning of the Γ should be clear.) In terms of $\smile$ and $\frown$, this gives for instance (if $x_1x_2x_3x_4$ is a simplex)

$$x_1x_2 \smile x_2x_3x_4 = x_1x_2x_3x_4, \qquad x_2x_3x_4 \frown x_1x_2x_3x_4 = x_1x_2.$$

(Γ_1) and (Γ_3) obviously hold. A simple calculation gives (Γ_2); see Alexander, [4].

For these products we clearly have

$$A \smile (B \smile C) = (A \smile B) \smile C, \quad A \frown (B \frown C) = (A \smile B) \frown C. \tag{6.2}$$

7. Products and simplicial maps

Having ordered the vertices of K', order those of K so that

$$\text{if } f(x_i) = x'_{\theta(i)}, \text{ then } \theta(i) < \theta(j) \text{ implies } i < j. \tag{7.1}$$

---

[17] H. Hopf, [7], Satz I.

Using the products of §6, we shall prove (5.13) and (5.14) in the form:

THEOREM 7. *In simplicial complexes, using the products of* §6,

$$f'(A' \smile B') = f'A' \smile f'B', \qquad f(f'A' \frown B) = A' \frown fB. \tag{7.2}$$

To prove the first equation, take any $\sigma_i'^p$ and $\sigma_j'^q$; say

$$\sigma_i'^p = x_0' \cdots x_p', \qquad \sigma_j'^q = x_p' \cdots x_{p+q}';$$

if the vertices of the simplexes are not related as shown, then clearly both sides of the equation (with these simplexes) vanish. By definition of f',

$$f'\sigma_i'^p = \sum x_{\alpha_0} \cdots x_{\alpha_p}, \qquad \theta(\alpha_h) = h,$$

$$f'\sigma_j'^q = \sum x_{\beta_p} \cdots x_{\beta_{p+q}}, \qquad \theta(\beta_h) = h.$$

Hence

$$f'\sigma_i'^p \smile f'\sigma_j'^q = \sum x_{\alpha_0} \cdots x_{\alpha_p} \cdots x_{\alpha_{p+q}}, \qquad \theta(\alpha_h) = h.$$

This is clearly also $f'(\sigma_i'^p \smile \sigma_j'^q)$.[18] To prove the second equation, we use the first, (4.6) and (5.4): For any cell σ' (of the proper dimension),

$$\sigma' \cdot f(f'A' \frown B) = f'\sigma' \cdot (f'A' \frown B) = (f'\sigma' \smile f'A') \cdot B = f'(\sigma' \smile A') \cdot B$$
$$= (\sigma' \smile A') \cdot fB = \sigma' \cdot (A' \frown fB).$$

8. Construction of the products

In a general complex, it is most convenient to construct the $\frown$ product. Take any integer γ. We shall construct all $\sigma_i^q \frown \sigma_j^{p+q}$ in succession for $p = 0, 1, 2, \cdots$. First consider $p = 0$. Set $\sigma_i^q \frown \sigma_j^q = 0$ for $j \neq i$. Let $\sigma_j^q \frown \sigma_j^q$ be any 0-chain in $\bar{\sigma}_j^q$ the sum of whose coefficients is γ; this is possible, by Lemma 2. The required properties hold so far. Suppose all $\sigma_i^q \frown \sigma_j^{r+q}$ are properly constructed for all q and all $r < p$; we must construct each $\sigma_i^q \frown \sigma_j^{p+q}$. If σ_i^q is not a face of σ_j^{p+q}, set $\sigma_i^q \frown \sigma_j^{p+q} = 0$. As $\mathrm{St}(\sigma_i^q) \cdot \bar{\sigma}_j^{p+q}$ has no cells, both sides of (Q_2) with σ_i^q and σ_j^{p+q} vanish. Now suppose σ_i^q is a face of σ_j^{p+q}. Set

$$C^{p-1} = (-1)^p \delta\sigma_i^q \frown \sigma_j^{p+q} + \sigma_i^q \frown \partial\sigma_j^{p+q}; \tag{8.1}$$

this is a chain in $\bar{\sigma}_j^{p+q}$.

Suppose first that $p = 1$. Then for some α, $\delta\sigma_i^q = \alpha\sigma_j^{p+q} + \cdots$, and

$$\begin{aligned} I \cdot C^{p-1} &= -I \cdot (\alpha\sigma_j^{p+q} \frown \sigma_j^{p+q}) + I \cdot (\sigma_i^q \frown \alpha\sigma_i^q) \\ &= -\alpha\gamma + \alpha\gamma = 0. \end{aligned} \tag{8.2}$$

Hence, by (K_4), we may choose the 1-chain $\sigma_i^q \frown \sigma_j^{p+q}$ in $\bar{\sigma}_j^{p+q}$ so that its boundary is C^{p-1}. Then (Q_1) through (Q_3) hold.

[18] By using joins, and the 0-chains $X_h = f'x_h'$, one could give a direct formal proof.

Suppose next that $p > 1$. Then (Q_2) gives

$$\begin{aligned}(8.3)\qquad \partial C^{p-1} &= (-1)^p[(-1)^{p-1}\delta\delta\sigma_i^q \frown \sigma_j^{p+q} + \delta\sigma_i^q \frown \partial\sigma_j^{p+q}] \\ &\quad + (-1)^{p-1}\delta\sigma_i^q \frown \partial\sigma_j^{p+q} + \sigma_i^q \frown \partial\partial\sigma_j^{p+q} = 0.\end{aligned}$$

Hence, by (K_4), we may choose $\sigma_i^q \frown \sigma_j^{p+q}$ in $\bar{\sigma}_j^{p+q}$ with the boundary C^{p-1}. The properties again hold.

We remark that any $\frown$ product may be constructed in this manner.

9. Uniqueness of the products

We first prove

THEOREM 8. *Let $\frown$ be any product with $\gamma = 0$. Then there is a bilinear operation $\wedge$ such that*

(R_1) $\sigma_i^q \wedge \sigma_j^{p+q}$ is a $(p + 1)$-chain in $\overline{\mathrm{St}(\sigma_i^q)\cdot\sigma_j^{p+q}}$.

(R_2) $\sigma_i^q \frown \sigma_j^q = \partial(\sigma_i^q \wedge \sigma_j^q)$.

(R_3) $\sigma_i^q \frown \sigma_j^{p+q} = \partial(\sigma_i^q \wedge \sigma_j^{p+q}) + (-1)^p\delta\sigma_i^q \wedge \sigma_j^{p+q} + \sigma_i^q \wedge \partial\sigma_j^{p+q}$ for $p > 0$.

We shall construct $\wedge$ for $p = 0, 1, \cdots$. As $\gamma = 0$, we can construct $\sigma_i^q \wedge \sigma_j^q$ so (R_2) holds. Suppose all $\sigma_i^q \wedge \sigma_j^{r+q}$ are constructed for all q and $r < p$; we shall construct $\sigma_i^q \wedge \sigma_j^{p+q}$. We make it 0 if σ_i^q is not a face of σ_j^{p+q}. Suppose it is. Set

$$C^p = \sigma_i^q \frown \sigma_j^{p+q} - (-1)^p\delta\sigma_i^q \wedge \sigma_j^{p+q} - \sigma_i^q \wedge \partial\sigma_j^{p+q}.$$

By (Q_2), and (R_3) solved for $\partial(\sigma_i^q \wedge \sigma_j^{p+q})$,

$$\begin{aligned}\partial C^p &= (-1)^p\delta\sigma_i^q \frown \sigma_j^{p+q} + \sigma_i^q \frown \partial\sigma_j^{p+q} - (-1)^p[\delta\sigma_i^q \frown \sigma_j^{p+q} - \delta\sigma_i^q \wedge \partial\sigma_j^{p+q}] \\ &\quad - [\sigma_i^q \frown \partial\sigma_j^{p+q} - (-1)^{p-1}\delta\sigma_i^q \wedge \partial\sigma_j^{p+q}] = 0.\end{aligned}$$

Hence C^p is a cycle in $\bar{\sigma}_j^{p+q}$, and (as $p > 0$) we may choose $\sigma_i^q \wedge \sigma_j^{p+q}$ with C^p as boundary.

We may form linear combinations of $\frown$ products (and also of $\smile$ products) by defining

$$(9.1)\qquad A(\alpha_1\frown_1 + \alpha_2\frown_2)B = \alpha_1(A\frown_1 B) + \alpha_2(A\frown_2 B).$$

By applying this to any $\sigma_i^q \frown \sigma_j^q$, we find

$$(9.2)\qquad \gamma(\alpha_1\frown_1 + \alpha_2\frown_2) = \alpha_1\gamma(\frown_1) + \alpha_2\gamma(\frown_2).$$

THEOREM 9. *For any two products $\frown$ and $\frown'$,*

$$(9.3)\qquad \gamma(\frown)(A^p \frown' B^q) \sim \gamma(\frown')(A^p \frown B^q) \quad \text{if} \quad \delta A^p = \partial B^q = 0.$$

As a consequence, if v^p and u^q denote cohomology and homology classes, then

$$(9.4)\qquad \gamma(\frown)(v^p \frown' u^q) = \gamma(\frown')(v^p \frown u^q).$$

To prove this, set $\frown'' = \gamma(\frown)\frown' - \gamma(\frown')\frown$. Then $\gamma(\frown'') = 0$. By Theorem 8, (R_2) or (R_3), $A^p \frown'' B^q \sim 0$, from which (9.3) follows.

Theorem 10. *Theorem 9 holds for the $\smile$ product. Thus*

$$\gamma(\smile)(A^p \smile' B^q) \backsim \gamma(\smile')(A^p \smile B^q) \quad \text{if} \quad \delta A^p = \delta B^q = 0. \tag{9.5}$$

As above, it is sufficient to show that if $\gamma(\smile) = 0$, then $A^p \smile B^q \backsim 0$ for all cocycles A^p, B^q. Let $\frown$ correspond to $\smile$; then $\gamma(\frown) = 0$, and we may construct $\wedge$. Construct a corresponding $\vee$ (using relations of the form (5.1), (5.2)); then, as in (5.4),

$$(D^{p+1} \vee E^q)\cdot F^{p+q} = D^{p+1}\cdot(E^q \wedge F^{p+q}). \tag{9.6}$$

Now for $p > 0$ and any cell $\sigma = \sigma_k^{p+q}$, as $A\cdot\partial C = \delta A\cdot C$,

$$\begin{aligned}(A \smile B)\cdot\sigma = A\cdot(B \frown \sigma) &= A\cdot[\partial(B \wedge \sigma) + (-1)^p\delta B \wedge \sigma + B \wedge \partial\sigma] \\ &= (\delta A \vee B)\cdot\sigma + (-1)^p(A \vee \delta B)\cdot\sigma + \delta(A \vee B)\cdot\sigma,\end{aligned}$$

and hence

$$A^p \smile B^q = \delta(A^p \vee B^q) + \delta A^p \vee B^q + (-1)^p A^p \vee \delta B^q; \tag{9.7}$$

if $p = 0$, then $A^0 \smile B^q = \delta A^0 \vee B^q$. Hence if A and B are cocycles, then $A \smile B \backsim 0$.

10. Proof of properties in §5

To prove (5.12), we define a new product, $\smile'$, by

$$A^p \smile' B^q = (-1)^{pq} B^q \smile A^p.$$

Clearly (P_1) holds for $\smile'$. To prove (P_2), (P_2) for $\smile$ gives

$$\begin{aligned}\delta(A^p \smile' B^q) &= (-1)^{pq}[\delta B^q \smile A^p + (-1)^q B^q \smile \delta A^p] \\ &= (-1)^{pq+(q+1)p} A^p \smile' \delta B^q + (-1)^{pq+q+q(p+1)} \delta A^p \smile' B^q,\end{aligned}$$

which reduces to the desired formula. (P_3) is a consequence of Theorem 3. (5.12) now follows from (9.5).

To prove (5.10) and (5.11), take a simplicial subdivision K' of K, and define new products in K, using those of §6 in K', by (15.3) and (15.6). The properties now follow at once from (6.2). (Compare the proof of Theorem 14.)

To prove Theorem 6, we apply (9.5) and (7.2): Using $\smile_0$ for the product of §6,[19]

$$f'(A' \smile B') \backsim \gamma(\smile) f'(A' \smile_0 B') = \gamma(\smile)(f'A' \smile_0 f'B') \backsim f'A' \smile f'B'.$$

The other relation is proved similarly.

---

[19] If $C' \backsim D'$, then $D' - C' = \delta E'$, and

$$f'D' - f'C' = f'\delta E' = \delta f'E', \qquad f'C' \backsim f'D'.$$

11. The products, general coefficient groups

Let G be an abelian group. Let $\mathrm{Ch}_Y(X)$ be the group of homomorphisms (characters) of the group X into the group Y. Then $H = \mathrm{Ch}_G(G)$ is a ring with the definitions (writing $h \cdot g$ for $h(g)$)

$$(11.1) \qquad (h + h')\cdot g = h\cdot g + h'\cdot g, \qquad (hh')\cdot g = h\cdot(h'\cdot g).$$

If we use the coefficient group G in chains with which we form cycles, then the most useful coefficients for chains from which cocycles are formed are those of H.[20]

EXAMPLES. Let I_0, I_m, R, $\mathfrak{R}$, R_1, $\mathfrak{R}_1$ denote the groups of integers, integers mod m, real numbers, rational numbers, reals mod 1, and rationals mod 1. All but the last two are rings. Then, using $\approx$ for isomorphic, examples of G and corresponding H are

(α) $G \approx I_0$, $H \approx I_0$.

(β) $G \approx I_m$, $H \approx I_m$ ($m \geqq 1$).

(γ) $G \approx R$ or $\mathfrak{R}$, $H \approx R$ or $\mathfrak{R}$.

(δ) $G \approx R_1$ or $\mathfrak{R}_1$, $H \approx I_0$.

(α) is a special case of (β). H is commutative in these cases. If G is a direct sum of such groups, we may find the corresponding H with the following rules:

$$\mathrm{Ch}_Y (X_1 + X_2) \approx \mathrm{Ch}_Y (X_1) + \mathrm{Ch}_Y (X_2).$$

$$\mathrm{Ch}_{Y_1+Y_2} (X) \approx \mathrm{Ch}_{Y_1} (X) + \mathrm{Ch}_{Y_2} (X).$$

Our object here is three-fold. (a) we point out how the products with G and H are defined in terms of the former products. (b) The products are discussed, assuming merely that they satisfy the (P_i) and the (Q_i). (c) In a simplicial complex, for certain groups G and H, when one of $\smile$, $\frown$ is known in the cohomology and homology groups, the other may be determined at once.

(a) If the products with integral coefficients are given, we may set

$$(11.2) \qquad (\textstyle\sum h_i\sigma_i^p)\cdot(\sum g_i\sigma_i^p) = \sum h_i\cdot g_i,$$

$$(11.3) \qquad h\sigma_i^p \smile h'\sigma_j^q = \textstyle\sum_k {}^{pq}\Gamma_k^{ij} hh'\sigma_k^{p+q},$$

$$(11.4) \qquad h\sigma_j^q \frown g\sigma_k^{p+q} = \textstyle\sum_l {}^{pq}\Gamma_k^{lj} h\cdot g\sigma_l^p.$$

The relations (P_2) and (Q_2), and hence (5.6) through (5.9), continue to hold, so the products are defined among the cohomology and homology groups. If u^p and v^p denote homology and cohomology classes, the explicit definitions are

[20] As $h\cdot(g + g') = h\cdot g + h\cdot g'$, if we consider H as an additive group, then H and G form a "group pair" with respect to G. We could replace H by any subring. Thus, in (γ) below, we could take $H \approx I_0$.

$$(11.5)\qquad v^pA^p\cdot u^pB^p = A^p\cdot B^p,$$

$$(11.6)\qquad v^pA^p \smile v^qB^q = v^{p+q}(A^p \smile B^q),$$

$$(11.7)\qquad v^qB^q \frown u^{p+q}C^{p+q} = u^p(B^q \frown C^{p+q}).$$

(5.4) holds, because of (11.1), and hence

$$(11.8)\qquad (v_i^p \smile v_j^q)\cdot u_k^{p+q} = v_i^p\cdot(v_j^q \frown u_k^{p+q}).$$

(P_3) and (Q_3) with $\gamma = 1$ are replaced by the relations

$$(11.9)\qquad hI \smile h'\sigma_j^q = hh'\sigma_j^q, \qquad I\cdot(h\sigma_j^q \frown g\sigma_j^q) = h\cdot g.$$

(We could leave out the h in the first.) Thus the integer $\gamma(\smile) = \gamma(\frown)$ is replaced by the operations $\gamma(\smile; h, h') = hh'$, $\gamma(\frown; h, g) = h\cdot g$; if G is the group of integers, then $\gamma = \gamma(\smile; 1, 1) = \gamma(\frown; 1, 1)$. To a general γ correspond general bilinear maps $\gamma(\smile; h, h')$ into H and $\gamma(\frown; h, g)$ into G, and (11.9) becomes

$$(11.10)\qquad hI \smile h'\sigma_j^q = \gamma(\smile; h, h')\sigma_j^q, \qquad I\cdot(h\sigma_j^q \frown g\sigma_j^q) = \gamma(\frown; h, g).$$

The relations (5.10) through (5.14) and (5.4) hold, if in (5.12) we assume that H is commutative. For the proofs, we need merely multiply by elements of H and G and sum, using (11.1). The same remark holds for Theorem 2. Suppose ${}^p\partial_i^j = \pm 1$ or 0 in the complex. With the above interpretation for γ, the proofs of Theorems 1 and 3 hold. (Note that only the revised (P_i) and (Q_i) are used here.)

(b) Suppose $\smile$ and $\frown$ products are defined, satisfying the analogues of (P_1) and (Q_1), (P_2) and (Q_2), and (11.10), which we call (P_3') and (Q_3'). That $\smile$ and $\frown$ are bilinear means that there are bilinear maps ${}^{pq}\Phi_k^{ij}$ of (H, G) into G and ${}^{pq}\Psi_k^{ij}$ of (H, H) into H such that

$$(11.11)\qquad h\sigma_i^p \smile h'\sigma_j^q = \textstyle\sum_k {}^{pq}\Psi_k^{ij}(h, h')\sigma_k^{p+q},$$

$$(11.12)\qquad h\sigma_j^q \frown g\sigma_k^{p+q} = \textstyle\sum_l {}^{pq}\Phi_k^{lj}(h, g)\sigma_l^p.$$

(P_2) and (Q_2) translate into (Γ_2) with ${}^{pq}\Gamma_k^{ij}$ replaced by ${}^{pq}\Psi_k^{ij}(h, h')$ and ${}^{pq}\Phi_k^{ij}(h, g)$ respectively. We suppose scalar products are defined by (11.2). We say $\smile$ and $\frown$ *correspond* if (5.4) holds. By (11.11) and (11.12), this is so if and only if

$$(11.13)\qquad {}^{pq}\Psi_k^{ij}(h, h')\cdot g = h\cdot{}^{pq}\Phi_k^{ij}(h', g).$$

(In the simplest case, $\Psi = \gamma hh'$ and $\Phi = \gamma h'\cdot g$ for some integer γ, and (11.13) follows from (11.1).) The products may not be derivable from the products in (a). For instance, if G is the group of real numbers, the ${}^{pq}\Gamma_k^{ij}$ may not be integers.

If $\frown$ *is given, and* $H = \mathrm{Ch}_G(G)$, *then* (11.13) *determines* $\smile$. For, given h and h', the right hand side is a homomorphism of G into itself, and thus determines an element of H; we call this element ${}^{pq}\Psi_k^{ij}(h, h')$. Clearly Ψ is bilinear, and (11.13) holds. If we take (Γ_2) with $\Phi(h', g)$ and apply h, using (11.13) gives a relation with terms acting on g. It being true for all g, we can drop it, and find (Γ_2) with $\Psi(h, h')$, i.e. (P_2). (Q_1) gives (P_1). By (Q_3'), i.e. the second relation in (11.10), and (5.4) (which follows from (11.13)),

$$(hI \smile h'\sigma_j^q)\cdot g\sigma_j^q = hI\cdot(h'\sigma_j^q \frown g\sigma_j^q) = h\cdot\gamma(\frown; h', g).$$

For fixed h and h', the last term is a map of G into itself, defining an element of H, which we call $\gamma(\smile; h, h')$; thus

$$\gamma(\smile; h, h')\cdot g = h\cdot\gamma(\frown; h', g). \tag{11.14}$$

Clearly $\gamma(\smile; h, h')$ is bilinear. If h^* is the coefficient of $hI \smile h'\sigma_j^q$ in σ_j^q, then the above relations give

$$h^*\cdot g = h^*\sigma_j^q\cdot g\sigma_j^q = \gamma(\smile; h, h')\cdot g, \qquad h^* = \gamma(\smile; h, h').$$

Hence, by (Q_1), the first relation in (11.10), i.e. (P_3'), holds.

Let 1 be the identity in H: $1\cdot g = g$ (all g). Consider the following *hypothesis* (H) on H, or on any subring containing 1: The only homomorphism ϕ of H into G (i.e. $\phi(h_1 + h_2) = \phi(h_1) + \phi(h_2)$) such that $\phi(1) = 0$ is $\phi(h) = 0$ (all h). This holds for instance if H is any of the rings given above. If $G = I_0 + I_0 = I_0^2$, we may use, in place of $H \approx I_0^4$, a subring $H' \approx I_0$.

If $\smile$ *is given, and* H *satisfies* (H), *then* $\frown$ *is determined.* For fixed h' and g, the left hand side of (11.13) is a homomorphism ϕ of H into G. Set $g^* = \phi(1)$. Then for all h in H, $\phi(h) = h\cdot g^*$; for if $\phi'(h) = \phi(h) - h\cdot g^*$, then $\phi'(1) = 0$. Set ${}^{pq}\Phi_k^{ij}(h', g) = g^*$; Φ is bilinear, and (11.13) holds. Applying (Γ_2) with $\Psi(h, h')$ to g and using (11.13) gives a relation for all h. Setting $h = 1$ gives (Q_2). (P_1) gives (Q_1). By (5.4) and the first relation in (11.10),

$$hI\cdot(h'\sigma_j^q \frown g\sigma_j^q) = \gamma(\smile; h, h')\sigma_j^q\cdot g\sigma_j^q = \gamma(\smile; h, h')\cdot g.$$

If we set $\gamma(\frown; h', g) = \gamma(\smile; 1, h')\cdot g$, then $\gamma(\frown; h', g)$ is bilinear and (Q_3') holds (as $1\, I\cdot g = I\cdot g$).

THEOREM 11. *Let K be a simplicial complex. Then with coefficient groups G and $H = \mathrm{Ch}_G(G)$, any two $\frown$ products satisfying the new (Q_i), with the same $\gamma(\frown; h, g)$, give the same product in the cohomology and homology groups. The same is true for $\smile$ if $H = \mathrm{Ch}_G(G)$ satisfies the hypothesis (H).*

Recall that (P_3') and (Q_3') are (11.10). First take a fixed vertex x_i^p in each simplex σ_i^p, and define the "join" $x_i^p\sigma_j^q$ for any σ_j^q in $\bar\sigma_i^p$ (which vanishes if σ_j^q contains x_i^p). Set $x_i^p \sum g_j\sigma_j^q = \sum g_j x_i^p\sigma_j^q$. Then if C^q is a q-G-cycle in $\bar\sigma_i^p$, $\partial(x_i^pC^q) = C^q$. (See for instance Lefschetz, [10], p. 111, (9').)

Now given $\frown_1$ and $\frown_2$ with the same γ, set $\frown = \frown_2 - \frown_1$. Then $\gamma(\frown; h, g) = 0$ (all h, g). We shall construct a bilinear product $h\sigma_i^q \wedge g\sigma_j^{p+q}$ as in Theorem 8. The proof there given holds, if we are careful to define the product for all h and g at each step. At a typical step, we have a p-G-cycle

$C^p = h\sigma_i^q \frown g\sigma_j^{p+q} + \cdots$ in $\bar{\sigma}_j^{p+q}$. To make $\partial(h\sigma_i^q \wedge g\sigma_k^{p+q}) = C^p$, we set $h\sigma_i^q \wedge g\sigma_j^{p+q} = x_j^{p+q} C^q$. Because of the new ($R_2$) and ($R_3$), the $\frown$ product vanishes in the cohomology and homology groups; it follows that $\frown_1$ and $\frown_2$ give the same products there.

To prove the last statement, we shall show that if $\gamma(\smile; h, h') = 0$, then $A \smile B \backsim 0$ for cocycles A and B. Construct $\frown$ corresponding to $\smile$; then by (11.14), $\gamma(\frown; h', g) = 0$. Hence, as we saw above, a bilinear $\wedge$ operation may be defined satisfying the revised (R_i). Next we shall construct $\vee$ so that (9.6) will hold. In terms of the coefficients defining $\wedge$ and $\vee$, this takes the form of the relation (11.13). The right hand side being given, we construct the coefficients on the left so the relation is satisfied, exactly as we constructed $\smile$ in terms of $\frown$. We can now show that $(A \smile B)\cdot g\sigma = \delta(A \vee B)\cdot g\sigma$ for cocycles A and B, as in the proof of Theorem 10. As this is true for all g, $A \smile B$ and $\delta(A \vee B)$ have the same coefficient in σ. As this is true for all σ, $A \smile B = \delta(A \vee B) \backsim 0$.

(c) For certain sets of groups G, H and Z, the homology and cohomology groups ${}^p\mathbf{H}^G$ and ${}^p\mathbf{H}_H$ satisfy

$$(11.15) \qquad {}^p\mathbf{H}^G \approx \mathrm{Ch}_Z({}^p\mathbf{H}_H), \qquad {}^p\mathbf{H}_H \approx \mathrm{Ch}_Z({}^p\mathbf{H}^G).$$

As shown in Whitney, [13], this is so whenever the following conditions are satisfied.

(1) G and H form a group pair with respect to Z.
(2) $G \approx \mathrm{Ch}_Z(H)$, $H \approx \mathrm{Ch}_Z(G)$.
(3) G and H resolve each other completely.
(4) Z is infinitely (better term: completely) divisible.

(11.15) holds for any of the examples (β), (γ), (δ) above, with $Z = G$. We can show this for (β) by first using $Z = R_1$, and noting that in the maps, only a subgroup $\approx G$ of R_1 is used.

THEOREM 12. *Let the products* (11.5) *be defined, and suppose* (11.15) *is satisfied. Then, for any* γ, *as soon as one of* $\smile$, $\frown$ *is known, the other is determined by* (11.8).

Recall that for any γ, $\smile$ and $\frown$ are uniquely defined. We shall show that, given $\smile$ or $\frown$, there is a unique corresponding $\frown$ or $\smile$ satisfying (11.8); as the correct products satisfy this relation, the theorem will be proved.

Suppose $\smile$ is defined. For fixed v_j^q and u_k^{p+q}, the left hand side of (11.8) is a homomorphism of ${}^p\mathbf{H}_H$ into Z, and hence, by (11.15), corresponds to a unique element of ${}^p\mathbf{H}^G$; we call this element $v_j^q \frown v_k^{p+q}$. Then $\frown$ is bilinear, and (11.8) is satisfied. We find $\smile$ in terms of $\frown$ in the same manner.

12. Construction of the products in low dimensional complexes

We shall construct all products $\sigma^q \frown \sigma^{p+q}$ for $p \leqq 2$ in a particular fashion; then $\sigma^p \smile \sigma^q$ is determined for $p \leqq 2$. We can then determine the $\smile$ products in the cohomology and homology groups of complexes of dimensions $\leqq 5$, with the help of Theorem 12 and (5.12); the remaining $\frown$ products are then given with the aid of Theorem 12.

We begin by ordering the vertices of K in a fixed fashion, $x_1, x_2, \cdots$. For each σ_i^p, let V_i^p be its first vertex. For each vertex x_h of σ_i^p, let $W_i^p x_h$ be a 1-chain in $\bar{\sigma}_i^p$ whose boundary is $x_h - V_i^p$. Set (using $\gamma = 1$)

$$\sigma_i^p \frown \sigma_i^p = V_i^p, \tag{12.1}$$

$$\sigma_j^{p-1} \frown \sigma_i^p = {}^p\partial_i^j W_i^p V_j^{p-1} \qquad (\sigma_j^{p-1} \text{ in } \bar{\sigma}_i^p). \tag{12.2}$$

Finally, take any σ_j^{p-2} of $\bar{\sigma}_i^p$; say

$$\partial\sigma_i^p = \sum \alpha_k \sigma_k^{p-1}, \qquad \delta\sigma_j^{p-2} = \sum \beta_k \sigma_k^{p-1} \text{ in } \bar{\sigma}_i^p.$$

As $\partial\partial\sigma_i^p = 0$, $\sum \alpha_k\beta_k = 0$. Now (see §8)

$$C^1 = \delta\sigma_j^{p-2} \frown \sigma_i^p + \sigma_j^{p-2} \frown \partial\sigma_i^p = \sum \beta_k W_i^p V_k^{p-1} + \sum \alpha_k W_k^{p-1} V_j^{p-2}$$

is a cycle in $\bar{\sigma}_i^p$, and we may choose a 2-chain $\sigma_j^{p-2} \frown \sigma_i^p$ bounded by it.

In the simplest case, we will have

$$\partial\sigma_i^p = \sigma_s^{p-1} + \sigma_t^{p-1} + \cdots, \qquad \delta\sigma_j^{p-2} = \sigma_s^{p-1} - \sigma_t^{p-1} + \cdots,$$

and

$$C^1 = W_i^p V_s^{p-1} - W_i^p V_t^{p-1} + W_s^{p-1} V_j^{p-2} + W_t^{p-1} V_j^{p-2},$$

and we can find a possible $\sigma_j^{p-2} \frown \sigma_i^p$ at once. For most σ_j^{p-2} on σ_i^p (for instance, for all in which $V_j^{p-2} = V_s^{p-1} = V_t^{p-1}$), C^1 will vanish; but there will in general be at least one for which it does not.

III. Invariance Theorems

13. Subdivision and consolidation

Let K' be a complex, satisfying (K_1), (K_2) and (K_3) of §2. Let $\{E_i^p\}$ be a set of distinct closed subcomplexes of K' which cover K', E_i^p being of dimension p. Let $F_i^p = F(E_i^p)$ be the union of all E_j^q ($q < p$) contained in E_i^p. Set $O_i^p = E_i^p - F_i^p$. Assume

(K_1') The common part $E_i^p \cdot E_j^q$ of any two of the subcomplexes is either void or the union of a subset of the subcomplexes.

(K_2') If E_j^q is in E_i^p, $E_j^q \neq E_i^p$, then $q < p$.

(K_3') With integer coefficients, O_i^p is monocyclic or acyclic[21] in the dimension p and is acyclic in all lower dimensions.

[21] O_i^p is *acyclic* in the dimension q if every q-cycle (which need not be boundary-like) is a boundary; O_i^p is *monocyclic* in the dimension p if there is a p-cycle $X^p \neq 0$ such that any p-cycle is a multiple of X^p. (All this is with integer coefficients.) Compare Tucker: "cell-like," "null-like." In [4], Ch. VI, a similar assumption, using any coefficient group G, is made. Note that O_i^p is a subcomplex of K; a chain A in O_i^p is a cycle if, considered as a chain in K, ∂A^p has no part in O_i^p.

Define a complex K as follows. With each O_i^p which is monocyclic we associate a p-cell σ_i^p. Say σ_j^q is a face of σ_i^p if and only if E_j^q is in E_i^p. We shall define the ${}^p\partial_j^i$ in K and prove (K_1), (K_2) and (K_3) for K below. We call K a *consolidation*[22] of K', and K', a *subdivision* of K. If K may be formed from a simplicial K' in this manner, we say K *admits a simplicial subdivision.*

14. Structure of K and K'

Denote the cells of K' by τ_i^p. Let J^p be the union of all E_i^q for $q \leqq p$. We prove:

(α_1) O_i^p contains all the p-cells of E_i^p; for F_i^p is of dimension $<p$.

(α_2) $E_i^p \cdot E_j^p$ is in J^{p-1} if $j \neq i$; for the common part is a union of E_k^q, and as each such E_k^q is in E_i^p, $q < p$, by (K_2').

(α_3) Each τ_i^p is in a unique O_j^q; then E_j^q is the smallest subcomplex containing τ_i^p. To prove this, choose E_j^q as stated. E_j^q is uniquely determined. For if τ_i^p is also in E_k^r, then it is in their common part and hence in an E_l^s contained in both, by (K_1'); hence $E_l^s = E_j^q$, and E_j^q is in E_k^r. As τ_i^p is in no E_k^{q-1}, it is in O_j^q. Suppose τ_i^p were also in $O_k^r \neq O_j^q$. Then E_k^r contains E_j^q, and by (K_2'), $r > q$. Hence E_j^q is in F_k^r, and τ_i^p is not in O_k^r, a contradiction.

In (α_3), we define $E_j^q = E(\tau_i^p)$, $O_j^q = O(\tau_i^p)$. If O_j^q is monocyclic (in the dimension q), we set $\sigma_j^q = \sigma(\tau_i^p)$.

(α_4) If τ is in O_i^p and in E_j^q, then E_j^q contains E_i^p, by (α_3).

For each O_i^p which is monocyclic, let

$$X_i^p = \sum\nolimits_j {}^pa_i^j \tau_j^p \tag{14.1}$$

be the corresponding p-cycle; for other O_i^p, set $X_i^p = 0$. Map $L^p(K)$ into $L^p(K')$ by

$$Sd\sigma_i^p = X_i^p. \tag{14.2}$$

Define $Sd \sum \alpha_i \sigma_i^p$ by linearity. Then

(α_5) $SdA^p = 0$ implies $A^p = 0$. For if $A^p = \sum \alpha_i \sigma_i^p$, then $\sum \alpha_i X_i^p = 0$, and (α_2) shows that each $\alpha_i X_i^p = 0$; as $X_i^p \neq 0$ in this case, $\alpha_i = 0$.

LEMMA 4. *Any cycle A'^p in J^p (with integer coefficients) is SdA^p for a uniquely defined A^p in K.*

First, write $A'^p = \sum A_i'^p$, $A_i'^p$ in E_i^p. As $A'^p - A_i'^p$ is in $\sum_{j \neq i} E_j^p$, by (α_2), so is $-\partial(A'^p - A_i'^p) = \partial A_i'^p$. By ($\alpha_4$), $\partial A_i'^p$ has no part in O_i^p, that is (using (α_1)), $A_i'^p$ is a p-cycle in O_i^p. Hence, if $A_i'^p \neq 0$, then O_i^p is monocyclic, and for some α_i,

$$A_i'^p = \alpha_i X_i^p; \qquad \text{then } A'^p = Sd \sum \alpha_i \sigma_i^p.$$

The uniqueness follows from (α_5). We remark that "in J^p" may be replaced by "in $J^p - J^{p-1}$."

---

[22] "*Zellenzerspaltung*" in [4], Ch. VI. But see the last foot-note.

We now define the boundary relations in K. Given σ_i^p, $\partial Sd\sigma_i^p$ is a cycle in F_i^p, which is in J^{p-1}; hence it may be written uniquely as SdB^{p-1}. *We define* $\partial\sigma_i^p = B^{p-1}$. Then

$$\partial SdA^p = Sd\partial A^p. \tag{14.3}$$

Clearly (K_1) holds for K. Also $\partial Sd\sigma_i^p$ is in E_i^p, hence $\partial\sigma_i^p$ is in $\bar{\sigma}_i^p$, and (K_2) holds. For any σ_i^p, $Sd\partial\partial\sigma_i^p = \partial\partial Sd\sigma_i^p = 0$; hence, by ($\alpha_5$), $\partial\partial\sigma_i^p = 0$, and ($K_3$) holds.

THEOREM 13. *If K' is a subdivision of K, then there are homomorphisms ϕ of $L^p(K')$ into $L^p(K)$ and ψ of $L^p(K')$ into $L^{p+1}(K')$ such that*

(a) *$\phi\tau^p$ is in $\bar{\sigma}(\tau^p)$; $\psi\tau^p$ is in $E(\tau^p)$.*

(b) $\partial\phi A'^p = \phi\partial A'^p$.

(c) $\phi SdA^p = A^p$.

(d) $Sd\phi A'^p = A'^p - \psi\partial A'^p - \partial\psi A'^p$.

If we take duals of these relations and (14.3), using:

$\phi' = D(\phi)$, mapping $L^p(K)$ into $L^p(K')$,

$Sd' = D(Sd)$, mapping $L^p(K')$ into $L^p(K)$,

$\psi' = D(\psi)$, mapping $L^{p+1}(K')$ into $L^p(K')$,

we obtain

(b′) $\delta\phi' A^p = \phi'\delta A^p$,

(c′) $Sd'\phi' A^p = A^p$,

(d′) $\phi' Sd' A'^p = A'^p - \psi'\delta A'^p - \delta\psi' A'^p$,

$$\delta Sd' A'^p = Sd'\delta A'^p. \tag{14.3′}$$

We begin by constructing homomorphisms θ_p and ψ_p as follows. θ_p and ψ_p are defined in J^p, and map $L^r(K')$ into $L^r(K')$ and into $L^{r+1}(K')$ respectively. For τ^q in J^{p-1}, set $\theta_p\tau^q = \tau^q$, $\psi_p\tau^q = 0$. If τ^q is in O_i^p, then we shall have:

$$\psi_p\tau^q \text{ is in } O_i^p, \qquad \theta_p\tau^q \text{ is in } \begin{cases} F_i^p \text{ if } q < p, \\ E_i^p \text{ if } q = p. \end{cases}$$

Also

$$\partial\psi_p\tau^q = \tau^q - \psi_p\partial\tau^q - \theta_p\tau^q. \tag{14.4}$$

For τ^0 in J^0, set $\theta_0\tau^0 = \tau^0$, $\psi_0\tau^0 = 0$. Suppose all θ_r and ψ_r are constructed in J^{p-1}; we shall construct them in J^p. We need merely consider θ_p and ψ_p; for (14.4) holds for θ_r and ψ_r, $r > p$, by their definitions. For τ^0 in O_i^p, we may choose $\psi_p\tau^0$ in O_i^p and $\theta_p\tau^0$ in F_i^p so that

$$\partial\psi_p\tau^0 = \tau^0 - \theta_p\tau^0,$$

by (K_3'); then (14.4) holds.

Now suppose ψ_p and θ_p are defined for all cells of dimension $< q$ in O_i^p ; we shall define them for τ^q. If $\partial_1\tau^q$ is the part of $\partial\tau^q$ in O_i^p, $\psi_p\partial_1\tau^q$ is in O_i^p ; if $\partial_2\tau^q$ is the part in F_i^p, $\psi_p\partial_2\tau^q = 0$. Hence $\psi_p\partial\tau^q$ is in O_i^p. Also, using (14.4) applied to $\partial\tau^q$,

$$\partial(\tau^q - \psi_p\partial\tau^q) = \partial\tau^q - (\partial\tau^q - \theta_p\partial\tau^q) = \theta_p\partial\tau^q, \tag{14.5}$$

which is in F_i^p. Hence $\tau^q - \psi_p\partial\tau^q$, as a chain in the complex O_i^p, is a cycle.

Suppose first that $q < p$. Then, by (K_3'), we may find a chain $\psi_p\tau^q$ in O_i^p and a chain $\theta_p\tau^q$ in F_i^p so that (14.4) holds. Suppose next that $q = p$. Then if $\tau^q = \tau_j^p$,

$$\tau_j^p - \psi_p\partial\tau_j^p = \alpha_j X_i^p, \tag{14.6}$$

by (K_3'). If $X_i^p = 0$, we take $\alpha_j = 0$. Set $\psi_p\tau_j^p = 0$, $\theta_p\tau_j^p = \alpha_j X_i^p$; again (14.4) follows.

We prove some properties of ψ_p and θ_p. Taking the boundary of (14.4) gives

$$\begin{aligned} 0 &= \partial\tau^q - (\partial\tau^q - \psi_p\partial\partial\tau^q - \theta_p\partial\tau^q) - \partial\theta_p\tau^q, \\ \partial\theta_p\tau^q &= \theta_p\partial\tau^q. \end{aligned} \tag{14.7}$$

Next, by (14.1) and (14.6), summing over all p-cells in O_i^p,

$$\textstyle\sum_j \alpha_j{}^p a_i^j X_i^p = \sum_j {}^p a_i^j(\tau_j^p - \psi_p\partial\tau_j^p) = X_i^p - \psi_p\partial X_i^p.$$

But ∂X_i^p is in F_i^p, and hence $\psi_p\partial X_i^p = 0$. Therefore

$$\textstyle\sum_j \alpha_j{}^p a_i^j = 1 \qquad \text{if } X_i^p \neq 0. \tag{14.8}$$

It follows that, even if $X_i^p = 0$,

$$\theta_p X_i^p = \textstyle\sum_j {}^p a_i^j \theta_p\tau_j^p = \sum_j {}^p a_i^j \alpha_j X_i^p = X_i^p. \tag{14.9}$$

Now define a homomorphism θ of $L^q(K')$ into $L^q(J^q)$ by

$$\theta\tau^q = \theta_q\theta_{q+1}\cdots\theta_n\tau^q, \tag{14.10}$$

supposing K' is of dimension n. By the definition of $\theta_p\tau^q$ for $p = q$, we may define ϕ in $L^q(K')$ so that

$$\theta\tau^q = Sd\phi\tau^q. \tag{14.11}$$

By (14.3), (14.9) and (14.7),

$$\partial\theta\tau^q = Sd\partial\phi\tau^q = \theta_{q-1}Sd\partial\phi\tau^q = \theta_{q-1}\partial\theta\tau^q = \theta_{q-1}\theta_q\cdots\theta_n\partial\tau^q = \theta\partial\tau^q; \tag{14.12}$$

hence

$$Sd\partial\phi\tau^q = \partial Sd\phi\tau^q = \partial\theta\tau^q = Sd\phi\partial\tau^q,$$

and (b) follows from (α_5).

Finally, define ψ by

$$\psi\tau^q = \textstyle\sum_{k=q}^n \psi_k\theta_{k+1}\cdots\theta_n\tau^q, \tag{14.13}$$

using $\theta_{n+1} \cdots \theta_n \tau^q = \tau^q$. As $k + 1 > q$ in each term of the sum, $\theta_{k+1} \cdots \theta_n \tau^q$ is in J^k, so that $\psi_k \theta_{k+1} \cdots \theta_n \tau^q$ is defined. Note that $\psi_{q-1}\theta_q \cdots \theta_n \partial \tau^q = 0$, as $\theta_q \cdots \theta_n \partial \tau^q$ is of dimension $q - 1$. Hence, by (14.4) and (14.7),

$$\begin{aligned} \partial\psi\tau^q &= \sum_{k=q}^{n} \partial\psi_k \theta_{k+1} \cdots \theta_n \tau^q \\ &= \sum_{k=q}^{n} \theta_{k+1} \cdots \theta_n \tau^q - \sum_{k=q}^{n} \psi_k \partial\theta_{k+1} \cdots \theta_n \tau^q \\ &\quad - \sum_{k=q}^{n} \theta_k \theta_{k+1} \cdots \theta_n \tau^q = \tau^q - \psi\partial\tau^q - \theta\tau^q; \end{aligned}$$

this, with (14.11), gives (d).

To prove (c), (14.9) gives

$$Sd\phi Sd\sigma_i^p = Sd\phi X_i^p = \theta X_i^p = \theta_p X_i^p = X_i^p = Sd\sigma_i^p;$$

(c) now follows from (α_5).

15. Combinatorial invariance

We can now prove (compare Theorem 13)

THEOREM 14. *Let K' be a subdivision of K, and let G be an abelian group. Then both ϕ and Sd induce isomorphisms between ${}^p\mathbf{H}^G(K)$ and ${}^p\mathbf{H}^G(K')$, and both ϕ' and Sd' induce isomorphisms between ${}^p\mathbf{H}_G(K)$ and ${}^p\mathbf{H}_G(K')$. If K and K' each admits a product theory, then these isomorphisms preserve products.*[23]

The meaning of the last phrase is seen from (15.4), (15.5), (15.7) and (15.8) below.

Set $Sd \sum g_i\sigma_i^p = \sum g_i Sd\sigma_i^p$ etc. Using u and v for homology and cohomology classes as in §11, set

(15.1) $\qquad \phi u'A'^p = u\phi A'^p, \qquad SduA^p = u'SdA^p,$

(15.2) $\qquad \phi' vA^p = v'\phi' A^p, \qquad Sd'v'A'^p = vSd'A'^p,$

the chains being cycles in (15.1) and cocycles in (15.2). The proof that these are isomorphisms follows at once from Theorem 13. Consider for instance ϕ'. To show that $\phi' v$ is uniquely determined, suppose $vA^p = vB^p$. Then $A^p - B^p = \delta C^{p-1}$, and

$$v'\phi'A^p - v'\phi'B^p = v'\phi'\delta C^{p-1} = v'\delta\phi' C^{p-1} = 0.$$

Suppose $\phi' vA^p = \phi' vB^p$. Then $\phi' A^p - \phi' B^p = \delta C'^{p-1}$, and

$$A^p - B^p = Sd'\phi'(A^p - B^p) = Sd'\delta C'^{p-1} = \delta Sd' C'^{p-1},$$

so that $vA^p = vB^p$. Given a $v'A'^p$, to find a vA^p mapping into it, set $A^p = Sd'A'^p$. Then $\delta A^p = Sd'\delta A'^p = 0$, and

$$\phi' vA^p = v'\phi' Sd'A'^p = v'(A'^p - \psi'\delta A'^p - \delta\psi' A'^p) = v'A'^p.$$

[23] If topological coefficient groups are used, the isomorphisms are continuous. The part of this theorem relating to the homology groups has been proved by Tucker, [12], for integral coefficients, and by Alexandroff-Hopf, [4], Ch. VI, using the stronger condition noted in footnote 21.

To prove that the $\smile$ products agree in the cohomology groups of K and K' (if they are defined), consider first integer coefficients. Take a definite $\smile$ product in K', and define a new one in K by

$$A \smile B = Sd'(\phi' A \smile \phi' B). \tag{15.3}$$

We must prove (P_1) through (P_3). To prove (P_1), say

$$\sigma_i^p \smile \sigma_j^q = \alpha\sigma_k^r + \cdots, \qquad \alpha \neq 0 \qquad (r = p + q).$$

Then

$$\phi'\sigma_i^p \smile \phi'\sigma_j^q = \beta\tau^r + \cdots, \qquad Sd'\tau^r = \gamma\sigma_k^r, \qquad \beta \neq 0, \gamma \neq 0.$$

As then $Sd\sigma_k^r = \gamma\tau^r + \cdots$, τ^r is in O_k^r, and $\sigma(\tau^r) = \sigma_k^r$. Say

$$\phi'\sigma_i^p = \epsilon\tau^p + \cdots, \qquad \phi'\sigma_j^q = \zeta\tau^q + \cdots, \qquad \tau^p \smile \tau^q = \eta\tau^r + \cdots$$
$$(\epsilon, \zeta, \eta \neq 0).$$

As then $\phi\tau^p = \epsilon\sigma_i^p + \cdots$, σ_i^p is in $\bar{\sigma}(\tau^p)$. As τ^p is in $\bar{\tau}^r$, $E(\tau^p)$ is in $E(\tau^r)$, and σ_i^p is in $\bar{\sigma}(\tau^r) = \bar{\sigma}_k^r$. Similarly σ_j^q is in $\bar{\sigma}_k^r$.

To prove (P_2), (b′) and (14.3′) with (P_2) for $\smile$ in K' give

$$\delta(A \smile B) = Sd'(\phi'\delta A \smile \phi' B \pm \phi' A \smile \phi'\delta B) = \delta A \smile B \pm A \smile \delta B.$$

To prove (P_3), note that $\phi' I = I'$; hence, using (P_3) in K',

$$I \smile \sigma_j^q = Sd'(\phi' I \smile \phi'\sigma_j^q) = Sd'\phi'\sigma_j^q = \sigma_j^q.$$

Therefore we may use the product of (15.3) in K. To prove that ϕ' and Sd' preserve products in the cohomology groups, note that (for cocycles), using (d′) and (5.6), (5.7),

$$\phi'(A \smile B) = \phi' Sd'(\phi' A \smile \phi' B) \backsim \phi' A \smile \phi' B, \tag{15.4}$$

$$Sd'A' \smile Sd'B' = Sd'(\phi' Sd'A' \smile \phi' Sd'B') \backsim Sd'(A' \smile B'). \tag{15.5}$$

We may similarly define $\frown$ in K in terms of $\frown$ in K' by

$$A \frown B = \phi(\phi' A \frown SdB). \tag{15.6}$$

This corresponds to the $\smile$ product in K, and hence is a $\frown$ product. To show this, (4.6) gives

$$(A \smile B)\cdot C = Sd'(\phi' A \smile \phi' B)\cdot C = (\phi' A \smile \phi' B)\cdot SdC,$$
$$A\cdot(B \frown C) = A\cdot\phi(\phi' B \frown SdC) = \phi' A\cdot(\phi' B \frown SdC).$$

Applying (5.4) in K' shows that these are equal; hence $\smile$ and $\frown$ correspond in K, by (5.4).

We map cycles [cocycles] from K into K' and from K' into K with Sd and ϕ [with ϕ' and Sd']. The invariance of the $\frown$ product is given by

$$(15.7) \qquad Sd(A \frown B) = Sd\phi(\phi' A \frown SdB) \sim \phi' A \frown SdB,$$

$$(15.8) \qquad Sd' A' \frown \phi B' = \phi(\phi' Sd' A' \frown Sd\phi B') \sim \phi(A' \frown B'),$$

where A, A' are cocycles, B, B' are cycles.

Now consider the coefficient group G as in §11, (a). The $\smile$ products in the cohomology groups in both K and K' are formed by (11.3). As (15.3) holds with A and B replaced by ${}^p h\sigma_i^p$ and ${}^q h\sigma_j^q$, (15.3) etc. hold with any coefficient group. (15.4) and (15.5) prove the invariance. Similarly for the $\frown$ product.

16. Topological invariance

We shall show how to associate homology and cohomology groups and a product theory with a polyhedron P by means of any simplicial subdivision. By Theorem 14 we may find these groups and products, using any complex K which admits, as a subdivision, a simplicial triangulation of P, and admits a product theory.

The theorem and proof extend at once to prove the existence of groups and products in a bicompact space; compare Steenrod (see footnote 8), §9.

The proof is based on Theorem 13. However, if we restrict ourselves to simplicial complexes, Theorem 13 becomes much more simple. Hence we give it again for this case, as a lemma. We may then prove, as before, the first part of Theorem 14.

LEMMA 6. *Let K' be a simplicial subdivision of the simplicial complex K, and let ϕ be a pseudo-identical map*[24] *of K' into K. Then there is a map ψ as in Theorem* 13 *such that the conclusions of Theorem* 13 *hold.*

The statements about ϕ and Sd are well-known; we shall construct ψ. For each vertex x' of K', let $\psi x'$ be any 1-chain in the subdivision of the smallest cell $\sigma(x')$ of K containing x', which is bounded by $x' - \phi x'$. Suppose ψ is constructed in $L^0(K'), \cdots, L^{p-1}(K')$. Then applying (d) of Theorem 13 to the $(p - 1)$-chain $\partial\tau^p$, we find

$$\partial(Sd\phi\tau^p - \tau^p + \psi\partial\tau^p) = \partial Sd\phi\tau^p - \partial\tau^p + \partial\tau^p - Sd\phi\partial\tau^p = 0.$$

Hence $Sd\phi\tau^p - \tau^p + \psi\partial\tau^p$ is a p-cycle in the subdivision of $\sigma(\tau^p)$ $(p > 0)$, and therefore, as is well-known, we can find a chain $-\psi\tau^p$ there bounded by it. The ψ as thus constructed clearly has the required properties.

THEOREM 15. *Let K and K' be simplicial triangulations of homeomorphic polyhedra P and P'. Then there are isomorphisms between ${}^p\mathbf{H}^G(K)$ and ${}^p\mathbf{H}^G(K')$ and between ${}^p\mathbf{H}_G(K)$ and ${}^p\mathbf{H}_G(K')$ which preserve the $\smile$ and $\frown$ products.*

[24] That is, a simplicial map such that each vertex x' of K' goes into a vertex of a cell of K containing it.

By identifying corresponding points of P and P', we may suppose K and K' are triangulations of P. We may subdivide K' into K_1', then K into K_1, and then K_1' into K_2', so that the following conditions hold.[25] There are simplicial maps ψ_2, ϕ_1, ψ_1, of K_2' into K_1, K_1 into K_1', and K_1' into K, such that $\Psi = \phi_1\psi_2$ and $\Phi = \psi_1\phi_1$ are pseudo-identical. The duals ψ_1', ϕ_1', ψ_2' of ψ_1, ϕ_1, ψ_2 induce homomorphisms of the cohomology groups of K, K_1', K_1 into those of K_1', K_1, K_2'. Combining pairs of these homomorphisms gives homomorphisms of ${}^p\mathbf{H}_G(K)$ into ${}^p\mathbf{H}_G(K_1)$ and of ${}^p\mathbf{H}_G(K_1')$ into ${}^p\mathbf{H}_G(K_2')$. These are induced by the duals Φ' and Ψ' of Φ and Ψ, and are isomorphisms, by Theorem 14.[26] Set $\psi^* = (\Phi')^{-1}\phi_1'$; then ψ_1' and ψ^* map ${}^p\mathbf{H}_G(K)$ into ${}^p\mathbf{H}_G(K_1')$ and vice versa. Further, using E for the identity, we have, in the cohomology groups,

$$\begin{aligned}\psi^*\psi_1' &= (\phi_1'\psi_1')^{-1}\phi_1'\psi_1' = E,\\ \psi_1'\psi^* &= (\Psi'^{-1}\Psi')\psi_1'(\phi_1'\psi_1')^{-1}\phi_1'\\ &= (\psi_2'\phi_1')^{-1}\psi_2'(\phi_1'\psi_1')(\phi_1'\psi_1')^{-1}\phi_1' = E.\end{aligned}$$

It follows that ψ_1' and ψ^* are isomorphisms (see for instance [4], p. 558). Combining this isomorphism with an isomorphism χ between ${}^p\mathbf{H}_G(K_1')$ and ${}^p\mathbf{H}_G(K')$ (Theorem 14) gives an isomorphism θ between ${}^p\mathbf{H}_G(K)$ and ${}^p\mathbf{H}_G(K')$. By the same process it is seen that ψ_1 induces an isomorphism between ${}^p\mathbf{H}^G(K_1')$ and ${}^p\mathbf{H}^G(K)$. As ψ_1 is simplicial, the isomorphisms induced by ψ_1' and ψ_1 preserve products (see §7). The same is true of χ and its dual, and hence of θ and its dual. This completes the proof.

IV. Manifolds

17. Dual complexes in a manifold

Let K be a subdivision of a closed oriented combinatorial manifold,[27] and let K' be the first derived (simplicial) subdivision of K. Order the vertices of K' by choosing first the vertices of K, next the centers of 1-cells of K, etc. Say K is of dimension n. For each cell σ_i^p of K, let E_i^{n-p} be the subcomplex of K' containing all $(n - p)$-cells of K' which have the center of σ_i^p as their first vertex, and all faces of these cells. The hypotheses on K show that the complex K^* thus formed, the "dual" of K, admits K' as a simplicial subdivision (see §13), and admits a product theory. The maps Sd and Sd^* of $L^p(K)$ and $L^p(K^*)$ into $L^p(K')$ are defined in the natural manner; for the latter, see (19.4). Define ϕ and ϕ^* as in Part III.

[25] See J. W. Alexander, *Combinatorial Analysis Situs*, Trans. Am. Math. Soc., vol. 28 (1926), pp. 308–310.

[26] Hence $\Phi' = \phi_1'\psi_1'$ has an inverse; but this alone does not imply that ϕ_1' and ψ_1' have inverses.

[27] Compare Seifert-Threlfall, *Topologie*, Ch. X, or Lefschetz, [10], Ch. III.

18. The Poincaré duality theorem

Using K and K^*, we may find at once the form of the duality theorem given by Kolmogoroff, [8], and Čech, [5]. Let $\tau_i^{n-p} = \mathfrak{D}(\sigma_i^p)$ be the cell of K^* dual to σ_i^p ; set $\sigma_i^p = \mathfrak{D}^*(\tau_i^{n-p})$. Recall that

$$(18.1) \qquad \partial\mathfrak{D}(\sigma^{p-1}) = (-1)^p\mathfrak{D}\sigma^p + \cdots \text{ if and only if } \partial\sigma_i^p = \sigma^{p-1} + \cdots ;$$

it follows that

$$(18.2) \qquad \partial\mathfrak{D}(A^p) = (-1)^{p+1}\mathfrak{D}(\delta A^p), \qquad \delta\mathfrak{D}(A^p) = (-1)^p\mathfrak{D}(\partial A^p).$$

Similar relations hold for $\mathfrak{D}^*$. From (18.2) we may conclude at once that $\mathfrak{D}$ establishes an isomorphism between ${}^p\mathbf{H}_G(K)$ and ${}^{n-p}\mathbf{H}^G(K^*)$. But ϕSd^* establishes an isomorphism between ${}^{n-p}\mathbf{H}^G(K^*)$ and ${}^{n-p}\mathbf{H}^G(K) = {}^{n-p}\mathbf{H}^G$ (see Theorem 13); hence

$$(18.3) \qquad {}^p\mathbf{H}_G \approx {}^{n-p}\mathbf{H}^G.$$

Suppose that, as in §11, (c), ${}^p\mathbf{H}_G(K) \approx \mathrm{Ch}_Z({}^p\mathbf{H}^H(K))$; then

$$(18.4) \qquad {}^{n-p}\mathbf{H}^G \approx \mathrm{Ch}_Z({}^p\mathbf{H}^H).$$

19. Products and intersections[28]

Supposing K *is simplicial,* we shall (a) define intersections in terms of the $\frown$ product, (b) give a relation defining $\mathfrak{D}$, or rather $Sd^*\mathfrak{D}$, (c) find the relation between $\smile$ and $\frown$ in the cohomology and homology groups, and (d) relate $\smile$ and intersections.

(a) The intersection of a chain A^p of K and a chain B^{*q} of K^* is the following chain[29] of K':

$$(19.1) \qquad A^p \circ B^{*q} = \phi^{*\prime}\mathfrak{D}A^p \frown Sd^*B^{*q} \qquad (\frown \text{ product from §6}).$$

We may deduce the ordinary boundary relation:

$$\begin{aligned} \partial(A^p \circ B^{*q}) &= (-1)^{q-(n-p)}\delta\phi^{*\prime}\mathfrak{D}A^p \frown Sd^*B^{*q} + \phi^{*\prime}\mathfrak{D}A^p \frown \partial Sd^*B^{*q} \\ (19.2) \qquad &= (-1)^{n-q}\phi^{*\prime}\mathfrak{D}\partial A^p \frown Sd^*B^{*q} + \phi^{*\prime}\mathfrak{D}A^p \frown Sd^*\partial B^{*q} \\ &= (-1)^{n-q}\partial A^p \circ B^{*q} + A^p \circ \partial B^{*q}. \end{aligned}$$

Note that, by (7.2) and Theorem 13, (c),

$$(19.3) \qquad \phi^*(A^p \circ B^{*q}) = \mathfrak{D}A^p \frown B^{*q}.$$

---

[28] Compare Čech, [5]; Freudenthal, [6].

[29] We wish $A \circ B^*$ to be a cycle if A and B^* are cycles. To apply the $\smile$ or $\frown$ product, we must turn at least one of them into a cocycle; we use $\mathfrak{D}A$. It is best to use a fixed $\frown$ product, which we may do in K'. We map a cocycle of K^* into a cocycle of K' with $\phi^{*\prime}$, and a cycle of K^* into a cycle of K' with Sd^*. In this manner the form of (19.1) is determined.

(b) Order the vertices of K' in the *opposite* manner from that used above. Define ϕ by mapping each vertex of K' into the last vertex of the simplex of K containing it. If Z^n is the fundamental n-cycle of K, then $Z'^n = SdZ^n$ is the fundamental n-cycle of K'. The map $Sd^*\mathfrak{D}$ is given by

$$(19.4) \qquad Sd^*\mathfrak{D}A^p = \phi'A^p \frown Z'^n \qquad (\frown \text{ from §6}).$$

Further, as $\phi Z'^n = \phi SdZ^n = Z^n$, (7.2) gives

$$(19.5) \qquad \phi Sd^*\mathfrak{D}A^p = A^p \frown \phi Z'^n = A^p \frown Z^n;$$

it is this map that Čech uses in place of $\mathfrak{D}$.

(c) Set $\theta A^p = A^p \frown Z^n$. Then, by (6.2),

$$(19.6) \qquad \theta(A \smile B) = A \frown \theta B.$$

As Sd^* and ϕ induce isomorphisms in the homology groups, (19.5) shows that θ induces the same isomorphism of ${}^p\mathbf{H}_G$ into ${}^{n-p}\mathbf{H}^G$ that $\mathfrak{D}$ does. Hence θ^{-1} exists in these groups, and (19.6) gives

$$(19.7) \qquad v_1 \smile v_2 = \theta^{-1}(v_1 \frown \theta v_2), \qquad v_1 \frown u_1 = \theta(v_1 \smile \theta^{-1}u_1).$$

(d) By (19.1), (19.4) and (6.2),

$$\mathfrak{D}^*A^* \circ \mathfrak{D}B = \phi^{*\prime}A^* \frown Sd^*\mathfrak{D}B = (\phi^{*\prime}A^* \smile \phi'B) \frown Z'^n.$$

By (c), applying this to the cohomology groups gives

$$(19.8) \qquad \theta v_1 \circ \theta v_2 = \theta(v_1 \smile v_2),$$

and hence

$$(19.9) \qquad u_1 \circ u_2 = \theta(\theta^{-1}u_1 \smile \theta^{-1}u_2).$$

A final remark. For a positively oriented σ^n, $\sigma^n \frown Z^n$ is the first vertex of σ^n. Hence

$$(19.10) \qquad I \cdot (A^n \frown Z^n) = A^n \cdot Z^n.$$

20. On intersections of chains and complexes

Let M^n be a manifold, and let K and K' be singular complexes (i.e. continuous maps of complexes into M^n) in "general position" in M^n; that is, so that no σ^p intersects any σ'^{n-p-1}. (A slight deformation of K' will bring this about.) Then all Kronecker indices $(\sigma_i^p \circ \sigma_j'^{n-p})$ have meaning, and $(A^p \circ \partial B'^{n-p+1}) = (-1)^p(\partial A^p \circ B'^{n-p+1})$. (See for instance Lefschetz, [10], p. 169, (20).) Set

$$(20.1) \qquad g\sigma_i^p = \sum_j (\sigma_i^p \circ \sigma_j'^{n-p})\sigma_j'^{n-p}, \qquad g'\sigma_i'^{n-p} = \sum_j (\sigma_j^p \circ \sigma_i'^{n-p})\sigma_j^p,$$

and hence define gA^p, $g'A'^p$. These are dual. We may call the chain gA^p of K' *the intersection of the chain A^p of K with K'.*

For A^p in K and B'^{n-p} in K', we have the Kronecker index

$$(A^p \circ B'^{n-p}) = gA^p \cdot B'^{n-p} = A^p \cdot g'B'^{n-p}. \tag{20.2}$$

As

$$\delta g A^p \cdot \sigma' = gA^p \cdot \partial \sigma' = (A^p \circ \partial \sigma') = (-1)^p (\partial A^p \circ \sigma') = (-1)^p g \partial A^p \cdot \sigma',$$

we have $\delta g = (-1)^p g \partial$. Taking duals also, we have[30]

$$\delta g A^p = (-1)^p g \partial A^p, \qquad \delta g' A'^p = (-1)^p g' \partial A'^p. \tag{20.3}$$

In particular, cycles of K or K' map into cocycles of K' or K.

We may let K be a subdivision of M^n and let K' be a deformed position of K; then chains of K are mapped into chains of K', which may be considered as chains of K again. Then g takes the place of the $\mathfrak{D}$ in §18.

21. Dual bases

In this section, *we shall use the group R of real or of rational numbers* as coefficient group. Say a set of p-cycles $X_1^p, \cdots, X_s^p$ forms a *base* if they are linearly independent with homology, i.e.

$$\alpha_1 X_1^p + \cdots + \alpha_s X_s^p \sim 0 \qquad \text{implies} \qquad \alpha_1 = \cdots = \alpha_s = 0,$$

and if any p-cycle is homologous to a linear combination of them with real or rational coefficients. In other words, their homology classes form a base for $\mathbf{H}^p = {}^p\mathbf{H}^R$. Define a base for p-cocycles similarly. *Bases exist in any complex.* To show this, note first that $\mathbf{H}^p$ is a vector group; for it has a finite number of generators, (using elements of R as coefficients), and no elements of finite order. (If $kX^p \sim 0$, $k \neq 0$, then $kX^p = \partial Y^{p+1}$, $X^p = \partial(Y^{p+1}/k) \sim 0$.) Hence we may choose independent generators $u_1, \cdots, u_s$. Let X_i^p be a cycle in the class u_i; then $X_1^p, \cdots, X_s^p$ form a base. Say a base $X_1^p, \cdots, X_s^p$ for p-cycles and a base $C_1^p, \cdots, C_t^p$ for p-cocycles (then $t = s$) are *dual* if

$$C_i^p \cdot X_j^p = \delta_{ij} \quad (= 1 \text{ if } i = j, \text{ and } = 0 \text{ if } i \neq j). \tag{21.1}$$

Dual bases exist in any complex. First, let $X_1^p, \cdots, X_s^p$ be a base for p-cycles. As $\mathbf{H}_p = {}^p\mathbf{H}_R \approx \mathrm{Ch}_R \mathbf{H}^p$, the group of characters of $\mathbf{H}^p$ into R, (see for instance Whitney, [13], Theorems 7 and 8), we may choose cocycles $C_1^p, \cdots, C_s^p$ such that (21.1) holds. Let v_i be the cohomology class of C_i^p. Clearly any character of $\mathbf{H}^p$ may be expressed uniquely as a linear combination of $v_1, \cdots, v_s$; hence $C_1^p, \cdots, C_s^p$ form a base for p-cocycles.

Now consider a closed orientable manifold M^n. Dual bases for n-cycles and n-cocycles are formed by the fundamental n-cycle Z^n and a single n-cell σ^n, oriented so that $\sigma^n \cdot Z = 1$; similarly a vertex x and the cocycle I form dual

[30] The converse relations $\partial g = \pm g\delta$ etc. are false in general.

bases. Say bases $C_1^p, \cdots, C_s^p$ and $D_1^{n-p}, \cdots, D_t^{n-p}$ form *dual bases* for p-cocycles and $(n - p)$-cocycles (then $t = s$) if

$$C_i^p \smile D_j^{n-p} \backsim \delta_{ij}\sigma^n, \qquad \text{i.e.} \qquad (C_i^p \smile D_j^{n-p})\cdot Z^n = \delta_{ij}. \tag{21.2}$$

We may have $n - p = p$. Then if

$$X_i^{n-p} = C_i^p \frown Z^n, \qquad Y_i^p = D_i^{n-p} \frown Z^n, \tag{21.3}$$

we find, using (5.4) and (5.12),

$$C_i^p \cdot Y_j^p = \delta_{ij}, \qquad D_i^{n-p}\cdot X_j^{n-p} = (-1)^{p(n-p)}\delta_{ij}.$$

Hence the C_i^p and Y_i^p, also the $(-1)^{p(n-p)}D_i^{n-p}$ and X_i^{n-p}, form dual bases.

Finally, the X_i^{n-p} and Y_i^p form dual bases in the ordinary sense. For, using simple properties of intersections and (19.8) and (19.10),

$$\begin{aligned}(X_i^{n-p}\circ Y_j^p) &= I'\cdot(\theta C_i^p \circ \theta D_j^{n-p}) = I\cdot\theta(C_i^p \smile D_j^{n-p})\\ &= I\cdot[(C_i^p \smile D_j^{n-p}) \frown Z^n] = (C_i^p \smile D_j^{n-p})\cdot Z^n = \delta_{ij}.\end{aligned}$$

V. Products in Product Complexes

22. Definition of the products

Let K_1 and K_2 be two complexes (simplicial or not), with cells σ_i^p and τ_k. Then we have (properly oriented) $(p + r)$-cells $\sigma_i^p \times \tau_k$ and $(p + r)$-chains $A^p \times B^r$ in the product complex $K^* = K_1 \times K_2$. We recall that

$$\partial(\sigma^p \times \tau^r) = (\partial\sigma^p \times \tau^r) + (-1)^p(\sigma^p \times \partial\tau^r). \tag{22.1}$$

It follows that

$$\delta(\sigma^p \times \tau^r) = (\delta\sigma^p \times \tau^r) + (-1)^p(\sigma^p \times \delta\tau^r). \tag{22.2}$$

Choose products in K_1 and K_2. We define products in $K_1 \times K_2$ in terms of these by

$$(\sigma^p \times \tau^r) \smile (\sigma^q \times \tau^s) = (-1)^{qr}(\sigma^p \smile \sigma^q) \times (\tau^r \smile \tau^s), \tag{22.3}$$

$$(\sigma^p \times \tau^r) \frown (\sigma^q \times \tau^s) = (-1)^{p(s-r)}(\sigma^p \frown \sigma^q) \times (\tau^r \frown \tau^s). \tag{22.4}$$

We shall show that[31] the $\smile$ product has the required properties in $K_1 \times K_2$. (P_1) of §5 clearly holds. Also,

$$I^* \smile I^* = (I_1 \times I_2) \smile (I_1 \times I_2) = (I_1 \smile I_1) \times (I_2 \smile I_2) = I_1 \times I_2 = I^*,$$

[31] Of course $K_1 \times K_2$ as above defined admits a simplicial subdivision, with the proper geometric interpretation. Hence, by Theorem 14, we obtain the correct products from $K_1 \times K_2$.

so that $\gamma = 1$. Now let us calculate the terms in (P_2). We find

$$\delta[(\sigma^p \times \tau^r) \smile (\sigma^q \times \tau^s)]$$
$$= (-1)^{qr}[(\delta\sigma^p \smile \sigma^q) \times (\tau^r \smile \tau^s) + (-1)^p(\sigma^p \smile \delta\sigma^q) \times (\tau^r \smile \tau^s)]$$
$$+ (-1)^{p+q(r+1)}[(\sigma^p \smile \sigma^q) \times (\delta\tau^r \smile \tau^s) + (-1)^r(\sigma^p \smile \sigma^q) \times (\tau^r \smile \delta\tau^s)],$$

$$\delta(\sigma^p \times \tau^r) \smile (\sigma^q \times \tau^s) = (-1)^{qr}(\delta\sigma^p \smile \sigma^q) \times (\tau^r \smile \tau^s)$$
$$+ (-1)^{p+q(r+1)}(\sigma^p \smile \sigma^q) \times (\delta\tau^r \smile \tau^s),$$

$$(\sigma^p \times \tau^r) \smile \delta(\sigma^q \times \tau^s) = (-1)^{(q+1)r}(\sigma^p \smile \delta\sigma^q) \times (\tau^r \smile \tau^s)$$
$$+ (-1)^{q(r+1)}(\sigma^p \smile \sigma^q) \times (\tau^r \smile \delta\tau^s).$$

From these equations (P_2) follows at once. We may prove similarly that the $\frown$ product has the required properties, or that it corresponds to the $\smile$ product.

In the product K of three complexes, the $\smile$ and $\frown$ products come out the same whether we write K in the form $(K_1 \times K_2) \times K_3$ or $K_1 \times (K_2 \times K_3)$. The signs are as in (23.8) and (23.9) below.

23. The products in Euclidean space

We may subdivide Euclidean n-space $E^n = (u_1, \cdots, u_n)$ by means of the planes $u_i =$ an integer ($i = 1, \cdots, n$). (We could subdivide similarly any small portion of a differentiable manifold.) We shall work out explicitly a $\smile$ product in E^n by using the product of §6 and writing $E^n = E^1 \times \cdots \times E^1$.

First consider E^1. We may denote its cells by

(23.1) $$(\alpha, \beta), \qquad \alpha \text{ an integer}, \beta = 0 \text{ or } 1;$$

it is either the vertex $u = \alpha$ or the 1-cell $\alpha \leqq u \leqq \alpha + \beta$; its dimension is β. Set $(\alpha, \beta) = 0$ for any other β. If we order the vertices by letting $(\alpha, 0)$ precede $(\alpha', 0)$ if $\alpha < \alpha'$, then the $\smile$ product of §6 may be written

(23.2) $$(\alpha, \beta) \smile (\alpha + \beta, \gamma) = (\alpha, \beta + \gamma),$$

the product vanishing in all other cases.

We turn now to the n-dimensional case. The cells of E^n are

(23.3) $$(\alpha_1, \beta_1; \cdots; \alpha_n, \beta_n), \qquad \alpha_i \text{ integral}, \beta_i = 0 \text{ or } 1;$$

the dimension of such a cell is $\sum \beta_i$. Any such symbol with some $\beta_i \neq 0$ or 1 is set $= 0$. Set

(23.4) $$S_i(\beta) = S_i(\beta_1, \cdots, \beta_n) = (-1)^{\beta_1+\cdots+\beta_{i-1}};$$

in particular, $S_1(\beta) = 1$. The boundary relations are

(23.5) $$\partial(\alpha_1, \beta_1; \cdots; \alpha_n, \beta_n) = \textstyle\sum_i S_i(\beta)(\alpha_1, \beta_1; \cdots; \alpha_i + 1, \beta_i - 1; \cdots)$$
$$- \textstyle\sum_i S_i(\beta)(\alpha_1, \beta_1; \cdots; \alpha_i, \beta_i - 1; \cdots),$$

$$(23.6)\quad \delta(\alpha_1, \beta_1; \cdots; \alpha_n, \beta_n) = -\sum_i S_i(\beta)(\alpha_1, \beta_1; \cdots; \alpha_i, \beta_i + 1; \cdots) + \sum_i S_i(\beta)(\alpha_1, \beta_1; \cdots; \alpha_i - 1, \beta_i + 1; \cdots).$$

Set also

$$(23.7)\quad P(\beta; \gamma) = P(\beta_1, \cdots, \beta_n, \gamma_1, \cdots, \gamma_n) = (-1)^\epsilon, \qquad \epsilon = \sum_{i>j} \beta_i \gamma_j.$$

Suppose each β_i and γ_i is 0 or 1, and $\beta_i + \gamma_i \leqq 1$. Then, dropping out all zeros, $P(\beta; \gamma)$ is the sign of the permutation which carries the β's and γ's from their positions shown to their positions in $(\beta_1 + \gamma_1, \cdots, \beta_n + \gamma_n)$. From (22.3) and (23.2) we now find

$$(23.8)\quad \begin{aligned}(\alpha_1, \beta_1; \cdots; \alpha_n, \beta_n) \smile (\alpha_1 + \beta_1, \gamma_1; \cdots; \alpha_n + \beta_n, \gamma_n) \\ = P(\beta; \gamma)(\alpha_1, \beta_1 + \gamma_1; \cdots; \alpha_n, \beta_n + \gamma_n),\end{aligned}$$

the product vanishes in all other cases. As an example,

$$(\alpha_1, 1; \alpha_2, 0) \smile (\alpha_1 + 1, 0; \alpha_2, 1) = (\alpha_1, 1; \alpha_2, 1),$$

$$(\alpha_1, 0; \alpha_2, 1) \smile (\alpha_1, 1; \alpha_2 + 1, 0) = -(\alpha_1, 1; \alpha_2, 1).$$

By taking the dual of (23.8) as in (5.4), we find at once

$$(23.9)\quad \begin{aligned}(\alpha_1 + \gamma_1 - \beta_1, \beta_1; \cdots; \alpha_n + \gamma_n - \beta_n, \beta_n) \frown (\alpha_1, \gamma_1; \cdots; \alpha_n, \gamma_n) \\ = P(\gamma - \beta, \beta)(\alpha_1, \gamma_1 - \beta_1; \cdots; \alpha_n, \gamma_n - \beta_n).\end{aligned}$$

It is not hard to prove the formulas for $\delta(A \smile B)$ and $\partial(A \frown B)$ from the above formulas, and thus show directly that they are the required products.

24. On the maps of products of manifolds into a complex

We shall use only chains with real or rational coefficients; then dual bases may be defined as in §20. We begin by proving, after Hopf,

THEOREM 16.[32] *Let M^n be a closed orientable manifold with fundamental n-cycle Z^n, and let f be a simplicial map of M^n into a complex K'. Let $C_1^p, \cdots, C_s^p$ and $D_1^q, \cdots, D_s^q$ be dual bases for p-cocycles and q-cocycles in M^n ($p + q = n$; p may $= q$), and let $C_1'^p, \cdots$, and $D_1'^q, \cdots$ be bases in K'. Say*

$$(24.1)\qquad f'C_i'^p \backsim \sum_j \lambda_{ij} C_j^p, \qquad f'D_j'^q \backsim \sum_i \mu_{ij} D_i^q.$$

If $fZ^n \sim 0$, then

$$(24.2)\qquad N_{ij} = \sum_{k=1}^{s} \lambda_{ik}\mu_{jk} = 0 \qquad \text{(all } i, j).$$

[32] This theorem was communicated to me by H. Hopf in a letter of September 9, 1937, together with part of Theorem 17. The results (with the complex replaced by a manifold) are corollaries of his paper [7]; we base them on the corresponding relations (7.2).

First we shall find another interpretation of the λ_{ij} and μ_{ij}. Let $\{X_i^p\}$, $\{Y_i^q\}$, $\{X_i'^p\}$, $\{Y_i'^q\}$ be bases for cycles dual to $\{C_i^p\}$, $\{D_i^q\}$, $\{C_i'^p\}$, and $\{D_i'^q\}$ respectively. Say

$$(24.3)\qquad fX_i^p \sim \sum_j \xi_{ij} X_j'^p, \qquad fY_i^q \sim \sum_j \eta_{ij} Y_j'^q;$$

then

$$(24.4)\qquad C_i'^p \cdot fX_j^p = \xi_{ji} = f'C_i'^p \cdot X_j^p = \lambda_{ij},$$

$$(24.5)\qquad D_i'^q \cdot fY_j^q = \eta_{ji} = f'D_i'^q \cdot Y_j^q = \mu_{ij},$$

so that $N_{ij} = \sum \xi_{ki}\eta_{kj}$.

To prove the theorem, we need merely note that

$$0 = (C_i'^p \smile D_j'^q)\cdot fZ^n = f'(C_i'^p \smile D_j'^q)\cdot Z^n = (f'C_i'^p \smile f'D_j'^q)\cdot Z^n$$
$$= \sum_{k,l} \lambda_{ik}\mu_{jl}(C_k^p \smile D_l^q)\cdot Z^n = \sum_{k,l} \lambda_{ik}\mu_{jl}\delta_{kl} = N_{ij}.$$

THEOREM 17. *Let M^p be a closed orientable p-manifold, and let S^q be a q-sphere, $q \geqq p$. Let $\bar{X}^p$ and $\bar{Y}^q$ be corresponding fundamental cycles. Let M^{p+q} be a simplicial subdivision of $K^{p+q} = M^p \times S^q$, with fundamental cycle Z^{p+q}. If x, y are vertices of M^p, S^q, then*

$$(24.6)\qquad X^p = Sd(\bar{X}^p \times y) \quad \text{and} \quad Y^q = Sd(x \times \bar{Y}^q)$$

are cycles of M^{p+q}. Let f be a simplicial map of M^{p+q} into a complex K', and suppose $fZ^{p+q} \sim 0$.

(a) *If $q > p$ or $q = p$ is even, then either $fX^p \sim 0$ or $fY^q \sim 0$.*

(b) *If $q = p$ is odd, then one of fX^p, fY^q is homologous to a multiple of the other.*

If $q > p$, then as $\mathbf{H}^r(S^q) = 0$ $(r = 1, \cdots, q-1)$, it follows that X^p and Y^q generate $\mathbf{H}^p(M^{p+q})$ and $\mathbf{H}^q(M^{p+q})$ respectively. Let I and J be the sums of vertices, and let σ^p and τ^q be cells, in M^p and S^q. Let ϕ and Sd be the maps of Theorem 13, and set

$$(24.7)\qquad C^p = \phi'(\sigma^p \times J), \qquad D^q = \phi'(I \times \tau^q), \qquad \sigma^{p+q} = \sigma^p \times \tau^q.$$

These are cocycles. Then, using Theorem 13, (c).

$$(24.8)\qquad C^p\cdot X^p = (\sigma^p \times J)\cdot \phi Sd(\bar{X}^p \times y) = 1, \qquad D^q\cdot Y^q = 1,$$

so that X^p and C^p, also Y^q and D^q, form dual bases. Further, as $\bar{Z}^{p+q} = \phi Z^{p+q}$ is the fundamental cycle of K^{p+q}, (15.4) gives

$$(24.9)\qquad \begin{aligned}(C^p \smile D^q)\cdot Z^{p+q} &= \phi'[(\sigma^p \times J) \smile (I \times \tau^q)]\cdot Z^{p+q}\\ &= (\sigma^p \times \tau^q)\cdot\phi Z^{p+q} = \sigma^{p+q}\cdot \bar{Z}^{p+q} = 1,\end{aligned}$$

so that C^p and D^q form dual bases. Hence, defining the $X_i'^p$ and $Y_j'^q$ as before, and defining ξ_i, η_i by

$$(24.10)\qquad fX^p \sim \sum \xi_i X_i'^p, \qquad fY^q \sim \sum \eta_i Y_i'^q,$$

Theorem 16 gives, with (24.4) and (24.5),

$$\xi_i \eta_j = 0 \qquad \text{(all } i, j). \tag{24.11}$$

It follows that all the $\xi_i = 0$ or all the $\eta_i = 0$, so that $fX^p \sim 0$ or $fY^q \sim 0$.

Now suppose $p = q$. Then X^p and Y^p together generate $\mathbf{H}^p(M^{2p})$ and C^p and D^p together generate $\mathbf{H}_p(M^{2p})$. The relations (24.6) through (24.9) still hold. Also,

$$\begin{aligned} &D^p \smile C^p \backsim (-1)^{p^2} C^p \smile D^p, \qquad (D^p \smile C^p)\cdot Z^{2p} = (-1)^p, \\ &(C^p \smile C^p)\cdot Z^{2p} = (D^p \smile D^p)\cdot Z^{2p} = 0, \end{aligned} \tag{24.12}$$

so that $\{C^p, D^p\}$ and $\{D^p, (-1)^p C^p\}$ form dual bases for p-cocycles. Further,

$$C^p \cdot Y^p = (\sigma^p \times J)\cdot(x \times \bar{Y}^p) = 0, \qquad D^p \cdot X^p = 0, \tag{24.13}$$

so that $\{X^p, Y^p\}$ and $\{C^p, D^p\}$, and also $\{Y^p, (-1)^p X^p\}$ and $\{D^p, (-1)^p C^p\}$, form dual bases. Comparing (24.10) with (24.3), in which $Y_i'^p = X_i'^p$, we see that (ξ_i, η_i) takes the place of[33] $(\xi_{1i}, \xi_{2i}, \cdots)$, while $(\eta_j, (-1)^p \xi_j)$ takes the place of $(\eta_{1j}, \eta_{2j}, \cdots)$. Hence, by Theorem 16,

$$\xi_i \eta_j + (-1)^p \xi_j \eta_i = 0 \qquad \text{(all } i, j). \tag{24.14}$$

Suppose that $p = q$ is even. If not all the ξ_i are $= 0$, say $\xi_\nu \neq 0$; then setting $i = j = \nu$ in (24.14) shows that $2\xi_\nu \eta_\nu = 0$, $\eta_\nu = 0$, and then setting $i = \nu$, any j, shows that all other $\eta_j = 0$. Thus the theorem is proved for this case.

Finally, if $p = q$ is odd, (24.14) shows that all determinants of the matrix

$$\begin{Vmatrix} \xi_1 & \xi_2 & \cdots \\ \eta_1 & \eta_2 & \cdots \end{Vmatrix}$$

vanish, so the two sets of numbers are proportional. Thus (b) of the theorem holds.

COROLLARY.[34] *If n is even and f is a map of $S^n \times S^n$ into a complex K' of dimension $< 2n$, then one of $S^n \times y$, $x \times S^n$ is mapped into a cycle ~ 0.*

We shall consider briefly what becomes of Theorem 17 when we consider the product $M^{p+q+r} = M^p \times S^q \times S^r$, $p \leqq q \leqq r$. Using the fundamental cycles of the three manifolds, define the three corresponding cycles, X^p, Y^q, Z^r as before. If $p < q < r$, we find by the above methods that $f(Z^{p+q+r}) \sim 0$ implies that one of $f(X^p)$, $f(Y^q)$, $f(Z^r)$ is ~ 0. Consider the case $p = q = r$. Say

[33] For a direct treatment, compare (24.18) below.

[34] For $K' = S^n$, the theorem was proved by H. Hopf, Fund. Math., vol. 25 (1935), pp. 427–440, part of Satz V. In the present form (with K' a manifold), it was communicated to me by Hopf in the letter referred to.

(24.15) $fX^p \sim \sum \xi_i X_i'^p, \qquad fY^p \sim \sum \eta_i X_i'^p, \qquad fZ^p \sim \sum \zeta_i X_i'^p.$

Let $C_1'^r, \cdots, C_s'^p$ form a base for p-cocycles in K'. Then if $fZ^{3p} \sim 0$, we have

$$M_{ijk} = ((C_i'^p \smile C_j'^p) \smile C_k'^p) \cdot fZ^{3p} = 0 \qquad \text{(all } i, j, k). \tag{24.16}$$

Define C^p, D^p, E^p as before. Then

$$\begin{gathered} C^p \smile C^p \backsim D^p \smile D^p \backsim E^p \smile E^p \backsim 0, \\ ((C^p \smile D^p) \smile E^p) \cdot Z^{3p} = 1. \end{gathered} \tag{24.17}$$

As $f'C_i'^p \backsim \xi_i C^p + \eta_i D^p + \zeta_i E^p$ etc., the above relations give

$$\begin{aligned} M_{ijk} &= \xi_i \eta_j \zeta_k (C^p \smile D^p \smile E^p) \cdot Z + \xi_i \zeta_j \eta_k (C^p \smile E^p \smile D^p) \cdot Z + \cdots \\ &= \xi_i(\eta_j \zeta_k + \epsilon \eta_k \zeta_j) + \xi_j(\eta_k \zeta_i + \epsilon \eta_i \zeta_k) + \xi_k(\eta_i \zeta_j + \epsilon \eta_j \zeta_i), \end{aligned} \tag{24.18}$$

where $\epsilon = (-1)^p$. Suppose p is even. Then by considering such quantities as M_{iii}, M_{iij}, we see that all the ξ_i or all the η_i or all the ζ_i vanish, so that one of fX^p, fY^p, fZ^p is ~ 0. If p is odd, then all determinants of the matrix of ξ_i, η_i and ζ_i vanish, so that fX^p, fY^p and fZ^p are linearly dependent under homology.

APPENDIX

MISCELLANEOUS QUESTIONS

25. The products in terms of other operations

We shall show how the products of §6 may be defined in terms of two other operations; one acts on single chains, and the other, on two chains of the same dimension.

Writing simplexes in their normal form §6, we define two homomorphisms of L^p into L^q, $q \leqq p$, by means of

$$\xi_p^q(x_{i_0} \cdots x_{i_p}) = x_{i_0} \cdots x_{i_q}, \qquad \zeta_p^q(x_{i_0} \cdots x_{i_p})\, x_{i_{p-q}} \cdots x_{i_p}. \tag{25.1}$$

Let ξ_q^p and ζ_q^p ($q \leqq p$) be the dual maps of L^q into L^p. Define also

$$(\textstyle\sum \alpha_i \sigma_i^p) \circ (\sum \beta_i \sigma_i^p) = \sum \alpha_i \beta_i \sigma_i^p. \tag{25.2}$$

Then (6.1) gives

$$A^p \smile B^q = \xi_p^{p+q} A^p \circ \zeta_q^{p+q} B^q, \qquad A^q \frown B^{p+q} = \xi_{p+q}^p (\zeta_q^{p+q} A^q \circ B^{p+q}). \tag{25.3}$$

Let I^p be the sum of all (oriented) p-cells: $I^p = \sum x_{i_0} \cdots x_{i_p}$. Then clearly $I^0 = I$. Some relations following at once from the definitions are

$$I^p \cdot (A \circ B) = A \cdot B, \qquad (A \circ B) \cdot C = A \cdot (B \circ C), \tag{25.4}$$

$$\xi_q^p A^q \cdot B^p = A^q \cdot \xi_p^q B^p, \qquad (A \smile B) \cdot C = A \cdot (B \frown C). \tag{25.5}$$

$$I^p \circ A^p = A^p \circ I^p = A^p, \qquad \xi_p^q I^p = \zeta_p^q I^p = I^q (q \geqq p), \tag{25.6}$$

$$\delta I^{2p} = 0, \qquad \delta I^{2p+1} = I^{2p+2}. \tag{25.7}$$

For any subcomplex K' of K, let us interpret K' also as a chain of mixed dimension, namely, the sum of all its cells. Then $K = \sum I^p$, $K' \circ K' = K'$, and

(25.8) $$A^p \circ K' = K' \circ A^p = \text{the ``part of } A^p \text{ in } K'\text{''}.$$

A^p is in K' if and only if $A^p \circ K' = A^p$. Set $K'' = K - K'$. Then all the following twelve conditions are equivalent:
K' is closed, K'' is open, $\bar{K}' = K'$, $\mathrm{St}(K'') = K''$,

$$\sigma_i^p \circ K' = \sigma_i^p \quad \text{implies} \quad \partial\sigma_i^p \circ K' = \partial\sigma_i^p,$$

$$\sigma_i^p \circ K'' = \sigma_i^p \quad \text{implies} \quad \delta\sigma_i^p \circ K'' = \delta\sigma_i^p,$$

$$\partial(A \circ K') \circ K'' = 0, \qquad \delta(A \circ K'') \circ K' = 0,$$

$$\partial(A \circ K'') \circ K'' = \partial A \circ K'', \qquad \delta(A \circ K') \circ K' = \delta A \circ K',$$

$$\partial(A \circ K') \circ K' = \partial(A \circ K'), \qquad \delta(A \circ K'') \circ K'' = \delta(A \circ K'').$$

We may obtain certain chains related to vertices:

$$\zeta_0^p x_i \smile \xi_0^q x_i = \sum \text{all } (p+q)\text{-cells with } x_i \text{ as } (p+1)\text{th vertex},$$

$$\sum_k \zeta_0^{p-k} x_i \smile \xi_0^k x_i = \sum \text{all } p\text{-cells with } x_i \text{ as a vertex}.$$

26. Resolution of chains into boundaries and cocycles

By considering the ranks of the incidence matrices, we may prove

THEOREM 18. *Using real or rational numbers as coefficients, any p-chain may be written uniquely as a p-boundary plus a p-cocycle, or as a p-coboundary plus a p-cycle.*

We shall prove the first statement. To prove uniqueness, suppose

$$\partial A_1 + B_1 = \partial A_2 + B_2 \qquad (\delta B_1 = \delta B_2 = 0).$$

Then setting $A = A_2 - A_1$, $B = B_1 - B_2 = \sum \beta_i \sigma_i^p$, (4.7) gives

$$B \cdot B = B \cdot \partial A = \delta B \cdot A = 0, \qquad \sum \beta_i^2 = 0, \qquad \text{each } \beta_i = 0.$$

Hence $B_1 = B_2$, and $\partial A_1 = \partial A_2$, as required. To prove the existence of the decomposition, consider the rank ρ^p and the numbers of rows and columns α^{p-1} and α^p of $\| {}^p\partial_j^i \|$; α^p is the number of p-cells in K. Clearly

(26.1) $$\rho^{p+1} = \text{number of independent } p\text{-boundaries},$$

(26.2) $$\alpha^p - \rho^{p+1} = \text{number of independent } p\text{-cocycles}.$$

With real coefficients, L^p is a linear vector space of dimension α^p. By the above relations, the p-boundaries and p-cocycles form linear subspaces of dimensions ρ^{p+1} and $\alpha^p - \rho^{p+1}$. We saw that the subspaces were orthogonal; hence they generate L^p.

With integral coefficients, the p-boundaries and p-cocycles generate a subgroup M^p of L^p of rank α^p. Hence the difference group $L^p - M^p$ is finite, and for some integer $m \neq 0$, $m\theta = 0$ for all θ in $L^p - M^p$. This means that *any chain mA^p is (uniquely) a boundary plus a cocycle.* From this the theorem with rational coefficients follows at once.

EXAMPLE. Let K consist of a, b, c, ab, bc, ca. Then using integer coefficients, a 0-chain A^0 is a boundary plus a cocycle if and only if $I \cdot A^0 \equiv 0 \pmod 3$, and a 1-chain A^1 is a coboundary plus a cycle if and only if $I^1 \cdot A^1 \equiv 0 \pmod 3$.

27. Dual maps and changes of base

We shall prove

THEOREM 19. *Let G and H be free groups, and let dual homomorphisms be defined in terms of the bases $x_1, \cdots, x_m$ in G and $y_1, \cdots, y_n$ in H. Then the bases $x_1', \cdots, x_m'$ in G and $y_1', \cdots, y_n'$ in H give the same definition of duality if and only if the new bases are formed from the original ones by permutations and changes of sign.*

Clearly any such changes of base are allowable. To prove the other half of the theorem, suppose

$$x_i' = \sum_j \alpha_{ij} x_j, \qquad x_i = \sum_j \alpha_{ij}^{-1} x_j', \qquad |\alpha_{ij}| = \pm 1,$$
$$y_i' = \sum_j \beta_{ij} y_j, \qquad y_i = \sum_j \beta_{ij}^{-1} y_j', \qquad |\beta_{ij}| = \pm 1.$$

Suppose both pairs of bases give the same definition of dual homomorphisms. Then if ϕ and ψ are dual, we may write

$$\phi x_i = \sum_j \phi_{ij} y_j, \qquad \psi y_i = \sum_j \phi_{ji} x_j,$$
$$\phi x_i' = \sum_j \phi_{ij}' y_j', \qquad \psi y_i' = \sum_j \phi_{ji}' x_j'.$$

These equations with the former ones give

$$\phi_{ij}' = \sum_{k,l} \alpha_{ik} \phi_{kl} \beta_{lj}^{-1}, \qquad \phi_{ji}' = \sum_{k,l} \beta_{ik} \phi_{lk} \alpha_{lj}^{-1}.$$

As these hold for all ϕ_{kl}, we may set $\phi_{kl} = 1$, and all other $\phi_{pq} = 0$. This gives

$$\alpha_{ik} \beta_{lj}^{-1} = \alpha_{ki}^{-1} \beta_{jl} \qquad \text{(all } i, j, k, l). \tag{27.1}$$

Multiplying by $\alpha_{pk} \beta_{jq}$ and summing over j and k gives

$$\delta_{lq} \sum_k \alpha_{pk} \alpha_{ik} = \delta_{pi} \sum_j \beta_{jl} \beta_{jq} \qquad \text{(all } i, l, p, q).$$

Giving p and q different values shows that

$$\sum_k \alpha_{pk} \alpha_{ik} = c\delta_{pi}, \qquad \sum_k \beta_{kp} \beta_{ki} = c\delta_{pi}, \tag{27.2}$$

for some number c. The left hand member is an element of the matrix product of $\| \alpha_{ij} \|$ by its transposed; hence its determinant is $| \alpha_{ij} |^2 = 1$. Therefore $c^m = 1$, and $c = \pm 1$; clearly $c = 1$. It follows that $\| \alpha_{ij} \|$ (and also $\| \beta_{ij} \|$) has a ± 1 in each column, the rest of the column being zeros. This proves the theorem.

Suppose we replace *free group* by *linear vector space*. Again we find (27.2); this time we know merely that $c > 0$. Hence *the bases may be altered by applying to each base an orthogonal transformation, and then multiplying each vector of each set by the same constant* $\neq 0$; *these are the only allowable alterations.*

BIBLIOGRAPHY

1. J. W. Alexander, *On the chains of a complex and their duals*, Proc. Nat. Ac. Sci., vol. 21 (1935), pp. 509–511.
2. J. W. Alexander, *On the ring of a compact metric space*, Proc. Nat. Ac. Sci., vol. 21 (1935), pp. 511–512.
3. J. W. Alexander, *On the connectivity ring of an abstract space*, Annals of Math., vol. 37 (1936), pp. 698–708.
4. P. Alexandroff-H. Hopf, *Topologie I*, Berlin, 1936.
5. E. Čech, *Multiplications on a complex*, Annals of Math., vol. 37 (1936), pp. 681–697.
6. H. Freudenthal, *Über Mannigfaltigkeiten und ihre Abbildungen*, Kon. Ak. v. Wet., Amsterdam, vol. 40 (1937), pp. 54–60. Also Annals of Math., vol. 38(1937), pp. 647–655 and 847–853. See also Comp. Math. vol. 5 (1937), pp. 315–318 (added in proof).
7. H. Hopf, *Zur Algebra der Abbildungen von Mannigfaltigkeiten*, Jour. f. Math., vol. 163 (1930), pp. 71–88.
8. A. Kolmogoroff, *Über die Dualität im Aufbau der kombinatorischen Topologie*, Recueil Math. Moscow, vol. 1 (43) (1936), pp. 97–102.
9. A. Kolmogoroff, *Homologierung des Komplexes und des local-bikompakten Raumes*, Recueil Math. Moscow, vol. 1 (43) (1936), pp. 701–705.
10. S. Lefschetz, *Topology*, Colloquium lectures, New York, 1930.
11. S. Lefschetz, *The rôle of algebra in topology*, Bull. Am. Math. Soc., vol. 43 (1937), pp. 345–359.
12. A. W. Tucker, *An abstract approach to manifolds*, Annals of Math., vol. 34 (1933), pp. 191–243.
13. H. Whitney, *On matrices of integers and combinatorial topology*, Duke Math. Jour., vol. 3 (1937), pp. 35–45.

Harvard University.

Reprinted from DUKE MATHEMATICAL JOURNAL
Vol. 4, No. 3, September, 1938

TENSOR PRODUCTS OF ABELIAN GROUPS

BY HASSLER WHITNEY

1. **Introduction.** Let G and H be Abelian groups. Their direct sum $G \oplus H$ consists of all pairs (g, h), with $(g, h) + (g', h') = (g + g', h + h')$. If we consider G and H as subgroups of $G \oplus H$, with elements $g = (g, 0)$ and $h = (0, h)$, then we may form $g + h$, and the ordinary laws of *addition* hold. Our object in this paper is to construct a group $G \circ H$ from G and H, in which we can form $g \cdot h$, with the properties of *multiplication*; that is, the distributive laws

$$(1.1) \qquad (g + g')\cdot h = g\cdot h + g'\cdot h, \qquad g\cdot(h + h') = g\cdot h + g\cdot h'$$

hold. Clearly $G \circ H$ must contain elements of the form $\sum g_i \cdot h_i$; it turns out (Theorem 1) that with these elements, *assuming only* (1.1), *we obtain an Abelian group,* which we shall call the *tensor product* of G and H.[1]

The tensor product is known in one important case; namely, in tensor analysis (see §4, (b), and §11), though the definition in the form here given does not seem to have been used. Certain other cases are known (see §4). We refer to the examples there given for further indications of the scope of the theory. A direct product of algebras has been constructed by J. L. Dorroh,[2] by methods closely allied to those of the present paper.

As is to be expected, we see in Part I that when we multiply several groups together, the associative and commutative laws hold; also the distributive laws with respect to direct sums and difference groups. The group of integers plays the rôle of a unit group.[3] The rest of Part I is devoted largely to a study of the relation between groups with operator rings and tensor products; in particular, divisibility properties are considered.

In Part II, a detailed study of tensor products of linear spaces is made; we now assume $rg\cdot h = g\cdot rh$ (r real). With any element α of $G \circ H$ are associated subspaces $G(\alpha)$ of G and $H(\alpha)$ of H; their dimensions equal the minimum number of terms in an expression $\sum g_i \cdot h_i$ for α, and in this expression the g_i and h_i form bases in $G(\alpha)$ and $H(\alpha)$. The elementary operations of tensor algebra are derived, and a direct manner of introducing covariant differentiation is indicated.[4] If the linear spaces are topological, a topology may be introduced into

Received February 23, 1938; presented to the American Mathematical Society, February 26, 1938. See Proceedings of the National Academy of Sciences, vol. 23(1937), p. 290.

[1] This is so even if G and H are not Abelian; see Theorem 11. If G and H are linear or topological, we use a slightly different definition.

[2] J. L. Dorroh, *Concerning the direct product of algebras*, Annals of Mathematics, vol. 36 (1935), pp. 882–885. The author is indebted to the referee for pointing out this paper to him.

[3] In linear spaces, the group of real numbers also is a unit.

[4] Some of these results have been derived independently by H. E. Robbins.

the tensor product. If the spaces are not of finite dimension, there are of course various topologies possible in the product; the one we give is probably at an extreme end, in that a neighborhood of 0 in any topology will contain a neighborhood of the sort here given. The topology has certain defects in that the associative and distributive laws seem not to hold in general with topology preserved. In the case of Hilbert spaces, there is a natural definition of the topology in the product (see Murray and von Neumann, reference in §4, (c)). In the intermediate case of Banach spaces, probably the norm $|\alpha|$ may be defined as the lower bound of numbers $\sum |g_i| |h_i|$ for expressions $\sum g_i \cdot h_i$ of α.[5]

In topological groups which contain denumerable dense sets, the product may be given a topology, as is shown in Part III; it agrees with that in Part II when both are defined. Again, in complicated groups, other topologies are possible and perhaps preferable. Finally, for a more complete theory, one must allow infinite sums $\sum g_i \cdot h_i$.

2. **Notations.** Write $G \approx H$ if G and H are isomorphic. The symbol 0 means the zero in any group, or the group with only the zero element. $A \cap B$ is the set of elements in both A and B. ag (a an integer > 0) means $g + \cdots + g$ (a terms); $(-a)g = a(-g)$, $0g = 0$. $g + A$ is the set of all $g + g'$, g' in A; similarly for $A + B$. $g \cdot B$ is the set of all $g \cdot h$, h in B, etc. $aA =$ all ag, g in A. Note that $2A \subset A + A$, etc. Write $a \mid g$ if there is a g' with $ag' = g$; g is then "divisible" by the integer a. $a \mid A$ means $a \mid g$ for all g in A. G is "completely divisible" if for every $a \neq 0$, $a \mid G$, i.e., $aG = G$. The "nullifier" of H in G (of G in H) is the set of all g (all h) such that $g \cdot h = 0$ for all h in H (all g in G).

Let $\sum^* A$ denote the set of all finite sums $a_1 + \cdots + a_k$, a_i in A, any k; this is a subgroup of G (if $A \subset G$). $\sum^*_i A_i$ is the set of all $a_1 + \cdots + a_k$ (a_i in A_i, any k).

Let $G \oplus H$ and $G \ominus G'$ denote direct sums and difference groups. There is a "natural homomorphism" of G into $G \ominus G'$. Some particular groups we shall use are: $I_0 =$ group of integers; $I_\mu = I_0 \ominus \mu I_0 =$ integers mod μ (with elements a_μ for integral a); $Rt =$ rational numbers; $Rl =$ real numbers. Set $G_\mu = G \ominus \mu G$.

I. Discrete groups

3. **Discrete tensor products.** Let G and H be groups (not necessarily Abelian), with the operation $+$. Let $\mathfrak{S}$ be the set of all symbols

$$(g_1, h_1; \cdots; g_n, h_n) \qquad (g_i \text{ in } G,\ h_i \text{ in } H,\ n \text{ any integer} > 0).$$

We add two symbols by the rule

$$(g_1, h_1; \cdots) + (g_{n+1}, h_{n+1}; \cdots) = (g_1, h_1; \cdots; g_{n+1}, h_{n+1}; \cdots).$$

[5] This definition was suggested to me by H. E. Robbins.

Clearly + is associative. We may put any element of $\mathfrak{S}$ in the normal form $(g_1, h_1) + \cdots + (g_n, h_n)$; if we write

$$g_i \times h_i = (g_i, h_i),$$

we obtain

$$(g_1, h_1; \cdots; g_n, h_n) = g_1 \times h_1 + \cdots + g_n \times h_n.$$

Define two equivalence relations as follows:

$$\cdots + (g + g') \times h + \cdots \sim \cdots + g \times h + g' \times h + \cdots, \tag{3.1}$$

$$\cdots + g \times (h + h') + \cdots \sim \cdots + g \times h + g \times h' + \cdots. \tag{3.2}$$

Any succession $s_1 \sim s_2 \sim \cdots \sim s_p$ we shall call an *equivalence sequence* between s_1 and s_p. If two elements s, s' are joined by an equivalence sequence, we say they are *equivalent*, $s \sim s'$. Let also $s \sim s$. The elements of $\mathfrak{S}$ fall into subsets under this relation; these form the elements of the *discrete tensor product* $G \circ H$. In case G and H are discrete, we call this the *tensor product*, in agreement with the definition in Part III. Let $\sum g_i \cdot h_i = g_1 \cdot h_1 + \cdots$ be the element of $G \circ H$ containing the element $\sum g_i \times h_i$ of $\mathfrak{S}$.

To define the group operation, which we temporarily call $\oplus$, in $G \circ H$, take any α and α', and let $\sum g_i \times h_i$ and $\sum g_i' \times h_i'$ be corresponding elements of $\mathfrak{S}$; we set

$$\alpha \oplus \alpha' = \sum g_i \cdot h_i + \sum g_i' \cdot h_i'. \tag{3.3}$$

We must show that this is independent of the choices of $s = \sum g_i \times h_i$ and $s' = \sum g_i' \times h_i'$. If we had chosen t and t', then there are equivalence sequences joining s to t and s' to t'; applying these sequences to $\sum g_i \times h_i + \sum g_i' \times h_i'$ shows that the same element $\alpha \oplus \alpha'$ is determined. Henceforth we use + instead of $\oplus$. Note that + is associative, and (1.1) holds.

We prove in succession the following facts.

(a)
$$\begin{aligned} g \cdot 0 = (g + g - g) \cdot 0 &= g \cdot 0 + g \cdot 0 + (-g) \cdot 0 = g \cdot (0 + 0) + (-g) \cdot 0 \\ &= g \cdot 0 + (-g) \cdot 0 = (g - g) \cdot 0 = 0 \cdot 0; \end{aligned}$$

similarly $0 \cdot h = 0 \cdot 0$.

(b)
$$g \cdot h + 0 \cdot 0 = g \cdot h + g \cdot 0 = g \cdot (h + 0) = g \cdot h,$$

and hence $0 \cdot 0 = g \cdot 0 = 0 \cdot h$ plays the rôle of the identity.

(c)
$$\begin{aligned} g \cdot h = g \cdot h + 0 \cdot (-h) &= g \cdot h + g \cdot (-h) + (-g) \cdot (-h) \\ &= g \cdot 0 + (-g) \cdot (-h) = (-g) \cdot (-h). \end{aligned}$$

Next, we may operate with the product as if G and H were Abelian. For

(d)
$$\begin{aligned} g \cdot (h + h') = g \cdot h + g \cdot h' &= (-g) \cdot (-h) + (-g) \cdot (-h') \\ &= (-g) \cdot (-h - h') = g \cdot (h' + h); \end{aligned}$$

similarly $(g + g')\cdot h = (g' + g)\cdot h$. Also

$$g\cdot(h + h' + h'') = g\cdot h + g\cdot(h' + h'') = g\cdot h + g\cdot(h'' + h')$$
$$= g\cdot(h + h'' + h'), \text{ etc.}$$

Finally, the operation in $G \circ H$ is commutative. For [6]

$$\alpha = (g + g')\cdot(h' + h) = g\cdot(h' + h) + g'\cdot(h' + h)$$
$$= g\cdot h' + g\cdot h + g'\cdot h' + g'\cdot h,$$

also

$$\alpha = (g + g')\cdot h' + (g + g')\cdot h = g\cdot h' + g'\cdot h' + g\cdot h + g'\cdot h,$$

and hence

$$\text{(e)} \qquad g\cdot h + g'\cdot h' = (-g)\cdot h' + \alpha + (-g')\cdot h = g'\cdot h' + g\cdot h.$$

Remark. We would have obtained the same results if we had replaced the elementary equivalence relations by

$$\cdots + (g + g') \times h + \cdots \sim \cdots + g' \times h + g \times h + \cdots, \text{ etc.}$$

THEOREM 1. *$G \circ H$ is an Abelian group; the identity is $0\cdot 0 = g\cdot 0 = 0\cdot h$, and the inverse of $g\cdot h$ is*

$$(3.4) \qquad -(g\cdot h) = (-g)\cdot h = g\cdot(-h).$$

The distributive laws (1.1) *hold.*

This follows from the above results. Because of (d), *we henceforth assume G and H are Abelian*, except in Theorem 11.

THEOREM 2. *In any $G \circ H$, for any integer a,*

$$(3.5) \qquad a(g\cdot h) = ag\cdot h = g\cdot ah.$$

For instance,

$$(-2)g\cdot h = (-g - g)\cdot h = -[(g + g)\cdot h] = -[g\cdot h + g\cdot h] = (-2)(g\cdot h).$$

Using the distributive laws, we may use summation signs as usual; for instance,

$$\sum_i \left(\sum_j a_{ij}g_j\right)\cdot h_i = \sum_i \sum_j (a_{ij}g_j\cdot h_i) = \sum_j \sum_i (g_j\cdot a_{ij}h_i) = \sum_j \left(g_j\cdot \sum_i a_{ij}h_i\right).$$

4. **Examples.** A system with both "addition" and "multiplication" may in general be defined by starting with a system or systems, using addition alone,

[6] For a direct proof, we have

$$g\cdot h + g'\cdot h' = g\cdot h + g\cdot h' + (-g + g')\cdot h' = g\cdot(h + h') + (g' - g)\cdot(h + h')$$
$$+ (g' - g)\cdot(-h) = (g + g' - g)\cdot(h + h') + g'\cdot(-h) + (-g)\cdot(-h)$$
$$= g'\cdot(h + h' - h) + g\cdot h = g'\cdot h' + g\cdot h.$$

forming a tensor product, and defining new equality relations. Specifically, any group pair is an example.

(a) The Abelian groups G and H form a *group pair* with respect to the group Z if a multiplication $g \times h = z$ is given, satisfying both distributive laws. *Any such group pair may be defined by choosing a homomorphism of $G \circ H$ into Z.* Clearly

$$\phi(\sum g_i \cdot h_i) = \sum g_i \times h_i$$

has the required properties. Practically all further examples come under this head.

(b) The most important example of a true tensor product (and the example from which we chose the word "tensor") is probably the following. If G is the tangent vector space at a point of a differentiable manifold, then $G \circ G$ is the space of contravariant tensors of order 2 at the point. (Here $G \circ G$ is not the discrete, but the reduced, or topological, tensor product; see Part II or Part III. The same remark applies to other examples below.) Using also the "conjugate space" $L(G)$ and iterated tensor products gives tensors of all orders (see §11). Of course these spaces are usually defined in terms of coördinate systems in G.

Note that in terms of a fixed coördinate system, $G \circ G$ gives: vector times vector equals matrix. For a generalization, see (i) below.

(c) If G in (b) is replaced by Hilbert space, $G \circ G$ is a Hilbert space,[7] except for the completeness postulate (which could be obtained by completing the space or allowing certain infinite sums in $G \circ G$).

(d) The true tensor product $G \circ H$ has also been used in case one of G, H has a finite number of generators, and has been applied in topology.[8] From the examples (j) and Theorems 3 and 5 below, we may at once determine $G \circ H$ if both G and H have finite sets of generators.

The remaining examples are in general not true tensor products, but come under the heading (a). The general case $G \circ H \to Z$ does not often occur. The case $G \circ G \to Z$ appears in (b). The cases $G \circ H \to H$ and $G \circ G \to G$ appear in (e) and (g) below.

(e) If G is a group, with "operators" from the group R, i.e., $r \cdot g = g'$, the distributive laws are generally assumed; we have $R \circ G \to G$. Here one generally lets R be a ring (see §6).

(f) If G is a group and R is a ring, and we wish to form from G a group G^*

[7] See F. J. Murray and J. von Neumann, *On rings of operators*, Annals of Mathematics, vol. 37(1936), pp. 116–229, Chapter I. As a bounded operator A in G corresponds uniquely to an element f in G: $A(g) = (f, g)$, their space $G \otimes G$ corresponds to our $G \circ G$. M. H. Stone and J. W. Calkin have also considered a direct definition of $G \circ G$ such as we give. Compare also M. Kerner, *Abstract differential geometry*, Compositio Mathematica, vol. 4 (1937), pp. 308–341.

[8] See Alexandroff-Hopf, *Topologie* I, pp. 585–586 and p. 233, (15), and H. Freudenthal, Fundamenta Mathematicae, vol. 29(1937). The definition of $G \circ H$ is indirect. The case that one of G, H is a free group has been studied by H. Freudenthal, Compositio Mathematica, vol. 4(1937), pp. 145–234, Chapter III.

which "admits" R as operator ring, we need merely use $G^* = R \circ G$ (see Theorem 12 below). If we wish to replace G by a group G^* in which division by any integer $\neq 0$ is possible and unique, we use $G^* = Rt \circ G$ (see §8).

(g) If G is a group, any choice of $G \circ G \to G$ makes G a ring (in general non-associative), and conversely.

(h) Let V_p, V_q and V_r be linear spaces (= vector spaces) of dimensions p, q and r. Set $G = Ch_{V_q}(V_p)$ (= group of linear maps of V_p into V_q), $H = Ch_{V_r}(V_q)$, $Z = Ch_{V_r}(V_p)$. Obviously, we have $G \circ H \to Z$. G, H, Z, and $G \circ H$ are vector spaces of dimensions pq, qr, pr, and pq^2r. Hence $Z \approx G \circ H$ is possible only if $q = 1$, i.e., $V_q \approx Rl$. In this case it is true, as shown by (10.7) and (10.11) below. If we choose fixed coördinate systems in V_p, V_q and V_r, then G, H and Z may be interpreted as groups of matrices.

(i) If $G = H$ is the (additive) group of continuous functions $g(x)$, $0 \leqq x \leqq 1$, we may interpret $G \circ H$ as a subgroup of the group of continuous functions $z(x, y)$, $0 \leqq x \leqq 1$, $0 \leqq y \leqq 1$, with $g \cdot h$ corresponding to $z(x, y) = g(x)h(y)$. As is well known from the theory of integral equations, if we allow infinite sums, we may obtain all continuous functions $z(x, y)$.

(j) Finally, we give some examples of tensor products, using the groups most commonly used as coefficient groups in topology. Let Rt_1 and Rl_1 be Rt and Rl reduced mod 1.

$$I_0 \circ G \approx G, \qquad I_\mu \circ G \approx G_\mu \qquad \text{(Theorems 7, 8)},$$

$$I_\mu \circ I_\nu \approx I_{(\mu,\nu)},$$

$$I_\mu \circ Rt \approx I_\mu \circ Rl \approx I_\mu \circ Rt_1 \approx I_\mu \circ Rl_1 \approx 0 \qquad (\mu > 0),$$

$$Rt \circ Rt \approx Rt, \qquad Rt \circ Rl \approx Rl \circ Rl \approx Rl,$$

$$Rt \circ Rt_1 \approx Rt \circ Rl_1 \approx Rt_1 \circ Rt_1, \text{ etc.}, \approx 0.$$

5. General properties. We first consider commutative and associative properties.

THEOREM 3. *There is a natural isomorphism $G \circ H \approx H \circ G$, given by*

$$\phi(\textstyle\sum g_i \cdot h_i) = \sum h_i \cdot g_i. \tag{5.1}$$

THEOREM 4. *There are natural isomorphisms*

$$F \circ (G \circ H) \approx F \circ G \circ H \approx (F \circ G) \circ H,$$

where $F \circ G \circ H$ is the group of all $\sum f_i \cdot g_i \cdot h_i$, using the three distributive laws. The isomorphisms are given by

$$\phi(\textstyle\sum f_i \cdot g_i \cdot h_i) = \sum (f_i \cdot g_i) \cdot h_i, \qquad \psi(\sum f_i \cdot g_i \cdot h_i) = \sum f_i \cdot (g_i \cdot h_i). \tag{5.2}$$

The first theorem is evident; we prove the second, using ϕ. The definition of ϕ is unique, as any equivalence relation in the $\sum f_i \cdot g_i \cdot h_i$ corresponds to one in the $\sum (f_i \cdot g_i) \cdot h_i$. If $\phi(\sum f_i \cdot g_i \cdot h_i) = 0$, then an equivalence sequence carries

$\sum(f_i \cdot g_i) \cdot h_i$ into 0; a corresponding sequence carries $\sum f_i \cdot g_i \cdot h_i$ into 0; hence ϕ is an isomorphism into a subgroup of $(F \circ G) \circ H$. Finally, given any

$$\sum_i z_i \cdot h_i = \sum_i \left(\sum_j f_{ij} \cdot g_{ij}\right) \cdot h_i = \sum_{i,j} (f_{ij} \cdot g_{ij}) \cdot h_j$$

in $(F \circ G) \circ H$, ϕ carries $\sum f_{ij} \cdot g_{ij} \cdot h_i$ into it. This completes the proof.

Next we prove the distributive laws with respect to direct sums and difference groups.

THEOREM 5. *There is a natural isomorphism*

$$(F \oplus G) \circ H \approx F \circ H \oplus G \circ H,$$

given by

$$\text{(5.3)} \quad \begin{aligned} &\phi[(f_1, g_1) \cdot h_1 + \cdots + (f_n, g_n) \cdot h_n] \\ &\qquad = (f_1 \cdot h_1 + \cdots + f_n \cdot h_n ; g_1 \cdot h_1 + \cdots + g_n \cdot h_n). \end{aligned}$$

To show that ϕ is uniquely defined, we have, for instance, as $(f, g) + (f', g') = (f + f', g + g')$,

$$\begin{aligned} \phi[\cdots + (f, g) \cdot h + (f', g') \cdot h + \cdots] \\ &= (\cdots + f \cdot h + f' \cdot h + \cdots, \cdots + g \cdot h + g' \cdot h + \cdots) \\ &= (\cdots + (f + f') \cdot h + \cdots, \cdots + (g + g') \cdot h + \cdots) \\ &= \phi[\cdots + \{(f, g) + (f', g')\} \cdot h + \cdots]. \end{aligned}$$

ϕ maps the first group into the whole of the second; for

$$\text{(5.4)} \quad \phi[(f_1, 0) \cdot h_1 + \cdots + (0, g_1) \cdot h_1' + \cdots] = (f_1 \cdot h_1 + \cdots, g_1 \cdot h_1' + \cdots).$$

Clearly ϕ is a homomorphism. Now suppose $\phi(\alpha) = 0$; let α be given as in (5.3). First, we may transform α into the form of the left side of (5.4). For each half of the right side of (5.3), there is an equivalence sequence carrying it into 0. There are corresponding sequences acting on the left side of (5.4), which shows that $\alpha = 0$. Hence ϕ is an isomorphism.

THEOREM 6. *If G' is a subgroup of G, there is a natural isomorphism*

$$(G \ominus G') \circ H \approx G \circ H \ominus \sum{}^* (G' \cdot H),$$

given as follows. If ψ and Ψ are the natural homomorphisms of G into $G \ominus G'$ and of $G \circ H$ into $G \circ H \ominus \sum^*(G' \cdot H)$, we set

$$\text{(5.5)} \qquad \phi[\psi(g_1) \cdot h_1 + \cdots + \psi(g_n) \cdot h_n] = \Psi(g_1 \cdot h_1 + \cdots + g_n \cdot h_n).$$

By Theorem 3, there is a similar relation with G and H interchanged.

To show that ϕ is uniquely defined, suppose first that $\psi(g_1) = \psi(\bar{g}_1)$. Then $\bar{g}_1 = g_1 + g'$ (g' in G'), and

$$\Psi(\bar{g}_1 \cdot h_1 + \cdots) = \Psi(g_1 \cdot h_1 + \cdots) + \Psi(g' \cdot h_1) = \Psi(g_1 \cdot h_1 + \cdots).$$

The rest of the proof is like previous proofs. For instance, if the element (5.5) vanishes, then $\sum g_i \cdot h_i$ is in $\sum^*(G' \cdot H)$, and hence may be transformed into the form $\sum g_i' \cdot h_i'$ (g_i' in G'). The same transformations may be carried out on the left side of (5.5); as $\psi(g_i') = 0$, this gives $\sum \psi(g_i) \cdot h_i = 0$.

Remark. $\sum^*(G' \cdot H)$ is perhaps "smaller" than $G' \circ H$; for instance, if $G = I_0$, $G' = 2G$, $H = I_2$, then $G' \circ H \approx I_2$, $\sum^*(G' \cdot H) \approx 0$. But *there is a natural homomorphism of $G' \circ H$ onto the whole of $\sum^*(G' \cdot H)$*, clearly. Compare Theorem 28, Part II.

THEOREM 7. *There is a natural isomorphism $I_0 \circ G \approx G$, given by*

$$\phi(\sum a_i \cdot g_i) = \sum a_i g_i. \tag{5.6}$$

The proof is like previous proofs. Note that we have a normal form for elements of $I_0 \circ G$: if we use Theorem 2,

$$\sum a_i \cdot g_i = \sum 1 \cdot a_i g_i = 1 \cdot \sum a_i g_i = 1 \cdot g'. \tag{5.7}$$

The expression of an element in the normal form is unique, by the theorem.

THEOREM 8. *There is a natural isomorphism $I_\mu \circ G \approx G_\mu$, given by*[9]

$$\phi(\sum_i a_\mu^i \cdot g^i) = \sum_i a^i g_\mu^i. \tag{5.8}$$

Using Theorems 6 and 7, we see easily that the following isomorphism is the one given by the theorem:

$$I_\mu \circ G = (I_0 \ominus \mu I_0) \circ G \approx I_0 \circ G \ominus \sum^* (\mu I_0 \cdot G)$$
$$= I_0 \circ G \ominus \sum^* (I_0 \cdot \mu G) \approx I_0 \circ (G \ominus \mu G) \approx G_\mu .$$

THEOREM 9. *If G is completely divisible and every element of H is of finite order, then $G \circ H \approx 0$.*

For if $mh = 0$, then $g \cdot h = mg' \cdot h = g' \cdot mh = 0$.

THEOREM 10. *If G' and H' are subgroups of the nullifiers of H and G in G and H, respectively, then there are natural isomorphisms*

$$G \circ H \approx (G \ominus G') \circ H \approx G \circ (H' \ominus H) \approx (G \ominus G') \circ (H \ominus H');$$

if ϕ and ψ are the natural isomorphisms of G into $G \ominus G'$ and of H into $H \ominus H'$, these are given by

$$\sum g_i \cdot h_i \approx \sum \phi(g_i) \cdot h_i \approx \sum g_i \cdot \psi(h_i) \approx \sum \phi(g_i) \cdot \psi(h_i).$$

First, applying Theorem 6, we find, as $G' \cdot H = 0$,

$$G \circ H \approx G \circ H \ominus \sum^* (G' \cdot H) \approx (G \ominus G') \circ H, \text{ etc.}$$

Next, for any h' in H', $\phi(g) \cdot h'$ corresponds to $g \cdot h' = 0$ in the first isomorphism above; hence $(G \ominus G') \cdot H' = 0$, and

$$(G \ominus G') \circ H \approx (G \ominus G') \circ H \ominus \sum^* ((G \ominus G') \cdot H') \approx (G \ominus G') \circ (H \ominus H').$$

[9] g_μ is the element of G_μ corresponding to g in G.

We end by showing that the discrete tensor product of any two groups, not necessarily Abelian, is isomorphic to the discrete tensor product of the two groups "made Abelian".

THEOREM 11. *Let G and H be any two groups, and let G' and H' be their commutator subgroups. Then there is a natural isomorphism*

$$G \circ H \approx (G \ominus G') \circ (H \ominus H').$$

Because of Theorem 10, we need merely show that any commutator is in the nullifier of the other group; this follows at once from §3, (d).

6. **Sets, groups, rings, operators.** If A and B are two sets of elements, we may define their (discrete) tensor product as the set of all symbols $\pm a_1 \cdot b_1 \pm \cdots \pm a_n \cdot b_n$, with the obvious definition of $+$, which we assume commutative. This is a free group, generated by all $a \cdot b$; if A and B have m and n elements, respectively, then $A \circ B$ has mn generators.

If G is an Abelian group and A is a set of elements, their tensor product is the set of all $\sum g_i \cdot a_i$, with the distributive law as in (3.1), postulating that $+$ is commutative, and $0 \cdot a + g \cdot a' = g \cdot a'$. This is the "group of all linear forms over elements of A, with coefficients in G". An example is given by the groups of chains used in topology.

We shall say an Abelian group G *admits the ring R as operator ring*, or *admits* R simply, if R has a unit 1, and $rg = g'$ is defined satisfying

$$\begin{aligned} &r(g + g') = rg + rg', \qquad (r + r')g = rg + r'g, \\ &r(r'g) = (rr')g \text{ or } (r'r)g, \qquad 1g = g. \end{aligned} \tag{6.1}$$

We call R a *left* or *right* operator according as we use $(rr')g$ or $(r'r)g$ in the third relation. In the second case, we might write gr in place of rg, obtaining $(gr')r = g(r'r)$. Suppose, for definiteness, we write $r[g]$ instead of rg. Then a ring can operate on itself in both ways, using

$$r[r'] = rr' \quad \text{and} \quad r[r'] = r'r. \tag{6.2}$$

The associative law $r[r'[r'']] = (r[r'])[r'']$ holds in either case.

If G and H both admit R, to left or right, we say an isomorphism ϕ between G and H is an *operator isomorphism* if $\phi(rg) = r\phi(g)$; we use $\approx$ again, and say ϕ *preserves* the operator.

THEOREM 12. *If R is a ring with unit, and we define $R \circ G$, considering R as a group under addition, then $R \circ G$ admits R to left or right, under the definitions*

$$r\left(\sum r_i \cdot g_i\right) = \sum rr_i \cdot g_i \quad \text{or} \quad \sum r_i r \cdot g_i . \tag{6.3}$$

The proof is simple. The following theorem is a generalization.

THEOREM 13. *If G admits R to left or to right, then so does any tensor product $G \circ H$ or $H \circ G$, under the definition*

$$r\left(\sum g_i \cdot h_i\right) = \sum rg_i \cdot h_i , \qquad r\left(\sum h_i \cdot g_i\right) = \sum h_i \cdot rg_i . \tag{6.4}$$

Suppose G and H both admit R, each to one side. Then we define the *reduced tensor product* $G \circ' H$ with respect to R as follows. Take the tensor product $G \circ H$, and define a new relation

$$rg \cdot h = g \cdot rh. \tag{6.5}$$

$G \circ' H$ is the group thus formed; it is the difference group of $G \circ H$ with the group generated by all $rg \cdot h - g \cdot rh$.

THEOREM 14. *If G admits R to the left, then there is a natural operator isomorphism*

$$R \circ' G \approx G,$$

letting R act on itself to the right and on $R \circ' G$ to the left, given by

$$\phi(\sum r_i \cdot g_i) = \sum r_i g_i . \tag{6.6}$$

Here, (6.5) is replaced by

$$rr' \cdot g = r'[r] \cdot g = r \cdot r'[g] = r \cdot r'g. \tag{6.5'}$$

To show that ϕ is uniquely defined, we have for instance

$$\phi(rr' \cdot g) = (rr')g = r(r'g) = \phi(r \cdot r'g).$$

ϕ is a homomorphism into the whole of G; for $\phi(1 \cdot g) = 1g = g$. It preserves the operator, for

$$\phi(r(\sum r_i \cdot g_i)) = \phi(\sum rr_i \cdot g_i) = \sum (rr_i)g_i = \sum r(r_i g_i)$$
$$= r(\sum r_i g_i) = r\phi(\sum r_i \cdot g_i).$$

Finally, ϕ is (1-1). For if $\phi(\sum r_i \cdot g_i) = \sum r_i g_i = 0$, then

$$\sum r_i \cdot g_i = \sum 1 \cdot r_i g_i = 1 \cdot \sum r_i g_i = 1 \cdot 0 = 0.$$

The theorem clearly holds with "right" and "left" interchanged.

Suppose R and S are rings.[10] Then we can make $R \circ S$ a ring in four different ways, namely,

$$(r \cdot s)(r' \cdot s') = rr' \cdot ss' \quad \text{or} \quad rr' \cdot s's, \text{ etc.},$$
$$(\sum r_i \cdot s_i)(\sum r'_j \cdot s'_j) = \sum \sum (r_i \cdot s_i)(r'_j \cdot s'_j). \tag{6.7}$$

The uniqueness of the definition is easily established. The associative and distributive laws hold. If R and S have units 1_R and 1_S, then so has $R \circ S$, namely, $1_R \cdot 1_S$.

We shall not discuss the questions of zero-divisors or of fields.

7. Rational multipliers and tensor products.

Definition. For any rational number r, $r = a/b$, $(a, b) = 1$, and any $A \subset G$

[10] Compare J. L. Dorroh, loc. cit.

(including $A = g$), we let rA be the set of all elements g' such that $bg' = ag$, g in A. This agrees with the definition of aA and with the natural definition of $(1/a)A$. Then some of the formal properties of rational numbers as multipliers hold. In particular, some elements can be divided by certain integers. *Division by integers, when it exists, is unique if and only if G has no elements $\neq 0$ of finite order.* For if g' and g'' are in $(1/a)g$, $g' \neq g''$, then $a(g' - g'') = g - g = 0$, so that $g' - g''$ is of finite order; if $g \neq 0$ is of finite order a, then $(1/a)0$ is not unique. We shall say G has *unique division* if it is completely divisible and has no elements $\neq 0$ of finite order. Because of Theorem 15 below, we may then multiply by rational numbers in such a group, and all formal laws will hold.

The only theorem we will need in §8 is the following.

THEOREM 15. *The following three statements are equivalent:*

(a) *G admits Rt as operator ring; we shall write $r[g]$.*

(b) *G has unique division.*

(c) *For each rational r and each g in G, rg is a unique element of G.*

Further, G can admit Rt in at most one way; if it does, then $rg = r[g]$.

First, if G admits Rt, then G has no elements of finite order. For, note first that (for $a > 0$, and hence for $a \leqq 0$),

$$(*) \qquad a[g] = (1 + \cdots + 1)[g] = 1[g] + \cdots + 1[g] = ag.$$

Now if $ag = 0$, $a \neq 0$, then $a[g] = ag = 0 = a0 = a[0]$; hence

$$g = 1[g] = \left(\frac{1}{a}\, a\right)[g] = \frac{1}{a}[a[g]] = \frac{1}{a}[a[0]] = 1[0] = 0.$$

Next, if (a) holds, then for each integer $a \neq 0$ and each g in G, $g' = (1/a)[g]$ exists, and $ag' = a[g'] = g$; hence (b) holds. (b) clearly implies (c). If (c) holds, then setting $r[g] = rg$ gives (a).

Finally, if two operations $r[g]$ and $r\{g\}$ are defined, then they agree; for by (*),

$$b\left(\frac{a}{b}[g]\right) = \left(b\,\frac{a}{b}\right)[g] = a[g] = ag = b\left(\frac{a}{b}\{g\}\right);$$

as G can have no elements of finite order, $(a/b)[g] = (a/b)\{g\}$. Also

$$b\left(\frac{a}{b}[g]\right) = a[g] = ag = b\left(\frac{a}{b}\, g\right),$$

and hence $r[g] = rg$.

Before considering tensor products, we consider some divisibility properties in general groups. Let δ_r denote the denominator of r; $\delta_r = b$ if $r = a/b$, $(a, b) = 1$.

LEMMA 1. *If rg is not void, then $\delta_r \mid g$, and conversely.*

For if $r = a/b$, $bg' = ag$, and $pa + qb = 1$, then

$$b(qg + pg') = qbg + pag = g.$$

The converse is clear.

LEMMA 2. *If* $(a, b) = 1$, *then*

$$\frac{a}{b} A = a\left(\frac{1}{b} A\right) = \frac{1}{b} (aA). \tag{7.1}$$

To prove the first relation, the elements of $a((1/b)A)$ are all g', $g' = ag^*$, g^* in $(1/b)A$, i.e., $bg^* = g$ in A; then $bg' = ag$, and as $(a, b) = 1$, g' is in $(a/b)A$. Conversely, if g' is in $(a/b)A$, then $bg' = ag$ (g in A). Choose p, q so that $pa + qb = 1$, and set $g^* = qg + pg'$. Then

$$bg^* = qbg + pag = g, \qquad ag^* = qbg' + pag' = g',$$

so that g^* is in $(1/b)A$ and g' is in $ag^* \subset a((1/b)A)$. The second relation is clear.

LEMMA 3. *For any integers a and b,*

$$\frac{1}{a}\left(\frac{1}{b} A\right) = \frac{1}{ab} A, \qquad a\left(\frac{1}{a} A\right) \subset A, \qquad \frac{1}{a} (aA) \supset A. \tag{7.2}$$

The proof is simple.

We turn now to tensor products.

LEMMA 4. *If* $\delta_r \mid g$ *and* $\delta_r \mid h$, *then*

$$g' \cdot h = g \cdot h' \quad \text{for any } g' \text{ in } rg \text{ and any } h' \text{ in } rh. \tag{7.3}$$

Set $r = a/b$, $(a, b) = 1$. If

$$bg' = ag, \qquad g = bg^*, \qquad bh' = ah, \qquad h = bh^*,$$

then

$$g \cdot h' = bg^* \cdot h' = g^* \cdot bh' = g^* \cdot ah = g^* \cdot abh^* = abg^* \cdot h^* = ag \cdot h^*$$
$$= bg' \cdot h^* = g' \cdot bh^* = g' \cdot h.$$

Example. If $\delta_r \mid h$ is false, $rg \cdot h$ may not be uniquely defined. For if $G = H = I_2$, $g = 0_2$, $h = 1_2$, then $G \circ H \approx I_2$, and $\frac{1}{2}g \cdot h$ contains both 0_2 and 1_2.

THEOREM 16. *If* $\delta_r \mid A$ *and* $\delta_r \mid B$, *then*

$$rA \cdot B = A \cdot rB; \tag{7.4}$$

if A and B are single elements, so is $rA \cdot B$.

This follows from Lemmas 1 and 4.

Remark. $r(g \cdot h)$ may be $\neq rg \cdot h$. For example, if $G = H = I_2$, $g = h = 0_2$, $r = \frac{1}{2}$, then $rg \cdot h = 0_2$, while $r(g \cdot h)$ contains both 0_2 and 1_2. However,

$$r(A \cdot B) \supset rA \cdot B; \tag{7.5}$$

for if $r = a/b$, $(a, b) = 1$, g in A, h in B, $bg' = ag$, so that $g' \cdot h$ is in $rA \cdot B$, then

$$b(g' \cdot h) = bg' \cdot h = ag \cdot h = a(g \cdot h) \text{ is in } a(A \cdot B),$$

so that $g' \cdot h$ is in $r(A \cdot B)$.

LEMMA 5. *If $b \mid A$ and $b \mid B$, then*

$$\frac{1}{b}(aA)\cdot B = \frac{a}{b}A\cdot B = a\left(\frac{1}{b}A\right)\cdot B = A\cdot\frac{1}{b}(aB) = A\cdot\frac{a}{b}B = A\cdot a\left(\frac{1}{b}B\right); \tag{7.6}$$

if A and B are single elements, so is the above.

Say $(a, b) = k$, $a = a'k$, $b = b'k$; then $(a', b') = 1$. To prove the first relation, we use Lemmas 2 and 3 and Theorem 16, and the fact $b \mid aA$:

$$\frac{1}{b}(aA)\cdot B = \frac{1}{b'}\left(\frac{1}{k}(k(a'A))\right)\cdot B \supset \frac{1}{b'}(a'A)\cdot B = \frac{a'}{b'}A\cdot B = \frac{a}{b}A\cdot B,$$

$$\frac{1}{b}(aA)\cdot B = A\cdot a\left(\frac{1}{b}B\right) = A\cdot a'\left(k\left(\frac{1}{k}\left(\frac{1}{b'}B\right)\right)\right) \subset A\cdot a'\left(\frac{1}{b'}B\right) = \frac{a}{b}A\cdot B.$$

From these the relation follows. The other relations are consequences of this one or are easily proved. The last statement follows from Theorem 16.

THEOREM 17. *If $\delta_r\delta_{r'} \mid A$ and $\delta_r\delta_{r'} \mid B$,*[11] *then*

$$r(r'A)\cdot B = (rr')A\cdot B = A\cdot(rr')B, \text{ etc.}; \tag{7.7}$$

if A and B are single elements, so is the above.

Say $r = a/b$, $r' = c/d$, $(a, b) = (c, d) = 1$. As $bd \mid cA$, etc.,

$$r(r'A)\cdot B = a\left(\frac{1}{b}\left(\frac{1}{d}(cA)\right)\right)\cdot B = \frac{1}{bd}(cA)\cdot aB = ac\left(\frac{1}{bd}A\right)\cdot B$$

$$= \frac{ac}{bd}A\cdot B = (rr')A\cdot B, \text{ etc.}$$

8. **The tensor product** $Rt \circ G$. First note that, if F is any completely divisible group (in particular, Rt), then in studying $F \circ G$, we could assume that G has no elements $\neq 0$ of finite order. For otherwise, let G' be the subgroup of elements of finite order of G. As G' is in the nullifier of F, $\sum^*(F\cdot G') = 0$ (see Theorem 9); hence, by Theorem 10,

$$F \circ G \approx F \circ (G \ominus G').$$

Thus we may replace G by $G \ominus G'$, which has no elements $\neq 0$ of finite order.

THEOREM 18. *In $Rt \circ G$, each element may be written in the form $(1/a)\cdot g$. If G has no elements $\neq 0$ of finite order, then $r\cdot g = 0$ if and only if $r = 0$ or $g = 0$.*

First,

$$\sum r_i\cdot g_i = \sum\frac{a_i}{a}\cdot g_i = \frac{1}{a}\cdot\sum a_i g_i = \frac{1}{a}\cdot g.$$

Next, suppose we have an equivalence sequence reducing $r\cdot g$ to $0\cdot 0$. In all terms occurring, there is a least common denominator c. Multiplying every-

[11] Possibly this hypothesis can be weakened.

thing by c gives an equivalence sequence, which may be interpreted as a sequence in $I_0 \circ G$, or again, in G itself. Hence, if $r = a/b$, we have $(ca/b)g = 0$. If $r \neq 0$, then $ca/b \neq 0$, and as G has no elements of finite order, $g = 0$.

THEOREM 19. *$Rt \circ G$ has unique division.*

This follows from Theorems 12 and 15.

THEOREM 20. *There is an isomorphism $G \approx Rt \circ G$, given by $\phi(\sum r_i \cdot g_i) = \sum r_i g_i$, if and only if G has unique division.*

This is an extension of Theorem 15. One half follows from Theorem 19; the other half is clear.

THEOREM 21. *If G has no elements $\neq 0$ of finite order, then $Rt \circ G$ is the smallest completely divisible group containing G. That is, if H is completely divisible and contains a subgroup $H_1 \approx G$, then it contains a subgroup $H_2 \approx Rt \circ G$.*

Let H' be the subgroup of elements of finite order of H. Clearly H' is completely divisible; hence we may write $H = H' \oplus H''$.[12] For any $h = h' + h''$, write $h' = \phi(h)$, $h'' = \psi(h)$; then ϕ and ψ are homomorphisms. Set $H_1'' = \psi(H_1)$; then $H_1'' \approx G$. For if $\psi(h_1) = 0$ (h_1 in H_1), then h_1 is in H', and hence is of finite order; but h_1 is in $H_1 \approx G$, which gives $h_1 = 0$.

Let H_2 be the subgroup of H'' containing all elements with multiples in H_1''. H_2 is completely divisible. For given h in H_2 and an integer $a \neq 0$, choose h^* in H so that $ah^* = h$, and set $h_1 = \psi(h^*)$. Then h_1 is in H'', and as h is in H'',

$$ah_1 = a\psi(h^*) = \psi(ah^*) = \psi(h) = h;$$

hence h_1 is in H_2. As H'' has no elements $\neq 0$ of finite order, neither has H_2; hence H_2 has unique division.

Let θ be the isomorphism of G into H_1''. As rh is uniquely defined for h in the group H_2 (Theorem 15), and clearly obeys $(r + r')/h = rh + r'h$, $r(h + h') = rh + rh'$, we may set

$$\Theta(\sum r_i \cdot g_i) = \sum r_i \theta(g_i),$$

defining a homomorphism of $Rt \circ G$ into H_2. Suppose $\Theta(\alpha) = 0$. If $\alpha = (1/a) \cdot g$ (Theorem 18), then $\Theta(\alpha) = (1/a)\theta(g) = 0$. Multiplying by a gives $\theta(g) = 0$, and hence $g = 0$, and $\alpha = 0$, as θ is an isomorphism. Hence Θ is (1-1). For any h in H_2, we may take a so that ah is in H_1''; then for some g, $ah = \theta(g) = \Theta(1 \cdot g)$, and $h = \Theta((1/a) \cdot g)$; hence Θ is an isomorphism, and the theorem is proved.

9. **Tensor products and character groups.** In some cases, the group $Ch_H(G)$ of homomorphisms of G into H can be expressed in terms of the two groups H and $Ch_{I_0}(G)$, by (9.1). See also Theorem 25 of Part II. We remark in passing that $Ch_H(G)$ and G form a group pair with respect to H, with the definition $\Phi(\sum \phi_i \cdot g_i) = \sum \phi_i(g_i)$ (ϕ_i in $Ch_H(G)$, g_i in G).

[12] See R. Baer, *The subgroup of elements of finite order of an Abelian group*, Annals of Mathematics, vol. 37(1936), pp. 766–781, (1; 1).

THEOREM 22.[13] *There is a natural isomorphism*

$$(9.1) \qquad Ch_{I_0}(G) \circ H \approx Z \subset Ch_H(G),$$

defined as follows. For u_i *in* $Ch_{I_0}(G)$ *and* h_i *in* H,

$$(9.2) \qquad \Phi(\sum u_i \cdot h_i ; g) = \sum u_i(g)h_i .$$

If either G or H is a free group with a finite number of generators, then $Z = Ch_H(G)$.

It is clear that the definition of Φ is unique, and Φ is a homomorphism. We must show that it is (1-1). Suppose the element (9.2) equals 0. Say the sum contains n terms. Let $A = I_0 \oplus \cdots \oplus I_0$ be the group of all n-tuples $(a_1, \cdots, a_n)$ of integers, and let A' be the subgroup of all $(a_1, \cdots, a_n)$ in A for which $\sum a_i h_i = 0$. We may choose a base

$$\alpha_1, \cdots, \alpha_n ; \qquad \alpha_i = (a_{i1}, \cdots, a_{in})$$

in A and integers $p_1, \cdots, p_m$ $(m \leqq n)$ such that

$$p_1\alpha_1, \cdots, p_m\alpha_m$$

form a base in A'.[14] For each g, let $u(g)$ be the element $(u_1(g), \cdots, u_n(g))$ of A; as $\sum u_i(g)h_i = 0$, $u(g)$ is in A'. Hence, for each g, there is a uniquely defined set of numbers $\rho_1(g), \cdots, \rho_m(g)$ such that

$$u(g) = \sum_{j=1}^{m} \rho_j(g) p_j \alpha_j ;$$

hence

$$u_i = \sum_{j=1}^{m} p_j a_{ji} \rho_j .$$

As the $u_i(g)$ are homomorphisms, so are $u(g)$ and the $\rho_i(g)$; the $\rho_i(g)$ are in $Ch_{I_0}(G)$. Set

$$\bar{h}_i = \sum_{k=1}^{n} a_{ik} h_k \qquad (i = 1, \cdots, m);$$

then

$$p_i \bar{h}_i = \sum_{k=1}^{n} p_i a_{ik} h_k = 0 \qquad (i = 1, \cdots, m),$$

by the choice of the α_i and p_i. Hence, using the distributive laws in $Ch_{I_0}(G) \circ H$,

$$\sum_{i=1}^{n} u_i \cdot h_i = \sum_{i=1}^{n} \left(\sum_{j=1}^{m} p_j a_{ji} \rho_j \right) \cdot h_i = \sum_{j=1}^{m} \left(\rho_j \cdot p_j \sum_{i=1}^{n} a_{ji} h_i \right)$$

$$= \sum_{j=1}^{m} (\rho_j \cdot p_j \bar{h}_j) = \sum_{j=1}^{m} (\rho_j \cdot 0) = 0,$$

as required.

[13] Compare Theorem 25.

[14] See, for example, Alexandroff-Hopf, loc. cit., p. 566.

Now suppose H has a base $\bar{h}_1, \cdots, \bar{h}_n$, so any h may be written uniquely $\sum a_i\bar{h}_i$. Let ϕ be any homomorphism of G into H; then we may write

$$\phi(g) = \sum u_i(g)\bar{h}_i,$$

and the $u_i(g)$ are elements of $Ch_{I_0}(G)$. Also

$$\Phi(\sum u_i\cdot\bar{h}_i; g) = \sum u_i(g)\bar{h}_i = \phi(g),$$

so Φ maps $Ch_{I_0}(G) \circ H$ into the whole of $Ch_H(G)$.

Suppose finally that G has a base $\bar{g}_1, \cdots, \bar{g}_n$. Let $\bar{u}_i(g)$ be the element of $Ch_{I_0}(G)$ defined by $\bar{u}_i(\bar{g}_i) = 1$, $\bar{u}_i(\bar{g}_j) = 0$ $(j \neq i)$. Take any homomorphism ϕ of G into H. Then for any $g = \sum a_i\bar{g}_i$, $\bar{u}_i(g) = a_i$, and

$$\phi(g) = \sum a_i\phi(\bar{g}_i) = \sum \bar{u}_i(g)\phi(\bar{g}_i);$$

hence, setting $h_i = \phi(\bar{g}_i)$,

$$\Phi(\sum \bar{u}_i\cdot h_i; g) = \sum \bar{u}_i(g)\phi(\bar{g}_i) = \phi(g).$$

This completes the proof.

Examples. Suppose $G = H = I_2$. Then $Ch_H(G)$ has two elements, while $Ch_{I_0}(G) \circ H$ has only one. Again, let G be the additive group of triadic rational numbers (all numbers of the form $a/3^b$), and set $H = I_2$. There are two elements in $Ch_H(G)$, determined by $\phi(1) = 0_2$ and $\phi(1) = 1_2$; but there is only one element in $Ch_{I_0}(G) \circ H$.

II. Linear spaces

10. **Products, finite dimensional spaces.** A *linear space*, or *vector space*, G, is an Abelian[15] group which admits the real numbers Rl as operators (see §6). Let $G(g_1, \cdots, g_m)$ be the subspace of G generated by $g_1, \cdots, g_m$, i.e., all $\sum a_ig_i$ (a_i real). If such a set generates G itself, then let $g_1, \cdots, g_m$ be such a set with the least number of elements. Then these elements form a *base* for G, and G is of *dimension m*.

In any finite dimensional linear space G, with a base $g_1, \cdots, g_m$, we may introduce a *natural topology* by defining neighborhoods $U(\epsilon)$ of 0 for each $\epsilon > 0$, consisting of all $\sum a_ig_i$ with $\sum a_i^2 < \epsilon^2$. The topology is independent of the choice of a base. In this topology, the operation ag is continuous in both variables.

In the tensor product $G \circ H$, we clearly wish to have

$$a(g\cdot h) = ag\cdot h = g\cdot ah \qquad (a \text{ in } Rl); \tag{10.1}$$

hence we use the *reduced tensor product* (see (6.4)), but call it the *tensor product* simply. Without this, we would have for instance in Rl, $\sqrt{2}\cdot 1 \neq 1\cdot\sqrt{2}$. Further, *if we assume that $g\cdot h$ is continuous, then* (10.1) *follows.* To show this, the last statement in Theorem 15, and Theorem 16, show that $bg\cdot h = g\cdot bh$ for any rational b. Letting $b \to a$ gives the result.

[15] The group is necessarily Abelian. Compare §3, (e). If G is not linear, it can be made so by taking $Rl \circ G$; see Theorem 12, §6.

We assume in the rest of §10 that G and H have bases $\bar{g}_1, \cdots, \bar{g}_m$ and $\bar{h}_1, \cdots, \bar{h}_n$, respectively.

THEOREM 23. *An element of $G \circ H$ may be written uniquely in any one of the three normal forms*

$$\sum_{i=1}^{m} \sum_{j=1}^{n} a_{ij}(\bar{g}_i \cdot \bar{h}_j) = \sum_{i=1}^{m} \bar{g}_i \cdot h'_i = \sum_{j=1}^{n} g'_j \cdot \bar{h}_j. \tag{10.2}$$

For, if

$$g_k = \sum b_{ki}\bar{g}_i, \qquad h_k = \sum c_{ki}\bar{h}_i, \tag{10.3}$$

then the distributive laws give

$$\sum_k g_k \cdot h_k = \sum_k \left(\sum_i b_{ki}\bar{g}_i\right) \cdot \left(\sum_j c_{kj}\bar{h}_j\right) = \sum_k \sum_i \left(\bar{g}_i \cdot \sum_j b_{ki}c_{kj}\bar{h}_j\right)$$

$$= \sum_i \bar{g}_i \cdot \sum_j \sum_k b_{ki}c_{kj}\bar{h}_j, \text{ etc.};$$

thus (10.2) holds with

$$a_{ij} = \sum_k b_{ki}c_{kj}, \qquad h'_i = \sum_j a_{ij}\bar{h}_j, \qquad g'_i = \sum_j a_{ji}\bar{g}_j. \tag{10.4}$$

Given any expression $\sum g_i \cdot h_i$ for α in $G \circ H$, the above procedure gives the normal forms in a unique manner; we must show that if $\sum g_i \cdot h_i = \sum g_i^* \cdot h_i^*$, the two expressions give the same result. It is sufficient to prove this for $(g + g^*) \cdot h$ and $g \cdot h + g^* \cdot h$, for $g \cdot (h + h^*)$ and $g \cdot h + g \cdot h^*$, and for $ag \cdot h$ and $g \cdot ah$. In each case, the proof is simple.

Let $Ch_H(G)$ denote the group of linear maps (= continuous homomorphisms) of G into H; this is a linear space of dimension mn. In particular, $L(G) = Ch_{Rl}(G)$ is the group of linear real-valued functions in G, and is called the *conjugate space* of G. Here, isomorphism will mean continuous isomorphism = operator isomorphism. The following theorem is well known.

THEOREM 24. $L(G) \approx G$. *Further, there is a natural isomorphism*

$$L(L(G)) \approx G, \tag{10.5}$$

defined as follows. For any g in G, $\phi(g)$ is the element of $L(L(G))$ which, for any u in $L(G)$, has the value $u(g)$.

Let $\bar{u}_i(g)$ be the element of $L(G)$ such that $\bar{u}_i(\bar{g}_i) = 1$, $\bar{u}_i(\bar{g}_j) = 0$ $(j \neq i)$. Clearly $\bar{u}_1, \cdots, \bar{u}_m$ form a base in $L(G)$; hence $L(G) \approx G$. Next, ϕ is linear. It is (1-1); for if $\phi(g) = 0$, then $u(g) = 0$ (all u in $L(G)$), which implies $g = 0$. Given any v in $L(L(G))$, set $a_i = v(\bar{u}_i)$; then for any $u = \sum b_i \bar{u}_i$,

$$u\left(\sum a_i \bar{g}_i\right) = \sum_i a_i \sum_j b_j \bar{u}_j(\bar{g}_i) = \sum a_i b_i = \sum b_i v(\bar{u}_i) = v(u),$$

so that $\phi(\sum a_i \bar{g}_i) = v$. Clearly $\phi(ag) = a\phi(g)$; hence ϕ is an isomorphism.

THEOREM 25.[16] *There is a natural isomorphism*

$$Ch_H(G) \approx L(G) \circ H, \tag{10.6}$$

given by

$$\phi(\sum u_i \cdot h_i \,;\, g) = \sum u_i(g) h_i . \tag{10.7}$$

ϕ is clearly uniquely defined. If we write all elements of $L(G) \circ H$ in the third normal form $\sum u_i \cdot \bar{h}_i$, the properties of ϕ are easily established; for any element of $Ch_H(G)$ can be written uniquely as $\sum u_i(g)\bar{h}_i$, and if this is the zero element, i.e., it is equal to zero in H for all g, then all $u_i(g) = 0$.

COROLLARY I. *$G \circ H$ may be written in the form*

$$G \circ H \approx L(L(G)) \circ H \approx Ch_H(L(G)). \tag{10.8}$$

The isomorphism of the first group into the last is given as follows. For $\sum g_i \cdot h_i$ in $G \circ H$ and u in $L(G)$,

$$\phi(\sum g_i \cdot h_i \,;\, u) = \sum u(g_i) h_i . \tag{10.9}$$

COROLLARY II. *There is a natural isomorphism*

$$Ch_G(Rl) \approx G; \tag{10.10}$$

for u in $Ch_G(Rl)$, $\phi(u) = u(1)$.

For $L(Rl) \circ G \approx Rl \circ G \approx G$. (Moreover, a direct proof is obvious.)

THEOREM 26. *$G \circ H$ is a linear space of dimension mn, with a base $\bar{g}_1 \cdot \bar{h}_1, \cdots, \bar{g}_m \cdot \bar{h}_n$. If $\{U\}$ and $\{V\}$ are neighborhood systems in G and H, respectively, defining their natural topologies, then either of the following neighborhood systems, if we use $p = \min(m, n)$,*

$$N(U, V) = U \cdot V + \cdots + U \cdot V \qquad (p \text{ summands}), \tag{10.11}$$

$$N(U_1, U_2, \cdots ; V_1, V_2, \cdots) = \sum_k{}^*(U_k \cdot V_k) \tag{10.12}$$

defines the natural topology in $G \circ H$.[17] The multiplication $g \cdot h$ is continuous.

The first part of the theorem follows from Theorem 23. Let N, N', N'' denote natural neighborhoods and those of (10.11) and (10.12). Given an $N = N(\epsilon)$, consisting of all $\sum a_{ij} \bar{g}_i \cdot \bar{h}_j$ with $\sum a_{ij}^2 \leqq \epsilon^2$, set $\epsilon_1 = \epsilon/(mn)^{\frac{1}{2}}$, and let

$$U_k = U(\epsilon_1/2^k), \qquad V_k = V(1), \qquad (k = 1, 2, \cdots),$$

be natural neighborhoods in G and H. Then if $g_k = \sum b_{ki} \bar{g}_i$ is in U_k and $h_k = \sum c_{kj} \bar{h}_j$ is in V_k, (10.4) gives, if we use any finite number ν of summands in (10.12),

$$|a_{ij}| = \left| \sum_k b_{ki} c_{kj} \right| < \sum_{k=1}^{\nu} \epsilon_1/2^k < \epsilon_1 = \epsilon/(mn)^{\frac{1}{2}}.$$

[16] This holds if at least one of G, H is of finite dimension. Compare Theorem 22.

[17] If we map Rl into a curve everywhere dense on the torus, the topology of the torus gives an "unnatural" topology in Rl. In $Rl \circ Rl$, either type of neighborhood as here given then contains the whole space.

Hence $\sum a_{ij}^2 < \epsilon^2$, and $\sum_{k=1}^{\nu} g_k \cdot h_k$ is in N. Thus any N contains an N''.

Next, given an N'', take

$$U \subset U_1 \cap \cdots \cap U_p, \qquad V \subset V_1 \cap \cdots \cap V_p.$$

Then clearly $N' = N(U, V) \subset N''$.

Next, take any $N' = N(U, V)$. Suppose for definiteness that $p = m$. Take ϵ_1 so that $U(2\epsilon_1) \subset U$ and $V(\epsilon_1) \subset V$, and set $\epsilon = \epsilon_1^2$. Now take any α of $G \circ H$ in $N(\epsilon)$; then we can write $\alpha = \sum a_{ij}\bar{g}_i \cdot \bar{h}_j$, with $\sum a_{ij}^2 < \epsilon^2$. Also,

$$\alpha = \sum_{i=1}^{m} \left(\epsilon_1 \bar{g}_i \cdot \sum_{j=1}^{n} \theta a_{ij} \bar{h}_j\right), \qquad \theta = \frac{1}{\epsilon_1}.$$

As $m = p$ and $\epsilon_1 \bar{g}_i$ is in $U(2\epsilon_1) \subset U$, to show that $N(\epsilon) \subset N(U, V)$, it is sufficient to show that $\sum_j \theta a_{ij} \bar{h}_j$ is in $V(\epsilon_1)$. But

$$\sum_j \theta^2 a_{ij}^2 \leqq \theta^2 \sum_{i,j} a_{ij}^2 < \theta^2 \epsilon^2 = \epsilon_1^2,$$

and this proves the statement.

The continuity of $g \cdot h$ is clear from the relation

$$\sum (a_i + a_i')\bar{g}_i \cdot \sum (b_j + b_j')\bar{h}_j - \sum a_i \bar{g}_i \cdot \sum b_j \bar{h}_j = \sum (a_i' b_j + a_i b_j' + a_i' b_j')\bar{g}_i \cdot \bar{h}_j.$$

If G and H are metric, and hence scalar products $g \circ g'$ and $h \circ h'$ are defined, we may define scalar products and hence a metric in $G \circ H$ by

(10.13) $$\left(\sum_k g_k \cdot h_k\right) \circ \left(\sum_l g_l' \cdot h_l'\right) = \sum_{k,l} (g_k \circ g_l')(h_k \circ h_l').$$ [18]

11. **Tensor algebra.** Let G be a linear space of finite dimension n; in §12, it will be the "tangent space" at a point of a manifold. Any element of G we shall call a *contravariant vector*. An element of $H = L(G)$ we call a *covariant vector*. Any element of the linear space

(11.1) $$T(p, q) = G \circ \cdots \circ G \circ H \circ \cdots \circ H \qquad (p \text{ factors } G,\ q \text{ factors } H)$$

we shall call a *tensor of contravariant order p and covariant order q*. As $L(p, q)$ is a linear space, we may *add* two tensors of the same type, and *multiply* a tensor by a real number. Using Theorems 3 and 4 in Part I, we have

$$(G \circ \cdots \circ H \circ \cdots) \circ (G \circ \cdots \circ H \circ \cdots)$$
$$\approx G \circ \cdots \circ G \circ \cdots \circ H \circ \cdots \circ H \circ \cdots.$$

Hence a tensor of $T(p, q)$ and a tensor of $T(p', q')$ may be *multiplied*, giving a tensor of $T(p + p', q + q')$.

The process of *contraction* is as follows. To contract the element $g \cdot h$ of

[18] For a study of this metric in Hilbert spaces, see Murray and von Neumann, loc. cit.

$T(1, 1) = G \circ H$, recall that $H = L(G)$, and set $\phi(g \cdot h) = h(g)$, a real number. To contract the element

$$\alpha = \sum_k g_k^1 \cdots g_k^p \cdot h_k^1 \cdots h_k^q \quad \text{of } T(p, q)$$

with respect to the p-th g and the q-th h, for example, set

$$\phi(\alpha) = \sum_k h_k^q(g_k^p) g_k^1 \cdots g_k^{p-1} \cdot h_k^1 \cdots h_k^{q-1}; \tag{11.2}$$

this is an element of $T(p - 1, q - 1)$.

Let $\bar{g}_1, \cdots, \bar{g}_n$ form a base in G, and choose $\bar{h}^i$ so that $\bar{h}^i(\bar{g}_j) = \delta_j^i$; then $\bar{h}^1, \cdots, \bar{h}^n$ form a base in H. By the proof of Theorem 23, we may write any element of $T(p, q)$ uniquely in the normal form

$$\alpha = \sum_{i_r, j_s = 1}^{n} A_{j_1 \cdots j_q}^{i_1 \cdots i_p} \bar{g}_{i_1} \cdots \bar{g}_{i_p} \bar{h}^{j_1} \cdots \bar{h}^{j_q}; \tag{11.3}$$

there are n^{p+q} terms in the sum, and the $A_{j_1 \cdots j_q}^{i_1 \cdots i_p}$ are called the *components of* α in the coördinate system of the $\bar{g}_i$. Let us verify the *laws of transformation* of the components. Suppose we introduce the new base $g_1', \cdots, g_n'$. Say

$$\bar{g}_i = \sum_{k=1}^{n} a_i^k g_k', \qquad g_i' = \sum_{k=1}^{n} a_i'^k \bar{g}_k.$$

If $h'^i(g_j') = \delta_j^i$, then setting $h'^i = \sum b_k^i \bar{h}^k$ gives

$$\delta_j^i = h'^i(g_j') = \sum_k b_k^i \bar{h}^k (\sum_l a_j'^l \bar{g}_l) = \sum_{k,l} b_k^i a_j'^l \delta_l^k = \sum_k b_k^i a_j'^k.$$

Hence $b_j^i = a_j^i$, and $\bar{h}^i = \sum a_k'^i h'^k$. Putting in (11.3) and using the distributive laws gives

$$\alpha = \sum A_{j_1 \cdots j_q}^{i_1 \cdots i_p} a_{i_1}^{k_1} \cdots a_{i_p}^{k_p} a_{l_1}'^{j_1} \cdots a_{l_q}'^{j_q} g_{k_1}' \cdots g_{k_p}' \cdot h'^{l_1} \cdots h'^{l_q}.$$

Calling the new components $A'^{k_1 \cdots k_p}_{l_1 \cdots l_q}$, we have the ordinary laws of transformation. Note that

$$h(g) = \sum_i B_i \bar{h}^i (\sum_j A^j \bar{g}_j) = \sum_{i,j} A^j B_i \bar{h}^i(\bar{g}_j) = \sum_i A^i B_i,$$

so that the terms as here introduced agree with the usage in tensor algebra.

12. **Tensor analysis.** Let M be a differentiable manifold.[19] By a *parametrized curve* C starting at the point x_0 in M we shall mean a differentiable map ϕ of an interval $0 \leqq t \leqq \eta$ into M, with $\phi(0) = x_0$. Let us introduce a coördinate system into a neighborhood of M about x_0, i.e., a (1-1) differentiable map θ of a region of the space E of sets of n numbers $(x^1, \cdots, x^n)$ into M, with non-vanishing Jacobian; say $\theta(0, \cdots, 0) = x_0$. Then C translates into a curve C' in E,

[19] See, for instance, O. Veblen-J. H. C. Whitehead, *Foundations of Differential Geometry*, Cambridge Tracts in Mathematics, No. 29, 1933, or H. Whitney, *Differentiable manifolds*, Annals of Mathematics, vol. 37(1936), pp. 645–680.

given by $\theta^{-1}(\phi(t))$, if η is small enough. We say two parametrized curves starting at x_0 are *equivalent* if, when translated into E, they have the same tangent vector (in both magnitude and direction). Clearly the definition of equivalence is independent of the coördinate system chosen. Hence the classes of equivalent curves form a set of elements intrinsically defined in M; we call these *contravariant vectors* at x_0. Using a fixed coördinate system, we may obtain a (1-1) correspondence between contravariant vectors g at x_0 and vectors v in E at 0, merely by choosing, as an interval, the line segment of v, parametrized so that $t = 1$ at its end, and mapping it (or a portion of it, if it does not lie wholly in the region) into M with θ. We may *add* two contravariant vectors at x by taking the corresponding vectors in E, adding, and mapping back into M. Again the result is independent of the coördinate system chosen; hence *the contravariant vectors at x_0 form an intrinsically defined linear space*, the *tangent space $G(x_0)$ to M at x_0*.

We may obtain an intrinsic definition of $L(G(x_0)) = H(x_0)$ at x_0 by considering differentiable functions defined in a neighborhood of x_0, which vanish at x_0, and calling two functions equivalent if their partial derivatives at x_0 are the same in any coördinate system. To add covariant vectors, we need merely add the corresponding functions.

We shall consider briefly *covariant differentiation* in M. Suppose that to any two sufficiently near points x_0 and x_1 of M corresponds a linear map $\Psi_{x_1 x_0}$ of $G(x_1)$ into $G(x_0)$, so that certain simple continuity and linearity properties are satisfied, which we shall not make precise. This will define an *affine connection*[20] in M. Now let $A(x)$ be a differentiable tensor field, being, for each x, an element of $T(p, q; x)$ (using $G(x)$). Let g be any contravariant vector at x_0, and let C, given by $\phi(t)$, be a corresponding parametrized curve. Then if $x_t = \phi(t)$, we may define

$$\nabla_g A(x_0) = \lim_{t \to 0} \frac{1}{t} [\Psi_{x_t x_0} A(x_t) - A(x_0)]. \tag{12.1}$$

(Of course $\Psi_{x_t x_0}$ may be used to translate a tensor at x_t into a tensor at x_0.) For each g at x_0, $\nabla_g A(x_0)$ is a tensor of $T(p, q; x_0)$, and it depends linearly on g; hence we have a linear map of $G(x_0)$ into $T(p, q; x_0)$. By Theorem 25, there is a natural isomorphism

$$Ch_{T(p,q,x_0)}(G(x_0)) \approx T(p, q; x_0) \circ L(G(x_0)) \approx T(p, q + 1; x_0).$$

Hence, at each point x_0 we have a tensor of $T(p, q + 1; x_0)$, of the same contravariant order as A and of covariant order one greater; this is the *covariant derivative* of A at x_0. Again, the definition is intrinsic.

[20] By using a coördinate system about x_0 and letting $x_1 \to x_0$, we may use this connection to obtain an affine connection in the ordinary sense. Conversely, given an ordinary affine connection, we may define geodesics in M, and by following along them, define a connection as above. If we imbed M in a Euclidean space as in Whitney, loc. cit., Theorem 1, we may realize the tangent spaces by tangent planes of dimension n, and define an affine connection by projecting one tangent plane onto another.

13. **Products, general linear spaces.** A representation $\sum g_i \cdot h_i$ of an element α of $G \circ H$ is *minimal* if there is no representation with fewer summands. The *rank* $\rho(\alpha)$ of α is the number of summands in a minimal representation of α. We consider $0 = 0 \cdot 0$ as having no summands, and set $\rho(0) = 0$.

We collect some known results (at least for finite dimensional spaces) in the following theorem.

THEOREM 27. *$G \circ H$ is a linear space. For any α in $G \circ H$ there are corresponding linear subspaces $G(\alpha)$ and $H(\alpha)$ of G and H with the following properties.*

(a) *There is a representation $\sum g_i \cdot h_i$ for α with g_i in $G(\alpha)$, h_i in $H(\alpha)$. In any representation $\sum g_i' \cdot h_i'$ for α, $G(\alpha) \subset G(g_1', \cdots)$, $H(\alpha) \subset H(h_1', \cdots)$.*

(b) $\dim G(\alpha) = \dim H(\alpha) = \rho(\alpha)$; *$G(\alpha)$ and $H(\alpha)$ are $G(g_1, \cdots)$ and $H(h_1, \cdots)$ in any minimal representation $\sum g_i \cdot h_i$ of α.*

(c) *$\sum g_i \cdot h_i$ is minimal if and only if the sets $g_1, \cdots$ and $h_1, \cdots$ are each independent.*

(d) *If $g_1, \cdots, g_m$ and $h_1, \cdots, h_n$ are bases in subspaces G' of G and H' of H, and $\alpha = \sum a_{ij} g_i \cdot h_j$, then $\rho(\alpha) = \text{rank} \, \| a_{ij} \|$.*

(e) *If $g \cdot h = 0$, then either $g = 0$ or $h = 0$.*

(f) *$g \cdot h = g' \cdot h' \neq 0$ if and only if $g' = ag$, $h' = (1/a)g$ for some real a.*

The first statement follows from Theorems 1 and 13.

Suppose $\alpha = \sum g_i' \cdot h_i' = \sum g_i'' \cdot h_i''$, g_i' in G', g_i'' in G'', h_i' and h_i'' in H^*. Set $G^* = G' \cap G''$, and choose subspaces G_1 and G_2 (possibly containing 0 alone) such that

$$G' + G'' = G^* \oplus G_1 \oplus G_2, \qquad G_1 \subset G', \qquad G_2 \subset G''.$$

Choose bases $\{g_i^*\}$ in G^*, $\{g_i^1\}$ in G_1, $\{g_i^2\}$ in G_2; then all the g's form a base in $G' + G''$. By Theorem 23, we may write uniquely, for some h_i^*, etc., in H^*,

$$\alpha = \sum g_i^* \cdot h_i^* + \sum g_i^1 \cdot h_i^1 + \sum g_i^2 \cdot h_i^2.$$

Now $G' = G^* \oplus G_1$; hence, if we reduce $\sum g_i' \cdot h_i'$ to this normal form, the third group of terms will not appear. As the normal form is unique, the third sum $= 0$. Similarly, as $G'' = G^* \oplus G_2$, the second sum vanishes. Hence $\alpha = \sum g_i^* \cdot h_i^*$ can be expressed by using g's from $G' \cap G''$ alone. Hence there is a minimal subspace $G(\alpha)$ which may be used. Find similarly a minimal $H(\alpha)$. Now α can be expressed, by using $G(\alpha)$ and $H' \supset H(\alpha)$, and $G' \supset G(\alpha)$ and $H(\alpha)$. Choosing bases properly in G' and H' and using the first normal form, we see at once that α may be expressed, using $G(\alpha)$ and $H(\alpha)$. This proves (a).

Next we show that rank $\| a_{ij} \|$ depends on α alone. Suppose $\{g_i\}$ and $\{g_i'\}$ are bases in G', $\{h_i\}$ is a base in H', and $\alpha = \sum a_{ij} g_i \cdot h_j = \sum a_{kj}' g_k' \cdot h_j$. If $g_i = \sum_k b_{ki} g_k'$, then $a_{kj}' = \sum_i b_{ki} a_{ij}$, i.e., $A' = BA$. As B is non-singular, rank $A =$ rank A'. Similarly, a change of base in H' causes no change in the rank. If $G'' \supset G'$ and $H'' \supset H'$, and we choose bases in these spaces containing the above g_i and h_i, then $\sum a_{ij} g_i \cdot h_j$ is also a normal form for α, using G'' and H''. The new $\| a_{ij} \|$ is the old $\| a_{ij} \|$ with extra rows and columns of zeros; the ranks are therefore the same. Now given any two representations of α in normal form, using the pair G', H' and the pair G'', H'', we may also write α in normal form, using $G' + G''$ and $H' + H''$. The above proof shows that all ranks of matrices are the same.

If $\sum_{i=1}^{\rho(\alpha)} g_i \cdot h_i$ is minimal, then obviously the sets $\{g_i\}$ and $\{h_i\}$ are independent. They form bases in spaces G' and H', say, and the expression $\sum g_i \cdot h_i$ is then in normal form. The matrix is the unit matrix, and hence is of rank $\rho(\alpha)$. This proves (d). As $G(\alpha) \subset G'$, and dim $G(\alpha) < \rho(\alpha)$ is clearly impossible, $G(\alpha) = G'$ and dim $G(\alpha) = \rho(\alpha)$; similarly for $H(\alpha)$. (b) is now proved. If $\alpha = \sum_{i=1}^{r} g_i \cdot h_i$ and the sets $\{g_i\}$, $\{h_i\}$ are independent, then we have a representation in normal form, with matrix of rank r; hence $r = \rho(\alpha)$, and $\sum g_i \cdot h_i$ is minimal. This proves (c).

To prove (e), suppose $g \neq 0$, $h \neq 0$. Then $g \cdot h$ is minimal, by (c), hence $\rho(g \cdot h) = 1$, and $g \cdot h \neq 0$. (f) follows from the fact that for $\alpha = g \cdot h = g' \cdot h' \neq 0$, $G(\alpha) =$ all multiples of $g =$ all multiples of g'.

THEOREM 28. *If G' is a linear subspace of G, then there is a natural isomorphism* $G' \circ H \approx \sum^*(G' \cdot H)$.

Using $\sum g_i \times h_i$ in $G' \circ H$, set $\phi(\sum g_i \times h_i) = \sum g_i \cdot h_i$. Clearly ϕ is a uniquely defined homomorphism onto the whole of $\sum^*(G' \cdot H)$. Suppose $\phi(\sum g_i \times h_i) = \sum g_i \cdot h_i$, $\sum g_i \times h_i \neq 0$. We may suppose $\sum g_i \times h_i$ is minimal. Then the sets $\{g_i\}$ and $\{h_i\}$ are independent, and hence $\sum g_i \cdot h_i$ is minimal, by the last theorem, and $\sum g_i \cdot h_i \neq 0$. Hence ϕ is (1-1), and this completes the proof.

14. **On topological linear spaces.** We shall use the following definition. If $G' \subset G^*$, a *projection* of G^* into G' is a linear map of G^* into G' such that every element of G' is fixed.

Definition. We shall call a *topological linear space* G a linear space with sets $U, V, \cdots$, called neighborhoods (of 0), such that:

(1) 0 is in every U;

(2) given U, V, there is a $W \subset U \cap V$;

(3) given U, there is a V such that for $-1 \leqq a \leqq 1$, $aV \subset U$;

(4) given U, there is a V with $V + V \subset U$;

(5) for every U and every g in G there is an a with g in aU;

(6) for every finite dimensional subspace G' of G and every natural neighborhood U' in the space G' (see §10), there is a neighborhood U in G with the following property. If $G^* \supset G'$ is a finite dimensional subspace, then there is a projection of G^* into G' which carries $U \cap G^*$ into U'.

We shall relate this definition to Definition 2b of von Neumann.[21]

Note that (6) implies a separation postulate: If $g \neq 0$, then there is a U which does not contain g. The reason for using our (6) is that with it one may prove the same property, and hence the separation postulate, in tensor products.

[21] J. von Neumann, *On complete topological spaces*, Transactions of the American Mathematical Society, vol. 37 (1935), pp. 1–20. We refer to this paper as N.

THEOREM 29. *A topological linear space (even with* (6) *replaced by a separation postulate) is a regular Hausdorff space;* $g + g'$ *and* ag *are each continuous in both variables.*

As our definition has all the properties in N, Definition 2b, except his (2) and (7), and a separation postulate holds, his proof holds without change.[22] We may now use U_{cl} = closure of U and U_i = inner points of U, etc., as in N.

Preparatory to proving Theorem 30, we note the following facts.

(a) If a set of sets U satisfies the above properties, then so do the sets U_{cl}, the sets U_i, and the sets $U - U$ (= all $g - g'$, g and g' in U).

(b) The sets U_{cl}, $(U_{cl})_i$, and $U - U$ define the same topology (i.e., give the same definition of S_i for any S) as the sets U.

These hold also if N, Definition 2b is used. To prove these facts, first note that N, Theorem 3, in particular, $(aS)_{cl} = aS_{cl}$, holds for closures; the proof is essentially the same. (a) and (b) now follow easily, using especially: $U_{cl} \subset U + U$; $V + V \subset U$ implies $V \subset U_i$.

(c) In a convex space as in N, we may suppose that the U are convex, and either closed or open, and that $-U = U$.

For we may use either the U_{cl} or the $(U_{cl})_i$. The U_{cl} are convex, by N, Theorem 13. To prove this for $(U_{cl})_i = S_i$, take g and g' in S_i and $0 < a < 1$. Set $g^* = ag + (1 - a)g'$, and choose V so that $g + V \subset S$, $g' + V \subset S$. Then

$$g^* + V \subset a(g + V) + (1 - a)(g' + V) \subset S_{cl} = S,$$

and hence g^* is in S_i. Finally, replace each U by $U - U = U'$; then all former properties hold, and $-U' = U'$.

LEMMA 6. *Let G be a convex topological linear space as in* N, *satisfying our* (c). *Let $g_1, \cdots, g_\mu, g'$, be independent; let $g_1, \cdots, g_\mu$ determine the subspace G_1 of G, and the whole set, the subspace G^*. Let m be an integer $\leqq \mu$. Let U be a neighborhood such that*

$$\text{if } \sum_{i=1}^{\mu} a_i g_i \text{ is in } U, \quad \text{then} \quad |a_i| \leqq t_i \qquad (i = 1, \cdots, m). \tag{14.1}$$

Then there is a projection of G^* into G_1 such that the projection of $U \cap G^*$ satisfies the same inequalities.

It will not restrict the generality if we suppose that t_i are the smallest numbers such that (14.1) holds. As $-U = U$, no inequality in (14.1) can be bettered now.

First, suppose we have two elements

$$g_1'' = \sum a_i g_i + cg', \qquad g_2'' = \sum b_i g_i + cg', \qquad \text{in } U; \tag{14.2}$$

then as $U = -U$ is convex,

$$\tfrac{1}{2}(g_1'' - g_2'') = \sum \tfrac{1}{2}(a_i - b_i) g_i \qquad \text{is in } U,$$

[22] In Hausdorff, *Mengenlehre*, Berlin, 1927, there is an error on p. 229: (5) does not follow from (6), as shown by a space in which the only open sets are the null set and the whole space. In N, proof of Theorem 6, one should mention that a separation postulate holds, as a consequence of Definition 2b, (2).

and hence

$$(14.3) \qquad | a_i - b_i | \leqq 2t_i \qquad (i = 1, \cdots, m).$$

Now take any $c \geqq 0$ for which there is an element of the form (14.2) in U; let $\phi_i(c)$ and $\psi_i(c)$ $(i = 1, \cdots, m)$ be the greatest lower bound and least upper bound respectively of all numbers d such that for some numbers $a_1, \cdots, a_{i-1}, a_{i+1}, \cdots, a_\mu$,

$$\sum_{j \neq i} a_j g_j + (\pm t_i + d) g_i + cg' \qquad \text{is in } U \quad (- \text{ for } \phi, + \text{ for } \psi).$$

In other words, $\phi_i(c)$ and $\psi_i(c)$ show how much U sticks out beyond the rectangle of the t_i, in the g_i direction, at the height c, with respect to the direction of g'. By the choice of the t_i, $\phi_i(0) = \psi_i(0) = 0$.

By (14.3), $\phi_i(c) \geqq \psi_i(c)$. We now show that

$$(14.4) \qquad \text{if } \ 0 < c < c', \ \text{ then } \ \frac{\phi_i(c)}{c} \leqq \frac{\phi_i(c')}{c'}, \qquad \frac{\psi_i(c)}{c} \geqq \frac{\psi_i(c')}{c'}.$$

Suppose, for instance, the first inequality is false. Then there are numbers a_j $(j \neq i)$, e, such that

$$g_1' = \sum_{j \neq i} a_j g_j + (-t_i + d) g_i + c'g' \qquad \text{is in } U,$$

$$d = \frac{c'}{c} [\phi_i(c) - e], \qquad e > 0.$$

By the choice of the t_i, there is an element

$$g_2' = \sum_{j \neq i} b_j g_j + (-t_i + d') g_i \text{ in } U, \qquad d' < \frac{c'e}{c' - c}.$$

As U is convex,

$$\frac{c}{c'} g_1' + \frac{c' - c}{c'} g_2' = \sum_{j \neq i} a_j' g_j + \left(-t_i + \frac{c}{c'} d + \frac{c' - c}{c'} d' \right) g_i + cg'$$

is in U. But also

$$\frac{c}{c'} d + \frac{c' - c}{c'} d' = \phi_i(c) - e + \frac{c' - c}{c'} d' < \phi_i(c),$$

contradicting the definition of $\phi_i(c)$.

The inequalities show that we may define

$$(14.5) \qquad \phi_i' = \lim_{c \to 0+} \frac{\phi_i(c)}{c}, \qquad \psi_i' = \lim_{c \to 0+} \frac{\psi_i(c)}{c};$$

then

$$(14.6) \qquad \frac{\phi_i(c)}{c} \geqq \phi_i' \geqq \psi_i' \geqq \frac{\psi_i(c)}{c} \qquad (c > 0; i = 1, \cdots, m).$$

Set

$$g'' = g' + \sum_{i=1}^{m} \phi_i' g_i; \tag{14.7}$$

we shall show that if we project U along the direction of g'' into G_1, i.e., use $\phi(\sum a_i g_i + a'' g'') = \sum a_i g_i$, then (14.1) will hold for the projection. As $-U = U$, it will be sufficient to consider the part of U with $c > 0$. If this is false, say for i, then there is an element

$$\sum_{j=1}^{\mu} a_j g_j + c g'' \text{ in } U, \qquad c > 0, \qquad a_i > t_i \text{ or } a_i < -t_i. \tag{14.8}$$

Using (14.7), we have

$$\sum_{j \neq i}^{j \leqq m} (a_j + c\phi_j')g_j + \sum_{j=m+1}^{\mu} a_j g_j + (a_i + c\phi_i')g_i + cg' \text{ in } U. \tag{14.9}$$

Suppose first that $a_i < -t_i$. Then

$$a_i + c\phi_i' < -t_i + c\,\frac{\phi_i(c)}{c} = -t_i + \phi_i(c),$$

contradicting the definition of $\phi_i(c)$. Next, if $a_i > t_i$, then

$$a_i + c\phi_i' > t_i + c\psi_i' \geqq t_i + \psi_i(c),$$

contradicting the definition of $\psi_i(c)$. This completes the proof.

THEOREM 30. *Any convex topological linear space as in* N *is a topological linear space as here defined, even if his* (2) *is replaced by a separation postulate.*

We may suppose his neighborhoods satisfy our (c). We must prove our (6). Let $g_1, \cdots, g_m$ form a base for G', and choose $t_1, \cdots, t_m$ so that all points $\sum a_i g_i$, $|a_i| \leqq t_i$, lie in U'. Let R be the closed region $|a_i| \leqq t_i$, and let A be its boundary. For each g in A, we may choose a $U(g)$ not containing it, and a $V(g)$ so that $V(g) - V(g) \subset U(g)$. As the operations in G are continuous, $\sum a_k g_k$ is continuous in the a_k, so the $(V(g))_i \cap G'$ (S_i = inner points of S) are open in the natural topology in G'; hence a finite number of the sets $g + V(g) \cap G'$ cover A. Let U be a neighborhood in the corresponding set $V(g_1) \cap \cdots \cap V(g_\lambda)$. Now U contains no element of A. For suppose g is in $A \cap U$. Say g is in $g_k + V(g_k)$. Then as g is in $U \subset V(g_k)$, g_k is in $V(g_k) - V(g_k) \subset U(g_k)$, a contradiction. As U is convex, $U \cap G'$ is in the complement of A in R.

Let $g_{m+1}', \cdots, g_n'$ form, with $g_1, \cdots, g_m$, a base in G^* (if $G^* \neq G'$), and let G_i be the space generated by $g_1, \cdots, g_i'$ ($i = m + 1, \cdots, n$). We shall prove, by induction on i, that G_i can be projected into G' so that $U \cap G_i$ goes into R; as $R \subset U'$, the case $i = n$ gives the theorem. There will be elements $g_{m+1}, \cdots, g_n$ such that $g_1, \cdots, g_i$ also determine G_i, and the projection of G_i into G' is with respect to $g_{m+1}, \cdots, g_i$:

$$\phi\left(\sum_{k=1}^{i} a_k g_k\right) = \sum_{k=1}^{m} a_k g_k.$$

Suppose we have found the elements $g_{m+1}, \cdots, g_\mu$. As the projection of G_μ into G' carries $U \cap G_\mu$ into R, (14.1) is satisfied. Hence we may apply Lemma 6 with $g' = g'_{\mu+1}$, $G^* = G_{\mu+1}$; this gives a projection of $G_{\mu+1}$ into G_μ such that the projection of $U \cap G_{\mu+1}$ satisfies (14.1). Let $g_{\mu+1}$ give the direction of the projection; then projecting $G_{\mu+1}$ into G' with respect to $g_{m+1}, \cdots, g_{\mu+1}$ carries $U \cap G_{\mu+1}$ into R, as required.

15. **Products, topological linear spaces.** We prove

THEOREM 31. *If G and H are topological linear spaces, so is $G \circ H$, the topology being given by* (10.12). *(We may use either open or closed neighborhoods in G and in H; see* §14.) *The multiplication $g \cdot h$ is continuous. The topology in $G \circ H$ depends only on the topologies in G and in H, not on the neighborhood systems employed.*

First, $G \circ H$ is a linear space, by Theorem 27. We shall prove the postulates of §14. (1) is trivial. To prove (2), take any two neighborhoods $N(U_1, \cdots; V_1, \cdots) = N(U_i; V_i)$, and $N(U'_i; V'_i)$. Choose U''_i in $U_i \cap U'_i$ and V''_i in $V_i \cap V'_i$; then

$$N(U''_i; V''_i) \subset N(U_i; V_i) \cap N(U'_i; V'_i).$$

To prove (3), given $N(U_i; V_i)$, choose U'_i so that $aU'_i \subset U_i$ if $|a| \leqq 1$ ($i = 1, 2, \cdots$). Then

$$aN(U'_i; V'_i) = N(aU'_i; V'_i) \subset N(U_i; V_i).$$

To prove (4), take any $N(U_i; V_i)$. Choose U'_i and V'_i so that

$$U'_i \subset U_{2i-1} \cap U_{2i}, \qquad V'_i \subset V_{2i-1} \cap V_{2i}.$$

Now take any $\sum g_i \cdot h_i$ and $\sum g'_i \cdot h'_i$ in $N(U'_i; V'_i)$. As $g_1 \cdot h_1$ is in $U'_1 \cdot V'_1 \subset U_1 \cdot V_1$, $g'_1 \cdot h'_1$ is in $U'_1 \cdot V'_1 \subset U_2 \cdot V_2$, $g_2 \cdot h_2$ is in $U'_2 \cdot V'_2 \subset U_3 \cdot V_3$, etc., we see that $\sum g_i \cdot h_i + \sum g'_i \cdot h'_i$ is in $N(U_i; V_i)$.

To prove (5), take any $\sum_{i=1}^{s} g_i \cdot h_i$ and any $N(U_i; V_i)$. Choose U^*_i and V^*_i so that

$$cU^*_i \subset U_i, \qquad cV^*_i \subset V_i, \qquad (|c| \leqq 1; i = 1, \cdots, s),$$

take a_i and b_i so that

$$g_i \text{ is in } a_i U^*_i, \qquad h_i \text{ is in } b_i V^*_i,$$

and let a be the largest of the $|a_i|$ and $|b_i|$. Then

$$a_i U^*_i = a(a_i/a) U^*_i \subset a U_i, \qquad \text{etc.};$$

it follows that $\sum g_i \cdot h_i$ is in

$$a^2 N(U_i; V_i) = N(aU_i; aV_i).$$

We now prove (6). Let F' be a subspace of $F = G \circ H$, generated by $f_1, \cdots, f_s$. Set (see Theorem 27)

$$G' = G(f_1) + \cdots + G(f_s), \qquad H' = H(f_1) + \cdots + H(f_s).$$

Let $g_1, \cdots, g_m$ and $h_1, \cdots, h_n$ be bases in G' and H'; set $f_{ij} = g_i \cdot h_j$. Then the f_{ij} form a base (see Theorem 26) in a space $F'' \supset F'$. Take a fixed projection of F'' into F'. Given a natural neighborhood N' in F', we may choose a natural neighborhood N'' in F'' which projects into a subset of N'. As any projection of an F^* into F'' combines with the above projection to give a projection of F^* into F', it is sufficient to prove (6) with F', N' replaced by F'', N''.

Choose $\epsilon > 0$ so that any $\sum a_{ij} f_{ij}$ with each $| a_{ij} | \leqq \epsilon$ is in N''. Let A and B be the sets of elements $\sum a_i g_i$ and $\sum b_i h_i$ in G' and H' with $| a_i | \leqq \frac{1}{2}\epsilon$, $| b_i | \leqq 1$. Choose U_1 in G by (6) so that any $U_1 \cap G^*$ can be projected into A, and choose V_1 in H so that any $V_1 \cap H^*$ can be projected into B. Choose $U_2, U_3, \cdots$ so $2U_i \subset U_{i-1}$, and set $V_2 = V_3 = \cdots = V_1$. Set $N = N(U_i ; V_i)$.

Now take any $F^* \supset F''$. Choose a base $f_1^*, \cdots, f_t^*$ in F^*, and set $G^* = \sum G(f_i^*)$, $H^* = \sum H(f_i^*)$. Choose projections of G^* into G' and H^* into H' so $U_1 \cap G^*$ goes into A and $V_1 \cap H^*$ goes into B. If G_1 is the subset of G^* projecting into 0 in G', and $g_{m+1}, \cdots, g_\mu$ is a base in G_1, then $g_1, \cdots, g_\mu$ is a base in G^*; choose a base $h_1, \cdots, h_\nu$ in H^* similarly. Now any element of F^* can be written uniquely in the form

$$f = \sum_{(i,j)=(1,1)}^{(m,n)} a_{ij} f_{ij} + \sum{}' a_{ij} g_i \cdot h_j,$$

where in $\sum'$, either $i > m$ or $j > n$. (Not all such elements need be in F^*.) Dropping out the second group of terms defines a projection of F^* into F''.

We shall show by induction that any $(U_i \cdot V_i) \cap F^*$ projects into elements $\sum a_{kl} f_{kl}$ with $| a_{kl} | \leqq \epsilon/2^i$; it will follow that $N = \sum_i^* (U_i \cdot V_i)$ projects into N''.

Take first any α in $(U_1 \cdot V_1) \cap F^*$; we may suppose $\alpha \neq 0$. Then $\alpha = g \cdot h$, g in U_1, h in V_1. As α is in F^*, $G(\alpha) \subset G^*$. But also $G(\alpha) \subset G(g)$; hence $G(\alpha) \subset G^* \cap G(g)$. As $\alpha \neq 0$, $G(\alpha)$ contains elements $\neq 0$, which implies that g is in G^*. Similarly h is in H^*. Say

$$g = \sum_{i=1}^{\mu} a_i g_i, \qquad h = \sum_{j=1}^{\nu} b_j h_j.$$

Then as g projects into A and h into B, $g \cdot h$ projects into

$$\sum_{(i,j)=(1,1)}^{(m,n)} a_i b_j f_{ij}, \qquad | a_i b_j | < \tfrac{1}{2}\epsilon,$$

so that the statement holds for $(U_1 \cdot V_1) \cap F^*$. Supposing it holds for $k - 1$, we shall prove it for k. Take any g in U_k and h in V_k such that $g \cdot h$ is in F^*. Then

$$2g \text{ is in } 2U_k \subset U_{k-1}, \qquad h \text{ is in } V_{k-1},$$

so that $2(g\cdot h)$ is in $(U_{k-1}\cdot V_{k-1}) \cap F^*$, and hence projects into $\sum a_{ij}f_{ij}$ with $|a_{ij}| \leqq \epsilon/2^{k-1}$. Hence the required inequality on $g\cdot h$ holds. This completes the proof of (6).

To show that $g\cdot h$ is continuous, as $+$ is continuous in $G \circ H$ (Theorem 29), and

$$(15.1) \qquad (g + g')\cdot(h + h') - g\cdot h = g\cdot h' + g'\cdot h + g'\cdot h',$$

it is sufficient to show that $g\cdot h'$ is continuous in h' at $h' = 0$, $g'\cdot h$ is continuous in g' at $g' = 0$, and $g'\cdot h'$ is continuous in g' and h' at $g' = 0$, $h' = 0$. For the first case, given $N = N(U_i ; V_i)$, choose a so that g is in aU_1, and choose V so that $aV \subset V_1$ (N, Theorem 1, with $n = 1$). Then

$$g\cdot V \subset aU_1\cdot V = U_1\cdot aV \subset U_1\cdot V_1 \subset N.$$

The second case is similar. The third is clear, as $U_1\cdot V_1 \subset N$.

Finally, let $\{U\}$, $\{\bar{U}\}$ and $\{V\}$, $\{\bar{V}\}$ be equivalent neighborhood systems in G and H, respectively. Given an $N(U_i ; V_i)$, choose $\bar{U}_i \subset U_i$ and $\bar{V}_i \subset V_i$ $(i = 1, 2, \cdots)$; then $\bar{N}(\bar{U}_i ; \bar{V}_i) \subset N(U_i ; V_i)$. Similarly find an N in any $\bar{N}$. Hence the $\{N\}$ and $\{\bar{N}\}$ are equivalent. The theorem is now completely proved.

III. Topological groups

16. **The topological tensor product.** An Abelian *topological group* G is an Abelian group which is at the same time a Hausdorff space,[23] and in which $\phi(g, g') = g + g'$ and $\psi(g) = -g$ are continuous. If $U, U', \cdots$ are the neighborhoods of 0, we may let the sets $g + U$, $g + U'$, $\cdots$ be the neighborhoods of the element g, without altering the topology. If we assume that the separation postulate in Hausdorff, page 229, (4), holds, then the postulate (5) follows.

We shall say the space G is *sequence-separable* if it contains a finite or denumerable set of points forming a dense set.[24]

If G and H are sequence-separable topological groups, we define their *topological tensor product* $G \circ H$, or *tensor product* simply, as follows. Let $\bar{g}_1, \bar{g}_2, \cdots$ and $\bar{h}_1, \bar{h}_2, \cdots$ be sequences of points dense in G and H, respectively. Let $P_1, P_2, \cdots$ be a sequence of pairs of elements, $P_i = (\bar{g}_{\mu_i}, \bar{h}_{\nu_i})$, such that each pair $(\bar{g}_j, \bar{h}_k)$ occurs infinitely often among the P_i. Let T' be the discrete tensor product of G and H, with elements $\sum g_i \times h_i$. For each sequence U_1, $U_2, \cdots$ of neighborhoods (of 0) in G and each sequence $V_1, V_2, \cdots$ in H, set

$$(16.1) \qquad \begin{aligned} Q_i'(U, V) &= \bar{g}_{\mu_i} \times V + U \times \bar{h}_{\nu_i} + U \times V, \\ N'(U_1, \cdots ; V_1, \cdots) &= \sum_i{}^* Q_i'(U_i, V_i). \end{aligned}$$

Next, call two elements α, β of T' *equivalent*, $\alpha \sim \beta$, if every $\alpha + N$ contains β or vice versa, or if there is a succession $\alpha = \alpha_0, \alpha_1, \cdots, \alpha_n = \beta$, with α_i and

[23] See Hausdorff, loc. cit. Note that neighborhoods are open sets here.

[24] For metric spaces, this is the same as separability.

α_{i+1} equivalent as above. The sets of equivalent elements form the elements of the tensor product $T = G \circ H$. The *neighborhoods* N of 0 in $G \circ H$ are the images of the sets N' in T'; they are obtained by replacing $\times$ by $\cdot$ in (16.1). *Addition* in T is the image of addition in T'. The element $\sum g_i \cdot h_i$ in T is the image of $\sum g_i \times h_i$ in T'.

THEOREM 32. *$G \circ H$ is a sequence-separable Abelian topological group; the multiplication $g \cdot h$ satisfies* (1.1), *and is continuous. The topology in $G \circ H$ is independent of the sequences $\{\bar{g}_i\}$, $\{\bar{h}_i\}$, and of the neighborhood systems $\{U\}$, $\{V\}$, employed.*

We begin by showing that T' has all the properties of a topological group, except for the separation postulate. First we prove Hausdorff's postulates (B), (C) (loc. cit., p. 228). Given $N'(U_1, \cdots; V_1, \cdots) = N'(U_l; V_l)$ and $N'(U_l'; V_l')$, take $U_i'' \subset U_i \cap U_i'$ and $V_i'' \subset V_i \cap V_i'$ $(i = 1, 2, \cdots)$; then clearly

$$N'(U_l''; V_l'') \subset N'(U_l; V_l) \cap N'(U_l'; V_l'). \tag{16.2}$$

To prove (C), it is sufficient to show that, for any $N'(U_l; V_l)$ and any α in $N'(U_l; V_l)$, there is an $N'(U_l'; V_l')$ with

$$\alpha + N'(U_l'; V_l') \subset N'(U_l; V_l). \tag{16.3}$$

As α is in $N'(U_l; V_l)$, it is in $\sum_{i=1}^{s} Q_i'(U_i; V_i)$ for some s. Choose numbers $\phi(1) > s, \phi(2) > \phi(1), \phi(3) > \phi(2), \cdots$ so that $P_i = P_{\phi(i)}$, and set $U_i' = U_{\phi(i)}$, $V_i' = V_{\phi(i)}$. Then

$$\begin{aligned} Q_i'(U_i', V_i') &= \bar{g}_{\mu_i} \times V_i' + U_i' \times \bar{h}_{\nu_i} + U_i' \times V_i' \\ &= \bar{g}_{\mu_{\phi(i)}} \times V_{\phi(i)} + U_{\phi(i)} \times \bar{h}_{\nu_{\phi(i)}} + U_{\phi(i)} \times V_{\phi(i)} = Q_{\phi(i)}'(U_{\phi(i)}, V_{\phi(i)}); \end{aligned}$$

hence $\sum_i^* Q_i'(U_i', V_i') \subset \sum_{i>s}^* Q_i'(U_i, V_i)$, and (16.3) follows.

We show that the group operations in T' are continuous. Given $N'(U_l; V_l)$, take

$$U_i' \subset U_i \cap (-U_i), \qquad V_i' \subset -V_i;$$

then

$$-Q_i'(U_i', V_i') = \bar{g}_{\mu_i} \times (-V_i') + (-U_i') \times \bar{h}_{\nu_i} + U_i' \times (-V_i') \subset Q_i'(U_i, V_i);$$

hence

$$N'(U_l'; V_l') \subset -N'(U_l; V_l), \tag{16.4}$$

and $-\alpha$ is continuous in α. To show that $\alpha + \beta$ is continuous, we must find $N'(U_l'; V_l')$ corresponding to $N'(U_l; V_l)$ such that

$$N'(U_l'; V_l') + N'(U_l'; V_l') \subset N'(U_l; V_l). \tag{16.5}$$

Choose in succession integers

$$\phi(1), \qquad \psi(1) > \phi(1), \qquad \phi(2) > \psi(1), \qquad \psi(2) > \phi(2), \qquad \cdots$$

so that $P_i = P_{\phi(i)} = P_{\psi(i)}$. Take

$$U'_i \subset U_{\phi(i)} \cap U_{\psi(i)}, \qquad V'_i \subset V_{\phi(i)} \cap V_{\psi(i)}.$$

Then as $g_{\mu\phi(i)} = g_{\mu i}$, etc.,

$$Q'_i(U'_i, V'_i) \subset Q'_{\phi(i)}(U_{\phi(i)}, V_{\phi(i)}) \cap Q'_{\psi(i)}(U_{\psi(i)}, V_{\psi(i)}).$$

Hence

$$Q'_i(\cdots) + Q'_i(\cdots) \subset Q'_{\phi(i)}(\cdots) + Q'_{\psi(i)}(\cdots),$$

and (16.5) follows.

We now consider equivalent elements in T'. First we prove

(*) *If $\alpha \sim \beta$, then for every N', $\alpha + N'$ contains β.*

For suppose there is a succession $\alpha = \alpha_0, \alpha_1, \cdots, \alpha_n = \beta$, such that for each i, either every $\alpha_i + N'$ contains α_{i+1}, or every $\alpha_{i+1} + N'$ contains α_i. The latter condition implies the former. For given an N', choose $N'_1 \subset -N'$ by (16.4); then as α_i is in $\alpha_{i+1} + N'_1$, α_{i+1} is in $\alpha_i - N'_1 \subset \alpha_i + N'$. Next, given an N', choose N'_1 (using (16.5)) so that

$$N'_1 + N'_1 + \cdots + N'_1 \subset N' \qquad (n \text{ summands}).$$

Setting $A_k = N'_1 + \cdots + N'_1$ (k summands), we have

$$\alpha + N' \supset \alpha_0 + A_n \supset \alpha_1 + A_{n-1} \supset \cdots \supset \alpha_{n-1} + N'_1 \supset \beta,$$

as required.

Next we prove that T is a topological group. Let $\theta(\alpha)$ be the element of T containing the element α of T'. We must show that addition in T is uniquely defined; this is so if $\alpha \sim \alpha'$ and $\beta \sim \beta'$ imply $\alpha + \beta \sim \alpha' + \beta'$. Given any N', choose N'_1 so that $N'_1 + N'_1 \subset N'$. By the property (*), $\alpha + N'_1 \supset \alpha'$ and $\beta + N'_1 \supset \beta'$; hence

$$(\alpha + \beta) + N' \supset (\alpha + N'_1) + (\beta + N'_1) \supset \alpha' + \beta',$$

and $\alpha + \beta \sim \alpha' + \beta'$. Further,

$$\theta(\alpha + \beta) = \theta(\alpha) + \theta(\beta),$$

so that θ is a homomorphism of T' into T (which is clearly an Abelian group). To prove that T is a Hausdorff space, suppose $N_1 = \theta(N'_1)$ and $N_2 = \theta(N'_2)$ are given; take $N' \subset N'_1 \cap N'_2$; then $N = \theta(N') \subset N_1 \cap N_2$. Next, suppose α^* is in $N = \theta(N')$; then $\alpha^* = \theta(\alpha)$, α in N'. Choose N'_1 so that $\alpha + N'_1 \subset N'$; then $\alpha^* + \theta(N'_1) = \theta(\alpha + N'_1) \subset N$. To prove the separation postulate, suppose $\alpha^* \neq 0$. Say $\alpha^* = \theta(\alpha)$. As $\theta(0) = 0$, α is not ~ 0, and there is an N' not containing α. Set $N = \theta(N')$; then α^* is not in N. For if it were, then we would have $\alpha^* = \theta(\beta)$, β in N' and $\beta \sim \alpha$; but taking N'_1 by (16.3) so that $\beta + N'_1 \subset N'$, the property (*) gives $\alpha \subset \beta + N'_1 \subset N'$, a contradiction. To prove that the operations in T are continuous, given $N = \theta(N')$, take $N'_1 \subset -N'$; then $N_1 = \theta(N'_1) \subset -N$; also given $N = \theta(N')$, choose N'_1 so that $N'_1 + N'_1 \subset N'$; then $\theta(N'_1) + \theta(N'_1) \subset N$.

Next, (1) holds, as it holds for $\times$. We shall now show that $g \cdot h$ is continuous. First we show that it is continuous in h at $h = 0$; given $N = N(U_l ; V_l)$, we must find V so that $g \cdot V \subset N$. As the $\bar{g}_i$ are dense in G, we may choose j so that $g - \bar{g}_{\mu_j}$ is in U_1. Choose $V \subset V_1 \cap V_j$; then

$$g \cdot V \subset (g - \bar{g}_{\mu_j}) \cdot V + \bar{g}_{\mu_j} \cdot V \subset U_1 \cdot V_1 + \bar{g}_{\mu_j} \cdot V_1 \subset N.$$

Similarly $g \cdot h$ is continuous in g at $g = 0$. Further, $g \cdot h$ is continuous in both variables at $g = 0$, $h = 0$; for $U_1 \cdot V_1 \subset N(U_l ; V_l)$. Finally, as addition is continuous in T, (15.1) shows that $g \cdot h$ is continuous.

Next we show that T is sequence-separable; in fact, that the set of all finite sums $\sum \bar{g}_{p_i} \cdot \bar{h}_{q_i}$ is dense in T. As each element of T is a finite sum $\sum g_i \cdot h_i$ and $+$ is continuous, it is sufficient to show that for any $g \cdot h$ and any $N = N(U_l ; V_l)$ there is a $\bar{g}_i \cdot \bar{h}_j$ in $g \cdot h + N$. As $\cdot$ is continuous, we may choose U and V so that $(g + U) \cdot (h + V) \subset g \cdot h + N$; we need now merely take $\bar{g}_i$ in $g + U$ and $\bar{h}_j$ in $h + V$.

That the topology in T is independent of the neighborhood systems chosen is trivial; see the end of the proof of Theorem 31. We must still show that the topology is independent of the choice of the $\bar{g}_i$ and $\bar{h}_i$. By symmetry, it is sufficient to show that if $\{\bar{g}_i\}$ is replaced by the dense sequence $\{g_i^*\}$, then any $N(U_l; V_l)$ contains an $N^*(U_l'; V_l')$, defined in terms of the g_i^*. Let ξ_i, η_i replace μ_i, ν_i. Given $N(U_l ; V_l) = N_0$, choose $N_1, N_2, \cdots$ in succession so that $N_{i+1} + N_{i+1} \subset N_i$. As $\cdot$ and $+$ are continuous, we can choose U_i' and V_i' so that

$$Q_i^* = \bar{g}_{\xi_i} \cdot V_i' + U_i' \cdot \bar{h}_{\eta_i} + U_i' \cdot V_i' \subset N_i ;$$

then for any s,

$$\sum_{i=1}^{s} Q_i^* \subset N_1 + \cdots + N_{s-1} + N_s \subset N_1 + \cdots + N_{s-2} + N_{s-1} + N_{s-1}$$

$$\subset N_1 + \cdots + N_{s-2} + N_{s-2} \subset \cdots \subset N_1 + N_1 \subset N_0,$$

and hence $N^*(U_l'; V_l') \subset N_0$, as required. This completes the proof of the theorem.

THEOREM 33. *Let $g_1^*, g_2^*, \cdots$ and $h_1^*, h_2^*, \cdots$ be (finite or infinite) sequences such that the sets $\sum a_i g_i^*$ and $\sum a_i h_i^*$ (integral a_i) are dense in G and H, respectively. Then we may use these sequences in place of dense sequences in defining the topology in $G \circ H$.*

Let $h_1, h_2, \cdots$ be either the above sequence $h_1^*, h_2^*, \cdots$, or a dense sequence in H. Arrange all $\sum a_i g_i^*$ in a sequence $\bar{g}_1, \bar{g}_2, \cdots$. Let $N(U_l; V_l)$ be defined in terms of the sets $\{\bar{g}_i\}$, $\{h_i\}$, and $N^*(U_l; V_l)$, in terms of the sets $\{g_i^*\}$, $\{h_i\}$. It is sufficient to show that these two sets of neighborhoods give the same topology in T. As the g_i^* occur among the $\bar{g}_i$, it is clear that any N contains an N^*; we must prove the converse.

Let $P_i = (\bar{g}_{\mu_i}, h_{\nu_i})$ and $P_i^* = (g_{\xi_i}^*, h_{\eta_i})$ define the sequences of pairs defining the N and the N^*. If $\bar{g}_{\mu_i} = \sum_j a_{ij} g_j^*$, set $m(i) = \sum_j |a_{ij}|$. Then $\bar{g}_{\mu_i} \cdot V$ is contained in $m(i)$ terms of the form $g_j^* \cdot (\pm V)$, and $Q_i(U_i', V_i')$ is contained in $m(i) + 2$ terms of forms appearing (except for the $\pm$) in $N^*(U_l ; V_l)$. For each i we may choose $m(i) + 2$ numbers $\phi_1(i), \cdots, \phi_{m(i)+2}(i)$ such that $\phi_k(i) \neq \phi_l(j)$ whenever $i \neq j$, and the k-th part into which Q_i is split corresponds to part of $Q^*_{\phi_k(i)}$. Then if the U_i' and V_i' are chosen small enough, we will have

$$Q_i(U_i', V_i') \subset \sum_{k=1}^{m(i)} Q^*_{\phi_k(i)}(U_{\phi_k(i)}, V_{\phi_k(i)}),$$

and hence $N(U_l'; V_l') \subset N^*(U_l ; V_l)$, as required.

17. Relation to linear spaces; examples. We shall show that whenever the definitions of tensor products in Parts II and III both apply, they coincide.

THEOREM 34. *If G and H are sequence-separable topological linear spaces, then their topological tensor product T is the same as their topological reduced tensor product T^*.*

Let $\sum g_i \times h_i$ and $\sum g_i \cdot h_i$ denote elements of T^* and T, respectively. Set $\phi(\sum g_i \times h_i) = \sum g_i \cdot h_i$; we shall show that ϕ is a topological isomorphism. To show that ϕ is uniquely defined, we must show that $ag \cdot h = g \cdot ah$ for any real a; but this follows from the continuity of $g \cdot h$ (see §10). ϕ is a homomorphism; we shall show that it is continuous. Use $N(U_l ; V_l)$ in T and $N^*(U_l ; V_l)$ in T^*. Given $N(U_l; V_l)$, we wish to find $N^*(U_l'; V_l')$ mapping into it. From (10.12) and (16.1) it is apparent that we may use $U_i' = U_i$, $V_i' = V_i$.

Next we show that for any $N^* = N^*(U_l ; V_l)$, there is an $N = N(U_l'; V_l') \subset \phi(N^*)$. Say $\bar{g}_1, \bar{g}_2, \cdots$ and $\bar{h}_1, \bar{h}_2, \cdots$ are the dense sequences used in G and H, and $P_1, P_2, \cdots, P_i = (\bar{g}_{\mu_i}, \bar{h}_{\nu_i})$, the pairs. By §14, (5), we may choose for each i a number a_i such that $\bar{g}_{\mu_i}$ is in $a_i U_{3i-2}$, and a number b_i such that $\bar{h}_{\nu_i}$ is in $b_i V_{3i-1}$. By von Neumann, loc. cit., Theorem 1 (with $n = 1$), there is a V_i'' such that $a_i V_i'' \subset V_{3i-2}$, and a U_i'' such that $b_i U_i'' \subset U_{3i-1}$. Choose

$$U_i' \subset U_i'' \cap U_{3i}, \qquad V_i' \subset V_i'' \cap V_{3i}.$$

Now

$$\bar{g}_{\mu_i} \cdot V_i' + U_i' \cdot \bar{h}_{\nu_i} + U_i' \cdot V_i' \subset a_i U_{3i-2} \cdot \frac{1}{a_i} V_{3i-2} + \frac{1}{b_i} U_{3i-1} \cdot b_i V_{3i-1} + U_{3i} \cdot V_{3i}$$
$$= \phi(U_{3i-2} \times V_{3i-2} + U_{3i-1} \times V_{3i-1} + U_{3i} \times V_{3i});$$

hence $N \subset \phi(N^*)$.

Clearly ϕ maps T^* into the whole of T. When we have shown that ϕ is (1-1), the proof will be completed. Let T' be the discrete tensor product of G and H; use $g \circ h$ here. Take any $\alpha^* = \sum g_i \times h_i \neq 0$ in T^*; then $\alpha' = \sum g_i \circ h_i$ is a corresponding element of T'. There is an $N^* = N^*(U_l ; V_l)$ which does not contain α^*. Construct $U_1', U_2', \cdots$ and $V_1', V_2', \cdots$ by the method given above, and set

$$N' = \sum_i{}^* (\bar{g}_{\mu_i} \circ V_i' + U_i' \circ \bar{h}_{\nu_i} + U_i' \circ V_i').$$

The map $\psi(\sum g_i' \circ h_i') = \sum g_i' \times h_i'$ of T' into T^* is uniquely defined. By the choice of the U_i' and V_i', $\psi(N') \subset N^*$; hence α' is not in N'. Therefore, by the property (*) in §16, α' is not ~ 0, so that the corresponding element $\sum g_i \cdot h_i$ of T is $\neq 0$. Consequently $\alpha^* \neq 0$ implies $\phi(\alpha^*) \neq 0$, and ϕ is (1-1).

Examples. That the topology in (10.12) cannot be used in the general case is shown by the example $I_0 \circ Rl$. A neighborhood U of 0 in I_0 is 0 itself; hence $U \cdot V = 0 \cdot V = 0$, and $I_0 \circ Rl$ would be discrete; the multiplication $a \cdot g$ would not be continuous. However, the sets $1 \cdot V$ form a neighborhood system in $I_0 \circ Rl$. In fact, *if G has a finite number of generators $\bar{g}_1, \cdots, \bar{g}_n$, then the sets*

$$\bar{g}_1 \cdot V_1 + \cdots + \bar{g}_n \cdot V_n \qquad (V_1, \cdots, V_n \textit{ neighborhoods in } H) \tag{17.1}$$

form a neighborhood system in $G \circ H$. This is an easy consequence of Theorem 33 (compare Theorem 26).

HARVARD UNIVERSITY.

Reprinted from the Proceedings of the NATIONAL ACADEMY OF SCIENCES,
Vol. 26, No. 2, pp. 143–153. February, 1940.

SOME COMBINATORIAL PROPERTIES OF COMPLEXES

BY HASSLER WHITNEY

DEPARTMENT OF MATHEMATICS, HARVARD UNIVERSITY

Communicated January 13, 1940

1. *Introduction.*—The purpose of this note is twofold. First, we give a definition of the "dual" of a cell in a larger cell, and use this to define products in a complex.[1] Second, we discuss "locally isomorphic" complexes, and products in such complexes. The results are needed in the following note.

2. *Dual Systems in a Complex.*—For each pair of oriented cells σ^r, σ^{r+s} in a geometric complex K, let

$$D(\sigma^r, \sigma^{r+s}) = \text{dual of } \sigma^r \text{ in } \sigma^{r+s}$$

be a singular s-chain whose cells are in σ^{r+s}, and each cell of the closure of which is in either σ^{r+s} or σ^r. Let it be 0 if σ^r and σ^{r+s} are not incident. Set

$$D(\Sigma\alpha_i\sigma_i^r, \Sigma\beta_j\sigma_j^{r+s}) = \Sigma\alpha_i\beta_j D(\sigma_i^r, \sigma_j^{r+s}).$$

If these satisfy the two conditions

$$D(x_i, x_i) = x_i \qquad \text{(for each vertex } x_i\text{)},$$

$$\partial D(\sigma^r, \sigma^{r+s}) = (-1)^{r+1}[D(\delta\sigma^r, \sigma^{r+s}) - D(\sigma^r, \partial\sigma^{r+s})],$$

we say these form a *system dual to K.*

3. *Intersections in a Cell.*—Let E'^n be an oriented half-plane bounded by E^{n-1}: $\partial E'^n = E^{n-1}$. Let

$$\sigma^r = x_{\lambda_0}\ldots x_{\lambda_r}, \ \sigma^s = x_{\mu_0}\ldots x_{\mu_s}, \ (r + s = n),$$

be simplexes in $E'^n + E^{n-1}$. With a simple definition of *general position relative to* E^{n-1}, we may define their Kronecker index $\{\sigma^r, \sigma^s, E'^n\}$ as usual;

we omit anything in E^{n-1}. (The cells may be singular, if any one is wholly in E'^n or E^{n-1}.) Let their index in any E^* be 0 if they are not both in E^*. Then

$$\{A^r, \partial B^{r+s}, E'^n\} - (-1)^r\{\partial A^r, B^s, E'^n\} + \{\partial A^r, \partial B^s, \partial E'^n\} = 0.$$

Applying the same theory to a cell and its faces gives the same formula with E'^n replaced by a cell or chain C^n.

4. *Products in Terms of Intersections.*—Say two systems D, D' dual to K are in *general position* if their sets of singular cells are in general position in each σ^r relative to $\partial\sigma^r$.

If this holds, and we set

$$(\sigma^r \smile \sigma^s)\cdot\sigma^{r+s} = \{D(\sigma^r, \sigma^{r+s}), D'(\sigma^s, \sigma^{r+s}), \sigma^{r+s}\},$$

then the $\smile$ and corresponding $\frown$ products satisfy the conditions in PC.

5. *Special Products.*—The products of PC, §6, may be obtained as follows: Let K be simplicial, with ordered vertices. Take $0 < a < 1$, and set

$$y_{\lambda_0} = x_{\lambda_0}, \quad y_{\lambda_0\ldots\lambda_r} = (1-a)x_{\lambda_0} + ay_{\lambda_1\ldots\lambda_r} \quad (\lambda_0 < \ldots < \lambda_r).$$

$y_{\lambda_0\ldots\lambda_r}$ is interior to $x_{\lambda_0}\ldots x_{\lambda_r}$. Let

$$[\lambda_0' \ldots \lambda_r'] = +1 \text{ or } -1$$

according as $(\lambda_0',\ldots,\lambda_r')$ is an even or odd permutation of the natural order of these numbers. If $(\lambda_0',\ldots,\lambda_r')$ is a permutation of the natural order $\lambda_0 < \ldots < \lambda_r$, we use

$$x_{\lambda_0}\ldots x_{\lambda_r} = [\lambda_0'\ldots\lambda_r']x_{\lambda'_0}\ldots x_{\lambda'_r}, \quad y_{\lambda'_0\ldots\lambda'_r} = y_{\lambda_0\ldots\lambda_r}.$$

Now take any cells

$$\sigma^r = x_{\lambda_0}\ldots x_{\lambda_r}, \quad \sigma^{r+s} = [\lambda_0\ldots\lambda_r\ldots\lambda_{r+s}]x_{\lambda_0}\ldots x_{\lambda_r}\ldots x_{\lambda_{r+s}}, \quad \lambda_0 < \ldots < \lambda_r, \quad \lambda_{r+1} < \ldots < \lambda_{r+s},$$

and set

$$D_a(\sigma^r, \sigma^{r+s}) = \sum_{(\mu)}[\lambda_0\ldots\lambda_r\mu_1\ldots\mu_s]y_{\lambda_0\ldots\lambda_r}y_{\lambda_0\ldots\lambda_r\mu_1}\ldots y_{\lambda_0\ldots\lambda_r\mu_1\ldots\mu_s},$$

summed over all permutations $(\mu_1,\ldots,\mu_s)$ of $(\lambda_{r+1},\ldots,\lambda_{r+s})$. Then *this is a system dual to* K. *If* $0 < a' < a < 1$, *then the systems* $D_a, D_{a'}$ *are in general position, and the definition in* §4 *gives the products of* PC, §6. (This is easy to verify for $r + s \leq 2$.) If we replace a' by 0 and a by 1, we find interesting combinatorial formulas.

6. *A New Particular* $\smile$ *Product.*—Writing $A^r = \Sigma a_{\lambda_0 \ldots \lambda_r} x_{\lambda_0} \ldots x_{\lambda_r}$, etc., then $A^r \smile B^s = C^{r+s}$ may be defined as follows, for various r and s:

$$c_\lambda = a_\lambda b_\lambda, \quad c_{\lambda_0 \ldots \lambda_r} = a_{\lambda_0 \ldots \lambda_r} b_{\lambda_0}, \quad c_{\lambda_0 \ldots \lambda_s} = a_{\lambda_1} b_{\lambda_0 \lambda_1 \ldots \lambda_s}, \quad c_{\lambda_0 \ldots \lambda_{r+s}} = (-1)^{r+1} a_{\lambda_1 \ldots \lambda_{r+1}} (b_{\lambda_1 \lambda_{r+1} \ldots \lambda_{r+s}} - b_{\lambda_0 \lambda_{r+1} \ldots \lambda_{r+s}}).$$

This product is used in the following note, §6.

7. *Locally Isomorphic Complexes.*—Let K be an abstract complex. In place of (K_4) of PC, we assume: (*a*) Each σ^1 is incident with exactly two vertices; $\partial\sigma^1 = \pm a \pm b$. (*b*) Each $\partial\sigma^r (r > 1)$ is combinatorially connected (through paths with non-zero incidence numbers). (*c*) 0- and 1-cycles mod 2 in any closed cell $\bar{\sigma}$ bound in $\bar{\sigma}$, if the 0-cycles have an even number of vertices. Two complexes K, K' are *locally isomorphic* if there is a (1–1) correspondence ϕ between their non-oriented cells which preserves incidences, and such that for each σ, $\bar{\sigma}$ and $\phi\bar{\sigma}$ are isomorphic complexes. Examples:

$$\begin{aligned} K:\quad & \partial(ab) = b - a,\ \partial(ac) = c - a,\ \partial(bc) = c - b, \\ K':\quad & \partial'(ab) = b - a,\ \partial'(ac) = c - a,\ \partial'(bc) = -c - b. \end{aligned}$$

If $\sigma = bc$, then reorienting c to c' in K' gives $\partial'(bc') = \partial(bc)$. K' has 1-dimensional torsion.

Note that chains mod 2 may be transferred to locally isomorphic complexes at will.

A *unit* 0-*chain* I is a sum $\Sigma(\pm x_i)$ of all oriented vertices. Let a_2 be the integer a mod 2; let $A|_2 = (A)_2$ be the chain A reduced mod 2.

For any I, $W = (\frac{1}{2}\delta I)_2$ exists and is a cocycle; its cohomology class $\mathbf{W}$ is independent of the I chosen. If K and K' are locally isomorphic, then they are isomorphic if and only if $\mathbf{W} = \mathbf{W}'$.

K is *normal* if $\mathbf{W} = 0$.[2] Any geometric complex is normal. Given any normal K and any 1-cocycle X mod 2 in K, there is a locally equivalent K' with $W' = X$.

8. *Particular Orientations in Simplicial Complexes.*—Set

$$\rho(0_2) = 1,\ \rho(1_2) = -1,\ \rho(\Sigma\alpha_i\sigma_i) = \Sigma\rho(\alpha_i)\sigma_i.$$

Let σ_i^r denote oriented cells in either of the locally isomorphic complexes K, K', K being normal; the incidence numbers may differ in sign. Define new incidences in K' as follows. If σ^r, σ^{r+1} have the same first vertex, let $[\sigma^r : \sigma^{r+1}]' = [\sigma^r : \sigma^{r+1}]$. If their first vertices x_i, x_j are different, set

$$[\sigma^r : \sigma^{r+1}]' = \rho(W' \cdot (x_i x_j))[\sigma^r : \sigma^{r+1}].$$

The K' thus constructed is isomorphic with the given one. (The cells may be reoriented.) Thus, in the example, we may use $W' = (bc)_2$, $\rho(W' \cdot bc) =$

$\rho(1_2) = -1$, and $[c:bc]' = -[c:bc] = -1$. We may also write, with the $\frown$ product of PC, §6,

$$[\sigma^r : \sigma^{r+1}] = \rho[W' \cdot (\sigma^r \frown [\sigma^{r+1})] [\sigma^r : \sigma^{r+1}].$$

9. *Products in General Complexes.*—The theory in PC holds only in normal complexes. It may be extended as follows. Let K', K'', K''' be locally isomorphic complexes, with

$$\mathbf{W}' + \mathbf{W}'' + \mathbf{W}''' = 0.$$

Then we may define products of the form

$$X' \smile Y'' = Z''', \qquad X'' \frown A''' = B',$$

as follows. Orient the vertices x_i', etc., so that if $I' = \Sigma x_i'$, etc., then

$$\left(\frac{1}{2}\delta I'\right)_2 + \left(\frac{1}{2}\delta I''\right)_2 + \left(\frac{1}{2}\delta I'''\right)_2 = 0.$$

Let K be the corresponding normal complex, in which $\frown_0$ is defined. Take any face σ_1 of any σ_2; choose a vertex x in σ_1. Let Φ', etc., be isomorphisms of $\bar\sigma_2$ into $\bar\sigma_2'$, etc., such that $\Phi'(x) = x'$, etc. Then set

$$\sigma_1'' \frown \sigma_2''' = \Phi'[(\Phi'')^{-1}(\sigma_1'') \frown_0 (\Phi''')^{-1}(\sigma_2''')].$$

If K is simplicial, with ordered vertices, we may use, if $\lambda_0 < \ldots < \lambda_p < \ldots < \lambda_{p+q}$,

$$(x_{\lambda_p} \ldots x_{\lambda_{p+q}})'' \frown (x_{\lambda_0} \ldots x_{\lambda_p} \ldots x_{\lambda_{p+q}})''' = \rho[W''(x_{\lambda_0} x_{\lambda_p})](x_{\lambda_0} \ldots x_{\lambda_p})'.$$

The products are unique just as in PC.

10. *Subdivisions; Chain-Mappings.*—Any complex K', in which each $\bar\sigma$ has the homology groups of a closed simplex, is locally isomorphic with a similar normal complex K; the latter has a "regular" subdivision, with geometric meaning. The theory of chain-mappings[3] carries over to the present case. Products are preserved under subdivision and chain-mappings. If K_1' is the "regular" subdivision of K', then $W' = Sd'W_1'$.

11. *Coboundary Operators.*—Set $\omega(0_2) = 0$, $\omega(1_2) = -1$ (we could use $+1$). Set $\omega(\Sigma \alpha_i \sigma_i^r) = \Sigma \omega(\alpha_i) \sigma_i^r$ ($\alpha_i = 0_2$ or 1_2). Then for any p-I_2-cocycle X, $\frac{1}{2}\,\delta\omega X$ is a $(p + 1)$-I_0-cocycle whose cohomology class depends only on that of X. The dual operation $\frac{1}{2}\,\partial\omega A$ is well known.

If K' and K'' are locally isomorphic, and we let σ_i^r denote cells in either, then $\frac{1}{2}\,(\delta''A' - \delta'A')$ is defined for all chains A' in K'; as a chain in K'', if $\delta'A' = 0$, it is a cocycle whose class depends only on that of A'. If $\delta'A' = 0$, we may write it as $\frac{1}{2}\,\delta''\omega(A\,|_2)$.

Corresponding to the pair K', K'' is a characteristic 2-I_0-cocycle in the corresponding complex K''',

$$X''' = \frac{1}{4}\delta''\delta'I' \qquad (I' = \Sigma x_i').$$

If $\mathfrak{B}'$, $\mathfrak{B}''$ are 0-sphere-bundles with these as classes, then their product has the 2-class X''' (see the following note). Similarly, to K', K'', K''' corresponds $\frac{1}{8}\delta'''\delta''\delta'I'$, etc.

12. *Relation to Products.*—Using K', K'', K''' again, and $\omega A = A$ if A has integer coefficients, set

$$X' \cup Y'' = \omega X' \smile \omega Y''.$$

This has topological significance if each chain either is integral or equals W' or W''. To show this, we use, if σ_2 is a p-face of $\sigma_1 = \sigma_1^{p+1}$, and x_2,x_1 are their first vertices,

$$\frac{1}{2}(\delta'''\sigma_2 - \delta''\sigma_2)\cdot\sigma_1 = \rho[W''(x_1x_2)]\omega W'(x_1x_2).$$

We find

$$W' \cup Y'' = \frac{1}{2}(\delta'''Y'' - \delta''Y''),$$

$$W' \cup W'' = \frac{1}{2}(\delta'''\omega W'' - \delta''\omega W'') = \text{the } X''' \text{ above.}$$

13. *On Non-Orientable Manifolds.*[4]—Let K be a subdivision of the non-orientable manifold M. For each oriented σ, let ϵ be an orientation of the part of M about σ; set

$$\sigma' = (\sigma, \epsilon) = (-\sigma, -\epsilon),\ -\sigma' = (-\sigma, \epsilon) = (\sigma, -\epsilon),\ [\sigma_1':\sigma_2'] = [\sigma_1:\sigma_2] \text{ or} - [\sigma_1:\sigma_2]$$

according as the orientations ϵ_1 and ϵ_2 agree or disagree. The new complex K' is locally isomorphic with K; its class has the property that any 1-circuit A^1 in K preserves orientation if and only if $A^1\cdot W' = 0_2$. The Poincaré duality theorem (compare PC, §18) says:

$$^p\mathbf{H}_G(K) \approx {}^{n-p}\mathbf{H}^G(K'), \qquad ^p\mathbf{H}_G(K') \approx {}^{n-p}\mathbf{H}^G(K).$$

Note that in K', the sum of the n-cells forms a cycle with integral coefficients.

[1] We recall the notations ∂ for boundary, δ for coboundary; I_μ = integers mod μ, I_0 = integers. An r-I_μ-chain is an r-chain with coefficients in I_μ. The reference PC will mean Whitney, "On Products in a Complex," *Ann. Math.*, **39**, 397–432 (1938). We note some corrections to this paper: In footnote 9, 3-sphere should read 3-complex.

In Theorem 12, we mean the (uniquely determined) products of 11 (a). In place of the following paragraphs, read: As knowing all $v_i^p \cdot u^p$ for all i uniquely determines u^p, and the correct $\frown$ satisfies (11.8), we have found the correct $\frown$. (NOTE: Theorem 11 is not used.) The end of the first paragraph in 12 should read: Find $u^p \smile u^q$ over I_0, then over R_1, ($p \leqq 2$), then for all p, by (5.12); then find $\frown$ over I_0, by Theorem 12. (We must know (11.15).) In Theorem 13, (a), add: $\phi\tau^p = 0$ if $O(\tau^p)$ is acyclic. Relation (14.9) follows directly from (14.4). After (25.7), add: $I^p \smile I^q = I^{p+q}$.

[2] *Augmentable* in Tucker, *Ann. Math.*, **34**, 191–243 (1933).

[3] See Tucker, these PROCEEDINGS, **25**, 371–374 (July, 1939). The theory was developed independently by S. Lefschetz and myself.

[4] This case of the theory is due to de Rham; see *Comm. Math. Helv.*, **4**, 151–157 (1933).

ON REGULAR FAMILIES OF CURVES[1]

HASSLER WHITNEY

A family F of non-intersecting curves filling a metric space is called *regular* if, in a neighborhood of any point p, it is homeomorphic with a family of straight lines. We have given in another paper[2] a necessary and sufficient condition, which we shall call (A') (to be described below), that a family F be regular. We shall prove in this note that the following condition is sufficient:

(A) Given any point p, and a direction on the curve through p, there is an arc pq in this direction with the following property. For every $\epsilon>0$ there is a $\delta>0$ such that for any p', with $\rho(p', p)<\delta$, there is an arc $p'q'$ of $C(p')$ such that

$$(1) \qquad p'q' \subset V_\epsilon(pq), \qquad q' \subset V_\epsilon(q).$$

The condition (A') is the same, except that after (1), we add:

(2) If r' and s' are on $p'q'$ and $\rho(r', s')<\delta$, then $\delta(r's')<\epsilon$.

From the present theorem it is clear that the families of curves recently defined by Niemitzki[3] are regular.

To prove the theorem, suppose (A) holds, but (A') does not. Then the following is true:

(B) There is a point p, and a direction of the curve $C(p)$, such that for any arc pq on $C(p)$ in this direction, there is an $\epsilon>0$, such that for any $\delta>0$, there is a point p', with $\rho(p', p)<\delta$, such that for any q' on $C(p')$,

(3) either $p'q' \not\subset V_\epsilon(pq)$, or $q' \not\subset V_\epsilon(q)$,

---

[1] Presented to the Society, April 27, 1940.

[2] Annals of Mathematics, (2), vol. 34 (1933), pp. 244–270. We refer to this paper as RF. By RF, Theorem 7A, F is regular as there defined. The converse is proved as follows. By Theorem 17A, there is a cross-section S through p. In a neighborhood of p, the curves are orientable (this is easily seen, for instance, with the help of Theorem 9B). Choose an open subset S' of S, and let U be all points $q'=g'(q, \alpha)$, q in S', $|\alpha|<\epsilon$ (see RF, §15); U is a neighborhood of p, expressed as the product of S' and the open line segment $-\epsilon<\alpha<\epsilon$.

By a curve, we shall mean here the topological image of an open line segment or of a circle. We shall use $\rho(p, q)$ for distance, $\delta(A)$ for the diameter of the set A, and $V_\epsilon(A)$ for the set of all points p, $\rho(p, A)<\epsilon$. Let $C(p)$ mean the curve of F through p.

[3] V. Niemytzki, Recueil Mathématique de Moscou, vol. 6 (48) (1939), pp. 283–292. We mention two further papers in the subject: H. Whitney, Duke Mathematical Journal, vol. 4 (1938), pp. 222–226, showing that if the curves fill a region in 3-space, a cross-section may be chosen so as to be a 2-cell; W. Kaplan, Duke Mathematical Journal, vol. 7 (1940), pp. 154–185, studying families filling the plane.

(4) or there are points r', s' on $p'q'$ such that $\rho(r', s') < \delta$, and $\delta(r's') \geqq \epsilon$.

Choose a point p and a direction on $C = C(p)$, by (B). Choose q on C in this direction, by (A). Choose $\epsilon > 0$ by (B). For each positive integer i, choose δ_i by (A), with ϵ replaced by ϵ/i. Choose p_i by (B), with δ replaced by δ_i. Choose q_i by (A), with p' replaced by p_i; then

$$(5) \qquad p_iq_i \subset V_{\epsilon/i}(pq), \qquad q_i \subset V_{\epsilon/i}(q).$$

By (B), as $\epsilon_i < \epsilon$, we may choose p_i' and q_i' on p_iq_i so that

$$(6) \qquad \rho(p_i', q_i') < \delta_i, \qquad \delta(p_i'q_i') \geqq \epsilon.$$

By (6), we may choose r_i on $p_i'q_i'$ so that $\rho(p_i', r_i) \geqq \epsilon/2$. By (5) and (6), we may choose a subsequence so that for some points p' and r on pq,

$$(7) \qquad p'_{\lambda_i} \rightarrow p', \qquad q'_{\lambda_i} \rightarrow p', \qquad r_{\lambda_i} \rightarrow r;$$

then $r \neq p'$. Say, for definiteness, that r is in the direction of q from p'. The set of such points r which are limits of such sequences $\{r_{\lambda_i}\}$ forms a closed set, which, by (5), is in $p'q$; we shall let r be the point furthest from p'. (It might be q.)

Assuming that (A) holds for the point r and the direction away from p', we shall arrive at a contradiction. Choose a point s on C in this direction from r, by (A). (If C is a closed curve, it might happen that s is on the arc pr.) Choose r' and s' on C just behind and just in front of r, so that r' is on neither pp' nor rs, and s' is not on $p'r$. We shall show that for any $\epsilon' > 0$ there is an integer j and a point s_j on $p_j'q_j'$ within ϵ' of s'; as s is in pq, by (5), this will contradict the definition of r, and thus prove the theorem.

Set

$$(8) \qquad 4\eta = \min\left[\rho(r', rs), 2\epsilon'\right].$$

Choose r_-, r_+, s_-, s_+ on C in the order $r_-r'r_+rs_-s's_+s$, so that r_- is not in rs and s_+ is not in pr (if C is closed), and so that

$$(9) \qquad r_-r'r_+ \subset V_\eta(r'), \qquad s_-s's_+ \subset V_\eta(s').$$

Set

$$(10) \qquad 2\epsilon'' = \min\left[\rho(pr_-, r_+q), \rho(rs_-, s_+s), \eta\right].$$

Using r, s, and ϵ'', choose $\delta'' > 0$ by (A). By (7), we may choose j so that

$$(11) \qquad \epsilon/j < \epsilon'', \quad \rho(p_j', p') < \epsilon'', \quad \rho(q_j', p') < \epsilon'', \quad \rho(r_j, r) < \delta''.$$

By the choice of δ'', we may choose s^* on $C(r_j)$ so that

$$(12) \qquad r_j s^* \subset V_{\epsilon''}(rs), \qquad s^* \subset V_{\epsilon''}(s).$$

As $r_j s^*$ is a connected set, (11), (12) and (10) show that there is a point s_j on it such that $\rho(s_j, rs_- + s_+ s) \geqq \epsilon''$; hence, by (12), $\rho(s_j, s_- s_+) < \epsilon''$, and by (9),

$$(13) \qquad \rho(s_j, s') < \epsilon'' + \eta < 2\eta \leqq \epsilon'.$$

By (12) and (8),

$$(14) \qquad \rho(r', r_j s_j) > 2\eta.$$

By (5) and (11),

$$(15) \qquad p_j' q_j' \subset V_{\epsilon''}(pq).$$

By (11), (10), (5) and (9), there are points p_j^* in $p_j' r_j$ and q_j^* in $r_j q_j'$ such that

$$(16) \qquad \rho(p_j^*, r') < 2\eta, \qquad \rho(q_j^*, r') < 2\eta.$$

By this and (14), the arc $r_j s_j$ is contained in the arc $p_j^* r_j q_j^* \subset p_j' q_j'$. Hence $s_j \subset p_j' q_j'$, which, with (13), gives the contradiction.

HARVARD UNIVERSITY

TOPICS IN THE THEORY OF ABELIAN GROUPS. I
DIVISIBILITY OF HOMOMORPHISMS

HASSLER WHITNEY

1. Introduction. The theory of character groups of Abelian groups has in recent years become of great importance, especially through applications to topology and algebra. The character group of G is the group H of homomorphisms of G into the real numbers mod 1. Extending this, we may consider the group $H = \text{Hom}\ (G, Z)$ of homomorphisms of G into a third group Z, or more generally, a "pairing" of groups H and G into Z: a multiplication $h \cdot g = z$, satisfying both distributive laws.

Of course the duality theorems for character groups will not hold in the more general cases; but, under certain conditions, substitutes may hold. We expect in later notes to give various facts about pairings, and the closely associated problem of divisibility by integers. In the present note, we answer the question of when a homomorphism of G into Z is divisible by an integer m; this has an immediate application to a theorem in combinatorial topology.

2. Divisibility theorems, discrete groups. We use the following notations for a group X and an integer m.

$$mX = \text{all } mx, \qquad x \in X,$$
$$_mX = \text{all } x \in X \text{ such that } mx = 0.$$

Note that $x \in mX$ means that x is divisible by m.

In the theorems below, we use

$$H = \text{Hom}\ (G, Z). \tag{1}$$

Let $H'(G')$ denote all $h(g)$, $h \in H'$, $g \in G'$.

LEMMA 1. *Suppose G' is a subgroup of G,*

$$h \in H, \qquad h' \in \text{Hom}\ (G', Z).$$

Then h' can be extended over G so that $h = mh'$ in G' if and only if:

(α) *$h(G) \subset mZ$.*

(β) *$h'(mg) = h(g)$ if $mg \in G'$.*

(γ) *If m' is a divisor of m, then $h'(m'g) \in m'Z$ if $m'g \in G'$.*

The necessity of the conditions is clear. We prove the sufficiency first for the case

Received by the editors November 19, 1943.

$$m = p^n, \qquad p \text{ prime.}$$

To begin with, extend h' over $mG+G'$ by setting

$$(2) \qquad h'(mg + g') = h(g) + h'(g') \qquad (g' \in G').$$

This definition is unique. For if $mg+g' = mg_1+g_1'$, then $m(g_1-g) \in G'$, and (β) gives

$$0 = h'(0) = h'[m(g_1 - g) + (g_1' - g')] = h(g_1 - g) + h'(g_1' - g'),$$

so that

$$h(g) + h'(g') = h(g_1) + h'(g_1').$$

Clearly (β) holds in $mG+G'$. To prove (γ), suppose $m = m'm''$. Take any

$$m'g = mg_1 + g' \in mG + G'.$$

Then $m'(g - m''g_1) = g' \in G'$, and hence, using (2), (α), and (γ) in G', we find

$$h'(m'g) = h(g_1) + h'[m'(g - m''g_1)] \in m'Z.$$

Let S be the set of all pairs (ϕ, G''), where

$$mG + G' \subset G'' \subset G,$$

$$\phi \in \text{Hom}\,(G'', Z), \qquad \phi = h' \text{ in } mG + G',$$

and such that (β) and (γ) hold with ϕ and G'' in place of h' and G'. Partially order these by setting

$$(\phi_1, G_1) < (\phi_2, G_2) \quad \text{if} \quad G_1 \subset G_2 \quad \text{and} \quad \phi_2 = \phi_1 \text{ in } G_1,$$

that is, ϕ_2 is an extension of ϕ_1. By[1] Zorn's lemma, there is a maximal simply ordered set S' of elements of S. Let G_0 be the set of all g in some G'' with $(\phi, G'') \in S'$, and set $\phi_0(g) = \phi(g)$ for such a (ϕ, G''). Clearly G_0 is a subgroup of G, $\phi_0(g)$ is independent of the (ϕ, G'') chosen, and $(\phi_0, G_0) \in S'$; hence it is the last element in S', and there is no later element in S. Therefore ϕ_0 cannot be further extended in G. To prove the lemma for $m = p^n$, it is sufficient to show that $G_0 = G$; for $\phi_0 = h'$ in G' by hypothesis, and (β) gives

$$m\phi_0(g) = \phi_0(mg) = h(g), \qquad m\phi_0 = h.$$

Suppose $G_0 \neq G$. Then choose an element $g_1 \notin G_0$ such that for a largest possible divisor m_1 of m, $g_1 \in m_1 G$. Since $mG \subset G_0$, it follows

[1] Compare Remark 2 in §3.

that $m_1 < m$, and $m_1 = p^\mu$, $\mu < n$. Choose g_3 and g_2 so that

$$(3) \qquad g_1 = m_1 g_3, \qquad g_2 = p g_1 = p m_1 g_3;$$

then $g_2 \in G_0$. Set $m_2 = p^{n-\mu-1}$; then $p m_1 m_2 = m$. Since $\phi_0 = h'$ in $mG + G'$, (2) gives

$$(4) \qquad m_2 \phi_0(g_2) = \phi_0(p m_1 m_2 g_3) = h(g_3).$$

Since $g_2 = p m_1 g_3 \in G_0$, we may choose z_2 by (γ) so that

$$(5) \qquad \phi_0(g_2) = p m_1 z_2.$$

Let G_1 be the subgroup of G generated by G_0 and g_1. Since p is prime, and pg_1 is in G_0 while g_1 is not, $ag_1 \in G_0$ if and only if $a \equiv 0 \pmod p$. Extend ϕ_0 through G_1 by setting

$$(6) \qquad \phi_0(ag_1 + g) = a m_1 z_2 + \phi_0(g) \qquad (a \text{ integral}, g \in G_0).$$

To show that this is unique, suppose

$$ag_1 + g = bg_1 + g', \qquad g, g' \in G_0.$$

Then

$$(a - b)g_1 = g' - g \in G_0, \qquad a - b = kp,$$

and

$$\begin{aligned}[a m_1 z_2 + \phi_0(g)] - [b m_1 z_2 + \phi_0(g')] &= kp m_1 z_2 + \phi_0(g - g') \\ &= k\phi_0(g_2) - \phi_0(kpg_1) \\ &= \phi_0(kg_2 - kg_2) = 0,\end{aligned}$$

as required. Clearly ϕ_0 is now a homomorphism in G_1.

Since $mG \subset G_0$, (β) holds in G_1. To prove (γ), take any

$$m'g = ag_1 + g_0 \in G_1 \qquad (g_0 \in G_0).$$

If $m'g \in G_0$, then $\phi_0(m'g) \in m'Z$, since (γ) holds in G_0. Suppose it is not. By the choice of g_1 and m_1, $m' \leqq m_1$; hence for some m'', $m'm'' = m_1$. Now by (3),

$$g_0 = m'g - a m_1 g_3 = m'(g - a m'' g_3),$$

and by (γ) in G_0, $\phi_0(g_0) = m' z_0$ for some z_0. Now by (6),

$$\phi_0(m'g) = a m_1 z_2 + \phi_0(g_0) = m'(a m'' z_2 + z_0),$$

as required.

We have now shown that $(\phi_0, G_1) \in S$. Since ϕ_0 has been extended from G_0, $(\phi_0, G_0) \prec (\phi_0, G_1)$. But this contradicts the fact that S' was maximal. It follows that $G_0 = G$, and the lemma is proved for the case $m = p^n$.

For a general m, we shall use induction on the number $\rho(m)$ of distinct prime divisors of m. Say $m=m_1m_2$, $(m_1, m_2)=1$, $\rho(m_1)$ and $\rho(m_2)$ being less than $\rho(m)$. Set

$$h_1 = m_2h', \qquad h_2 = m_1h', \qquad \text{in } G'.$$

Then the conditions of the lemma hold for both h_1 and h_2, with m_1 and m_2 respectively. Property (α) for m_1 and m_2 is clear. To prove (β), we have

$$h_1(m_1g) = m_2h'(m_1g) = h'(mg) = h(g).$$

To prove (γ), suppose m_1' is a divisor of m_1 and $m_1'g \in G'$; then m_1' is a divisor of m, and (γ) for h' gives

$$h_1(m_1'g) = m_2h'(m_1'g) \in m_2(m_1'Z) \subset m_1'Z.$$

Therefore h_1 and h_2 may be extended over G, so that

$$h = m_1h_1 = m_2h_2.$$

Now choose integers r, s so that $rm_1+sm_2=1$. Then

$$rh_2 + sh_1 = (rm_1 + sm_2)h' = h' \text{ in } G',$$
$$m(rh_2 + sh_1) = rm_1m_2h_2 + sm_2m_1h_1 = h \text{ in } G,$$

so that $h'=rh_2+sh_1$ is the required homomorphism.

THEOREM 1. *For any integer m, $h \in mH$ if and only if*
(a) $h(G) \subset mZ$,
(b) $h({}_mG)=0$.

Let G' be the identity alone, and set $h'(0)=0$. We need merely prove (β) and (γ) of Lemma 1. Clearly (β) reduces to (b); (γ) is trivial.

3. **Divisibility properties, topological groups.** We shall consider only topological groups with the following property:

(P) For each integer $m>0$, the mapping mg of G into itself is interior, that is, maps open sets into open sets.

An equivalent statement is the following: For every neighborhood U of 0 in G and every $m>0$ there is a neighborhood U' of 0 such that $U' \subset mU$; that is, all sufficiently small elements of G are divisible by m, a solution lying in U. The property holds in the commonly used groups.

REMARK 1. It follows that mG is open in $G(m>0)$; hence, for any subgroup G', $mG+G'$ is open in G.

EXAMPLE. If G is the group of dyadic rational numbers, using the topology of the real numbers, and $m=3$, then $3G$ contains no element of the form $1/2^n$. Thus G does not satisfy (P).

LEMMA 2. *Let G and Z be topological groups, and let* (P) *hold in G. Let h' be a homomorphism of G into Z, and set $h=mh'$. Then h' is continuous if and only if h is.*

That continuity of h follows from that of h' is simple (and does not use (P)). Suppose h is continuous. Given any neighborhood V of 0 in Z, choose a neighborhood U_1 of 0 in G so that $h(U_1)\subset V$, and choose $U\subset mU_1$. Now take any $g\in U$. Then $g=mg_1$, $g_1\in U_1$, and $h'(g)=h(g_1)\in V$, proving that h' is continuous.

By Hom (G, Z) we shall now mean the group of *continuous homomorphisms* of G into Z. We do not need to consider a topology in H here.

THEOREM 2. *If G and Z are topological groups, and* (P) *holds in G, then Lemma* 2 *and Theorem* 1 *hold.*

Using the former proofs, we need merely prove the continuity of h'. But this follows from the continuity of h and the last lemma.

REMARK 2. Since mG is open in G and h' is first extended over mG, if the proof in Lemma 1 is given in the equivalent form of extending h' over larger and larger subgroups of G (as from G_0 to G_1 in the proof given), only a finite number of steps will be required in the case of most groups G.

4. On the complete resolution of H by G. We use the following definitions and facts.[8] If H, G are paired into Z, (H, G') (the *nullifier* or *annihilator* of G' in H) is the set of all h with $h\cdot G'=0$. If $(H, G')=0$, then G' *resolves* H. (Then if $h_1\neq h_2$, there is an element $g\in G'$ with $h_1\cdot g\neq h_2\cdot g$.) If $(H, {}_mG')=mH$, G' *m-resolves* H. (Then if h is not divisible by m, there is an element $g\in {}_mG'$ such that $h\cdot g\neq 0$.) "Resolves" and "0-resolves" are equivalent. If G' m-resolves H for all integers $m\geqq 0$, G' *resolves H completely. Z is completely divisible* if $mZ=Z$ for each integer $m\neq 0$. The main application of the Pontrjagin duality theory to topology lies in the fact[8] that any group resolves completely and is resolved completely by its character group.

THEOREM 3. *If Z is completely divisible, then G resolves H* $=$Hom (G, Z) *completely.*

---

[8] Compare H. Whitney, *On matrices of integers and combinatorial topology*, Duke Math. J. vol. 3 (1937) pp. 35–45. The main statement in footnote 8 of this paper is not true.

If the groups are topological, we assume G satisfies (P).

If $h \in mH$, say $h = mh'$, then $mg = 0$ implies

$$h(g) = h'(mg) = h'(0) = 0,$$

and $h \in (H, {}_mG)$. Conversely, if $h \in (H, {}_mG)$, that is, $h({}_mG) = 0$, Theorem 1 (or Theorem 2) gives $h \in mH$.

5. **Application to combinatorial topology.** Let $\boldsymbol{H}^r(K, X)$ and $\boldsymbol{H}_r(K, X)$ denote the rth homology and cohomology groups, respectively, of K, with the coefficient group X.

THEOREM 4. *Let Z be completely divisible. Then for any G, and $H = \text{Hom}\ (G, Z)$, we have the isomorphisms*

$$\boldsymbol{H}_r(K, H) \approx \text{Hom}\ [\boldsymbol{H}^r(K, G), Z],$$

$$\boldsymbol{H}^r(K, H) \approx \text{Hom}\ [\boldsymbol{H}_r(K, G), Z].$$

If G and Z are topological, we assume G satisfies (P).

This is Theorem 8 of Whitney, loc. cit., with the hypothesis weakened; it is essentially the Pontrjagin Duality Theorem, slightly generalized. For the proof, we need merely apply Theorem 3 to obtain the remainder of the hypothesis.

HARVARD UNIVERSITY

COMPLEXES OF MANIFOLDS[1]

BY HASSLER WHITNEY

DEPARTMENT OF MATHEMATICS, HARVARD UNIVERSITY

Communicated December 10, 1946

1. *Introduction.* A smooth manifold M (smooth means continuously differentiable) is a well-defined concept. A bounded manifold may have, for boundary, anything from a smooth manifold to quite a general point set. For analytical purposes, the boundary should at least be made up of pieces of manifolds, joining together smoothly; that is, it should be formed of a "complex of manifolds," or "complifold" for short. Then $M = \sigma^n$, together with the boundary cells σ_i^r, forms a complifold K. If one tries to define K abstractly, one finds that all sorts of curious situations may occur locally. In this preliminary note, we wish to point out some of the properties that will bring K back to reasonableness. If K is imbeddable in a Euclidean space E^m, each cell being imbedded smoothly, this should be satisfactory.

2. *Complifolds.* Instead of giving a full definition, we shall point out some of the properties that are clearly necessary. The cells should attach to each other, in the point set sense, like the cells of an ordinary complex. The relation between coördinate systems in incident cells should be smooth, with Jacobian matrix of maximum rank. If a vector is tangent to faces σ_1, σ_2 of a cell σ at p, it should be tangent to a common face of σ_1 and σ_2.

Among the important subjects we cannot discuss here, we mention the following. The property of K being "locally flat" or "locally simplicial"; the property of K_1 and K_2 being in "general position" in K; the fact that, in this case, $K_1 \cap K_2$ is a complifold.

The local properties of K can be described largely in terms of tangent vectors, to which we now turn.

3. *The tangent space of K at a point.* For each (closed) cell σ, and point p in σ, there is a *tangent cone* $\Gamma(\sigma, p)$ of vectors (σ, v) which are tangent to σ at p; these generate a vector space $V(\sigma, p)$. Given p in K, let $\sigma_1, \ldots, \sigma_s$ be the cells containing p. Define (abstractly) the direct sum

$$V^*(p) = V(\sigma_1, p) \oplus \ldots \oplus V(\sigma_s, p).$$

The vectors

$$w_{ij}(v) = (\sigma_i, v) - (\sigma_j, v), \; v \text{ in } V(\sigma_i, p) \cap V(\sigma_j, p)$$

generate a subspace $V^{**}(p)$ of $V^*(p)$. The *tangent space*

$$V(K, p) = V^*(p) \ominus V^{**}(p)$$

of K at p is formed from $V^*(p)$ by setting all $w_{ij}(v) = 0$.

We say K is *consistent at* p if each $\Gamma(\sigma, p)$ lies in $V(K, p)$ in a one-one manner. This is obviously a necessary condition that K be smoothly imbeddable in E^m.

It may happen that a relation $a_1v_1 + \ldots + a_kv_k = 0$ is not true at p, yet arbitrarily near p a relation arbitrarily near this one holds. This might preclude smooth imbedding.

A necessary and sufficient condition for imbeddability of a general complifold seems highly difficult to obtain. Instead, we shall give a slight restriction on K.

4. *Cellwise homogeneity.* Let K be consistent at each p. If for each σ, and cells $\sigma_1, \ldots, \sigma_s$, with σ as face, the part of $V(K, p)$ over these cells is of constant dimension as p moves over σ, we say K is *cellwise homogeneous.*

THEOREM: *Any cellwise homogeneous complifold of dimension* n *may be smoothly imbedded in* E^{2n+1}, *and smoothly immersed in* E^{2n}.

[1] Presented to the American Mathematical Society Sept. 8, 1942.

ANNALS OF MATHEMATICS
Vol. 50, No. 1, January, 1949

RELATIONS BETWEEN THE SECOND AND THIRD HOMOTOPY GROUPS OF A SIMPLY CONNECTED SPACE

BY HASSLER WHITNEY

(Received March 29, 1948)

1. Introduction

In the study of the classification and extension of mappings into simply connected spaces R (see two forthcoming papers), certain special mappings ϕ of a 3-sphere S_0^3 into R turn out to be of importance. There is a set of "tubes" T_i in S_0^3; ϕ maps each cross section of a single tube in the same manner into R, and maps all of S_0^3 outside these tubes into a fixed point $\bar{p}$ of R.

The mapping ϕ, in each cross section of a tube, determines an element α of the second homotopy group $\pi_2 = \pi_2(R)$ of R; in S_0^3, it determines an element ξ of the third homotopy group $\pi_3 = \pi_3(R)$. It is the purpose in Part II of this paper to study the relation between these elements. The formulas are based on two fundamental operations on elements of π_2, giving elements of π_3; these operations have been defined by J. H. C. Whitehead [1]. (His operations are defined for more general dimensions.)

If π_2 has elements of finite order, a somewhat more complicated system is necessary. Suppose α is of order n. Then we may have cylinders, which we call "junctions", into each of which n tubes enter; the cylinders are mapped into R in some fixed fashion, which need not be specified. The resulting element of π_3 is studied in Part III.

The space R is assumed to be an arcwise connected and simply connected Hausdorff space. Any mapping f of an oriented sphere S^r into R determines an element of π_r, since R is simply connected. We assume the reader is acquainted with the elementary properties of homotopy groups. Some proofs are given in rather intuitive geometric terms; the reader can make them analytically precise if he desires.

PART I. TUBES

2. Definition of the tubes

Let E^3 be Euclidean 3-space. Let $\bar{E}^3$ denote E^3 plus the point at infinity; this is a 3-sphere, which we may use in place of S_0^3. We assume a fixed orientation of E^3 is given.

Let T_0 be a solid torus, that is, the Cartesian product $S_0^1 \times \sigma_0^2$ of a circle S_0^1 and a disc σ_0^2. If vectors (v_1, v_2) at $q \in \sigma_0^2$ and v_3 at $p \in S_0^1$ orient σ_0^2 and S_0^1 respectively, then (v_1, v_2, v_3) orients T_0 at $p \times q$.

A *tube* in E^3 is a mapping Φ of T_0 into E^3 with the following properties.

(a) Φ is of class C^2 (has continuous second partial derivatives), and is one–one, with non-vanishing Jacobian.

(b) Φ is linear in each $p \times \sigma_0^2$.

(c) Φ is orientation preserving.

We shall commonly use the term "tube," denoted by T, to mean the point set $\Phi(T_0)$, together with its correspondence with T_0.

Each set $\Phi(p \times \sigma_0^2)$ we call a *cross section* of T; each set $\Phi(S_0^1 \times q)$ we call a *filament* of T. If q_0 is the center of σ_0^2, and q_1 is a fixed ("top") point of σ_0^2, we call $\Phi(S_0^1 \times q_0)$ and $\Phi(S_0^1 \times q_1)$ the *center filament* and *top filament* of T respectively.

Note that the orientation of any cross section determines the orientation of any filament, and conversely (using the orientation of E^3). If we reverse the orientation of either the cross sections or E^3, the orientation of the filaments must be reversed; Φ must be changed correspondingly.

The *looping coefficient* of T is the looping coefficient of any two of its filaments ζ, ζ' in E^3:

$$LC(T) = LC(\zeta, \zeta'). \tag{2.1}$$

This is independent of the orientation of the tube (i.e. of its filaments).

If T_1 and T_2 are disjoint tubes in E^3, with filaments ζ_1, ζ_2, their *looping coefficient* is

$$LC(T_1, T_2) = LC(\zeta_1, \zeta_2); \tag{2.2}$$

again, any filaments of the tubes may be used.

Note that if T_1 is a short tube going around a piece of T_2, then $LC(T_1, T_2) = 1$ or -1 according as a filament of T_1 has the same or opposite orientation as the boundary of a cross section of T_2.

We remark that if v_1 and v_2 are tangent vectors to σ_0^2 at q_0, then a tube T is determined by naming Φ over $S_0^1 \times q_0$, thus determining the center filament ζ_0, and naming the images of v_1 and v_2 at each point of ζ_0.

Let us define a *canonical tube* with given looping coefficient l as follows. Let Φ_0^* be a mapping of T_0 into E^3 which gives a solid torus T_0^* in the usual manner; $LC(T_0^*) = 0$. Thinking of p as an angular coordinate in S_0^1, let ψ_θ denote a rotation of σ_0^2 through the angle θ. Set

$$\Phi_l^*(p \times q) = \Phi_0^*(p \times \psi_{lp}(q)).$$

The point set $\Phi_l^*(T_0)$ is the same as $\Phi_0^*(T_0)$; clearly a tube T_l^* is defined with $LC(T_l^*) = l$.

LEMMA 2a. *Any unknotted tube may be deformed into a canonical tube with the same looping coefficient.*

Let $T = \Phi_0(T_0)$ be unknotted; that is, $\zeta_0(T)$ is unknotted. We may then deform $\zeta_0(T)$ into $\zeta_0(T_0^*) = \zeta_0(T_l^*)$ (all l); that is, Φ_t is defined in $S_0^1 \times q_0$, and $\Phi_1 = \Phi_0^*$ there. We shall extend Φ_t through T_0.

For fixed vectors v_1, v_2 in σ_0^2 at q_0, we first tell how each Φ_t is to map these. Choose these arbitrarily, so they are continuous in $p \times q_0$ and t, and so that for each p and t they define a plane not tangent to $\Phi_t(S_0^1 \times q_0)$.

We may do this for instance as follows. First, deform Φ_0 to make the images of v_1 and v_2 perpendicular, and perpendicular to the filament, for each $p \times q_0$; keep the filament fixed during this process. Now if v_3 is the tangent vector to the filament, define v_1 and v_2 (for each p) as the solutions of the differential equations

$$\frac{dv_i}{dt} = -\left(v_i \cdot \frac{dv_3}{dt}\right) v_3, \qquad i = 1, 2;$$

then the vectors remain perpendicular. This defines a deformation of Φ_0, using (b).

The deformation is certainly of class C^1; if it is not of class C^2, we may obtain a deformation of class C^2 by an approximation. (This is not difficult to show, using for instance methods in H. Whitney, "*Differentiable Manifolds*," Annals of Math., vol. 37 (1936), pp. 645–680.)

If Φ_t is not now one-one for each t, it certainly will be if we alter it so that v_1 and v_2 are mapped into shorter vectors, making the tubes thinner (except at $t = 0$).

We must still fit Φ_1 to some Φ_l^*. That one may be deformed into the other is an easy consequence of the fact that the looping coefficients are the same, if we set $l = LC(T)$.

3. Mappings corresponding to tubes

Let $T = \Phi(T_0)$ be a tube, and let f be a mapping of σ_0^2 into R such that $f(\partial\sigma_0^2) = \bar{p}$. Then the corresponding mapping $\phi = [T, f]$ of $\bar{E}^3$ into R is defined as follows:

$$(3.1) \qquad \begin{cases} \phi(\Phi(p \times q)) = f(q), & \\ \phi(p') = \bar{p}, & p' \in \bar{E}^3 - T. \end{cases}$$

This is clearly a continuous mapping of $\bar{E}^3$ into R. We call it a *tube mapping*. If there is a set of disjoint tubes, mapped as above, the rest of $\bar{E}^3$ going into $\bar{p}$, we call the mapping a *tube system mapping*.

Suppose f_0 and f_1 define the same element α of π_2. Then f_t is defined, $0 \leqq t \leqq 1$; $f_t(\partial\sigma_0^2) = \bar{p}$. The mappings $[T, f_t]$ are then defined, showing that $[T, f_0]$ and $[T, f_1]$ are homotopic mappings of $\bar{E}^3$ into R. Hence, for any T and any $\alpha \in \pi_2$, we may define $[T, \alpha] \in \pi_3$.

Let Φ_t be a deformation of the mapping Φ_0 of T_0 in E^3; $\Phi_t(p \times q)$ is continuous, and each $T_t = \Phi_t(T_0)$ is a tube. Then $[T_t, f]$ is a deformation of the mapping $[T_0, f]$ of $\bar{E}^3$ in R (for any fixed f); hence, if f defines α, $[T_0, \alpha] = [T_1, \alpha]$.

LEMMA 3a. *For unknotted tubes,* $[T, \alpha]$ *depends on* $LC(T)$ *and* α *only.*

Suppose T_0 and T_1 are unknotted, with the same looping coefficient l, and f_0 and f_1 both define α; we must prove $[T_0, f_0] = [T_1, f_1]$. By the proof above, we may suppose $f_0 = f_1$. By Lemma 2a, we may deform both T_0 and T_1 into T_l^*; hence both elements of π_3 equal $[T_l^*, \alpha]$.

LEMMA 3b. *If T is an unknotted tube with looping coefficient* 0, *then* $[T, \alpha] = 0$ *for all* α.

By the last lemma, we may suppose that $T = T_0^*$. Let x_1, x_2, x_3 be axes in

E^3 such that the x_1, x_2-plane E_0 is a plane of symmetry of T_0^*, the center filament ζ_0 lying in the x_1, x_2-plane. Let f define α. Define a mapping ϕ_t, for all $t \geqq 0$, of $\bar{E}^3$ into R as follows. First, $\phi_0 = [T, f]$, and $\phi_t(\infty) = \bar{p}$. For other points of $\bar{E}^3$, set

$$\phi_t(x_1, x_2, x_3) = \begin{cases} \phi_0(x_1, x_2, x_3 + t), & x_3 \geqq 0, \\ \phi_0(x_1, x_2, x_3 - t), & x_3 \leqq 0. \end{cases}$$

Clearly ϕ_t is continuous in each closed half-space. Since $\phi_0(x_1, x_2, t) = \phi_0(x_1, x_2, -t)$, ϕ_t is continuous in E^3. For t large enough, $\phi_t(\bar{E}^3) = \bar{p}$.

Since ϕ_0 defines $[T, \alpha]$ and is homotopic to zero, $[T, \alpha] = 0$.

4. Splicing of tubes

Let ϕ be a tube system mapping of $\bar{E}^3$ into R. Let $[T_1, f]$ and $[T_2, f]$ be two of the tube mappings in ϕ, with the same f. We shall show how ϕ may be deformed into ϕ' so that, corresponding to ϕ', T_1 and T_2 are each broken apart at a cross section, and the resulting ends of T_1 are joined to those of T_2.

Let C be a cube in E^3 whose side is of length 2, bisected by the plane E_0 (see §3); let B^+ and B^- be the faces of C on opposite sides of E_0. By a deformation, we may carry T_1 around so that it enters C at B^+ and leaves through it again, $T_1 \cap B^+$ consisting of two cross sections; do the same with T_2, using B^-. We may carry this out so that, in C, T_2 is exactly a reflection of T_1 across E_0. (Besides being pushed into corresponding positions, the tubes are twisted, so that for instance their top filaments are each "on top" (have the largest x_2) in C; the positive senses of the filaments must disagree.) No other piece of any tube shall touch C.

For $p \in \bar{E}^3 - C$, set $\phi_t(p) = \phi(p)$. In C set

$$\phi_t(x_1, x_2, x_3) = \begin{cases} \phi(x_1, x_2, \min (x_3 + t, 1)), & x_3 \geqq 0, \\ \phi(x_1, x_2, \max (x_3 - t, -1)), & x_3 \leqq 0. \end{cases}$$

This has the effect of pulling the tubes further into C towards E_0, and letting the parts disappear as they hit E_0. Clearly ϕ_t is a deformation of ϕ. Under ϕ_1, there are two pieces of tubes in C, stretching from B^- to B^+, and joining corresponding parts of T_1 and T_2.

5. The product $\alpha\beta$

Let T_1 and T_2 be tubes in E^3, in the shape of tori, such that

$$(5.1) \qquad LC(T_i) = 0 \ (i = 1, 2), \qquad LC(T_1, T_2) = 1.$$

Take elements α, β of π_2, and let f, g be corresponding mappings of σ_0^2 into R. Then the mapping ϕ of $\bar{E}^3$ into R, containing the two tube mappings $[T_1, f]$, $[T_2, g]$, defines an element of π_3, which we call $\alpha\beta$. (This definition is easily seen to be equivalent to that of Whitehead, [1].)

Clearly $\alpha\beta$ is independent of the choice of f and g. Deforming the two tori together, we see that it is independent of their positions so long as (5.1) is preserved.

By a deformation of $\bar{E}^3$, we may interchange the two tubes; this shows that

$$\alpha\beta = \beta\alpha. \tag{5.2}$$

Suppose $\beta = 0$. Then we may represent β by the mapping $g(\sigma_0^2) = \bar{p}$. Using this in the definition of $\alpha\beta$, the tube T_2 has disappeared; Lemma 3b now gives, using (5.2),

$$\alpha 0 = 0\alpha = 0. \tag{5.3}$$

The product $\alpha\beta$ is bilinear:

$$\alpha(\beta + \gamma) = \alpha\beta + \alpha\gamma. \tag{5.4}$$

This is Theorem 1 of Whitehead, loc. cit. The proof may be given as follows. We have linked tubes T_1 and T_2 with cross sections defining α and $\beta + \gamma$ respectively; the whole defines $\alpha(\beta + \gamma)$. Now T_2 falls into two tubes T_3 and T_4, with cross sections defining β and γ respectively. If we pull these two tubes apart, and splice two parts of T_1 between them, we obtain a pair of linked tori, defining $\alpha\beta$ and $\alpha\gamma$ respectively.

Since $\alpha\beta + (-\alpha)\beta = (\alpha - \alpha)\beta = 0\beta = 0$,

$$(-\alpha)\beta = \alpha(-\beta) = -\alpha\beta. \tag{5.5}$$

Using (5.4) gives, for integers n,

$$(n\alpha)\beta = \alpha(n\beta) = n(\alpha\beta). \tag{5.6}$$

6. Pulling one tube through another

Let ϕ be a mapping of $\bar{E}^3$ into R such that two pieces T_1 and T_2 of tubes pass close by each other, in non-parallel directions. Let ϕ' be the mapping obtained from ϕ by making T_1 pass by T_2 on the other side of where it is under ϕ. Let ϕ, ϕ' define ξ, ξ' in π_3; we shall determine $\xi - \xi'$.

If we pull the part of T_1 which is not too close to T_2 towards the other side of T_2, then part of T_1 now wraps most of the way around T_2. Pull the ends of this latter part of T_1 towards each other, and splice (see §4); we have left a T_1-torus around T_2, plus a new T_1 on the other side of T_2.

Now start pulling T_2 away from the T_1-torus; part of T_2 wraps around it; perform a splice. What remains is a new T_1 and T_2, in the desired position, plus a pair of linked tori.

LEMMA 6a. *Let ϕ and ϕ' be tube system mappings in which pieces of tubes T_1, T_2 pass on opposite sides of each other in non-parallel directions, as described above. Let p_1 and p_2 be nearby points of T_1 and T_2 respectively (in the position defining ϕ). Let cross sections of T_1 and T_2 define α and β respectively, let ϕ and ϕ' define ξ and ξ' respectively. Let v_1 and v_2 be vectors in the direction of the center filaments of T_1 and T_2 at p_1 and p_2 respectively; set $v = p_2 - p_1$. Set $\epsilon = 1$ or -1*

according as (v_1, v, v_2) *defines the positive or negative orientation of* E^3. *Then*

$$\xi - \xi' = \epsilon\alpha\beta.$$

This is clear from the discussion above, if we note that the looping coefficient of the two tori is ϵ.

REMARKS. The conditions described and the formula are symmetrical; note that, if $v' = p_1 - p_2$, then (v_2, v', v_1) and (v_1, v, v_2) define the same orientation in E^3. We may find ϵ as follows. Take a piece of one tube going by the other; go, instead, all the way around the other; the resulting looping coefficient is ϵ.

7. The function $\eta(\alpha)$

Given $\alpha \epsilon \pi_2$, choose a mapping f of σ_0^2 defining α, and choose an unknotted tube T with looping coefficient 1. (The tube could be knotted; see §11.) Then $[T, f]$ defines an element of π_3 which we call $\eta(\alpha)$ (compare Whitehead, [1]). The result of course does not depend on f; by Lemma 3a, it does not depend on the choice of T (we could use the canonical T; see Lemma 2a).

REMARK. If $R = S_0^2$ and f is of degree 1, defining α, then $\eta(\alpha)$ is the well known Hopf mapping of S_0^3 into S_0^2, and is a generator of π_3.

We now prove various properties of the operations. First, the definition gives immediately

$$\eta(0) = 0. \tag{7.1}$$

Next we show that

$$\eta(-\alpha) = \eta(\alpha). \tag{7.2}$$

To this end, let $[T_1, f]$ define $\eta(\alpha)$. Let g be a mapping of σ_0^2 into R obtained from f by a reflection across a diameter in σ_0^2; then g defines $-\alpha$. Let Φ_1 be the mapping of $S_0^1 \times \sigma_0^2$ defining T_1. Choose a new mapping Φ_2 of $S_0^1 \times \sigma_0^2$ into the same point set T_1, so that as p runs around S_0^1, $\Phi_1(p \times q_0)$ and $\Phi_2(p \times q_0)$ run around the center filament in opposite directions, and so that in each cross section, $g\Phi_2^{-1} = f\Phi_1^{-1}$. This defines a tube mapping $[T_2, g]$, which (as a mapping of $\bar{E}^3$ into R) is identical with $[T_1, f]$. Since $LC(T_2)$ does not depend on the orientation of the filaments, $LC(T_2) = LC(T_1) = 1$. Hence $[T_2, g]$ defines $\eta(-\alpha)$, and (7.2) follows.

Now we prove

$$\eta(\alpha + \beta) = \eta(\alpha) + \eta(\beta) + \alpha\beta. \tag{7.3}$$

Note that this gives a definition of the product in terms of the η-operation.

Let f and g map patches in σ_0^2 into R to define α and β respectively. If f' is the resulting mapping, and T_1^* is the canonical tube with looping coefficient 1 (§2), then $[T_1^*, f']$ defines $\eta(\alpha + \beta)$. Clearly T_1^* falls into two tubes T_2 and T_3, mapped according to f and g respectively. Moreover,

$$LC(T_2) = LC(T_3) = LC(T_2, T_3) = 1.$$

Pull T_3 away from T_2; this subtracts $\alpha\beta$ from $\eta(\alpha + \beta)$, by Lemma 6a. There remain tube mappings $[T_2, f]$ and $[T_3, g]$ defining $\eta(\alpha)$ and $\eta(\beta)$ respectively, and (7.3) follows.

The relation

$$(7.4) \qquad 2\eta(\alpha) = \alpha\alpha$$

now follows at once. For

$$0 = \eta(0) = \eta(\alpha - \alpha) = \eta(\alpha) + \eta(-\alpha) + \alpha(-\alpha)$$
$$= 2\eta(\alpha) - \alpha\alpha.$$

This could be proved directly as follows. Take two linked tori, defining $\alpha\alpha$; take them as large and thin, lying close together all along, but oppositely directed. Perform two splices between them, so that they fall into two separated tubes, each with looping coefficient 1; thus $2\eta(\alpha)$ is defined.

Using induction, we may show that

$$(7.5) \qquad \eta(a\alpha) = a^2\eta(\alpha).$$

This is true if $a = 0$ or 1. For $a > 1$,

$$\eta(a\alpha) = \eta[(a-1)\alpha + \alpha] = \eta[(a-1)\alpha] + \eta(\alpha) + (a-1)\alpha\alpha$$
$$= (a-1)^2\eta(\alpha) + \eta(\alpha) + 2(a-1)\eta(\alpha) = a^2\eta(\alpha).$$

For $a < 0$,

$$\eta(a\alpha) = \eta(-a\alpha) = (-a)^2\eta(\alpha) = a^2\eta(\alpha).$$

We now give a formula combining (7.3) and (7.5):

$$(7.6) \qquad \eta\left(\sum_i a_i\alpha_i\right) = \sum_i a_i^2\eta(\alpha_i) + \sum_{i<j} a_i a_j \alpha_i \alpha_j.$$

This is true if there is but a single term in the sum. Using induction, we have

$$\eta\left(\sum_{i=1}^{n} a_i\alpha_i\right) = \eta\left(\sum_{i=1}^{n-1} a_i\alpha_i\right) + \eta(a_n\alpha_n) + \left(\sum_{i=1}^{n-1} a_i\alpha_i\right)(a_n\alpha_n)$$
$$= \sum_{i=1}^{n-1} a_i^2\eta(\alpha_i) + \sum_{i<j\leq n-1} a_i a_j\alpha_i\alpha_j + a_n^2\eta(\alpha_n) + \sum_{i<n} a_i a_n \alpha_i \alpha_n,$$

which gives the required formula.

If α is of finite order, so is $\eta(\alpha)$. In fact:

$$(7.7) \qquad \text{If} \quad n\alpha = 0, \quad \text{then} \quad 2n\eta(\alpha) = 0;$$

$$(7.8) \qquad \text{if} \quad n\alpha = 0,\ n \text{ odd}, \quad \text{then} \quad n\eta(\alpha) = 0.$$

To prove (7.7), we have simply

$$2n\eta(\alpha) = n[2\eta(\alpha)] = n(\alpha\alpha) = (n\alpha)\alpha = 0\alpha = 0.$$

To prove (7.8), suppose $n = 2k + 1$. Then

$$n\eta(\alpha) = (n^2 - n^2 + n)\eta(\alpha) = n^2\eta(\alpha) - n(n - 1)\eta(\alpha)$$
$$= \eta(n\alpha) - 2kn\eta(\alpha) = \eta(0) - kn\alpha\alpha = 0.$$

8. An example

One might suppose that $n\alpha = 0$ would imply $n\eta(\alpha) = 0$ in all cases. But this is not true if n is even. For example, let S^2 be a 2-sphere, let σ^3 be a 2-sphere plus interior, and let ψ be a mapping of $\partial\sigma^3$ onto S^2 with the degree n; performing all identifications $\psi(p) = p$ gives a space R_n. Let α be a generator of $\pi_2(S^2)$. Then, in R_n, $n\alpha = 0$, but $n\eta(\alpha) \neq 0$.

This follows from a theorem of J. H. C. Whitehead (loc. cit. Lemma 4); by this theorem, the kernel of the injection $\pi_3(S^2) \to \pi_3(R_n)$ is generated by the elements

$$\eta(n\alpha) = n^2\eta(\alpha), \qquad (n\alpha)\alpha = 2n\eta(\alpha);$$

for n even, this gives just the multiples of $2n\eta(\alpha)$. The result appears also in a note by Pontrjagin, [2]. The proof (not given there) consists in finding a numerical invariant (mod $2n$), with the help of a generalized product, and involves much less homotopy theory.

9. A pairing for certain cyclic subgroups of π_2

Let the cyclic subgroup π_2' of π_2 generated by α be either infinite or of odd order n. Then we may define a pairing of π_2' with itself into π_3, by the definition

$$(a\alpha)\circ(b\alpha) = ab\eta(\alpha). \tag{9.1}$$

For if $a\alpha = a'\alpha$ and $b\alpha = b'\alpha$, then $a \equiv a'$ and $b \equiv b'$ mod n ($n = 0$ or n odd), and

$$a'b'\eta(\alpha) - ab\eta(\alpha) = kn\eta(\alpha) = 0.$$

For α of order n, n even, this is in general impossible; for we would have to have $n\eta(\alpha) = (n\alpha)\circ\alpha = 0$.

The Whitehead pairing is half this pairing:

$$2(a\alpha\circ b\alpha) = ab(2\eta(\alpha)) = ab\alpha\alpha = (a\alpha)(b\alpha). \tag{9.2}$$

Part II. Closed tube systems

10. The reduction formula

Let T be a tube system, composed of a set of closed tubes. Choose a filament (properly oriented) in each of these; these, together, form a *filament* of T.

Let $[T, f]$ be a tube system mapping, where f defines an element $\alpha \in \pi_2$ in each cross section. Then if ζ is a filament of T, there is a *corresponding* 1-cycle $\alpha\zeta$ with coefficients in π_2; this is the singular 1-cycle formed by attaching the value α to ζ.

If $[T, f]$, $[T', f']$ are tube system mappings, whose tubes are disjoint, they form together a tube system mapping. If f, f' define α, α' respectively, we may define the looping coefficient

$$LC(\alpha\zeta, \alpha'\zeta') = LC(\zeta, \zeta')\alpha\alpha'. \tag{10.1}$$

Note that d times this element of π_3 is 0, if α, α' are of order a, a' respectively and d is the greatest common division of a, a'.

Let $[T, f]$ be a tube system mapping, f defining α. Let ζ, ζ' be disjoint filaments of T. Then we may define the looping coefficient

$$LC^0(\alpha\zeta, \alpha\zeta') = LC(\zeta, \zeta')\eta(\alpha). \tag{10.2}$$

Then

$$LC(\alpha\zeta, \alpha\zeta') = 2LC^0(\alpha\zeta, \alpha\zeta').$$

We now give a formula which reduces the problem of finding the element of π_3 corresponding to a set of disjoint tube system mappings to that of finding the elements of π_3 corresponding to the individual ones.

THEOREM 10A. *Let $[T_1, f_1]$, $\cdots$, $[T_m, f_m]$ be tube system mappings with disjoint tubes. Let the whole system define $\xi \in \pi_3$, and let the ith subsystem define $\xi_i \in \pi_3$. Let f_i define $\alpha_i \in \pi_2$, and let ζ_i be a filament of T_i. Then*

$$\xi = \sum_i \xi_i + \sum_{i<j} LC(\alpha_i\zeta_i, \alpha_j\zeta_j). \tag{10.3}$$

REMARK. The formula holds equally well if there are junctions present (see Part III). The proof requires no change.

We shall use induction on m. It is trivial if $m = 1$. Taking the general case, let ξ' be the element of π_3 defined by the tube systems, in which T_m has been carried into a position T'_m such that there is a plane P separating T'_m from the other T_i. Then, using induction,

$$\xi' = [\sum_{i<m} \xi_i + \sum_{i<j<m} LC(\alpha_i\zeta_i, \alpha_j\zeta_j)] + \xi_m.$$

To compare ξ with ξ', we must see what happens when T_m is carried into T'_m. We may do this so that, during the process, tubes of T_m cut through tubes of other T_i a finite number of times. Let ξ'' and ξ''' be the elements of π_3 defined just before and after a tube of T_m cuts through a tube of T_i; let l'' and l''' be the values of $\sum_{j<m} LC(\alpha_j\zeta_j, \alpha_m\zeta_m)$ just before and after respectively.

Define ϵ for the cutting as in Lemma 6a; then

$$\xi'' - \xi''' = \epsilon\alpha_m\alpha_i.$$

At the cutting, ζ_m cuts through ζ_i; hence, with proper consideration of the sign, $LC(\zeta_i, \zeta_m)$ changes by $-\epsilon$, and consequently

$$l'' - l''' = \epsilon\alpha_i\alpha_m = \xi'' - \xi'''.$$

At the end of the process, the two elements of π_3 under discussion are ξ' and 0 respectively; at the start, they are ξ and $l_0 = \sum_{j<m} LC(\alpha_j\zeta_j, \alpha_m\zeta_m)$. Hence $\xi - \xi' = l_0$, and using the expression above for ξ' gives (10.3).

11. The formula in terms of looping coefficients

We consider first a tube system in which each cross section determines the same $\alpha \in \pi_2$.

THEOREM 11A. *Let* $[T, f]$ *be a tube system mapping defining* $\xi \in \pi_3$, f *defining* $\alpha \in \pi_2$. *Then if* ζ, ζ' *are disjoint filaments of* T,

$$\xi = LC^0(\alpha\zeta, \alpha\zeta') = LC(\zeta, \zeta')\eta(\alpha). \tag{11.1}$$

Let $T_1, \cdots, T_m$ be the tubes of T. Suppose first that $m = 1$, and $T = T_1$ is in the shape of a torus. The formula is certainly true if $l = LC(T) = LC(\zeta, \zeta') = 0$ or 1. Suppose $l > 1$. Stretch T out like a rubber band, with two long straight pieces T', T'', close to each other, joined by curved pieces at their ends. Twist T so that, for short pieces $T_1', \cdots, T_l'$ of T', ζ' goes around ζ once in each, and so that ζ' lies above ζ in the rest of T. Now perform $l - 1$ splices, between pieces of T' separating T_i' from T_{i+1}' and nearby pieces of T''; this leaves l tubes, each with looping coefficient 1. By definition of η, the result defines $l\eta(\alpha)$; this proves (11.1) in this case.

Suppose that $l < 0$. Take a similar tube mapping $[T', f']$ with $LC(T') = -l$. Splice T' to T; this gives $[T'', f'']$ with $LC(T'') = 0$, hence defining $0 \in \pi_3$. Since $[T', f']$ defines $-l\eta(\alpha)$, $[T, f]$ defines $l\eta(\alpha)$.

Take now a general m. If the tubes are in the shape of tori, separated from each other by planes, applying the formula just proved to each tube separately gives the result. Now let the tubes have any position. Let us straighten out the tubes and separate them, to reduce the problem to that just solved. We may do this with a finite number of cuttings. We need now only show that at each cutting, the change in the element of π_3 defined equals the change in the formula.

Taking ϵ as before, the change in the element of π_3 is $-\epsilon\alpha\alpha$. Consider the change Δ in $LC(\zeta, \zeta')$. At the cutting, a piece of tube containing pieces of both ζ and ζ' cuts through another such piece. The cutting of ζ through ζ' and of ζ' through ζ each contribute to Δ; hence $\Delta = -2\epsilon$. Therefore the change in $LC'(\alpha\zeta, \alpha\zeta')$ is $-2\epsilon\eta(\alpha) = -\epsilon\alpha\alpha$, as required.

We now give a formula for the general case.

THEOREM 11B. *With the hypotheses of Theorem* 10A, *if* ζ_i' *is a filament of* T_i *disjoint from* ζ_i,

$$\begin{aligned} \xi &= \sum_i LC^0(\alpha_i\zeta_i, \alpha_i\zeta_i') + \sum_{i<j} LC(\alpha_i\zeta_i, \alpha_j\zeta_j) \\ &= \sum_i LC(T_i)\eta(\alpha_i) + \sum_{i<j} LC(T_i, T_j)\alpha_i\alpha_j. \end{aligned} \tag{11.2}$$

This follows at once from Theorems 10A and 11A.

REMARK. In the last term, we may use $LC(\alpha_i\zeta_i, \alpha_j\zeta_j')$. Also, using two commutative laws shows that $i < j$ may be replaced by $i > j$. Adding two expressions for ξ thus obtained gives

$$(11.3) \qquad 2\xi = \sum_{i,j} LC(\alpha_i \zeta_i, \alpha_j \zeta_j') = LC\left(\sum_i \alpha_i \zeta_i, \sum_i \alpha_i \zeta_i'\right).$$

If π_3 contains elements of order 2, we cannot divide this expression by 2 to recover an expression for ξ however.

12. Projections into a plane

Consider a tube system in E^3, composed of subsystems $T_1, \cdots, T_m$. Let ζ_i be a filament of T_i. Let ψ be the orthogonal projection of E^3 onto a plane P. Then at least after a slight deformation of the tube system, $\psi\zeta_1, \cdots, \psi\zeta_m$ are sets of smooth curves in P (ψ mapping tangent vectors to the ζ_i into non-vanishing vectors), and there are at most a finite number of crossing points.

Here, a *crossing point* of arcs A, B in P is a point where the arcs cross, at a non-zero angle. Crossing points can come from arcs of the same $\psi\zeta_i$ or from arcs of different ones.

We take P oriented, say by the pair of vectors (u_1, u_2). Then the positive normal to P is given by a vector u such that (u, u_1, u_2) defines the positive orientation of E^3. The side of P in the direction of u we call the *upper* side of P.

Let A, B be non-intersecting oriented arcs in E^3 such that ψA, ψB have a single crossing point p. Let p_A and p_B be the points of A and B projecting into p respectively. Let v_A and v_B be the tangent vectors to ψA and ψB at p respectively, and set $v_{AB} = p_B - p_A$. We call the crossing *positive* or *negative* according as the set of vectors

$$(v_A, v_{AB}, v_B)$$

defines the positive or negative orientation of E^3. Note that (v_B, v_{BA}, v_A) defines the same orientation; hence the definition is independent of the order of A, B.

LEMMA 12a. *The sign of the above crossing is positive or negative according as* (v^+, v^-) *defines the positive or negative orientation of* P, *where* v^+ *and* v^- *are the tangent vectors to the upper and lower curves at the crossing respectively.*

Suppose, for definiteness, that A is the upper and B is the lower crossing; then v_{BA} is a positive multiple of u. Then the ordered triples

$$(v_A, v_{AB}, v_B), \qquad (-v_{AB}, v_A, v_B), \qquad (v, v_A, v_B)$$

define the same orientation of E^3. Hence (v_A, v_{AB}, v_B) defines the positive or negative orientation of E^3 according as (v_A, v_B) defines the positive or negative orientation of P, as required.

13. The formula in terms of crossings

Let ψ project a tube system into P as above. Let N_i^+ be the number of crossing points of $\psi\zeta_i$ with itself of positive type, and N_i^-, the number of negative type.

For $i \neq j$, let N_{ij}^{+} and N_{ij}^{-} be the numbers of crossing points of $\psi\zeta_i$ with $\psi\zeta_j$ *in which* ζ_i *lies above* ζ_j at the point, of positive and negative types respectively. Set

$$N_i = N_i^{+} - N_i^{-}, \qquad N_{ij} = N_{ij}^{+} - N_{ij}^{-};$$

these are the algebraic numbers of self-crossings of $\psi\zeta_i$ and of crossings of $\psi\zeta_i$ with $\psi\zeta_j$ with ζ_i above ζ_j respectively.

We now find the formula for the element of π_3 determined by a tube system, in terms of crossing points under a projection, assuming that, seen from above, no tube twists. It would be easy to include the result of twists in the formula, but we do not need this.

THEOREM 13A. *Let* $[T_1, f_1], \cdots, [T_m, f_m]$ *be tube system mappings with disjoint tubes, defining* $\xi \in \pi_3$. *Let* ψ *be the projection of* E^3 *into a plane* P *such that filaments* $\zeta_1, \cdots, \zeta_m$ *of* $T_1, \cdots, T_m$ *respectively project into sets of curves with a finite number of crossings. Let the top filaments of the tubes lie on top of the tubes* (i.e. *be highest above* P). *Then if* f_i *defines* α_i,

$$\xi = \sum_i N_i \eta(\alpha_i) + \sum_{i<j} N_{ij} \alpha_i \alpha_j. \tag{13.1}$$

We may suppose ζ_i is the top filament of T_i. Let ζ_i' be a filament to the right of ζ_i. Let us determine $l_i = LC(\zeta_i, \zeta_i')$. To this end, pull ζ_i downwards till it is entirely below ζ_i'. Each time a piece of ζ_i cuts through a piece of ζ_i', an amount ϵ say is lost from l_i. Each such cutting corresponds to a crossing of $\psi\zeta_i$ with itself, and conversely; moreover, $\epsilon = 1$ or -1 according as the crossing is positive or negative. It follows that

$$LC(T_i) = LC(\zeta_i, \zeta_i') = N_i.$$

We next determine $l_{ij} = LC(\zeta_i, \zeta_j)$ by pulling ζ_i downwards. We lose ϵ from l_{ij} each time ζ_i cuts through ζ_j; each such cutting corresponds to a crossing of $\psi\zeta_i$ with $\psi\zeta_j$ such that ζ_i lies above ζ_j there, and conversely; moreover, the sign of the crossing is ϵ. Hence

$$LC(T_i, T_j) = LC(\zeta_i, \zeta_j) = N_{ij}.$$

Putting these equations into (11.2) gives (13.1).

PART III. TUBE SYSTEMS WITH JUNCTIONS

14. The junctions

Let α be an element of π_2 of finite order n. We shall define what we call an *α-junction*, which is a mapping of a cylindrical portion of E^3 into R. There will be two types, positive and negative.

Let f be a *fixed* mapping of σ_0^2 into R defining α. Let J be a cylinder, with top and bottom and a curved side. Let $Q_0, \cdots, Q_{n-1}$ be patches around the side, in cyclic order (in the sense of a right hand screw moving upwards), each being the portion of the side cut out by a small cylinder T_i piercing the side. Orient the inside of J like the orientation of E^3, and orient the Q_i so that each forms part of the boundary ∂J. Take a cross section of T_i as σ_0^2, with its top point on top, oriented like Q_i; projecting onto Q_i puts Q_i into correspondence with σ_0^2; map Q_i into R like the mapping f in σ_0^2. All points of ∂Q_i are mapped into $\bar{p}$; map the rest of ∂J into $\bar{p}$.

The mapping of ∂J just defined defines the element $n\alpha = 0$ of π_2; hence the mapping may be extended through J. Choose a *fixed* way of doing this; this defines the *positive junction defined by* α, which we also call J. Note that the mapping is determined when we know which is the top of the cylinder and which the first patch Q_0 is. Note also that if tubes are attached to the patches so that a cross section coincides with the patch in its correspondence with σ_0^2, then the positive direction of each filament is away from J, just like the positive normal to ∂J.

Let P be a vertical plane cutting both the top and bottom of J (and hence J itself) in half. We let P be perpendicular to the cylinder which cuts the side of J in Q_0. Reflecting across P gives a new mapping of the same point set into R; this defines the *negative junction defined by* α. We let the image of Q_0 be the first patch of the junction, the image of Q_1 the second, etc. Tubes attached to the patches have their filaments pointing into J.

15. Effect of rotating a junction

Let ϕ be a mapping of E^3 into R such that in a certain cylinder, it defines the α-junction J (either positive or negative). Suppose we take J and turn it, in either direction, so that each patch moves into the position of the next adjacent patch. Since all patches are mapped alike, this defines a new mapping ϕ' of E^3 into R. We say ϕ' is obtained from ϕ by rotating the junction one step.

LEMMA 15a. *Let ϕ and ϕ' be mappings of E^3 into R, one obtained from the other by rotating an α-junction J one step. Then the elements ξ, ξ' defined by ϕ, ϕ' respectively differ by $n\eta(\alpha)$, if α is of order n.*

REMARKS. Since $2n\eta(\alpha) = 0$, we need not specify the direction of rotation. Suppose n is even. Rotating J through two steps brings us back to the same element of π_3. Hence, if we insert J to fit a given set of patches, and these patches are called "even" and "odd" alternately, any manner of inserting J so that the even patches Q_{2k} of J go into even patches and the odd patches Q_{2k+1} go into odd patches gives the same element of π_3; any other insertion gives the same element, plus $n\eta(\alpha)$. If n is odd, then since $n\eta(\alpha) = 0$, the manner of insertion is immaterial.

To prove the lemma, we first deform the tubes so that in a reasonably large neighborhood U of J, the tube T_i attached to Q_i moves straight out without twisting.

We now deform ϕ into a mapping ϕ_1 which is equivalent to rotating the junction and adding a tube mapping. In the figure, the junction is in the center, $n = 4$, and the curved lines are the tubes. Take a little piece of each tube T_i, pull it out and stretch it around and under the next tube T_{i+1}, and bring it back close to T_{i+1}. This piece of T_i may now be spliced to the adjacent piece of T_{i+1}. In the resulting tube system, starting in towards J along T_i, we run into J in the next patch Q_{i+1}. We may straighten out these tubes, provided that we rotate the junction through an angle $-2\pi/n$. This part of the new tube system is now exactly like the original, except for the rotation of the junction. There is also a separate closed tube in the new system; it lies above the rest of the system, and, projected into a plane below, it has n self-crossings, all of the same type. Applying Theorem 13A to this tube shows that, by itself, it defines the element $\pm n\eta(\alpha) = n\eta(\alpha)$ of π_3. Since the result was obtained by a deformation, the new element of π_3 plus $n\eta(\alpha)$ equals the old element of π_3, and the conclusion of the lemma follows.

FIG. 1

16. Eliminating a pair of junctions

Let J and J' be junctions, joined by a short straight untwisted tube T; let T join the first patches of the tubes. Let $T_2, \cdots, T_n$ and $T'_2, \cdots, T'_n$ be the other tubes on J and J' respectively, in cyclic order. Suppose we remove J, J' and T. We may then join T_i to T'_i, without twisting ($i = 2, \cdots, n$). For if T is oriented from J towards J' for instance, the T'_i are oriented towards J', and the T_i, away from J; also, all top filaments near the junctions are on corresponding parts of the tubes. (Furthermore, J is positive and J' is negative.)

Note that *two junctions joined by a tube are always of opposite signs.*

LEMMA 16a. *Let $[T^*, f]$ be a tube system mapping, defining $\xi \in \pi_3$. Let J and J' be junctions, joined by a tube T, as above. Remove J and J' as described. Then the resulting tube system mapping defines the same element ξ.*

We shall obtain the new tube system mapping by means of a deformation of f. First, deforming f gives a situation as follows. There is a cube C containing

J and J'. A plane P bisecting C also bisects T; the parts of the tubes and the junctions in C are symmetric with respect to P. (Recall the definition of a negative junction.) Hence, if p and p' are symmetric points of C, $f(p) = f(p')$. Moreover, the tubes T_i and T'_i enter opposite faces of C, and are perpendicular to the faces there. Then, exactly as in §4, we get rid of the junctions, and find T_i joined to T'_i by a straight tube through C.

REMARK. There may be two tubes joining the same two junctions, one joining patches of the same parity (both even or both odd), and the other joining patches of opposite parity. In this case, when projected into a plane, there must be crossings.

17. Linear graphs in the plane

Let T be a tube system with junctions in E^3. Take a certain direction as upwards; we shall always assume the tops of the junctions are on top, in the sense of this direction. For each tube ending in junctions, extend its center filament to the center of the junctions (which are distinct, since the junctions are of opposite sign). After a slight deformation, we may project into a plane P, so that the center filaments and junctions go into the 1-cells and vertices (0-cells) of a linear graph G in P; the 1-cells are either closed curves or arcs joining distinct vertices; there are but a finite number of crossings, occuring at interior points of curves. At each vertex corresponding to an α-junction, α of order n, n even, the ends of 1-cells attaching to the vertex are named even and odd alternately. The 1-cells and vertices are oriented. (A vertex is oriented by naming it positive or negative.)

18. The coboundary Z

Let G be the graph in P corresponding to a tube system T (see §17) with a fixed α of order $n > 0$, n even. We define a corresponding 1-cochain Z in G (mod 2) as follows. Take any 1-cell A of G. Define

(18.1) $Z' \cdot A$ = number (mod 2) of crossings of A with other 1-cells;

(18.2) $Z'' \cdot A$ = 0 or 1 (mod 2) according as A joins patches of the same or opposite parity;

= 0 if A is a closed curve.

Setting $Z' \cdot \sum_i A_i = \sum_i Z' \cdot A_i$ and similarly for Z'' defines 1-cochains Z', Z''. Set

$$Z = Z' + Z''. \tag{18.3}$$

Note that, for any 1-chain A, (KI = Kronecker index mod 2, intersections at interior points only count)

$$Z' \cdot A = KI(A, G - A), \tag{18.4}$$

i.e., the number of crossings of arcs of A with arcs of $G - A$. For if A =

$A_1 + \cdots + A_k$, any crossing of A_i with A_j $(i \neq j)$ counts in both terms $KI(A_i, G - A_i)$ and $KI(A_j, G - A_j)$ appearing in (18.1), and hence does not contribute to the sum of such terms.

For any $A = \sum A_i$, since $Z'' \cdot A_i = 0$ or 1 according as whether there are an even or an odd number of ends of A_i named odd, we have

(18.5) $\quad Z'' \cdot A \equiv$ the number of ends of 1-cells in A named odd.

LEMMA 18a. *Z is a coboundary*: $Z = \delta Y$.

It is sufficient to show that for any cycle C forming a closed curve, $Z \cdot C = 0$. (C may have self-intersections.) Let the vertices and 1-cells making up C, in order, be $V_1, A_1, V_2, A_2, \cdots, V_m, A_m, V_1$. Since each 1-cell joins junctions of opposite sign, m is even. (The modification in the proof required if C is a single 1-cell is easy to give.)

Let $C' = G - C$ be the 1-chain (mod 2) of 1-cells not in C. Since each vertex is on an even number of 1-cells of G, $\partial G = 0$; therefore $\partial C' = \partial G - \partial C = 0$. Hence we may write

$$C' = C'_1 + \cdots + C'_q,$$

each C'_i being a closed curve.

Let B be a closed curve formed from C as follows. It contains all but a little of each end of each A_i; it replaces the ends of A_{i-1} and A_i at each V_i (setting $A_0 = A_m$) by a curve B''_i going around V_i (on either side). Let B'_i be the part of B on A_i; set

$$B' = \sum B'_i, \qquad B'' = \sum B''_i;$$

then $B = B' + B''$.

The crossings of C with C' are the same as of B' with C'; hence

$$Z' \cdot C = KI(C, C') = KI(B', C').$$

Since B and C'_i are closed curves,

$$KI(B, C') = \sum KI(B, C'_i) = 0;$$

therefore

$$\begin{aligned} Z' \cdot C &= KI(B, C') + KI(B', C') \\ &= KI(B'', C') = \sum KI(B''_i, C'). \end{aligned}$$

Now $KI(B''_i, C') = 0$ or 1 (mod 2) according as B''_i cuts an even or an odd number of 1-cells of C' at V_i. Since C' contains all 1-cells not in C, these cases happen according as the ends of A_{i-1} and A_i at V_i are named of the opposite or same parity. Thus $Z' \cdot C$ is the number k (mod 2) of vertices V_i of C at which the ends of the adjoining 1-cells of C are of the same parity.

Since m is even, $m - k \equiv k$ (mod 2). Therefore $Z' \cdot C$ is the number (mod 2) of vertices V_i of C at which the ends of the two adjoining 1-cells of C are of the opposite parity, that is, at which just one of these ends is named odd. Hence

$Z' \cdot C$ is (mod 2) the number of ends of 1-cells of C named odd. Therefore, by (18.5),

$$Z' \cdot C = Z'' \cdot C, \qquad Z \cdot C = 0,$$

completing the proof.

19. The formula using projections

We are now ready to extend Theorem 13A to the case that junctions are present.

THEOREM 19A. *Let $[T_1, f_1], \cdots, [T_m, f_m]$ be tube system mappings which may have junctions, with disjoint tubes, defining $\xi \in \pi_3$. With a certain direction in E^3 named upwards, let the top filament of each tube lie on top; then the top face of each junction is on top. Deform slightly so that, when projected into a plane P, the linear graph G is formed as in §17. Define N_i, N_{ij} as in §13. Let f_i define $\alpha_i \in \pi_2$, of order n_i. For each i with $n_i > 0$ even, define the coboundary Z_i* (mod 2) *in the subgraph G_i of G corresponding to T_i as in §18. Choose a* 1*-chain D_i* (mod 2) *in G_i such that*

$$\partial D_i = \text{sum of vertices of } G_i \pmod 2. \tag{19.1}$$

Then

$$\xi = \sum_i N_i \eta(\alpha_i) + \sum_{i<j} N_{ij} \alpha_i \alpha_j + \sum_{n_i > 0, \text{ even}} (Z_i \cdot D_i) n_i \eta(\alpha_i). \tag{19.2}$$

The meaning of the last term must be made clear, since $Z_i \cdot D_i$ is an integer mod 2. Replace $Z_i \cdot D_i$ simply by any corresponding integer (say 0 or 1); since $2n_i \eta(\alpha_i) = 0$, it does not matter which integer is chosen.

First we note that D_i exists. This follows easily from the fact that each 1-cell not a closed curve joins vertices of opposite sign. (A slightly stronger theorem will be proved in Lemma 21a.)

Next, note that the right hand side of (19.2) is independent of the choice of D_i. For suppose, instead, that D_i' were chosen. Then

$$Z_i \cdot D_i' - Z_i \cdot D_i = \delta Y_i \cdot (D_i' - D_i) = Y_i \cdot (\partial D_i' - \partial D_i) = 0.$$

We shall prove the formula by induction on the number of junctions in the whole system. If there are none, it is true, by Theorem 13A. In the general case, choose a junction J; let T be a tube ending at J; the other end of T is at a junction $J' \neq J$, since J and J' are of opposite signs. We shall prove (19.2) by apply ng the formula to a tube system like the given one, but without the junctions J and J'.

There are three main steps in the proof. The first step is to pull J along T towards J', until it is close to J'; we let T shrink into a very short straight tube, and let the other tubes ending at J stretch, and follow along the path where T was. This of course does not alter ξ; we must show that it does not alter the rest of (19.2). The second step is to rotate the junctions if necessary, so that T

joins the first patches of each. The last step is to eliminate the two junctions, as in §16.

We consider now the *first step*. During the deformation, the only alteration in G is the motion of the vertex V towards the vertex V', and the shrinking of the 1-cell A corresponding to T and the stretching of the other 1-cells on V. So long as V crosses no other 1-cells of G, the right hand side of (19.2) is unchanged. We must show that it is still unchanged when V crosses the 1-cell A' of G. Say T is in T_i, and T' (corresponding to A') is in T_j.

CASE I. $j \neq i$. Then the first and last sums in (19.2) are unchanged; we must show that the second sum is. Suppose first that $i < j$, and T is below T' in the crossing. Then this crossing does not count in the sum, nor do the $n_i - 1$ crossings after V has passed through A'; hence the sum is unchanged. The same is true if $i > j$ and T is above T'. Suppose now that $i < j$ and T is above T' at the crossing. Then before V has passed through A', the crossing contributes $\epsilon\alpha_i\alpha_j$ ($\epsilon = \pm 1$); afterwards, since the crossing of the single 1-cell A, with a certain orientation, with A', is replaced by a crossing of $n_i - 1$ 1-cells with the opposite orientation, the new crossings contribute $-(n_i - 1)\epsilon\alpha_i\alpha_j$. Hence the change in the sum is

$$\pm n_i\alpha_i\alpha_j = \pm(n_i\alpha_i)\alpha_j = \pm 0\alpha_j = 0.$$

The same is true if $i > j$ and T is below T'.

CASE II. $j = i$. Then the second sum in (19.2) is unchanged. Consider the first sum. The single crossing of A with A' (which may equal A) of one sign is replaced by $n_i - 1$ crossings of the opposite sign; hence the first sum changes by $n_i\eta(\alpha_i)$. If n_i is odd, then this quantity is 0, and the value i does not occur in the last sum; hence the whole is unchanged. Suppose now that $n_i > 0$ is even. We shall show that the change in $Z_i \cdot D_i$ is $\equiv 1 \pmod 2$; this will prove that the whole change is $\pm n_i\eta(\alpha_i) \pm n_i\eta(\alpha_i) = 0$, since $2n_i\eta(\alpha_i) = 0$.

Because of (19.1), an odd number p of 1-cells of D_i end at V. Since n_i is even, an odd number $q = n_i - p$ of 1-cells of $G_i - D_i$ also end at V. Suppose first that A' is not in D_i. We find the change in $Z'_i \cdot D_i$ by using (18.4). The change is due to the change in $KI(D_i, A')$. Since each 1-cell of D_i ending at V loses a crossing (if it is A) or adds one (if it is $\neq A$), the number of changes is $p \equiv 1 \pmod 2$. Suppose next that A' is in D_i. Then the change is due to the change in $KI(A', G_i - D_i)$, and the same argument applies, with p replaced by q. Hence $Z'_i \cdot D_i$ changes by 1 (mod 2) in all cases. Of course $Z''_i \cdot D_i$ is unchanged; hence the change in $Z_i \cdot D_i$ is 1 (mod 2), and the first step is completed.

We turn now to the *second step*. The junctions J, J' are now beside each other, joined by the short straight tube T. Rotate the junctions if necessary so that T joins the first patches of each. If n_i is odd, nothing is altered; assume n_i even. Say J and J' are rotated through p and p' steps respectively. Then by Lemma 15a, ξ is altered by $(p + p')n_i\eta(\alpha_i)$, which is 0 or $n_i\eta(\alpha_i)$ according as $p + p'$ is even or odd. In this process, only the last sum in (19.2) can change. Consider a rotation through one step of the junction J. For each 1-cell A^*

ending at V, $Z_i''\cdot A^*$ changes by 1 (mod 2); see (18.2). Since the number of 1-cells of D_i ending at V is odd, $Z_i''\cdot D_i$ changes by 1 (mod 2). Of course $Z_i'\cdot D_i$ is unchanged; hence $Z_i\cdot D_i$ changes by 1 (mod 2) for the rotation of one step. The same holds for V'. Consequently the $p + p'$ rotations alter the last term by $(p + p')n_i\eta(\alpha_i)$, which is the same as the alteration of ξ (recall that $2n_i\eta(\alpha_i) = 0$). Thus the rotations do not alter (19.2).

We are now ready for the *last step*. The elimination of the junctions as in §16 does not alter ξ. This turns the tube system T_i into a tube system T_i^* (which need not be connected, even if T_i is). Again the case n_i odd is trivial; assume n_i even. Choose D_i^* in the corresponding graph G_i^* so that ∂D_i^* is the sum of the vertices. Since D_i in G_i may be chosen arbitrarily, we choose it as follows. Each 1-cell of G_i^* in D_i^* which is also a 1-cell in G_i we put in D_i. Take any 1-cell $A_i \neq A$ ending on V or on V'; it becomes part of a 1-cell A_i^* in G_i^*; put A_i in D_i if A_i^* is in D_i^*. There are now the same number of p of 1-cells of D_i ending on V as there are ending on V'; if p is even, put A in D_i, while if it is not, do not. It is clear from the construction that ∂D_i is the sum of the vertices of G_i; hence D_i may be used in (19.2).

By induction, (19.2) holds in the new system. The first two sums on the right retain their meaning, and are unchanged; the same is true for the terms of the last sum with indices $\neq i$. Since ξ is unchanged, it remains to show simply that

$$Z_i\cdot D_i \equiv Z_i^*\cdot D_i^* \pmod 2.$$

The ends of 1-cells of D_i are exactly those of D_i^*, plus the ends on V and V'. Since T joins patches of the same parity in J and J', clearly the ends of D_i on V and V' are paired off, the two of a pair being of the same parity. Hence, by (18.5), $Z_i''\cdot D_i \equiv Z_i^{*\prime\prime}\cdot D_i^*$. It is clear that $Z_i'\cdot D_i \equiv Z_i^{*\prime}\cdot D_i^*$. Hence $Z_i\cdot D_i \equiv Z_i^*\cdot D_i^*$, completing the proof of the theorem.

20. A generalization

We shall show here that formula (19.2) holds, with a wider meaning given to the objects involved. It is the last sum in (19.2) we wish to generalize. Call a vertex of the corresponding graph G *special* if the corresponding junction has an even number of patches (i.e., the corresponding α is of even order). We shall pull pieces of 1-cells of G near special vertices into the vertices, forming the vertices of a new graph G^*.

Around each special vertex V_i of G choose a simple closed curve C_i, containing no other vertices, no crossings, and no arcs corresponding to tubes with a different α, and having itself a finite number of crossings with 1-cells of G. The vertices V_i^* of G^* correspond to the vertices V_i of G, and have the same sign; V_i^* is special if V_i is. The 1-cells of G^* correspond to the connected pieces of the 1-cells of G which are left when everything inside each C_i is dropped out; each is oriented like the corresponding 1-cell in G. Note that if the C_i are small enough, then G and G^* are isomorphic. Note also that G^* may contain 1-cells joining any two special vertices with the same α, distinct or not, and the 1-cell may be oriented in either direction. We may represent G^* in the plane by taking G and shrinking each C_i to a point.

From the special vertex V_i, follow along each 1-cell ending at V_i until we first reach C_i. It is clear that these arcs divide the interior of C_i into n regions (if α is of order n), which we call the *sectors* of C_i; we call the arcs we have followed the *sector edges*. Each sector edge is named even or odd according to how its end on V_i is named.

LEMMA 20a. *If V_i is special, the ends of 1-cells of G^* on V_i^* may be named even and odd alternately in a unique manner so that each 1-cell whose continuation in C_i is a sector edge is named the same as the sector edge.*

For each 1-cell of G entering a sector of C_i must leave it again, and hence the number of crossings of the part of C_i between two adjacent sector edges is even.

The subgraphs G_i^* of G^* correspond to the G_i in G (see §19). The numbers N_i^*, N_{ij}^* clearly have the same meaning as N_i, N_{ij} in G, and are equal to these respectively. The meanings of $Z_i^{*\prime}$, $Z_i^{*\prime\prime}$ are the same as of Z_i', Z_i''. Define the cochain $Z_i^{*\prime\prime\prime}$ by the following relation. For any 1-cell A^* of G_i^* (α_i of order $n_i > 0$, even),

$$Z_i^{*\prime\prime\prime}\cdot A^* = 0 \text{ or } 1 \text{ according as the ends of } A^* \text{ join vertices of the opposite or same sign.} \tag{20.1}$$

Set

$$Z_i^* = Z_i^{*\prime} + Z_i^{*\prime\prime} + Z_i^{*\prime\prime\prime}. \tag{20.2}$$

THEOREM 20A. *Let the conditions of Theorem 19A hold. Define G^* as above, and name ends of 1-cells on special vertices according to Lemma 20a. Define Z_i^* by (20.2), and choose D_i^* so that ∂D_i^* is the sum of the vertices in G_i^* (mod 2). Then (19.2) holds, using N_i^*, N_{ij}^*, Z_i^* and D_i^*.*

The existence of D_i^* will be proved below.

We begin the proof by showing that $Z^* = Z_i^*$ is a coboundary. We follow the proof in §18. We shall show that for any closed curve C in G_i^*, $Z^*\cdot C = 0$. In the present case, m need not be even. Define B, B', B'' as before. Again, $Z^{*\prime}\cdot C$ is the number k (mod 2) of vertices V_i^* of C at which the ends of the adjoining 1-cells of C are of the same parity.

The number of 1-cells of C which join vertices of opposite sign is clearly even, and the number of 1-cells joining vertices of the same sign is $\equiv Z^{*\prime\prime\prime}\cdot C$ (mod 2); hence $m \equiv Z^{*\prime\prime\prime}\cdot C$ (mod 2). Therefore

$$m - k \equiv Z^{*\prime}\cdot C + Z^{*\prime\prime\prime}\cdot C \qquad \text{(mod 2)}.$$

Now $m - k$ is the number of vertices of C at which the ends of the adjoining 1-cells of C are of opposite parity, as before. This is $Z^{*\prime\prime}\cdot C$. Putting in the above relation gives $Z^*\cdot C = 0$, as required.

It follows from this that (19.2) (with stars) is independent of the D_i^* chosen.

We choose D_i^* as follows. Take a fixed D_i for G_i; say $D_i = \sum A_k$. For each A_k, let $B_{k1}^*, \cdots, B_{k\lambda_k}^*$ be the 1-cells of G^* corresponding to the pieces B_{kj} of A_k. Set

$$B_k^* = \sum_j B_{kj}^*, \qquad D_i^* = \sum B_k^*.$$

If $\partial A_k = V_p + V_q$ (mod 2), clearly $\partial B_k^* = V_p^* + V_q^*$ (mod 2); hence D_i^* has the required boundary.

We now prove (19.2) (with stars), using this D_i^*. Since $N_i^* = N_i$, $N_{ij}^* = N_{ij}$, it suffices to prove that

$$Z_i^* \cdot D_i^* \equiv Z_i \cdot D_i \qquad \text{(mod 2)}.$$

Clearly $Z_i^{*\prime} \cdot D_i^* = Z_i' \cdot D_i$; if we show that for each k,

$$(18.3) \qquad (Z_i''^* + Z_i^{*\prime\prime\prime}) \cdot B_k^* \equiv Z_i'' \cdot A_k \qquad \text{(mod 2)},$$

the above relation will follow.

As we go from one end of A_k to the other, we enter and leave again, in turn, the interior of curves $C_{\gamma_1}, \cdots, C_{\gamma_s}$ say ($s = \lambda_k - 1$; C_{γ_j} around V_{γ_j}). (Some of these may be the same; there may be none.) The ends of $B_{k,j}$ and $B_{k,j+1}$ on C_{γ_j} are named of opposite parity, since the part of C_{γ_j} between these is cut an even number of times by 1-cells of G. Hence (recalling the naming in Lemma 20a),

$$Z_i'' \cdot A_k \equiv \sum Z_i^{*\prime\prime} \cdot B_{kj}^* + (\lambda_k - 1) \equiv Z_i^{*\prime\prime} \cdot B_k^* + (\lambda_k - 1) \quad \text{(mod 2)}.$$

Now the number of 1-cells B_{kj}^* is λ_k. The number of these which join vertices of G^* of opposite sign is $\equiv 1$ (mod 2), since A_k joins vertices of opposite sign. The number of B_{kj}^* which join vertices of the same sign is $Z_i^{*\prime\prime\prime} \cdot B_k^*$. Hence

$$\lambda_k \equiv 1 + Z_i^{*\prime\prime\prime} \cdot B_k^* \qquad \text{(mod 2)}.$$

Putting this in the relation above gives (20.3), completing the proof of the theorem.

21. A formula using looping coefficients

We shall take a general tube system mapping with junctions, and describe a process giving directly the corresponding element of π_3, without using projections. The formula will not be needed in the two forthcoming papers referred to above.

Because of the reduction formula (§10), we may consider a single tube system mapping $[T, f]$, f defining $\alpha \in \pi_2$. Say α is of order n. Let G be the corresponding graph in E^3; its 1-cells and vertices are oriented.

LEMMA 21a. *There is a 1-chain B in G with integer coefficients such that ∂B is the sum of the negative vertices of G minus the sum of the positive ones.*

We find the part B_i of B in each connected part G_i of G as follows. Say there are p 1-cells in G_i. Each one has two ends, one positive and one negative (unless G_i is a closed curve). There are therefore $q = p/n$ positive vertices and

q negative vertices in G_i. Pair off the positive and negative vertices in any fashion: V_j^+ and V_j^-, $j = 1, \cdots, q$. Since G_i is connected, there is a path of 1-cells joining V_j^+ to V_j^-; forming these into a 1-chain B_{ij} whose coefficients are ± 1, we obtain $\partial B_{ij} = V_j^- - V_j^+$. Then $B_i = \sum_j B_{ij}$ is the required 1-chain in G_i, and $B = \sum_i B_i$.

Let $C = \sum c_i A_i$ be any 1-cycle in G with integer coefficients. We define corresponding singular cycles $\zeta(C)$, $\zeta'(C)$, with coefficients in the subgroup of π_2 generated by α, as follows. In the disc σ_0^2, let τ be the diameter dividing the top and bottom halves, and let τ' be a segment lying above τ. In each tube T_i, corresponding to A_i, choose $|c_i|$ positively oriented filaments, corresponding to points of σ_0^2 on τ, if $c_i \neq 0$. To each filament let correspond the coefficient α or $-\alpha$ according as $c_i > 0$ or $c_i < 0$. This gives a singular chain whose boundary is in the junctions. In each junction J, since $\partial C = 0$, the number of filaments ending at J (all of which of course have the same orientation) with the coefficient α equals the number with the coefficient $-\alpha$; moreover, they lie in cyclic order around the curved face of J. Hence we may join these up in pairs by arcs in J which, viewed from above, have no crossings, and attach the coefficient α to each, so that the result is a singular cycle $\zeta(C)$. We define $\zeta'(C)$ by shifting the points in τ to corresponding points in τ', and lifting the arcs in each junction upwards correspondingly.

THEOREM 21A. *Let $[T, f]$ be a tube system mapping into R, which may have junctions, defining $\xi \in \pi_3$, let f define $\alpha \in \pi_2$; let α be of order $n \neq 0$. Let G be the corresponding graph in E^3. Let G^1 be the integral 1-chain consisting of the oriented 1-cells of G. Choose a 1-chain $B = \sum b_i A_i$ in G by Lemma 21a; let $\bar{B}$ be this chain reduced* mod 2. *Let $\zeta(G^1 - nB)$ and $\zeta'(G^1 - nB)$ be the singular cycles defined above. If n is even, define the 1-cochain Z'' in G as in* §18. *Then (see* §10 *for the definition of LC^0)*

$$\xi = LC^0(\zeta(G^1 - nB), \zeta'(G^1 - nB)) + (Z'' \cdot \bar{B}) n\eta(\alpha), \tag{21.1}$$

omitting the last term if n is odd.

First we note that ∂G^1 has the coefficient $-n$ on each positive vertex and n on each negative one; hence $G^1 - nB$ is a cycle.

We shall replace the tube system by one to which we may apply Theorem 19A. First, having chosen a plane P in E^3 and an upwards direction, deform the tube system so that each junction has its top face on top.

The next step is to get the top filament of each tube on top. To this end, cut through the tube; we may now untwist it, getting the top filament on top, and then attach the pieces together again. We must show that this alters both sides of (21.1) by the same amount.

Take the tube end which, in its given orientation, points towards the other end. It is sufficient to show that a single twist of this end in the positive sense (using the orientation of σ_0^2) alters both sides by the same amount. It adds $\eta(\alpha)$ to ξ. The change in the right hand side of (21.1) occurs only in LC^0. Each filament of ζ' gets an added turn about each filament of ζ; the corresponding contribution (each filament with its given orientation) is $\eta(\alpha)$. Hence the total change is, if the tube corresponds to A_i,

$$(1 - nb_i)(1 - nb_i)\eta(\alpha) = \eta(\alpha);$$

for $(1 - nb_i)^2 \equiv 1$, mod n if n is odd, and mod $2n$ if n is even. Thus the change in both sides of (21.1) is the same.

We may now project G into a graph ψG in the plane P, and apply Theorem 19A to ψG. We shall evaluate the right hand side of (21.1) in terms of the quantities in (19.2), and show that exactly the right hand side of (19.2) (for a single tube system) is obtained.

Pull $\zeta(G^1 - nB)$ downwards till it is wholly below $\zeta'(G^1 - nB)$; then LC^0 is the algebraic number of cuttings with ζ' on the way, times $\eta(\alpha)$. Since top filaments were on top throughout, the cuttings occur only where part of one tube is above part of another (or the same) tube; that is, they correspond exactly to the crossings of ψG. Take such a crossing; say a piece of the 1-cell ψA_i crosses a piece of ψA_j there, and the crossing is positive. The corresponding contribution to LC^0 is

$$(1 - nb_i)(1 - nb_j)\eta(\alpha) = \eta(\alpha) - (b_i + b_j)n\eta(\alpha),$$

since $n^2\eta(\alpha) = 0$. Write

$$LC^0(\zeta(G^1 - nB), \zeta'(G^1 - nB)) = L_1 + L_2,$$

where L_1 and L_2 are the contributions to LC^0 of the first and second terms in the relation above respectively. Clearly $L_1 = N\eta(\alpha)$, giving the first part of (19.2); the second part drops out.

To evaluate L_2, divide the crossings into two types, those in which $b_i + b_j$ is even and odd respectively. We may suppose n is even, since otherwise L_2 and the rest of (21.1) and (19.2) all drop out. Each crossing of the first [second] type contributes 0 [$n\eta(\alpha)$] to L_2. Hence L_2 is the number p of crossings of the second type (taken mod 2 is sufficient), times $n\eta(\alpha)$. Now $b_i + b_j$ is odd if and only if just one of the 1-cells A_i, A_j appears in $\bar{B}$; hence p is the number of crossings of arcs in $\psi\bar{B}$ with arcs not in $\psi\bar{B}$, which gives $p = Z'\cdot\bar{B}$. Therefore

$$L_2 = (Z'\cdot\bar{B})n\eta(\alpha).$$

This quantity combines therefore with the last term of (21.1) to give the last term of (19.2), completing the proof.

Harvard University

References

1. Whitehead, J. H. C., *On adding relations to homotopy groups*, Annals of Math., vol. 42 (1941), pp. 409–428.
2. Pontrjagin, L., *A classification of the mappings of the 3-dimensional complex into the 2-dimensional sphere*, Rec. Math. [Mat. Sbornik] N.S. 9 (51) (1941), pp. 331–363.

ANNALS OF MATHEMATICS
Vol. 50, No. 2, April, 1949

CLASSIFICATION OF THE MAPPINGS OF A 3-COMPLEX INTO A SIMPLY CONNECTED SPACE

HASSLER WHITNEY

(Received March 29, 1948)

1. Introduction

Let R be an arcwise connected and simply connected Hausdorff space, whose second homotopy group $\pi_2 = \pi_2(R)$ has a finite set of generators. (This is always the case if R is a finite complex, since π_2 is then isomorphic with the second homology group of R, by a well known Theorem of Hurewicz.) We shall classify the mappings of a 3-complex K into R. We take K as a simplicial complex, with ordered vertices. For the relation of this paper to the literature, and references in brackets, see the following paper.

It would be natural to deduce the classification theorem from the extension theorem of the following paper. We shall not do this however; for the proof of the classification theorem is simpler, and furthermore, it is not evident how to obtain the relatively simple formula of the present paper from the more complicated one of the extension theorem.

The case that π_2 has no elements of even order is especially simple; the classification for this case is obtained in §8. The rest of the paper is devoted to the special problem arising when elements of even order are present; the study of "junctions" in a preceding paper[1] is then needed. It will be noted that in applying the formulas of the paper mentioned, which we refer to as RH, there are no "crossings of tubes" when projected into the given plane; thus a simplification of RH could be used in the present paper (but not in the following one).

The results of §§2, 3 and 4 are largely well known; they are fundamental to the discussions of the present paper and the following one.

2. Preliminaries

Let $\alpha_1, \cdots, \alpha_\nu$ be independent generators of π_2. Let α_μ be of order n_μ, and generate the cyclic subgroup $\pi_{2\mu}$ of π_2 ($\pi_{2\mu}$ is infinite if $n_\mu = 0$). Then π_2 is the direct sum

$$\pi_2 = \pi_{21} \oplus \cdots \oplus \pi_{2\nu}. \tag{2.1}$$

Corresponding to this decomposition, any r-π_2-cochain X may be written as a sum of r-$\pi_{2\mu}$-cochains:

$$X = X_1 + \cdots + X_\nu. \tag{2.2}$$

Let $\boldsymbol{C}^r(K, G)$, $\boldsymbol{Z}^r(K, G)$, $\boldsymbol{B}^r(K, G)$, $\boldsymbol{H}^r(K, G)$ denote the groups of r-G-cochains, r-G-cocycles, r-G-coboundaries, and r-G-cohomology classes respec-

[1] H. WHITNEY, *Relations between the second and third homotopy groups of a simply connected space*, these Annals, 50 (1949), 180–202.

tively. Let G be a finite group, expressed as the image of the group of integers under a homomorphism ϕ; then G is isomorphic to the integers mod n for some n. (We shall use $G = \pi_{2\mu}$, $n = n_\mu$, $r = 2$). There is a well known operation from $H^r(K, G)$ to $H^{r+1}(K, J)$ (J = integers), defined as follows. (This will not be used until §11; for the definition with ∂ in place of δ, see for instance Alexandroff-Hopf, *Topologie* I, Berlin, 1935, pp. 222–223.)

Given an r-G-cocycle X, let $\omega_n X = \bar{X}$ be the corresponding integral cochain, with each coefficient g (mod n) replaced by a, where $\phi(a) = g$ and $0 \leqq a < n$. (We could omit these inequalities if desired.) Since $\delta X = 0$, the coefficients of $\delta\bar{X}$ are multiples of n. Set

$$\theta_n^{r+1}(X) = (1/n)\, \delta\omega_n X = (1/n)\, \delta\bar{X}. \tag{2.3}$$

The cohomology class of $\theta_n^{r+1}X$ depends on that of X only. (We do not need this fact.)

If E^3 is oriented Euclidean 3-space, and $\bar{E}^3$ is E^3 plus the point at infinity, the elements of π_3 are represented by the mappings f of $\bar{E}^3$ into R, with the point at infinity going into the fixed point $\bar{p}$ of R. Note that, after a small deformation, we may assume that only a bounded part of E^3 goes into $R - \bar{p}$. With a fixed coordinate system x_1, x_2, x_3 in E^3, we shall let E_t denote the plane $x_3 = t$, and E_t^+ and E_t^-, the half spaces $x_3 \geqq t$, $x_3 \leqq t$ respectively. Adding a bar to any of these symbols denotes adding the point at infinity.

Let $\sigma_0^r, \sigma_1^r, \sigma_2^r$ denote oriented r-cells homeomorphic in a definite fashion. Identify corresponding points of their boundaries. Then

$$S_{01}^r = \sigma_1^r - \sigma_0^r, \qquad S_{12}^r = \sigma_2^r - \sigma_1^r, \qquad S_{02}^r = \sigma_2^r - \sigma_0^r,$$

are oriented r-spheres. Let f_i be a mapping of σ_i^r into R such that all three mappings agree on the boundary of the cells. Then mappings of the S_{ik}^r are determined, which determine elements h_{ik} of π_r, and

$$h_{01} + h_{12} = h_{02}, \tag{2.4}$$

a well known lemma in homotopy theory.

The Cartesian product $I \times K$ of the unit interval I and K is subdivided into cells $I \times \sigma_i^r$, oriented so that

$$\partial(I \times \sigma) = 1 \times \sigma - 0 \times \sigma - I \times \partial\sigma.$$

A deformation f_t ($0 \leqq t \leqq 1$) of f_0 into f_1 (both defined in K) is equivalent to a mapping F of $I \times K$.

Let τ^r be a convex Euclidean cell, and let τ^{r-1} be a boundary cell; let τ' be $\partial\tau^r$ minus the interior of τ^{r-1}. If f and g are mappings of τ^r into a space which agree in τ', then they are homotopic, keeping the points of τ' fixed. (This follows from the fact that τ' is retract of τ^r.) In particular, if f is defined in τ^r and $f(p) = \bar{p}$ in τ^{r-1}, then f is homotopic to f' with $f'(p) = \bar{p}$ in τ^r.

Applying these facts to $I \times K$ shows that if a mapping f_0 is defined in K, and f_t ($0 \leqq t \leqq 1$) is defined over a (closed) subcomplex of K, f_t may be extended over the rest of K.

3. Cocycles corresponding to certain mappings

Let L be a complex, formed of Euclidean cells. Let f_0 be a mapping of L into R such that $f_0(L^{r-1}) = \bar{p}$. The corresponding r-π_r-cocycle $W^r_{f_0}$ in L is defined as follows. For each σ^r, $f_0(\partial\sigma^r) = \bar{p}$; hence f_0 defines an element $W^r_{f_0}(\sigma^r)$ of π_r in σ^r. That $\delta W^r_{f_0} = 0$ follows from the fact that f_0 is defined over all $(r+1)$-cells of L. Any W^r is obtainable from a mapping of L^r, which is extendable through L^{r+1} if $\delta W^r = 0$.

Let f_t be a deformation of the above f_0 such that $f_t(L^{r-2}) = \bar{p}$, $f_1(L^{r-1}) = \bar{p}$. Let this define F in $I \times L$. Then for each σ^{r-1}, $F(\partial(I \times \sigma^{r-1})) = \bar{p}$; hence F maps $I \times \sigma^{r-1}$ into R, defining an element $X^{r-1}(\sigma^{r-1})$ of π_r. By definition of W^r_F,

$$W^r_F \cdot (I \times A^{r-1}) = X^{r-1} \cdot A^{r-1} \tag{3.1}$$

for all chains A^{r-1}. Since F in $0 \times L$ and $1 \times L$ define f_0 and f_1 respectively, and $\delta W^r_F = 0$, we have, for each σ^r,

$$0 = \delta W^r_F \cdot (I \times \sigma^r) = W^r_f \cdot \partial(I \times \sigma^r)$$
$$= W^r_{f_1} \cdot \sigma^r - W^r_{f_0} \cdot \sigma^r - X^{r-1} \cdot \partial\sigma^r;$$

consequently

$$W^r_{f_1} - W^r_{f_0} = \delta X^{r-1}. \tag{3.2}$$

We may express this by saying that the change in the cocycle corresponding to the mapping is the coboundary of the deformation cochain.

Conversely, if F is defined in $I \times K^r$ except interior to the cells $I \times \sigma^r_i$, if F in $0 \times K$ and $1 \times K$ defines f_0 and f_1 in K respectively, if $F(\partial(I \times \sigma^{r-1}_i)) = 0$, and if X^{r-1} exists satisfying (3.2), then F is extendable through $I \times K^r$, i.e., f_0 is homotopic to f_1 in K^r. For $\delta W^r_F = 0$.

Let f and g be mappings of L into R which agree on L^{r-1}. We cannot at once define W^r_f and W^r_g; but we can define the *difference cochain* $\Delta^r_{f,g}$, an r-π_r-cochain, as follows. For each σ^r_i, choose a τ^r_i; identify $\partial\tau^r_i$ with $\partial\sigma^r_i$; mapping σ^r_i according to g and τ^r_i according to f gives a mapping h of $S^r_i = \sigma^r_i - \tau^r_i$, defining $\Delta^r_{f,g}(\sigma^r_i) \in \pi_r$. (This cochain appears both in Eilenberg [2] and in Pontrjagin [7].) Clearly $\Delta^r_{f,g} + \Delta^r_{g,h} = \Delta^r_{f,h}$ (see §2), and

$$\Delta^r_{f,g} = W^r_g - W^r_f \qquad \text{if the latter are defined.} \tag{3.3}$$

Let f_t be defined so that $f_1 = f_0$ in L^r and $f_t(L^{r-1}) = \bar{p}$; we define the *corresponding deformation cochain* $\Delta^r_{f_t}$, an r-π_{r+1}-cochain, as follows (compare Pontrjagin, [7]). Take any σ^r_i; let F, in $I \times \sigma^r_i$, correspond to f_t in σ^r_i. Define G in $I \times \sigma^r_i$, corresponding to g_t in σ^r_i, with $g_t = f_0$. Then $G = F$ on $(I \times \sigma^r)^r = 0 \times \sigma^r \cup 1 \times \sigma^r \cup I \times \partial\sigma^r$, so that we may set

$$\Delta^r_{f_t}(\sigma^r_i) = \Delta^{r+1}_{G,F}(I \times \sigma^r_i). \tag{3.4}$$

Let f_t be a deformation of L such that $f_1 = f_0$ in L^r and $f_t(L^{r-1}) = \bar{p}$. Then

$$\Delta^{r+1}_{f_0,f_1} = \delta\Delta^r_{f_t}\,; \tag{3.5}$$

this is a generalization of (3.2). To prove this, take any σ_i^{r+1}; let f_t in σ_i^{r+1} define F in $I \times \sigma_i^{r+1}$. Let $g_t = f_0$ in $\partial\sigma_i^{r+1}$; let g_t define G in $I \times \partial\sigma_i^{r+1}$, and let $G = F$ in $0 \times \sigma_i^{r+1} \cup 1 \times \sigma_i^{r+1}$. Since g_t is constant, G defines the same element of π_{r+1} as would a mapping G' of the set obtained from $\partial(I \times \sigma_i^{r+1})$ by contracting each $I \times p$ ($p \in \partial\sigma_i^{r+1}$) to a point, and using $G = F$ in the two faces left. Since G' defines $\Delta_{f_0,f_1}^{r+1}(\sigma_i^{r+1})$, so does G in $\partial(I \times \sigma_i^{r+1})$.

Say $\partial\sigma_i^{r+1} = \sum \epsilon_j \sigma_j^r$; then $\partial(I \times \sigma_i^{r+1}) = -\sum \epsilon_j(I \times \sigma_j^r) + \cdots$. Choose σ_j^r, and change G to F in $I \times \sigma_j^r$; using (2.4) (applied to two mappings on $I \times \sigma_j^r$ and one on $\partial(I \times \sigma_i^{r+1}) - I \times \sigma_j^r$), this alters the element of π_{r+1} by $-\epsilon_j \Delta_{G,F}^{r+1}(I \times \sigma_j^r) = -\epsilon_j \Delta_{f_t}^r(\sigma_j^r)$. Changing G to F in each face in turn shows that the total alteration is

$$-\sum \epsilon_j \Delta_{f_t}^r(\sigma_j^r) = -\Delta_{f_t}^r \cdot \partial\sigma_i^{r+1} = -\delta\Delta_{f_t}^r(\sigma_i^{r+1}).$$

Thus F, in $\partial(I \times \sigma_i^{r+1})$, defines $\Delta_{f_0,f_1}^{r+1}(\sigma_i^{r+1}) - \delta\Delta_{f_t}^r(\sigma_i^{r+1})$. But F is defined in $I \times \sigma_i^{r+1}$; hence this element is 0, giving (3.5).

4. The 2-cocycle of a mapping

Let f_0 and f_1 be homotopic mappings of K (or of K^2) into R, such that $f_0(K^1) = f_1(K^1) = \bar{p}$. Then, as is well known, $W_{f_0}^2 \backsim W_{f_1}^2$. For let the deformation f_t of f_0 into f_1 define F in $L = I \times K$. Since R is simply connected, we may deform F over the 1-cells $I \times \sigma_i^0$ so that it maps them into $\bar{p}$; extend the deformation over L (keeping F fixed in $0 \times K$ and $1 \times K$), obtaining F'. This gives a new homotopy f_t' of f_0 into f_1 with $f_t'(K^0) = \bar{p}$. The result now follows from (3.2).

Conversely, if $f_0(K^1) = f_1(K^1) = \bar{p}$ and $W_{f_1}^2 \backsim W_{f_0}^2$, then f_0 and f_1 are homotopic over K^2; see the converse statement following (3.2). *The 2-cohomology class W_f^2 corresponding to f is a homotopy invariant.*

Choose a *fixed* cocycle W^2 from each cohomology class W^2; then each f is homotopic to some f' with $W_{f'}^2$ the chosen cocycle of W_f^2.

5. Standard mappings

Let σ_0^2 be a fixed oriented 2-cell, in the shape of a circular disc. A fixed point of σ_0^2 will be called its *top*. For each $\mu = 1, \cdots, \nu$, let ϕ_μ be a fixed mapping of σ_0^2 into R with $\phi_\mu(\partial\sigma_0^2) = \bar{p}$, defining the element α_μ of π_2.

Given any 2-simplex σ^2 with ordered vertices p_0, p_1, p_2, and any element α of π_2, we define a corresponding *standard mapping* of σ^2 into R, as follows. Write

$$(5.1) \qquad \alpha = a_1\alpha_1 + \cdots + a_\nu\alpha_\nu, \qquad 0 \leqq a_\mu < n_\mu \quad \text{if} \quad n_\mu > 0;$$

this representation of α is unique. Let p_1^* and p_2^* be the mid points of p_0p_1 and p_0p_2 respectively. Let Q be a rectangle interior to σ^2, bisected by the segment $p_1^*p_2^*$. Cut Q into rectangles $Q_1, \cdots, Q_\nu$, each bisected by $p_1^*p_2^*$, ordered in the direction from p_1^* to p_2^*. In each Q_μ, choose $k_\mu = |a_\mu|$ disjoint discs $Q_{\mu 1}, \cdots, Q_{\mu k_\mu}$, each bisected by $p_1^*p_2^*$. The *top* of $Q_{\mu i}$ is the point of the disc furthest from p_1p_2.

Take first, for a given μ, the case that $n_\mu > 0$; then $k_\mu = a_\mu$. There is a unique linear mapping of σ_0^2 into each $Q_{\mu i}$ which preserves orientation and sends the top of σ_0^2 into the top of $Q_{\mu i}$. Using ϕ_μ, this gives a mapping of $Q_{\mu i}$ into R, defining α_μ. Map $Q_\mu - \sum_i Q_{\mu i}$ into $\bar{p}$. The resulting mapping of Q_μ into R defines $a_\mu \alpha_\mu$.

Suppose now that $n_\mu = 0$. If $a_\mu \geqq 0$, we proceed as before. Suppose $a_\mu < 0$. Then there is a unique linear mapping of σ_0^2 into each $Q_{\mu i}$ which reverses orientation and sends the top of σ_0^2 into the top of $Q_{\mu i}$. Using ϕ_μ gives a mapping of $Q_{\mu i}$ defining $-\alpha$. Hence the mapping in Q_μ defines $-k_\mu \alpha = a_\mu \alpha$.

Map $\sigma^2 - \sum Q_{\mu i}$ into $\bar{p}$. This defines a *standard* mapping f of σ^2 into R which defines $\sum a_\mu \alpha_\mu = \alpha$.

A mapping f of K into R is *standard* if it is standard in each 2-simplex of K. It is *completely standard* if, further, the corresponding cocycle W_f^2 is the chosen cocycle of $\mathbf{W}_f^2$. Using §4, we see that any mapping is homotopic to a completely standard one.

6. A particular deformation

Let f_0 be a standard mapping of K into R, and let X^1 be a 1-π_2-cocycle in K. Corresponding to X^1 we shall define a deformation f_t of f_0 into f_1 such that $f_1 = f_0$ in K^2, and $\Delta^3_{f_0,f_1}$ bears a certain relation to X^1.

To define f_t, we define F in $I \times K$. First, set $F(I \times K^0) = \bar{p}$. Next, take any $\tau_i^2 = I \times \sigma_i^1$, $\sigma_i^1 = p_0 p_1$. Set $p_i^* = 1/2 \times p_i$; define rectangles Q_μ and discs $Q_{\mu i}$ in τ_i^2 exactly as in §5. We may then define a *standard mapping* F of τ_i^2 into R, defining the element $X^1(\sigma_i^1)$, as in §5.

We remark for future reference that if a mapping F' of $I \times K^1$ is given such that $F'(\partial(I \times \sigma_i^1)) = 0$ for each i, then letting $X^1(\sigma_i^1)$ be the element of π_2 defined by F' in $I \times \sigma_i^1$, and defining F in $I \times K^1$ as described, we may deform F' into F, keeping $F' = F$ fixed in each $\partial(I \times \sigma_i^1)$.

We now choose a *standard extension* of F through each $\tau^3 = I \times \sigma^2$, $\sigma^2 = p_0 p_1 p_2$. Choose a small circular disc C in σ^2, near the center of the edge $p_1 p_2$. Set $B = I \times C$. Take first $\mu = 1$. To each disc Q_{1i} in $I \times p_0 p_1$ we attach a tube; let it go into τ^3, then run down close to the face $I \times p_0 p_1$ towards $I \times p_1$, then run close to the face $I \times p_1 p_2$ towards B, and run into B; keep it always at the height $t = 1/2$, and keep the top filament on top. (See the figure in §7.)

Run tubes from the discs (with $\mu = 1$) in the face $I \times p_0 p_2$ into B in the same manner, moving close by $I \times p_2$. Run short tubes from the discs in $I \times p_1 p_2$ into B. We may do this so that, if projected into $0 \times \sigma^2$, no tubes cross.

We must now join the tubes in B. If first $n_\mu = 0$ ($\mu = 1$), the algebraic number of tubes (with an obvious definition) entering B is 0, since $\delta X^1 = 0$ and hence $\delta X_\mu^1 = 0$ (all μ); we may therefore insert short tubes in B, joining the above tubes in pairs. We do this so that top filaments are on top, and no tubes cross when projected into $0 \times \sigma^2$. Let F map each tube into R according to the mapping ϕ_1 defining α_1; then F is continuous so far.

Suppose now that $n_\mu > 0$ ($\mu = 1$). If the algebraic number of tubes entering

B is 0, join the tubes as before. Otherwise, this number is $\pm n_\mu$ ($\mu = 1$), as is easily seen (see §11). We then join any pairs of opposite sign, and insert a junction to which the rest are attached; we may clearly do this so that, when projected into $0 \times \sigma^2$, no tubes cross, and so that top filaments and the top of the junction are on top. The exact manner of inserting the junction will be given in §10.

We carry out the same process with $\mu = 2$, with the following change. As soon as each tube has left a face $I \times p_i p_j$, it is carried to a slightly higher level $t > 1/2$, so that it will intersect no tubes with $\mu = 1$. Insert the tubes with $\mu = 3$, bringing them to a higher level yet, etc.

If Q is the rectangle in σ^2 as defined in §5, then, since $f_0 = f_1$ in K^2, we may define F in $I \times Q$ so that $f_t = f_0$ in Q. Setting $F(p) = \bar{p}$ in the rest of $I \times \sigma^2$ makes F continuous.

So far, F is defined in $0 \times K \cup I \times K^2$. Extend F arbitrarily through $I \times K$ (see §2).

7. The corresponding $\Delta^3_{f_0, f_1}$, special case

In the special case that π_2 has no elements of even order, we now determine $\Delta^3_{f_0, f_1}$ for the mappings considered above. Take any $\sigma^3 = p_0p_1p_2p_3$; choose τ^3 and identify $\partial\tau^3$ with $\partial\sigma^3$; map σ^3 according to f_1 and τ^3 according to f_0; the resulting mapping f^* of $S^3 = \sigma^3 - \tau^3$ determines $\xi = \Delta^3_{f_0, f_1}(\sigma^3)$.

Define the oriented 3-sphere

$$S' = 0 \times \sigma^3 + I \times \partial\sigma^3 - 1 \times \sigma^3 = -\partial(I \times \sigma^3).$$

Using the mapping F of §6, define the mapping F' of S' into R by setting

$$F' = F \quad \text{in} \quad 0 \times \sigma^3 \cup I \times \partial\sigma^3, \qquad F'(1 \times p) = F(0 \times p);$$

we shall show that this determines ξ.

First, f_1 maps σ^3 the way F maps $1 \times \sigma^3$. Since F is defined throughout $I \times \sigma^3$, we may replace F in $1 \times \sigma^3$ by $F = F'$ in $0 \times \sigma^3 \cup I \times \partial\sigma^3$. Next, f_0 in τ^3 is the same as f_0 in σ^3, i.e., the same as F' in $1 \times \sigma^3$. Hence f^* in S^3 is equivalent to F' in S'.

Let us represent $t \times \partial\sigma^3$ by the subset $\bar{E}_t$ of $\bar{E}^3$, $0 \leqq t \leqq 1$, letting the point at infinity correspond to a point p' in $p_0p_2p_3$ near p_0, for which $F(t \times p') = \bar{p}$, $0 \leqq t \leqq 1$. We may now consider F in $I \times \partial\sigma^3$ as a mapping of $\bar{E}_{0,1} = \bar{E}_0^+ \cap \bar{E}_1^-$. The tube systems $T_1, \cdots, T_\nu$ which we have inserted in each $I \times \sigma_i^2$, for $\sigma_i^2 \subset \partial\sigma^3$, are now tube systems in $E_{0,1}$, approximately at the level $x_3 = 1/2$, and with top filaments on top.

Suppose the tube system T_μ is pulled away from the other tubes as described below. We then lose an amount $\Delta_\mu\xi$ say from ξ. The tube system T_μ, taken by itself, defines an element ξ_μ' of π_3. Then if ξ' is the mapping defined by F', but with $T_1, \cdots, T_\nu$ omitted,

$$\xi = \xi' + \sum \Delta_\mu\xi + \sum \xi_\mu'.$$

If $T_1, \cdots, T_\nu$ are omitted, then $F(t \times p) = F(0 \times p)$, $0 \leqq t \leqq 1$, $p \in \partial\sigma^3$. Now ξ' is defined by F' in S', but using this new F. With the new F, we may omit the interior of $I \times \partial\sigma^3$ and identify $0 \times \partial\sigma^3$ and $1 \times \partial\sigma^3$. The new F' is now a mapping of the cells $1 \times \sigma^3$ and $0 \times \sigma^3$, whose boundaries are identified. But in each of these, F' is the same as F in $0 \times \sigma^3$; hence $\xi' = 0$, and

$$\Delta^3_{f_0, f_1}(\sigma^3) = \sum \Delta_\mu \xi + \sum \xi'_\mu . \tag{7.1}$$

First we determine the $\Delta_\mu \xi$. Each T_λ lies wholly below T_μ, if $\lambda < \mu$. Hence we may bring the tube systems to separate levels, T_μ at the level $t = \mu/(\nu + 1)$. Keeping T_μ at this level, pull it towards the 2-cell $I \times p_2p_3$. In this process,

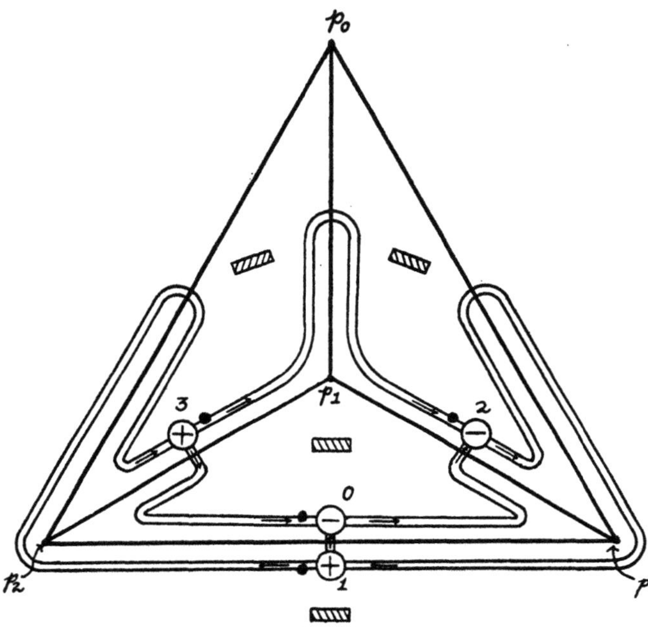

FIG. 1

the tubes through $I \times p_0p_1$ must cut through the vertical tubes in $I \times Q$, $Q \subset p_1p_2p_3$, but no other cuttings are necessary. (In the figure, Q appears as a shaded rectangle in $p_1p_2p_3$.) We may now shrink T_μ into a small region separated by a sphere from any other tube system. Hence $\Delta_\mu \xi$ is defined by the above cuttings. Since T_μ cuts through no T_λ, the order in which the T_μ are pulled away is immaterial.

Each tube of T_μ through $I \times p_0p_1$ has a cross section defining α_μ; the set of these tubes defines $X^1_\mu(p_0p_1)$. The set of tubes in $I \times Q$ defines $W^2_{f_0}(p_1p_2p_3)$. Pulling each tube of the first set in turn through all the tubes of the second set, a total change of $X^1_\mu(p_0p_1)W^2_{f_0}(p_1p_2p_3)$ is caused in ξ (see RH, §6). Hence

$$\Delta_\mu \xi = \pm X^1_\mu(p_0p_1)W^2_{f_0}(p_1p_2p_3) = \pm(X^1_\mu \smile W^2_{f_0})(\sigma^3), \tag{7.2}$$

using the usual formula for the $\smile$ product. The correct sign is determined in (12.1); we do not need it here.

The tube system T_μ in its new position defines ξ_μ'. If there are no junctions, RH Theorem 13A shows that $\xi_\mu' = 0$. If there are, then n_μ is odd, and RH Theorem 19A gives the same result. Therefore (7.1) and (7.2) give

$$\Delta^3_{f_0,f_1} = \pm X^1 \smile W^2_{f_0} \quad \text{in the special case.} \tag{7.3}$$

8. The classification theorem, special case

We give first a condition for the homotopy of two completely standard mappings, then an enumeration of the homotopy classes.

THEOREM 8A. *Let π_2 have no elements of even order. Then two completely standard mappings f, g of K into R are homotopic if and only if*

(a) $W^2_f = W^2_g$,

(b) *There is a 1-π_2-cocycle X^1 and a 2-π_3-cochain Y^2 such that*

$$\Delta^3_{f,g} = X^1 \smile W^2_f + \delta Y^2. \tag{8.1}$$

Suppose first that (a) and (b) hold; we must show that f and g are homotopic. Because of (a), $\Delta^3_{f,g}$ is defined. We first define a deformation f_t of $f = f_0$ into f_1, by the construction of §6, using X^1 or $-X^1$ so that (7.3) becomes

$$\Delta^3_{f,f_1} = X^1 \smile W^2_f.$$

Next define a deformation f_t' of $f_0' = f_1$ into f_1', as follows. Using F' in $I \times K$, set $F'(I \times K^1) = \bar{p}$. Define F' in each $I \times \sigma^2_i$ so that $f_1' = f_0'$ there, and $\Delta^2_{f_t'}(\sigma^2_i) = Y(\sigma^2_i)$ (see §3). Extend F' arbitrarily over the rest of $I \times K$. Then (3.5) gives

$$\Delta^3_{f_0',f_1'} = \delta\Delta^2_{f_t'} = \delta Y.$$

Therefore

$$\Delta^3_{f,f_1'} = \Delta^3_{f,f_1} + \Delta^3_{f_1,f_1'} = X^1 \smile W^2_f + \delta Y^2 = \Delta^3_{f,g},$$

and it follows that $\Delta^3_{f_1',g} = 0$.

Now f_1' and g agree on each $\partial\sigma^3_i$, and $\Delta^3_{f_1',g}(\sigma^3_i) = 0$; therefore f_1' may be deformed into g in σ^3_i (keeping it fixed in $\partial\sigma^3_i$). Thus f_1' and g, and hence f and g, are homotopic.

Suppose next that f and g are homotopic. Let f_t be the deformation, using the interval I': $0 \leqq t \leqq 2$, and let F be the corresponding mapping of $I' \times K$. Let I'' be the interval $1 \leqq t \leqq 2$. We shall carry out certain deformations on F, keeping it fixed at $t = 0$ and $t = 2$.

First, deform F so that $F(I' \times K^0) = \bar{p}$. Next, deform the new F so that $F(I'' \times K^1) = \bar{p}$. We now deform it so that it is standard in $I \times K^1$ (see §6). Finally, deform the new F in each $I \times \sigma^2_i$, keeping it fixed in $0 \times \sigma^2_i \cup I \times \partial\sigma^2_i$, so that it is the standard extension as defined in §6. (For the existence of the first two deformations, see the end of §2.)

Since f and g are homotopic, and $f = g$ in K^1, $W_g^2 \backsim W_f^2$ (see §4). Since f and g are completely standard, $W_g^2 = W_f^2$, proving (a).

Because of the last deformation of F, if we let $\pm X^1$ denote the deformation cocycle corresponding to F over the $I \times \sigma_i^1$, using the sign which occurs in (7.3), (7.3) gives

$$\Delta_{f,f_1}^3 = X^1 \smile W_f^2.$$

Since $W_{f_1}^2 - W_f^2 = \delta X^1 = 0$, $W_g^2 = W_f^2 = W_{f_1}^2$, and $g = f_2 = f_1$ in K^2. Also $f_t = f_1 (1 \leqq t \leqq 2)$ in K^1. Hence $Y^2 = \Delta_{f_t}^2 (t \in I'')$ is defined. Now (3.5) gives

$$\Delta_{f_1,g}^3 = \delta Y^2.$$

These two relations give (8.1), proving (b).

THEOREM 8B. *Let π_2 have no elements of even order. For each 2-π_2-cohomology class $\mathbf{W}^2$, let W^2 be the chosen cocyle in it, and let $\Psi(\mathbf{W}^2)$ be the set of all 3-π_3-cocycles*

$$(8.2) \qquad \psi(X^1, \mathbf{W}^2) = X^1 \smile W^2, \qquad X^1 \in \mathbf{Z}^1(K, \pi_2).$$

The set $\mathfrak{S}$ of homotopy classes of K into R falls into disjoint subsets $\mathfrak{S}(\mathbf{W}^2)$. The classes in $\mathfrak{S}(\mathbf{W}^2)$ are in one–one correspondence with the elements of the quotient group

$$(8.3) \qquad \mathbf{Z}^3(K, \pi_3) \bmod [\mathbf{B}^3(K, \pi_3) + \Psi(\mathbf{W}^2)].$$

Of course $\mathbf{Z}^3 = \mathbf{C}^3$, since $\dim(K) = 3$. If $\mathbf{W}^2 = 0$, (8.3) is simply $H^3(K, \pi_3)$; in this case the correspondence may be set up in a natural fashion, but apparently not so in the general case.

The partition of $\mathfrak{S}$ into subsets $\mathfrak{S}(\mathbf{W}^2)$ was noted in §4. Let us show that each $\mathfrak{S}(\mathbf{W}^2)$ is non void. Define f as follows. Set $f(K^1) = \bar{p}$. For each σ_i^2, choose the standard mapping f of σ_i^2 into R defining $W^2(\sigma_i^2)$. Since $\delta W^2 = 0$, we may extend f through $K = K^3$. Then the homotopy class of f is in $\mathfrak{S}(\mathbf{W}^2)$.

In each $\mathfrak{S}(\mathbf{W}^2)$, choose a fixed homotopy class $h_0 = h(\mathbf{W}^2)$, and let $f_0 = f(\mathbf{W}^2)$ be a completely standard mapping in this class. We shall set up a correspondence u between homotopy classes in $\mathfrak{S}(\mathbf{W}^2)$ and elements of (8.3). Set $u(h(\mathbf{W}^2)) = 0$. Now take any other homotopy class $h' \in \mathfrak{S}(\mathbf{W}^2)$; let f' be a corresponding completely standard mapping. Since $W_{f'}^2 = W^2 = W_{f_0}^2$, $f' = f_0$ in K^2. Hence we may define

$$\Delta_{f_0,f'}^3 \in \mathbf{Z}^3(K, \pi_3),$$

thus determining an element $u(h')$ of (8.3). Suppose we had chosen f'' instead of f'. Then f' and f'' are homotopic; hence, by Theorem 8A, $\Delta_{f',f''}^3 = \Delta_{f_0,f''}^3 - \Delta_{f_0,f'}^3$ lies in $\mathbf{B}^3 + \Psi$, and $u(h')$ is independent of the choice of f'.

If h' and h'' are distinct homotopy classes in $\mathfrak{S}(\mathbf{W}^2)$, and f', f'' are corresponding completely standard mappings, then by Theorem 8A, $\Delta_{f',f''}^3$ is not in $\mathbf{B}^3 + \Psi$; hence u is one–one into.

To show that u is onto, take any 3-π_3-cocycle X^*. Set $f = f_0$ in K^2. In each σ_i^3, define f so that $\Delta_{f_0,f}^3(\sigma_i^3) = X^*(\sigma_i^3)$. Then if h is the homotopy class of f, $u(h) = X^* \bmod (B^3 + \Psi)$. This completes the proof.

9. The case $R = S_0^2$

If R is a 2-sphere S_0^2, then π_2 is infinite cyclic. By RH, §9, we may define a new pairing $\alpha \cdot \beta$ of π_2 with itself into π_3 such that $2(\alpha \cdot \beta) = \alpha\beta$. Let $X \smile Y$ denote products, using this pairing; then

$$2(X \smile Y) = X \cup Y.$$

Making this substitution in Theorems 8A and 8B, we find the Theorem of Pontrjagin, [7]; see also Steenrod, [10].

10. Insertion of the junctions

In §6, we were not explicit about how junctions were to be inserted. For any μ with $n_\mu > 0$ and even, we must make the insertion more precise.

Take any $\sigma^2 = p_0p_1p_2$. Let the point p start at p_0, and move around $\partial\sigma^2$, going to p_1, then to p_2, then back to p_0. The segment $I \times p$ then moves around $I \times \partial\sigma^2$, crossing all discs (with this μ) to which tubes were attached. The first such disc (which is in $I \times p_0p_1$ if there are any discs there) will be named even for σ^2; the discs (with this μ) are named even and odd alternately. The tubes entering the cylinder B in $I \times \sigma^2$ are named according to the discs they end on, and if there is a junction J in B, the tubes ending at J are named even and odd correspondingly (compare RH, §20). We insert J so that its even discs attach to even tubes. (Note that the boundary of B corresponds to the curve C_i in RH, §20.)

In the figure in §7, let each pair of curves from an edge p_ip_j to a circle represent any number of tubes from $I \times p_ip_j$ to the corresponding cylinder B. The first tube, attached to the first disc, named even, is marked by a heavy dot.

11. What junctions are possible

Given any 1-π_2-cocycle $X^1 = \sum X_\mu^1$, recall that $\bar{X}_\mu^1$ is the integral cochain corresponding to X_μ^1, whose coefficients a_i satisfy $0 \leqq a_i < n_\mu$ (we are taking $n_\mu > 0$ even).

Take any $\sigma^2 = p_0p_1p_2$. Since $\delta X_\mu^1 = 0$,

$$\bar{X}_\mu^1(p_0p_1) + \bar{X}_\mu^1(p_1p_2) - \bar{X}_\mu^1(p_0p_2) \equiv 0 \pmod{n_\mu}. \tag{11.1}$$

Call the three terms a_2, a_0, a_1 respectively. Since $0 \leqq a_2 + a_0 < 2\, n_\mu$, we see that the left hand side of (11.1) equals 0 or n_μ according as $a_2 + a_0 < n_\mu$ or $\geqq n_\mu$; this is the condition that $\theta_{n_\mu}^2 X_\mu^1(\sigma^2)$ be 0 or 1 respectively.

Let the mapping F correspond to W_f^2 and X^1. Then for any σ^2 and μ, the algebraic number of tubes entering $I \times \partial\sigma^2$ equals the left hand side of (11.1). Hence *the number of junctions in $I \times \sigma^2$ equals $\theta_{n_\mu}^2 X_\mu^1(\sigma^2)$, and is 0 or 1 according as $\bar{X}_\mu^1(p_0p_1) + \bar{X}_\mu^1(p_1p_2)$ is $< n_\mu$ or $\geqq n_\mu$.*

We now consider orientations. Since

$$\partial(I \times \sigma^2) = -I \times \partial\sigma^2 + \cdots = -I \times p_0p_1 + \cdots,$$

the discs in $I \times p_0p_1$ and $I \times p_1p_2$ are oriented oppositely to $\partial(I \times \sigma^2)$; hence tubes in $I \times \sigma^2$ attached to these discs are oriented into $I \times \sigma^2$, towards the cylinder B. Therefore *if there is a junction in $I \times \sigma^2$, it is negative relative to the natural orientation of $I \times \sigma^2$.*

12. The corresponding $\Delta^3_{f_0,f_1}$, general case

Given $\sigma^3 = p_0p_1p_2p_3$, we determine $\xi = \Delta^3_{f_0,f_1}(\sigma^3)$ by (7.1). As to $\Delta_\mu\xi$, it was found before, except for the proper sign. To determine the sign, we shall determine some orientations.

Since

$$I \times \partial\sigma^3 = -I \times p_0p_1p_2 + \cdots, \qquad \partial(-I \times p_0p_1p_2) = I \times p_0p_1 + \cdots,$$

$I \times p_0p_1$ lies positively on the boundary of the part of S' in $I \times p_0p_1p_2$. Therefore positive tubes T'_μ cutting $I \times p_0p_1$ go out of $I \times p_0p_1p_2$, into $I \times p_0p_1p_3$. (See the arrows in the figure.)

The vertical tubes T^* in $I \times p_1p_2p_3$ have cross sections whose boundaries are oriented in the direction from p_1 to p_2 to p_3 to p_1. If the tubes T' are continued around the tubes T^*, they would go in the opposite direction. Therefore (see RH, §§2, 6)

$$\Delta_\mu\xi = -X^1_\mu(p_0p_1)W^2_{f_0}(p_1p_2p_3),$$

$$\sum \Delta_\mu\xi = -X^1(p_0p_1)W^2_{f_0}(p_1p_2p_3). \tag{12.1}$$

As before, $\xi'_\mu = 0$ unless $n_\mu > 0$ is even. Suppose this is the case. Let σ^2_i be the face of σ^3 opposite p_i. Since $S' = I \times \partial\sigma^3 + \cdots = \sum(-1)^i I \times \sigma^2_i + \cdots$, if there is a junction J_i in $I \times \sigma^2_i$, it must have the sign $-(-1)^i$; see the end of §11. Hence the possible junctions in $I \times \partial\sigma^3$ are:

$$J_0 : -, \qquad J_1 : +, \qquad J_2 : -, \qquad J_3 : +.$$

If there are no junctions, of course $\xi'_\mu = 0$. If there are, then there are the same number of negative as of positive junctions.

Let B_i be the vertical cylinder in $I \times \sigma^2_i$ in which the tubes are joined, containing J_i if it exists. Then each tube through a face $I \times \sigma^1_i$ runs from one B_j to another B_k, corresponding to cells of $I \times K$ with $I \times \sigma^1_i$ as a face. The tubes entering a given B_j are named even and odd alternately, as in §10.

Let Case (23) denote the case that just the junctions J_2 and J_3 are present, etc.

Case (23). Since J_2 is present, $\bar{X}^1_\mu(p_0p_1) > 0$ (see §11), so that there are tubes through $I \times p_0p_1$, which run from B_3 to B_2. Let T be the first one. Since p_0p_1 is the first edge of both σ^2_2 and σ^2_3, T is the first tube entering both B_2 and B_3, and hence is named even at both ends.

We now apply RH Theorem 20A; the formula (19.2) reads in the present case

$$\xi'_\mu = (Z_\mu \cdot D_\mu) n_\mu \eta(\alpha_\mu). \tag{12.2}$$

Here, the B_i which actually contain junctions, i.e., B_2 and B_3, are thought of as vertices of a linear graph G^*, and the parts of the tubes outside B_2 and B_3 are thought of as 1-cells of G^*. Each vertex B_i is oriented according to the sign of J_i. By D_μ is meant any 1-chain mod 2 whose boundary is the sum of the junctions; we may let D_μ correspond to the tube T. Here and in the later cases, the tubes employed will always join cylinders B_i, B_j of opposite signs; hence $Z_\mu''' \cdot D_\mu = 0$ (see RH, (20.1)). Since there are no crossings, when G^* is projected into $0 \times \partial\sigma^3$ (see the figure in §7), $Z'_\mu = 0$; hence we need consider only Z''_μ. Now $Z''_\mu \cdot D_\mu$ counts the number of 1-cells (mod 2) of D_μ whose ends are of opposite parity. Since T is even at both ends, $Z''_\mu \cdot D_\mu = 0$ in this case, and $\xi'_\mu = 0$.

Case (12). First suppose $X^1_\mu(p_0p_3) \neq 0$. Let T be the first tube through $I \times p_0p_3$; it runs from B_2 to B_1. In the given ordering of tubes entering B_2, T is the last tube; hence, cyclically, the next tube is the first, named even; hence T is named odd at B_2. For the same reason, T is odd at B_1. Hence, again using T for D_μ, we find $\xi'_\mu = 0$.

Now suppose $X^1_\mu(p_0p_3) = 0$; then there are no corresponding tubes through $I \times p_0p_3$. We then insert a closed tube T^* with cross section defining α_μ, which runs from B_2 through $I \times p_0p_3$ into B_1, then out again, through $I \times p_0p_3$ nearer $I \times p_3$ than before, into B_2 again and joining its starting point. We do this so that when projected into $0 \times \partial\sigma^3$, there is no crossing, and so that the top filament is on top. Let T be the part of T^* running from B_2 to B_1, and let T' be the part running back.

If we name the ends of T odd and the ends of T' even, the ends of tubes around B_2 and B_3 alternate in parity. Hence again we may apply RH Theorem 20A, with T for D_μ; again $\xi'_\mu = 0$.

Case (03). Since J_0 is present, $X^1_\mu(p_1p_2) \neq 0$. Hence there is a tube through $I \times p_1p_2$; call the first one T. At B_0, T is the first tube, and hence is named even. At B_3, all tubes through $I \times p_0p_1$ come before T; hence T is named even or odd according as the number of these, which is $\bar{X}^1_\mu(p_0p_1)$ (see §11), is even or odd. Thus

$$Z_\mu \cdot D_\mu = \bar{X}^1_\mu(p_0p_1) \quad (\text{mod } 2).$$

Putting in (12.2) gives

$$\xi'_\mu = \bar{X}^1_\mu(p_0p_1) n_\mu \eta(\alpha_\mu). \tag{12.3}$$

Case (01). There is a tube through $I \times p_2p_3$ in this case; call the first one T. At B_0, the end of T is named even or odd according to whether $\bar{X}^1_\mu(p_1p_2)$ is even or odd; at B_1, according to whether $\bar{X}^1_\mu(p_0p_2)$ is even or odd. Hence the ends of T are named in the same or opposite manner according to whether (using (11.1))

$$\bar{X}^1_\mu(p_1p_2) + \bar{X}^1_\mu(p_0p_2) \equiv \bar{X}^1_\mu(p_0p_1) \equiv 0 \quad (\text{mod } 2)$$

or not. Hence the formula for ξ'_μ is the same as in Case (03).

Case (0123). An example of this case, with $n_\mu = 4$, is given by ($X_{01} = \bar{X}^1_\mu(p_0p_1)$ etc.).

$$X_{01} = 3, \quad X_{02} = 2, \quad X_{03} = 1, \quad X_{12} = 3, \quad X_{13} = 2, \quad X_{23} = 3.$$

There are tubes through both $I \times p_0p_1$ and $I \times p_2p_3$; call the first tubes of each kind T and T' respectively. We let D_μ correspond to $T + T'$. As before, the ends of T are both named even, and the ends of T' are named the same or opposite according to whether $\bar{X}^1_\mu(p_0p_1)$ is even or odd. Consequently the formula for ξ'_μ is the same as in cases (03) and (01).

We may summarize the cases as follows. If J_0 is absent, $\xi'_\mu = 0$. If J_0 is present, ξ'_μ is given by (12.3). Hence, referring to §11,

$$\xi'_\mu = \theta^2_{n_\mu} X^1_\mu(p_1p_2p_3)\bar{X}^1_\mu(p_0p_1)n_\mu\eta(\alpha_\mu) \tag{12.4}$$

in all cases.

Putting this relation and (12.1) into (7.1) gives

$$\Delta^3_{f_0,f_1} = -X^1 \smile W^2_{f_0} + \sum_{n_\mu>0,\ \text{even}} (\bar{X}^1_\mu \smile \theta^2_{n_\mu}X^1_\mu)n_\mu\eta(\alpha_\mu). \tag{12.5}$$

Note that the last term may be written

$$\sum_{n_\mu>0,\ \text{even}} (\bar{X}^1_\mu \smile \delta\bar{X}^1_\mu)\eta(\alpha_\mu). \tag{12.6}$$

We shall find another way of writting this. Write X^1_μ in the form

$$X^1_\mu = \sum (a_i\alpha_\mu)\sigma^1_i, \qquad 0 \leqq a_i < n_\mu. \tag{12.7}$$

Then $\bar{X}^1_\mu = \sum a_i\sigma^1_i$, and if $[\sigma^1_j : \sigma^2_i]$ denotes incidence numbers,

$$\delta\bar{X}^1_\mu = \sum_i b_i\sigma^2_i, \qquad b_i = \sum_j [\sigma^1_j : \sigma^2_i]a_j.$$

We define θ^* by

$$\theta^*_{n_\mu}X^1_\mu = \sum_i (b_i/n_\mu)\alpha_\mu\sigma^2_i; \tag{12.8}$$

note that

$$\theta^*_{n_\mu}X^1_\mu = (\theta^2_{n_\mu}X^1_\mu)\alpha_\mu. \tag{12.9}$$

Now

$$\begin{aligned} X^1_\mu \smile \theta^*_{n_\mu}X^1_\mu &= \sum_i (a_i\alpha_\mu)\sigma^1_i \smile \sum_j (b_j/n_\mu)\alpha_\mu\sigma^2_j \\ &= (\sum_i a_i\sigma^1_i \smile \sum_j (b_j/n_\mu)\sigma^2_j)\alpha_\mu\alpha_\mu, \end{aligned}$$

$$X^1_\mu \smile \theta^*_{n_\mu}X^1_\mu = (\bar{X}^1_\mu \smile \theta^2_{n_\mu}X^1_\mu)\alpha_\mu\alpha_\mu. \tag{12.10}$$

Thus (12.5) may be written

$$\Delta^3_{f_0,f_1} = -X^1 \smile W^2_{f_0} + \sum_{n_\mu>0,\ \text{even}} n_\mu/2(X^1_\mu \smile \theta^*_{n_\mu}X^1_\mu). \tag{12.11}$$

13. The classification theorem, general case

The homotopy and classification theorems are the obvious generalizations of Theorems 8A and 8B; the proofs of the present theorems are like those of the earlier ones.

THEOREM 13A. *Two completely standard mappings f, g of K into R are homotopic if and only if:*
(a) $W_f^2 = W_g^2$,
(b) *there is* 1-π_2-*cocycle* X^1 *and a* 2-π_3-*cochain* Y^2 *such that*

$$\Delta_{f,g}^3 = \delta Y - X^1 \smile W_f^2 + \textstyle\sum_{n_\mu>0,\,\text{even}} (n_\mu/2)(X_\mu^1 \smile \theta_{n_\mu}^* X_\mu^1). \tag{13.1}$$

THEOREM 13B. *For each* $\mathbf{W}^2$, *with chosen cocycle* W^2, *let* $\Psi(\mathbf{W}^2)$ *be the set of all* 3-π_3-*cocycles*

$$\psi(X^1, \mathbf{W}^2) = -X^1 \smile W^2 + \textstyle\sum_{n_\mu>0,\,\text{even}} (n_\mu/2)(X_\mu^1 \smile \theta_{n_\mu}^* X_\mu^1), \tag{13.2}$$

for 1-π_2-*cocycles* X^1. *The set* $\mathfrak{S}$ *of homotopy classes of K into R falls into disjoint subsets* $\mathfrak{S}(\mathbf{W}^2)$. *The classes in* $\mathfrak{S}(\mathbf{W}^2)$ *are in one-one correspondence with the elements of the quotient group*

$$\mathbf{Z}^3(K, \pi_3) \bmod [\mathbf{B}^3(K, \pi_3) + \Psi(\mathbf{W}^2)]. \tag{13.3}$$

We must show that the subset $\mathbf{Z}'$ of $\mathbf{Z}^3$ enclosed in brackets in (13.3) is a subgroup of $\mathbf{Z}^3$; in the former case it was clear, since ψ was linear in X^1. It is sufficient to show that

$$\psi(X^1, \mathbf{W}^2) + \psi(Y^1, \mathbf{W}^2) \backsim \psi(X^1 + Y^1, \mathbf{W}^2), \tag{13.4}$$

$$\psi(-X^1, \mathbf{W}^2) \backsim -\psi(X^1, \mathbf{W}^2). \tag{13.5}$$

Use (12.6) in (13.2), and write x, y η for $\bar{X}_\mu^1$, Y_μ^1, $\eta(\alpha_\mu)$. Then

$$\begin{aligned} u &= \psi(X^1, \mathbf{W}^2) + \psi(Y^1, \mathbf{W}^2) - \psi(X^1 + Y^1, \mathbf{W}^2) \\ &= \{\textstyle\sum(x \smile \delta x) + \sum(y \smile \delta y) - \sum[(x + y) \smile \delta(x + y)]\}\eta \\ &= -\textstyle\sum(x \smile \delta y + y \smile \delta x)\eta. \end{aligned}$$

Now (see RH, (7.7))

$$2(x \smile \delta y)\eta = (\bar{X}_\mu^1 \smile \theta_{n_\mu}^2 Y_\mu^1)\cdot 2n_\mu\eta(\alpha_\mu) = 0;$$

hence we may replace $x \smile \delta y$ by $-\, x \smile \delta y$. For the same reason, we may, in the expression

$$(y \smile \delta x)\eta = (\bar{Y}_\mu^1 \smile \theta_{n_\mu}^2 X_\mu^1)n_\mu\eta(\alpha_\mu),$$

replace $\bar{Y}_\mu^1$ and $\theta_{n_\mu}^2 X_\mu^1$ by the corresponding cochains y' and x^* reduced mod 2. But

$$\delta\bar{Y}_\mu^1 = n_\mu\theta_{n_\mu}^2 Y_\mu^1 = 2(n_\mu/2)\theta_{n_\mu}^2 Y_\mu^1;$$

hence $\delta y' = 0$. Since $\delta\theta_{n_\mu}^2 X_\mu^1 = 0$, we may apply the commutative law, giving

$$(y' \smile x^*)n_\mu\eta \backsim (x^* \smile y')n_\mu\eta = (\delta x \smile y)\eta.$$

Therefore

$$-\, u \backsim \textstyle\sum(\delta x \smile y - x \smile \delta y)\eta = \delta[\sum(x \smile y)\eta],$$

proving (13.4). The proof of (13.5) is similar; we must note that

$$\theta^2_{n_\mu}(-X^1_\mu) \smile -\theta^2_{n_\mu}(X^1_\mu). \tag{13.6}$$

REMARK. We could prove that Z' is a subgroup of Z^3 as follows. Let f_0 be a completely standard mapping defining W^2. Choose

$$V^3_i = \delta Y^2_i + \psi(X^1_i, W^2) \in Z', \qquad i = 1, 2;$$

we wish to show that $V^3_1 + V^3_2$ and $-V^3_1$ are in Z'. To V^3_1 and V^3_2 correspond deformations $f_t(0 \leqq t \leqq 1)$ and $f_t(1 \leqq t \leqq 2)$; then

$$\Delta^3_{f_0,f_1} = V^3_1, \qquad V^3_{f_1,f_2} = V^3_2.$$

Now f_0 and f_2 are homotopic. Hence, by Theorem 13A, there exist cochains Y^2_3 and X^1_3 such that

$$\Delta^3_{f_0,f_2} = \delta Y^2_3 + \psi(X^1_3, W^2).$$

This gives $V^3_1 + V^3_2 = \Delta^3_{f_0,f_2} \in Z'$. Next, define $f_t(0 \leqq t \leqq 1)$ as before, and set $f_t = f_{2-t}$ $(1 \leqq t \leqq 2)$. The latter deformation corresponds to some V^3_4 in Z'. Using f_t, $0 \leqq t \leqq 2$, shows that $V^3_1 + V^3_4 = 0$; hence $-V^3_1 = V^3_4 \in Z'$.

HARVARD UNIVERSITY

ANNALS OF MATHEMATICS
Vol. 50, No. 2, April, 1949

AN EXTENSION THEOREM FOR MAPPINGS INTO SIMPLY CONNECTED SPACES

HASSLER WHITNEY

(Received March 29, 1948)

1. Introduction

The modern work on the classification and extension of mappings had its starting point in a paper[1] [5] of H. Hopf, classifying the mappings of an n-dimensional complex into an n-sphere S^n. A simple formulation and proof of this theorem were given by the author [11] as one of the first applications of the new cohomology theory. Hurewicz [6] generalized the theorem, replacing S^n by any space whose homotopy groups below the nth vanish.

An early theorem of the same sort where this condition failed was that of Hopf [4], on the mappings of S^3 into S^2. The author in 1936 (in an unpublished manuscript) gave algebraic conditions for the homotopy of mappings of an n-complex into projective n-space for $n = 2, 3$. This was generalized by Robbins [9], considering mappings of a 2-complex into any space. A great simplification and generalization of this work has recently been obtained by Paul Olum.

Pontrjagin [7] classified the mappings of a 3-complex into S^2, and gave a corresponding extension theorem for a 4-complex; this was one of the first applications of the new "cup products." (The same theorem was obtained independently by the author; see an abstract in the Bull. Am. Math. Soc., vol. 42 (1936), p. 338. The method used is contained in the first half of CM.) In another paper [8], the 3-dimensional homotopy group of a simply connected space was determined, with the help of algebraic methods. (Proofs were not given.) A new type of cup product, $X * Y$, was made use of.

This product of Pontrjagin furnished the clue to the general $\cup_i$ products of Steenrod [10]; using them, Steenrod classified the mappings of an $(n + 1)$-complex into an n-sphere, and gave the corresponding extension theorems.

The object of the present paper is to give an extension theorem in the case $n = 2$. Given any simplicial complex K of dimension 4 (whose vertices we always take as given in a fixed order), we give an algebraic criterion that a mapping f of the 2-dimensional part K^2 of K into a simply connected space R be extendable over K. We assume first that f is deformed into a "standard mapping", which may always be done. In the case that $\pi_2(R)$ has no elements of finite order, the formula was obtained by the author in 1940 (see the abstract in the Bull. Am. Math. Soc., vol. 46 (1940), p. 897; the formula was mistakenly stated for the general case).

[1] Further historical notes and references are given in Steenrod [10]. We shall make use of the preceding paper, which we denote by CM, and the paper RH, referred to in CM, §1. We note a correction to RH: the paper of Pontrjagin referred to in §8 should be to the paper [8] (not to [7]) in the References below.

The proof is straightforward and geometric in character; it lies rather purely in the domain of Hurewicz's homotopy methods. As a result, the nature of the terms in our formula becomes reasonably clear. On the other hand, it is not clear to what degree topological significance can be ascribed to parts of the formula. There is a great similarity between the formula and Pontrjagin's formula for $X * Y$ mentioned above; it may be possible to combine parts of our formula in a significant manner. Moreover, since our formula contains $\pi_3(R)$ implicitly, which group is the subject of Pontrjagin's formulas in [8], it may be possible to combine the results so as to obtain theorems using more purely the algebraic properties of both spaces K and R.

Two products, $\smile_1$ and $\smile_2$, of Steenrod will be used. In §4 we give a brief description of the $\smile_i$ products (mod 2). The simplicity of the description, together with the fact that the products occur exactly in this manner in our formula, is good evidence of the naturalness of Steenrod's definition.

Besides the results of the paper RH, various facts and notations from the preceding paper CM will be used. We recall in particular the decomposition of π_2 given in CM §2, and the relation between r-$\pi_{2\mu}$-cochains X^r_μ, integral cochains $\bar{X}^r_\mu$, and cochains $\theta^{r+1}_{n_\mu} X^r_\mu$, given in CM (2.3). For further orientation, see Steenrod [10].

As in CM, the formula is simpler if π_2 has no elements of even order; it will be obtained for this case in §9.

2. The nature of the problem

Let f be any mapping of K^2 into R. Then for each σ^3_i, the mapping f in $\partial\sigma^3_i$ determines an element $W^3_f(\sigma^3_i)$ of $\pi_2 = \pi_2(R)$, since R is simply connected. This determines a 3-π_2-cochain W^3_f, the *first obstruction* of f. There is an extension of f over K^3 if and only if $W^3_f = 0$.

Suppose this condition holds. Let g be an extension of f over K^3. Then for each σ^4_i, g in $\partial\sigma^4_i$ determines $W^4_g(\sigma^4_i)$ in π_3; W^4_g is the *g-obstruction* of f; g is extendable over $K = K^4$ if and only if $W^4_g = 0$.

Let g' be another extension of f over K^3. For each σ^3_i, since $g = g'$ in $\partial\sigma^3_i$, the *difference cochain* $\Delta^3_{g,g'}(\sigma^3_i)$ is defined (see CM, §3). Since

$$W^4_{g'} - W^4_g = \delta\Delta^3_{g,g'}, \tag{2.1}$$

altering g alters W_g by a coboundary. Hence the 4-π_3-cohomology class W^4_g of W^4_g depends on f alone, and we may name it the *second obstruction* W^4_f of f. Since W_g may be altered by any coboundary, *f is extendable over $K = K^4$ if and only if $\mathsf{W}^4_f = 0$.* If this holds, one may define the next higher obstruction, etc.

The precise object of this paper is to find a formula for W^4_f in terms of W^2_f, supposing that f is "standard" in K^2. We explain this term. We say that f is *standard in* K^2 if $f(K^1) = \bar{p}$ and each σ^2_i is mapped in a fixed way, depending only on the element $W^2_f(\sigma^2_i)$ of π_2 defined by f in σ^2_i. Of course whenever W^3_f is defined,

$$W^3_f = \delta W^2_f. \tag{2.2}$$

Since any mapping is homotopic to a standard one, we lose no generality in restricting ourselves to such mappings.

Suppose that f is standard and $\delta W_f^3 = 0$. Consider now any σ_i^3. As soon as we know the part of W_f^3 in $\partial\sigma_i^3$, we know f in $\partial\sigma_i^3$; choose a definite extension of f through σ_i^3. Doing this for each σ_i^3 gives a *standard mapping* of K^3 into R.

Note that with a given definition of standard mappings, W_f^4 is determined by W_f^3; if the definition of standard mappings in 3-simplexes is altered, W_f^4 is changed by a coboundary.

3. The abstract formulation

The above deformations are purely local in character. This enables one to formulate the situation in more abstract terms, as follows. (This formulation is due essentially to Eilenberg-MacLane, [2] and [3]; see also Robbins [9].)

Corresponding to the simply connected space R, construct an abstract complex K^*, as follows. For each r, the r-cells τ^r of K^* are the 2-π_2-cocycles (considered as abstract entities) in a fixed simplex σ_0^r with ordered vertices. (We do not need to discuss the case $r < 2$ here.) An r-G-cochain X^{*r} is a function $X^{*r}(\tau^r)$ defined over the τ^r and with values in the group G. We define δX^{*r} as follows. Take any τ^{r+1}, given by the cocycle Z^2 in σ_0^{r+1}. Say $\partial\sigma_0^{r+1} = \sum(-1)^i\sigma_i'$, σ_i' being the r-face of σ_0^{r+1} opposite the vertex in the position i. Since each σ_i' has ordered vertices, we may identify it with σ_0^r; the part of Z^2 in σ_i' becomes a cocycle Z_i^2 in σ_0^r. Set

$$\delta X^{*r}(\tau^{r+1}) = \sum(-1)^i X^{*r}(Z_i^2).$$

Define standard mappings of 2- and 3-simplexes with ordered vertices into R as above. In terms of these, there is a definite 4-π_3-cochain W^{*4} in K^*, defined as follows. Take any τ^4 in K^*, defined by the cocycle Z^2 in σ_0^4. Map the 1-simplexes of σ_0^4 into $\bar{p}$, and map each 2-simplex σ_j^2 of σ_0^4 into R by the standard mapping defining $Z^2(\sigma_j^2)$. Since $\delta Z^2 = 0$, we may extend this mapping over the 3-simplexes σ_i^3 of σ_0^4; do this in the standard manner. The mapping is now defined in $\partial\sigma_0^4$, and defines $W^{*4}(\tau^4)$.

Suppose the definition of standard mappings in 3-simplexes was changed. Then a new W'^4 results, with $W'^4 \backsim W^4$ (see (2.1)). Hence the cohomology class $\mathbf{W}^{*4}$ is independent of the standard mappings in the 3-simplexes. (We take W^{*4} in K^{*4} for the moment.)

Suppose the standard mappings in the 2-simplexes were altered. For each $\alpha \in \pi_2$, let $f_t(p; \alpha)$ denote a deformation of the old standard mapping correspond to α into the new one. (We may keep $\partial\sigma_0^2$ at $\bar{p}$.) For each τ^3, defined by Z^2 in σ_0^3, using $f_t(p; Z^2(\sigma_i^2))$ in each face σ_i^2 of σ_0^3 and extending this deformation of $\partial\sigma_0^3$ in a definite manner over σ_0^3 gives a new mapping of σ_0^3, which we call the new standard mapping. For any τ^4, the new mapping in $\partial\sigma_0^4$ is obtained from the old one by a deformation; hence W^{*4} is unaltered. This proves that $\mathbf{W}^{*4}$ *is an invariant of R.*

We now show that $\delta W^{*4} = 0$ in K^*. Take any τ^5, defined by Z^2 in σ_0^5. Let f be the corresponding standard mapping of $(\sigma_0^5)^3$. Set $f_0((\sigma_0^5)^3) = \bar{p} \in R$. Say

$Z^2 = \delta X^1$ in σ_0^5; deforming f_0 in each 1-simplex σ_k^1 according to the element $X^1(\sigma_k^1)$ of π_2 and extending over $(\sigma_0^5)^3$ gives an f_1 with $W_{f_1}^2 = Z^2$ in σ_0^5; hence we may deform f_1 into f_2 with $f_2 = f$ in $(\sigma_0^5)^2$. Now $\Delta_{f_2,f}^3$ is defined, and

$$W_f^4 - W_{f_2}^4 = \delta\Delta_{f_2,f}^3.$$

Since f_0 is extendable over σ_0^5, $W_{f_0}^4 = 0$. Since f_2 is homotopic to f_0 in $(\sigma_0^5)^3$, $W_{f_2}^4 = W_{f_0}^4 = 0$. Hence $\delta W_f^4 = \delta\delta\Delta_{f_2,f}^3 = 0$ in σ_0^5, as required.

We now relate this discussion to §2. Let K^* have one vertex τ^0 and one 1-cell τ^1; let all incidence numbers $[\tau^0 : \tau^1]$ and $[\tau^1 : \tau_i^2]$ vanish. Take any K, and any standard mapping f of K^2 into R. We then define a chain-mapping κ of K into K^*, as follows. Send each vertex into the chain $0\tau^0$ and each 1-simplex into the chain $0\tau^1$. For any σ_i^r, $r \geqq 2$, let $\kappa(\sigma_i^r)$ be the cell τ^r defined by the part of W_f^2 in σ_i^r. This chain transformation is clearly boundary preserving; hence the dual κ^* satisfies $\kappa^*\delta = \delta\kappa^*$. Clearly

$$W_f^4 = \kappa^* W^{*4}; \tag{3.1}$$

hence f is extendable over $K = K^4$ if and only if $\kappa^* W^{*4} = 0$.

4. The $\cup_i$ products mod 2

The reader interested only in the special case of the theorem, given in §9, can skip this section.

Let K be a simplicial complex with ordered vertices. We shall give a definition of the product $X^r \cup_i Y^s = Z^n$, $n = r + s - i$, with integers mod 2 as coefficients, which is easily seen to be equivalent to that of Steenrod [10]. It is sufficient to define the product for simplexes. This is 0 unless both simplexes are faces of some σ_0^n. For a given σ_0^n, the set of all products $\sigma^r \cup_i \tau^s = \sigma_0^n$ (mod 2) is in one-one correspondence with the set of all symbols of the following nature. We enclose between a pair of brackets $n + 1$ symbols in order, each being either a hyphen or a star; for instance, with $n = 7$,

$$[- * * - * - - *]. \tag{4.1}$$

Each position corresponds to a vertex of σ_0^n. For each hyphen, the corresponding vertex of σ_0^n is in both σ and τ. For each star, the corresponding vertex is in exactly one of σ, τ, according to the following rules:

(a) The first star belongs to σ or τ according as the number of hyphens preceding it is even or odd.

(b) A given star belongs to the same or opposite simplex (σ or τ) as the last preceding star according as the number of hyphens between them is even or odd.

Replacing stars in the above example by the corresponding simplexes gives

$$[- \tau\tau - \sigma - - \sigma]. \tag{4.2}$$

The corresponding product is, if $\sigma_0^n = p_0 \cdots p_7$,

$$p_0p_3p_4p_5p_6p_7 \cup_3 p_0p_1p_2p_3p_5p_6 = \sigma_0^n \qquad \text{(mod 2)}.$$

In a "filled in" symbol as above, $i + 1$ is the number of hyphens, $r + 1$ is the number of symbols $\neq \tau$, and $s + 1$ is the number of symbols $\neq \sigma$.

Let us find all products with

$$r = 3, \quad s = 2, \quad i = 1, \quad n = 4.$$

There are five places, two hyphens, two symbols σ, and one symbol τ. The full list of such symbols is

$$[\sigma\sigma - \tau -], \quad [\sigma - \tau - \sigma], \quad [- \tau - \sigma\sigma].$$

For a shorthand notation, write

$$\bar{3} \text{ for } p_0p_1p_2p_4, \qquad \overline{01} \text{ for } p_2p_3p_4,$$

etc. Then the corresponding products giving $\sigma_0^4 = p_0 \cdots p_4$ are

$$\bar{3} \smile_1 \overline{01}, \quad \bar{2} \smile_1 \overline{04}, \quad \bar{1} \smile_1 \overline{34}. \tag{4.3}$$

Now we list all products with

$$r = 3, \quad s = 3, \quad i = 2, \quad n = 4:$$

$$[\sigma - \tau - -], \quad [\sigma - - - \tau], \quad [- - \sigma - \tau], \quad [- \tau - \sigma -];$$

$$\bar{2} \smile_2 \bar{0}, \quad \bar{4} \smile_2 \bar{0}, \quad \bar{4} \smile_2 \bar{2}, \quad \bar{1} \smile_2 \bar{3}. \tag{4.4}$$

5. Standard mappings of K^3

We have defined standard mappings of K^2 in CM, §5. Given any standard mapping f of K^2, with $\delta W_f^2 = 0$, we now define the standard extension over K^3. Given any $\sigma^3 = p_0p_1p_2p_3$, let σ_i^2 be the face opposite the vertex p_i; the definition of f will depend simply on the values of $W_f^2(\sigma_i^2)$. We may picture σ^3 simply in E^3 by letting σ_0^2 lie in the x_1, x_2-plane E_0, and letting p_0 lie above the center of σ_0^2, with $x_3 = 1$.

In each σ_i^2 there are discs, mapped according to elements α_μ of π_2; the rest of σ_i^2 is mapped into $\bar{p}$. For each μ, we join the discs corresponding to α_μ by tubes inside σ^3. We describe these tubes first for a fixed μ.

Take a point p^* of $\sigma_0^2 = p_1p_2p_3$ fairly close to the center of p_2p_3, and let B_μ be a small cylinder in σ^3 a little above p^*, with top and bottom faces and curved side. We order the μ-discs in $\partial\sigma^3$ as follows. Let the point p move from p_1 to p_2, to p_3, and back to p_1; the segment p_0p sweeps out the faces σ_3^2, σ_1^2, σ_2^2 in order; order the μ-discs in these faces according to when they were swept over. Now let p run from p_3 to p_2; the discs in σ_0^2 swept out by p_1p follow the above discs in order. Attach α_μ-tubes to the discs, and carry them into the curved face of B_μ, preserving this order.

We must be more explicit. Let ψ be the projection of $\sigma^3 - p_0$ from the point p_0 into E_0. If we look at $\sigma^3 - p_0$ from p_0, we see it as though it were projected. Now the discs in σ_0^2 appear near the center. (In the figure below, the position of these discs is marked by the symbol 123. The edges p_1p_3 etc. are drawn curved, to give more room. The cylinder B_μ is marked by the figure 4.) The discs in the other faces appear projected into corresponding sides of σ_0^2 (and are marked 012 etc. in the figure); the order of the discs defined above is clear. To each disc in σ_i^2, $i > 0$, there is a unique manner of attaching a tube (with cross section mapped into R by the chosen mapping ϕ_μ defining α_μ); the top filament is on top at the disc; the orientation of the tube is determined by that of the disc.

We carry this tube into B_μ, keeping the top filament on top. We may do this for all the discs in the σ_i^2, $i > 0$, so that no two tubes cross (seen from p_0).

The discs in σ_0^2 have their tops away from p_2p_3. Attach tubes to these, rising upwards towards p_0; then bend the tubes over towards B_μ; the top filaments now appear on top. They run into B_μ in the correct order.

We now attach the tubes inside B_μ. This is done exactly as in CM, §§6 and 10. If $n_\mu = 0$, no junctions are needed. If $n_\mu \neq 0$, one junction may be needed; it may be of either sign (see §10); it is inserted as in CM, §10. (The "first" tube, named even, is marked by a heavy dot.) The mapping of the part of σ_0^3 in the tubes, and in the junction if present, is now determined; the rest of σ_0^3 is mapped into $\bar{p}$.

For the various values of μ, we must be sure that the different tubes do not intersect. To this end, for $\lambda < \mu$, we let the μ-tubes, after leaving their respective discs, rise slightly higher than the λ-tubes; B_μ lies above B_λ.

6. Position of the tubes in $\partial\sigma^4$

The definition of f having been extended through K^3, we are ready to start work on the determination of W_f^4. Take any $\sigma^4 = p_0 \cdots p_4$. We shall represent $\partial\sigma^4$ as $\bar{E}$ (see CM §2) in the following manner. Let σ_i^3 be the face of σ^4 opposite p_i. We let σ_0^3 correspond to $\bar{E}_0^-$. Now $\partial\sigma_0^3 = \bar{E}_0$; we let the point at infinity correspond to a point n $p_1p_3p_4$ near p_1 (this is not essential); the other faces of $\partial\sigma_0^3$ are in a bounded part of E_0.

The point p_0 we take in E_0^+, above p_2, at $x_3 = 1$. The 2-faces $p_0p_jp_k$ are formed by drawing segments from p_0 to p_jp_k. Then σ_1^3, σ_3^3 and σ_4^3 are bounded, while σ_2^3 is unbounded.

The correspondence can clearly be chosen so that seen from p_0, the tubes in σ_1^3, σ_3^3 and σ_4^3 have the same appearance (no tubes crossing, top filaments on top) as in the original simplexes. This is true also of σ_2^3, except that the tubes from $B_{2\mu}$ (the cylinder in σ_2^3 corresponding to μ) go way around before they can enter $p_0p_1p_3$ and $p_0p_1p_4$. For $\lambda < \mu$, each $B_{i\mu}$ lies above $B_{i\lambda}$ if $i > 0$.

In the figure, the tubes mentioned so far are those drawn in solid lines.

The tubes in σ_0^3 lie in $\bar{E}_0^-$. In the next section we shall deform these tubes so that, seen from p_0, their top filaments are on top.

7. Deformation of the tubes

Let us alter the correspondence of $\partial\sigma^4$ with $\bar{E}^3$ as follows. The part $p_1p_3p_4$ of $\partial\sigma_0^3$ which is mapped into the unbounded part of $\bar{E}_0$ is pulled into a bounded part of $\bar{E}_0^-$, until $\partial\sigma_0^3$ has somewhat the shape of a cylinder, the top face containing $\partial\sigma_0^3 - p_1p_3p_4$, and the bottom face A lying in E_0^- at $x_3 = -1$. Note that, seen from p_0, the tubes from $B_{4\mu}$ (marked 4 in the figure) to the discs (marked 123) in $p_1p_2p_3$ move towards p_1, with top filaments on top; on the other side of E_0, they start back in a direction towards a point underneath $B_{4\mu}$; similarly for the other tubes running into σ_0^3. We may suppose they lie below tubes in E_0^+ (seen from p_0), with each $B_{0\mu}$ below each $B_{1\mu}$. Let us stretch the tubes in E_0^- downwards, till most of each tube lies fairly close to A.

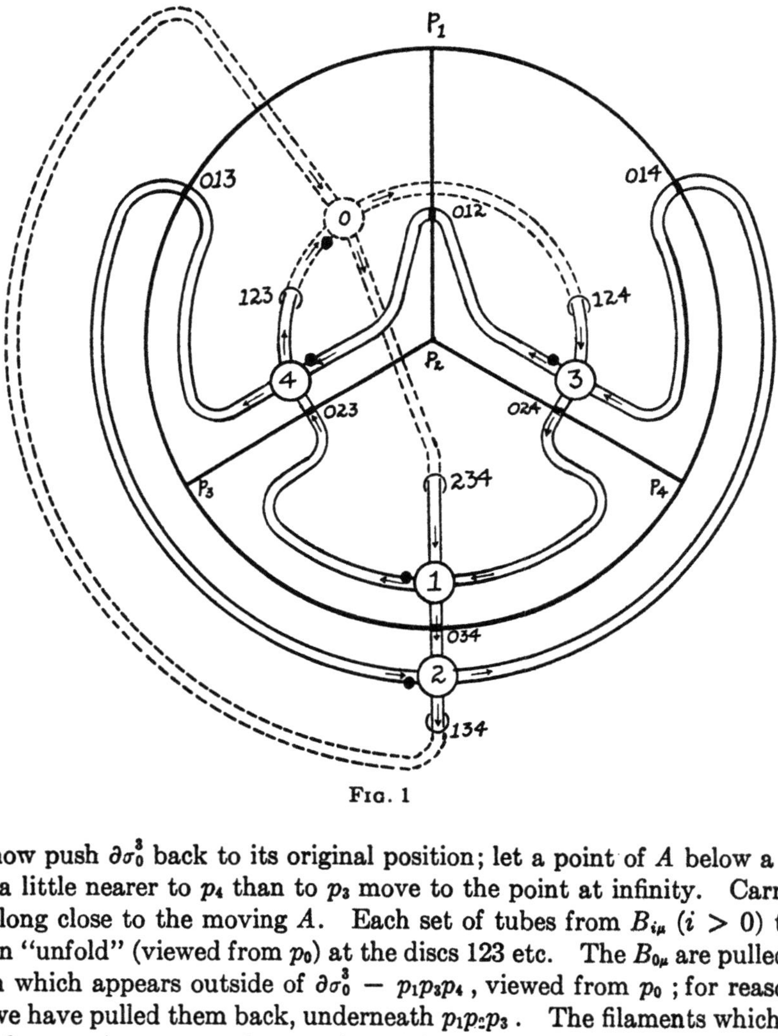

FIG. 1

We now push $\partial\sigma_0^3$ back to its original position; let a point of A below a point of p_3p_4 a little nearer to p_4 than to p_3 move to the point at infinity. Carry the tubes along close to the moving A. Each set of tubes from $B_{i\mu}$ $(i > 0)$ to $B_{0\mu}$ will then "unfold" (viewed from p_0) at the discs 123 etc. The $B_{0\mu}$ are pulled to a position which appears outside of $\partial\sigma_0^3 - p_1p_3p_4$, viewed from p_0; for reasons of space, we have pulled them back, underneath $p_1p_2p_3$. The filaments which were below, close to A, are now close to the original position of A, and hence are on top. Each cylinder $B_{0\mu}$ (and any junction in it) has turned right side up. For $\lambda < \mu$, the λ-tubes were formerly above the μ-tubes in σ_0^3, but are now below them there. We may arrange matters so that, viewed from p_0, the only crossings between μ-tubes, for given μ, are between tubes through $p_0p_1p_2$ and tubes through $p_2p_3p_4$. For $\lambda < \mu$, λ-tubes are below μ-tubes throughout, except that λ-tubes through $p_0p_1p_2$ are above μ-tubes through $p_2p_3p_4$.

Another way of seeing that the tube system T in E_0^- may be replaced by the new position T' of T as described above is the following. Let T^* denote the tube system in E_0^+ obtained from T' by a reflection across E_0. Through a reflection in a vertical plane, or from RH Theorem 21A, we see that the pair T, T^* gives a mapping of $\bar{E}^3$ into R which defines the element 0 of π_3. Consider three mappings of a closed 3-cell (E_0^+ or E_0^-) into R, defined by the tube system T_0 in $\partial\sigma^4 - \sigma_0^3$, by T, and by T'; these agree on the boundary (i.e. on E_0). By CM (2.4) (with an obvious meaning of $\xi(\cdots)$),

$$\xi(T_0, T) + \xi(T, T') = \xi(T_0, T').$$

The first term is $W_f^4(\sigma^4)$, the second is 0 (as is seen by using the T^* above), and hence the third is $W_f^4(\sigma^4)$.

8. Separation of the tube systems

We now separate the tube systems for the different values of μ. If π_2 has no elements of even order, this will give the final formula for W^4; in the general case, there will be terms corresponding to the individual tube systems to evaluate.

To begin with, pull the last tube system T_ν upwards in E^3, away from the other systems T_μ. In this process, all ν-tubes through $p_2p_3p_4$ cut through all μ-tubes through $p_0p_1p_2$, for all $\mu < \nu$; no other cuttings are needed. A ν-tube through a positive patch in $p_2p_3p_4$ is oriented away from p_0 in the natural orientation of $\sigma_1^3 = p_0p_2p_3p_4$, since $\partial\sigma_1^3 = p_2p_3p_4 + \cdots$; since σ_1^3 lies negatively in $\partial\sigma^4 = \bar{E}^3$, the tube is oriented towards p_0 relative to $\bar{E}^3$ (see the figure). (We could note simply, using $p_1p_2p_3p_4$, that $\partial\sigma^4$ orients $\bar{E}^3$ like a right handed system, and that a right hand screw at 234 turning in the direction $p_2 \to p_3 \to p_4 \to p_2$ moves upward.) If we carry such a tube around $\partial(p_0p_1p_2)$, it goes around in the positive sense. The tube would therefore have a looping coefficient 1 with a positive tube through $p_0p_1p_2$. Hence, by RH, §6, the element of π_3 lost in pulling T_ν away is

$$\textstyle\sum_{\mu<\nu} \overline{W}_{f\mu}^2(p_0p_1p_2)\overline{W}_{f\nu}^2(p_2p_3p_4)\alpha_\mu\alpha_\nu .$$

Now pull $T_{\nu-1}$ away; an element of π_3 is lost which is given by a similar formula. Continuing, we see that the element lost on separating all the systems is

$$\textstyle\sum_{\lambda<\mu} \overline{W}_{f\lambda}^2(p_0p_1p_2)\overline{W}_{f\mu}^2(p_2p_3p_4)\alpha_\lambda\alpha_\mu .$$

Let ξ_μ be the element determined by the tube system T_μ by itself. Then

$$W_f^2(\sigma^4) = \textstyle\sum_{\lambda<\mu} \overline{W}_{f\lambda}^2(p_0p_1p_2)\overline{W}_{f\mu}^2(p_2p_3p_4)\alpha_\lambda\alpha_\mu + \sum_\mu \xi_\mu . \tag{8.1}$$

Suppose that π_2 has no elements of even order. We may then determine ξ_μ by RH (19.2), or by RH (13.1) if π_2 has no elements of finite order. Only a single value of i occurs in (19.2), namely, μ; we have $\xi_\mu = N_\mu\eta(\alpha_\mu)$. From RH

§12 we see that crossings of μ-tubes through positive patches in $p_0p_1p_2$ with μ-tubes through positive patches in $p_2p_3p_4$ are of positive type; hence, counting the crossings,

$$N_\mu = \overline{W}^2_{f\mu}(p_0p_1p_2)\overline{W}^2_{f\mu}(p_2p_3p_4). \tag{8.2}$$

9. The theorem in a special case

The last two relations give

THEOREM 9A. *If π_2 has no elements of even order, then*

$$W^4_f = \textstyle\sum_{\lambda<\mu}(\overline{W}^2_{f\lambda} \smile \overline{W}^2_{f\mu})\alpha_\lambda\alpha_\mu + \sum_\mu(\overline{W}^2_{f\mu} \smile \overline{W}^2_{f\mu})\eta(\alpha_\mu). \tag{9.1}$$

If π_2 is cyclic, infinite or of odd order, then we can use the pairing of RH, §9: $a\alpha \circ b\alpha = ab\eta(\alpha)$. Define in this case

$$\textstyle\sum a_i\alpha\sigma^r_i \smile_0 \sum b_j\alpha\sigma^s_j = \sum a_ib_j\eta(\alpha)(\sigma^r_i \smile \sigma^s_j). \tag{9.2}$$

Then (9.1) becomes

$$W^4_f = W^2_f \smile_0 W^2_f. \tag{9.3}$$

This holds, for example, if R is a 2-sphere (see [7], p. 357, and [10], Theorem 21.1).

10. Study of the general case

We must still determine ξ_μ in case $n_\mu > 0$ is even. For a shorthand notation, we set

$$w(ijk) = \overline{W}^2_{f\mu}(p_ip_jp_k).$$

There are different cases to consider, just as in CM, §12, according to what junctions are present. As in CM, §11, we see that there is at most one junction in each cylinder $B_{i\mu}$; but it may be either positive or negative, since two sets of tubes enter $B_{i\mu}$ with each orientation.

If there are no junctions, of course $\xi_\mu = N_\mu\eta(\alpha_\mu)$, as before. Suppose there are just two; then they are of opposite sign. We may suppose there is a tube joining them through the face common to their 3-simplexes; if not, such a tube can be inserted, exactly as in CM, §12, Case (12).

In any given Case (ij), where junctions J_i and J_j are present, in $B_{i\mu}$ and $B_{j\mu}$ respectively, we let T be the first tube joining the cylinders through the face common to σ^3_i and σ^3_j. Let A be the corresponding 1-cell of G^* (see CM, and RH §20). Then (RH, Theorem 20A)

$$\xi_\mu = N_\mu\eta(\alpha_\mu) + (Z_\mu \cdot D_\mu)n_\mu\eta(\alpha_\mu). \tag{10.1}$$

The first term is given by (8.2). As to the second, using $D = A$,

$$Z_\mu \cdot D = Z'_\mu \cdot A + Z''_\mu \cdot A + Z'''_\mu \cdot A. \tag{10.2}$$

The first term on the right is the number (mod 2) of crossings of A with other arcs of G^*, i.e. the number of μ-tubes which, projected into E_0 from p_0, cross T, and the second term is 0 or 1 (mod 2) according as the ends of T on $B_{i\mu}$ and $B_{j\mu}$ are of the same or opposite parity; see RH §18. The third term will always vanish.

CASE (34). Since T is the first tube for both $B_{3\mu}$ and $B_{4\mu}$, $Z''_\mu \cdot A = 0$. Now T crosses (when projected) just those tubes through $p_2p_3p_4$; hence $Z'_\mu \cdot A = w(234)$ (mod 2). We have therefore:

$$(34): \qquad Z_\mu \cdot D \equiv w(234) \qquad (\text{mod } 2).$$

CASE (14). Here, $Z'_\mu \cdot D = 0$. To bring the first tube entering $B_{4\mu}$ to T, which is the first tube entering through the face $p_0p_2p_3$, we must cross all tubes through $p_0p_1p_2$, $w(012)$ in number. Since T is the first tube entering $B_{1\mu}$,

$$(14): \qquad Z_\mu \cdot D \equiv w(012) \qquad (\text{mod } 2).$$

CASE (24). Again $Z'_\mu \cdot D = 0$. To get from the first tube entering $B_{4\mu}$ to T, we must make one more step than the number of tubes through $p_1p_2p_3$. Hence

$$(24): \qquad Z_\mu \cdot D \equiv w(123) + 1 \qquad (\text{mod } 2).$$

CASE (12). To get from the first tubes entering $B_{1\mu}$ and $B_{2\mu}$ to T, we must cross $w(023)$ and $w(013)$ tubes respectively. Since the number of tubes entering $B_{4\mu}$ is $\equiv 0$ (mod 2), (equivalently, since $\delta W^2_f \cdot \sigma^3_4 = 0$), $w(023) + w(013) \equiv w(012) + w(123)$ (mod 2). Hence

$$(12): \qquad Z_\mu \cdot D \equiv w(012) + w(123) \qquad (\text{mod } 2).$$

For the other cases with two junctions, we find similarly (omitting the mod 2):

(04):	1,	(23):	$w(123) + w(234)$,
(13):	$w(012) + w(234) + 1$,	(03):	$w(234)$,
(02):	$w(123) + 1$,	(01):	$w(012)$.

REMARK. The sum of the values of $Z_\mu \cdot D$ in cases (ij), (ik) and (jk) is always 1.

Consider now the cases where there are more than two junctions. There are then four, two positive, say J_i and J_j, and two negative, say J_k and J_l. Choose tubes T_1 joining $B_{i\mu}$ to $B_{k\mu}$ and T_2 joining $B_{j\mu}$ to $B_{l\mu}$, as was done in the cases above. Let these correspond to 1-cells A_1, A_2 in G^*; set $D = A_1 + A_2$. Since Z_μ is linear (being a cochain),

$$Z_\mu \cdot D = Z_\mu \cdot A_1 + Z_\mu \cdot A_2,$$

and hence is found by adding the value for the cases (ik) and (jl).

In case $(ijkl)$, the manner of splitting up depends on the signs of the junctions. But on finding the values of $Z_\mu \cdot D$ for each of the two cases obtained from any splitting, it is seen that the answer always comes out the same. We find

$$(1234): \qquad w(012) + w(123) + w(234),$$

$$(0234): \qquad w(123) + w(234) + 1,$$

(0134): $w(012) + w(234)$,

(0124): $w(012) + w(123) + 1$,

(0123): $w(012) + w(123) + w(234)$.

11. The formula for W_f^4

Let us write

(11.1) $$Z_\mu \cdot D = z_1 + z_2,$$

where z_1 is, in each case, the part of the expression given by w's, and z_2 is the part (if any) given by the number 1.

We now summarize the results of the above cases. We see that in any Case (ij), $z_2 \equiv 1 \pmod 2$ if and only if i and j are both even or both odd. Recall from CM §11 that

(11.2) $$v(\sigma_i^3) = \theta_{n_\mu}^3 W_{f\mu}^2(\sigma_i^3)$$

is the number of junctions (mod 2) in σ_i^3. Consulting (4.4) shows that, in all cases,

(11.3) $$z_2 \equiv (v \smile_2 v) \qquad (\text{mod } 2).$$

For instance, in Case (02),

$$(v \smile_2 v)(\sigma^4) = [(\bar{0} + \bar{2}) \smile_2 (\bar{0} + \bar{2})](\sigma^4) = (\bar{2} \smile_2 \bar{0})(\sigma^4) = 1.$$

Consulting (4.3), we see that if we set

(11.4) $$\begin{cases} \gamma(\sigma_1^3) = w(012), \gamma(\sigma_2^3) = w(123), \gamma(\sigma_3^3) = w(234), \\ \qquad \gamma(\sigma_0^3) = \gamma(\sigma_4^3) = 0, \end{cases}$$

(11.5) $$\gamma(\sum a_i \sigma_i^3) = \sum a_i \gamma(\sigma_i^3),$$

all mod 2, then

$$(\sigma_i^3 \smile_1 \overline{W}_{f\mu}^2)(\sigma^4) = \gamma(\sigma_i^3),$$

and hence

(11.6) $$(X^3 \smile_1 \overline{W}_{f\mu}^2)(\sigma^4) = \gamma(X^3).$$

For instance, recalling that $\overline{34}$ stands for $p_0p_1p_2$,

$$(\sigma_1^3 \smile_1 \overline{W}_{f\mu}^2)(\sigma^4) = [\sigma_1^3 \smile_1 w(012)p_0p_1p_2](\sigma^4) = w(012).$$

Referring to the various cases above, we see that $z_1 = \gamma(\bar{\imath} + \bar{\jmath})$ in Case (ij), for all i, j, and hence $z_1 = \gamma(\bar{\imath} + \bar{\jmath} + \bar{k} + \bar{l})$ in Case $(ijkl)$. Therefore

(11.7) $$z_1 \equiv \gamma(v) \equiv (v \smile_1 \overline{W}_{f\mu}^2)(\sigma^4).$$

Combining this with (11.3) and (11.1) gives

(11.8) $$Z_\mu \cdot D \equiv [v \smile_1 \overline{W}_{f\mu}^2 + v \smile_2 v](\sigma^4) \qquad (\text{mod } 2).$$

Let us now put this in (10.1), and insert the result in (8.1) (compare (8.2) and (9.1)). This gives the theorem in the general case:

THEOREM 11A. *Given the standard mapping f of the 2-dimensional part* K^2 *of the complex K with ordered vertices into the arcwise connected and simply connected Hausdorff space R such that* $\pi_2 = \pi_2(R)$ *has a finite number of generators, let the corresponding cochain* W_f^2 *be a cocycle. Then in terms of a given decomposition of* π_2 *as in* CM §2, *the second obstruction* W_f^4 *is given by*

$$W_f^4 = \textstyle\sum_{\lambda<\mu}(\overline{W}_{f\lambda}^2 \smile \overline{W}_{f\mu}^2)\alpha_\lambda\alpha_\mu + \sum_\mu(\overline{W}_{f\mu}^2 \smile \overline{W}_{f\mu}^2)\eta(\alpha_\mu) + \sum_{n_\mu>0,\ \text{even}}[\theta_{n_\mu}^3 W_{f\mu}^2 \smile_1 \overline{W}_{f\mu}^2 + \theta_{n_\mu}^3 W_{f\mu}^2 \smile_2 \theta_{n_\mu}^3 W_{f\mu}^2]n_\mu\eta(\alpha_\mu). \tag{11.9}$$

We recall that $\overline{W}$ and θ were defined in CM, §2.

HARVARD UNIVERSITY

REFERENCES

1. S. EILENBERG, *Cohomology and continuous mappings*. Ann. of Math. 41 (1940): 231–251. See also Lectures in Topology, Univ. of Mich. Press, 1941, 57–100.
2. S. EILENBERG AND S. MACLANE, *Relations between homology and homotopy groups of spaces*. Ann. of Math. 46 (1945): 480–509.
3. S. EILENBERG AND S. MACLANE, *Determination of the second homology and cohomology groups of a space by means of homotopy invariants*. Proc. Nat. Acad. Sci. 32 (1946): 277–280.
4. H. HOPF, *Über die Abbildungen der dreidimensionalen Sphäre auf die Kugelfläche*. Math. Ann. 104 (1931): 637–665.
5. H. HOPF, *Die Klassen der Abbildungen der n-dimensionalen Polyeder auf die n-dimensionale Sphäre*. Comment. Math. Helv. 5 (1933): 39–54.
6. W. HUREWICZ, *Beiträge zur Topologie der Deformationen, III, Klassen und Homologietypen von Abbildungen*. Proc. Neder. Akad. Wetensch. 39 (1936): 117–126.
7. L. PONTRJAGIN, *A classification of mappings of the 3-dimensional complex into the 2-dimensional sphere*. Rec. Math. [Mat. Sbornik] N.S. 9 (51) (1941): 331–363.
8. L. PONTRJAGIN, *Mappings of the 3-dimensional sphere into an n-dimensional complex*. C. R. Acad. Sci. URSS. 34 (1942): 35–37.
9. H. E. ROBBINS, *On the classification of the mappings of a 2-complex*. Trans. Am. Math. Soc. 49 (1941): 308–324.
10. N. E. STEENROD, *Products of cocycles and extensions of mappings*. Ann. of Math. 48 (1947): 290–320.
11. H. WHITNEY, *The maps of an n-complex into an n-sphere*. Duke Math. J. 3 (1937): 51–55.

Chapter 4

Geometric Integration Theory

Reprinted from the Proceedings of the NATIONAL ACADEMY OF SCIENCES,
Vol. 33, No. 1, pp. 1–11. January, 1947

ALGEBRAIC TOPOLOGY AND INTEGRATION THEORY

BY HASSLER WHITNEY

DEPARTMENT OF MATHEMATICS, HARVARD UNIVERSITY

Communicated December 10, 1946

1. *Introduction.* We give here a preliminary account of a study of that side of integration theory which is useful particularly in differential geometry and fields centering around Stokes's theorem and the like. The subject is integration over r-dimensional domains (or, more generally, over r-chains) in an n-dimensional manifold or more general space (see below); the methods are direct, and are carried out largely without using coördinate systems. There are connections with real variable theory, but the problems are in general of a different nature. The methods are based on the fundamentals of algebraic topology; in fact, the work may be looked upon as the study of algebraic topology with real numbers as coefficients.

The integral $\int_A X$ of a function X over a domain A corresponds to the "scalar" or "Kronecker" product $X \cdot A$ of a cochain X and a chain A; we shall write $X \cdot A$ in place of $\int_A X$. The values of integrals are real numbers, but could equally well be elements of a Banach space. The work is largely a matter of proving inequalities; to make this possible, Lipschitz conditions are used throughout. In particular, cells σ used for domains of integration will be images of Euclidean cells σ_0 under "Lipschitz mappings" f:

$$\text{dist}[f(p), f(q)] \leqq N\ \text{dist}(p, q) \text{ for some } N.$$

The smallest possible such N is the *expansion* $\xi(f, \sigma_0)$ of f in σ_0. (Though the inequalities depend on the metric in the space, the results are independent of Lipschitz changes of metric.) With this notion, a definition of the size $|A|$ of a chain A is given; for usual domains, this is the r-dimensional volume of the domain. The integral corresponds to that of a bounded function: $|\int_A X| \leqq N|A|$ for some N. The extension to the unbounded case will not be considered here.

The most general space to which our results apply is a "Lipschitz space" R, defined as follows. It is a metric space (assumed compact for simplicity), for which there is an imbedding f into some Euclidean space E^n, both f and f^{-1} being Lipschitz. Further, $f(R)$ is supposed "locally Lipschitz connected" in each dimension r. That is, for some $\eta > 0$ and some N_r, any singular $(r - 1)$-sphere S^{r-1} of diameter $< \eta$ in $f(R)$ bounds a singular r-cell Q^r in $f(R)$ such that

$$\text{diam }(Q^r) \leqq N_r \text{ diam }(S^{r-1}).$$

There is then a retraction ρ of a neighborhood U of $f(R)$ in E^n into $f(R)$ such that ρ is Lipschitz.

2. *Lipschitz chains.* By vol (σ_0^r) we mean the usual r-volume of a Euclidean simplex σ_0^r in metric Euclidean space. We define $|\sigma^r|$ for a Lipschitz singular simplex $\sigma^r = f(\sigma_0^r)$, i.e., image of a Euclidean simplex σ_0^r under a Lipschitz mapping f, as follows. Consider subdivisions $\sigma_0^r = \Sigma\sigma_{0i}^r$ of σ_0^r into Euclidean simplexes σ_{i0}^r, and affine mappings g_i of Euclidean simplexes τ_i^r into σ_{0i}^r; set $f_i(p) = f(g_i(p))$ in τ_i^r; then

$$|\sigma^r| = GLB\sum_i \xi^r(f_i, \tau_i) \text{ vol }(\tau_i),$$

taking the greatest lower bound over all such expressions. For a Euclidean simplex or smooth simplex σ, $|\sigma| = \text{vol }(\sigma)$. To extend the definition to Lipschitz singular chains $A = \Sigma a_i\sigma_i^r = \Sigma a_i f(\sigma_{0i}^r)$, consider simultaneous subdivisions $\Sigma_j\sigma_{0ij}^r$ of the σ_{0i}^r, and proceed as above, using $|a_i|$; coefficients must be collected over the singular simplexes occurring.[1]

3. *Lipschitz cochains. A Lipschitz r-cochain* X^r in R is a real-valued function $X \cdot A^r$ of Lipschitz singular chains A^r in R, with the following properties: For some numbers N, N',

(*a*) $X\cdot(-\sigma) = -X\cdot\sigma,$

(*b*) $|X\cdot\sigma| \leqq N|\sigma|,$

(*c*) $|X\cdot\partial\sigma| \leqq N'|\sigma|.$

Here, $\partial\sigma$ denotes the boundary of an $(r + 1)$-simplex σ. We let $|X|$ and $|\delta X|$ denote the least possible values of N and N', respectively. X, defined for simplexes, becomes a linear function of Lipschitz singular r-chains by setting $X\cdot\Sigma a_i\sigma_i^r = \Sigma a_i X\cdot\sigma_i^r$. Now $|X\cdot A| \leqq |X|\,|A|$.

The *coboundary* δX of X is defined by

$$\delta X^r\cdot A^{r+1} = X^r\cdot\partial A^{r+1}.$$

Written in the form $\int_A\delta X = \int_{\partial A}X$, this becomes Stokes's theorem.

If f is a Lipschitz mapping of R into R', it carries cochains $X' = X'^r$ of R' into cochains f^*X' of R, defined by $f^*X'\cdot A = X'\cdot fA$. We have $\delta f^*X' = f^*\delta X'$, and

$$|f^*X'| \leqq \xi^r(f)\,|X'|,\ |\delta f^*X'| \leqq \xi^{r+1}(f)|\delta X'|.$$

A 0-cochain X is simply a point function: $X(p) = X\cdot p$. Its coboundary is given by $\delta X\cdot(pq) = X\cdot\partial(pq) = X\cdot q - X\cdot p$.

4. *Definition for polyhedral chains.* Let $A = fA_0$ be a Lipschitz chain in R, expressed as the image of the chain A_0 in the simplicial complex K_0. Let K_1 be a simplicial subdivision of K_0. Let $\mathfrak{S}\,A_0$ be a subdivision of A_0. Construct the cartesian product $I \times K_0$ of the unit interval $I = (0, 1)$ and K_0; consider K_0 as $0 \times K_0$ and K_1 as $1 \times K_0$. We may subdivide it simplicially so that

$$\partial\mathfrak{S}(I \times A_0) = \mathfrak{S}A_0 - A_0 - \mathfrak{S}(I \times \partial A_0).$$

Map $I \times K_0$ into R by setting $F(t, p) = f(p)$. Then for each τ of $\mathfrak{S}(I \times A_0)$ not in $0 \times K_0$ or $1 \times K_0$, $|F(\tau)| = 0$. It follows from (*b*) and (*c*) that

$$X\cdot\mathfrak{S}A = X\cdot F\mathfrak{S}A_0 = X\cdot FA_0 = X\cdot A.$$

Thus $X\cdot A$ is independent of subdivisions of A, and may be defined for "polyhedral singular chains" A.

Essentially the same method (with $K_1 = K_0$) shows that $X\cdot A$ is continuous under deformations of A.

For n-cochains X in E^n (or in an n-manifold) we may define $X\cdot Q$ for open sets Q; thus, X gives rise to a Lebesgue integral $X\cdot Q = \int_Q D_X(p)dp$, with the ordinary Lebesgue measure. Such a theorem is unknown for X^r in E^n, $1 < r < n$.

5. *Special cochains; the de Rham Theorem.* For an oriented n-simplex σ^n in metric oriented E^n, let $[\sigma^n]$ be $\pm$ vol (σ^n) according as σ^n and E^n are oriented alike or unlike. Write $I^n\cdot\sigma^n = [\sigma^n]$; then I^n is the *unit n-cochain* of metric oriented E^n. For any 0-cochain W^0 and r-cochain X^r, we may define the r-cochain W^0X^r by letting $(W^0X^r)\cdot\sigma^r = (W^0\cdot p)(X^r\cdot\sigma^r)$ approximately, for p in σ^r, if σ^r is small; the definition is completed by a passage to the limit (using subdivisions of σ). Now, with a notation from §4, $X^n = D_XI^n$ in E^n.

A cochain X is a *cocycle* if $\delta X = 0$; X and Y are *cohomologous* if $Y - X = \delta Z$ for some Z. The classes of cohomologous cocycles are the elements of the rth *Lipschitz cohomology group* of the space.

In a simplicial complex K, to each Lipschitz cochain X corresponds an algebraic cochain ϕX, defined by $\phi X\cdot\sigma_i^r = X\cdot\sigma_i^r$ for each r-simplex σ_i^r of K. *This defines an isomorphism between the Lipschitz and the algebraic cohomology groups.* This is the counterpart of the de Rham Theorem[2] (originally expressed in terms of homology theory). See also §§7 and 8.

In the proof, we define for each algebraic cochain X a Lipschitz cochain ψX as follows. It is sufficient to define $\psi X\cdot\tau$ for simplexes τ lying in simplexes of K (compare §6). If we define $\psi\sigma^r$ for each σ^r, ψX for all

X will be defined. Take $\sigma^r = p_0 \dots p_r$, and any $\tau^r = q_0 \dots q_r$ in a simplex of K. If τ is not in the star of σ, set $\psi\sigma \cdot \tau = 0$. Otherwise, say τ is in $\sigma^s = p_0 \dots p_r p_{r+1} \dots p_s$. Set

$$\psi\sigma \cdot \tau = [q_0 \dots q_r p_{r+1} \dots p_s]/[\sigma].$$

This is independent of the (Euclidean) metric employed. We may prove: $\phi\psi X = X$, $\delta\psi X = \psi\delta X$.

6. *Lipschitz skeleton cochains.* A Lipschitz r-cochain X must have a value for all Lipschitz r-chains A; these chains form a very large class of objects. We shall show that Lipschitz cochains may be derived from "Lipschitz skeleton cochains," defined for a much more restricted class of objects.

A *skeleton r-simplex* σ^r in a metric space R is a set of $r + 1$ points $p_0, \dots, p^r$ in R. It is oriented like an abstract simplex. A *defining set of sides* of σ^r is a set $p_{\lambda_i} p_{\mu_i}$ such that if the p_i are put in E^r, the vectors $v_i = p^{\mu_i} - p_{\lambda_i}$ $(i = 1, \dots, r)$ are independent. The *potentiality* of σ is

$$\text{Pot}(\sigma) = \frac{1}{r!} \min \text{dist}(p_{\lambda_1}, p_{\mu_1}) \dots \text{dist}(p_{\lambda_r}, p_{\mu_r}),$$

taking the minimum over defining sets of sides of σ.

A *Lipschitz skeleton cochain* x is a function $x \circ \sigma$ such that (a′) $x \circ (-\sigma) = -x \circ \sigma$, and

$$\text{(b}') \quad |x \circ \sigma| \leqq N \,\text{Pot}(\sigma), \quad \text{(c}') \quad |x \circ \partial\sigma| \leqq N' \,\text{Pot}(\sigma),$$

for some N, N', as before. Again $x \circ \Sigma a_i \sigma_i^r = \Sigma a_i x \circ \sigma_i^r$.

We shall use a *standard* sequence of subdivisions[3] $K_1, K_2, \dots$, of a complex K_0: only a finite number of shapes of simplexes may occur in all the K_i. If $A = fA_0$, A_0 in K_0, and $\mathfrak{S}_n A_0$ is the subdivided chain in K_n, etc., as in §4, then

$$\bar{x} \cdot A = \lim_{n \to \infty} x \circ f \mathfrak{S}_n A_0$$

exists, and $\bar{x}$ is a Lipschitz cochain X, depending on x only. Conversely, any X gives rise to an x for which $\bar{x} = X$.

Skeleton cochains x, y are *equivalent* if $\bar{x} = \bar{y}$. This holds if for each $\epsilon > 0$ there is a $\delta > 0$ such that

$$|y \circ \sigma - x \circ \sigma| \leqq \epsilon \,\text{Pot}(\sigma) \text{ if diam}(\sigma) < \delta.$$

7. *Tensor cochains.* For simplicity, we shall remain in E^n. A *Lipschitz tensor cochain* is a function $T(p;\ v_1, \dots, v_r)$ of the point p and vectors $v_1, \dots, v_r$, which (a″) is linear and alternating in the vectors, and such that

$$\text{(b}'') \quad |T(p;\ v_1, \dots, v_r)| \leqq N|v_1| \dots |v_r|,$$

$$\text{(c}'') \quad |T(p';\ v_1, \dots) - T(p;\ v_1, \dots)| \leqq N'|p' - p|\, |v_1| \dots |v_r|.$$

We may define $T(p;\ V)$ where V is a contravariant r-vector, i.e., alternating product, $v_1\ldots v_r$ (or a sum of such). For fixed p, T is a covariant r-vector.

For a point function $u(p)$ and a vector v, set

$$\nabla_v u(p) = \lim_{t\to 0}\,[u(p + tv) - u(p)]/t.$$

The covariant derivative of T is, if T is differentiable,

$$\delta T(p;\ v_1\ldots v_{r+1}) = \sum (-1)^{i-1}\nabla_{v_i} T(p;\ v_1\ldots\hat{v}_i\ldots v_{r+1});$$

$\hat{v}_i$ denotes that this term is to be omitted.

To any oriented Euclidean simplex $\sigma^r = p_0\ldots p_r$ in E^n corresponds a contravariant r-vector $\{\sigma^r\}$, defined by

$$\{\sigma^r\} = \frac{1}{r!}\,(p_1 - p_0)\ldots(p_r - p_0);$$

if $v_1, \ldots, v_r$ is any defining set of sides of σ^r, $\{\sigma^r\} = \pm v_1\ldots v_r$. The center of σ^r is $p_\sigma = \Sigma p_i/(r + 1)$.

To each T corresponds a Lipschitz skeleton cochain $x = \Theta T$ defined by

$$\Theta T\cdot\sigma = T(p_\sigma;\ \{\sigma\}).$$

In the proof that (c′) holds and for other matters, the following combinatorial theorem (corresponding to Stokes's theorem) is needed. Let $L_0, \ldots, L_r$ be covariant r-vectors. Let $\sigma^r = p_0\ldots p_r$ be a simplex, with a defining set of sides $v_i = p_{\mu_i} - p_{\lambda_i} (i = 1, \ldots, r)$. Let σ_i be the face of σ opposite p_i. Then

$$\sum_{i=0}^{r} (-1)^i L_i\,\{\sigma_i\} = \frac{1}{(r-1)!}\sum_{j=1}^{r} (-1)^j [L_{\mu_j}(\ldots\hat{v}_j\ldots) - L_{\lambda_j}(\ldots\hat{v}_j\ldots)].$$

Though $\underline{\delta\Theta T} \neq \underline{\Theta\delta T}$ in general, they are equivalent (if T is differentiable); hence $\overline{\delta\Theta T} = \delta\overline{\Theta T} = \overline{\Theta\delta T}$, as Lipschitz cochains. Using $\overline{\Theta T}$ defines an integration process without using coördinate systems.[4] With a coördinate system, the theory becomes ordinary integration; the theory of the Jacobian, Stokes's theorem, etc., follows simply.

The "tensor cohomology groups" again are isomorphic to the algebraic cohomology groups. (The de Rham Theorem is for the case that the tensors are differentiable.)

8. *Products of cochains.* The product of tensor cochains is well known. The corresponding product among Lipschitz cochains satisfies the following conditions:

(α) X^rY^s is an $(r + s)$-cochain, linear in both X^r and Y^s.

(β) X^0Y^s is as defined in §5.

(γ) $|X^r Y^s| \leqq N_{rs}|X^r|\,|Y^s|$.

(δ) $\delta(X^r Y^s) = \delta X^r Y^s + (-1)^r X^r \delta X^s$.

For general Lipschitz cochains, the products with these properties exist and are uniquely determined. (This holds in a Lipschitz space.) The usual properties hold:

$$(XY)Z = X(YZ); \quad Y^s X^r = (-1)^{rs} X^r Y^s;$$

for Lipschitz mappings f, $f^*(XY) = (f^*X)(f^*Y)$.

The product corresponds (under the operation ϕ of §5) to the product[5] $X \smile Y$ of algebraic cochains. Hence the cohomology rings in the algebraic, Lipschitz and tensor cases are isomorphic, completing the de Rham Theorem for these cases.

The fundamental tool for this section is the approximation to a cochain X by a set of coboundaries Z_i. Taking R in E^n, let ρ be the retraction of §1. Cut U into small pieces Q_i; take p_i in Q_i, and set

$$Z_i \cdot \sigma^{r-1} = X \cdot \rho J(p_i, \sigma^{r-1})$$

for any σ^{r-1} in Q_i; J denotes the join. Then $|X - \delta Z_i|$ is small in Q_i.

9. *The continuity theorem.* A continuity theorem was mentioned in §4. We state a much more general (and difficult) theorem. Let A be a Lipschitz chain in R. For any $\epsilon > 0$ and N, there is an $\eta > 0$ with the following property. For any Lipschitz chain B satisfying $|B| < N$, $|\partial B| < N$, and one further condition described below,

$$|X \cdot B - X \cdot A| \leqq \max. (|X|, |\delta X|)\epsilon.$$

The further condition is: there exist chains (not necessarily Lipschitz) C within η of ∂A and D within η of A such that

$$B - A = \partial D + C.$$

Letting $\eta \to 0$ is interesting: two chains A and B are "homology equivalent," i.e., C and D exist for all $\eta > 0$, if and only if $X \cdot A = X \cdot B$ for all X.

10. *Stokes's theorem.* With the help of methods described above (including the continuity theorem), a very general form of Stokes's theorem may be proved. We shall not give details here.

[1] See, for instance, Seifert-Threlfall, *Lehrbuch der Topologie*, Leipzig, 1934, Chapter IV.

[2] See G. de Rham, *Jour. de Math.* (10), **9**, 115–200 (1931).

[3] See H. Freudenthal, *Annals Math.*, **43**, 580–582 (1942).

[4] In an earlier version of this work, a large part of the theory was worked out by Paul Olum (1940, unpublished). See in this connection H. Whitney, *Duke Math. Jour* **4**, 495–528 (1938), §§11, 12.

[5] See, for instance, H. Whitney, *Annals Math.*, **39**, 397–432 (1938); see also H. Whitney, *Bull. Am. Math. Soc.*, **43**, 785–805 (1937), p. 803.

GEOMETRIC METHODS IN COHOMOLOGY THEORY[1]

BY HASSLER WHITNEY

DEPARTMENT OF MATHEMATICS, HARVARD UNIVERSITY

Communicated December 10, 1946

1. *Introduction.* In algebraic topology, the concepts of chains and cycles have considerable geometric content. Singular chains and cycles carry this content over to more general spaces such as polyhedra and absolute neighborhood retracts (*ANR*'s). Cochains and cocycles, algebraically as simple as chains and cycles, did not appear until about 1935, in part because of the lack of a geometric interpretation. We give such an interpretation here;[2] it is, we believe, the natural one. Geometric cochains will be functions of singular chains in "general position."

The theory should apply to *ANR*'s R; at present, some of the basic theorems are proved only for polyhedra P. Full details will be given elsewhere. The coefficient group H will be supposed discrete. If H is continuous, for example, if it is the group of reals, the "nuclear system" (see below) may be discarded in whole or in part.[3]

2. *r-thin subsets of R.* A subset N of R is *r-thin* in R if any singular r-cell $\sigma^r = f(\sigma_0^r)$ in R may be pulled away from N by an arbitrarily small deformation f_t; $f_1(\sigma_0^r) \cap N = 0$. For example, a k-plane in n-space E^n is $(n-k-1)$-thin in E^n. The set N and its complement may be very complicated.

Let f map the polyhedron P into P'; let N be r-thin in P'. Then there is an arbitrarily small deformation f_t of f such that $f_1^{-1}(N)$ is r-thin in P. (This has not been proved for *ANR*'s).

Let N, N' be point sets in R. The singular chain A in R is in *general position* relative to (N, N') if A and ∂A do not intersect N' and N, respectively.

3. *Geometric cochains.* A *geometric r-H-cochain* $X = X^r$ in R is a system of the following sort. It has a *nucleus* $\bar{X}$ and a *nuclear boundary* $\bar{X}' \subset \bar{X}$; $\bar{X}$ and $\bar{X}'$ are closed, and are $(r-1)$-thin and r-thin, respectively. For any singular r-chain A, with integer coefficients, in general position relative to $(\bar{X}, \bar{X}')$, $X \cdot A$ is defined, and is in H; this is a homomorphism of the group of these chains into H. Further,

(a) $X^r \cdot A^r = 0$ if A^r does not intersect $\bar{X}$,

(b) $X^r \cdot \partial A^{r+1} = 0$ if A^{r+1} does not intersect $\bar{X}'$.

The *coboundary* δX of X, with nuclear system $(\bar{X}', 0)$, is defined by $\delta X^r \cdot A^{r+1} = X^r \cdot \partial A^{r+1}$.

Clearly X is determined by its values for arbitrarily small singular cells near points of $\bar{X} - \bar{X}'$. Keeping all chains in general position, $X \cdot A$ is easily proved independent of subdivisions and deformations of A.

If $X \cdot \sigma = 0$ for all σ in some neighborhood U of p in $\bar{X}$, the points of $\bar{X}$ in U may be dropped out. In this way, we obtain the *irreducible nucleus* of X. Note now that $X \cdot A$ is always defined if $\delta X = 0$, $\partial A = 0$.

For a simple example, let $\bar{X}$ be a square in E^3; let $\bar{X}'$ be its boundary. Let u be a normal vector to $\bar{X}$. Then if A^1 is a segment cutting through $\bar{X}$, let $X^1 \cdot A^1$ be 1 or -1 according as A^1 is directed like u or unlike u. If A^2 is a disc cutting $\bar{X}'$, $\delta X^1 \cdot A^2 = X^1 \cdot \partial A^2 = \pm 1$.

Without further conditions, a cochain may be very complicated. For example, in E^n ($n \geqq 1$), $X^n \cdot \sigma^n$ may have arbitrarily large values for arbitrarily small σ^n near a given p in X.

4. *Standard cochains in a complex.* In a complex K the "dual" ξ^r_i of a cell σ^r_i of K consists of simplexes of the first derived of K, of various dimensions in general; it forms a point set which is $(r - 1)$-thin in K. It forms the nucleus of a cochain $X_i^r = \phi\sigma_i^r$ such that $X_i^r \cdot \sigma_j^r = \delta_{ji}$. Setting $\phi\Sigma a_i\sigma_i^r = \Sigma a_i X_i^r$, we have $\delta\phi X = \phi\delta X$ for algebraic cochains X of K. These *standard cochains* operate on the algebraic chains of K, thought of as singular chains, just as the algebraic cochains do.

5. *Mappings and deformations.* Let f map P_1 into P_2. Then given X_2 in P_2, we may deform f into f_1 (see §2) to make $(f_1^{-1}(\bar{X}_2), f^{-1}(\bar{X}_2'))$ the nuclear system of a cochain $X_1 = f_1^*X_2$ in P_1, defined by

$$f_1^*X_2 \cdot A_1 = X_2 \cdot f_1 A_1.$$

Then $\delta f_1^*X_2 = f_1^*\delta X_2$. If $f_t(p)$ (p in P_1, $0 \leqq t \leqq 1$) is a deformation, such that $f_0^*X_2$ and $f_1^*X_2$ are defined, we may "deform" the deformation for $0 < t < 1$, and then define an $(r - 1)$-cochain $f_I^*X_2$, such that

$$\delta f_I^*X_2 = f_1^*X_2 - f_0^*X_2 - f_I^*\delta X_2.$$

6. *A local property of cochains.* Given X^r in R ($r > 0$), and p in $\bar{X} - \bar{X}'$, let f_t be a deformation of the identity in R such that $f_1(p)$ is not in $\bar{X}$. Say p is in U, $f_1(U) \cap \bar{X} = 0$, and $f_t(U) \cap \bar{X}' = 0$ for all t. Now $f_0^* X = X$, $f_1^*X = 0$ in U, and $f_I^*\delta X = 0$ in U; hence, by the relation above, $X = \delta f_I^*X$ in U. Thus, *every geometric cochain of dimension* > 0 *is locally a coboundary.*

7. *Geometric cohomology groups in a complex.* Let X^r ($r > 0$) be a geometric cochain in the complex K such that δX is standard (§4). With methods above, and other tools, we may show that there is a standard Y and a geometric Z such that $X - Y = \delta Z$. It follows that the geometric and algebraic cohomology groups of K are isomorphic.

8. *Properties of singular chains in terms of geometric cochains.* Conditions that a cochain X be 0, or a cocycle, or a coboundary, are expressible in a simple manner in terms of $X \cdot A$ for various A; we shall state some corresponding theorems for singular chains. Say A_1 and A_2 are *essentially equal:* $A_1 =_e A_2$, if, for arbitrary neighborhoods U of $A_1 \cup A_2$ and U'

of $\partial A_1 \cup \partial A_2$, there are singular chains $B \subset U$ and $C \subset U'$ such that $A_1 - A_2 = \partial B + C$. (This relation is not transitive, unless some assumption to eliminate chains with space-filling characteristics is made.) Now the conditions $A = {}_e0$, $\partial A = {}_e0$, $A = {}_e\partial B$ for cycles A, are equivalent to $X \cdot A = 0$, $\delta X \cdot A = 0$, $X \cdot A = 0$ if $\delta X = 0$, for all cochains X of the proper dimension in general position relative to A.

9. *Products of geometric cochains.* We first give an example. Let $\bar{X}$ and $\bar{Y}$ be squares in E^3, as in §3; let them intersect in a segment $\bar{Z}$. (Do not let $\bar{X}'$ and $\bar{Y}'$ intersect.) Take p in $\bar{Z}$; let u and v be vectors at p normal to $\bar{Z}$, and lying in $\bar{Y}$ and $\bar{X}$, respectively. If A^1 is a segment cutting $\bar{X}$ in the direction of u, say $X \cdot A^1 = \alpha$; define β similarly for $\bar{Y}$ and v. Then if A^2 is a disc cutting $\bar{Z}$ at p and containing u and v, and A^2 is oriented by the ordered pair (u, v), set $Z^2 \cdot A^2 = \alpha\beta$. We set

$$Z = X \smile Y; \; \bar{Z} = \bar{X} \cap \bar{Y}, \bar{Z}' = (\bar{X} \cap \bar{Y}') \cup (\bar{X}' \cap \bar{Y}).$$

We now discuss the general situation. Say X^r and Y^s are in *general position* if $\bar{Z}$ and $\bar{Z}'$ defined above are $(r + s - 1)$-thin and $(r + s)$-thin, respectively, and $\bar{X}' \cap \bar{Y}'$ is $(r + s + 1)$-thin. (Given arbitrary X and Y, slight deformations f_t and g_t of the identity $f_0 = g_0$ will bring $f_1{}^*X$ and $g_1{}^*Y$ into general position.) Given any p in $\bar{Z} - \bar{Z}'$, define Z in a small neighborhood U of p as follows. First, if $r = 0$,

$$Z^s \cdot \sigma^s = (X^0 \cdot p)(Y^s \cdot \sigma^s), \; \sigma^s \text{ in } U.$$

For $r > 0$, say $X = \delta X_1$, $\delta Y = 0$, in U (see §6); using induction, set

$$Z^{r+s} \cdot \sigma^{r+s} = (X_1{}^{r-1} \smile Y^s) \cdot \partial\, \sigma^{r+s}, \; \sigma^{r+s} \text{ in } U.$$

The proof of existence and uniqueness, together with the usual properties of the product, is carried out with the help of induction.

10. *The usual products in a simplicial complex.* If K is a simplicial complex with ordered vertices, there is a simple way of defining twosystems of duals $\xi_i{}^r$, $\eta_j{}^s$ (see §4) so that any corresponding $\varphi_\xi X^r$, $\varphi_\eta Y^s$ are always in general position.[4] Now for any algebraic X^r and Y^s, and any simplex σ^{r+s} of K, which we may think of as a singular simplex,

$$(\varphi_\xi X^r \smile \varphi_\eta Y^s) \cdot \sigma^{r+s}$$

is defined, and is the same as the usual algebraic product $(X \smile Y^s) \cdot \sigma^{r+s}$.

[1] Presented to the Am. Math. Society, Feb. 21, 1941, under the title "Geometric methods in combinatorial topology."

[2] For earlier work along these lines, see S. Wylie, *Proc. London Math. Soc.*, **46**, 174–198 (1940). An application will be found in H. Whitney, *Annals Math.* (2), **45**, 220–246 (1944), §5.

[3] The theory then takes on a completely different form. See the preceding paper in these PROCEEDINGS.

[4] Compare H. Whitney, these PROCEEDINGS, **26**, 143–148 (1940), §5.

LA TOPOLOGIE ALGÉBRIQUE ET LA THÉORIE DE L'INTÉGRATION;

PAR HASSLER WHITNEY.
(Harvard University, Cambridge, Mass.)

1. Introduction. — Dans cette conférence, je me propose d'exposer des méthodes pour étudier la théorie de l'intégration, méthodes qui s'appliquent surtout à la géométrie différentielle, et aux théories où entre le théorème de Stokes. Le but le plus important est d'utiliser des éléments de caractère géométrique plutôt qu'analytique; en particulier, nous ferons très peu usage des systèmes de coordonnées (1).

Remarquons que la théorie de Cartan-De Rham (2) peut être présentée comme suit. Une forme X^r de degré r (élément d'intégration r-uple) dans une variété différentiable M sera dénommée *r-cochaîne différentiable;* l'intégrale $\int_{A^r} X^r$ de X^r sur un champ d'intégration, ou r-chaîne, A^r est bien définie.

Désignons respectivement par δX^r et ∂A^{r+1} la forme dérivée de X^r, et la frontière de A^{r+1}; on a alors la formule de Stokes,

$$\int_{\partial A^{r+1}} X^r = \int_{A^{r+1}} \delta X^r.$$

L'intégrale $\int_A X$ correspond au produit algébrique XA; en utilisant cette notation, la formule peut être écrite

$$X^r \partial A^{r+1} = \delta X^r A^{r+1},$$

(1) Les résultats principaux de cette théorie ont été présentés dans une Note : *Algebraic topology and integration theory* [Proceedings of the National Academy of Sciences (U. S. A.), vol. 33, 1947, p. 1-6].

(2) *Voir* G. DE RHAM, *Sur l'Analysis situs des variétés à n dimensions* (*Journal de Math., pures et appl.*, 1931, p. 115-200); *Relations entre la Topologie et la Théorie des intégrales multiples* (*L'Enseignement Mathématique*, 35e année, 1936, p. 213-228).

formule bien connue en topologie algébrique. Une cochaîne X^r est un *cocycle* si $\delta X^r = 0$; X^r et Y^r sont *cohomologues* si $Y^r - X^r = \delta Z^{r-1}$; les cocycles cohomologues forment les éléments du *groupe de cohomologie de dimension* r de la variété M. Le théorème fondamental de M. de Rham nous apprend que ce *groupe est isomorphe au groupe de cohomologie algébrique de* M *formé avec une subdivision polyédrale de* M; de plus, *les deux anneaux de cohomologie sont isomorphes.*

La théorie des formes de Cartan-De Rham donne ainsi un aspect géométrique à la topologie algébrique sur une variété différentiable M. Or, la théorie de l'intégration ne dépend pas du fait que les formes sont différentiables. Nous allons remplacer cette hypothèse par des conditions de Lipschitz; il en résultera une théorie de l'intégration qui équivaut à la théorie de la cohomologie algébrique, et qui est valable dans des espaces bien plus généraux, dans des *espaces de Lipschitz*, qui contiennent en particulier les polyèdres (*voir* ci-dessous). Tous les espaces considérés seront métriques; nous les supposerons compacts, quoique ce ne soit pas indispensable.

2. Représentations de Lipschitz. — Une représentation f d'un espace R dans un autre R′ sera dite *de Lipschitz* s'il existe un nombre N tel que

$$\rho[f(p), f(q)] \leqq N\rho(p, q) \qquad (p, q \in R).$$

Nous désignerons par *extension* $\mathcal{E}(f, R) = \mathcal{E}(f)$ de f la borne inférieure des nombres N.

Si f est une représentation d'un sous-ensemble de l'espace euclidien E^n dans R, nous définirons l'*extension réduite* $\mathcal{E}_*(f)$ comme suit. Soit Φ une représentation affine de E^n en lui-même, avec $\mu_n[\Phi(Q)] = \mu_n(Q)$; $\mathcal{E}_*(f)$ est alors la borne inférieure de $\mathcal{E}(f\Phi)$. [$\mu_n(Q) =$ mesure de Q.]

Un espace (métrique compact) R sera dit *espace de Lipschitz* s'il existe une homéomorphie f de R dans un espace euclidien E, telle que f et f^{-1} soient des représentations de Lipschitz, et telle qu'il existe une rétraction Φ de Lipschitz d'un voisinage $U \subset E$ de $R_1 = f(R)$ sur R_1; c'est-à-dire, Φ est une représentation de Lipschitz de U dans R_1, et $\Phi(p) = p$, $p \in R_1$. Pour ceci, il faut et il suffit que R_1 soit un espace *localement connexe de Lipschitz* dans toutes les dimensions.

3. Chaînes de Lipschitz. — Nous désignerons par $\mu_r(Q)$ la mesure r-dimensionnelle de Q, définie par exemple si Q est un ensemble mesurable dans E^r, ou si Q est un polyèdre métrique r-dimensionnel.

Nous désignerons par *simplexe de Lipschitz* dans R un couple (τ, f), τ étant un simplexe euclidien, et f une représentation de Lipschitz de τ dans R; s'il existe une représentation affine Φ de τ' sur τ telle que $f'(p') = f[\Phi(p')]$ $(p' \in \tau')$, nous poserons $(\tau, f) = (\tau', f')$. Nous écrirons aussi $(\tau, f) = f\tau$. Au moyen de ces simplexes de Lipschitz on peut construire des *simplexes singuliers de Lipschitz* par la méthode classique (*voir* par exemple Seifert-Threlfall, *Topologie*).

Nous appellerons *masse* $|f\tau^r|$ de τ^r la borne inférieure

$$|f\tau^r| = \inf \sum_i \mathcal{E}^r_*(f, \tau^r_i)\mu_r(\tau^r_i),$$

où $\mathfrak{S}\tau^r = \Sigma\tau^r_i$ est une subdivision arbitraire de τ^r en simplexes euclidiens τ^r_i. On définit de même la masse $|A|$ d'une chaîne $A = \Sigma a_i f\tau^r_i$ de Lipschitz (les a_i étant des nombres réels), et au moyen des identifications définissant les chaînes singulières, la masse $|A|_*$ d'une chaîne singulière de Lipschitz.

Si l'on utilisait l'égalité $\mathfrak{S}\tau^r = \Sigma Q_i$, les Q_i étant des ensembles mesurables de τ^r, on trouverait une définition de $|f\tau|$ de type de Lebesgue; mais l'étude des cochaînes de Lipschitz présenterait alors des difficultés.

4. Cochaînes de Lipschitz. — Nous appellerons *r-cochaîne de Lipschitz* dans l'espace R une fonction linéaire $X^r(A^r)$ (nous écrirons aussi XA) des r-chaînes A^r de Lipschitz, telle qu'il existe des nombres N, N' satisfaisant aux conditions suivantes :

$$(a) \qquad |X^r A^r| \leqq N|A^r|,$$

$$(b) \qquad |X^r \partial A^{r+1}| \leqq N'|A^{r+1}|.$$

Les valeurs de XA peuvent être des nombres réels, ou plus généralement, des éléments d'un espace de Banach. Nous désignerons par δX la cochaîne de Lipschitz définie par la formnle du paragraphe 1. Nous désignerons respectivement par $|X|$ et $|\delta X|$ les bornes inférieures des nombres N et N'.

Si f est une représentation de Lipschitz de l'espace R dans l'espace R', à chaque cochaîne de Lipschitz X' dans R' correspond une cochaîne de Lipschitz $f^* X'$ dans R, définie par la formule

$$f^* X' A = X' f A.$$

Nous pouvons démontrer les formules suivantes :

$$\delta f^* X' = f^* \delta X', \qquad |f^* X'| \leqq \mathcal{E}^r(f)\,|X'|.$$

Soit K un complexe, $\mathfrak{S}K$ une subdivision de K, et soient A_0, $\mathfrak{S}A_0$ respectivement une chaîne de K et la subdivision correspondante de A_0. Soit f une représentation de Lipschitz de K dans R. Construisons le produit cartésien $I \times K$ de l'intervalle-unité I par K, et posons $F(t, p) = f_t(p)$. Considérons aussi une subdivision $\mathfrak{S}(I \times K)$ telle que

$$\mathfrak{S}(O \times K) = K, \qquad \mathfrak{S}(I \times K) = \mathfrak{S}K.$$

Posons $A = f A_0$, $\mathfrak{S}A = f \mathfrak{S} A_0$. On a alors la formule

$$\partial F \mathfrak{S}(I \times A_0) = \mathfrak{S}A - A - F\mathfrak{S}(I \times \partial A_0).$$

D'où

$$|X\mathfrak{S}A - XA| \leqq |\delta X|\,|F\mathfrak{S}(I \times A_0)| + |X|\,|F\mathfrak{S}(I \times \partial A_0)|.$$

D'après la définition de F, on a $|\tau| = 0$ pour chaque simplexe entrant dans $\mathfrak{S}(I \times A_0)$ ou $\mathfrak{S}(I \times \partial A_0)$. Par suite $X\mathfrak{S}A = XA$. Il en résulte facilement qu'on peut définir XA pour une chaîne polyédrale A par la formule $XA = X\mathfrak{S}A$, puisque $X\mathfrak{S}A$ ne dépend pas de la subdivision choisie.

Si X^n est une cochaîne de Lipschitz dans l'espace euclidien E^n, on peut définir $X^n Q$ pour les ensembles ouverts et bornés Q, donc pour les ensembles mesurables. Il existe par suite une fonction bornée mesurable $D_X(p)$ telle que

$$X^n Q = \int_Q D_X(p)\, dp, \qquad |X^n| = \sup |D_X(p)|.$$

5. Cochaînes squelettes de Lipschitz. — Un *r-simplexe squelette* $\sigma^r = p_0 \ldots p_r$ dans l'espace métrique R est un ensemble de $r+1$ points $p_0, \ldots, p_r$ dans R. Choisissons r arêtes $p_{\lambda_1}p_{\mu_1}, \ldots, p_{\lambda_r}p_{\mu_r}$ de σ^r, de sorte que chaque paire de sommets de σ^r soient réunis par une suite de ces arêtes. Nous définissons alors

$$\mathrm{Pot}(\sigma^r) =' \frac{1}{r!} \min \rho(p_{\lambda_1}, p_{\mu_1}) \ldots \rho(p_{\lambda_r}, p_{\mu_r}).$$

Les *chaînes-squelettes* de R sont les formes linéaires $\Sigma a_i \sigma_i^r$. Une *cochaîne-squelette de Lipschitz* est une fonction linéaire $x \circ \mathrm{A}$ des chaînes squelettes de Lipschitz telle qu'il existe des nombres N, N' satisfaisant aux conditions suivantes :

$$(a) \qquad | x \circ \sigma^r | \leqq \mathrm{N}\, \mathrm{Pot}(\sigma^n),$$

$$(b') \qquad | x \circ \partial\sigma^{r+1} | \leqq \mathrm{N}'\, \mathrm{Pot}(\sigma^{r+1}).$$

On définit $|x|$, $|\delta x|$ comme précédemment.

En utilisant une *suite standard* $\mathfrak{S}_1\mathrm{K}$, $\mathfrak{S}_2\mathrm{K}$, ... de subdivisions de K, c'est-à-dire une suite telle que [3]

$$\frac{\mu_r(\tau^r)}{[\mathrm{diam}(\tau^r)]^r} \geqq \alpha > 0,$$

pour tous les simplexes τ^r de chaque $\mathfrak{S}_i\mathrm{K}$, on peut définir

$$\bar{x}\, \mathrm{A}^r = \lim_{i \to \infty} x \circ \mathfrak{S}_i \mathrm{A}^r.$$

Nous pouvons démontrer que *cette limite existe, indépendamment du choix des subdivisions;* de plus, *$\bar{x}$ est une cochaîne de Lipschitz, telle que*

$$|\bar{x}| \leqq |x|, \qquad |\delta\bar{x}| \leqq |\delta x|.$$

Nous pouvons remplacer A^r dans cette formule par une *chaîne de Lipschitz-Vietoris*. Inversement, *pour chaque* X *dans un espace de Lipschitz il existe une x telle que $\bar{x} = \mathrm{X}$.*

Nous dirons que x, y sont *équivalentes* si $\bar{x} = \bar{y}$. Pour ceci il suffit qu'il existe pour chaque $\varepsilon > 0$, un nombre $\eta > 0$ tel que

$$|y \circ \sigma - x \circ \sigma| \leqq \varepsilon\, \mathrm{Pot}(\sigma) \quad \text{si} \quad \mathrm{diam}(\sigma) < \eta.$$

---

[3] *Voir* H. Freudenthal, *Annals of Math.*, vol. 43, 1942, p. 580-582.

6. Cochaînes tensorielles de Lipschitz. — Pour simplifier les idées, nous étudierons seulement l'espace euclidien E^n. Une *r-cochaîne tensorielle de Lipschitz* est une fonction $T(p; v_1, \ldots, v_r)$ du point p et des vecteurs $v_1, \ldots, v_r$, qui est une fonction linéaire et alternée des vecteurs, telle qu'il existe des nombres N, N' satisfaisant aux conditions suivantes :

$$(a'') \qquad |T(p; v_1, \ldots, v_r)| \leqq N |v_1| \ldots |v_r|,$$
$$(b'') \qquad |T(p'; v_1, \ldots) - T(p; v_1, \ldots)| \leqq N' |p' - p| |v_1| \ldots |v_r|.$$

Nous faisons correspondre à T une cochaîne squelette de Lipschitz x comme suit. Soit $\sigma^r = p_0 \ldots p_r$ un simplexe euclidien dans E^n. Soient $v_1 = p_{\mu_1} - p_{\lambda_1}, \ldots, v_r = p_{\mu_r} - p_{\lambda_r}$ des vecteurs sur des arêtes de σ^r tels que $(v_1, \ldots, v_r)$ définissent l'orientation positive de σ^r. Soit p_σ le centre de gravité des sommets de σ^r. Nous posons alors

$$x \circ \sigma^r = \frac{1}{r!} T(p_\sigma; v_1, \ldots, v_r).$$

Si T est différentiable, on peut définir la dérivée alternée (différentielle extérieure de Grassman) δT; la cochaîne squelette correspondante est équivalente à δx.

On peut définir

$$\int_Q T = \int_Q \bar{x} = \bar{x} Q,$$

ce qui donne la théorie de l'intégration ordinaire.

7. Produits de cochaînes. — On définit pour commencer le produit $X^0 Y^s$ comme suit. Si $f\tau^s$ est un simplexe de Lipschitz dans R, alors $f^* Y^s$ est une cochaîne de Lipschitz dans τ^s, et $D_{f^*Y^s}$ est défini comme au paragraphe 4; nous posons alors

$$X^0 Y^s f\tau^s = \int_{\tau^s} X^0 f(p) D_{f^*Y^s}(p)\, dp.$$

Ce produit est une cochaîne de Lipschitz.

En général, il est possible de définir d'une manière unique le produit $X^r Y^s$ dans un espace de Lipschitz, de manière que les conditions suivantes soient satisfaites :

α. $X^r Y^s$ est une $(r+s)$-cochaîne de Lipschitz, linéaire dans X^r et Y^s.

β. Il existe un nombre N_{rs} tel que

$$|X^r Y^s| \leq N_{rs} |X^r| |Y^s|.$$

γ. On a la formule

$$\delta(X^r Y^s) = \delta X^r Y^s + (-1)^r X^r \delta Y^s.$$

On peut alors démontrer que

$$(XY)Z = X(YZ), \qquad Y^s X^r = (-1)^{rs} X^r Y^s,$$
$$f^*(XY) = f^* X f^* Y.$$

8. Le théorème de G. de Rham. — Dans un complexe K, nous pouvons faire correspondre à chaque cochaîne de Lipschitz X^r une cochaîne algébrique ΦX, par la formule

$$\Phi X(\sigma^r) = X \sigma^r.$$

On peut démontrer que la transformation Φ établit un isomorphisme entre les anneaux de cohomologie algébrique et les anneaux de cohomologie de Lipschitz.

De même la correspondance Φ définie par la formule

$$\Phi T(\sigma^r) = \int_{\sigma^r} T,$$

établit un isomorphisme entre les anneaux de cohomologie algébrique et les anneaux de cohomologie tensorielle.

Reprinted from Vol. I, PROCEEDINGS OF THE
INTERNATIONAL CONGRESS OF MATHEMATICIANS, 1950
Printed in U.S.A.

r-DIMENSIONAL INTEGRATION IN n-SPACE

HASSLER WHITNEY

1. Introduction. There are various elementary and fundamental questions in integration theory as applied to geometry and physics that are not covered by the modern theory of the Lebesgue integral. In particular, basic problems concerning integrals over domains of dimension less than that of the containing space, as functions of the domain, are largely untouched. We shall present here a general approach to this type of problem.[1]

For an example from physics, consider the flux through a surface S in Euclidean 3-space E^3. Cut S into small pieces $\sigma_1, \sigma_2, \cdots$; we find the flux through each σ_i, and add. Take a typical small piece σ, in the form of a parallelogram with vectors v_1, v_2 along two of its sides, and containing the point p. The flux $X(\sigma)$ through σ depends on p, the area of σ, and the direction of σ. If we change the direction of σ, $X(\sigma)$ varies; $X(\sigma)$ is approximately proportional (for small σ) to the cosine of the angle between the normal $v_1 \times v_2$ to σ and the direction of the flux through p. In particular, for a rotation through π, $X(\sigma)$ is replaced by its negative. Thus σ must be taken as oriented, say by ordering the vectors (v_1, v_2).

We may represent the flux by a vector function $\omega(p)$; then for small σ as above, $X(\sigma)$ is approximately $\omega(p) \cdot (v_1 \times v_2)$, and thus is linear and skew-symmetric in the vectors v_1, v_2 determining σ. More generally, for surface integrals in E^n, an integral $\int_S$ may be determined by naming a function $\omega(p)$ which, for each p, is a linear skew-symmetric function of v_1, v_2; that is, ω is a differential form, or a "2-form" for short. An r-dimensional integral may be determined similarly by an r-form.

Given $\omega(p)$, say with $r = 2$, the integral $\int_S \omega$ may be defined as follows. Cut S into small triangles (approximately) $\sigma_1, \cdots, \sigma_m$. S being oriented, let v_{i1}, v_{i2} be the vectors of two of the sides of σ_i, so that (v_{i1}, v_{i2}) orient σ_i like S. Let p_i be a point of σ_i. Then

$$\int_S \omega = \lim \frac{1}{2} \sum_i \omega(p_i; v_{i1}, v_{i2}),$$

taking the limit as the mesh of the subdivision of S approaches 0. The factor 1/2 is due to the use of triangles instead of parallelograms. The usual Riemann type theory may be set up in this fashion, essentially without use of coordinate systems.[2]

[1] Some of the theory given here has appeared in notes by the author, *Algebraic topology and integration theory*, Proc. Nat. Acad. Sci. vol. 33 (1947) pp. 1–6, and *La topologie algébrique et la théorie de l'intégration*, in the volume *Topologie algébrique*, Colloques Internationaux du Centre National de la Recherche Scientifique, Paris, 1949.

[2] This was carried out by P. Olum in a Senior Thesis at Harvard, 1940. See also H. Federer, *An introduction to differential geometry*, mimeographed notes, Stenographic Bureau, Brown University, Providence, R. I. (see Math. Reviews vol. 10 (1949) p. 264), and A. Lichnerowicz, *Algèbre et analyse linéaires*, Masson, Paris, 1947.

2. The general problem. We now ask what a general theory of integration should look like. For a given "integrand" X, the integral is a function of the domain A; we shall write it as $X \cdot A$. The basic property in the usual theory is additivity:

$$\int_{P \cup Q} = \int_P + \int_Q \quad \text{if } P \cap Q = 0.$$

In the general case, we may wish to integrate over domains with self-intersections. For instance, we may desire the line integral over a curve C consisting, in succession, of the oriented arcs C_1, C_2, C_3, C_2, C_4. We clearly should have

$$\int_C = \int_{C_1} + 2\int_{C_2} + \int_{C_3} + \int_{C_4}.$$

Thus a general r-dimensional domain is an "r-chain" $A = \sum a_i \sigma_i^r$, with oriented r-cells σ_i^r, each having a coefficient a_i (which we may take as any real number). Now $\int_A = \sum a_i \int_{\sigma_i^r}$, i.e.,

$$X \cdot \sum a_i \sigma_i^r = \sum a_i (X \cdot \sigma_i^r), \tag{2.1}$$

and the integral is a linear function of r-chains, generalizing the additivity above.

Recall the Theorem of Stokes: an r-form ω with differentiable coefficients has an exterior derivative $\delta\omega$; given an $(r + 1)$-chain A, with boundary ∂A,

$$\int_A \delta\omega = \int_{\partial A} \omega. \tag{2.2}$$

If X is an integral (abstractly defined) over r-chains, we may *define* an integral δX over $(r + 1)$-chains by the corresponding formula

$$\delta X \cdot A = X \cdot \partial A. \tag{2.3}$$

If we wish an integral $X \cdot A$ to be expressible in terms of A and local properties of X, such as is the case if X is defined by a differential form, we cannot let A be too general. For instance, for $r = 1$, there is a curve C and a differentiable function f in the plane, such that the partial derivatives of f vanish at all points of C, and yet f is not constant in C.[3] The natural assumption here is that C should be rectifiable. A similar assumption is in order for $r > 1$; see §8 below.

When we have chosen what domains should be allowable for setting up r-chains, we must decide how general the allowable linear functions X should be. Without some restrictions, little theory could be obtained. For instance, for $r = 0$, a typical domain is a point p; then $X \cdot p = \phi(p)$ would be an arbitrary function. In the usual Lebesgue theory, bounded measurable functions as integrands form a special class of summable functions; yet they illustrate much of the basic theory. We might invent some kind of corresponding bound, which we

[3] See H. Whitney, *A function not constant on a connected set of critical points*, Duke Math. J. vol. 1 (1935) pp. 514–517.

call the "mass" $|X|$, of X. It turns out that assuming both $|X|$ and $|\delta X|$ are bounded leads to a very satisfactory theory, to be described below; see (4.2). In particular, any X can be represented by a differential form (§6). An important problem is to obtain general results under weaker hypotheses than the boundedness of $|X|$ and $|\delta X|$.

3. Polyhedral chains. We introduce integration theory in a space R as follows. First choose a set of elements, "r-chains", forming a linear space; introduce a norm in this space; when completed, this gives a Banach space $\mathbf{C}^r$. Then the space of "r-cochains" X is the conjugate space $\bar{\mathbf{C}}^r$ of $\mathbf{C}^r$. The function $X(A) = X \cdot A$ is the "integral" of X over A. By making the norm in $\mathbf{C}^r$ small, we enlarge the set of elements which can be in $\mathbf{C}^r$ (i.e., for which the norm exists); the corresponding norm in $\bar{\mathbf{C}}^r$ is then large, which restricts the elements occurring in $\bar{\mathbf{C}}^r$.

We shall remain mostly in Euclidean space E^n; later we consider briefly the case of manifolds and more general spaces. The norms in $\mathbf{C}^r$ and $\bar{\mathbf{C}}^r$ will depend on the metric of E^n; but the sets of elements in these spaces are independent of the metric.

Among r-chains in E^n we must certainly include polyhedral r-chains $\sum a_i\sigma_i^r$, each σ_i^r and its boundary cells being flat. Define the *mass* by

$$(3.1) \qquad |\sum a_i\sigma_i^r| = \sum |a_i||\sigma_i^r|$$

(the σ_i^r non-overlapping), $|\sigma_i^r|$ denoting the r-dimensional volume of σ_i^r.

The mass is too large a norm for our purposes. For instance, for $r = 0$, using $|p| = 1$ shows that any bounded function would define an element of $\bar{\mathbf{C}}^0$. Consider the case $r = n$ for a moment. If we take a domain P and translate it by a small vector v, giving $Q = T_vP$, then $\int_P$ and $\int_Q$ are nearly the same. For a general r, we might require $\int_{\sigma^r}$ and $\int_{T_v\sigma^r}$ to be nearly the same, even though v need not lie in the r-plane of σ^r. In terms of chains, we may require the norm $|A|^\#$ to satisfy

$$(3.2) \qquad |T_v\sigma^r - \sigma^r|^\# \leqq |v||\sigma^r|/(r+1).$$

This does not relate r-cells in different r-planes. To take care of this, assume that for any $(r+1)$-cell σ^{r+1}, whose boundary is an r-chain $\partial\sigma^{r+1}$, we have $|\partial\sigma^{r+1}|^\# \leqq N|\sigma^{r+1}|$ for some fixed N. There is a uniquely defined largest norm satisfying these conditions, and also $|A|^\# \leqq |A|$. We call it the *tight norm*. It is independent of N for N large.

With this norm the conjugate space corresponds exactly to what we formerly called the set of "tensor cochains" in E^n. (See §5 below.) Though these cochains X are easy to work with, they are not general enough for some purposes. For instance, f^*X, defined in §9, is not tightly Lipschitz.

The most important norm for our purposes is the *Lipschitz norm* $|A|^*$, intermediate in size between $|A|$ and $|A|^\#$. It is the largest norm satisfying the conditions

(3.3) $$|\sigma^r|^* \leqq |\sigma^r|, \qquad |\partial\sigma^{r+1}|^* \leqq |\sigma^{r+1}|.$$

It is easy to find an explicit expression for $|A|^*$, as follows. Considering all polyhedral $(r+1)$-chains D, $|A|^*$ is the greatest lower bound

(3.4) $$|A|^* = \operatorname*{GLB}_D (|A - \partial D| + |D|).$$

For example, let σ be a segment of length a, and let σ' be $T_v\sigma$, $|v| = b$, b small. Let D be the parallelogram formed by carrying σ into σ', oriented so that $\partial D = \sigma' - \sigma + C$, where C is composed of the remaining short sides of D. Then $|D| \leqq ab$, $|C| = 2b$, and hence

$$|\sigma' - \sigma|^* \leqq |C| + |D| \leqq (a+2)b.$$

Of course $|\sigma' - \sigma| = 2a$; also $|\sigma' - \sigma|^\# = ab/2$ for b small.

Given A, taking $D = 0$ in (3.4) shows that

(3.5) $$|A|^* \leqq |A|.$$

If $A^r = \partial B^{r+1}$, choosing D_2^{r+2} so that

$$|B - \partial D_2| + |D_2| < |B|^* + \epsilon$$

and setting $D_1 = B - \partial D_2$ gives

$$|A - \partial D_1| + |D_1| = |D_1| < |B|^* + \epsilon;$$

this proves

(3.6) $$|\partial B|^* \leqq |B|^* \leqq |B|.$$

In particular, (3.3) is proved.

For some special dimensions, considering $aq - bp$ as a 0-chain and $q - p$ as a vector, we have

$$r = 0:\ |1q - 1p|^* = |1q - 1p|^\# \leqq |q - p|,$$

with the sign $=$ if $|q - p| \leqq 2$, and

$$r = n:\ |A^n|^* = |A^n|,$$

since there are no nontrivial $(n+1)$-chains in E^n.

If $A \neq 0$, then $|A|$, $|A|^*$, and $|A|^\#$ are all >0.

For any σ^r, $|\sigma^r| = |\sigma^r|^* = |\sigma^r|^\#$. In fact, $|A| = |A|^* = |A|^\#$ if $A = \sum a_i\sigma_i^r$, the σ_i^r are parallel and similarly oriented, and the a_i are $\geqq 0$.

4. Lipschitz cochains. A Lipschitz r-cochain in E^n (or in a subset R of E^n) is an element of the conjugate space $\bar{\mathbf{C}}^r$ of $\mathbf{C}^r$ (or of $\mathbf{C}^r(R)$, using chains in R only). The Lipschitz norm $|X|^*$ is the norm of X in $\bar{\mathbf{C}}^r$; the mass $|X|$, corresponding to $|A|$, is defined. The definitions are

(4.1) $$|X|^* = \operatorname{LUB}_{|A|^*=1} |X \cdot A|, \qquad |X| = \operatorname{LUB}_{|A|=1} |X \cdot A|.$$

Though $\mathbf{C}^r$ is separable, $\bar{\mathbf{C}}^r$ is not; hence $\mathbf{C}^r$ is not reflexive.

Note that $\delta\delta X = 0$; for $\delta\delta X \cdot A^{r+2} = \delta X \cdot \partial A = X \cdot \partial\partial A = 0$.

The following relation shows that $X \in \bar{\mathbf{C}}^r$ if and only if $|X|$ and $|\delta X|$ are finite:

(4.2) $$|X|^* = \max(|X|, |\delta X|).$$

To prove this, note first that $|X| \leqq |X|^*$, because of (3.5) and (4.1). Next, (2.3), (4.1), and (3.6) give, using any B^{r+1},

$$|\delta X \cdot B| = |X \cdot \partial B| \leqq |X|^* |\partial B|^* \leqq |X|^* |B|^*;$$

hence δX is a Lipschitz $(r + 1)$-cochain, and $|\delta X| \leqq |X|^*$. This proves $\geqq$ in (4.2). To prove the reverse inequality, we need merely show that for any polyhedral r-chain A,

(4.3) $$|X \cdot A| \leqq \max(|X|, |\delta X|)\, |A|^*.$$

Given $\epsilon > 0$, choose $D = D^{r+1}$ so that

$$|A - \partial D| + |D| < |A|^* + \epsilon.$$

Then if $C = A - \partial D$,

$$|X \cdot A| \leqq |X \cdot C| + |X \cdot \partial D| \leqq |X|\,|C| + |\delta X|\,|D|$$
$$\leqq \max(|X|, |\delta X|)\,(|A|^* + \epsilon),$$

which gives (4.3).

Consider the case $r = 0$. For any Lipschitz 0-cochain X, $w(p) = X \cdot p$ is a real-valued function. The relations

$$|w(p)| = |X \cdot p| \leqq |X|\,|p| = |X|,$$
$$|w(q) - w(p)| = |X \cdot (1q - 1p)| = |\delta X \cdot (pq)| \leqq |\delta X|\,|q - p|$$

show that w is bounded and satisfies a Lipschitz condition. Conversely, any such w defines an X.

Now take $r = n$. With E^n oriented, $X \cdot Q$ is defined for polyhedral regions Q oriented like E^n. Since $|X \cdot Q| \leqq |X|\,|Q|$, X is extendable to be an additive set function over measurable sets, satisfying the same inequality. Define the "fullness" of a set Q by

(4.4) $$\Theta_n(Q) = |Q|/[\operatorname{diam} Q]^n.$$

By standard Lebesgue theory, there is a measurable function $D_X(p)$, $|D_X(p)| \leqq |X|$, such that

(4.5) $$X \cdot Q = \int_Q D_X(p)\, dp \qquad (Q \text{ oriented like } E^n).$$

Moreover, using sequences of cells $\sigma_1, \sigma_2, \cdots$ containing p and with

$$\Theta_n(\sigma_i) \geqq \eta > 0$$

for all i,

$$D_X(p) = \lim_{i\to\infty} X\cdot\sigma_i/|\sigma_i| \qquad \text{a. e. in } E^n. \tag{4.6}$$

5. The λ-norms. Suppose, instead of (3.3), we require

$$|\sigma^r|_\lambda^* \leqq |\sigma^r|, \qquad |\partial\sigma^{r+1}|_\lambda^* \leqq |\sigma^{r+1}|/\lambda. \tag{5.1}$$

We then obtain the *Lipschitz λ-norm*, with the property

$$|A|_\lambda^* = \operatorname*{GLB}_D (|A - \partial D| + |D|/\lambda). \tag{5.2}$$

The corresponding λ-norm for cochains satisfies

$$|X|_\lambda^* = \max(|X|, \lambda|\delta X|). \tag{5.3}$$

We obtain $|A|_\lambda^\#$ similarly, using $|v|\,|\sigma^r|/(r+1)\lambda$ in (3.2). Define the *Lipschitz constant* of X by

$$\mathfrak{L}(X) = \mathrm{LUB}\,\frac{|X\cdot(T_v\sigma - \sigma)|}{|\sigma|\,|v|}; \tag{5.4}$$

call X *tightly Lipschitz* (a "tensor cochain" in[1]) if this is finite. If this holds for the Lipschitz r-cochain X, then

$$|\delta X| \leqq (r+1)\mathfrak{L}(X), \tag{5.5}$$

$$|X|_\lambda^\# = \max[|X|, (r+1)\lambda\mathfrak{L}(X)]. \tag{5.6}$$

The sets of elements in the spaces $\overline{\mathbf{C}}_\lambda^r$ for various λ are the same; only the norms differ; similarly for $\overline{\mathbf{C}}_\lambda^{r\#}$.

Given σ^r, it is not hard to construct a tightly Lipschitz X vanishing outside an arbitrary neighborhood of σ, such that $|X| = 1$, $X\cdot\sigma = |\sigma|$. Then for λ small enough, $|X|_\lambda^\# = 1$. Using this, we may prove, for polyhedral A,

$$\lim_{\lambda\to 0} |A|_\lambda^\# = \lim_{\lambda\to 0} |A|_\lambda^* = |A|. \tag{5.7}$$

From this we may prove that $|A|$ is a lower semicontinuous function of polyhedral chains A in $\mathbf{C}^r$ (in fact, in any $\mathbf{C}_\lambda^r$ or $\mathbf{C}_\lambda^{r\#}$). For a general A in $\mathbf{C}^r$, *define* $|A|$ as the lower bound of $\liminf |A_i|$ for sequences of polyhedral chains A_i with $|A_i - A|^* \to 0$; it may be infinite. $|A|$ is still lower semicontinuous.

For $r = 0$, $|A|_\lambda^\# = |A|_\lambda^*$.

6. Lipschitz cochains and differential forms. We present here the theorem of Wolfe,[4] that the Lipschitz r-cochains in an open subset R of E^n correspond to

[4] J. H. Wolfe, *Tensor fields associated with Lipschitz cochains*, Harvard Thesis, 1948.

the differential forms in R satisfying certain conditions. Suppose X is given. At any point p we wish to find a corresponding linear skew-symmetric function $\omega(p;\, v_1, \cdots, v_r)$, or equivalently, a linear function $\omega(p)\cdot\alpha$ of contravariant r-vectors α. The natural definition is the following. The set of points

$$p + \sum_{i=1}^{r} t_i v_i \qquad (0 \leqq t_i \leqq t,\, i = 1, \cdots, r)$$

forms a parallelopiped σ_t, oriented by the ordered set $(v_1, \cdots, v_r)$. Set

$$\omega(p;\, v_1, \cdots, v_r) = \lim_{t\to 0} X\cdot\sigma_t/t^r, \tag{6.1}$$

if this exists. (Cf. (4.6).) Then for almost all p in R, this exists and defines a linear function $\omega(p)$; for fixed α, $\omega(p)\cdot\alpha$ is measurable. Also $|\,\omega(p;\, v_1, \cdots, v_r)\,| \leqq |\,X\,|\,|\,v_1\,| \cdots |\,v_r\,|$. The same is true for $\delta\omega$, defined from δX. Moreover, for all σ,

$$X\cdot\sigma = \int_\sigma \omega(p;\, e_1, \cdots, e_r)\, dp, \tag{6.2}$$

$(e_1, \cdots, e_r)$ being an orthonormal set in the plane of σ, oriented like σ.

The theorem is proved by first smoothing X by an averaging process (giving a tightly Lipschitz cochain), in which case the corresponding form is more easily found, and then passing to the limit. The reason $\omega(p)$ turns out to be linear may be illustrated for the case $r = 1$ as follows. Given vectors u, v, set

$$q_t = p + tu, \qquad q_t' = q_t + tv = p + t(u + v);$$

set $\sigma_t = pq_tq_t'$. For small t, $\omega(p)\cdot tu = \int_{pq_t}\omega$ approximately, etc.; thus

$$\begin{aligned} |\,\omega(p)\cdot tu + \omega(p)\cdot tv - \omega(p)\cdot t(u + v)\,| &=_{ap} |\,X\cdot\partial\sigma_t\,| \\ &= |\,\delta X\cdot\sigma_t\,| \leqq t^2\,|\,\delta X\,|\,|\,\sigma_1\,|\,; \end{aligned}$$

dividing by t and letting $t \to 0$ gives the result.

Conversely, suppose $\omega(p)$ is defined and linear a. e. in R, with $\omega(p)\cdot\alpha$ measurable for each α; suppose $|\,\omega(p)\,|$ is bounded as above, and $|\int_{\partial\sigma}\omega\,| \leqq N\,|\,\sigma\,|$ for each $(r + 1)$-cell σ for which the integral is defined. Such a form we call a *Lipschitz r-form* in R. (The conditions can be weakened; for instance, we need merely continuity in α of $\omega(p)\cdot\alpha$ for simple unit r-vectors α, not linearity.) Then there is a corresponding Lipschitz r-cochain X in R, for which (6.2) holds whenever the integral is defined.

Say ω_1, ω_2 are *equivalent* if for each α, $\omega_1(p)\cdot\alpha = \omega_2(p)\cdot\alpha$ a. e. in R. Then *the Lipschitz r-cochains in R correspond exactly to the classes of equivalent Lipschitz r-forms in R.*

Note that for any Lipschitz r-form ω in R, we may find the corresponding Lipschitz r-cochain X, take δX, and find the corresponding $\delta\omega$, even if the components of ω are not differentiable; whenever they are, this definition of $\delta\omega$ agrees with the analytic one.

A "simple" contravariant r-vector is one expressible as a product of r vectors. Define the "simple norm" $|\xi|_s$ of a covariant r-vector ξ by

$$(6.3) \qquad |\xi|_s = \text{LUB}\,|\xi\cdot\alpha|, \qquad \alpha \text{ simple}, \qquad |\alpha| = 1.$$

Then for any X and corresponding ω,

$$(6.4) \qquad |X| = \text{essential}\ \underset{p}{\text{LUB}}\ |\omega(p)|_s\,.$$

7. General r-chains. Recall that $\mathbf{C}^r$ was the completion of the space of polyhedral r-chains; the new elements of $\mathbf{C}^r$ we call "general Lipschitz r-chains".

Consider first the case $r = 0$. Let R be a locally compact separable metric space. An additive set function Φ in R assigns to each Borel set Q a number $\Phi(Q)$, such that $\Phi(\sum Q_i) = \sum\Phi(Q_i)$ if the Q_i are disjoint. Any such set function defines uniquely a general Lipschitz 0-chain A_Φ, as follows. Let $R = R_1 \cup \cdots \cup R_m$ be a partition of R into disjoint Borel sets; choose $p_i \in R_i$, and set

$$(7.1) \qquad B = \sum\Phi(R_i)p_i\,;$$

this is an approximation to A_Φ by a polyhedral 0-chain, if the R_i are "small" enough (i.e. cut up Φ finely enough).

The mass $|A_\Phi|$ is the total variation of Φ. The expression for $|A_\Phi|^*$ is more complicated. Not all 0-chains are expressible in this manner; but the 0-chains described are dense in $\mathbf{C}^0$. For any Lipschitz 0-cochain X, corresponding to the function $\omega(p)$,

$$(7.2) \qquad X\cdot A_\Phi = \int_R \omega(p)\, d\Phi(p),$$

using the Lebesgue Stieltjes integral. The integral is defined for more general functions ω; on the other hand, $X\cdot A$ is defined for more general A.

Of course $\Phi(Q)$ is defined without regard to orientation properties. This corresponds to the fact that a 0-cell, i.e., a point, has a natural orientation. We may consider the theory of additive set functions and the Lebesgue Stieltjes integral as 0-dimensional integration; r-dimensional integration for $r > 0$ requires orientation properties.

Consider next n-chains in E^n. Recall the L^1-norm for summable functions F:

$$(7.3) \qquad |F| = \int_{E^n} |F(p)|\, dp.$$

If F is cellwise constant for some subdivision of E^n, it defines an n-chain A with $|A|^* = |A| = |F|$; this shows that summable functions define elements of $\mathbf{C}^n$; in fact, they give the whole of $\mathbf{C}^n$.

Take the case $n = 1$. If F is not only summable but differentiable, it is easy to show that $-dF/dx$ defines a 0-chain which is exactly the boundary (see below) of A_F. On the other hand, suppose

$$F_a(x) = \log |x|, \qquad x < 0; \qquad F_a(x) = \log x + a, \qquad x > 0;$$

then $-dF_a(x)/dx = -1/x$, all $x \neq 0$. For each a, F_a defines a 1-chain A_a, and the same function $-1/x$ (which is not summable) "corresponds" to ∂A_a.

The boundary ∂A of a general chain A may always be defined. For if A is defined by the sequence $A_1, A_2, \cdots$ of polyhedral chains A_i, then

$$|\partial A_j - \partial A_i|^* \leqq |A_j - A_i|^*$$

by (3.6), and hence $\partial A_1, \partial A_2, \cdots$ is a Cauchy sequence, and defines ∂A.

Now take $0 < r < n$, in E^n. Let $\alpha(p)$ be any field of contravariant r-vectors, the components being summable functions. It may be shown that α defines uniquely a general Lipschitz r-chain A_α, with the property that for any X, corresponding to ω,

$$X \cdot A_\alpha = \int_{E^n} \omega(p) \cdot \alpha(p)\, dp. \tag{7.4}$$

Such A_α we call "spread out" r-chains. They are dense in $\mathbf{C}^r$; this holds even if we require the components to be analytic functions. (This holds in open subsets of E^n.)

If the components $\alpha^{\lambda_1 \cdots \lambda_r}$ of α are differentiable, and

$$\beta^{\lambda_1 \cdots \lambda_{r-1}} = \sum_k \partial \alpha^{\lambda_1 \cdots \lambda_{r-1} k}/\partial x_k, \tag{7.5}$$

then

$$\partial A_\alpha = (-1)^r A_\beta. \tag{7.6}$$

Define the "mass" of a contravariant r-vector α by

$$|\alpha|_m = \text{GLB} \sum |\alpha_i|, \tag{7.7}$$

for expressions $\alpha = \sum \alpha_i$ of α as a sum of simple r-vectors. (Then the norms $|\alpha|_m$ and $|\xi|_s$ of (6.3) correspond, considering the spaces of covariant and contravariant r-vectors as conjugate spaces of each other.) Then

$$|A_\alpha| = \int |\alpha(p)|_m\, dp. \tag{7.8}$$

There seems to be no simple expression for $|A_\alpha|^*$.

8. Lipschitz r-chains. In the applications, one integrates commonly over domains which are not polyhedral (curved surfaces, etc.). The domains are generally expressible as images of flat domains, under mappings f which are differentiable or piecewise differentiable. We shall assume merely that f satisfies a Lipschitz condition: If ρ denotes distance,

$$\mathfrak{L}_f = \text{LUB}\, \rho(f(p), f(q))/\rho(p, q) \text{ is finite.} \tag{8.1}$$

If f is a Lipschitz mapping of an r-dimensional polyhedron P, in which an r-chain A is given, into a metric space R, we call (A, f), or fA for short, a *Lip-*

schitz r-chain in R. We may as well take the cells of P as disjoint cells in E^r. With E^r oriented, A may be replaced by a summable function φ (see §7); φ defines a general r-chain $\bar{\varphi}$ in E^r; we call $(\bar{\varphi}, f) = f\bar{\varphi}$ a *Lipschitz-Lebesgue r-chain* in R.

If R is a metric space satisfying certain conditions, Lipschitz chains in R can be used to set up integration theory. The mass $|A|$ and norm $|A|^*$ can be defined; see below. Particularly important is the case that R is a smooth manifold M; if a metric is not given, one may be introduced, and the spaces of chains and cochains (though not the norms) are independent of the metric (at least in compact subsets of M).

A Lipschitz-Lebesgue r-chain $f\bar{\varphi}$ in E^n is a general Lipschitz chain, given by approximating φ by a cellwise constant function φ' and f by a simplicial mapping f' (thus defining an approximating polyhedral chain). The following continuity theorem holds: Given $f\bar{\varphi}$ and numbers L and $\epsilon > 0$, there is a $\zeta > 0$ with the following property. For any φ' with $|\bar{\varphi} - \bar{\varphi}'|^* \leqq \zeta$ and for any f' with $\mathfrak{L}_{f'} \leqq L$ and $|f'(p) - f(p)| \leqq \zeta$ (all p), we have $|f'\bar{\varphi}' - f\bar{\varphi}|^* < \epsilon$. For given polyhedral $A = \bar{\varphi}$, and L, keeping $\varphi' = \varphi$, we may take $\zeta = c\epsilon$ for some c ($c = L^r|A| + L^{r-1}|\partial A|$); but this is not possible in the general case.

A Lipschitz chain A, being a general chain, has a boundary ∂A; a Lipschitz $(r-1)$-form ω defines X and hence δX. General theory now gives Stokes' Theorem:

$$\int_{\partial A} \omega = X \cdot \partial A = \delta X \cdot A = \int_A \delta\omega.$$

It is always possible to represent a Lipschitz-Lebesgue chain in E^n as $f\bar{\varphi}$ with f one-one in the carrier $\mathrm{Car}(\varphi)$ of φ, i.e., the set of points p where $\varphi(p) \neq 0$. For such a representation, the mass is given by the formula

$$(8.2) \qquad |f\bar{\varphi}| = \int |\varphi(p)|\,|J_f(p)|\,dp,$$

$J_f(p)$ being the Jacobian, which exists a.e. since f is Lipschitz (Rademacher's Theorem). For any representation, (8.2) holds with $\leqq$. In the proof, we make use of the existence of a tightly Lipschitz cochain X such that

$$(8.3) \qquad |X| = 1, \qquad |f\bar{\varphi}| - \epsilon < X \cdot f\bar{\varphi} \leqq |f\bar{\varphi}|,$$

for arbitrary $\epsilon > 0$.

Let $S_1, S_2, \cdots$ be a sequence of subdivisions of E^r with mesh $\to 0$. Then given $f\bar{\varphi}$, if $\bar{\varphi}_{ki}$ is the part of $\bar{\varphi}$ in the cell σ_{ki} of S_k, we have $\sum_i \bar{\varphi}_{ki} = \bar{\varphi}$, and

$$(8.4) \qquad \lim_{k\to\infty} \sum_i |f\bar{\varphi}_{ki}|^* = |f\bar{\varphi}|.$$

Given $f\bar{\varphi}$ in E^n, with f one-one in $\mathrm{Car}(\varphi)$, and given $\eta > 0$ and $\epsilon > 0$, there is a $\zeta > 0$ with the following property. Let S be any simplicial subdivision of P of mesh $< \zeta$, whose simplexes σ_i satisfy $\Theta_r(\sigma_i) \geqq \eta$. Let φ' in σ_i be the average of φ in σ_i, and let f' be the simplicial mapping which coincides with f at the vertices of S. Then

$$(8.5) \qquad \big|\, |f\bar{\varphi}| - |f'\bar{\varphi}'| \,\big| < \epsilon.$$

Given a Lipschitz mapping f of a subset Q of E^r into a metric space R, define the "reduced Lipschitz constant"

$$(8.6) \qquad \bar{\mathfrak{L}}_{f,Q} = \operatorname*{GLB}_{\varphi} \mathfrak{L}_{f\varphi^{-1}(Q)},$$

where φ is an affine volume-preserving mapping of E^r into itself, using f in Q and φ in $\varphi^{-1}(Q)$. Define the "local constant" $\mathfrak{L}_f^*(p)$ of f at p as follows. Let $\rho_r a^r$ be the volume of the set $x_1^2 + \cdots + x_r^2 \leqq a$. Let $U_\zeta(p)$ denote the ζ-neighborhood of p. Given ζ and η such that $\eta < \rho_r \zeta^r$, set

$$(8.7) \qquad \mathfrak{L}_f^*(p, \zeta, \eta) = \operatorname*{GLB}_{Q} \bar{\mathfrak{L}}_{f,Q}, \qquad Q \subset U_\zeta(p), \qquad |Q| > \eta,$$

$$(8.8) \qquad \mathfrak{L}_f^*(p) = \lim_{\mu \to 0} \ \liminf_{\zeta \to 0} \ \mathfrak{L}_f^*[p, \zeta, (1 - \mu)\rho_r \zeta^r].$$

If f is a Lipschitz mapping of a measurable subset Q of E^r into E^n, then

$$(8.9) \qquad |J_f(p)| = [\mathfrak{L}_f^*(p)]^r \text{ a. e. in } Q.$$

Now the set of $(\bar{\varphi}, f)$, with certain equivalence relations, forms the space of Lipschitz r-chains A in R. For any such A, considering the various expressions $(\bar{\varphi}, f)$ of A, define

$$(8.10) \qquad |A| = \operatorname{GLB} \int |\varphi(p)| \, [\mathfrak{L}_f^*(p)]^r \, dp.$$

Actually (with some restrictions on R) we may choose $(\bar{\varphi}, f)$ so that f is one-one in $\operatorname{Car}(\varphi)$; then $|A|$ equals the integral above. In particular, taking $\varphi(p) = 1$ gives the "r-volume" of the "rectifiable" manifold $f(\operatorname{Car}(\varphi))$ in R. The definitions agree with the previous ones if R is an open subset of E^n.

The norm $|A|^*$ is definable from $|A|$; then cochains may be introduced as before.

9. Properties of Lipschitz mappings. Let f be a Lipschitz mapping of E^n into E^m. Then each polyhedral r-chain A in E^n is carried into a Lipschitz r-chain fA in E^n. The definition may be extended to general Lipschitz r-chains in E^n. The usual properties hold. In particular,

$$(9.1) \qquad |fA| \leqq \mathfrak{L}_f^r |A|, \quad |fA|^* \leqq \max(\mathfrak{L}_f^r, \mathfrak{L}_f^{r+1}) \, |A|^*.$$

Given X in E^m, f^*X in E^n is defined by $f^*X \cdot A = X \cdot fA$. Then $|f^*X| \leqq \mathfrak{L}_f^r |X|$, etc.

If ω is a Lipschitz r-form in E^m, it defines a Lipschitz r-cochain X, this defines f^*X, and this gives ω^* in E^n; call this $f^*\omega$. Recall that X defines δX and hence $\delta\omega$; also $\delta f^*X = f^*\delta X$; hence $\delta f^*\omega = f^*\delta\omega$. The usual analytic theory requires the differentiability of both f and ω for any parts of this last formula even to be defined.

Nevertheless, with the "Lipschitz" hypotheses used here, an analytic treatment of $f^*\omega$ can be given. We first note a fundamental difficulty: ω may fail to be defined in a set of measure 0 in E^m, and f might even map the whole of E^n into this set. Thus the first necessity is to enlarge or improve the definition of ω. This is done by determining the corresponding X, and finding a new ω' from X; then $\omega' = \omega$ a.e. (see §6). Then the usual analytic formula defines $f^*\omega'$ a.e. in E^n.

The proof requires the following approximation theorem. Let X be a Lipschitz r-cochain in E^m, and let p be any point at which the corresponding $\omega(p)$ exists and is linear (see §6). Then for any $\eta > 0$ and $\epsilon > 0$ there is a $\zeta > 0$ with the following property. Let σ be any oriented r-simplex with

$$\sigma \subset U_\zeta(p), \qquad \Theta_r((p) \cup \sigma) \geqq \eta: \tag{9.2}$$

if $\{\sigma\}$ denotes the r-vector defined by σ, then

$$| X \cdot \sigma - \omega(p) \cdot \{\sigma\} | \leqq \epsilon | \sigma |. \tag{9.3}$$

As a particular case, suppose $r = 1$, and X is the coboundary δY of a Lipschitz 0-cochain Y, Y corresponding to the real-valued function $w(p)$. Then the theorem gives the Rademacher Theorem on the total differentiability a.e. of the Lipschitz function w.

HARVARD UNIVERSITY,
CAMBRIDGE, MASS., U. S. A.

GEOMETRIC INTEGRATION THEORY

Introduction

The purpose of this preliminary chapter is first of all to provide motivation for the methods and tools appearing in the book, and secondly to illustrate some of the general considerations through a study of special cases, particularly in three dimensions. The Introduction and the body of the book are independent of each other; however, the meaning of the full theory will become clearer if the Introduction is read in conjunction with the rest of the book.

In Part A (which is rather abstract in character), we ask what a theory of r-dimensional integration in n-space should look like. An "integral" $X \cdot \sigma$ is defined over an oriented r-cell σ for instance, and changes sign if the orientation is reversed. We may now define $X \cdot (a\sigma) = a(X \cdot \sigma)$ for real numbers a, and $X \cdot (A + B) = X \cdot A + X \cdot B$; assuming that a subdivision of cells does not affect the integral, we now have a linear function defined over "polyhedral r-chains." To give this linear function analytical properties, we introduce some continuity hypotheses. Next we study the local nature of integration. Near any point, in any r-dimensional direction, the integral over a cell is approximately proportional to the r-dimensional volume of the cell (with the strongest continuity hypothesis); this fact leads to the construction of a point function $D_X(p)$ which acts on "r-vectors" $\{\sigma\}$ of oriented r-cells σ. These r-vectors must have certain simple properties, which in turn lead directly to the construction of Grassmann algebra. Finally, D_X becomes a differential r-form, whose integral $\int_\sigma D_X$ over any σ equals the original $X \cdot \sigma$.

In Part B, we start with the elements of Grassmann algebra as derived above, and work out from a geometric point of view some of the fundamentals of calculus. We consider vector analysis in three dimensions, differentials, Jacobians, transformation of "multiple integrals," manifolds, and the Theorems of Stokes and de Rham.

The purpose of Part C is to introduce the reader to some of the general methods in the later parts of the book. In the last two sections, we touch on the way some particular modes of integration may be considered as r-dimensional for different r.

A. The General Problem of Integration

1. The integral as a function of the domain. For an integration theory there must certainly be various possible "domains of integration."

Whatever kind of process integration (with real values) is, a definite "integrand" X will give a real number when applied to a permissible domain A. Thus, for a fixed X, we have a real valued function of domains A; we denote the value on A by $X \cdot A$. We consider integration in Euclidean space.

If we are to call the integration r-dimensional, we must certainly include among the permissible domains the simplest r-dimensional figures. An r-cell σ, consisting of a closed bounded part of an r-plane, bounded by a finite number of pieces of $(r-1)$-planes, is such a figure. We make our first hypothesis:

HYPOTHESIS (H_1). The integral over σ depends on the orientation of σ; a reversal of orientation reverses the sign of the integral.

We discuss the meaning of and reason for this hypothesis. A line segment σ^1 has two end points p and q; the two orientations of σ^1 are the two directions along σ^1, from p to q and from q to p; we may denote the two oriented cells by $pq = -qp$ and by $qp = -pq$ respectively. They may be defined by a choice of the vector $q - p$ (from p to q) or $p - q$ (from q to p). We may orient a triangle $p_0p_1p_2$ by choosing an ordered pair of independent vectors in it; for instance, the pair $(p_1 - p_0, p_2 - p_0)$. Interchanging these or reversing the direction of either would reverse the orientation. Similarly, an r-cell σ is oriented by the choice of an ordered set of r independent vectors in it. A 0-cell, i.e. a single point, has no orientation properties.

The triangle $\sigma = p_0p_1p_2$, oriented as above, has a boundary $\partial\sigma$, consisting of the oriented segments p_0p_1, p_1p_2, and p_2p_0. The boundary $\partial(pq)$ of the oriented segment pq consists of the point q counted positively, and the point p counted negatively. The boundary $\partial\sigma$ of an r-cell σ contains its $(r-1)$-faces, properly oriented (App. II, 5).

For any oriented cell σ, let $-\sigma$ denote the oppositely oriented cell; then (H_1) may be written in the form

$$X \cdot (-\sigma) = -X \cdot \sigma. \tag{1}$$

We consider some examples. Let ϕ be a real valued function defined in 3-space E^3, and let C be an oriented curve, from the point p to the point q. If we integrate the rate of change of ϕ along C, we obtain $\phi(q) - \phi(p)$. If the orientation of C were reversed, we would obtain $\phi(p) - \phi(q)$. Next, consider any 1-dimensional integral $\int_A \omega$. Let $\sigma = p_0p_1p_2$ be an oriented triangle, cut into the two oriented triangles $\sigma' = p_0p'p_2$, $\sigma'' = p'p_1p_2$, by the segment $p'p_2$, with p' in p_0p_1. Then

$$\int_{\partial\sigma'} \omega + \int_{\partial\sigma''} \omega = \int_{\partial\sigma} \omega; \tag{2}$$

for

$$\int_{p_0p'} \omega + \int_{p'p_1} \omega = \int_{p_0p_1} \omega, \qquad \int_{p'p_2} \omega + \int_{p_2p'} \omega = 0,$$

the latter since $p'p_2$ and p_2p' are oppositely oriented. This type of relation is fundamental for geometric properties of the integral; it would be impossible if orientation properties were disregarded.

Of course integration over non-oriented domains is possible and sometimes of importance; in this case, much more general types of domains are permissible, but the geometric properties are largely lost. We prefer to think of such integration as 0-dimensional; see § 26 below.

The most typical requirement of integration theory is additivity, or in the present exposition, *invariance under subdivisions* (used in the example above):

HYPOTHESIS (H_2). If the oriented r-cell σ is cut into similarly oriented r-cells $\sigma_1, \cdots, \sigma_m$, then

$$X\cdot\sigma = X\cdot\sigma_1 + \cdots + X\cdot\sigma_m. \tag{3}$$

2. Polyhedral chains. We wish to write the boundary $\partial\sigma$ of the triangle $\sigma = p_0p_1p_2$ as a domain of integration. With the point p' in p_0p_1 as in § 1, we would like to write $\partial\sigma$ in various ways, such as

$$\partial\sigma = p_0p_1 + p_1p_2 + p_2p_0 = p_0p' + p'p_2 - p_0p_2 + p'p_1 + p_1p_2 - p'p_2$$

etc. This suggests the definition of a *polyhedral r-chain A* as being a linear combination of oriented r-cells, with real numbers as coefficients, together with the properties

$$1\sigma = \sigma, \qquad 0\sigma = 0, \qquad a(-\sigma) = (-a)\sigma = -(a\sigma), \tag{1}$$

and invariance under subdivision: if the oriented cell σ is cut into $\sigma_1, \cdots, \sigma_m$, then σ and $\sigma_1 + \cdots + \sigma_m$ are the same polyhedral chain. The definitions of aA (for real numbers a) and $A + B$ are obvious; the set of polyhedral r-chains now forms a linear space.

Because of (H_1) and (H_2), it is possible to define $X\cdot A$ for any polyhedral r-chain $A = \sum a_i\sigma_i$, by the relation

$$X\cdot\sum a_i\sigma_i = \sum a_i(X\cdot\sigma_i); \tag{2}$$

X is now a *linear function of polyhedral r-chains*. For this reason, we call X an *r-cochain*. That X is a cochain is equivalent to assuming that $X\cdot\sigma$ is defined, with the properties (H_1) and (H_2).

The boundary of $A = \sum a_i\sigma_i$ is defined to be $\partial A = \sum a_i\,\partial\sigma_i$; this is easily seen to be a well defined polyhedral $(r-1)$-chain. A polyhedral 0-chain is an expression $A^0 = \sum a_ip_i$, the p_i being points; we set $\partial A^0 = 0$.

3. Two continuity hypotheses. For a satisfactory integration theory, the permissible domains must include oriented curved r-cells for instance. One should be able to obtain these as limits of polyhedral r-chains, and the integral should be definable as the limit of the integrals over the approximating polyhedral chains. This requires some continuity hypotheses on the integral. We give two hypotheses in this section with which a satisfactory general theory may be obtained; if we include also the hypothesis of the next section, the integral has simpler analytical properties.

Let σ be an oriented triangle, of area $|\sigma|$. If we cut the plane containing σ into small rectangles and let $\tau_1, \cdots, \tau_m$ be those contained in σ, then these τ_i fill up most of σ, and it is natural to require that $\sum X\cdot\tau_i$ be near $X\cdot\sigma$. This will follow from:

HYPOTHESIS (H_1'). Given the r-cochain X, there is a number N_1 such that

$$|X\cdot\sigma| \leqq N_1 |\sigma|, \qquad \text{all oriented } r\text{-cells } \sigma, \tag{1}$$

where $|\sigma|$ is the r-dimensional volume of σ.

This is of course a stronger hypothesis than needed for the above requirement. We assume it largely for the sake of the analytical methods described in § 6 below.

Take the case $r = 0$. We may consider the "0-dimensional volume" of a point p to be 1. Any 0-cochain X corresponds to a real function ϕ: $\phi(p) = X\cdot p$ for all points p. Now (H_1') says that $|\phi(p)| \leqq N_1$, all p; that is, ϕ is bounded. These functions are too general; we must restrict them further. We look for a hypothesis suggested by the case $r = 1$.

Take the above triangle σ again; let τ be the union of the τ_i. We may consider the boundary $\partial\tau$ of τ as an approximation to $\partial\sigma$, even though it is made up of segments not parallel to the sides of σ in general. Taking $r = 1$, we may require that $X\cdot\partial\tau$ be near $X\cdot\partial\sigma$, i.e. that $X\cdot\partial(\sigma - \tau)$ be small, as a result of the area of $\sigma - \tau$ being small.

HYPOTHESIS (H_2'). Given the r-cochain X, there is a number N_2 such that

$$|X\cdot\partial\sigma^{r+1}| \leqq N_2 |\sigma^{r+1}|, \qquad \text{all oriented } (r+1)\text{-cells } \sigma^{r+1}. \tag{2}$$

Note that (H_2') is trivially satisfied if $r = n$.

For $r = 0$, (H_2') says that for any oriented segment pq,

$$|X\cdot q - X\cdot p| = |X\cdot\partial(pq)| \leqq N_2 |q - p|, \tag{3}$$

$|q - p|$ being the length of pq; that is, the function $\phi(p) = X\cdot p$ satisfies a Lipschitz condition.

Any cochain satisfying (H_1') and (H_2') we call a *flat* cochain.

4. A further continuity hypothesis. If the cell σ is moved into a nearby position σ', we may assume that $X\cdot\sigma'$ is near $X\cdot\sigma$. We shall consider rigid motions without turning; that is, translations by means of vectors v. Let $T_v\sigma$ denote the new cell. Our hypothesis is that $X\cdot T_v\sigma$ differs from $X\cdot\sigma$ by at most some fixed multiple of the r-volume $|\sigma|$ times the distance $|v|$ of translation.

HYPOTHESIS (H_3'). Given the r-cochain X, there is a number N_3 such that for any oriented r-cell σ and vector v,

$$(1) \qquad |X \cdot T_v\sigma - X \cdot \sigma| \leq N_3 |\sigma| |v|.$$

In the case $r = 0$, this hypothesis is equivalent to (H_2'). For $r = n$, it is non-trivial, whereas (H_2') is trivial.

Any cochain satisfying all three hypotheses we call *sharp*. We shall see that sharp cochains correspond to differential forms. (This holds also in the flat case; see Chapter IX.)

5. Some examples. We help elucidate some of the hypotheses through the study of a steady flow of fluid in oriented 3-space E^3. Take any oriented 2-cell σ. Let (v_1, v_2) define its orientation, and choose a vector v_3 so that (v_1, v_2, v_3) (or equivalently, (v_3, v_1, v_2)) defines the given orientation of E^3; then the *positive* direction through σ is the direction in the sense of v_3. Let $X \cdot \sigma$ be the quantity (positive or negative) of fluid flowing through σ in the positive direction in unit time; this is the *flux* across σ. Clearly (H_1) and (H_2) hold; hence X is a 2-cochain. Of course $X \cdot S$ for any oriented surface S is definable.

If the density of fluid and the velocity of flow are bounded, then clearly (H_1') holds. Now take any 3-cell τ. If fluid is being neither created nor destroyed, then (we assume the density constant in time) the total rate of flow out of τ, which equals $X \cdot \partial\tau$, must be 0. In general, $X \cdot \partial\tau$ equals the total rate of creation of fluid in τ. Hence (H_2') is equivalent to assuming that the total rate of creation per unit volume is bounded.

With the same flow of fluid, consider the *circulation* along an oriented curve C. At a point p of C, if $u(p)$ is the unit tangent vector at p in the positive direction along C, $v(p)$ is the velocity vector of the fluid at p, and $\rho(p)$ is the density at p, then the circulation is

$$(1) \qquad Y \cdot C = \int_C \rho v \cdot u$$

(compare (18.3) below). Again (H_1) and (H_2) hold.

Hypothesis (H_1') will follow from the boundedness of v and ρ. Given an oriented 2-cell σ, $Y \cdot \partial\sigma$ is the circulation around the boundary $\partial\sigma$. Taking arbitrarily small cells σ near a point p, we see that (H_2') will follow if curl $(v) = \nabla \times v$ exists and is finite; compare (21.4).

Suppose the flow is through a pipe. Then the above cochains X and Y are not defined throughout E^3, but only in the region of flow; the hypotheses need be assumed only in this region.

6. The case $r = n$. For an n-cochain X which satisfies (H_1') in oriented E^n ((H_2') is satisfied trivially), it is standard Lebesgue theory that there is a bounded measurable function Φ such that (using the Lebesgue integral)

$$(1) \qquad X \cdot \sigma = \int_\sigma \Phi, \qquad \text{all } n\text{-cells } \sigma^n \text{ oriented like } E^n.$$

We consider briefly the simpler case when (H_3') is also satisfied. Given the point p, let $\sigma_1, \sigma_2, \cdots$ be a sequence of n-cells oriented like E^n, in smaller and smaller neighborhoods of p. Set

$$(2) \qquad \Phi(p) = \lim_{i \to \infty} \frac{X \cdot \sigma_i}{|\sigma_i|}.$$

We indicate the proof of existence and uniqueness of the limit. Let $\tau_1, \tau_2, \cdots$ be a similar sequence of cubes. Suppose i_0 and k_0 are such that for $i \geqq i_0$ and $k \geqq k_0$, σ_i and τ_k are in a neighborhood of p of diameter $<\epsilon$. We may take k so large that translations of τ_k nearly fill up σ_i:

$$\sigma_i = \sigma' + R, \qquad \sigma' = T_{v_1}\tau_k + \cdots + T_{v_s}\tau_k, \qquad |R| \text{ small.}$$

By (H_3'),

$$\left| \frac{X \cdot \sigma'}{|\sigma'|} - \frac{X \cdot \tau_k}{|\tau_k|} \right| \leqq \frac{1}{s} \sum_{j=1}^{s} \left| \frac{X \cdot T_{v_j}\tau_k - X \cdot \tau_k}{|\tau_k|} \right| \leqq N_3 \epsilon,$$

and the statement follows, using (H_1').

Using (H_3') again, it is clear that Φ satisfies

$$(3) \qquad |\Phi(p + v) - \Phi(p)| \leqq N_3 |v|,$$

and that (1) holds (using the Riemann integral). Moreover,

$$(4) \qquad |X \cdot \sigma - |\sigma| \Phi(p_0)| = \left| \int_\sigma [\Phi(p) - \Phi(p_0)]\, dp \right| \leqq N_3 \zeta |\sigma|$$

if all points of σ are within ζ of p_0.

7. The r-vector of an oriented r-cell. Let X be a sharp r-cochain in E^n. Then for each oriented r-plane P in E^n, we may consider $X \cdot \sigma$ for r-cells in P, and hence find a function Φ_P in P as in § 6. For any point p, the values of $\Phi_P(p)$ for the various oriented r-planes P through p are of interest.

We shall use a closely related function. Given σ and p, let P be the r-plane through p parallel to σ and oriented like σ, and set

$$(1) \qquad D_X(p) \cdot \{\sigma\} = |\sigma| \Phi_P(p).$$

We must give meaning to $D_X(p)$, to $\{\sigma\}$, and to their combination. With the σ_i as in § 6, (1) and (6.2) give

$$(2) \qquad D_X(p) \cdot \{\sigma\} = \lim_{i \to \infty} \frac{|\sigma|}{|\sigma_i|} X \cdot \sigma_i.$$

As a function of σ, the right hand side is known as soon as we know the set of r-planes parallel with σ, the orientation of σ, and $|\sigma|$; we call this triple the *r-vector* $\{\sigma\}$ of σ. We may now define $D_X(p)$ to be that real valued function defined on all r-vectors of oriented r-cells which is given by (1). (Later $D_X(p)$ will be taken to be defined on more special spaces $\mathbf{T}_r$.)

We remark that $D_X(p)\cdot\{\sigma\}$ is unchanged if we alter the metric in E^n; for using (2), we see that $|\sigma|/|\sigma_i|$ is unchanged, as is $X\cdot\sigma_i$.

Given $\{\sigma\}$ and the real number $a \neq 0$, let $a\{\sigma\}$ denote $\{\sigma'\}$ for any oriented σ' parallel with σ, oriented like or opposite to σ according as $a > 0$ or $a < 0$, and such that $|\sigma'| = |a||\sigma|$. If σ_1 and σ_2 are parallel, say $\{\sigma_2\} = a\{\sigma_1\}$; set

$$\{\sigma_1\} + \{\sigma_2\} = (1 + a)\{\sigma_1\},$$

for $a \neq -1$. If we include a "zero r-vector" 0, then with these definitions the r-vectors associated with a fixed set of parallel r-planes form a linear space isomorphic with the real numbers. In any such space of r-vectors, we see easily from (2) that $D_X(p)$ is linear:

$$(3) \qquad D_X(p)\cdot(a\{\sigma\} + b\{\sigma'\}) = aD_X(p)\cdot\{\sigma\} + bD_X(p)\cdot\{\sigma'\}.$$

8. On r-vectors and boundaries of $(r + 1)$-cells. Let us immerse the set of all r-vectors of oriented r-cells in a linear space $\mathbf{S}_r$ (of infinite dimension for $0 < r < n$) as follows. Take a fixed point p_0. An element α of $\mathbf{S}_r$ is a finite set of distinct r-planes $P_1, \cdots, P_m$ through p_0, together with an r-vector α_i associated with each P_i; we may include extra planes P_j, if we associate the zero r-vector with them. We form $a\alpha$ by replacing each α_i by $a\alpha_i$. Given α and β, we may take enough planes $P_1, \cdots, P_m$ so that α and β are defined by α_i and β_i in P_i respectively $(i = 1, \cdots, m)$; let $\alpha + \beta$ be defined by $\alpha_i + \beta_i$ in P_i. Clearly $\mathbf{S}_r$ is independent of the choice of p_0. The linear spaces described in the last section are linear subspaces of $\mathbf{S}_r$.

We may extend $D_X(p)$ to be a linear function in $\mathbf{S}_r$ by defining

$$(1) \quad D_X(p)\cdot\alpha = \sum_{i=1}^{m} D_X(p)\cdot\alpha_i \quad \text{if} \quad \alpha \text{ is defined by } \alpha_i \text{ in } P_i \ (i = 1, \cdots, m).$$

Now take any oriented $(r + 1)$-cell τ. Its boundary is an r-chain $\sigma_1 + \cdots + \sigma_m$. We wish the sum of the corresponding r-vectors to be 0:

$$(2) \quad \{\sigma_1\} + \cdots + \{\sigma_m\} = 0 \quad \text{if} \quad \sigma_1 + \cdots + \sigma_m = \partial\tau \text{ for some } \tau.$$

Requiring all such relations to hold turns $\mathbf{S}_r$ into a linear space $\mathbf{T}_r$. We shall call any element of $\mathbf{T}_r$ an *r-vector*. (Strictly speaking, let $\mathbf{S}_r'$ denote the linear subspace of $\mathbf{S}_r$ generated by all such elements $\{\sigma_1\} + \cdots + \{\sigma_m\}$; then $\mathbf{T}_r$ is the quotient space.) We now turn $D_X(p)$ into a linear function in the space $\mathbf{T}_r$, by letting its value on an element of $\mathbf{T}_r$ be its value on any corresponding element of $\mathbf{S}_r$. To show that this is possible, we must prove the following relation:

$$(3) \quad D_X(p)\cdot\{\sigma_1\} + \cdots + D_X(p)\cdot\{\sigma_m\} = 0 \quad \text{if} \quad \sigma_1 + \cdots + \sigma_m = \partial\tau.$$

We may suppose p is in τ. Given $\lambda > 0$, let us contract E^n towards p by the factor λ; then τ becomes τ_λ and σ_i becomes $\sigma_{\lambda i}$, and we have

$$\partial\tau_\lambda = \sum \sigma_{\lambda i}, \qquad |\tau_\lambda| = \lambda^{r+1}|\tau|, \qquad |\sigma_{\lambda i}| = \lambda^r |\sigma_i|.$$

Take a sequence $\lambda_1, \lambda_2, \cdots \to 0$. We may use $\sigma_{\lambda_1 i}, \sigma_{\lambda_2 i}, \cdots$ in (7.2); with the help of (7.1), we find

$$D_X(p)\cdot\{\sigma_i\} = \lim_{k\to\infty} \frac{|\sigma_i|}{|\sigma_{\lambda_k i}|} X\cdot\sigma_{\lambda_k i} = \lim_{k\to\infty} \frac{1}{\lambda_k^r} X\cdot\sigma_{\lambda_k i}.$$

Also, by (H_2'),

$$\left|\frac{1}{\lambda_k^r}\sum_i X\cdot\sigma_{\lambda_k i}\right| = \frac{1}{\lambda_k^r}|X\cdot\partial\tau_{\lambda_k}| \leqq \frac{1}{\lambda_k^r} N_2 |\tau_{\lambda_k}| = \lambda_k N_2 |\tau|.$$

These relations give (3).

9. Grassmann algebra. We shall find a special manner of writing elements of $\mathbf{T}_r$. Take any ordered set $(v_1, \cdots, v_r)$ of independent vectors. Let σ be the parallelepiped with a point p as vertex and with these vectors along the edges from p, oriented by the v_i. We define the symbol $v_1 \vee \cdots \vee v_r$ by

(1) $$v_1 \vee \cdots \vee v_r = \{\sigma\}.$$

Now any element of $\mathbf{T}_r$ can be written as a sum of such elements.

From the properties of § 7, we see that the "product" $v_1 \vee \cdots \vee v_r$ is skew symmetric:

(2) $$v_2 \vee v_1 = -v_1 \vee v_2, \quad \text{and hence} \quad v \vee v = 0.$$

Also

(3) $$v_1 \vee \cdots \vee (av_i) \vee \cdots \vee v_r = a(v_1 \vee \cdots \vee v_i \vee \cdots \vee v_r).$$

We shall prove that it is linear in v_1:

(4) $$(v_1 + v_1') \vee v_2 \vee \cdots \vee v_r = v_1 \vee v_2 \vee \cdots \vee v_r + v_1' \vee v_2 \vee \cdots \vee v_r;$$

hence it is linear in all the v_i.

If v_1' is in the r-plane determined by $v_1, \cdots, v_r$, this is a simple geometric fact about addition of volumes. We assume this is not the case, and consider a few values of r.

For $r = 1$, a 1-vector is now represented by a vector; the 1-vector $\{pq\}$ of the oriented segment pq is represented by the vector $q - p$. Relation (4) says that addition of 1-vectors (appearing on the right) is equivalent to addition of vectors (appearing on the left). We show this as follows. Choose a point p_0, and define the points and triangle

$$p_1 = p_0 + v_1, \qquad p_2 = p_1 + v_1' = p_0 + (v_1 + v_1'), \qquad \sigma = p_0p_1p_2.$$

By (8.2),

$$\{p_0p_1\} + \{p_1p_2\} + \{p_2p_0\} = 0, \quad \text{hence} \quad \{p_0p_2\} = \{p_0p_1\} + \{p_1p_2\};$$

this is the required relation.

For $r = 2$, take the above points p_i, and also the points $q_i = p_i + v_2$ $(i = 0, 1, 2)$. Set $\sigma' = q_0q_1q_2$. The p_i and q_i are vertices of a 3-cell τ whose faces are σ and σ', and also three parallelograms σ_{01}, σ_{12}, σ_{02}, where σ_{ij} contains p_ip_j. We have clearly

$$\{\sigma\} = \{\sigma'\} = \tfrac{1}{2}v_1 \vee v_1',$$

$$\{\sigma_{01}\} = v_1 \vee v_2, \qquad \{\sigma_{12}\} = v_1' \vee v_2, \qquad \{\sigma_{02}\} = (v_1 + v_1') \vee v_2.$$

With proper regard to orientations, we see that

$$\partial\tau = \sigma' - \sigma + \sigma_{01} + \sigma_{12} - \sigma_{02};$$

applying (8.2) gives (4) for $r = 2$. The general case is similar.

The set of all vectors in E^n forms a vector space $V = V(E^n)$. Let $e_1, \cdots, e_n$ be a base in V. Then any vector v can be written uniquely as $\sum v^i e_i$; the v^i are the components of v. Since we consider 1-vectors and vectors as the same, the e_i form a base in $\mathbf{T}_1$. For $r = 2$, using (2), (3) and (4) gives

$$v_1 \vee v_2 = \sum_{i,j=1}^{n} v_1^i v_2^j (e_i \vee e_j) = \sum_{i<j} \begin{vmatrix} v_1^i & v_2^i \\ v_1^j & v_2^j \end{vmatrix} e_i \vee e_j. \tag{5}$$

It may be shown that the $e_{ij} = e_i \vee e_j$ $(i < j)$ are independent in $\mathbf{T}_2$; by (5), they form a base in $\mathbf{T}_2$. Similarly, the $e_{\lambda_1 \cdots \lambda_r} = e_{\lambda_1} \vee \cdots \vee e_{\lambda_r}$ $(\lambda_1 < \cdots < \lambda_r)$ form a base in $\mathbf{T}_r$, and any r-vector α can be written uniquely in the form

$$\alpha = \sum_{\lambda_1 < \cdots < \lambda_r} \alpha^{\lambda_1 \cdots \lambda_r} e_{\lambda_1 \cdots \lambda_r}. \tag{6}$$

The $\alpha^{\lambda_1 \cdots \lambda_r}$ are the "components" of α.

It follows that $\mathbf{T}_r$ is of dimension $\binom{n}{r}$. In particular, $\mathbf{T}_n$ is of dimension 1, with base element $e_{1 \cdots n}$, and each $\mathbf{T}_k$ for $k > n$ contains the zero element only.

Through the definition

$$(v_1 \vee \cdots \vee v_r) \vee (v_{r+1} \vee \cdots \vee v_{r+s}) = v_1 \vee \cdots \vee v_{r+s}, \tag{7}$$

we have a bilinear multiplication between $\mathbf{T}_r$ and $\mathbf{T}_s$, with product in $\mathbf{T}_{r+s}$. If we include the space $\mathbf{T}_0$ = the real numbers, the system of the $\mathbf{T}_r$ with these operations is the *Grassmann algebra* of V. Let us denote $\mathbf{T}_r$ by $V_{[r]}$.

For $n \leqq 3$, any r-vector α (for any $r > 0$) equals $\{\sigma\}$ for some oriented r-cell σ. This is not the case for $n \geqq 4$; for instance, $e_{12} + e_{34}$ cannot be written in this form. Any α of the form $\{\sigma\}$ is a *simple* r-vector.

10. The dual algebra. The set of linear functions f in a vector space V forms a vector space, with the following definitions: the function af has the value $af(v)$ at v; the function $f+g$ has the value $f(v)+g(v)$ at v. This space is the *conjugate space* $\bar{V}$ of V.

Since $D_X(p)$ is a linear function in the vector space $\mathbf{T}_r = V_{[r]}$, we may consider it as being an element of the conjugate space of $V_{[r]}$; we denote this conjugate space by $V^{[r]}$. We call its elements *r-covectors*; the elements of $\bar{V} = V^{[1]}$ are *covectors*. We shall find a special manner of representing elements of $V^{[r]}$.

Let $e_1, \cdots, e_n$ be a base in V. Setting

$$e^i \cdot v = e^i \cdot (v^1 e_1 + \cdots + v^n e_n) = v^i \tag{1}$$

defines a linear function e^i in V, i.e. an element of $\bar{V}$. The elements $e^1, \cdots, e^n$ are easily seen to form a base in the base *dual* to the e_i. Now any element f of $\bar{V}$ may be written uniquely as $\sum f_i e^i$; the f_i are the components of f. Hence also $\bar{V}$ and V are of the same dimension; since $V^{[r]}$ and $V_{[r]}$ are conjugate, they are also of the same dimension.

The e^i may be defined by the relation

$$\begin{aligned} e^i \cdot e_j = \delta^i_j &= 1 \quad \text{if} \quad i = j, \\ &= 0 \quad \text{if} \quad i \neq j. \end{aligned} \tag{2}$$

Let us define base elements $e^{\lambda_1 \cdots \lambda_r}$ in $V^{[r]}$ by the same formula:

$$e^{\lambda_1 \cdots \lambda_r} \cdot e_{\lambda_1 \cdots \lambda_r} = 1 \qquad (\lambda_1 < \cdots < \lambda_r), \tag{3}$$

and $e^{\lambda_1 \cdots \lambda_r} \cdot e_{\mu_1 \cdots \mu_r} = 0$ for other μ_i $(\mu_1 < \cdots < \mu_r)$.

As a consequence of (2), we have

$$f(v) = f \cdot v = \sum_{i,j=1}^{n} f_i v^j e^i \cdot e_j = \sum_{i=1}^{n} f_i v^i; \tag{4}$$

similarly, writing

$$\xi = \sum_{\lambda_1 < \cdots < \lambda_r} \xi_{\lambda_1 \cdots \lambda_r} e^{\lambda_1 \cdots \lambda_r}, \tag{5}$$

the $\xi_{\lambda_1 \cdots \lambda_r}$ are the components of ξ, and we have, using (9.6),

$$\xi(\alpha) = \xi \cdot \alpha = \sum_{\lambda_1 < \cdots < \lambda_r} \xi_{\lambda_1 \cdots \lambda_r} \alpha^{\lambda_1 \cdots \lambda_r}. \tag{6}$$

Of course a change of base results in a change in components. For $n = 3$, $r = 2$, (5), (9.6) and (6) read

$$\xi = \xi_{12} e^{12} + \xi_{13} e^{13} + \xi_{23} e^{23}, \qquad \alpha = \alpha^{12} e_{12} + \alpha^{13} e_{13} + \alpha^{23} e_{23}, \tag{7}$$

$$\xi \cdot \alpha = \xi_{12} \alpha^{12} + \xi_{13} \alpha^{13} + \xi_{23} \alpha^{23}. \tag{8}$$

We wish to define expressions like $f^1 \vee \cdots \vee f^r$, the f^i being in $\bar{V}$ and the result being in $V^{[r]}$. We wish this multiplication to be skew symmetric and linear in each variable, and we wish to have $e^{\lambda_1}\vee\cdots\vee e^{\lambda_r} = e^{\lambda_1\cdots\lambda_r}$. This determines the multiplication. For instance, for $r = 2$,

$$f\vee g = \sum_{i,j=1}^{n} f_i g_j e^i\vee e^j = \sum_{i<j} \begin{vmatrix} f_i & f_j \\ g_i & g_j \end{vmatrix} e^{ij}. \tag{9}$$

The value of $(f^1\vee\cdots\vee f^r)\cdot(v_1\vee\cdots\vee v_r)$ is the determinant with the elements $f^i\cdot v_j$. For instance, for $r = 2$,

$$(f\vee g)\cdot(v\vee w) = \begin{vmatrix} f\cdot v & f\cdot w \\ g\cdot v & g\cdot w \end{vmatrix}. \tag{10}$$

To show this, we note that both sides of (10) are skew symmetric and linear in the quantities v, w; hence it is sufficient to prove this in the particular case $v = e_k$, $w = e_l$, $k < l$. But this follows at once from (9) and (3). Note that if we use (9) and (9.5) to obtain $(f\vee g)\cdot(v\vee w)$, comparing with (10) gives the Lagrange identity

$$\begin{vmatrix} \sum f_i v^i & \sum f_i w^i \\ \sum g_i v^i & \sum g_i w^i \end{vmatrix} = \sum_{i<j} \begin{vmatrix} f_i & f_j \\ g_i & g_j \end{vmatrix} \begin{vmatrix} v^i & v^j \\ w^i & w^j \end{vmatrix}. \tag{11}$$

11. Integration of differential forms. A *differential r-form* ω in E^n is a function whose values $\omega(p)$ are r-covectors. Hence, for any p and r-vector α, $\omega(p)\cdot\alpha$ is a real number. If ω is continuous, we may define its integral over any oriented r-cell σ in E^n as follows. Take a fine subdivision of σ into oriented r-cells $\sigma_1, \cdots, \sigma_s$; choose a point p_i in each σ_i; form the sum

$$\sum \omega(p_i)\cdot\{\sigma_i\}; \tag{1}$$

take the limit of this, using a sequence of subdivisions with diameters of cells approaching 0. *It is not necessary that E^n be metric or oriented.*

For a sharp r-cochain, D_X is a continuous differential form, and

$$X\cdot\sigma = \int_\sigma D_X = \lim \sum D_X(p_i)\cdot\{\sigma_i\}. \tag{2}$$

For if P is the r-plane of σ, oriented like σ, and we choose a metric in E^n, we may apply (7.2). See (V, 10).

B. Some Classical Topics

12. Grassmann algebra in metric oriented n-space. In metric E^n, scalar products $u\cdot v$ of vectors are defined. Take any vector u. The function $\phi_u(v) = u\cdot v$ (we write also $\phi_u\cdot v$) of vectors v is linear; hence ϕ_u is a definite element of the conjugate space $\bar{V}$ of V. This is easily shown to define an isomorphism between V and $\bar{V}$. Thus any linear function ψ in V can be written in the form $\psi(v) = u\cdot v$ for a unique vector u. Since $V_{[r]}$

and $V^{[r]}$ are conjugate for each r, a definite choice of metric in $V_{[r]}$ (see (I, 12.1)) establishes a definite isomorphism between $V_{[r]}$ and $V^{[r]}$.

Now take E^n to be oriented also. Then there is a definite "unit n-vector" α_0, which equals $\{\sigma\}$ for any n-cell σ of unit volume, oriented like E^n. For any orthonormal base $(e_1, \cdots, e_n)$ (consisting of perpendicular unit vectors defining the given orientation of E^n), we clearly have

$$\alpha_0 = e_{1\cdots n} = e_1 \vee \cdots \vee e_n. \tag{1}$$

There is also a definite "unit n-covector" ω_0 such that

$$\omega_0 = e^{1\cdots n} = e^1 \vee \cdots \vee e^n, \qquad \omega_0 \cdot \alpha_0 = 1. \tag{2}$$

Given any r-vector α, set

$$\Phi(\beta) = \omega_0 \cdot [\alpha \vee \beta], \qquad \text{all } (n-r)\text{-vectors } \beta. \tag{3}$$

Now Φ is a linear function in $V_{[n-r]}$ and hence is a definite element of $V^{[n-r]}$. Thus we have a definite isomorphism between $V_{[r]}$ and $V^{[n-r]}$.

13. The same, $n = 3$. Applying the results of the last section shows that in metric oriented E^3, all spaces $V_{[r]}$ and $V^{[r]}$ are isomorphic in a definite way either to $V_{[1]} = V$ or to $V_{[0]} =$ the reals. The isomorphism between $V = V_{[1]}$ and $\bar{V} = V^{[1]}$ is given by the scalar product; we consider the isomorphisms between V and $V^{[2]}$ and between V and $V_{[2]}$.

Take any vector v; set

$$\Psi_v(\alpha) = \omega_0 \cdot (v \vee \alpha), \qquad \text{all 2-vectors } \alpha. \tag{1}$$

Since Ψ_v is a linear function of α, it is a 2-covector.

Next, take any 2-vector α; set

$$\Theta_\alpha \cdot w = \omega_0 \cdot (\alpha \vee w), \qquad \text{all vectors } w. \tag{2}$$

Since this function of w is linear, it is given by the scalar product of w by a definite vector Θ_α.

We use this to define the *vector product* of vectors:

$$u \times v = \Theta_{u \vee v}; \tag{3}$$

thus $u \times v$ is defined by

$$(u \times v) \cdot w = \omega_0 \cdot (u \vee v \vee w), \qquad \text{all vectors } w. \tag{4}$$

Since all operations on the right of (4) are linear, $u \times v$ is a bilinear function of u and v. Since $v \vee u = -u \vee v$ and $u \vee u = 0$, we have

$$v \times u = -u \times v, \qquad u \times u = 0. \tag{5}$$

Let (e_1, e_2, e_3) be an orthonormal base giving the orientation chosen in E^3. Let each of u, v, w be each of e_1, e_2, e_3 in (4) in turn. As one example, we have

$$(e_1 \times e_3) \cdot e_2 = \omega_0 \cdot (e_1 \vee e_3 \vee e_2) = -\omega_0 \cdot \alpha_0 = -1. \tag{6}$$

In this manner we see that

(7) $$e_1 \times e_2 = e_3, \qquad e_1 \times e_3 = -e_2, \qquad e_2 \times e_3 = e_1.$$

Working out $u \times v = \sum_{i,j} u^i v^j e_i \times e_j$ gives

(8) $$u \times v = \begin{vmatrix} u_2 & u_3 \\ v_2 & v_3 \end{vmatrix} e_1 - \begin{vmatrix} u_1 & u_3 \\ v_1 & v_3 \end{vmatrix} e_2 + \begin{vmatrix} u_1 & u_2 \\ v_1 & v_2 \end{vmatrix} e_3.$$

We shall find a geometric interpretation of the vector product. Given u and v, choose an orthonormal base (e_1, e_2, e_3) orienting E^3 properly so that u is in the e_1-direction and v is in the plane of e_1 and e_2, on the same side of e_1 that e_2 is. Say

$$u = ae_1, \qquad v = b'e_1 + be_2; \qquad \text{then} \quad a \geqq 0, \qquad b \geqq 0.$$

Then applying (7) gives $u \times v = abe_3$. Thus, if u and v are independent, in which case a and b are $\neq 0$, then $u \times v$ is a vector perpendicular to both, and oriented so that $(u, v, u \times v)$ gives the orientation of E^3. Otherwise, $u \times v = 0$. Let σ be the parallelogram with a point p as vertex and sides along u and v. Then $|\sigma| = ab$, and hence the length of $u \times v$ is

(9) $$|u \times v| = |\sigma|.$$

We shall find the components of the 2-covector Ψ_v in (1) in an orthonormal coordinate system. Using (10.5) and (10.3), we have for instance

$$(\Psi_v)_{13} = \Psi_v(e_{13}) = \omega_0 \cdot \left(\sum v^i e_i \vee e_{13} \right) = \omega_0 \cdot v^2 e_2 \vee e_1 \vee e_3 = -v^2.$$

We find

(10) $$\Psi_v = v^3 e_{12} - v^2 e_{13} + v^1 e_{23}.$$

Similarly, for the vector Θ_α in (2),

(11) $$\Theta_\alpha = \alpha^{23} e_1 - \alpha^{13} e_2 + \alpha^{12} e_3.$$

14. The differential of a mapping. Let f be a mapping of E^n (or of an open set in E^n) into E^m which is smooth; that is, with coordinate systems in the spaces, the first partial derivatives of each component of the mapping function exist and are continuous. The *differential* ∇f of f is a concept which may be used in place of these partial derivatives, as follows. Take any point p and any vector v in E^n; then for each real number $t > 0$, $p + tv$ is a point in E^n. We set

(1) $$\nabla_v f(p) = \nabla f(p, v) = \lim_{t \to 0+} \frac{1}{t} [f(p + tv) - f(p)].$$

This is a vector in E^m, tangent to the curve which is the image under f of the line through p in the direction of v. For each p, we have a function $\nabla f(p)$ mapping vectors in E^n into vectors in E^m; it is elementary to show that this function is linear.

If $m = 1$ and E^m is the space $\mathfrak{R}$ of real numbers, we have a real valued function ϕ in E^n; with coordinates $(x^1, \cdots, x^n)$ in E^n and corresponding vectors $e_1, \cdots, e_n$ along the axes at the point p, we clearly have

$$(2) \qquad \nabla_{e_i}\phi(p) = \frac{\partial\phi(p)}{\partial x^i}, \qquad i = 1, \cdots, n.$$

In this case, $\nabla_v\phi(p)$ is a real linear function of vectors v, and thus $\nabla\phi(p)$ is a covector. Hence $\nabla\phi$ is a differential 1-form in E^n, called the *gradient* of ϕ.

We return to the general case. Given p and vectors $v_1, \cdots, v_r$ in E^n, set

$$(3) \qquad \nabla f(p, v_1\vee\cdots\vee v_r) = \nabla f(p, v_1)\vee\cdots\vee\nabla f(p, v_r);$$

this defines a linear transformation of r-vectors in E^n into r-vectors in E^m, which we also call $\nabla f(p)$.

Now let ω be any r-form in E^m. Take any point $q = f(p)$ in E^m. Then $\omega(q)\cdot\alpha'$ is a linear function of r-vectors α' in E^m; hence $\omega(q)\cdot\nabla f(p, \alpha)$ is a linear function of r-vectors α in E^n, and is thus an r-covector in E^n, which we call $(f^*\omega)(p)$. Now $f^*\omega$ is an r-form in E^n. The definition is given by

$$(4) \qquad (f^*\omega)(p)\cdot\alpha = \omega(f(p))\cdot\nabla f(p, \alpha), \qquad \text{all } r\text{-vectors } \alpha \text{ in } E^n.$$

Because of (3), we find, for differential forms ω and ξ in E^m,

$$(5) \qquad f^*(\omega\vee\xi) = f^*\omega\vee f^*\xi.$$

15. Jacobians. Let f be a smooth mapping of E^n (assumed metric and oriented) into E^m. Then the image under $\nabla f(p)$ of the unit n-vector α_0 of E^n is an n-vector in E^m, which we call the *Jacobian* $J_f(p)$ of f at p. Take for instance $n = 2$, $m = 3$, and let (e_1, e_2) be an orthonormal base in E^2. The images from p of these vectors are

$$(1) \qquad w_1(p) = \nabla f(p, e_1), \qquad w_2(p) = \nabla f(p, e_2),$$

and the Jacobian at p is the 2-vector

$$(2) \qquad J_f(p) = \nabla f(p, \alpha_0) = \nabla f(p, e_1\vee e_2) = w_1(p)\vee w_2(p).$$

If $J_f(p)$ is $\neq 0$, then $w_1(p)$ and $w_2(p)$ are independent, which clearly implies that f is one-one in a neighborhood of p; in this case the image of a neighborhood of p is a smooth piece S of surface in E^3.

Suppose f maps E^n into itself. Then $J_f(p)$ is a multiple $a\alpha_0$ of α_0; we call this number a the *algebraic Jacobian* $\mathsf{J}_f(p)$. Thus, with the unit n-covector ω_0 of E^n,

$$(3) \qquad J_f(p) = \mathsf{J}_f(p)\alpha_0, \qquad \omega_0\cdot J_f(p) = \mathsf{J}_f(p).$$

The term "Jacobian" is commonly used to denote $\mathsf{J}_f(p)$; note that $\mathsf{J}_f(p)$ is independent of the metric or orientation of E^n.

Take the case $n = 2$, $m = 3$ again. Using orthonormal coordinates (s, t) in E^2 and (x, y, z) in E^3, write the components of $\nabla_{e_1} f(p) = \partial f(p)/\partial s$ etc. as a set of three numbers. We have

$$(4) \qquad \frac{\partial f(p)}{\partial s} = \left(\frac{\partial x}{\partial s}, \frac{\partial y}{\partial s}, \frac{\partial z}{\partial s}\right), \qquad \frac{\partial f(p)}{\partial t} = \left(\frac{\partial x}{\partial t}, \frac{\partial y}{\partial t}, \frac{\partial z}{\partial t}\right).$$

Define the "Jacobian determinants"

$$\frac{\partial(x, y)}{\partial(s, t)} = \begin{vmatrix} \partial x/\partial s & \partial y/\partial s \\ \partial x/\partial t & \partial y/\partial t \end{vmatrix}, \text{ etc.}$$

Then by (9.5),

$$(5) \qquad J_f(p) = \frac{\partial(x, y)}{\partial(s, t)} e_{12} + \frac{\partial(x, z)}{\partial(s, t)} e_{13} + \frac{\partial(y, z)}{\partial(s, t)} e_{23}.$$

Define the "Jacobian vector" $\bar{J}_f(p)$ at p as that vector corresponding to the 2-vector $J_f(p)$; see (13.2). By (2) and (13.3), we may write it in the form (using orthonormal coordinates)

$$(6) \qquad \bar{J}_f(p) = \frac{\partial f(p)}{\partial s} \times \frac{\partial f(p)}{\partial t} = w_1(p) \times w_2(p).$$

By (13.11) or (13.8),

$$(7) \qquad \bar{J}_f(p) = \left(\frac{\partial(y, z)}{\partial(s, t)}, -\frac{\partial(x, z)}{\partial(s, t)}, \frac{\partial(x, y)}{\partial(s, t)}\right).$$

16. Transformation of the integral. Let ω be a uniformly continuous differential 2-form in the bounded open set R of the oriented plane E^2. We consider a smooth mapping f of the bounded open set R_0 of the space E'^2 onto R, with Jacobian $J_f \neq 0$ at all points. We wish to express $\int_R \omega$ as an integral over R_0.

Cut E'^2 into small rectangles, and let $\sigma_1, \cdots, \sigma_s$ be those lying in R_0. The image $\tau_i = f(\sigma_i)$ of σ_i is a small "curvilinear parallelogram" in R. Let p_i be a corner of σ_i, and let v_{i1}, v_{i2} be vectors on adjacent sides of σ_i, so that

$$\{\sigma_i\} = v_{i1} \vee v_{i2}.$$

Now

$$w_{i1} = \nabla f(p_i, v_{i1}), \qquad w_{i2} = \nabla f(p_i, v_{i2})$$

are side vectors of a parallelogram τ_i' which is a good approximation to τ_i if σ_i is small. By § 11, clearly

$$(1) \qquad \sum \omega(q_i) \cdot \{\tau_i'\}, \qquad q_i = f(p_i),$$

is a good approximation to $\int_R \omega$. By (14.3),

$$\{\tau_i'\} = w_{i1} \vee w_{i2} = \nabla f(p_i, v_{i1} \vee v_{i2}) = \nabla f(p_i, \{\sigma_i\}),$$

and by (14.4),

$$\sum \omega(q_i)\cdot\{\tau_i'\} = \sum \omega(f(p_i))\cdot\nabla f(p_i, \{\sigma_i\}) = \sum (f^*\omega)(p_i)\cdot\{\sigma_i\},$$

which is a good approximation to $\int_{R_0} f^*\omega$. Thus we see that

$$\int_R \omega = \int_{R_0} f^*\omega. \tag{2}$$

(The detailed proof is given in (III, 8).)

From this we shall derive the usual formula, using Jacobians, and taking $E'^2 = E^2$. With the unit 2-vector α_0 in E^2,

$$\{\tau_i'\} = \nabla f(p_i, \{\sigma_i\}) = |\sigma_i| \nabla f(p_i, \alpha_0) = |\sigma_i| J_f(p_i).$$

Let $\bar{\omega}$ be the real function corresponding to ω; that is, with the unit 2-covector ω_0 in E^2,

$$\omega(q) = \bar{\omega}(q)\omega_0, \qquad \bar{\omega}(q) = \omega(q)\cdot\alpha_0. \tag{3}$$

Now by (15.3),

$$\begin{aligned}\bar{\omega}(q_i)\,|\tau_i'| &= \omega(q_i)\cdot\{\tau_i'\}\\ &= \bar{\omega}(q_i)\omega_0\cdot|\sigma_i|\,J_f(p_i) = \bar{\omega}(q_i)J_f(p_i)\,|\sigma_i|\,.\end{aligned}$$

Summing and taking limits, we have

$$\int_R \bar{\omega} = \int_{R_0} \bar{\omega}(f(p))J_f(p)\,dp, \tag{4}$$

using the Riemann integral in both cases.

The same formulas hold in any number of dimensions. Note that neither side of (2) depends on a choice of metric.

17. Smooth manifolds. As a typical example, we take a piece S of smooth surface in E^3. At each point p of S there is a tangent plane $T(p)$. Some neighborhood U of p in $T(p)$ projects in a one-one way into S. A coordinate system in $T(p)$ projects into a coordinate system in S; this coordinate system is a smooth mapping of part of the space $\mathfrak{A}^2$ of pairs of real numbers into S. Where two such coordinate systems overlap, they are related by a smooth mapping of part of $\mathfrak{A}^2$ into itself, with non-vanishing Jacobian. This suggests the general definition of a smooth manifold, using such coordinate systems, without reference to any containing space, as in (II, 10).

With S as above, let v be a vector in $T(p)$. The points $p_t = p + tv$ in $T(p)$ project into points q_t in S (for t not too large). These points q_t form a "parametrized curve" C in S, which we may take as defining a corresponding vector "of S" at p. In place of the points p_t, we could use any function p_t' in $T(p)$ with $p_0' = p$ which has the same tangent vector at $t = 0$: $(\partial p_t'/\partial t)_{t=0} = v$; this would project into a parametrized curve

C' with points q_i' in S equivalent to C. We may use the definitions of av and of $v + w$ in $T(p)$ to give corresponding definitions in S at p. The vectors in S at p now form a vector space $V(p)$. Using this vector space, we may define r-vectors and r-covectors in S at p; hence we may define differential forms in S. In a general smooth manifold (not in Euclidean space), we may define vectors at a point by means of parametrized curves, and define addition of vectors with the help of coordinate systems.

To define r-dimensional integration in an n-dimensional manifold M, one must first choose simple r-dimensional domains of integration; (rectilinear) r-cells are not defined here. One may choose oriented pieces of r-dimensional submanifolds of M for such domains. We have now the problem of defining the integral of an r-form ω over an oriented piece of an r-manifold.

Take first $n = 1$; then we have an oriented curve C, abstractly defined. It is of course equivalent to an interval of the real numbers. Now $\int_C \omega$ is defined for a 1-form ω. Suppose C is in E^3 and ω is defined in a neighborhood of C. Let $p_0, p_1, \cdots, p_m$ be a division of C into short arcs. Then the vectors $p_{i+1} - p_i$ are nearly tangent to C, and

$$(1) \qquad \sum_{i=0}^{m-1} \omega(p_i)\cdot\{p_i p_{i+1}\} = \sum_{i=0}^{m-1} \omega(p_i)\cdot(p_{i+1} - p_i)$$

is an approximation to the integral $\int_C \omega$.

Take an oriented piece S of surface in E^3 again. Supposing S is a small piece, we may cut it up into small curvilinear pieces $\tau_1, \cdots, \tau_s$, for instance images $f(\sigma_1), \cdots, f(\sigma_s)$ of rectangles in a coordinate system (see the τ_i in § 16). With tangent vectors w_{i1}, w_{i2} at the vertex q_i of τ_i, we may form

$$(2) \qquad \sum \omega(q_i)\cdot(w_{i1} \vee w_{i2})$$

and use this as an approximation to $\int_S \omega$; compare (16.1). Because of (16.2), we see that we may equivalently define $\int_S \omega$ as being $\int_{S_0} f^*\omega$, if $S = f(S_0)$, S_0 being part of the Euclidean plane.

The latter definition is independent of the fact that S is in E^3. With S in E^3, if ω is defined throughout a neighborhood of S in E^3, we could use polyhedral approximations to define the integral; see Chapter X.

18. Particular forms of integrals in 3-space. Here we have r-dimensional integration for $r = 0$, 1, 2 and 3. Letting E^3 be metric and oriented, an r-covector in E^3 corresponds either to a real number or to a vector (see § 13); this gives us special forms of the integrals, which we discuss.

$r = 0$. A 0-cell is a point p (without orientation properties). A 0-covector is a real number; hence a 0-form is a real function. The integral of the 0-form ϕ over the 0-cell p is the number $\phi(p)$. A polyhedral 0-chain is an expression $A = \sum a_i p_i$, and the integral of ϕ over A is

$$\int_A \phi = \sum a_i \phi(p_i). \tag{1}$$

In particular if σ is the oriented 1-cell pq, then

$$\partial(pq) = 1q + (-1)p, \qquad \int_{\partial(pq)} \phi = \phi(q) - \phi(p). \tag{2}$$

$r = 1$. With an oriented curve C and a 1-form ω^1, an approximation to $\int_C \omega^1$ was given in (17.1). We replace ω^1 by a vector function v in E^3, as follows. For each p, $v(p)$ is the vector such that the scalar product $v(p)\cdot w$ equals $\omega^1(p)\cdot w$ for all vectors w; see § 12. With a common notation, a tangent vector along C is called the "line element" ds, and the integral is written (using (17.1))

$$\int_S \omega^1 = \int_S v\cdot ds = \lim \sum v(p_i)\cdot(p_{i+1} - p_i). \tag{3}$$

This is the "circulation" of v along C.

$r = 2$. In this case, an approximation to $\int_S \omega^2$ was given in (17.2). The 2-form ω^2 corresponds to a vector function v, as follows. For each p, $\omega^2(p) = \Psi_{v(p)}$, as in (13.1). This relation gives

$$\omega^2(p)\cdot\alpha = \omega_0\cdot[v(p)\vee\alpha], \qquad \text{all 2-vectors } \alpha.$$

Hence, by (13.4),

$$\omega^2(q_i)\cdot(w_{i1}\vee w_{i2}) = \omega_0\cdot[w_{i1}\vee w_{i2}\vee v(q_i)] = (w_{i1}\times w_{i2})\cdot v(q_i); \tag{4}$$

with the notations of (17.2), the sum of these quantities is an approximation to $\int_S \omega^2$. The vector

$$\mathbf{N}_i = w_{i1}\times w_{i2} \tag{5}$$

is called the "surface element" at q_i, and in analogy with (3), the integral is written

$$\int_S \omega^2 = \int_S v\cdot d\mathbf{N} = \lim \sum v(q_i)\cdot\mathbf{N}_i. \tag{6}$$

This is the "flux" of v across S (see § 5).

$r = 3$. The integral $\int_R \omega^3$ is defined for regions R in oriented E^3 and a 3-form ω^3. With the unit 3-covector ω_0, $\omega^3(p) = \bar{\omega}(p)\omega_0$ for a real function $\bar{\omega}$ as in (16.3), and

$$\int_R \omega^3 = \int_R \bar{\omega}, \tag{7}$$

the last integral being the Riemann integral. If we cut R into small cells σ_i oriented like E^3 and choose a point p_i in each σ_i, then

$$\sum \omega^3(p_i)\cdot\{\sigma_i\} = \sum \bar{\omega}(p_i)\,|\,\sigma_i\,| \tag{8}$$

is an approximation to the integral.

19. The Theorem of Stokes. Let ω be a smooth r-form in E^n. Then the general Theorem of Stokes says that there is an $(r + 1)$-form $\omega' = d\omega$ in E^n such that for any oriented piece M of a smooth manifold of dimension $r + 1$ in E^n with boundary ∂M,

$$\int_M d\omega = \int_{\partial M} \omega. \tag{1}$$

From general considerations we will see that ω' must exist; we deduce its form in the next section. The usual formulas in E^3 are given in § 21.

Define the r-cochain X in E^n by

$$X\cdot\sigma = \int_\sigma \omega, \qquad \text{all oriented } r\text{-cells } \sigma. \tag{2}$$

We define an $(r + 1)$-cochain $Y = dX$ by the formula

$$Y\cdot\tau = X\cdot\partial\tau = \int_{\partial\tau} \omega, \qquad \text{all oriented } (r+1)\text{-cells } \tau; \tag{3}$$

compare (1.2). It may be proved that X satisfies (H_2'), or equivalently, that Y satisfies (H_1'); see (V, 10). That Y satisfies (H_2') is trivial: For any oriented $(r + 2)$-cell τ',

$$Y\cdot\partial\tau' = X\cdot\partial\partial\tau' = X\cdot 0 = 0. \tag{4}$$

Thus Y is a flat $(r + 1)$-cochain. We call it the *coboundary* dX of X. If we assume that the partial derivatives of ω satisfy a Lipschitz condition, then we may show that Y is sharp. There is then (§ 11) a corresponding $(r + 1)$-form $\omega' = D_Y$ such that

$$\int_\tau \omega' = Y\cdot\tau = \int_{\partial\tau} \omega,$$

giving (1) for cells and hence for M. (Of course ω' exists under much more general conditions; see for instance (III, 16) and (IX, 12).)

20. The exterior differential. Given the smooth r-form ω in E^n, the $(r + 1)$-form $\omega' = d\omega$ of § 19 is called the *exterior differential* of ω. Using an affine coordinate system $(x^1, \cdots, x^n)$ in E^n, we wish to find the components of ω' in terms of those of ω.

Take first $r = 0$. Given the point p, take any i, set $p_t = p + te_i$, and consider the segment pp_t $(t > 0)$. Since

$$\int_0^t \omega_i' \, dx^i = \int_{pp_t} \omega' = \int_{\partial(pp_t)} \omega$$
$$= \omega(p_t) - \omega(p) = \int_0^t \frac{\partial \omega}{\partial x^i} \, dx^i,$$

we must have

$$\text{(1)} \qquad r = 0: \qquad \omega_i' = \frac{\partial \omega}{\partial x^i}.$$

Take next $r = 1$. Given p and i, j, let us write x, y for x^i, x^j, and omit the other coordinates. Take any $h, k > 0$, and let σ be the oriented rectangle with corners

$$p = (x, y), \quad p' = (x + h, y), \quad p'' = (x + h, y + k), \quad p''' = (x, y + k).$$

Since $\int_{p'}^{p''} \omega = \int_y^{y+k} \omega_j(x + h, t) \, dt$ etc., we have

(2)

$$\int_{\partial\sigma} \omega = \int_y^{y+k} [\omega_j(x + h, t) - \omega_j(x, t)] \, dt - \int_x^{x+h} [\omega_i(s, y + k) - \omega_i(s, y)] \, ds.$$

Now

$$\omega_j(x + h, t) - \omega_j(x, t) = \int_x^{x+h} \frac{\partial}{\partial x} \omega_j(s, t) \, ds,$$

and similarly for the other integrand. Hence we must have

$$\text{(3)} \qquad \int_\sigma \omega' = \int_{\partial\sigma} \omega = \int_x^{x+h} \int_y^{y+k} \left[\frac{\partial}{\partial x} \omega_j(s, t) - \frac{\partial}{\partial y} \omega_i(s, t)\right] dt \, ds.$$

Taking h and k arbitrarily small, this shows that

$$\text{(4)} \qquad r = 1: \qquad \omega_{ij}' = \frac{\partial \omega_j}{\partial x^i} - \frac{\partial \omega_i}{\partial x^j}.$$

The same method may be used in the general case. In particular,

$$\text{(5)} \qquad r = 2: \qquad \omega_{ijk}' = \frac{\partial \omega_{jk}}{\partial x^i} - \frac{\partial \omega_{ik}}{\partial x^j} + \frac{\partial \omega_{ij}}{\partial x^k}.$$

There are similar formulas relating ω and ω' directly. For instance,

$$\text{(6)} \qquad r = 1: \qquad d\omega(p) \cdot (v \vee w) = \nabla_v[\omega(p) \cdot w] - \nabla_w[\omega(p) \cdot v].$$

We mention some general properties. For any r-form ω and s-form ξ,

$$\text{(7)} \qquad dd\omega = 0, \qquad d(\omega \vee \xi) = d\omega \vee \xi + (-1)^r \omega \vee d\xi.$$

If f is a smooth mapping of one Euclidean space into another and ω is an r-form in the second, then

$$\text{(8)} \qquad df^* \omega = f^* \, d\omega.$$

If ω is an r-form defined in a manifold, we may define $d\omega$ with the help of a coordinate system; because of (8), the definition is independent of the coordinate system employed. The above properties continue to hold, except for (6).

21. Some special formulas in metric oriented E^3. We give examples of the exterior differential and Stokes' Theorem.

$r = 0$. Here, the exterior differential is simply the differential, i.e. the gradient: $d\phi = \nabla\phi$. (This formula fails for $r > 0$.) With the "del" operator

$$\nabla = e_1 \frac{\partial}{\partial x} + e_2 \frac{\partial}{\partial y} + e_3 \frac{\partial}{\partial z},$$

the vector function corresponding to $d\phi$ (see § 18) is

$$(1) \qquad \nabla\phi = \operatorname{grad}(\phi) = \frac{\partial\phi}{\partial x} e_1 + \frac{\partial\phi}{\partial y} e_2 + \frac{\partial\phi}{\partial z} e_3.$$

For an oriented curve C from p to q, (19.1) gives, with the notations of § 18,

$$(2) \qquad \int_C \nabla\phi\cdot d\mathbf{s} = \int_C \operatorname{grad}(\phi)\cdot d\mathbf{s} = \int_C d\phi = \phi(q) - \phi(p).$$

If C is an interval of the real axis, this is the "fundamental theorem of the integral calculus."

$r = 1$. Given the smooth vector function v in E^3, there is a corresponding 1-form ω, this has an exterior differential $\omega' = d\omega$, and this 2-covector valued function corresponds to a vector function v', called the *curl* of v. Because of (10.2), $\omega_i = v^i$, and because of (13.10),

$$(3) \qquad \omega'_{12} = v'^3, \qquad \omega'_{13} = -v'^2, \qquad \omega'_{23} = v'^1.$$

Hence, by (20.4),

$$(4) \quad \nabla \times v = \operatorname{curl}(v)$$
$$= \left(\frac{\partial v^3}{\partial y} - \frac{\partial v^2}{\partial z}\right) e_1 + \left(-\frac{\partial v^3}{\partial x} + \frac{\partial v^1}{\partial z}\right) e_2 + \left(\frac{\partial v^2}{\partial x} - \frac{\partial v^1}{\partial y}\right) e_3.$$

For an oriented piece S of surface with boundary curve C, (19.1) gives

$$(5) \qquad \int_S (\nabla \times v)\cdot d\mathbf{N} = \int_S \operatorname{curl}(v)\cdot d\mathbf{N} = \int_C v\cdot d\mathbf{s}.$$

This is the commonly called "Theorem of Stokes."

$r = 2$. Given the smooth vector function v, there is a corresponding 2-form ω, and $d\omega = \omega'$ is a 3-form corresponding to a real function $\bar{\omega}'$

as in (16.3), called the *divergence* of v. The relation between v and ω is given by (3), and $\bar{\omega}' = \omega'_{123}$. Hence, by (20.5),

$$\nabla \cdot v = \operatorname{div}(v) = \frac{\partial v^1}{\partial x} + \frac{\partial v^2}{\partial y} + \frac{\partial v^3}{\partial z}. \tag{6}$$

For a region R (oriented like E^3), with boundary S, (19.1) gives, using the volume element dR,

$$\int_R \nabla \cdot v \, dR = \int_R \operatorname{div}(v)\, dR = \int_S v \cdot d\mathbf{N}. \tag{7}$$

By taking particular vector functions in (5) or (7), we obtain special standard formulas. Recall that they are all cases of (19.1), which formula holds independently of a metric in the space.

The formula $dd\omega = 0$ gives, in the cases $r = 0$ and $r = 1$,

$$\nabla \times \nabla \phi = \operatorname{curl}(\operatorname{grad}(\phi)) = 0, \tag{8}$$

$$\nabla \cdot (\nabla \times \phi) = \operatorname{div}(\operatorname{curl}(\phi)) = 0; \tag{9}$$

these are easily verified directly.

We may start with a real function ϕ, find its gradient, interpret this as a vector function, and find its divergence, giving a real function again; this is the *Laplacian* of ϕ. By (1) and (6), this is

$$\nabla \cdot \nabla \phi = \Delta \phi = \frac{\partial^2 \phi}{\partial x^2} + \frac{\partial^2 \phi}{\partial y^2} + \frac{\partial^2 \phi}{\partial z^2}. \tag{10}$$

Because of its construction, the right hand side is independent of the choice of orthonormal coordinates.

22. An existence theorem. In a convex subset of E^3, any vector function is a gradient provided its curl is 0 (it is then commonly called an exact differential), any vector function is a curl provided its divergence is 0, and any real function is a divergence. (The functions are assumed smooth.) These are cases of the general theorem, that a smooth r-form ω ($r \geqq 1$) in a convex subset of E^n équals $d\xi$ for some $(r-1)$-form ξ if and only if $d\omega = 0$. Note that for a 0-form (real function) ϕ, $d\phi = 0$ if and only if ϕ is constant. The theorem fails for more general subsets of E^n or for manifolds; see the next section.

We give a formula denoted by $(n;\ r)$, for the components of a possible ξ in terms of those of the r-form ω in E^n, for various values of n and r, in a cube containing the origin. For $n = r = 1$, set

$$\xi(x) = \int_0^x \omega_1(t)\, dt; \tag{1;1}$$

then $(d\xi)_1(x) = d\xi(x)/dx = \omega_1(x)$, and hence $d\xi = \omega$.

For $n = 2$, $r = 1$, we carry out the idea above by integrating along the path from $(0, 0)$ to $(0, y)$ and thence to (x, y):

$$\xi(x, y) = \int_0^y \omega_2(0, t)\, dt + \int_0^x \omega_1(s, y)\, ds. \tag{2;1}$$

Then $\partial\xi(x, y)/\partial x = \omega_1(x, y)$, and since $\partial\omega_2/\partial x - \partial\omega_1/\partial y = 0$,

$$\frac{\partial\xi(x, y)}{\partial y} = \omega_2(0, y) + \int_0^x \frac{\partial\omega_1(s, y)}{\partial y}\, ds$$

$$= \omega_2(0, y) + [\omega_2(x, y) - \omega_2(0, y)] = \omega_2(x, y).$$

We write down some further cases:

$$\xi_1(x, y) = 0, \qquad \xi_2(x, y) = \int_0^x \omega_{12}(s, y)\, ds; \tag{2;2}$$

$$\xi(x, y, z) = \int_0^z \omega_3(0, 0, u)\, du + \int_0^y \omega_2(0, t, z)\, dt + \int_0^x \omega_1(s, y, z)\, ds; \tag{3;1}$$

$$\begin{cases} \xi_1(x, y, z) = 0, \qquad \xi_2(x, y, z) = \int_0^x \omega_{12}(s, y, z)\, ds, \\ \xi_3(x, y, z) = \int_0^y \omega_{23}(0, t, z)\, dt + \int_0^x \omega_{13}(s, y, z)\, ds; \end{cases} \tag{3;2}$$

$$\xi_{12}(x, y, z) = \xi_{13}(x, y, z) = 0, \qquad \xi_{23}(x, y, z) = \int_0^x \omega_{123}(s, y, z)\, ds. \tag{3;3}$$

A general proof of existence of ξ will be given in (IV, 25).

23. De Rham's Theorem. The Theorem of de Rham relates differential forms in a manifold to homology and cohomology properties of the manifold; see (IV, Theorem 29A). We shall discuss the theorem briefly for the case of a torus M, in the dimension 1.

The points of M are described by naming a pair of angular coordinates (θ, ψ), $0 \leqq \theta \leqq 2\pi$, $0 \leqq \psi \leqq 2\pi$, where $\theta = 0$ is identified with $\theta = 2\pi$, and similarly for ψ. Thus we have

$$(0, \psi) = (2\pi, \psi), \qquad (\theta, 0) = (\theta, 2\pi). \tag{1}$$

The points $(\theta, 0)$ as θ goes from 0 to 2π form an oriented closed curve C_1; similarly, the points $(0, \psi)$ form C_2. Now let C be any oriented closed curve. By an elementary argument we may show that C can be deformed into an oriented closed curve C' which follows C_1 a number μ_1 of times, and then C_2 μ_2 times; μ_1 and μ_2 are any integers. For $\mu_1 = \mu_2 = 0$, C' reduces to a point.

We call a differential form ω *closed* if $d\omega = 0$, and *derived* if $\omega = d\xi$ for some ξ. Since $dd\xi = 0$, any derived form is closed. Call ω and ω' *cohomologous* if $\omega - \omega'$ is derived. A closed r-form, together with all cohomologous forms, forms an r-dimensional *differential cohomology class* in M (with real coefficients); the set of these classes form the r-dimensional *differential cohomology space* $\mathbf{H}^r$. By the Theorem of de Rham, this is isomorphic to the r-dimensional algebraic cohomology space with real coefficients.

The *fundamental periods* of the closed 1-form ω in the torus are the numbers

$$P_1(\omega) = \int_{C_1} \omega, \qquad P_2(\omega) = \int_{C_2} \omega. \tag{2}$$

For any cohomologous ω', Stokes' Theorem gives

$$P_i(\omega') - P_i(\omega) = \int_{C_i} (\omega' - \omega) = \int_{C_i} d\xi = \int_{\partial C_i} \xi = 0,$$

since C_i is closed. Hence we may speak of the fundamental periods of a differential cohomology class. We shall show that there is a unique class with any given pair of numbers as periods. This is the original formulation of the theorem in the present situation.

Define the closed 1-forms

$$\omega_1^* = d\theta, \qquad \omega_2^* = d\psi. \tag{3}$$

Though θ and ψ are not single valued in M, because of (1), clearly ω_1^* and ω_2^* are. The fundamental periods of ω_1^* and of ω_2^* are $(2\pi, 0)$ and $(0, 2\pi)$ respectively; hence

$$P_1(a\omega_1^* + b\omega_2^*) = 2\pi a, \qquad P_2(a\omega_1^* + b\omega_2^*) = 2\pi b. \tag{4}$$

Thus there are closed forms with any given fundamental periods.

We next note that if the oriented closed curve C is deformed into C' as above, then for any closed 1-form ω,

$$\int_C \omega = \int_{C'} \omega = \mu_1 P_1(\omega) + \mu_2 P_2(\omega). \tag{5}$$

isomorphic to the r-dimensional algebraic cohomology space.

For we may find a succession of curves $C_0 = C, C_1, \cdots, C_m = C'$, as follows. For each i, there is an oriented 2-cell σ_i with boundary $\partial\sigma_i = B_i - A_i$, such that C_{i+1} contains B_i, C_i contains A_i, and C_i and C_{i+1} are the same otherwise. Now

$$\int_{C_{i+1}} \omega - \int_{C_i} \omega = \int_{\partial\sigma_i} \omega = \int_{\sigma_i} d\omega = 0,$$

and the first relation in (5) follows; the second is clear.

We must show still that if ω and ω' are closed 1-forms with the same fundamental periods, then their difference $\eta = \omega' - \omega$ is derived: $\eta = d\phi$. We may define ϕ by the formula (2;1) in § 22:

$$\phi(\theta, \psi) = \int_0^\psi \eta_2(0, t)\, dt + \int_0^\theta \eta_1(s, \psi)\, ds. \tag{6}$$

To show that ϕ is single-valued, apply (6) to give

$$\phi(\theta, 2\pi) - \phi(\theta, 0) = \int_0^{2\pi} \eta_2(0, t)\, dt = \int_{C_2} \eta = 0.$$

Similarly $\phi(2\pi, \psi) = \phi(0, \psi)$. That $d\phi = \eta$ follows as in § 22.

We remark that for any oriented closed curve C, we can find the associated numbers μ_1, μ_2 through ω_1^*, ω_2^*. With C' as before,

$$(7) \qquad \int_C \omega_i^* = \int_{C'} \omega_i^* = \mu_1 P_1(\omega_i^*) + \mu_2 P_2(\omega_i^*) = 2\pi\mu_i.$$

The Theorem of de Rham also relates products of forms to products of algebraic cohomology classes or intersections of cycles. We give one formula. Clearly $\omega_1^* \vee \omega_2^*$ is the unit 2-covector function in M; hence

$$(8) \qquad \int_M \omega_1^* \vee \omega_2^* = 4\pi^2.$$

It follows easily, for any closed ω_1, ω_2, that

$$(9) \qquad \int_M \omega_1 \vee \omega_2 = 4\pi^2[P_1(\omega_1)P_2(\omega_2) - P_2(\omega_1)P_1(\omega_2)] \quad ;$$

we use the fact that ω_i is cohomologous to a form $\omega_i' = a_i\omega_1^* + b_i\omega_2^*$.

C. Indications of General Theory

24. Normed spaces of chains and cochains. Given the flat r-cochain X in E^n (§ 3), let $|X|$ and $|dX|$ be the smallest N_1 and N_2 satisfying the hypotheses (H_1') and (H_2') respectively. Then the function

$$(1) \qquad |X|^\flat = \sup\{|X|, |dX|\}$$

of flat cochains is a norm, the *flat norm*, and the linear space of flat r-cochains with this norm is a Banach space $\mathbf{C}^{\flat r}$.

Let the *mass* of the polyhedral r-chain $A = \sum a_i\sigma_i$ be

$$(2) \qquad |A| = \sum |a_i||\sigma_i| \qquad \text{(if the } \sigma_i \text{ are non-overlapping);}$$

then the definition of $|X|$ gives $|X\cdot A| \leqq |X||A|$, and

$$|X\cdot\partial D| \leqq |dX||D|, \qquad \text{all polyhedral } (r+1)\text{-chains } D.$$

Given X and A, take any such D. Then

$$|X\cdot A| = |X\cdot(A - \partial D) + X\cdot\partial D|$$
$$\leqq |X||A - \partial D| + |dX||D| \leqq |X|^\flat(|A - \partial D| + |D|).$$

This suggests defining the *flat norm* of the polyhedral r-chain A by

$$(3) \quad |A|^\flat = \inf\{|A - \partial D| + |D|\}, \qquad \text{all polyhedral } (r+1)\text{-chains } D.$$

This is a norm in the space of polyhedral r-chains, and is the smallest norm such that, for all flat X,

$$(4) \qquad |X\cdot A| \leqq |X|^\flat |A|^\flat.$$

For an example, take an oriented 1-cell σ, and let $\sigma' = T_v\sigma$ denote σ translated by the vector v. Now σ and σ' are two sides (of length $|\sigma|$) of a parallelogram τ; the other two sides A, B are of length $|v|$, and $|\tau| \leqq |\sigma||v|$. Hence

$$(5) \qquad |T_v\sigma - \sigma|^\flat \leqq |T_v\sigma - \sigma - \partial\tau| + |\tau| \leqq (2 + |\sigma|)|v|.$$

Since

$$|X\cdot(T_v\sigma - \sigma)| \leqq |X|^\flat |T_v\sigma - \sigma|^\flat \leqq (2 + |\sigma|)\,|X|^\flat |v|,$$

$X\cdot T_v\sigma$ is near $X\cdot\sigma$ if v is small.

Each flat r-cochain X has a *coboundary* dX, which is a flat $(r + 1)$-cochain, defined by

$$(6) \qquad dX\cdot B = X\cdot\partial B, \qquad \text{all polyhedral } (r+1)\text{-chains } B.$$

This relation is the abstract counterpart of Stokes' Theorem (19.1).

The space of polyhedral r-chains in the flat norm is not complete; if we complete it, we obtain the space $\mathbf{C}_r^\flat$ of *flat* r-chains. Now $\mathbf{C}^{\flat r}$ is the conjugate space of $\mathbf{C}_r^\flat$.

We may define the *sharp norms* $|X|^\sharp$ of cochains and $|A|^\sharp$ of polyhedral r-chains similarly. Completing the space of r-chains in the sharp norm gives the space $\mathbf{C}_r^\sharp$ of sharp r-chains, which includes $\mathbf{C}_r^\flat$.

Now integration theory may be set up as follows: Define a norm (for instance the sharp or flat norm) in the linear space of polyhedral r-chains; then an "integrand" (cochain) X is a bounded linear function in this space; it satisfies continuity conditions, depending on the norm chosen. Now $X\cdot A = \int_A D_X$ is defined not only over polyhedral r-chains, but also over all elements of the completed space of chains; with properly chosen norms, this space will include curved pieces of r-dimensional manifolds, etc. This is the program we carry out starting with Chapter V.

25. Continuous chains. We return for the moment to the torus M of § 23. Let $C_{1\psi}$ be the oriented closed curve of points (θ, ψ) as θ runs from 0 to 2π. Since $C_{1\psi}$ may be deformed into C_1, we have, for any closed 1-form ω in M,

$$(1) \qquad \int_0^{2\pi} [\omega(\theta, \psi)\cdot e_1]\, d\theta = \int_0^{2\pi} \omega_1(\theta, \psi)\, d\theta = \int_{C_{1\psi}} \omega = P_1(\omega),$$

where e_1 is the unit vector in the θ-direction. Integrating with respect to ψ and dividing by 2π gives

$$(2) \qquad \frac{1}{2\pi}\int_M \omega\cdot e_1 = \frac{1}{2\pi}\int_0^{2\pi}\int_0^{2\pi} [\omega(\theta, \psi)\cdot e_1]\, d\theta\, d\psi = P_1(\omega).$$

This expresses $P_1(\omega)$ by means of the Riemann integral of $\omega\cdot e_1$ over the whole of M. We have here a 1-dimensional integral $P_1(\omega)$ in the guise of a 2-dimensional integral.

We ask now, *dropping the restriction that ω be closed*, can the left hand side of (2) be interpreted as a 1-dimensional integral? We may do this as follows. First, the curves $C_{1\psi}$ may be considered as simple 1-dimensional figures, and

$$X\cdot C_{1\psi} = \int_{C_{1\psi}} \omega = F(\psi)$$

is defined. Now take a fine subdivision $\mathfrak{S}$ of $(0, 2\pi)$ by the points $\psi_0 = 0$, $\psi_1, \cdots, \psi_m = 2\pi$. Corresponding to $\mathfrak{S}$, define the 1-chain (similar to a polyhedral 1-chain)

$$(3) \qquad A_{\mathfrak{S}} = \sum_{i=0}^{m-1} (\psi_{i+1} - \psi_i) C_{1\psi_i}.$$

Then

$$X\cdot A_{\mathfrak{S}} = \sum (\psi_{i+1} - \psi_i) X\cdot C_{1\psi_i} = \sum F(\psi_i)(\psi_{i+1} - \psi_i).$$

Taking a sequence $\mathfrak{S}_1, \mathfrak{S}_2, \cdots$ of such subdivisions, with mesh $\to 0$, this shows that

$$(4) \qquad \lim_{k\to\infty} X\cdot A_{\mathfrak{S}_k} = \int_0^{2\pi} F(\psi)\, d\psi,$$

which is the desired integral over M. Moreover, we show below that in the flat norm (§ 24), the sequence $A_{\mathfrak{S}_1}, A_{\mathfrak{S}_2}, \cdots$ is a Cauchy sequence; hence it has a limit A:

$$(5) \qquad \lim_{k\to\infty}{}^\flat A_{\mathfrak{S}_1} = A.$$

Now $X\cdot A = \int_0^{2\pi} F(\psi)\, d\psi$, so that the integral in (2) is simply the value of X on the flat 1-chain A.

To illustrate the convergence in (5), take the square $0 \leqq \theta \leqq 2\pi$, $0 \leqq \psi \leqq 2\pi$, in $\mathfrak{A}^2$ (see § 17) instead of in M; the $C_{1\psi_i}$ are now segments σ_i. Suppose $\mathfrak{S}_l$ is a refinement of $\mathfrak{S}_k$. Then each term $(\psi_{i+1} - \psi_i)\sigma_i$ in $\mathfrak{S}_k$ is replaced in $\mathfrak{S}_l$ by a sum of terms $(\psi_{i,j+1} - \psi_{ij})\sigma_{ij}$; if $\psi_{i+1} - \psi_i < \epsilon$ (all i), then (24.5) gives

$$\Big| \sum_j (\psi_{i,j+1} - \psi_{ij})\sigma_{ij} - (\psi_{i+1} - \psi_i)\sigma_i \Big|^\flat = \Big| \sum_j (\psi_{i,j+1} - \psi_{ij})(\sigma_{ij} - \sigma_i) \Big|^\flat$$
$$\leqq \sum_j (\psi_{i,j+1} - \psi_{ij}) \left| \sigma_{ij} - \sigma_i \right|^\flat \leqq (\psi_{i+1} - \psi_i)(2 + 2\pi)\epsilon.$$

Therefore

$$(6) \qquad | A_{\mathfrak{S}_l} - A_{\mathfrak{S}_k} |^\flat \leqq \sum_i (\psi_{i+1} - \psi_i)(2 + 2\pi)\epsilon = 2\pi(2 + 2\pi)\epsilon,$$

and (5) follows for this case (and hence for the above case).

Note that it is the e_1 in (2) which determined the flat chain A. In general, in an n-dimensional Riemannian manifold M, we may have a continuous point function $\alpha(p)$, whose values are r-vectors; the function α defines a flat r-chain A, called a *continuous r-chain*. An r-form $\omega = D_X$ corresponds to a flat (in fact, sharp) cochain X (see (11.2)), and using the Riemann integral,

$$(7) \qquad X \cdot A = \int_M \omega \cdot \alpha.$$

Suppose the function α is smooth. Then it may be shown (VI, 9) that the boundary ∂A of the corresponding chain A is defined by a function α' formed from the partial derivatives of α.

26. On 0-dimensional integration. We may divide kinds of integration into two types: (a) those in which orientation properties of the domains are used, and (b) those where these properties are not used. Under (a), there is of necessity a certain dimension number r attached to the integral. This might even be the case under (b), but the essential geometric properties are missing; we show below how to set up such an integral as a 0-dimensional integral.

Whenever a domain of integration can be turned in space (thus moving in part in directions not parallel to its original position), orientation properties are probably needed; witness the discussion of flux in § 5. Where this type of motion is not used, orientation can in general be disregarded. Thus, as in (18.7), the integral $\int_R \omega$ of a 3-form ω in oriented E^3 over a domain R oriented like E^3 can also be written as the Riemann integral $\int_R \bar{\omega}$ of the corresponding real function $\bar{\omega}$ (assuming E^3 is metric); in the latter, orientation properties are not used. In the transformation formula (16.2) or (16.4) however, one must pay attention to orientation.

A typical example of an integral of type (b) is the evaluation of the total mass of a mass distribution. We might have mass spread out over curves or over surfaces in E^3; but these figures need not be oriented in determining the mass. A more general integral of the same sort is the integral of a real function ϕ in a metric space S which has a (positive) measure function μ attached. We may consider ϕ as a 0-form; then $X \cdot p = \phi(p)$ is a 0-cochain in S. We show how to consider any compact measurable subset Q of S as a "flat" 0-chain A_Q, such that

$$(1) \qquad \int_Q \phi \, d\mu = X \cdot A_Q;$$

we assume ϕ is continuous, so a Riemann type definition may be used.

For each subdivision $\mathfrak{S}$ of Q into measurable sets $Q_1, \cdots, Q_m$, together with a point p_i chosen in each Q_i, we have a 0-chain

$$(2) \qquad A_{\mathfrak{S}} = \sum \mu(Q_i) p_i;$$

for a sequence of such sets with mesh $\to 0$, the limit of the corresponding chains is a "flat" chain A. The proof rests on the following fact: Suppose each Q_i is of diameter $<\epsilon$, and $\mathfrak{S}'$ is a refinement of $\mathfrak{S}$. Say Q_i is divided into sets Q_{ij}, and p_{ij} is in Q_{ij}. Then considering $p_i p_{ij}$ as an abstract 1-cell and noting from (24.3) that $|\,\partial(pq)\,|^\flat \leqq$ distance from p to q,

$$|\,A_{\mathfrak{S}'} - A_{\mathfrak{S}}\,|^\flat = \Big|\sum_{i,j} \mu(Q_{ij})(p_{ij} - p_i)\Big|^\flat = \Big|\sum_{i,j} \mu(Q_{ij})\,\partial(p_i p_{ij})\Big|^\flat$$
$$\leqq \epsilon \sum_{i,j} \mu(Q_{ij}) = \epsilon\mu(Q).$$

The usual proof now shows that the flat 0-chain A exists. Since

$$X\cdot A_{\mathfrak{S}} = \sum_i \mu(Q_i) X\cdot p_i = \sum_i \phi(p_i)\mu(Q_i),$$

it is clear that (1) holds.

We remark that if we set

$$\phi A_{\mathfrak{S}} = \sum \phi(p_i)\mu(Q_i) p_i,$$

the flat limit gives a 0-chain ϕA_Q; there is a "unit" 0-cochain X_0, defined by $X_0 \cdot p = 1$ (all p), with corresponding real function $D_{X_0}(p) = 1$, and $X_0\cdot\phi A = \int_Q \phi\, d\mu$. This type of operation is studied in (VII, 1).

Chapter 5

Other Subjects

A LOGICAL EXPANSION IN MATHEMATICS*

BY HASSLER WHITNEY†

1. *Introduction*. Suppose we have a finite set of objects, (for instance, books on a table), each of which either has or has not a certain given property A (say of being red). Let n, or $n(1)$, be the total number of objects, $n(A)$ the number with the property A, and $n(\bar{A})$ the number without the property A (with the property not-A or $\bar{A}$). Then obviously

$$n(\bar{A}) = n - n(A). \tag{1}$$

Similarly, if $n(A\ B)$ denote the number with both properties A and B, and $n(\bar{A}\ \bar{B})$ the number with neither property, that is, with both properties not-A and not-B, then

$$n(\bar{A}\bar{B}) = n - n(A) - n(B) + n(AB), \tag{2}$$

which is easily seen to be true.

The extension of these formulas to the general case where any number of properties are considered is quite simple, and is well

---

* Presented to the Society, October 25, 1930.

† National Research Fellow.

known to logicians. It should be better known to mathematicians also; we give in this paper several applications which show its usefulness.

The notation is that used in a paper by the author, *Characteristic functions and the algebra of logic*,* which we shall refer to as C. F. It should cause no confusion if we use the same symbols for characteristic functions, sets, and properties. The paper named is not essential for an understanding of the present paper.

2. *The Logical Expansion.* We prove now the general formula

$$\begin{aligned} n(\bar{A}_1\bar{A}_2\cdots\bar{A}_m) &= n - [n(A_1) + n(A_2) + \cdots + n(A_m)] \\ &+ [n(A_1A_2) + n(A_1A_3) + \cdots + n(A_{m-1}A_m)] \\ &- [n(A_1A_2A_3) + \cdots] + \cdots + (-1)^m n(A_1A_2\cdots A_m), \end{aligned} \tag{3}$$

which gives the number of objects *without* certain properties in terms of the numbers of objects *with* various of these properties. The formula is exactly what we should get if we multiplied out the expression in brackets in $n[(1-A_1)(1-A_2)\cdots(1-A_m)]$ by ordinary algebra and applied the general formulas

$$n(F+G) = n(F) + n(G), \quad n(-F) = -n(F).$$

We assume the formula holds if there are two properties involved, and shall prove it for the case that three properties are involved. The proof obviously holds if the numbers two and three are replaced by i and $i+1$, and hence the formula is true in general, by mathematical induction.

Consider the objects counted in $n(\bar{A}_1\bar{A}_2)$, that is, those with neither of the properties A_1 and A_2. We wish to know how many of these have the property $\bar{A}_3$. Applying (1) to this set, we have

$$n(\bar{A}_1\bar{A}_2\bar{A}_3) = n(\bar{A}_1\bar{A}_2) - n(\bar{A}_1\bar{A}_2A_3).$$

We know, by hypothesis, that

$$n(\bar{A}_1\bar{A}_2) = n - [n(A_1) + n(A_2)] + n(A_1A_2).$$

To find $n(\bar{A}_1\bar{A}_2A_3)$ we need merely consider those objects with the property A_3, and apply the expansion to this set. Thus

$$n(\bar{A}_1\bar{A}_2A_3) = n(A_3) - [n(A_1A_3) + n(A_2A_3)] + n(A_1A_2A_3).$$

* To appear in the Annals of Mathematics.

Hence

$$(4)\quad n(\bar{A}_1\bar{A}_2\bar{A}_3) = n - [n(A_1) + n(A_2) + n(A_3)] + [n(A_1A_2) + n(A_1A_3) + n(A_2A_3)] - n(A_1A_2A_3),$$

as required.

3. *The Measure of Characteristic Functions.* This section relates this paper to the paper C. F. We shall find a general class of formulas which contains the logical expansion as a special case, making use of *characteristic functions*. With each element x of a set of n objects R we associate an integer $A(x)$, positive, negative, or zero. We define the *measure of* A by the equation

$$(5)\qquad n(A) = \sum_{x \text{ in } R} A(x).$$

Suppose A were one for certain elements of R (which elements form the set A'), and zero for the rest. Then A is the characteristic function of A', and $n(A)$ is just the number of elements in A'. In particular, if $A \equiv 1$, that is, A' is R, then $n(A)$ is

$$(6)\qquad n(1) = n,$$

and if $A \equiv 0$, that is, A' contains no elements, then $n(A)$ is

$$(7)\qquad n(0) = 0.$$

If A and B are any functions and p and q are any numbers, then

$$(8)\qquad \begin{aligned} n(pA + qB) &= \sum_{x \text{ in } R} [pA(x) + qB(x)] = p\sum A(x) + q\sum B(x) \\ &= pn(A) + qn(B). \end{aligned}$$

The logical expansion now follows at once if we expand $\bar{A}_1\bar{A}_2 \cdots \bar{A}_m$ into the second normal form as in C. F.*

4. *On Prime Numbers.* Let

$$p_1, p_2, \cdots, p_m$$

be any set of positive integers. We wish to find N, the number of numbers less than or equal to a given number x, which are not divisible by any of these numbers. If we let R be the set of all numbers $\leqq x$ and let P_i be those divisible by p_i, then

* Note that, for characteristic functions, $\bar{A}_i = 1 - A_i$. We make this substitution, multiply out, and use (8).

$$
\begin{aligned}
N &= n(\overline{P}_1\overline{P}_2 \cdots \overline{P}_m) \\
&= n - [n(P_1) + n(P_2) + \cdots] + [n(P_1P_2) + \cdots] \qquad (9)\\
&\quad - \cdots + (-1)^m n(P_1P_2 \cdots P_m),
\end{aligned}
$$

by the logical expansion (where $n = x$). This expansion is fundamental in prime number theory.* Its importance lies in the fact that the terms of the form $n(P_{i_1}P_{i_2} \cdots P_{i_k})$ are easier to calculate directly than the desired quantity.

Suppose we wish to find the number $\phi(x)$ of prime numbers $\leqq x$. Let $p_1, p_2, \cdots, p_m$ be the primes $<\sqrt{x}$. Then the primes $\leqq x$ are just those numbers $\leqq x$ which are not divisible by any of the numbers $p_1, p_2, \cdots, p_m$, except for the number 1, together with the m numbers $p_1, p_2, \cdots, p_m$. Thus

$$\phi(x) = N + m - 1. \qquad (10)$$

5. *A Problem in Probability.* Suppose a pack of cards is lying in a row on a table. If we lay out another pack of cards on top of these, what is the probability that no card of the second pack will lie on the same card of the first? If a pack contains m cards, there are $m!$ arrangements of the cards in the second pack. If there are N arrangements of the second pack such that no card of the second pack falls on the same card of the first, then the required probability is $p = N/m!$. If $n(A_i)$ denotes the number of arrangements such that the ith card of the second pack falls on the same card of the first, etc., then $N = n(\overline{A}_1 \cdot \overline{A}_2 \cdots \overline{A}_m)$. If we use the logical expansion, a typical term of the result is $(-1)^k n(A_{i_1} \cdot A_{i_2} \cdots A_{i_k})$. This is the number of arrangements in which the cards numbered $i_1, i_2, \cdots, i_k$ fall on the same cards of the first pack. These k cards being fixed in position, there are $(m-k)!$ arrangements of the remaining cards. The term is thus $(-1)^k(m-k)!$. There are $\binom{m}{k}$ terms with k factors, contributing together $(-1)^k\binom{m}{k}(m-k)! = (-1)^k m!/k!$. Summing over k and dividing by $m!$, we have

$$p = \sum_{k=0}^{m} \frac{(-1)^k}{k!} = 1 - \frac{1}{1!} + \frac{1}{2!} - \cdots + (-1)^m \frac{1}{m!},$$

which is the sum of the first $m+1$ terms in the expansion of $1/e$.

* See for instance E. Landau, *Primzahlen*, Leipzig, Tuebner, pp. 67 ff.

6. *On the Number of Ways of Coloring a Graph.* Consider any set of objects $a, b, c, \cdots, f$, and any set of pairs of these objects, as $ab, bd, \cdots, cf$. We call the whole collection a *graph*, containing the *vertices* $a, b, c, \cdots, f$, and the *arcs* $ab, bd, \cdots, cf$. It can be visualized simply by letting the vertices be points in space, and letting each arc be a curve joining the two vertices involved.

Suppose we have a fixed number λ of colors at our disposal. Any way of assigning one of these colors to each vertex of the graph in such a way that any two vertices which are joined by an arc are of different colors, will be called an admissible coloring of the graph. We wish to find the number $M(\lambda)$ of admissible colorings, using λ or fewer colors.

As a special case of this general problem we have the four-color map problem: To know if we can assign to each region of a map on a sphere one of four colors in such a way that no two regions with a common boundary are of the same color, that is, to see if $M(4)>0$. Our graph (which is just the dual graph of the map) is constructed by placing a vertex in each region of the map, and joining two vertices by an arc if the corresponding regions have a common boundary. We shall deduce a formula for the number $M(\lambda)$ of ways of coloring a graph *

If there are V vertices in the graph G, then there are λ^V possible colorings, formed by giving each vertex in succession any one of the λ colors. Let R be this set of colorings. Let A_{ab} denote those colorings with the property that a and b are of the same color, etc. Then the set of admissible colorings is

$$\bar{A}_{ab}\bar{A}_{bd}\cdots\bar{A}_{cf},$$

and the number of colorings is, if there are E arcs in the graph,

* *A determinant formula for the number of ways of coloring a map*, Annals of Mathematics, (2), vol. 14 (1912), pp. 42–46. The formula was discovered independently by the author by the method here given. See in this connection a paper by Birkhoff, *On the number of ways of coloring a map*, Proceedings of the Edinburgh Mathematical Society, (2), vol. 2 (1930), pp. 83–91; also a paper by the author, *The coloring of graphs*, to appear in the Annals of Mathematics. In regard to the four-color map problem, see references in Birkhoff's last mentioned paper, also a paper by the author, *A theorem on graphs*, Annals of Mathematics, (2), vol. 32 (1931), p. 379.

$$
\begin{aligned}
M(\lambda) &= n(\overline{A}_{ab}\overline{A}_{bd}\cdots\overline{A}_{cf}) \\
&= n - [n(A_{ab}) + n(A_{bd}) + \cdots + n(A_{cf})] \qquad (11)\\
&\quad + [n(A_{ab}A_{bd}) + \cdots] - \cdots \\
&\quad + (-1)^E n(A_{ab}A_{bd}\cdots A_{cf}).
\end{aligned}
$$

With each property A_{ab} is associated an arc ab of G. In the logical expansion, there is a term corresponding to every possible combination of the properties A_{pq}; with this combination we associate the corresponding arcs, forming a *subgraph* H of G. In particular, the first term corresponds to the subgraph containing no arcs, and the last term corresponds to the whole of G. *We let H contain all the vertices of G.*

Let us evaluate a typical term, such as $n(A_{ab}A_{ad}\cdots A_{ce})$. This is the number of ways of coloring G in λ or fewer colors in such a way that a and b are of the same color, a and d are of the same color, $\cdots$, c and e are of the same color. In the corresponding subgraph H, any two vertices that are joined by an arc must be of the same color, and thus all the vertices in a single connected piece in H are of the same color. If there are p connected pieces in H, the value of this term is therefore λ^p. If there are s arcs in H, the sign of the term is $(-1)^s$. Thus

$$(-1)^s n(A_{ab}A_{ad}\cdots A_{ce}) = (-1)^s\lambda^p.$$

If there are (p, s) (this is Birkhoff's symbol) subgraphs of s arcs in p connected pieces, the corresponding terms contribute to $M(\lambda)$ an amount $(-1)^s(p, s)\lambda^p$. Therefore, summing over all values of p and s, we find the polynomial in λ:

$$M(\lambda) = \sum_{p,s}(-1)^s(p, s)\lambda^p. \qquad (12)$$

Consider a subgraph H of G of s arcs, in p connected pieces. Let us define its rank i and its nullity j by the equations $i = V - p$, $j = s - i = s - V + p$. Then $p = V - i$, $s = i + j$, and putting $(p, s) = m_{V-p,s-V+p} = m_{ij}$, (thus, m_{ij} is the number of subgraphs of G of rank i, nullity j), we have

$$
\begin{aligned}
M(\lambda) &= \sum_{i,j}(-1)^{i+j}m_{ij}\lambda^{V-i} = \sum_i m_i\lambda^{V-i}, \qquad (13)\\
m_i &= \sum_j(-1)^{i+j}m_{ij}.
\end{aligned}
$$

7. *The m_i in Terms of the Broken Circuits of G.* We shall find in this section an interpretation of the coefficients of $M(\lambda)$ directly in terms of properties of the graph. Consider for instance the graph G containing the vertices a, b, c, d and the arcs ab, ac, bc, bd, cd, the arcs being given in this definite order. Make a list of the *circuits* in G, naming the arcs of a circuit in the order in which they occur above. Here, the circuits are ab, ac, bc, and bc, bd, cd, and ab, ac, bd, cd. From each circuit we now form the corresponding ***broken circuit*** by dropping out the last arc of the circuit. The broken circuits here are ab, ac, and bc, bd, and ab, ac, bd. Then ***the number*** $(-1)^i m_i$ ***is the number of subgraphs of G of i arcs which do not contain all the arcs of any broken circuit.***

To show this, we arrange the broken circuits of G in a definite order, where we put a broken circuit P_i before a broken circuit P_j if, in naming the arcs of G one by one in the given order, all the arcs of P_i are named before all those of P_j are named; otherwise, the ordering is immaterial. Suppose there are σ broken circuits, $P_1, P_2, \cdots, P_\sigma$. We now divide the subgraphs of G into $\sigma+1$ sets (some of which may be empty), putting in the first set, S_1, all those subgraphs containing all the arcs of P_1; in the second, S_2, all those not containing P_1, but containing P_2; in the third, S_3, all those containing neither P_1 nor P_2, but containing P_3; $\cdots$; in the last set, $S_{\sigma+1}$, all those containing none of these broken circuits.

Consider now all the terms in (11) corresponding to the first set of subgraphs. Suppose α_1 is the arc we dropped out of a circuit to form the first broken circuit P_1. To each subgraph in S_1 not containing α_1 corresponds a subgraph in S_1 containing α_1 and conversely, as α_1 is not in P_1. The subgraphs of S_1, and hence the corresponding terms of (11), are thus paired off. But the two terms of each pair cancel. For let H and H' be the two corresponding subgraphs. If H is in p connected pieces, so is H', as the arc α_1 joins two vertices already connected by the broken circuit P_1. The terms each contribute λ^p therefore; but they are of opposite sign, as H' contains one more arc than H.

Consider now the terms corresponding to S_2 (if there are any such). If α_2 is the arc dropped out in forming P_2, α_2 is in neither P_1 nor P_2, on account of the way we have ordered the broken circuits. Thus to each subgraph in S_2 not containing α_2 corresponds a subgraph in S_2 containing α_2, and conversely. The corresponding terms of (11) are thus paired off, and they cancel, exactly as before.

Continuing, we cancel all terms in S_3, S_4, $\cdots$, S_σ. We are left only with terms in $S_{\sigma+1}$, that is, those corresponding to subgraphs not containing all the arcs of any broken circuit, and none of these have been canceled.

Consider any such term containing i arcs. The corresponding subgraph H contains no circuit, as it contains no broken circuit. If we build up H arc by arc, each arc we add joins two vertices formerly not connected therefore, and the number of connected pieces is decreased by one each time. Thus the number of connected pieces in H is $V-i$, and the corresponding term contributes $(-1)^i\lambda^{V-i}$ to $M(\lambda)$. If there are l_i such subgraphs, they together contribute an amount $(-1)^i l_i\lambda^{V-i}$. Hence, summing over i, we have $P(\lambda)=\sum_i(-1)^i l_i\lambda^{V-i}$. Comparing with (13), we see that $l_i=(-1)^i m_i$, as required.

EXAMPLES. Let G contain the vertices a, b, c, and the arcs ab, ac, bc. There is one broken circuit: ab, ac. There is one subgraph of no arcs, and $m_0=1$. There are three subgraphs of a single arc, and $-m_1=3$. There are three subgraphs of two arcs; but one of them contains the broken circuit, so $m_2=2$. The subgraph of three arcs contains the broken circuit. Hence, as $V=3$,

$$M(\lambda) = \lambda^3 - 3\lambda^2 + 2\lambda = \lambda(\lambda - 1)(\lambda - 2).$$

This is easily verified. For we can color a in λ ways; there are $\lambda-1$ colors left for b; there are now $\lambda-2$ colors left for c.

Let G be the graph named at the beginning of this section. If a subgraph contains the last broken circuit, it contains the first also, so we can forget the last. We find

$$M(\lambda) = \lambda^4 - 5\lambda^3 + 8\lambda^2 - 4\lambda = \lambda(\lambda - 1)(\lambda - 2)^2$$

which again is easily verified.

PRINCETON UNIVERSITY

CHARACTERISTIC FUNCTIONS AND THE ALGEBRA OF LOGIC.[1]

By Hassler Whitney.[2]

Introduction. The algebra of logic differs from ordinary algebra in that there are two formulas,

$$A \odot A = A, \qquad A \oplus A = A,$$

which have no analogues in ordinary algebra. On this account the rules of the algebra of logic must be learned afresh; this usually involves studying a set of axioms and deducing formulas from them.[3] We shall show in this paper how, by means of "characteristic functions",[4] the formulas of the algebra of logic can be expressed as equations of ordinary algebra; with this representation, the common operations on sets are easily understood.

In Part II "generalized sets" are taken up; these are useful in various mathematical theories.

I. Real sets.

1. **Characteristic functions.** Let R' (the universe of discourse) be any set of objects (elements) $a, b, c, \cdots$, (for instance, the books on a table, or the points of a plane), and let A' be any subset of these elements (say the red books, or the points inside a certain circle). The *characteristic function* $A(x)$ of the set A' is a function which associates either the number 1 or the number 0 with each element x of R': 1 if x is in A', and 0 otherwise. Thus

$$(1) \qquad A(x) = \begin{cases} 1 \text{ if } x \text{ is in } A', \\ 0 \text{ if } x \text{ is not in } A'. \end{cases}$$

For instance, if R' consists of two red books a and b and one book c not red, then $A(a) = A(b) = 1$ and $A(c) = 0$.

[1] Received May 31 and July 7, 1932.—Presented to the American Mathematical Society. December 28, 1931.

[2] National Research Fellow.

[3] For a treatment of this algebra, see C. I. Lewis, A Survey of Symbolic Logic, University of California Press, 1918, Ch. II, or D. Hilbert and W. Ackermann, Grundzüge der theoretischen Logik, Springer, Berlin 1928, Ch. I and II.

[4] See de la Vallée Poussin, Intégrales de Lebesgue, Fonctions d'Ensembles, Classes de Baire, Paris 1916, p. 7.

Any set is determined by its characteristic function. Two sets A' and B' are the same if and only if their characteristic functions are identically equal:

$$(2') \qquad A(x) = B(x) \text{ for all } x \text{ in } R'.$$

We shall write this simply

$$(2) \qquad A = B.$$

If A' contains all the elements of R', that is, A' is R', then the characteristic function A is identically 1:

$$(3) \qquad A = 1.$$

If A' contains no elements, A is 0:

$$(4) \qquad A = 0.$$

We must keep in mind that *all equations in characteristic functions are identities, holding true for each element x in R'.* In such equations, 1 or 0 is the function equal to one or zero for each element in R'.

2. **Combinations of sets.** A mathematical operation $\dot{+}$ (read: plus dot) will be useful here; it is defined by the equations

$$(5) \qquad 1\dot{+}1 = 1\dot{+}0 = 0\dot{+}1 = 1, \qquad 0\dot{+}0 = 0.$$

If we make the general definition, for any two numbers X and Y,

$$(6) \qquad X\dot{+}Y = X+Y-XY,$$

the above formulas are seen to hold.

Suppose A' and B' are any two sets, and $A'\odot B'$ denotes that set whose elements are in both A' and B' (the common part of A' and B'). Its characteristic function must be 1 for elements in both A' and B', and 0 otherwise; that is, it must be 1 when both A and B are 1, and 0 when either A or B is 0. But this is just

$$(7) \qquad AB,$$

the product of the two characteristic functions A and B.

If $A'\oplus B'$ denotes the set whose elements are either in A' or in B' or in both (the union, or sum, of A' and B'), its characteristic function must be 1 when either A or B is 1, and 0 when both A and B are 0. This is just

$$(8) \qquad A\dot{+}B = A+B-AB.$$

Note that

$$A \dot{+} B = A + B \quad \textit{if and only if} \quad AB = 0.$$

If $\tilde{A}'$ is the set whose elements are not in A' (the complementary set of A'), then its characteristic function, which we call $\bar{A}$, is evidently

$$\bar{A} = 1 - A.\text{[5]} \tag{9}$$

3. Operations with sets. With the above arithmetical representation of sets, we can perform all the common operations on sets with ease. Some examples will illustrate this.

The two most typical relations in the algebra of logic are: $A' \odot A'$ is A', and $A' \oplus A'$ is A', i. e. the elements in both A' and A' are just those in A', as are those in either A' or A'. The first relation, in characteristic functions, is

$$AA = A, \tag{10}$$

which is easily seen to be true. For take any element x of R'; $A(x) \cdot A(x) = A(x)$ because $A(x) =$ either 1 or 0, and $1 \cdot 1 = 1$, $0 \cdot 0 = 0$. To prove the second, we have

$$A \dot{+} A = A + A - AA = A + A - A = A. \tag{11}$$

The associative and commutative laws are evident, when we use (8). Some fundamental identities are

$$\begin{aligned} A \cdot 0 = 0, \qquad A \cdot 1 = A, \\ A \dot{+} 0 = A + 0 - A \cdot 0 = A, \qquad A \dot{+} 1 = A + 1 - A = 1, \\ A(B \dot{+} C) = A(B + C - BC) = AB + AC - ABC \\ = AB + AC - AB\,AC = AB \dot{+} AC, \\ (A \dot{+} B)(A \dot{+} C) = A \dot{+} BC. \end{aligned} \tag{12}$$

Some identities involving complementary sets are

$$\begin{aligned} \bar{1} = 1 - 1 = 0, \quad \bar{0} = 1 - 0 = 1, \\ \bar{\bar{A}} = 1 - \bar{A} = 1 - (1 - A) = A, \\ A\bar{A} = A(1 - A) = A - A = 0, \\ A \dot{+} \bar{A} = A + \bar{A} - A\bar{A} = A + (1 - A) = 1, \\ \bar{A} \dot{+} \bar{B} = 1 - AB = \overline{AB}, \\ \overline{A \dot{+} B} = 1 - (A \dot{+} B) = (1 - A)(1 - B) = \bar{A}\bar{B}. \end{aligned} \tag{13}$$

[5] Note that although equations like this have been used by Boole and others, they were taken in an entirely different sense. It seemed as though the formulas (10) and (11) could not possibly occur in ordinary algebra.

An important relation between sets is that of *inclusion*. A set A' is *contained in* a set B' if every element of A' is an element of B'. (A' may be identical with B'.) In characteristic functions, if $A(x) = 1$ for a given element x, then $B(x) = 1$ also. This is equivalent to stating that for each element x, either $A(x) = 0$ or $\bar{B}(x) = 0$; or,

$$A\bar{B} = 0, \quad \text{or,} \quad \bar{A}\dot{+}B = 1. \tag{14}$$

As $A - AB = A\bar{B}$, and as $(A\dot{+}B) - B = A + B - AB - B = A\bar{B}$, two equivalent equations are

$$AB = A, \quad A\dot{+}B = B. \tag{15}$$

Another way of writing this is

$$A \leqq B. \tag{16}$$

4. **Properties and propositions.** Given a set of elements R', suppose A_1 is a property which each element either has or has not. Corresponding to this property A_1 we can define a set A', the set of elements with the property A_1. The characteristic function corresponding to the elements with both properties A_1 and B_1, or with either property, or without the property A_1, or with any property which holds for all the elements, or with any property which holds for none of the elements, is AB, or $A\dot{+}B$, or $\bar{A}$, or 1, or 0. The statement "every element with the property A_1 has also the property B_1" translates into "A' is contained in B'", or "$A\bar{B} = 0$".

We can reduce the logic of propositions to characteristic functions as follows. Consider a single imaginary element a. If P^* is any proposition, let P be the corresponding characteristic function, putting $P(a) = 1$ if P^* is true, and $P(a) = 0$ if P^* is false. Corresponding to the proposition P^* *and* Q^*, or *either* P^* *or* Q^*, or *not*-P^*, or P^* *implies* Q^*, is the characteristic function PQ, or $P\dot{+}Q$, or $\bar{P}$, or $\bar{P}\dot{+}Q$ (see equation (14)).

5. **First normal form.** We begin by explaining an important process, that of *expanding a characteristic function in terms of other characteristic functions*. To expand A in terms of B, we write

$$A = A\cdot 1 = A(B+\bar{B}) = AB+A\bar{B}. \tag{17}$$

The interpretation is simple: The elements in $A' \odot B'$ together with the elements in $A' \odot \bar{B}'$ make up the elements in A'. As another example, to expand 1 in terms of A, or in terms of A and B, we write

$$1 = A+\bar{A}, \tag{18}$$

or

$$1 = A(B+\bar{B})+\bar{A}(B+\bar{B}) = AB+A\bar{B}+\bar{A}B+\bar{A}\bar{B}. \tag{19}$$

In general, 2^n terms are formed in expanding a function in terms of n other functions. We give now the first normal form.

To reduce any function F to the first normal form, we first remove any $\dotplus$ signs there may be by (8), *next multiply out, next expand each term in terms of all other characteristic functions appearing in F, and finally collect terms.*

For instance,

$$\begin{aligned} F &= A(A \dotplus B) - A\bar{B} = A(A+B-AB) - A\bar{B} \\ &= A + AB - AB - A\bar{B} = A - A\bar{B} = (AB + A\bar{B}) - A\bar{B} = AB. \end{aligned}$$

As another example,

$$F = A + B = (AB + A\bar{B}) + (AB + \bar{A}B) = 2AB + A\bar{B} + \bar{A}B.$$

Suppose the functions appearing in F were $A_1, A_2, \cdots, A_n$. Then each term in the normal form will be of the form $A_1^* A_2^* \cdots A_n^*$, where each A_k^* is either A_k or $\bar{A}_k$, and will have an integral coefficient, positive, negative or zero. There are 2^n possible terms of this sort, which may be found by expanding 1 in terms of all the functions. Note that *if H and I are any two of these terms, then $HI = 0$*; for for some k, H contains A_k (or $\bar{A}_k$) and I contains $\bar{A}_k$ (or A_k); but $A_k \bar{A}_k = 0$.

If G is F in the first normal form, G is determined by naming the coefficient of each of these 2^n terms (some coefficients may be zero). For an explanation of what values these coefficients can take on, see § 8.

Any characteristic function F has a unique expression in the first normal form (provided, of course, that we make no use of equations we may happen to know).

For suppose G_1 and G_2 were two expressions for F in the normal form, and suppose that for some one of the terms, say $A_1^* A_2^* \cdots A_n^*$, G_1 had the coefficient k_1, G_2 had the coefficient k_2, and $k_1 \neq k_2$. As $G_1 = F = G_2$, $G_1 - G_2 = 0$ for all x in R'. Thus

$$(k_1 - k_2) A_1^* A_2^* \cdots A_n^* + \cdots = 0, \qquad \text{all } x \text{ in } R'.$$

Let us interpret R' as being a set containing but a single element a, such that $A_1^*(a) = A_2^*(a) = \cdots = A_n^*(a) = 1$. This we may do, as the reduction to the normal form is independent of the interpretation of the sets involved. Then, as $\bar{A}_1^*(a) = \bar{A}_2^*(a) = \cdots = \bar{A}_n^*(a) = 0$, the above equation reduces to

$$(k_1 - k_2) A_1^*(a) A_2^*(a) \cdots A_n^*(a) = k_1 - k_2 = 0,$$

contradicting the hypotheses that $k_1 \neq k_2$.

6. **Application of the normal form.** In § 3 we gave a number of identities in characteristic functions. If we reduce each side of an identity

to the first normal form, the coefficient of each term in the result must be the same on both sides, by the uniqueness theorem. This gives a simple method of checking any identity. As an example, each side of the last equation of (12) reduces to $ABC + AB\bar{C} + A\bar{B}C + A\bar{B}\bar{C} + \bar{A}BC$ (or, $A + \bar{A}BC$).

Suppose we know a number of things, which we write in the form of equations, and we wish to deduce consequences. The simplest way to get a view of the equations as a whole is to write each equation $F = G$ in the form $F - G = 0$, and expand $F - G$ into the first normal form, expanding in terms of all functions involved. A typical equation is now

$$a_1 H_1 + a_2 H_2 + \cdots + a_m H_m = 0, \tag{20}$$

where each a_i is $\neq 0$. Here, if $A_1, A_2, \cdots, A_n$ are the functions involved, each H_i is of the form $A_1^* A_2^* \cdots A_n^*$, where each A_k^* is either A_k or $\bar{A}_k$. We can then write

$$H_1 = H_2 = \cdots = H_m = 0. \tag{21}$$

For suppose not. Then one of these H_i is not identically 0, that is, there is an element x_1 in R' such that $H_i(x_1) = 1$. But as $H_i H_j = 0$ for $j \neq i$, $H_j(x_1) = 0$ for $j \neq i$. As (20) is true for every x in R' it is true for x_1; but for x_1, it becomes

$$a_i = 0,$$

a contradiction.

Examples. Let us show that "If the proposition P^* is true and the proposition P^* implies Q^* is true, then the proposition Q^* is true". The premises are $P = 1$ and $\bar{P} \dot{+} Q = 1$; we must show that $Q = 1$. The first premise we write in the form $1 - P = \bar{P} = 0$, or expanding in terms of Q, $\bar{P}Q + \bar{P}\bar{Q} = 0$. As $1 - (\bar{P} \dot{+} Q) = P\bar{Q}$, the premises become

$$\bar{P}Q = \bar{P}\bar{Q} = P\bar{Q} = 0.$$

To compare Q and 1, we expand them:

$$Q = PQ + \bar{P}Q = PQ,$$
$$1 = PQ + P\bar{Q} + \bar{P}Q + \bar{P}\bar{Q} = PQ.$$

Thus $Q = 1$, as required.[6]

Let us solve a problem of Venn.[7] "A certain club has the following rules: (a) The financial committee shall be chosen from the general com-

[6] Of course we could note simply that as $P = 1$ and hence $\bar{P} = 0$, $\bar{P} \dot{+} Q = 1$ only if $Q = 1$.

[7] See Lewis, loc. cit., p. 201. We are following the procedure given in this book.

mittee. (b) No one shall be a member of both the general and library committees unless he be also on the financial committee. (c) No member of the library committee shall be on the financial committee. Simplify the rules."

Letting R' be the set of people in the club, and putting F', G' and L' for the financial, general and library committees respectively, the premises become

$$F\bar{G} = 0, \qquad GL\bar{F} = 0, \qquad LF = 0.$$

Expanding, we find that

$$F\bar{G}L = F\bar{G}\bar{L} = \bar{F}GL = FGL = 0.$$

Combining the first two terms and the last two terms, we have the equivalent equations

$$F\bar{G} = 0, \qquad GL = 0.$$

Thus the simplified rules will be (a), together with: (b′) No member of the general committee shall be on the library committee.

7. **Relation to the algebra of logic.** The essential difference between the present method and the algebra of logic is that here, one does not lose sight of the elements making up a set—any equation $A = B$ is an equation holding for every element under consideration. When one considers a set as consisting of certain specified elements, one can define the sum of two sets etc. directly, and hence the relations holding between combinations of sets can be deduced immediately; the arithmetic method furnishes a convenient tool for doing this. In the algebra of logic, a set is considered as an abstract quantity; the sum of two sets etc. must be defined abstractly, and one knows nothing about it except what one postulates. For this reason axioms must be proposed, enough so that all ordinary formulas can be derived therefrom.

We can, if we wish, consider a characteristic function as an abstract quantity; in this case, we can deal with sets just as described above, but without thinking of each equation as being an identity for all elements of some basic set. That we can deduce all the standard relations between sets from the relations given above follows from the fact that we can prove a set of axioms for the algebra of logic.[8]

II. Generalized Sets.

8. **Real and generalized sets.** Suppose we associate with each element of a set R' any integer, positive, negative or zero, instead of

[8] For instance those of Lewis, loc. cit. His "elements" are our "characteristic functions".

merely one or zero. The resulting function will not in general be the characteristic function of a real set; but we may consider it as the characteristic function of a *generalized set,* where each element is counted any number of times. (This is done for example with chains in analysis situs.) We shall discuss here the question as to whether any given set is real or generalized, and shall then take up two more normal forms.

Note that formulas previously deduced which depend on (10) or (11) do not hold for generalized sets.

Given any function F, let us expand it into the first normal form—this expansion holds of course whether the corresponding set is real or generalized. We prove now several facts.

If each coefficient in the first normal form is either 1 *or* 0, F' *is real.* For each element of R' falls into exactly one of the 2^n sets corresponding to the 2^n characteristic functions $A_1^* A_2^* \cdots A_n^*$ described in § 5; F has therefore the value 1 or 0 for each element, and corresponds to a real set.

Conversely, if F' *is real, each coefficient in the first normal form is* 1 *or* 0, *at least after we have dropped out any of the* 2^n *terms which are* 0 (i. e. 0 for all elements in R').

For suppose not; then there is a term $A_1^* A_2^* \cdots A_n^*$ in the first normal form of which the coefficient k is neither 0 nor 1, while $A_1^* A_2^* \cdots A_n^*$ is not 0, that is, for some element x_1, $A_1^*(x_1) A_2^*(x_1) \cdots A_n^*(x_1) = 1$. Now from the normal form we see that the value of F for x_1 is just k; hence the element x_1 is counted in F' a number k of times not one or zero, and F' is not real, a contradiction.

As an example, if we know the function $F = A + B = 2AB + A\bar{B} + \bar{A}B$ to be real, then necessarily $AB = 0$; after we have dropped out this term, each coefficient is 1 or 0.

9. **Second and third normal forms.** *In the second normal form, we remove all* $\dot{+}$ *signs, replace all characteristic functions with dashes by functions without, multiply out and collect terms.* Thus

$$\begin{aligned} F = \bar{B} - A\bar{B}C + ABC &= (1-B) - A(1-B)C + ABC \\ &= 1 - B - AC + 2ABC. \end{aligned}$$

In the third normal form, we remove all $\dot{+}$ *signs, replace all functions without dashes by functions with (but we do not alter the numerical term, if there is one), multiply out, and collect terms.* Thus, for the same F,

$$\begin{aligned} F &= \bar{B} - (1-\bar{A})\bar{B}(1-\bar{C}) + (1-\bar{A})(1-\bar{B})(1-\bar{C}) \\ &= 1 - \bar{A} - \bar{B} - \bar{C} + 2\bar{A}\bar{B} + \bar{A}\bar{C} + 2\bar{B}\bar{C} - 2\bar{A}\bar{B}\bar{C}. \end{aligned}$$

If there are n functions $A_1, A_2, \cdots, A_n$ involved in F, and G is F in the second (or third) normal form, then G is determined by naming the coefficient of each of the 2^n terms $A_{i_1} A_{i_2} \cdots A_{i_k}$ (or $\bar{A}_{i_1} \bar{A}_{i_2} \cdots \bar{A}_{i_k}$), where $i_1, i_2, \cdots, i_k$ are any set of distinct numbers chosen from the set of numbers $1, 2, \cdots, n$. In the above example, $n = 3$ and $2^n = 8$.

Any function F has a unique expression in either normal form. We prove this for the second normal form; the third is exactly alike.

Suppose F contains the n characteristic functions $A_1, A_2, \cdots, A_n$. Let G and G' be two expressions for F in the second normal form. If we write $q_1, q_2, \cdots, q_{2^n}$ for the 2^n quantities $A_{i_1} A_{i_2} \cdots A_{i_k}$, and if $k_1, k_2, \cdots, k_{2^n}$, and $k'_1, k'_2, \cdots, k'_{2^n}$ are the coefficients of these terms in G and G' respectively, then as $G = G'$, or $G - G' = 0$, we must have for each x in R',

$$(k_1 - k'_1)\, q_1(x) + (k_2 - k'_2)\, q_2(x) + \cdots + (k_{2^n} - k'_{2^n})\, q_{2^n}(x) = 0.$$

We shall choose a definite interpretation of the functions $A_1, A_2, \cdots, A_n$, by letting R' be a set containing just 2^n elements, one lying in each of the 2^n sets corresponding to the 2^n functions $A_1^* A_2^* \cdots A_n^*$ of § 5. Call these functions $p_1, p_2, \cdots, p_{2^n}$ for short, and let x_i lie in the set corresponding to p_i, $i = 1, 2, \cdots, 2^n$. As then $p_i(x_i) = 1$ and $p_i(x_j) = 0$, $j \neq i$, the 2^n functions $p_i(x)$ are linearly independent.[9] Expanding these into the second normal form, they are expressed in terms of the 2^n functions $q_i(x)$; hence these are linearly independent also. Therefore the determinant of the $q_i(x_j)$ is $\neq 0$, and the above equation is only possible for every x if each coefficient vanishes, i. e. if $k_i = k'_i$, $i = 1, 2, \cdots, 2^n$.

It follows from the uniqueness theorems that *a necessary and sufficient condition that two functions $F(A_1, A_2, \cdots, A_n)$ and $G(A_1, A_2, \cdots, A_n)$ be the same for all determinations of $A_1, A_2, \cdots, A_n$, is that they be identical in any normal form.*

10. **Relation between the second and third normal forms.** Let F be a function and let F_2 and F_3 be F expanded into the second and third normal forms respectively. Consider in F_2 all terms containing i factors (e. g. AB contains two factors, 1 contains none), and let α_i be the sum of the coefficients of these terms. Define β_i similarly for F_3. Thus, in the last example,

$$\alpha_0 = 1, \quad \alpha_1 = -1, \quad \alpha_2 = -1, \quad \alpha_3 = 2;$$
$$\beta_0 = 1, \quad \beta_1 = -3, \quad \beta_2 = 5, \quad \beta_3 = -2.$$

We wish to find a relation between the α_i and the β_i.

To do this, consider any term of the first normal form, such as ABC (taking the coefficient 1), and expand it into the second normal form:

[9] That is, if $c_1 p_1(x) + c_2 p_2(x) + \cdots + c_{2^n} p_{2^n}(x) = 0$ for every x in R', then

$$c_1 = c_2 = \cdots = c_{2^n} = 0.$$

$$\begin{aligned} ABC &= (1-\bar{A})(1-\bar{B})(1-\bar{C}) \\ &= 1-(\bar{A}+\bar{B}+\bar{C})+(\bar{A}\bar{B}+\bar{A}\bar{C}+\bar{B}\bar{C})-\bar{A}\bar{B}\bar{C}. \end{aligned}$$

This term contributes $1 = \binom{3}{0}$ to β_0, $-3 = -\binom{3}{1}$ to β_1, $3 = \binom{3}{2}$ to β_2, and $-1 = -\binom{3}{3}$ to β_3. In general, a term (with coefficient 1) in the first normal form with k factors contributes $(-1)^i\binom{k}{i}$ to β_i. Hence the total contributed to β_i by terms containing k factors is $(-1)^i\binom{k}{i}\alpha_k$, and

$$\beta_i = (-1)^i \sum_k \binom{k}{i} \alpha_k. \tag{22}$$

Of course α_i is given by the same formula in terms of the β_k:

$$\alpha_i = (-1)^i \sum_k \binom{k}{i} \beta_k. \tag{23}$$

These formulas are easily verified in the above examples.

PRINCETON UNIVERSITY.

(ADDED IN PROOF). P. Delens has observed (Comptes Rendus, vol. 195 (1932), p. 686–689) that with the help of (8), (9), and (10), the formulas of the algebra of logic take on an arithmetical form. However, he does not employ characteristic functions.

Some recent literature may be noted: C. I. Lewis and C. H. Langford, Symbolic Logic, New York 1932; E. V. Huntington, New sets of independent postulates for the algebra of logic, Transactions of the American Mathematical Society, vol. 35 (1933), pp. 274–304.

AN INEQUALITY RELATED TO THE ISOPERIMETRIC INEQUALITY

L. H. LOOMIS AND H. WHITNEY

In this note we shall prove the following theorem.

THEOREM 1. *Let m be the measure of an open subset O of Euclidean n-space, and let $m_1, \cdots, m_n$ be the $(n-1)$-dimensional measures of the projections of O on the coordinate hyperplanes. Then*

$$m^{n-1} \leqq m_1 m_2 \cdots m_n. \tag{1}$$

Note that for n-dimensional intervals with faces parallel to the coordinate hyperplanes, (1) holds with the equality sign.

With any reasonable definition of the $(n-1)$-dimensional measure s of the boundary of O, $s \geqq 2m_i$ for each i, so that (1) gives

$$m^{n-1} \leqq s^n/2^n; \tag{2}$$

this is the isoperimetric inequality, without the best constant. Since the proof of the isoperimetric inequality with the best constant is difficult,[1] and since its applications do not necessarily require the best constant, our elementary proof of the theorem may be of some interest.

We first reduce the problem to a combinatorial one, in the following theorem.

THEOREM 2. *Let S be a set of cubes from a cubical subdivision of n-space; let S_i be the set of $(n-1)$-cubes obtained by projecting the cubes of S onto the ith coordinate hyperplane. Let N and N_i be the numbers of cubes in S and S_i respectively. Then*

$$N^{n-1} \leqq N_1 N_2 \cdots N_n. \tag{3}$$

Assuming Theorem 2, we prove Theorem 1 as follows. Given $\epsilon > 0$, choose a cubical subdivision of n-space into cubes of side δ, with δ so small that if S is the set of cubes interior to O forming the set $\bar{S}$, $\mu(O-\bar{S}) < \epsilon$ (μ = measure). Then

$$[\mu(\bar{S})]^{n-1} = N^{n-1}\delta^{n(n-1)} \leqq (N_1\delta^{n-1}) \cdots (N_n\delta^{n-1}) \leqq m_1 \cdots m_n,$$

and since ϵ is arbitrary, (1) follows.

Received by the editors July 31, 1948.

[1] See E. Schmidt, *Über das isoperimetrische Problem in Raum von n Dimensionen*, Math. Zeit. vol. 44 (1939) pp. 689–788.

PROOF OF THEOREM 2.[3] If $n=2$, the theorem is clear; we shall use induction on n. Each cube of S projects into an interval on the first coördinate axis; let $I_1, \cdots, I_k$ be the intervals thus obtained. Let T_i be the set of cubes projecting onto I_i, and let T_{ij} be the set of $(n-1)$-cubes obtained by projecting the cubes of T_i into the jth coordinate hyperplane $(j=2, \cdots, n)$. Let a_i and a_{ij} be the numbers of cubes in T_i and T_{ij} respectively. Clearly

(4) $$\sum_{i=1}^{k} a_i = N, \qquad a_i \leqq N_1 \; (i = 1, \cdots, k),$$

(5) $$\sum_{i=1}^{k} a_{ij} = N_j \qquad (j = 2, \cdots, n).$$

Also, by induction,

(6) $$a_i^{n-2} \leqq a_{i2} \cdots a_{in} \qquad (i = 1, \cdots, k).$$

From (6) and the second part of (4) we obtain

$$a_i^{n-1} \leqq N_1 a_{i2} \cdots a_{in} \qquad (i = 1, \cdots, k).$$

Now using successively the first part of (4), the above inequality, Hölder's inequality, and (5), we see that

$$\begin{aligned} N = \sum_{i=1}^{k} a_i &\leqq \sum_{i=1}^{k} N_1^{1/(n-1)} \prod_{j=2}^{n} a_{ij}^{1/(n-1)} \\ &\leqq N_1^{1/(n-1)} \prod_{j=2}^{k} \left(\sum_{i=1}^{m} a_{ij} \right)^{1/(n-1)} = \prod_{j=1}^{n} N_j^{1/(n-1)}, \end{aligned}$$

as required.

HARVARD UNIVERSITY

[3] The authors are indebted to M. R. Demers for a simplification in the proof of this theorem.

THE EXTENSION OF LINEAR FUNCTIONALS DEFINED ON H^∞

ANDREW M. GLEASON AND HASSLER WHITNEY

Dedicated to Marston Morse

1. Introduction. We consider the Banach space L^∞ of (classes of) bounded measurable complex functions on the unit circle Γ_1. It has a subspace $H = H^\infty$ consisting of those functions h which are the boundary value functions (existing almost everywhere by Fatou's theorem) of bounded analytic functions $\hat{h}$ in the interior S_1 of Γ_1. By the Hahn-Banach theorem, any bounded linear functional ϕ defined on H^∞ can be extended over L^∞ with no increase in norm. It is the primary purpose of this paper to prove that this extension is unique, provided that ϕ is defined (over H) by an integral with kernel in L^1. Without this hypothesis, uniqueness may fail,[1] as we shall see in 10.1.

The simplest case of the theorem occurs when $\phi(h)$ is defined to be $\hat{h}(0)(h \in H)$; then, for the unique norm preserving extension ϕ_e of ϕ, $\phi_e(f)$ is the average value of f on Γ_1.

If we assume that the extended functional is also defined through a kernel in L^1, then the uniqueness follows easily. The key to the proof (with no such assumption) is contained in lemma 3.1, asserting that any real nonnegative function f in L^∞ may be approximated from below in the L^1 norm by a nonnegative function f' whose harmonic extension through S_1 has a bounded conjugate function. At first sight this might seem surprising; for f' must in general be discontinuous, and the conjugate of a harmonic function whose boundary value function has a simple discontinuity is unbounded.

The proof, in the case of a nonnegative kernel, follows quickly from the lemma. For a general $K \in L^1$ which is extremal over H, i.e. such that $\sup \int hKd\theta = \int |K| \, d\theta$ (requiring $h \in H$ and $\|h\| \leqq 1$), an extremal function $h_0 \in H$(i.e. $\|h_0\| = 1, \int h_0 K d\theta = \int |K| \, d\theta$) is made use of; with its help, the proof of uniqueness is reduced to the previous case. If ϕ is defined through an arbitrary kernel $J \in L^1$, we use a lemma of Rogosinski and Shapiro to replace J by an extremal kernel K.

In the second half of the paper we consider the analogous problem on a compact bordered Riemann surface $\bar{S}$, with border Γ. In general, uniqueness of the extension fails; a simple example is given in 10.2. The extension of the given functional ϕ is unique at least as far as a

Received March 16, 1961. The first named author was supported during the preparation of this paper by the Office of Ordnance Research at the Institute for Advanced Study.

[1] An example of this was first given us by R. R. Phelps.

certain subspace of finite codimension ρ in L^∞, where ρ is the first Betti number of $\bar{S}$. If ϕ is defined in H by a real nonnegative kernel K, then the subspace mentioned is the annihilator G of a space P defined through considering the periods of conjugates of harmonic functions (see 6.16 and 6.17). The set of all norm preserving extensions through L^∞ forms a closed convex cell of dimension at most ρ; in particular, the extension is unique if $\rho = 0$. For $\rho > 0$, the extension may or may not be unique, depending on the nature of the kernel K. Some examples of this are given in § 10. The principal facts are summarized in Theorem 9.1.

2. **Preliminaries.** A function $f \in L^\infty$ has a unique harmonic extension $\hat{f}$ over S_1 such that the boundary value function of $\hat{f}$ equals f a.e. (almost everywhere). If f is continuous, the boundary values are taken on continuously. If $f = f_1 + if_2$, where f_1 and f_2 are real functions, then $\hat{f} = \hat{f}_1 + i\hat{f}_2$; now $f \in H$ if and only if $\hat{f}_2$ is a harmonic conjugate of $\hat{f}_1$; then $\hat{f}$ is the analytic function with boundary values f.

Furthermore, H is the set of all $h \in L^\infty$ such that

$$\int_{\Gamma_1} h(\theta)e^{in\theta}d\theta = 0 , \qquad n = 1, 2, \cdots .$$

It follows that, on considering L^∞ as the conjugate space of L^1 (setting $f \cdot K = \int_{\Gamma_1} fKd\theta$ for $f \in L^\infty$, $K \in L^1$), H is closed in the weak* topology. Since the bounded analytic functions in S_1 form an algebra, passing to boundary values shows that H is an algebra.

Denote by $\|\ \|$, $\|\ \|_1$, the norms in L^∞ and L^1 respectively. For a subspace E of L^∞, $\|\phi\|_E$ denotes the norm of the functional ϕ considered in E only.

The first (well known) lemma applies to $L^\infty(X)$ for any measure space X.

2.1. LEMMA. *Suppose ϕ is a bounded linear functional on L^∞ such that*

$$\phi(1) = \|\phi\| . \tag{2.2}$$

Then ϕ is real and nonnegative; that is, $\phi(f)$ is real if f is real a.e., and $\phi(f) \geqq 0$ if $f \geqq 0$ a.e. Moreover,

$$\phi(Re\, f) = Re\, \phi(f) , \qquad f \in L^\infty . \tag{2.3}$$

Proof. The case $\phi = 0$ being trivial, we suppose $\|\phi\| > 0$; we may normalize so that $\phi(1) = \|\phi\| = 1$. Take any $f \in L^\infty$. Now

$$Re\, \phi(f) = Re\, \phi(f + A) - A \leqq \|f + A\| - A$$

for any real constant A; letting $A \to \infty$ gives

$$Re\, \phi(f) \leqq \text{ess sup}\, \{Re\, f(\theta)\}\,.$$

Replacing f by $-f$ gives

$$Re\, \phi(f) \geqq \text{ess inf}\, \{Re\, f(\theta)\}\,.$$

Hence if f is purely imaginary, $Re\, \phi(f) = 0$, and $\phi(f)$ is purely imaginary; it follows that if f is real then $\phi(f)$ is real. The last inequality shows that ϕ is nonnegative. Finally, if $f = f_1 + if_2$ then $\phi(f) = \phi(f_1) + i\phi(f_2)$, and (2.3) follows.

To state that a function has bounded real part is to state that its values lie in a vertical strip. We approximate to such functions by bounded functions by means of the following lemma.

We consider a half closed vertical strip and half the unit disk;

$$T^+ = \left\{\lambda\colon 0 \leqq Re\, \lambda < \frac{\pi}{4}\right\}, \qquad S_1^+ = \{z\colon Re\, z \geqq 0, |z| < 1\}\,.$$

2.4. LEMMA. *The map* $\lambda \to \tan \lambda$ *carries* T^+ *conformally onto* S_1^+. *Moreover,*

(2.5) $\qquad Re \arctan(\alpha \tan \lambda) \leqq Re\, \lambda$ *if* $\lambda \in T^+$, α *real,* $0 \leqq \alpha \leqq 1$.

Proof. The first statement is standard. To prove the second, it is sufficient to show that Re arctan $\alpha\beta$, regarded as a function of the real variable α, is monotone increasing provided that $Re\, \beta > 0$. Since

$$c = [(1 + \alpha^2\beta^2)(1 + \alpha^2\bar{\beta}^2)]^{-1} > 0\,,$$

we have

$$\frac{\partial}{\partial\alpha} Re \arctan \alpha\beta = Re \frac{\partial}{\partial\alpha} \arctan \alpha\beta = Re \frac{\beta}{1 + \alpha^2\beta^2}$$
$$= cRe[\beta(1 + \alpha^2\bar{\beta}^2)] = cRe[\beta + \alpha^2 |\beta|^2\bar{\beta}] > 0\,,$$

as required.

3. Approximation from below.

We can now prove

3.1. LEMMA. *Given the real nonnegative function f in L^∞, there exists a sequence $\{h_n\}$ of functions in H such that*

(3.2) $\qquad$ *for all* n, $0 \leqq Re\, h_n \leqq f$ *a.e.* ,

(3.3) $\qquad$ $Re\, h_n \to f$ *in* L^1 .

Proof. We may assume $f(\theta) < \pi/4$ a.e. The harmonic extension $\hat{f}$ is the real part of an analytic function g in S_1 where $g(0)$ is real. Since $0 \leqq Re\, g(z) < \pi/4$, g maps S_1 into T^+. Set

$$g_n(z) = \arctan [(1 - 1/n) \tan g(z)] . \tag{3.4}$$

Clearly g_n is a bounded analytic function in S_1. Let h_n be the boundary value function of g_n. By (2.5), $Re\, g_n(z) \leqq Re\, g(z)$, and (3.2) follows. Moreover,

$$\begin{aligned} \| f - Reh_n \|_1 &= \int_{\Gamma_1} (f - Reh_n) d\theta \\ &= 2\pi[Reg(0) - Reg_n(0)] = 2\pi[g(0) - g_n(0)] \end{aligned}$$

by the mean value theorem for harmonic functions. Since arctan is continuous in the real interval [0, 1], this tends to 0, and (3.3) is proved.

4. The theorem for nonnegative kernels. The theorem in this case is:

4.1. THEOREM. *Suppose ϕ is a bounded linear functional on L^∞ such that*

$$\| \phi \| = \| \phi \|_H \tag{4.2}$$

and

$$\phi(f) = \int_{\Gamma_1} f(\theta) K(\theta) d\theta \tag{4.3}$$

for all f in H, where K is a real nonnegative function in L^1. Then (4.3) holds for all $f \in L^\infty$.

Proof. By (4.3), $| \phi(h) | \leqq \| h \| \cdot \| K \|_1$, $h \in H$; hence (since $1 \in H$)

$$\| \phi \| = \| \phi \|_H \leqq \| K \|_1 = \phi(1) \leqq \| \phi \| .$$

Therefore $\phi(1) = \| \phi \|$, and Lemma 2.1 applies.

Suppose $f \in L^\infty$ is real and nonnegative. Choose the sequence $\{h_n\}$ by Lemma 3.1. By Lemma 2.1, $\phi(f) \geqq \phi(\mathrm{Re}\, h_n) = \mathrm{Re}\, \phi(h_n)$, and hence

$$\int fK - \phi(f) \leqq \int fK - Re \int h_n K = \int (f - Re\, h_n) K .$$

Now $K \in L^1$, the functions $f - Re\, h_n$ are uniformly essentially bounded, and $f - Re\, h_n \to 0$ in L^1; it is elementary to show therefore that $\int (f - Re\, h_n) K \to 0$. Thus $\phi(f) \geqq \int fK$ for real nonnegative $f \in L^\infty$.

If $f \in L^\infty$ is real, pick a real number α so that $\alpha < f(\theta)$ a.e. Then

$$\phi(f) = \phi(f - \alpha) + \alpha\phi(1) \geqq \int (f - \alpha)K + \alpha \int K = \int fK .$$

Replacing f by $-f$ gives the opposite inequality; hence (4.3) holds for real f. Since ϕ is linear, (4.3) holds also for complex f.

5. The general theorem for the unit circle. Given any linear functional ψ on L^∞ and any measurable subset Z of Γ_1 with characteristic function χ_Z, define

(5.1) $$\psi_Z(f) = \psi(\chi_Z f) .$$

5.2. LEMMA. *Let ψ be a bounded linear functional on L^∞, and let X and Y be complementary measurable subsets of Γ_1. Then*

(5.3) $$\| \psi \| = \| \psi_X \| + \| \psi_Y \| .$$

Proof. Since $\psi = \psi_X + \psi_Y$, $\| \psi \| \leqq \| \psi_X \| + \| \psi_Y \|$. Conversely, given $\varepsilon > 0$, choose g_1 and g_2 in L^∞ so that $\psi_X(g_1)$ and $\psi_Y(g_2)$ are real, $\| g_1 \| = \| g_2 \| = 1$, and

$$\psi_X(g_1) \geqq \| \psi_X \| - \varepsilon , \qquad \psi_Y(g_2) \geqq \| \psi_Y \| - \varepsilon .$$

Set $g = \chi_X g_1 + \chi_Y g_2$. Then $\| g \| \leqq 1$, and

$$\psi(g) = \psi_X(g_1) + \psi_Y(g_2) \geqq \| \psi_X \| + \| \psi_Y \| - 2\varepsilon ,$$

giving the opposite inequality.

5.4. THEOREM. *Suppose that ψ is a linear functional on L^∞, K is in L^1,*

(5.5) $$\| \psi \| = \| \psi \|_H = \| K \|_1 ,$$

and

(5.6) $$\psi(f) = \int_{\Gamma_1} f(\theta)K(\theta)d\theta$$

for all $f \in H$. Then (5.6) *holds for all $f \in L^\infty$.*

Proof. Since the unit ball B in $L^\infty = (L^1)^*$ is weak* compact (see for instance [2], p. 37) and H is weak* closed, $H \cap B$ is weak* compact. Because of (5.6), ψ is weak* continuous in H; hence $\psi(H \cap B)$ is compact, and there is an $h_0 \in H$ such that

(5.7) $$\| h_0 \| \leqq 1 , \qquad \psi(h_0) = \| \psi \|_H .$$

The equality, with (5.6) and (5.5), gives

$$\int_{\Gamma_1} h_0(\theta)K(\theta)d\theta = \int_{\Gamma_1} |K(\theta)|\, d\theta\,, \tag{5.8}$$

from which we conclude that

(5.9) $h_0(\theta)K(\theta)$ is real and nonnegative a.e., and

(5.10) for almost all θ, either $|h_0(\theta)| = 1$ or $K(\theta) = 0$.

We remark aside that h_0 is unique a.e. (unless $\psi = 0$); for a second such element h_1 would agree a.e. with h_0 wherever $K \neq 0$, by (5.9) and (5.10), hence $h_1 = h_0$ on a set of positive measure, and it is known that this implies that $h_1 = h_0$ a.e. (see for instance [6], vol. II, p. 203).

Set

$$\phi(f) = \psi(h_0 f)\,, \qquad f \in L^\infty\,. \tag{5.11}$$

Since H is a ring,

$$\phi(h) = \psi(h_0 h) = \int h h_0 K\,, \qquad h \in H\,.$$

Also $\|h_0\| \leqq 1$, $\|h_0 f\| \leqq \|f\|$, and hence $\|\phi\| \leqq \|\psi\|$. Now

$$\|\phi\| \leqq \|\psi\| = \|K\|_1 = \phi(1) \leqq \|\phi\|_H \leqq \|\phi\|\,,$$

giving $\|\phi\| = \|\phi\|_H$. By Theorem 4.1,

$$\psi(h_0 f) = \phi(f) = \int f h_0 K \text{ for all } f \in L^\infty\,.$$

If h_0 had a bounded inverse h_1, we could use this to give $\psi(f) = \psi(h_0 h_1 f) = \int fK$; but this need not be the case, as we shall see by an example in 10.4.

By (5.10), $h_0 \bar{h}_0 K = K$ a.e.; hence replacing f by $\bar{h}_0 f$ in the above equation gives

$$\psi(h_0 \bar{h}_0 f) = \int fK \text{ for all } f \in L^\infty\,. \tag{5.12}$$

Let X be the subset of Γ_1 where K vanishes, and let Y be the complementary subset. Then $h_0 \bar{h}_0 = 1$ a.e. on Y. In the notation of Lemma 5.2,

$$\psi_Y(f) = \psi(\chi_Y f) = \psi(\chi_Y h_0 \bar{h}_0 f) = \int \chi_Y f K = \int fK$$

for all $f \in L^\infty$, whence, with the aid of (5.5) and (5.3),

$$\|\psi_Y\| = \|K\|_1 = \|\psi\| = \|\psi_X\| + \|\psi_Y\|\,.$$

Therefore $\psi_X = 0$, and

$$\psi(f - h_0\bar{h}_0 f) = \psi(\chi_x(f - h_0\bar{h}_0 f)) = \psi_x(f - h_0\bar{h}_0 f) = 0$$

for all $f \in L^\infty$. Thus (5.12) gives (5.6) for all $f \in L^\infty$, as claimed.

We wish to give a theorem without the hypothesis on the norm of the kernel. For this, the given kernel must be replaced by an "extremal kernel". This is possible, by the following lemma of Rogosinski and Shapiro, [4], Theorem 8, p. 303. (A proof will be given in 8.9.)

5.13. LEMMA. *Suppose ψ is a linear functional defined on H by $\psi(h) = \int hJ\ (h \in H)$, where $J \in L^1$. Then there is a function $K \in L^1$, unique a.e., such that $\psi(h) = \int hK\ (h \in H)$ and $\| K \|_1 = \| \psi \|_H$.*

Combining the last two results gives

5.14. THEOREM. *Suppose ψ is a bounded linear functional on H such that*

$$\psi(h) = \int_{\Gamma_1} h(\theta)J(\theta)d\theta \text{ for all } h \in \mathrm{H}\ , \tag{5.15}$$

where $J \in L^1$. Then ψ has a unique norm preserving extension ψ_e over L^∞. Moreover, there is a $K \in L^1$, unique a.e., such that

$$\psi_e(f) = \int_{\Gamma_1} f(\theta)K(\theta)d\theta \text{ for all } f \in L^\infty\ , \tag{5.16}$$

we have $\| K \|_1 = \| \psi \|_H = \| \psi_e \|$. If $\| J \|_1 = \| \psi \|_H$, then $K = J$ a.e.

6. **Bordered Riemann surfaces.** We now turn our attention to the analogous problem for a compact bordered Riemann surface $\bar{S}$. The theorems are similar to those of the preceding sections and the methods of proof are essentially the same, but we must take account of the fact that not every harmonic function is the real part of an analytic function. In this section we recall some of the facts about harmonic and analytic functions on $\bar{S}$.

6.1. A bordered Riemann surface $\bar{S}$ is a connected surface carrying an oriented conformal structure which is everywhere locally isomorphic to a relatively open subset of the closed upper half plane, these isomorphisms being given by the local coordinate functions (see for instance [3]). Let S denote the interior of $\bar{S}$, and Γ, the border; S is a Riemann surface in the ordinary sense. We assume always that $\bar{S}$ is compact and Γ is not empty; then Γ is a real analytic manifold with a natural orientation, consisting of a finite number of disjoint Jordan curves.

6.2. For any real continuous function f on Γ there is a unique harmonic function $\hat{f}$ in S which extends f continuously over $\bar{S}$ (see also 6.7). For each point $p \in S$ there is a real and everywhere positive analytic differential ω_p on Γ such that, for all f as above,

$$\hat{f}(p) = \int_\Gamma f\omega_p \, . \tag{6.3}$$

Choose $p_0 \in S$; this point will be fixed throughout the paper. The differential ω_{p_0} defines a measure μ on Γ (called the harmonic measure for p_0). Expressing ω_p in terms of ω_{p_0}, (6.3) becomes

$$\hat{f}(p) = \int_\Gamma f(q)k(p, q)d\mu(q) \, . \tag{6.4}$$

For each $p \in S$, $k(p, q)$ is an everywhere positive real analytic function on Γ. Furthermore, k is continuous in $S \times \Gamma$, and for each fixed q, is harmonic in S.

6.5. Measure theoretic statements will always refer to the measure μ. We form as usual the spaces L^1 and L^∞ of integrable and of bounded complex valued measurable functions respectively, and use $\|\ \|_1$ and $\|\ \|$ for the corresponding norms. Let L^1_{Re} and L^∞_{Re} denote the spaces of real functions in L^1 and L^∞ respectively. To reduce the notation we shall not distinguish between measurable functions and their equivalence classes modulo null sets. With the pointwise operations of addition and multiplication, the Banach space L^∞ becomes a Banach algebra.

As in § 2, L^∞ is the conjugate space of L^1, under the definition $f \cdot K = \int_\Gamma fKd\mu$. Again, the closed unit ball B in L^∞ is weak* compact, as is $B_{Re} = B \cap L^\infty_{Re}$ in L^∞_{Re}. Since L^1 is separable, the weak* topology of B is metrizable ([1], p. 426), so that questions of convergence in B can be handled with sequences and the Bolzano-Weierstrass theorem applies.

6.6. The space L^1, defined through the choice of p_0, is not intrinsically related to $\bar{S}$. This defect may be remedied as follows. Let C be the space of continuous functions on Γ. An analytic differential ω on Γ determines a linear functional ϕ_ω on C by the formula

$$\phi_\omega(f) = \int_\Gamma f\omega \, .$$

The norm is given by $\|\phi_\omega\| = \int_\Gamma |\omega|$. Let M be the closure, in the conjugate C^* of C, of the set of all ϕ_ω. Clearly, M is intrinsically related to $\bar{S}$. It is easy to identify L^1 with M, since any differential form on Γ may be written as a multiple of ω_{p_0} (see 6.2). The choice of

p_0 enables us to interpret elements of M as functions on Γ; this is valuable because of the corresponding identification of M^* with L^∞ and because the analytic properties of k may be used. Since the null sets on Γ do not depend on the choice of p_0, the space L^∞ is intrinsic a priori; it is easy to check that its identification with M^* is independent of the choice of p_0.

6.7 For each complex function $f \in L^\infty$, its harmonic extension $\hat{f}$ through S may be defined by (6.4); it is bounded by $\|f\|$, and has the boundary value functions f a.e. on Γ.

6.8. LEMMA. *Let f and the sequence $\{f_n\}$ be in L^∞. If $f_n \to f$ weak*, then $\hat{f}_n \to \hat{f}$ pointwise. If the $\hat{f}_n$ are uniformly bounded and $\hat{f}_n \to \hat{f}$ pointwise, then $f_n \to f$ weak*.*

The first statement follows from (6.4). Conversely, suppose the $\hat{f}_n$ are uniformly bounded, $\hat{f}_n \to \hat{f}$ in S, but $f_n \to f$ weak* is false. Since λB in the weak* topology is metrizable and compact for any real λ, we may find a subsequence $\{f_{m_i}\}$ converging weak* to g, where $\|f - g\| > 0$. Then $\hat{f}_{m_i} \to \hat{g}$ in S, hence $\hat{g} = \hat{f}$ in S, and hence $g = f$ a.e. on Γ, a contradiction.

6.9. As in the case of the unit circle, H is the space of boundary values of bounded analytic functions in S. Again, for any $f \in L^\infty$, $f \in H$ if and only if $\hat{f}$ is analytic in S, or (setting $f = f_1 + if_2$ with f_1 and f_2 real) $\hat{f}_2$ is a conjugate of $\hat{f}_1$. As before, H is a subalgebra of L^∞.

6.10. LEMMA. *H is weak* closed in L^∞.*

Proof. By [2], p. 39, it is sufficient to show that $H \cap B$ is weak* closed. As noted in 6.5, convergence in B is sequential. Suppose that $\{h_n\}$ is a sequence in $H \cap B$ which converges to $f \in L^\infty$ weak*. By Lemma 6.8, $\hat{h}_n \to \hat{f}$ in S. Since the $\hat{h}_n$ are uniformly bounded, $\hat{f}$ is analytic, by a general convergence theorem; hence $f \in H$, as required.

6.11. Let $f \in L^\infty$. We form the harmonic extension $\hat{f}$, its differential $d\hat{f}$, and the conjugate harmonic differential $(d\hat{f})^*$. Suppose $\hat{f}$ had a conjugate function $\hat{f}^*$. Then if ν is a Jordan arc from p_0 to p_1 in S, with normal vector n, we would have

$$\hat{f}^*(p_1) - \hat{f}^*(p_0) = \int_\nu \frac{\partial \hat{f}^*(p)}{\partial s} ds = \int_\nu \frac{\partial \hat{f}(p)}{\partial n} ds ;$$

substituting in (6.4) and changing the order of integration gives

$$\hat{f}^*(p_1) - \hat{f}^*(p_0) = \int_\Gamma f(q)\Phi_\nu(q)d\mu(q) \text{ if } \hat{f}^* \text{ exists,} \tag{6.12}$$

where

$$\Phi_\nu(q) = \int_\nu \frac{\partial k(p, q)}{\partial n} ds , \tag{6.13}$$

the integral being defined with the help of the local conformal structure of S.

6.14. In the general case, $\hat{f}^*$ exists if and only if $\int_\Gamma f\Phi_\delta d\mu = 0$ for each closed curve δ in S, or equivalently, for each closed curve of a homology basis. Let $\delta_1, \cdots, \delta_\rho$ be such a basis. Here ρ is the first Betti mumber of S or $\bar{S}$, and

$$\rho = 1 - \chi = 2g + m - 1 ,$$

where χ is the Euler characteristic of $\bar{S}$, g is its genus, and m is the number of components of Γ.

Now $\hat{f}^*$ exists if and only if the periods

$$\int_{\delta_i} (d\hat{f})^* = \int_\Gamma f(q)\alpha_i(q)d\mu(q) , \qquad \alpha_i(q) = \Phi_{\delta_i}(q) , \tag{6.15}$$

all vanish. The α_i are real analytic on Γ and are therefore in L^1.

6.16. Let P and P_{Re} be the complex and real spaces respectively spanned by the α_i; then $P_{Re} = P \cap L^1_{Re}$.

Let $c_1, \cdots, c_\rho$ be given real numbers. There exists an analytic differential ω on $\bar{S}$ which can be extended analytically over the double of $\bar{S}$ and which has the periods $ic_1, \cdots, ic_\rho$ (see [3], p. 172). Since the periods of $Re\ \omega$ vanish there is a harmonic function $\hat{f}$ in $\bar{S}$ with $d\hat{f} = Re\,\omega$; now $\hat{f}$ is bounded in $\bar{S}$, and $(d\hat{f})^* = Im\,\omega$ has the periods $c_1, \cdots, c_\rho$, It follows from (6.15) that $\alpha_1, \cdots, \alpha_\rho$ are linearly independent over the reals. Therefore P_{Re} has real dimension ρ, and P has complex dimension ρ.

6.17. Considering $P \subset L^1$, let $G \subset L^\infty$ be its annihilator space. Then G is weak* closed and has codimension ρ in L^∞. Hence every linear functional on L^∞ which vanishes on G has the form $f \to \int_\Gamma f\alpha$ for some $\alpha \in P$. Moreover $G_{Re} = G \cap L^\infty_{Re}$ is the annihilator of P_{Re} in L^∞_{Re}, and $G = G_{Re} + iG_{Re}$. The definitions of P and G show that for any $f \in L^\infty$, $\hat{f}$ has a harmonic conjugate (not necessarily bounded) if and only if $f \in G$. It follows that $H \subset G$.

7. The theorem with an extremal kernel. Given the linear functional ϕ on H, defined by means of an extremal kernel K, we show here that any norm preserving extension over L^∞ is given by means of a modified kernel, and we study the set of norm preserving extensions. As before, the proof is based on an approximation lemma. Recall the definition of G_{Re} in 6.17.

7.1. LEMMA. *Given the nonnegative function $f \in G_{Re}$, there exists a sequence $\{h_n\}$ of functions in H such that $0 \leqq Re h_n \leqq f$ a.e. in Γ and $Re h_n \to f$ in L^1.*

Proof. As in § 3, assume $f(q) < \pi/4$ a.e. in Γ. Since $f \in G_{Re}$, we may find g, g_n and h_n as before. Then $0 \leqq Re\, h_n \leqq f$, and that $Re\, h_n \to f$ in L^1 follows from

$$\| f - Re\, h_n \|_1 = \int_\Gamma (f - Re\, h_n) d\mu = Re\, g(p_0) - Re\, g_n(p_0) \to 0 .$$

7.2. THEOREM. *Suppose ϕ is a linear functional on L^∞ such that $\| \phi \| = \| \phi \|_H$ and*

$$\phi(f) = \int_\Gamma f K d\mu \tag{7.3}$$

for all $f \in H$, where $K \in L^1$ is real and nonnegative. Then (7.3) *holds for all $f \in G$. Moreover, there exists a function $\alpha \in P_{Re}$ such that $K + \alpha$ is nonnegative and*

$$\phi(f) = \int_\Gamma f(K + \alpha) d\mu \text{ for all } f \in L^\infty . \tag{7.4}$$

Proof. The first statement follows as in the proof of Theorem 4.1, with Lemma 7.1 replacing Lemma 3.1.

To derive the second statement, set

$$\psi(f) = \phi(f) - \int_\Gamma f K d\mu , \qquad f \in L^\infty ;$$

then $\psi(g) = 0$ for all $g \in G$, and (see 6.17) there is an $\alpha \in P$ such that

$$\psi(f) = \int_\Gamma f \alpha d\mu \qquad f \in L^\infty .$$

These two relations give (7.4). Since $\phi(1) = \| \phi \|$ (compare § 4), ϕ is real and nonnegative (Lemma 2.1); hence $K + \alpha$ is real and nonnegative.

7.5. THEOREM. *Let ϕ be a linear functional defined on H by* (7.3), *where $K \in L^1$ is real and nonnegative. Then the set of norm preserving* *extensions of* ϕ *over* L^∞ *is in one to one correspondence with the set* W *of functions* $\alpha \in P_{Re}$ *such that* $K+\alpha$ *is nonnegative. The set* W *is a compact convex cell of real dimension at most* ρ, *the first Betti number of* $\bar{S}$.

Proof. By the last theorem, every norm preserving extension corresponds to some $\alpha \in W$; clearly this correspondence is one to one. It is onto W. For suppose $\alpha \in W$. Then if $\phi_e(f) = \int_\Gamma f(K+\alpha)d\mu$, $\phi_e = \phi$ over H, since $H \subset G$. Since $1 \in G$,

$$\| K+\alpha \|_1 = \int (K+\alpha)d\mu = \int K d\mu = \| K \|_1 ;$$

therefore $\| \phi_e \| = \| \phi \|$, and ϕ_e is a norm preserving extension of ϕ corresponding to α.

Clearly W is a closed convex subset of P_{Re}, and hence is of dimension $\leqq \rho$. Since $\| K+\alpha \|_1 = \| K \|_1$, W is bounded and hence compact.

7.6. THEOREM. *Suppose that* ψ *is a linear functional defined on* H *by*

$$\psi(h) = \int_\Gamma hK d\mu , \tag{7.7}$$

where $K \in L^1$ *and* $\| K \|_1 = \| \psi \|_H$. *Then there is a function* $h_0 \in H$ *such that every norm preserving extension* ψ_e *of* ψ *over* L^∞ *has the form*

$$\psi_e(f) = \int_\Gamma f(K+\alpha\bar{h}_0)d\mu, \text{ where } K+\alpha\bar{h}_0 \in L^1 , \tag{7.8}$$

for some $\alpha \in P_{Re}$. *(See also* (7.14), (7.15) *and* (7.12).)

Proof. We follow the proof of Theorem 5.4 (using 6.5 and 6.10), finding the extremal function h_0 for ψ such that

$$\| h_0 \| = 1 , \qquad \psi(h_0) = \| \psi \|_H , \qquad \int h_0 K d\mu = \int | K | d\mu . \tag{7.9}$$

Again we conclude that for almost all $q \in \Gamma$,

$$h_0(q)K(q) \geqq 0 \text{ and either } | h_0(q) | = 1 \text{ or } K(q) = 0 . \tag{7.10}$$

Given ψ_e, set

$$\phi(f) = \psi_e(h_0 f) , \qquad f \in L^\infty .$$

Then

$$\phi(h) = \psi_e(h_0h) = \int_\Gamma hh_0Kd\mu\,, \qquad h\in H\,,$$

since $h_0h\in H$. Also, as in § 5, $\|\phi\| = \|\psi\|_H = \|\phi\|_H$. Therefore for some $\alpha\in P_{Re}$,

$$\psi_e(h_0f) = \phi(f) = \int_\Gamma f(h_0K+\alpha)d\mu\,, \qquad f\in L^\infty\,, \tag{7.11}$$

by Theorem 7.2, and

$$h_0K+\alpha \geqq 0 \text{ a.e.}, \quad \|h_0K+a\|_1 = \|\phi\| = \|\psi\|_H\,. \tag{7.12}$$

Since $h_0\not\equiv 0$, $h_0=0$ at most in a set a measure 0 (see [6], vol. II, p. 203); hence there is a measurable function g on Γ such that $gh_0=1$ a.e. on Γ. Set

$$E_n = \{q\in\Gamma : |g(q)|\leqq n\}\,,$$

let χ_n be the characteristic function of E_n, and set $\psi_n(f)=\psi_e(f\chi_n)$. By Lemma 5.2, which extends to the present case,

$$\|\psi\|_H = \|\psi_e\| = \|\psi_n\| + \|\psi_e-\psi_n\|\,. \tag{7.13}$$

Since $gf\chi_n\in L^\infty$, (7.11) gives

$$\begin{aligned}\psi_n(f) &= \psi_e(gh_0f\chi_n) = \int_\Gamma gf\chi_n(h_0K+\alpha)d\mu\\ &= \int_{E_n} f(K+\alpha g)d\mu\,, \qquad f\in L^\infty\,.\end{aligned}$$

Hence

$$\int_{E_n}|K+\alpha g|\,d\mu = \|\psi_n\| \leqq \|\psi_e\|\,,$$

and letting $n\to\infty$ gives

$$K+\alpha g\in L^1\,, \qquad \|K+\alpha g\|_1 \leqq \|\psi_e\|\,.$$

Therefore, with the help of (7.9) and (7.11),

$$\begin{aligned}\|\psi_e\| &= \|\psi\|_H = \psi(h_0) = \int_\Gamma (h_0K+\alpha)d\mu\\ &= \int_\Gamma h_0(K+\alpha g)d\mu \leqq \int_\Gamma |K+\alpha g|\,d\mu \leqq \|\psi_e\|\,,\end{aligned}$$

and

$$\lim_{n\to\infty}\|\psi_n\| = \int_\Gamma |K+\alpha g|\,d\mu = \|\psi_e\|\,.$$

Now (7.13) gives $\lim\|\psi_e-\psi_n\|=0$; consequently

$$\psi_e(f) = \lim_{n\to\infty} \psi_n(f) = \int_\Gamma f(K + \alpha g)d\mu\ , \qquad f \in L^\infty\ .$$

From the equalities $\|h_0\| = 1$ and $\int h_0(K + \alpha g)d\mu = \int |K + \alpha g|\, d\mu$ we conclude that for almost all $q \in \Gamma$, either $|h_0(q)| = 1$ or $K(q) + \alpha(q)g(q) = 0$. Comparing this with (7.10) shows that

(7.14) $\qquad$ a.e. in Γ, either $|h_0(q)| = 1$ or $K(q) = \alpha(q) = 0$.

Since $gh_0 = 1$ a.e., we have $\alpha g = \alpha\bar{h}_0$ a.e., and the theorem is proved.

Note that $Kh_0\bar{h}_0 = K$ a.e., by (7.10); hence (7.8) may be written in the form

(7.15) $$\psi_e(f) = \int_\Gamma f\bar{h}_0(h_0K + \alpha)d\mu \text{ for all } f \in L^\infty\ .$$

8. **Existence of extremal kernels.** In order to obtain a result analogous to Theorem 5.14 for bordered Riemann surfaces we must extend the lemma of Rogosinski and Shapiro 5.13 to the more general setting. Our first object is to prove that, if A is the space of complex valued, real-analytic functions on Γ, then $H \cap A$ is weak* dense in H. Note that $H \cap A$ is the space of restrictions to Γ of functions holomorphic on $\bar{S}$.

8.1. LEMMA. *Let E be any real or complex Banach space and let B be the closed unit ball in its conjugate space E^*. Suppose G is a weak* closed linear subspace of E^* of finite codimension. Suppose $0 < \lambda \leqq 1$ and that A is a linear subspace of E^* such that $A \cap B$ is weak* dense in λB. Then $A \cap G \cap B$ is weak* dense in $G \cap \lambda B$.*

Proof. Since G is weak* closed, G is the annihilator of a certain closed linear manifold $P \subset E$. Since G has finite codimension in E^*, say m, P has dimension m. Elements of A are linear functionals on P by restriction, and indeed a total set of linear functionals, for if not there would be a $p \in P$ such that $p \neq 0$ and $a(p) = 0$ for all $a \in A$ and $A \cap B$ would not be weak* dense in λB. Hence we can find an m-dimensional linear subspace $F \subset A$ which is total on P. Then F is a linear complement of G in E^*.

Now let g_0 be any element of $G \cap \lambda B$. We must show that g_0 is the weak* limit of some directed system in $A \cap G \cap B$. By hypothesis there is a directed system $\{a_\alpha\}$ in $A \cap B$ such that $a_\alpha \to g_0$ weak*. Write $a_\alpha = f_\alpha + g_\alpha$ where $f_\alpha \in F$, $g_\alpha \in G$. Then $g_\alpha \in A \cap G$. For any $p \in P$, $a_\alpha(p) \to g_0(p) = 0$; hence $f_\alpha(p) \to 0$. Since F is finite dimensional and the elements of P determine a total set of linear functionals on F, this implies that $f_\alpha \to 0$ in the norm. Hence $g_\alpha \to g_0$ weak*. Now put

$\xi_\alpha = (1 + \|f_\alpha\|)^{-1}$. Then $\xi_\alpha \to 1$ so $\xi_\alpha g_\alpha \to g_0$ weak*. Moreover, $\|\xi_\alpha g_\alpha\| = \xi_\alpha \|a_\alpha - f_\alpha\| \leqq \xi_\alpha(1 + \|f_\alpha\|) = 1$, so $\xi_\alpha g_\alpha \in A \cap G \cap B$. This completes the proof.

For the next lemma, let A be the space defined at the beginning of § 8, and set $A_{Re} = A \cap L^\infty_{Re}$.

8.2. LEMMA. *Let $g \in G_{Re}$ (defined in 6.17). Then there exists a sequence $\{h_n\}$ in $H \cap A$ such that*

$$\|Re\, h_n\| \leqq \|g\|, \tag{8.3}$$

$$Re\, h_n \to g \text{ weak* in } L^\infty . \tag{8.4}$$

Proof. L^∞_{Re} is the real conjugate space of L^1_{Re} and G_{Re} is the annihilator of the finite dimensional subspace P_{Re} of L^1_{Re}. It is well known that $A_{Re} \cap B_{Re}$ is weak* dense in B_{Re}, where B_{Re} is the unit ball of L^∞_{Re}. By Lemma 8.1, $A_{Re} \cap G_{Re} \cap B_{Re}$ is weak* dense in $G_{Re} \cap B_{Re}$.

To prove the lemma we may clearly suppose that $\|g\| = 1$. As noted in 6.5, the weak* topology of B_{Re} is metrizable, so there is a sequence $\{f_n\}$ in $A_{Re} \cap G_{Re} \cap B_{Re}$ such that $f_n \to g$ weak*. Since $f_n \in G_{Re}$, $\hat{f}_n$ has a conjugate function, and $\hat{f}_n = Re\, \hat{h}_n$ where $\hat{h}_n$ is a holomorphic function on S. On the other hand, since $\hat{f}_n$ has real analytic boundary values, $\hat{h}_n$ is analytic on $\bar{S}$. Thus $h_n \in H \cap A$ and $Re\, h_n = f_n$. Now (8.4) has already been established and (8.3) follows because $f_n \in B_{Re}$ so that $\|f_n\| \leqq 1 = \|g\|$.

We can immediately deduce the following fact which may be useful in other connections.

8.5. THEOREM. *Let $\bar{H} = \{\bar{h} : h \in H\}$. Then $H + \bar{H}$ is weak* dense in G.*

Proof. Let M be the weak* closure of $H + \bar{H}$. Since $H \subset G$ and $\bar{H} \subset G$ while G is weak* closed, $M \subset G$. On the other hand, by the last lemma, every $g \in G_{Re}$ is the weak* limit of a sequence $\{1/2(h_n + \bar{h}_n)\}$ in $H + \bar{H}$, so $g \in M$. Thus $M \supset G_{Re}$. By linearity $M \supset G_{Re} + iG_{Re} = G$.

8.6. LEMMA. *$H \cap A$ is weak* dense in H. Indeed, $H \cap A \cap B$ is sequentially weak* dense in $H \cap B$.*

Proof. Given $h \in H \cap B$ we must find a sequence in $H \cap A \cap B$ which converges weak* to h. If h is constant, this is trivial, so we suppose not. Then arctan $\hat{h}$ maps S holomorphically into the open strip $T = \{\lambda : -\pi/4 < Re\, \lambda < \pi/4\}$. Now Re arctan $\hat{h}$ has a boundary value function $g \in G_{Re}$; by Lemma 8.2 we can choose a sequence $\{h_n\}$ in $H \cap A$ so that

$$(8.7)\qquad \begin{cases} \| Re\, h_n \| \leq \pi/4\,, & Re\, h_n \to g \text{ weak}^*\,, \\ Im\, \hat{h}_n(p_0) = Im \arctan \hat{h}(p_0). \end{cases}$$

Since $h_n \in H \cap A$ and tan is holomorphic in $\bar{T}$, tan $h_n \in H \cap A \cap B$.

Given $p \in S$, let ν be an arc from p_0 to p; now (6.12) and (8.7) give

$$Im\, \hat{h}_n(p) - Im\, \hat{h}_n(p_0) = \int_\Gamma Re\, h_n \Phi_\nu d\mu$$

$$\to \int_\Gamma g\Phi_\nu d\mu = \hat{g}^*(p) - \hat{g}^*(p_0)\;;$$

using (8.7) again gives $Im\, \hat{h}_n(p) \to Im$ arctan $\hat{h}(p)$. Also, by Lemma 6.8, $Re\, \hat{h}_n(p) \to Re \arctan \hat{h}(p)$; thus $\hat{h}_n(p) \to \arctan \hat{h}(p)$. Therefore $\tan \hat{h}_n(p) \to \hat{h}(p)$ in S, and by Lemma 6.8, $\tan h_n \to h$ weak*, completing the proof.

We shall need the following extension of a theorem of F. and M. Riesz.

8.8. LEMMA. *Let C be the space of continuous complex-valued functions on Γ. A Borel measure on Γ which is orthogonal to $H \cap C$ is absolutely continuous with respect to any harmonic measure.*

For a proof see Wermer [5] Lemma 3. Because of the context in which he is working, Wermer's statement includes the hypothesis that Γ consists of a single Jordan arc, but this hypothesis is not used in the proof.

We are finally in a position to extend the lemma of Rogosinski and Shapiro.

8.9. LEMMA. *Suppose ψ is a linear functional defined on H by*

$$\psi(h) = \int_\Gamma hJd\mu \qquad \textit{for all } h \in H\,,$$

where $J \in L^1$. Then there exists a function K in L^1 such that $\| K \|_1 = \| \psi \|_H$ and

$$(8.10)\qquad \psi(h) = \int_\Gamma hKd\mu \qquad \textit{for all } h \in H\,.$$

Proof. Following [4], we consider ψ on $H \cap C$ and take a norm preserving extension over C. The extension can be represented by a Borel measure ν. Now $\| \nu \| = \| \psi \|_{H \cap C}$ and $\psi(h) = \int h d\nu$ for $h \in H \cap C$. Hence $\int h(d\nu - Jd\mu) = 0$ for $h \in H \cap C$. By Lemma 8.8, we may write $d\nu - Jd\mu = Md\mu$ where $M \in L^1$. Put $K = J + M \in L^1$ and we have $d\nu = Kd\mu$ and $\| \nu \| = \| K \|_1$.

The linear functional $h \to \int h(K - J)d\mu$ vanishes for $h \in H \cap C$. Since $H \cap A \subset H \cap C$ is weak* dense in H (Lemma 8.6), $\int h(K - J)d\mu = 0$ for all $h \in H$, giving 8.10. Moreover, $\| \psi \|_H \leqq \| K \|_1 = \| \nu \| = \| \psi \|_{H \cap C} \leqq \| \psi \|_H$. This gives $\| K \|_1 = \| \psi \|_H$ and completes the proof.

9. The final theorem. Combining Lemma 8.9 with the theorems of § 7 gives

9.1. THEOREM. *Let $\bar{S}$ be a compact bordered Riemann surface with non-empty boundary Γ and first Betti number ρ. Let H be the space of boundary value functions of bounded analytic functions defined on S. Suppose ψ is a bounded linear functional defined on H by*

$$\psi(h) = \int_\Gamma h(q)J(q)d\mu(q) \qquad \text{for all } h \in H, \tag{9.2}$$

where μ is the harmonic measure on Γ associated with some point $p_0 \in R$, and J is in $L^1(\mu)$. Then every norm preserving extension ψ_e of ψ over L^∞ has the form

$$\psi_e(f) = \int_\Gamma f(q)K(q)d\mu(q) \qquad \text{for all } f \in L^\infty, \tag{9.3}$$

where $K \in L^1(\mu)$. The set of all such extensions forms a compact convex cell of dimension at most ρ; the extension is unique if $\rho = 0$.

The first part of the theorem follows directly from Lemma 8.9 and Theorem 7.6.

By 7.6 again, the set of norm preserving extensions corresponds to the set W of those functions $\alpha \in P_{Re}$ such that insertion in (7.8) or in (7.15) gives such an extension. As in the proof of Theorem 7.5, W is a compact subset of P_{Re}; this space has dimension ρ.

To prove convexity, suppose

$$\alpha_1, \alpha_2 \in W, \qquad 0 < t < 1, \qquad \alpha_3 = (1 - t)\alpha_1 + t\alpha_2;$$

let ψ_i be determined by (7.15), using α_i. That ψ_3 is an extension of ψ follows at once. Moreover, by (7.12), $h_0K + \alpha_i \geqq 0$ a.e. ($i = 1, 2$); hence $h_0K + \alpha_3 \geqq 0$ a.e., and (7.15) and (7.14) give

$$\| \psi_3 \| = \int_\Gamma | \bar{h}_0(h_0K + \alpha_3) | d\mu = \int_\Gamma (h_0K + \alpha_3)d\mu;$$

the similar formulas hold for $\| \psi_1 \|$ and $\| \psi_2 \|$. Since $\| \psi_1 \| = \| \psi_2 \| = \| \psi \|_H$, we have $\| \psi_3 \| = \| \psi \|_H$, which completes the proof.

10. Examples. We give here some particular cases where non-uniqueness or uniqueness holds, and an example concerning the proof of Theorem 5.4.

10.1. We give first the example promised in the introduction.

Suppose ψ is a nonzero bounded linear functional on L^∞ which vanishes on H and takes real values on L^∞_{Re}. By the Hahn decomposition theorem ψ may be represented as the difference $\psi^+ - \psi^-$ of two nonnegative linear functionals; here, ψ^+ is defined for nonnegative f by

$$\psi^+(f) = \sup\{\psi(g) : 0 \leqq g \leqq f\} ,$$

and is extended over the rest of L^∞ by linearity. Since ψ^+ and ψ^- are real and nonnegative, $\| \psi^+ \| = \psi^+(1)$ and $\| \psi^- \| = \psi^-(1)$; since $1 \in H$,

$$\| \psi^+ \|_H = \| \psi^+ \| , \qquad \| \psi^- \|_H = \| \psi^- \| .$$

Furthermore, $\psi^+ = \psi^-$ over H, since ψ vanishes in H; thus ψ^+ and ψ^- are distinct norm preserving extensions over L^∞ of the same linear functional over H.

To complete the example we must construct ψ with the given properties. Let ν be an arc of the unit circle, and let ν' be its complement. Let f_0 equal 1 on ν and equal -1 on ν'. Let $Re\,H$ be the set of real parts of functions in H; we shall show that f_0 is at a distance 1 from $Re\,H$. If this were not so, we could find $f \in Re\,H$ with $\| f - f_0 \| = 1 - \varepsilon$, $\varepsilon > 0$. Now $f \geqq \varepsilon$ on ν and $f \leqq -\varepsilon$ on ν'; the expression, in terms of f, of the conjugate $\hat{f}^*$ of the harmonic extension $\hat{f}$ of f over the unit disk shows that $\hat{f}^*$ is unbounded near the ends of ν, so $f \notin Re\,H$, a contradiction. By the Hahn-Banach extension theorem there is a bounded real linear functional on L^∞_{Re} which vanishes on $Re\,H$ but not at f_0. This extends by complex linearity over L^∞ to give the required functional ψ.

10.2. To illustrate the theorem for Riemann surfaces, let $\bar{S}$ be the annulus $\{z : a \leqq |z| \leqq b\}$, where $0 < a < b$. Here Γ consists of two oppositely oriented circles, and the first Betti number ρ of $\bar{S}$ is 1. A function $f \in L^\infty_{Re}$ is in G_{Re} if and only if $\int_\Gamma f d\theta = 0$; using Lebesgue measure μ on Γ, this shows that P is spanned by the single function α on Γ, defined by

$$\alpha(z) = \begin{cases} 1/b \text{ if } |z| = b , \\ -1/a \text{ if } |z| = a . \end{cases}$$

Let K_0 be the constant function 1, defining the linear functional ϕ

on H. Then for any real t,

$$K_t(z) = K_0(z) + t\alpha(z) \geqq 0 \text{ on } \Gamma \text{ if } -b \leqq t \leqq a \ ,$$

but this fails for other values of t. By Theorem 7.5, the set of norm preserving extensions of ϕ corresponds to this set of values of t, and is thus one dimensional.

On the other hand, if for instance $K_1(z) = |Re\, z|$ on Γ, then K_1 is nonnegative but $K_1 + t\alpha$ takes on negative values if t is real and $\neq 0$; by Theorem 7.5, the linear functional on H determined by K_1 has a unique norm preserving extension over L^∞.

10.3. More generally, let $\bar{S}$ be any compact bordered Riemann surface with first Betti number $\rho > 0$. Suppose ϕ is the linear functional defined on H by a real kernel K_0 such that $K_0 \geqq a > 0$ on Γ. Then for all α in P_{Re} such that $\|\alpha\| \leqq a$, $K_0 + \alpha$ is nonnegative; so that (Theorem 7.5) the set of extensions has real dimension ρ.

On the other hand, we shall show that (with measure μ as in 6.2) there is an $\varepsilon > 0$ such that if Q is a measurable subset of Γ with $\mu(Q) < \varepsilon$ and K is the characteristic function of Q, then the linear functional over H defined by K has a unique norm preserving extension over L^∞.

To this end, note that every nonzero α in P_{Re} is continuous and takes on both positive and negative values (since $1 \in H$ and hence $\int 1 \cdot \alpha d\mu = 0$). Let $\alpha_1, \cdots, \alpha_\rho$ be a basis for P_{Re}, and let D denote the set of n-tuples $v = (v_1, \cdots, v_\rho)$ with $\Sigma v_i^2 = 1$. For each $v \in D$, set

$$U_v = \{q \in \Gamma : \Sigma v_i \alpha_i(q) < 0\} \ , \qquad \mu_v = \mu(U_v) \ .$$

Clearly the function μ_v on D is lower semicontinuous; since D is compact, μ_v takes on its lower bound $\varepsilon > 0$. Now if $Q \subset \Gamma$, $\mu(Q) < \varepsilon$, then every $\alpha \neq 0$ in P takes on negative values outside of Q; hence (see Theorem 7.5) the characteristic function K of Q has the required property.

10.4. We show that there is a kernel on the unit circle Γ_1 whose extremal function has a zero on Γ_1 and hence does not have a bounded inverse; this was promised in the proof of Theorem 5.4. Through conformal mapping, we can replace the unit disk by the half disk $S^+ = \{z : Re\, z \geqq 0, |z| \leqq 1\}$, its boundary Γ replacing Γ_1. (We are now in the setting of Theorem 7.6.)

Define K on Γ by

$$K(z) = \begin{cases} \bar{z} & \text{if } |z| = 1 \ , \ Re\, z > 0 \ , \\ 0 & \text{if } Re\, z = 0 \ , \ |Im\, z| \leqq 1 \ . \end{cases}$$

Set $h_0(z) = z$ in S^+. If ψ is the linear functional on L^∞ defined by K, using Lebesgue measure μ on Γ,

$$\psi(h_0) = \int_\Gamma zKd\mu = \pi = \int_\Gamma Kd\mu = \|K\|_1 ;$$

hence h_0 is the extremal function for K. But $0 \in \Gamma$ and $h_0(0) = 0$.

References

1. N. Dunford and J. T. Schwartz, *Linear Operators I*, Interscience Publishers, New York, 1958.
2. E. Hille and R. S. Phillips, *Functional analysis and semi-groups*, Amer. Math. Soc. New York, 1957.
3. A. Pfluger, *Theorie der Riemannschen Flächen*, Springer, Berlin, 1957.
4. W. W. Rogosinski and H. S. Shapiro, *On certain extremum problems for analytic functions*, Acta Math., **90** (1953), 287-318.
5. John Wermer, *Subalgebras of the algebra of all complex-valued continuous functions on the circle*, Amer. J. Math., **78** (1956), 225-242.
6. A. Zygmund, *Trigonometric series*, vols. I and II, University Press, Cambridge, 1959.

HARVARD UNIVERSITY
INSTITUTE FOR ADVANCED STUDY

THE MATHEMATICS OF PHYSICAL QUANTITIES

PART I: MATHEMATICAL MODELS FOR MEASUREMENT

HASSLER WHITNEY, Institute for Advanced Study

INTRODUCTION

1. Purpose of this paper. To set up a physical theory, one constructs a mathematical model, and considers its relation to certain aspects of the physical world. There is a variety of models associated with the concept of measurement. Certain systems of numbers are important; for instance:

N, the natural numbers;	R^+, the positive reals;
Q^+, the positive rationals;	R, the reals.

Commonly one takes R^+ or R as a model for measurement. There are disadvantages in this, however. These models contain a specific number 1, and there is no natural way of putting this number in correspondence with a particular measurement; moreover, the models contain an operation of multiplication, with no natural physical counterpart.

Let us consider the problem of choosing a model M for masses. An object A has a certain property which we call its "mass"; why not let this property itself be an element of the model? As far as the structure of the model is concerned, we need not theorize on what "mass" really is; we need merely give it certain properties in the model. For instance, if we have distinct objects A and B, with masses m_A and m_B, we may think of the objects as forming a single object C; its mass m_C should then be m_A+m_B. Therefore M should contain an operation of addition, and any further properties we choose.

The two types of models that best fit in most situations we shall call "rays" and "birays." A ray (like a half line) is used for positive measurements, and a biray (like an oriented line with starting point), for measurements of quantities both positive and negative. It turns out that numbers appear in a natural way as operators on the model (see the next section).

In this paper we set up the theory of rays and birays; the real number system is constructed along with the models in a natural manner. In fact, this approach gives a simple and elegant way of introducing the reals and finding their basic properties.

We shall not give applications of the models in the body of the paper; some remarks on the subject will be made in this introduction. In this connection, see Part II, which will appear in the next issue of this MONTHLY.

2. Numbers as operators on the models. If we choose a stick of length l, and wish to use it to measure another stick, we lay out the first stick along the second several times; we thus form $l+l$, $l+l+l$, and so on. We also call these lengths $2l$, $3l$, $\cdots$. Thus N appears as a natural set of operators on our model. For masses, there is a different physical process of addition; but again we may use $2m=m+m$ and so on, with the same set N of numbers.

If we have a stick, whose length we call l, and we find a shorter stick, of length l', such that $3l' = l$, then we wish to give an expression of l' in terms of l. It is natural to set $l' = (1/3)l$. We may now set $2l' = (2/3)l$, and thus introduce Q^+ as operators. Finally, if our model has a certain completeness property, we may enlarge Q^+ to R^+ as operator system, and if we have negative quantities, we may enlarge R^+ to R.

3. Working in the model. Various properties of a model and its operations have obvious meaning in the applications. For instance we have distributive and associative laws:

$$5 \text{ cakes} + 2 \text{ cakes} = (5 + 2)\text{cakes} = 7 \text{ cakes}$$
$$2 \text{ yd} = 2(3 \text{ ft}) = (2 \times 3)\text{ft} = 6 \text{ ft}.$$

The fact that "2 yd" and "6 ft" name the same element of the model enables us to say they are equal; there is no need for such mysterious phrases as "2 yd measures the same as 6 ft."

4. The use of units. If we wish to use R^+ as a model for measuring (positive) lengths, we must decide which length 1 corresponds to; this length will then be called our "unit length." The more natural model is a ray L (on which R^+ operates); since the elements of L themselves are "lengths," the above question does not arise.

If we choose a length $l_0 \in L$, and compare other lengths with it, we may call l_0 our "unit"; this serves merely to remind us that l_0 is being kept fixed for a period. Suppose we now find certain other lengths, for instance $5l_0$, $2l_0$, $7l_0$. If we wish to shorten our notations, and call these lengths 5, 2, 7, we are then replacing L by R^+. We can then say "the length 5 really means the length $5l_0$." More awkwardly, one could say "the length is 5 when measured in terms of l_0."

Suppose we wish to "change units," say from ft to in. Then since, for any $a \in R^+$, a ft $= a(12$ in$) = 12a$ in, we would replace "the length a" by "the length $12a$." If any problems about units arise, they are at once resolved by going back to the explicit phrase "a ft."

5. The postulational treatment. Though all rays (and all birays) have the same structure, one may wish to use several rays in a single investigation. For instance, in mechanics, one uses separate rays M, L, T for measurement of mass, length and time. (We study structures containing several rays in Part II.) Hence we introduce our models postulationally; the definitions show whether or not a given structure is a ray or a biray. However, only a single structure R (or one of its subsets) is needed for operators; hence we introduce R constructively. (We give the characterization of R as a complete ordered field at the end.)

A basic theorem in the subject is an *isomorphism theorem*; a homomorphism of one ray into another is necessarily an isomorphism onto, and has certain additional properties (and similarly for birays). This theorem is a great aid in setting up the theory; in particular, with its use, multiplication in R^+ and in R is introduced and its properties derived with a minimal effort.

The postulates used for rays and birays are few in number and simple in character, and correspond to simple experimental phenomena.

6. Other models. We introduce only the most important models. With the real numbers at our disposal, and the facts about rays and birays, other similar models are easily studied. For instance, in measuring masses, one wishes to allow the mass zero (not present in a ray). This extra element may be introduced and related to the remaining elements in the obvious manner.

A model of a somewhat different nature is an oriented affine one-dimensional space T^*; this is the natural model for instance for moments in time (or positions on a line). There is a corresponding biray T of translations of T^*; this is the natural model for intervals of time (or directed lengths). We do not consider models including for instance 3-dimensional space; the term "measurement" is not the best term here.

If the measures of some type of quantity form a progression, as in counting, it is natural to use N for a model. However, if several such types of quantities are considered together, it is better to use several isomorphic models. For a plebeian illustration, suppose there will be six children at a party. We wish each to have two balloons and three cookies. What is the total supply needed? The answer is:

$$6(2\text{ bl} + 3\text{ ck}) = 6(2\text{ bl}) + 6(3\text{ ck}) = 12\text{ bl} + 18\text{ ck}.$$

7. A finite model. We have no infinite sets available in our environment. What happens if we have a set G with a large number n of elements, called $1'$, $2', \cdots, n'$, and wish it to approximate to N? We could define $a'+b'$ to be $(a+b)'$, or n' if $n<a+b$. Note that we may set $a'b'=ab'$; now G operates on itself, thus defining multiplication in G. We find that these operations are commutative and associative, and the distributive laws hold. However, the cancellation laws fail.

We give an instance from everyday life. Helen is setting the table for lunch for four; she places two spoons at each place. Mother answers the doorbell; it is Mr. and Mrs. Jones. Perhaps they will stay; Helen needs 4 more spoons. There are only two left in the drawer, so Helen puts them out. (She thus makes $8_s+4_s=10_s$.) Hearing the visitors say goodbye, Helen thinks, take away four spoons. She then realizes that, actually, she must take away only two. In her model, $8_s+4_s=8_s+2_s$.

CHAPTER I. DIVISIBLE SEMI-GROUPS

This chapter is preliminary in nature; certain basic properties of rather general structures are derived. It is shown how N appears as a system of operators on any commutative semi-group, and Q^+, if the semi-group is "uniquely divisible." To save space and help in the grasping of concepts, the proofs pertaining to N are given in rather intuitive fashion. However, the Peano postulates are seen to hold for N; hence one may replace the proofs by the usual more formal proofs where desired.

8. Commutative semi-groups. We begin with the definition.

DEFINITION 8A. *A* semi-group $(G, +)$ *is a nonvoid set G and an associative binary operation + in G. The semi-group is* commutative *if addition is commutative. Only the commutative case will be considered here.*

We may define $x+y+z$ to mean either $(x+y)+z$ or $x+(y+z)$, since these are the same. More generally, as is well known and easy to see, in any sum, the terms may be written in any order, and parentheses may be inserted or removed at will.

We wish to introduce a shorthand notation for such expressions as $x+x$, $x+x+x$, etc. Since the letter x plays no role here, let us think of it as replaced by a dot. This suggests the expressions

$$\begin{array}{lcccc} & x & x+x & x+x+x & x+x+x+x\cdots \\ (8\cdot 1) & (\cdot) & (\cdot\quad\cdot) & (\cdot\quad\cdot\quad\cdot) & (\cdot\quad\cdot\quad\cdot\quad\cdot) \end{array}$$

We now consider the new expressions as being names for new objects, forming a set N. It does not matter what these objects are; only their relation to G (or any other semi-group) counts. We also set $1=(\cdot)$, $2=(\cdot\cdot)$, $3=(\cdot\cdot\cdot)$.

We next let N operate on G as follows:

$$(8.2) \qquad (\cdot)x = x, \quad (\cdot\cdot)x = x+x, \quad (\cdot\cdot\cdot)x = x+x+x, \text{etc.}$$

We give an elementary property of the operation:

$$\begin{aligned} (\cdot)(x+y) &= x+y = (\cdot)x + (\cdot)y, \\ (\cdot\cdot)(x+y) &= (x+y)+(x+y) = (x+x)+(y+y) = (\cdot\cdot)x+(\cdot\cdot)y, \\ (\cdot\cdot\cdot)(x+y) &= (x+y)+(x+y)+(x+y) \\ &= (x+x+x)+(y+y+y) = (\cdot\cdot\cdot)x+(\cdot\cdot\cdot)y, \end{aligned}$$

and clearly, in general,

$$(8.3) \qquad a(x+y) = ax+ay \qquad (a \in N; \quad x, y \in G).$$

With the obvious definition of "successor function" σ in N, the Peano postulates are clear:

(N_1) *The element* $(\cdot)=1$ *is not a successor*: $\sigma x \neq 1$ *for all* $x \in N$.
(N_2) *For all* $x \neq y$, $\sigma x \neq \sigma y$; *that is, the function* σ *is one-one.*
(N_3) *For any* $N' \subset N$, *if* $1 \in N'$, *and* $x \in N'$ *implies* $\sigma x \in N'$, *then* $N' = N$.

We could now give proofs involving N in the classical manner, using mathematical induction, defining addition in N, and showing how to give definitions by induction. We shall, however, continue to give intuitive derivations.

If we look at the pattern

$$\begin{array}{ccccc} (x+x+x) & + & (x+x) & = & x+x+x+x+x \\ (\cdot\quad\cdot\quad\cdot) & & (\cdot\quad\cdot) & & (\cdot\quad\cdot\quad\cdot\quad\cdot\quad\cdot) \end{array}$$

this suggests defining

$$(\cdots) + (\cdot\cdot) = (\cdots\cdot\cdot).$$

The general rule is clear: To add two elements of N, place the corresponding symbols beside each other, and remove the inner parentheses. The commutative and associative laws are evident; hence $(N, +)$ is a commutative semi-group. Moreover, because of our definition,

$$(\cdots)x + (\cdot\cdot)x = (\cdots\cdot\cdot)x = [(\cdots) + (\cdot\cdot)]x,$$

and more generally,

$$(8.4) \qquad (a + b)x = ax + bx \qquad (a, b \in N, x \in G).$$

Since $(N, +)$ is a commutative semi-group, we may operate on it by N itself, thus defining *multiplication* in N. This gives, for instance,

$$(\cdot\cdot)(\cdots) = (\cdots) + (\cdots), \qquad (\cdots)(\cdots) = (\cdots) + (\cdots) + (\cdots).$$

As a consequence,

$$[(\cdot\cdot)(\cdots)]x = [(\cdots) + (\cdots)]x = (\cdots)x + (\cdots)x = (\cdot\cdot)[(\cdots)x],$$

and more generally,

$$(8.5) \qquad (ab)x = a(bx) \qquad (a, b \in N, x \in G).$$

Letting N operate on itself, (8.5) and (8.4) give the associative law for multiplication and the distributive law. The commutative law for multiplication must be proved separately. The cancellation law: $x+u=y+u$ implies $x=y$, is obvious from the representation of the elements with dots; a proof with the Peano postulates is easily given.

9. Order in N. If we think of the elements of N as laid out in their natural order, then $x<y$ means that x comes before y. The usual properties of order are clear. For later purposes, we show how to derive the properties from the two following properties of addition in N:

(a) For all $a, b \in N$, $a+b \neq a$.
(b) If $a \neq b$, then either $a+c=b$ or $b+c=a$, for some c.

Now write $a<b$ if $a+c=b$ for some c. The following properties follow at once from the definition:

$$(9.1) \qquad a < a + b.$$

$$(9.2) \qquad \text{If } a < b \text{ and } b < c \text{ then } a < c.$$

$$(9.3) \qquad \text{If } a_i < b_i (i = 1, \cdots, n) \text{ then } \Sigma a_i < \Sigma b_i.$$

The *trichotomy* property is: For all $a, b \in N$, exactly one of the following is true:

$$a < b, \qquad a = b, \qquad b < a.$$

For if $a \neq b$, then (b) shows that $a<b$ or $b<a$. That at most one of these is true follows from (a). (If $a+c=b$, $b+d=a$, then $a+(c+d)=a$, contradicting (a).)

Order is related to addition by the property (writing "iffif" for "if and only if")

$$a < b \quad \text{iffif} \quad a + c < b + c. \tag{9.4}$$

This is evident, using the cancellation law. Finally,

$$a < b \quad \text{iffif} \quad na < nb; \tag{9.5}$$

for by (9.3), $a<b$, $a=b$, $b<a$ imply respectively $na<nb$, $na=nb$, $nb<na$.

Rather than discuss *subtraction* here, we give the equivalent discussion when studying rays, in section 14.

10. Some examples of semi-groups. We give first a general kind of example:

Example 10A. Let G contain the elements $1, 2, 3, \cdots, m, \cdots, n$ from N. Let the successor function in G be as in N, except that we set $\sigma n = m$. Now addition is defined in terms of this function; we require

$$x + 1 = \sigma x, \qquad x + \sigma y = \sigma(x + y). \tag{10.1}$$

(These relations are used in defining addition in N through the Peano postulates.) For instance, with $m=5$, $n=7$, the elements are 1, 2, 3, 4, 5, 6, 7; adding 3 and adding 4 gives respectively the sequences

$$4, 5, 6, 7, 5, 6, 7; \qquad 5, 6, 7, 5, 6, 7, 5.$$

Example 10B. Take $m=1$ in the last example. Then we have the ring of integers mod n; n is the zero element.

Example 10C. Take $m=n$ in Example 10A. This gives the example of Section 7.

If G and G' are semi-groups, their *direct sum* consists of all pairs (x, x') with $x \in G$, $x' \in G'$. Define addition componentwise:

$$(x, x') + (y, y') = (x + y, x' + y').$$

This is clearly a semi-group, commutative if G and G' are.

Example 10D. There are ten teaspoons and six dessert spoons in a drawer. This gives us semi-groups G_t and G_d as in section 7. In the direct sum G^*, we have for instance

$$(5_t, 5_d) + (3_t, 3_d) = (8_t, 6_d).$$

If now we do not need to differentiate between teaspoons and dessert spoons, we have sixteen spoons, forming a semi-group G'. There is a natural mapping of G^* into G', in which for instance

$$(5_t,\ 5_d) \to 10_s, \qquad (3_t,\ 3_d) \to 6_s, \qquad (8_t,\ 6d) \to 14s.$$

Note that this mapping is not a homomorphism: $10_s+6_s \neq 14_s$.

11. Divisible semi-groups. In the following definition, we make use of the fact that N acts on any semi-group.

DEFINITION 11A. *We say that the commutative semi-group* $(G, +)$ *is* divisible *if for each* $x \in G$ *and* $n \in N$ *there is a* $y \in G$ *such that* $ny = x$. *We say that* $(G, +)$ *is* uniquely divisible *if the above* y *is unique.*

Example 11B. The group N_m of integers mod $m (m > 1)$ is not divisible; nor is N itself. The semi-groups Q^+, R^+, R are uniquely divisible. The group of dyadic rationals, containing all $k/2^n$ (k an integer, $n \in N$) is not divisible. The group of rationals mod 1 is divisible, but not uniquely; the same is true of the reals mod 1, or equivalently, of the multiplicative group of complex numbers of absolute value 1.

The direct sum of a finite set of uniquely divisible commutative semi-groups is uniquely divisible; in particular, any vector space over the reals is so.

DEFINITION 11C. *An element* x *of* G *is* idempotent *if* $x + x = x$. *We say* G *is* idempotent *if all its elements are idempotent.*

In a group, only the identity element is idempotent.

THEOREM 11D. *If* G *is idempotent, it is uniquely divisible.*

First, induction shows at once that $nx = x$ for all n and x. Now given x and n, set $y = x$; then $ny = x$. If also $ny' = x$, then $y' = ny' = x = ny = y$.

Note that, therefore, the operation of N on an idempotent semi-group is trivial.

Example 11E. Let U be a set, and let S be a set of subsets of U such that if A and B belong to S, so does their union $A \cup B$. Then $(S, \cup)$ is an idempotent commutative semi-group.

REMARK 11F. It is a theorem that every idempotent commutative semi-group is of the form of Example 11E.

DEFINITION 11G. *Let us say that* $(G, +)$ separates N *if for any two distinct natural numbers* m, n *there is an* $x \in G$ *such that* $mx \neq nx$.

We cannot then reduce N to a smaller set of operators on G, as was done for instance in Section 7.

THEOREM 11H. *Any divisible commutative semi-group* $(G, +)$ *which is not idempotent separates* N.

Suppose not. Then for some m and n, $n = m + h$ and $mx = nx$ for all x. For some a and k, $m + a = kh$. Now take any $y \in G$. Choose x so that $khx = y$. Since $mx = (m+h)x$, adding ax gives $khx = (k+1)hx$. Adding hx gives $(k+1)hx = (k+2)hx$. Continuing gives $khx = 2khx$, i.e. $y = 2y$. Thus G is idempotent, a contradiction.

12. Introduction of Q^+. We assume here that $(G, +)$ is a uniquely divisible commutative semi-group which is not idempotent. Then, using Theorem 11H, we have the properties

(12.1) $ax = ay$ implies $x = y \qquad (a \in N;\ x, y \in G)$,

(12.2) $ax = bx$ for all x implies $a = b \qquad (a, b \in N;\ x \in G)$.

Given x and a, there is a unique y such that $ay=x$; let us denote this y by the expression $\bar{a}x$. Now, by definition,

(12.3) $$a(\bar{a}x) = x \qquad (a \in N;\ x \in G).$$

We may now form $a(\bar{b}x)$. We would like to write this in the form $(a\bar{b})x$. To this end, we must define new elements $\bar{b}$, and give the whole expression meaning.

First, we show that (writing $\bar{b}'$ to denote $\overline{b'}$)

(12.4) $$a(\bar{b}x) = a'(\bar{b}'x) \quad (\text{all } x \in G) \quad \text{iffif} \quad ab' = a'b.$$

For, using (8.5) and (12.3) gives

$$(bb')(a(\bar{b}x)) = (ab')(b(\bar{b}x)) = (ab')x,$$
$$(bb')(a'(\bar{b}'x)) = (a'b)(b'(\bar{b}'x)) = (a'b)x.$$

Now if the left hand side of (12.4) holds, then the above equations give $(ab')x = (a'b)x$ (all x), and $ab'=a'b$, by (12.2). The converse follows also, using (12.1).

Now consider the expressions $a\bar{b}$ as denoting new objects. Because of (12.4), we wish an equivalence relation between these objects:

(12.5) $$a\bar{b} \sim a'\bar{b}' \quad \text{iffif} \quad ab' = a'b.$$

(To prove that the relation is transitive, multiply $ab'=a'b$ by $a'b''=a''b'$, giving $ab'a'b''=a'ba''b'$, and apply the cancellation law in N, giving $ab''=a''b$.) Denote the equivalence class of $a\bar{b}$ by a/b. Now let Q^+ be the set of equivalence classes thus obtained. (We could shorten $1\bar{b}$ to $\bar{b}$.) Note that $ac/bc=a/b$.

We now let Q^+ operate on G, by setting

(12.6) $$\frac{a}{b}x = a(\bar{b}x) \qquad (a, b \in N;\ x \in G).$$

Because of (12.4), the result is independent of the name a/b chosen for the given element of Q^+.

Note that $\bar{1}x=x$; hence $(a/1)x=a(\bar{1}x)=ax$. We therefore *identify* the element $a/1$ of Q^+ with the element a of N, thus imbedding N in Q^+, and the operation of N on G is preserved. (This is permissible; for (12.5) shows that $a/1\neq b/1$ if $a\neq b$.)

If $\bar{b}x=\bar{b}y$, then $x=b(\bar{b}x)=b(\bar{b}y)=y$. Hence also, using (12.1),

(12.7) $rx = ry$ implies $x = y \qquad (r \in Q^+;\ x, y \in G.)$

The relation (12.4) gives:

(12.8) $rx = sx$ for all $x \in G$ implies $r = s \ (r, s \in Q^+.)$

13. Properties of Q^+ and G. First we note that for any $r=a/b \in Q^+$, using (8.5), $b(rx)=b(a(\bar{b}x))=a(b(\bar{b}x))=ax$. Hence

$$b[r(x+y)] = a(x+y) = ax + ay = b(rx) + b(ry) = b(rx+ry),$$

and applying (12.1) gives

$$r(x+y) = rx + ry \qquad (r \in Q^+;\ x, y \in G). \tag{13.1}$$

Next, since

$$\frac{a}{b}x + \frac{c}{b}x = a(\bar{b}x) + c(\bar{b}x) = (a+c)(\bar{b}x) = \frac{a+c}{b}x,$$

we have

$$\frac{a}{b}x + \frac{c}{d}x = \frac{ad}{bd}x + \frac{bc}{bd}x = \frac{ad+bc}{bd}x. \tag{13.2}$$

Hence it is natural to define

$$\frac{a}{b} + \frac{c}{d} = \frac{ad+bc}{bd}. \tag{13.3}$$

Taking $b=d=1$ shows that this extends the definition in N. Now (13.2) gives the distributive law

$$(r+s)x = rx + sx \qquad (r, s \in Q^+;\ x \in G). \tag{13.4}$$

That the definition of $r+s$ is independent of the manner of writing r and s may be verified directly; it also follows from (13.2) and (12.8) (using some G).

Since addition and multiplication in N are commutative, (13.3) shows that addition in Q^+ is commutative. Similarly, addition is associative. (Using G, these properties follow easily, on applying (12.8).) Hence Q^+ is a commutative semi-group.

From the definition of addition in Q^+ we have

$$\left(\frac{a}{b} + \frac{a}{b}\right) + \frac{a}{b} = \frac{2a}{b} + \frac{a}{b} = \frac{3a}{b},$$

and similarly, in general,

$$n\,\frac{a}{b} = \frac{na}{b}. \tag{13.5}$$

We can solve $ns=r$ for s: If $r=a/b$, set $s=a/nb$. Thus Q^+ is divisible. If $n(a/b)=n(a'/b')$, then (13.5) and (12.5) give $na/b=na'/b'$, $nab'=na'b$, $ab'=a'b$, $a/b=a'/b'$; thus division (by elements of N) is unique. Clearly Q^+ is not idempotent.

Now Q^+ operates on itself; we call this operation multiplication. This extends the operation of multiplication in N. Because of (13.5),

$$bd\left(\frac{a}{b}\left(\frac{c}{d}\right)\right) = ad\left(\frac{b}{b}\left(\frac{c}{d}\right)\right) = ad\left(\frac{c}{d}\right) = ac = bd\frac{ac}{bd};$$

hence (12.1) gives

$$\frac{a}{b}\,\frac{c}{d} = \frac{ac}{bd}. \tag{13.6}$$

We now prove the general associative law:

$$r(sx) = (rs)x \qquad (r, s \in Q^+; x \in G). \tag{13.7}$$

Say $r=a/b$, $s=c/d$. Then, with the help of (12.6) and (12.3),

$$bd\left(\frac{a}{b}\left(\frac{c}{d}x\right)\right) = ad\left(\frac{c}{d}x\right) = acx = bd\left(\frac{ac}{bd}x\right);$$

this gives the result.

In particular, multiplication in Q^+ is associative; it is clearly commutative. The distributive laws follow from (13.1) and (13.4). The cancellation law for multiplication follows from (12.7). The existence and uniqueness of division is clear: The solution of $(a/b)u=c/d$ is

$$u = \frac{c}{d} \div \frac{a}{b} = \frac{cb}{da}. \tag{13.8}$$

In particular, $1/(a/b)=b/a$.

The subtraction property in Q^+ is:

(13.9) If $r\neq s$, then either $r+t=s$ or $s+t=r$, for some t.

For we may write $r=a/c$, $s=b/c$; then $a\neq b$, hence $b=a+d$ or $a=b+d$, and we may use $t=d/c$.

The cancellation law for addition is also easy to prove: Suppose $r+s=r+t$. Write $r=a/d$, $s=b/d$, $t=c/d$. Then $a+b=a+c$, hence $b=c$, and $s=t$.

The relation $r+s=r$ in Q^+ is impossible. Hence we may define order in Q^+ as in N (section 9), and Q^+ is now simply ordered.

CHAPTER II. RAYS

A "semi-ray" has certain properties needed for the measurement of positive quantities: addition and subtraction, order, and the existence of arbitrarily small elements. If the semi-ray is Archimedean, it may be completed to form a ray. The operation of Q^+ on a ray is extended to the operation by R^+; we find the basic properties of R^+ here.

14. Semi-rays. We shall not need divisibility in the definition; it will appear as a consequence of completeness in the next section.

DEFINITION 14A. *A* semi-ray *L is a commutative semi-group such that:*

(R_1) *For all x and y in L, $x+y\neq x$.*

(R_2) *For all x and y in L with $x\neq y$, we can find u and v in L such that $x+u+v=y$ or $y+u+v=x$.*

DEFINITION 14B. *$x<y$ means that $x+u=y$ for some u.*

Since we now have properties (a) and (b) of section 9, we may deduce the properties of *order* given there. We also have the *cancellation law*:

$$\text{(14.1)} \qquad \textit{If } x+u=y+u \textit{ then } x=y.$$

For if $x\neq y$, then $x<y$ or $y<x$, by trichotomy, and hence $x+u<y+u$ or $y+u<x+u$, contrary to $x+u=y+u$.

We now introduce *subtraction*. If $x<y$, then there is a unique element u such that $x+u=y$; we call this element $y-x$. Now

$$\text{(14.2)} \qquad (y-x)+x=y \qquad \text{if } x<y.$$

We give some simple properties:

$$\text{(14.3)} \qquad (x-y)-z=x-(y+z) \qquad \text{if } y+z<x,$$

$$\text{(14.4)} \qquad (x+y)-z=\begin{cases} x+(y-z) & \text{if } z<y, \\ x-(z-y) & \text{if } y<z<x+y. \end{cases}$$

To prove each one, add a quantity to each side which will remove the minus signs. Thus, for the last equality, add z:

$$\begin{aligned}[x-(z-y)]+z &= [x-(z-y)]+[(z-y)+y] \\ &= [(x-(z-y))+(z-y)]+y=x+y=[(x+y)-z]+z;\end{aligned}$$

now apply (14.1).

We now show that, owing to the particular form of (R_2), L contains arbitrarily small elements:

THEOREM 14C. *For each $x\in L$ and $n\in N$ there is some $y\in L$ with $ny<x$.*

First, $x+x\neq x$; also $x+x+u+v\neq x$ for all u and v, by (R_1); hence, by (R_2), we may find u and v so that

$$x+u+v=x+x.$$

By (14.1), $u+v=x$. Either $u\leqq v$ or $v\leqq u$; say $u\leqq v$. Then $u+u\leqq u+v=x$, and $u<2u\leqq x$; the statement holds with $n=1$. For the general case, we use induction: Say $nz<x$. Find y such that $2y\leqq z$. Then

$$(n+1)y=ny+y\leqq n(y+y)\leqq nz<x.$$

DEFINITION 14D. *The semi-ray L is* Archimedean *if for each x and y in L there is an $n\in N$ such that $y<nx$.*

Example 14E. Let L consist of all ordered pairs $(x, 0)$ with $x\in R^+$, and (x, y) with $x\in R$ and $y\in R^+$ (or $y\in N$); define addition component-wise. Then $(L, +)$ is clearly a commutative semi-group, and (R_1) holds. To prove (R_2), take $(x, y)\neq(x', y')$. If $y=y'$, then say $x<x'$; now $(x, y)+(x'-x, 0)=(x', y')$. If $y\neq y'$, say $y<y'$; then $(x, y)+(x'-x, y'-y)=(x', y')$. Now take $u=v=\frac{1}{2}(x'-x, 0)$ or $\frac{1}{2}(x'-x, y'-y)$. Since $n(1, 0)=(n, 0)<(0, 1)$ for all n, $(L, +)$ is not Archimedean.

15. Rays. We introduce completeness through Dedekind cuts.

DEFINITION 15A. *An* upper set *U in a semi-ray L is a set such that*

(U_1) *$U\neq\varnothing$ and $U\neq L$.*
(U_2) *If $x\in U$ and $x<y$ then $y\in U$.*

We say the upper set U is strict *if also*
(U_3) *If $x\in U$ then there is some $y\in U$ with $y<x$.*

DEFINITION 15B. *Given $S\subset L$, z is a* lower bound *for S if $z\leqq x$ for all $x\in S$. Also, z is a* greatest lower bound (g.l.b. for short) *for S is z is a lower bound, but no $z'>z$ is.*

THEOREM 15C. *Any subset of L has at most one* g.l.b.

This is trivial.

DEFINITION 15D. *The semi-ray L is* complete *if any strict upper set has a* g.l.b.

THEOREM 15E. *If L is complete, then any upper set has a* g.l.b.

Suppose the upper set U is not strict. Then for some $z\in U$, $x<z$ implies $x\notin U$. Clearly z is the g.l.b. for U.

DEFINITION 15F. *A* ray *is a complete semi-ray.*

THEOREM 15G. *A ray is Archimedean.*

For suppose not. Say $nx\leqq y$ for all n. Set

$$U = \{u: nx \leqq u \text{ for all } n \in N\}.$$

Since $x\notin U$ and $y\in U$, (U_1) holds. Clearly (U_2) holds also; hence U is an upper set. By Theorem 15E, U has a g.l.b., say z. We show now that $z<nx$ for some n. If $z\leqq x$, then $z<2x$. Otherwise, $x<z$, and we may write $x+y=z$; now $y<z$ also. Therefore $y\notin U$, and we have $y<mx$ for some m; hence $z=x+y<(m+1)x=nx$. But nx is a lower bound for U, contradicting the definition of z.

THEOREM 15H. *Any ray L is uniquely divisible.*

Take any x and any $n>1$; suppose that $ny\neq x$ for all y. Set

$$U = \{u \in L: x < nu\}.$$

By Theorem 14C, $U \neq L$. Since $x \in U$, (U_1) holds. Using (9.5) in L shows that (U_2) holds; thus U is an upper set, and has a g.l.b., say z.

Suppose $nz < x$; say $nz + y = x$. Find h by Theorem 14C so that $nh < y$; then $n(z+h) < nz + y = x$ and (9.5) shows that $z+h$ is a lower bound for U, a contradiction. Hence $nz > x$, and we may write $x + y = nz$. Choose h so that $nh < y$. Now $nh < nz$, hence $h < z$, and we may write $v + h = z$; now

$$x + y = nz = n(v + h) < nv + y,$$

hence $x < nv$, and $v \in U$, contradicting $v < z$. We have proved that L is divisible. Since $z' < z$ implies $nz' < nz$, division is unique.

Since $x < x + x$, L is not idempotent. Hence (section 12) *Q^+ operates on any ray L.*

We give two facts about inequalities. For $r, s \in Q^+$ and $x, y \in L$,

(15.1) $$r < s \quad \text{iffif} \quad rx < sx,$$

(15.2) $$x < y \quad \text{iffif} \quad rx < ry.$$

If $r < s$, say $r + t = s$; then $rx + tx = sx$, and $rx < sx$. Similarly, if $r \geqq s$, then $rx \geqq sx$; hence if $rx < sx$ then $r < s$. (15.2) follows similarly.

Finally, we prove a *density* theorem:

(15.3) *If $x, y, z \in L$, $x < y$, then for some $r \in Q^+$, $x < rz < y$.*

Say $x + u = y$. For some $n \in N$, $nu > z$. Set $h = (1/n)z$; then $h < u$. Also $h < y$, and $mh > y$ for some $m \in N$. Hence, for some $k \in N$, $kh < y \leqq (k+1)h$. Set $r = k/n$; then $rz = kh < y$. Also $x + u = y \leqq rz + h < rz + u$, and hence $x < rz$.

16. Completion of an Archimedean semi-ray. Such a completion is a ray; in particular, we use this later to construct R^+ from Q^+.

LEMMA 16A. *If L is a semi-ray and $x \in L$, then $U(x) = \{u \in L: x < u\}$ is a strict upper set.*

For $x \notin U(x)$, $2x \in U(x)$, and (U_1) holds. (U_2) is clear. Say $y \in U(x)$; then $x < y$. By (R_2), we may write $x + u + v = y$. Now $x + u \in U(x)$ and $x + u < y$, so that (U_3) holds.

LEMMA 16B. *If U is an upper set in the Archimedean semi-ray L and $h \in L$, we may find $z \in L$ such that $z \notin U$ and $z + h \in U$.*

Choose $x \notin U$, $y \in U$. Say $nh > y$; then $x + nh \in U$. Hence, for z, we may take one of the elements $x, x+h, x+2h, \cdots, x+(n-1)h$.

LEMMA 16C. *If U and V are strict upper sets in the semi-ray L, then*

(16.1) $$U + V = \{u + v: u \in U, v \in V\}$$

is a strict upper set; we have:

(16.2) *If $u \notin U$, $v \notin V$, then $u + v \notin U + V$.*

First, given $u \notin U$, $v \notin V$, take any $z \in U+V$, and write $z=u'+v'$, with $u' \in U$, $v' \in V$; then $u<u'$, $v<v'$, hence $u+v<z$, and $u+v \notin U+V$.

Because of this, $U+V \neq L$. Clearly $U+V \neq \varnothing$. Properties (U_2) and (U_3) for $U+V$ are clear.

THEOREM 16D. *Let* $(L, +)$ *be an Archimedean semi-ray, let* L^* *be the set of all strict upper sets in* L, *and define addition in* L^* *by* (16.1); *then* $(L^*, +)$ *is a ray. The mapping* $x \rightarrow U(x)$ *of* L *into* L^* (see Lemma 16A) *is one-one and preserves addition; it thus imbeds* L *in* L^*.

With addition in L^* defined by (16.1), we clearly have a commutative semigroup. To prove (R_1), take any strict upper sets U and V. Take $h \notin V$, and find z for U by Lemma 16B. By (16.2), $z+h \notin U+V$; since $z+h \in U$, $U \neq U+V$.

Next we introduce subtraction in L^*. Suppose $u_1 \in U$, $u_1 \notin V$. Then set

$$W = \{w: \text{for some } w' < w,\ w' + u \in V \text{ for all } u \in U\}.$$

We have $V \subset W$; hence $W \neq \varnothing$. We may write $u_1 = u+h$, with $u \in U$; then $h \notin W$, and $W \neq L$. Clearly (U_2) and (U_3) hold for W; hence W is a strict upper set.

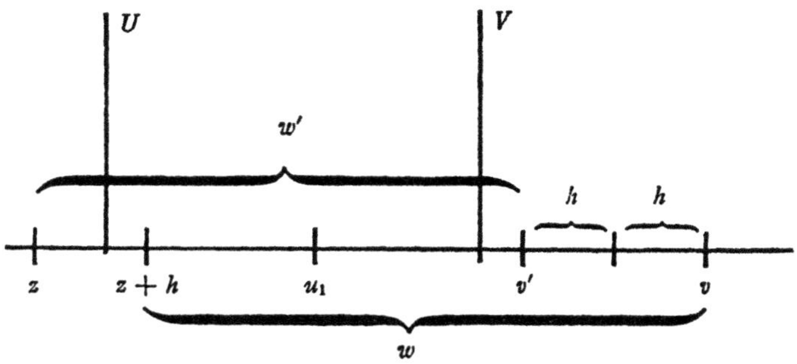

Clearly $U+W \subset V$. Conversely, take any $v \in V$. By (U_3) for V and Theorem 14C, we may write $v=v'+2h$, with $v' \in V$. By Lemma 16B, find $z \notin U$ such that $z+h \in U$. By (U_2) for U and for V, $z<u_1<v'$; hence we may write $z+w'=v'$. Set $w=w'+h$. Now if $u \in U$, then $u>z$, and hence $w'+u>w'+z=v' \in V$; thus $w'+u \in V$. This shows that $w \in W$. Also $z+h \in U$, $w \in W$, and $z+h+w=v$; thus $v \in U+W$, proving that $U+W=V$.

Next, with W as above, choose x so that $2x \notin W$. Since $2x \in U(x)$, we can find W' by the proof above so that $U(x)+W'=W$. Now we prove (R_2). Given U and V with $U \neq V$, either $U \not\subset V$ or $V \not\subset U$, say the former. The proof above gives $U+U(x)+W'=V$, as required. Thus $(L^*, +)$ is a semi-ray.

To prove that L^* is complete, take any strict upper set $U^* \subset L^*$. The elements of U^* are strict upper sets in L; let S be their union. Since $U^* \neq \varnothing$, $S \neq \varnothing$. Since $U^* \neq L^*$, there is a strict upper set $U \notin U^*$. Take $x \notin U$. For any $U' \in U^*$, $U<U'$ (in L^*), by (U_2) for U^*; hence (Definition 14B) $U+W=U'$

for some W, and thus $x \notin U'$. This shows that $x \notin S$, and $S \neq L$. Properties (U_2) and (U_3) for S are clear; thus S is a strict upper set in L, i.e. $S \in L^*$. Since $U \in U^*$ implies $U \subset S$ and hence $S \leqq U$ in L^*, S is a lower bound for U^* in L^*. If $S < S'$ in L^*, then $S + W = S'$, and there is some $x \in S$ with $x \notin S'$. Now $x \in U$ for some $U \in U^*$, and hence $U < S'$ in L^*; thus S' is not a lower bound for U^*. Therefore S is the g.l.b. of U^* in L^*, proving completeness.

Suppose $x < y$. Then say $x + u + v = y$; it follows that $x + u \in U(x)$, $x + u \notin U(y)$, and $U(x) \neq U(y)$. Thus the mapping $x \rightarrow U(x)$ is one-one.

We must show still that $U(x+y) = U(x) + U(y)$. Suppose $z \in U(x+y)$; then $x + y < z$, and we may write $x + y + u + v = z$. Now $x + u \in U(x)$ and $y + v \in U(y)$, and thus $z \in U(x) + U(y)$. Thus $U(x+y) \subset U(x) + U(y)$. The converse is clear, and the proof is complete.

Suppose we form the completion L^* of the ray L. Each element U of L^* is a strict upper set in L; it has a g.l.b., say x. The properties of strict upper sets show at once that $U = U(x)$ (compare Lemma 16A). Thus the mapping $x \rightarrow U(x)$ of L into L^* is onto, and we may identify L^* with L itself.

17. The isomorphism theorem. We shall show that there is essentially only one kind of ray; any two rays are isomorphic. All homomorphisms are isomorphisms; we find them all. For further information, see Section 19.

DEFINITION 17A. *Let $(G, +)$ and $(G', +)$ be commutative semi-groups. A mapping ϕ: $G \rightarrow G'$ is a* homomorphism *if $\phi(x+y) = \phi(x) + \phi(y)$ for all x and y in G. It is an* isomorphism *if it is also one-one, and is* onto G' *if the image of G is all of G'.*

THEOREM 17B. *Let ϕ be a homomorphism of the ray L into the ray L'. Then*

$$\text{(17.1)} \qquad \text{if } x < y \quad \text{then } \phi(x) < \phi(y),$$

$$\text{(17.2)} \qquad \text{if } r \in Q^+ \quad \text{then } \phi(rx) = r\phi(x).$$

If $x < y$, write $x + u = y$; then $\phi(x) + \phi(u) = \phi(y)$, and hence $\phi(x) < \phi(y)$. Next for any n, $\phi(x + \cdots + x) = \phi(x) + \cdots + \phi(x)$ (n terms in the sums); hence $\phi(nx) = n\phi(x)$. Also, if $y = (1/m)x$, i.e. $x = my$, then $\phi(x) = m\phi(y)$ and hence $\phi(y) = (1/m)\phi(x)$, from which (17.2) follows.

THEOREM 17C. *Let L and L' be rays, and suppose $w \in L$, $w' \in L'$. Then there is a unique homomorphism ϕ of L into L' such that $\phi(w) = w'$. Any homomorphism of L into L' is an isomorphism onto L'.*

For each $x \in L$, set

$$U'_x = \{y' \in L' : \text{for some } r \in Q^+, x < rw \text{ and } rw' < y'\}.$$

We see at once that U'_x is a strict upper set in L'; hence (see the end of Section 16) for a certain $x' \in L'$, $U'_x = U'(x')$, with $U'(x')$ defined as in Lemma 16A. Set $\phi(x) = x'$.

We show that

(17.4) $$U'_{x+y} = U'_x + U'_y.$$

If $z' \in U'_x + U'_y$, then $z' = x' + y'$, with $x' \in U'_x$, $y' \in U'_y$. For some r, $s \in Q^+$,

$$x < rw, \qquad rw' < x'; \qquad y < sw, \qquad sw' < y'.$$

Now $x+y<(r+s)w$, $(r+s)w'<z'$, and hence $z' \in U'_{x+y}$. Conversely, given $z' \in U'_{x+y}$, we may find r, u, v so that $x+y+u+v=rw$, $rw'<z'$. By (15.3) we may find s, $t \in Q^+$ so that

$$x < sw < x + u, \qquad y < tw < y + v.$$

Now $(s+t)w<rw$, and by (15.1), $s+t<r$, $(s+t)w'<rw'<z'$. We may write $(sw'+u')+(tw'+v')=z'$. Since $sw'+u' \in U'_x$ and $tw'+v' \in U'_y$, we have $z' \in U'_x + U'_y$, proving (17.4).

Recalling that $U'_z = U'(\phi(z))$ and $U'(x')+U'(y')=U'(x'+y')$ (see the end of Section 16), (17.4) gives

$$U'(\phi(x+y)) = U'_{x+y} = U'(\phi(x)) + U'(\phi(y)) = U'(\phi(x) + \phi(y));$$

hence $\phi(x+y)=\phi(x)+\phi(y)$, and ϕ is a homomorphism. Now we may apply (17.1), showing that ϕ is one-one.

Next we show that if ϕ' is any homomorphism of L' into L such that $\phi'(w')=w$, then $\phi(\phi'(x'))=x'$ for all $x' \in L'$. Set $x=\phi'(x')$, $y'=\phi(x)$. If $y'<x'$ we may find $r \in Q^+$ such that $y'<rw'<x'$. Then by Theorem 17B for ϕ', $rw = \phi'(rw')<\phi'(x')=x$, and hence, applying ϕ, $rw'<y'$, a contradiction. Similarly $x'<y'$ is false; hence $y'=x'$.

Because of this, ϕ maps L onto L'. Also, for another ϕ_1 like ϕ, we have $\phi_1(\phi'(x'))=x'=\phi(\phi'(x'))$, and since ϕ' is onto L, $\phi_1(x)=\phi(x)$ for all x; thus ϕ is uniquely determined, and the proof is complete.

18. Introduction of R^+. We saw in Section 13 that Q^+ is a completely divisible commutative semi-group; also, that $r+s=r$ is impossible in Q^+, so that (R_1) holds. Given r, $s \in Q^+$, choose $t \in Q^+$ by (13.9); set $u=v=t/2$; then either $r+u+v=s$ or $s+u+v=r$, proving (R_2). Therefore Q^+ is a semi-ray. Given r, $s \in Q^+$, we may write $r=a/c$, $s=b/c$; then since $ba \geqq b$, (13.5) shows that $br \geqq s$. Therefore Q^+ is Archimedean. We may now apply Theorem 16D, giving the completion R^+ of Q^+; R^+ is a ray, and we may consider Q^+ as imbedded in R^+, by the definition $r \rightarrow \{s \in Q^+ : r<s\}$.

19. The operation of R^+ on L. For each x in the ray L, let ϕ_x be the homomorphism of R^+ into L given by Theorem 17C, such that $\phi_x(1)=x$. By (17.2),

$$\phi_x(r) = \phi_x(r \cdot 1) = r\phi_x(1) = rx, \qquad (r \in Q^+);$$

hence, if we set

(19.1) $$ax = \phi_x(a) \qquad (a \in R^+, x \in L),$$

this extends the operation of Q^+ to an operation of R^+ on L. Note that

$$1x = x. \tag{19.2}$$

Since ϕ_x is a homomorphism, we have

$$(a+b)x = \phi_x(a+b) = \phi_x(a) + \phi_x(b) = ax + bx. \tag{19.3}$$

Given x and y in L, set $\psi(a) = \phi_x(a) + \phi_y(a)$ (all $a \in R^+$); then ψ is a homomorphism, and $\psi(1) = x + y$. By uniqueness in Theorem 17C, $\psi = \phi_{x+y}$. Hence

$$a(x+y) = \phi_{x+y}(a) = \psi(a) = \phi_x(a) + \phi_y(a) = ax + ay. \tag{19.4}$$

Since R^+ is a ray, R^+ operates on itself as above; let Φ_c be the corresponding functions. We call this operation *multiplication* in R^+. Thus

$$ab = \Phi_b(a); \qquad \Phi_b(1) = b. \tag{19.5}$$

Since the operation of R^+ on R^+ extends the operation of Q^+ on R^+ and hence of Q^+ on Q^+, and the latter operation is multiplication in Q^+, multiplication in R^+ extends that in Q^+.

The above distributive laws give, in R^+,

$$(a+b)c = ac + bc, \qquad a(b+c) = ab + ac. \tag{19.6}$$

Set $\Psi(a) = a$; Ψ is a homomorphism, and $\Psi(1) = 1$. By uniqueness, $\Psi = \Phi_1$. Hence $a1 = \Phi_1(a) = \Psi(a) = a$. Also, since $1 \in Q^+$ (or, by (19.5)), $1a = a$. Thus

$$1a = a1 = a. \tag{19.7}$$

We can *divide* in R^+: Given $a, b \in R^+$, since Φ_a is onto R^+, we can find $c \in R^+$ such that $ca = \Phi_a(c) = b$.

Given $b \in R^+$ and $x \in L$, let ψ be the composite mapping $\phi_x \circ \Phi_b$; this is a homomorphism of R^+ into L. Since

$$\psi(1) = \phi_x(\Phi_b(1)) = \phi_x(b) = bx,$$

uniqueness shows that $\psi = \phi_{bx}$. Hence we find the general associative law:

$$a(bx) = \phi_{bx}(a) = \psi(a) = \phi_x(\Phi_b(a)) = \phi_x(ab) = (ab)x. \tag{19.8}$$

In particular, taking $L = R^+$,

$$a(bc) = (ab)c. \tag{19.9}$$

If we set $\Psi_b(a) = \Phi_a(b)$, then the second part of (19.6) shows that Ψ_b is a homomorphism. Since $\Psi_b(1) = b1 = b = \Phi_b(1)$, $\Psi_b = \Phi_b$ and

$$ba = \Phi_a(b) = \Psi_b(a) = \Phi_b(a) = ab. \tag{19.10}$$

THEOREM 19A. *The operation of R^+ on L extends the operation of Q^+; we have properties* (19.1) *through* (19.10), *and also:*

(V_1) *For each x and y in L there is an $a \in R^+$ such that $ax = y$.*

(V_2) *If $ax = ay$ then $x = y$.*

(V_3) *If $ax = bx$ then $a = b$.*

To prove (V_1), recall that ϕ_x is onto L; hence for some a, $ax=\phi_x(a)=y$. To prove (V_2), set $c=1/a$; then $ax=ay$ gives $c(ax)=c(ay)$, and by (19.8) and (19.2), $x=y$. Since ϕ_x is one-one, (V_3) holds.

We now extend (17.2):

THEOREM 19B. *If ϕ is a homomorphism of the ray L into the ray L', then*

$$\phi(ax) = a\phi(x) \qquad (a \in R^+,\ x \in L). \tag{19.11}$$

To show this, take any fixed $x \in L$, and set $\psi(a)=\phi(ax)$, $\theta(a)=a\phi(x)$. Using (19.3) in L and in L' shows that ψ and θ are homomorphisms of R^+ into L'. Since $\psi(1)=\theta(1)$, uniqueness in Theorem 17C shows that $\psi=\theta$, as required.

CHAPTER III. BIRAYS

A biray is constructed from a ray by adjoining a zero element and negative elements. The biray constructed from the ray R^+ is the group of real numbers; multiplication in R is defined through the operation of R on itself.

20. Definition of birays. Essentially, a biray B is a commutative group containing a ray B^+, such that if $x\neq 0$ then x or $-x$ is in B^+.

DEFINITION 20A. *A* biray $(B, B^+, +)$ *is a set B, a subset B^+, and an operation of addition in B, such that:*

(B_1) *$(B, +)$ is a commutative semi-group.*

(B_2) *$(B^+, +)$ is a ray.*

(B_3) *For each $x, y \in B$ there is a $z \in B$ such that $x+z=y$.*

(B_4) *If $x\neq y$, $x+z=y$, and $y+z'=x$, then $z\in B^+$ or $z'\in B^+$.*

We prove first the *cancellation law*:

$$\text{If } x+z = x+z' \quad \text{then} \quad z = z'. \tag{20.1}$$

For suppose $z\neq z'$. By (B_3), we may write $z+u=z'$, $z'+v=z$. By (B_4), one of u, v is in B^+; say $u\in B^+$. Now use (B_3) to write $x+z+w=u$. These relations give

$$u+u = x+z+w+u = x+z'+w = x+z+w = u.$$

But $u\in B^+$, contrary to (B_2) and (R_1) (see Section 14).

We now find the *zero* element of B. Choose $x_0\in B^+$. Choose $0\in B$ so that $x_0+0=x_0$, by (B_3). We show that for all $x\in B$,

$$x + 0 = x. \tag{20.2}$$

By (B_3), we may write $x=x_0+y$. Now $x+0=y+x_0+0=y+x_0=x$. Moreover,

$$\text{if } x+u = x \quad \text{then} \quad u = 0, \tag{20.3}$$

by the cancellation law; hence the zero element 0 is uniquely defined by (20.2).

We now know that $(B, +)$ is a commutative group.

DEFINITION 20B. *For $x\in B$, $-x$ is the element such that*

$$x + (-x) = 0; \tag{20.4}$$

set also

$$(20.5) \qquad x - y = x + (-y).$$

We have

$$(20.6) \qquad (x - y) + y = x + [(-y) + y] = x + 0 = x.$$

Using this gives

$$(20.7) \qquad y = z - x \quad \text{iffif} \quad x + y = z$$

(add x in the first equation; add $-x$ in the second). Also

$$(20.8) \qquad -0 = 0, \qquad -(-x) = x;$$

for $0+0=0$; also $(-x)+x=0$, $(-x)+(-(-x))=0$, and the cancellation law gives the second part of (20.8). Some further elementary properties are:

$$(20.9) \qquad -(x + y) = (-x) + (-y) = -x - y,$$

$$(20.10) \qquad -(x - y) = y - x,$$

$$(20.11) \qquad (x - z) + (y - w) = (x + y) - (z + w).$$

We may prove each directly from (20.4) and (20.5). Or, we may add something to both sides and apply (20.1). For instance,

$$-(x - y) + x = -(x - y) + [(x - y) + y] = 0 + y = y,$$

and also $(y-x)+x=y$; hence (20.10) follows.

21. Order in birays. We first consider negative elements.

DEFINITION 21A. *B^- is the set of all x such that $-x \in B^+$. The elements of B^+ are* positive; *those of B^- are* negative.

We prove trichotomy: Each element of B is zero, positive or negative, and is only one of these.

Given x, at least one holds; for if $x \neq 0$, then since $x+(-x)=0$ and $0+x=x$, applying (B_4) shows that x is in B^+ or in B^-. At most one is true. For, since $(B^+, +)$ satisfies (R_1) (Section 14), and $0+0=0$, $0 \notin B^+$. Also, if $x \in B^- \cap B^+$, then $-x$ and x are in B^+, and by (B_2), $0=x+(-x) \in B^+$, a contradiction.

DEFINITION 21B. *$x<y$ means that $y-x \in B^+$.*

Because of (B_2), order is transitive. The trichotomy above is equivalent to trichotomy in terms of order (Section 9). Thus B is simply ordered. The properties (9.2), (9.3) and (9.4) clearly hold. Note that

$$(21.1) \qquad -x < -y \quad \text{iffif} \quad y < x.$$

One may introduce *absolute values* in terms of order. Note that if $x \leq y$ and $y \leq x$, then $x=y$. Hence the following definition makes sense:

$$(21.2) \qquad \text{If } x \leq y, \quad \text{then set } \inf\{x, y\} = x, \quad \sup\{x, y\} = y;$$

set

$$|x| = \sup\{x, -x\}. \tag{21.3}$$

The derivation of properties of absolute value is standard; we need not go into it here. The definitions of $\inf\{x_1, \cdots, x_n\}$, $\sup\{x_1, \cdots, x_n\}$ are clear.

22. The isomorphism theorem. We prove:

THEOREM 22A. *Let $(B, B^+, +)$ and $(B', B'^+, +)$ be birays. Take any $w \neq 0$ in B and any w' in B'. Then there is a unique homomorphism ϕ of $(B, +)$ into $(B', +)$ such that $\phi(w) = w'$. If $w' \neq 0$, then ϕ is an isomorphism onto B'. If w and w' are both positive or both negative, then ϕ carries B^+ onto B'^+ and B^- onto B'^-, and*

$$\phi(x) < \phi(y) \quad iffif \quad x < y; \tag{22.1}$$

if one of w, w' is positive and the other negative, ϕ has the opposite effect.

See also Theorem 24A.

Suppose first that w and w' are positive. By Theorem 17C, there is a unique homomorphism ϕ^+ of B^+ into B'^+ such that $\phi^+(w) = w'$; hence the restriction $\phi|B^+$ of ϕ to B^+ must be ϕ^+. We must have $\phi(x) = \phi(x+0) = \phi(x) + \phi(0)$, and hence $\phi(0) = 0$ (letting 0 denote the zero element in both birays). Also we must have $\phi(x) + \phi(-x) = \phi(x + (-x)) = \phi(0) = 0$, and hence $\phi(-x) = -\phi(x)$. Thus ϕ, if it exists, is unique. We may use these equations to define ϕ outside B^+.

To prove that ϕ is a homomorphism, we examine $\phi(\alpha+\beta)$ for any α, $\beta \in B$. The case that $\alpha = 0$ or $\beta = 0$ is clear, and the case α, $\beta \in B^+$ is known. The remaining cases are as follows: For x, $y \in B^+$,

$$\begin{aligned} \phi(-x + (-y)) &= \phi(-(x + y)) = -\phi(x + y) = -[\phi(x) + \phi(y)] \\ &= -\phi(x) - \phi(y) = \phi(-x) + \phi(-y); \end{aligned}$$

$$\phi(x + (-x)) = \phi(0) = 0 = \phi(x) - \phi(x) = \phi(x) + \phi(-x);$$

$$\text{if } y < x \text{ then } \phi(x + (-y)) = \phi(x - y) = \phi(x) - \phi(y) = \phi(x) + \phi(-y);$$

$$\begin{aligned} \text{if } x < y \text{ then } \phi(x + (-y)) &= \phi(-(y - x)) = -\phi(y - x) \\ &= -[\phi(y) - \phi(x)] = \phi(x) + \phi(-y); \end{aligned}$$

also $\phi((-x) + y) = \phi(y + (-x))$, and the above applies.

Since ϕ^+ is an isomorphism onto B'^+, ϕ carries B^+ onto B'^+ and B^- onto B'^-. Thus ϕ is onto B'. That ϕ is also one-one and is therefore an isomorphism is clear. If $x < y$, say $x + z = y$, $z \in B^+$. Expanding $\phi(x+z)$ shows that (22.1) holds.

Next suppose that w is positive and w' is negative. Set $w'' = -w'$, and let ϕ' be the homomorphism of B into B' with $\phi'(w) = w''$; then we see easily that $\phi(x) = -\phi'(x)$ is the unique required isomorphism.

If w is positive and $w' = 0$, we show that $\phi(w) = 0$ for all $x \in B$. If $x \in B^+$ and $\phi(x) \in B^+$, then by Theorem 17C, $\phi(y) \in B^+$ for all $y \in B^+$, contrary to $\phi(w) = 0$. If $x \in B^+$ and $\phi(x) \in B^-$, then $\phi' = -\phi$ is a homomorphism with $\phi'(w) = 0$ and

$\phi'(x) \in B^+$, and we again have a contradiction. Thus $\phi(x)=0$, all $x \in B^+$. The case that $x \in B^-$ is similar.

Finally, if w is negative, using the pair $(-w, -w')$ in place of (w, w') gives again the required properties.

23. Construction of a biray from a ray. Given a ray $(B^+, +)$, we construct a corresponding biray $(B, B^+, +)$. The elements of B are those of B^+, called positive, a new element 0, and for each element x of B^+, a new element x^*; the latter form B^-, the negative elements of B. We shall use $x, y, \cdots$ to denote elements of B^+, and $\alpha, \beta, \cdots$ to denote elements of B (including B^+). We define addition in B as follows:

$$\alpha + 0 = \alpha, \tag{23.1}$$

$$x + x^* = 0, \tag{23.2}$$

$$x + y^* = \begin{cases} x - y & \text{if } y < x, \\ (y - x)^* & \text{if } x < y, \end{cases} \tag{23.3}$$

$$x^* + y^* = (x + y)^*; \tag{23.4}$$

requiring addition to be commutative gives the remaining cases.

To prove properties of addition it is convenient to introduce new names for elements of B (corresponding to the "ordered pair" definition of negative numbers): For any $x, y \in B^+$, set

$$[x, y] = x + y^*. \tag{23.5}$$

As a consequence, for $x, y, z \in B^+$,

$$[y + z, y] = z, \qquad [x, x] = 0, \qquad [x, x + z] = z^*; \tag{23.6}$$

this shows that each element of B has new names.

We show now that

$$[x, y] = [x, y'] \quad \text{iffif} \quad y = y'. \tag{23.7}$$

Suppose $[x, y] = [x, y']$. From (23.5) and trichotomy we see first that whichever case in (23.2), (23.3) applies to x and y also must apply to x and y'; now if $y < x$, $x - y = x - y'$ implies $y = y'$, and the situation is similar in the other cases.

Applying (20.11) to (23.2) or (23.3) shows that

$$[x + z, y + z] = [x, y]. \tag{23.8}$$

Next, since $[x, y] = [x+u, y+u]$ and $[u, v] = [x+u, x+v]$, (23.7) gives

$$[x, y] = [u, v] \quad \text{iffif} \quad x + v = y + u. \tag{23.9}$$

We are now ready to prove that

$$[x, y] + [u, v] = [x + u, y + v]. \tag{23.10}$$

For $[x, y]$ there are three cases to consider: $[y+z, y]$, $[x, x]$ and $[x, x+z]$; similarly for $[u, v]$. Some of these cases are:

$$[x, x] + [u, v] = 0 + [u, v] = [u, v] = [x + u, x + v],$$
$$[y + z, y] + [v + w, v] = z + w = [y + z + v + w, y + v],$$
$$[x, x + z] + [u, u + w] = z^* + w^* = (z + w)^* = [x + u, x + z + u + w],$$

if $w<z$ then

$$[y + z, y] + [u, u + w] = z + w^* = z - w = [y + (z - w), y]$$
$$= [y + z + u, y + u + w],$$

if $w<z$ then

$$[y + z, y] + [u, u + w] = (w - z)^* = [y, y + (w - z)]$$
$$= [y + z + u, y + u + w];$$
$$[y + z, y] + [u, u + z] = z + z^* = 0 = [y + z + u, y + u + z];$$

the remaining cases are taken care of by commutativity.

Applying (23.10) twice gives

$$([x, y] + [u, v]) + [p, q] = [(x + u) + p, (y + v) + q];$$

now associativity in B^+ gives associativity in B. We now have (B_1) and (B_2) for $(B, B^+, +)$.

To prove (B_3), we need merely note that

$$(23.11) \quad [x, y] + [u + y, v + x] = [x + u + y, y + v + x] = [u, v].$$

Next we prove the cancellation law, through the implications

$$[x, y] + [u, v] = [x, y] + [p, q] \Rightarrow [x + u, y + v] = [x + p, y + q]$$
$$\Rightarrow x + u + y + q = y + v + x + p \Rightarrow u + q = v + p \Rightarrow [u, v] = [p, q].$$

To prove that (B_4) holds, suppose that

$$[x, y] \neq [u, v], \qquad [x, y] + \alpha = [u, v], \qquad [u, v] + \beta = [x, y].$$

By (23.11) and the cancellation law, we then have

$$\alpha = [u + y, v + x], \qquad \beta = [x + v, y + u].$$

Now by (23.9), $x+v \neq y+u$; hence either $x+v<y+u$ and $\alpha \in B^+$ or $x+v>y+u$ and $\beta \in B^+$.

We have now proved that $(B, B^+, +)$ is a biray.

REMARK: An alternative proof may be given as follows: First show that

$$\text{if } z + \alpha = z + \beta, \qquad z > |\alpha| + |\beta|, \quad \text{then } \alpha = \beta;$$

this is easy. Next, considering several cases, we find:

$$\text{If } z > |\alpha| + |\beta| \quad \text{then} \quad z + (\alpha + \beta) = (z + \alpha) + \beta.$$

Using these gives at once:

If $z > |\alpha| + |\beta| + |\gamma|$ then $z + [(\alpha + \beta) + \gamma] = z + [\alpha + (\beta + \gamma)]$;

this gives the associative law. To prove (B_3), use $x+(x^*+\beta)=\beta$, $x^*+(x+\beta)=\beta$. Finally, (B_4) is proved by considering several cases.

There is essentially only one biray containing a given ray:

THEOREM 23A. *If* $(B, B^+, +)$ *and* $(B', B^+, +)$ *are birays, then there is an isomorphism* ϕ *of the first onto the second, defined by* $\phi(0)=0'$, *and* $\phi(x)=x$, $\phi(-x)=-x$, $(x\in B^+)$.

Here, the first "$-x$" is interpreted in B, and the second, in B'; see Definition 20B. The theorem follows at once from Theorem 22A, choosing some $w=w'$ in B^+.

24. The operation of R on a biray. First, define $(R, R^+, +)$ to be the biray constructed from $(R^+, +)$ as in Section 23; by Theorem 23A, this is (up to isomorphisms) the only biray containing $(R^+, +)$ as its positive part.

We now define the operation of R on any biray $(B, B^+, +)$. For each $\alpha\in B$, let ϕ_α be the homomorphism of $(R, R^+, +)$ into $(B, B^+, +)$ given by Theorem 22A, with $\phi_\alpha(1)=\alpha$; set

(24.1) $$a\alpha = \phi_\alpha(a) \qquad (a \in R, \alpha \in B).$$

This extends the operation of R^+ on the part B^+ of B. Exactly as in Section 19, we find (always using $a, b, \cdots \in R$; $\alpha, \beta, \cdots \in B$)

(24.2) $$(a + b)\alpha = a\alpha + b\alpha, \qquad a(\alpha + \beta) = a\alpha + a\beta.$$

Since $(R, R^+, +)$ is a biray, it operates on itself; this operation we call *multiplication* in R. Now

(24.3) $$(a + b)c = ac + ac, \qquad a(b + c) = ab + ac.$$

There is no harm in identifying the zero element in different birays; in particular, in R and in B. Since ϕ_α is a homomorphism, $\phi_\alpha(0)=0$. We now have (compare Section 19)

(24.4) $$1\alpha = \alpha, \qquad 0\alpha = \alpha 0 = 0,$$

(24.5) $$1a = a1 = a, \qquad 0a = a0 = 0.$$

Again as in Section 19,

(24.6) $$a(b\alpha) = (ab)\alpha, \qquad a(bc) = (ab)c, \qquad ab = ba.$$

We now show that we can *divide* in R. Given $a, b\in R$ with $a\neq 0$, the homomorphism Φ_a of R into itself with $\Phi_a(1)=a$ is an isomorphism onto R (Theorem 22A); hence for some $c\in R$, $ca=\Phi_a(c)=b$.

THEOREM 24A. *If ϕ is a homomorphism of the biray B into the biray B', then*

$$\phi(a\alpha) = a\phi(\alpha) \qquad (a \in R, \alpha \in B). \tag{24.7}$$

The proof of Theorem 19B applies, using Theorem 22A (which holds also for the case that $\phi(\alpha)\equiv 0$).

THEOREM 24B. *$(R, +, \times)$ is a complete ordered field, unique up to isomorphisms.*

We have proved that we have a field; it is ordered. Since R^+ is complete, the proof that R is complete (using Dedekind cuts) is simple. Suppose $(R', +, \times)$ is another such field. Then $1 \in R^+$, $1' \in R'^+$. Let ϕ be the corresponding homomorphism of birays, with $\phi(1) = 1'$; this is an isomorphism onto R', by Theorem 22A.

Now take any fixed b, and set

$$\psi(a) = \phi(ab), \qquad \theta(a) = \phi(a)\phi(b).$$

Then

$$\psi(a_1 + a_2) = \phi((a_1 + a_2)b) = \phi(a_1b + a_2b)$$
$$= \phi(a_1b) + \phi(a_2b) = \psi(a_1) + \psi(a_2),$$

and similarly θ is a homomorphism. Since $\theta(1) = 1'\phi(b) = \phi(b) = \psi(1)$, Theorem 22A shows that $\psi = \theta$; that is, ϕ is a multiplicative homomorphism. Since $\phi(1) = 1'$, ϕ maps R^+ onto R'^+. Thus ϕ is an isomorphism of ordered fields.

Note that ϕ is the only such isomorphism; for we must have $\phi(1) = \phi(1 \cdot 1) = \phi(1)\phi(1)$, and since $\phi(1) \neq 0'$, $\phi(1) = 1'$.

THEOREM 24C. *For any biray B, we have:*
(V_1^*) *For each α and β in B, $\alpha \neq 0$, there is an $a \in R$ such that $a\alpha = \beta$.*
(V_2^*) *If $a\alpha = a\beta$, $a \neq 0$, then $\alpha = \beta$.*
(V_3^*) *If $a\alpha = b\alpha$, $\alpha \neq 0$, then $a = b$.*

The proofs are like the corresponding proofs in Theorem 19A; we use the fact that if $\alpha \neq 0$ then ϕ_α is onto B and is one-one.

THEOREM 24D. *Any biray $(B, B^+, +)$, with the operation of R, is an oriented one-dimensional vector space over R, and conversely.*

By (24.2), (24.6) and (24.4), the biray is a vector space; it is oriented by the choice of B^+ as "positive" part. Because of (V_1^*), it is one-dimensional. Conversely, any such space satisfies the postulates for a biray, and the operation by R is the same as that defined here.

THE MATHEMATICS OF PHYSICAL QUANTITIES

PART II: QUANTITY STRUCTURES AND DIMENSIONAL ANALYSIS

HASSLER WHITNEY, Institute for Advanced Study

INTRODUCTION

1. Mathematical models. In setting up a scientific theory, one chooses a framework in which the phenomena to be discussed may be expressed. This framework is given some kind of structure, which is then compared with various experiments. If this structure is given in a precise manner, so that one may draw rigorous conclusions about it, it is called a "mathematical model." The investigation then takes two separate directions:

(a) One studies the model itself, and draws various conclusions that are consequences of the chosen structure of the model.

(b) One tests the relation between the model and experiment.

The construction and employment of mathematical models is now one of the most important functions of applied mathematics, ever growing in scope. In fact, the use of models has been at the basis of scientific theories throughout history, though they have generally not been thought of quite in these terms. A very early model was the system of natural numbers, used in counting, and studied mathematically by Dedekind and Peano in the last century. The use of the "aether" as a model in which electromagnetic waves travel comes to mind; but this gave way to a more purely mathematical model to describe such phenomena.

Newton's laws of motion are expressed in terms of mass, length and time; these terms may be used to designate actual experimental findings, or concepts related to such findings. In writing equations, the latter meaning is best used; then we are building a mathematical model in which the laws take on precise meaning. Let us call a model of this nature a "quantity structure"; it is the kind of model studied in this paper. Though it is obviously basic to the expression of classical physics, it is curious that this sort of model seems never to have been set up in a complete and strict fashion. The essential features will be found in Drobot, [2].

The present Part II may be read independently of Part I (see [5]).

2. The kind of model studied. Classically, the results of measurements have been expressed in terms of numbers. The various sets of numbers that are important for us are:

N, the natural numbers; I, the integers (positive, zero or negative);

Q, the rationals; R^+, the positive reals; R, the real numbers.

Here is a typical statement in physics: The velocity of light is

$$c = 3 \times 10^{10} \frac{\text{cm}}{\text{sec}} \cdot \tag{2.1}$$

Such a statement is often interpreted as follows: The numbers express the "measure"; the "cm" and "sec" shows the units which are used. If such an interpretation is necessary, what does the equal sign mean? How does one work with equations, anyway? Obviously we cannot be on secure ground unless we give things precise meanings; we wish these meanings to fit the desired purpose, of course. It is said that you cannot "really" divide distance by time. Perhaps you cannot with actual experimental distance and time. But if we let "cm" and "sec" be elements of a mathematical model, we may introduce division and anything else we please; then the experimentalist may fit our model to the phenomena as he sees best.

Our first job is to choose the elements that will make up the model. Let us call these "quantities"; they correspond to physical quantities.

Next, we wish to add and subtract quantities; but only if they are of the same "type," corresponding to some physical dimension. (Then one of the quantities is a real multiple of the other.)

Certain physical quantities, like mass, are inherently positive; others, like time intervals, may be positive or negative, but may have a natural "positive direction." We suppose that each element of our model may be called positive, zero, or negative.

The positive quantities of a single type form a "ray"; R^+ is a typical example of a ray. All the quantities of a single type form a "biray"; this is simply an oriented one-dimensional vector space, with R as typical example. One may define the concepts of rays and birays from simple assumptions about addition. One may then define $3x$ to be $x+x+x$, and $(1/3)x$ to be y, where $3y=x$; continuing in this manner, one introduces R as a domain of operators on a biray; see [5]. In contrast with this, we assume a knowledge of R here, and addition is not mentioned in the definition of a quantity structure; instead, we *define* $ax+bx$ to be $(a+b)x$ (see Section 4).

We wish to be able to multiply any two quantities in our model; the result is a quantity, generally of a new type. Any nonzero quantity may be raised to any integral power; we may then for instance define cm ÷ sec to be $\text{cm}\times\text{sec}^{-1}$.

If we measure the height h upwards, a falling stone travels a distance

$$h = -(16\ \text{ft}/\text{sec}^2)t^2$$

in the time t. Solving for t gives

$$t = \sqrt{-h/16\ \text{ft}}\ \text{sec}.$$

Thus we must be able to take square roots of positive quantities. (Note that h must be negative in the above; this corresponds to the fact that the stone does not fall upwards.) More generally, we may allow any rational exponent of a positive quantity in our model, thus using Q as power operators; or we might allow R itself here.

We get in this manner certain mathematical quantities that might not appear experimentally; for example, $\text{cm}^{5/7}$, or $\sec^{\pi}$. In fact, 2 cm/cm = 2 is in our model, though it is not commonly thought of as a physical quantity.

The operations now introduced into our model must be subject to certain laws. These are in fact similar to those about numbers, and have always been used by scientists without worry. Some samples are:

$$3 \text{ cm} + 4 \text{ cm} = (3 + 4) \text{ cm}; \qquad 6 \text{ cm} \div 2 \sec = 3 \text{ cm}/\sec;$$

$$(9 \text{ cm}/\sec^2)^{1/2} = 9^{1/2} \text{ cm}^{1/2} (\sec^{-2})^{1/2} = 3 \text{ cm}^{1/2}/\sec.$$

3. Outline of the paper. In Chapter I, we define quantity structures, and study their general properties; in Chapter II, we illustrate their use through various examples, in particular, through "dimensional analysis."

Rays and birays are defined (Sections 4, 5) through the use of R^+ and R as operator domains. Quantity structures are defined postulationally (Sect. 6); that is, any given structure is by definition a quantity structure provided it satisfies certain given conditions. The elementary facts about these structures appear naturally.

Each element $x \neq 0$ of a given quantity structure S is on a biray $[x] = \{ax: a \in R\}$ (Sect. 7), consisting of all elements of the same type as x; the positive part $[x]^+$ of $[x]$ is a ray. Thus [in] = [ft] = [−100 ft], called the "dimension" of length. Considering each $[x]$ as an element of a new system gives us in a natural manner a vector space $[S]$ (Sect. 8) written multiplicatively; this is the space of possible "physical dimensions." If $[x_1], \cdots, [x_n]$ forms a base for $[S]$, then each element of S may be written in the form

$$x = ax_1^{\alpha_1} \cdots x_n^{\alpha_n}; \tag{3.1}$$

the representation is unique if $x \neq 0$ (see Sect. 9).

An isomorphism ϕ of S onto itself is a "similarity" if it preserves type: $[\phi(x)] = [x]$ (and preserves positiveness). With a base as above, the general similarity is defined by $\phi(x_i) = \lambda_i x_i$ for some positive numbers $\lambda_1, \cdots, \lambda_n$ (Sect. 11).

A quantity structure may be constructed from given base elements $x_1, \cdots, x_n$, or from the corresponding rays (Sections 12, 13). For example, choosing rays M, L, T (which represent mass, length and time), we build the model used in most problems in mechanics.

We end Chapter I by showing how algebra and calculus are carried out directly in a quantity structure; all terms in equations are elements of S, rather than just real numbers.

In Chapter II we give first some elementary applications of quantity structures, showing how one may "keep the units" in equations. Some remarks on "dropping units" are made; the end of Section 21 shows the danger in this. We then take up a powerful tool in many investigations, "dimensional analysis." This illustrates the fact that using a mathematical model (a quantity structure) enables one to be precise in statement, thus clearing up much of the confusion one normally sees in the subject. The key theorem is the "Pi Theorem" (Sect. 23): Let $f(x_1, \cdots, x_s)$ be a function for $x_i \in \Gamma_i$, where the Γ_i are rays in S, and let the values of f be in a ray Γ. Then if f is assumed "homogeneous," i.e. invariant under similarities of S, considerable knowledge about the form of f follows.

The use of dimensional analysis is quite standard; we illustrate it through a discussion of pendulums, of satellites, of the flow of water, and of torque of a wire.

CHAPTER I. QUANTITY STRUCTURES

4. Rays. We first consider operations on one set by another.

Definition 4A. Let A and X be sets. We say A *operates* on X if for each $a \in A$ and $x \in X$ there is a fixed corresponding element y of X. (Thus we have a function from $A \times X$ to X.) We say A is *multiplicative* if y is written as ax, and is a *power* operation if y is written as x^a. We say $x \in X$ *generates* X if each $y \in X$ is ax for some a; x generates X *simply* if this a is unique, for each y.

Definition 4B. A *ray* is a set X with an operation by R^+, such that:

(P_1) $a(bx) = (ab)x$ for all $a, b \in R^+$ and $x \in X$.

(P_2) There is an element $w \in X$ which generates X simply.

Note that R^+ is a ray; it operates on itself by multiplication.

We give some elementary properties of a ray X. First:

$$1x = x \qquad (x \in X). \tag{4.1}$$

For choosing w by (P_2), we may write $x = cw (c \in R^+)$. Now

$$1x = 1(cw) = (1c)w = cw = x.$$

Next:

(4.2) Each $x \in X$ generates X simply.

For we may write $x = cw$. Set $c' = 1/c$; then $c'x = c'(cw) = (c'c)w = 1w = w$. Now for any $y = dw$, $y = d(c'x) = (dc')x$. We must show still that

$$\text{If } a \neq b \text{ then } ax \neq bx \qquad (a, b \in R^+, x \in X). \tag{4.3}$$

With $x = cw$, we have $ax = (ac)w$, $bx = (bc)w$; since w generates X simply, $ac = bc$, and therefore $a = b$.

Definition of Addition. Choose w as before; we define

$$aw + bw = (a + b)w. \tag{4.4}$$

We now show that

$$(a + b)x = ax + bx \qquad (a, b \in R^+; x \in X). \tag{4.4'}$$

For, writing $x = cw$, we have

$$\begin{aligned}(a + b)x &= (a + b)(cw) = [(a + b)c]w = (ac + bc)w \\ &= (ac)w + (bc)w = a(cw) + b(cw) = ax + bx.\end{aligned}$$

Thus the definition of addition is independent of the w used.

The *commutative* and *associative* laws for addition follow at once. For instance, $aw+bw=(a+b)w=(b+a)w=bw+aw$. We give a second distributive law:

$$a(x+y)=ax+ay \qquad (a\in \mathrm{R}^+;\ x,y\in X). \tag{4.5}$$

For writing $x=cw$, $y=dw$, we have

$$a(x+y)=a[(c+d)w]=[a(c+d)]w=acw+adw=ax+ay.$$

There is a natural *order* in any ray, defined as follows:

$$x<y \text{ means } x+z=y \text{ for some } z. \tag{4.6}$$

Using x as generator in X gives: $ax<bx$ if $ax+cx=bx$ for some c; thus (writing "iffif" for "if and only if")

$$ax<bx \text{ iffif } a<b \qquad (a,b\in \mathrm{R}^+;\ x\in X). \tag{4.7}$$

We clearly have *trichotomy*: For each x, y, exactly one of $x<y$, $x=y$, $y<x$ is true. Also order is transitive: $x<y$ and $y<z$ imply $x<z$. Thus X is *simply ordered.*

We now show that *any two rays are isomorphic.* By definition, a *homomorphism* ϕ of a system X into a system X' of the same sort is a mapping which preserves all the properties of the systems. It is an *isomorphism* if it is one-one, and is an isomorphism *onto* X' if the image of X is all of X'.

THEOREM 4C. *Let X and X' be rays, and suppose $w\in X$, $w'\in X'$. Then there is a unique mapping ϕ of X into X' such that*

$$\phi(ax)=a\phi(x) \qquad (a\in \mathrm{R}^+,\ x\in X), \tag{4.8}$$

$$\phi(w)=w. \tag{4.9}$$

This ϕ is an isomorphism of X onto X', and hence preserves addition and order:

$$\phi(x+y)=\phi(x)+\phi(y), \tag{4.10}$$

$$\phi(x)<\phi(y) \text{ iffif } x<y. \tag{4.11}$$

First, since ϕ must satisfy $\phi(aw)=a\phi(w)=aw'$, ϕ is unique. Use this to define ϕ. Then for any $x=cw$ and any $a\in\mathrm{R}^+$,

$$\phi(x)=\phi(cw)=c\phi(w)=cw', \quad \phi(ax)=\phi(acw)=ac\phi(w)=acw'=a\phi(x);$$

thus ϕ is a homomorphism.

If $\phi(aw)=\phi(bw)$, then $aw'=bw'$, and hence $a=b$, by (4.3) in X'; thus ϕ is one-one. Since w' generates X', ϕ is onto X'. Since addition and order are defined in terms of the operation by R^+, these are preserved; a direct verification is also immediate.

COROLLARY 4D. *For each $x\in X$ there is a unique isomorphism ϕ of R^+ onto X such that $\phi(1)=x$; then $\phi(a)=ax$.*

COROLLARY 4E. *The isomorphisms of X into itself are in one-one correspondence with the numbers $\lambda \in \mathrm{R}^+$; they are given by $\phi_\lambda(x) = \lambda x$.*

For, taking a fixed $w \in X$, any $w' \in X$ may be written uniquely as λw for some λ; now apply Theorem 4C.

5. Birays. We give first a simple characterization of one-dimensional vector spaces.

THEOREM 5A. *Let V be a set on which* R *operates. Suppose* (P_1) *and* (P_2) *of Section* 4 *hold, with* R^+ *replaced by* R. *Then V, with addition defined as in* (4.4), *is a one-dimensional vector space over* R.

Choose $w \in V$ by (P_2), and set $O = 0w$. Now given $x \in V$, we may write $x = cw$, and

$$0x = 0(cw) = (0c)w = 0w = O.$$

Later we shall write O for the zero in any ray.

Also, for $a \in \mathrm{R}$, $aO = a(0w) = (a0)w = 0w = O$. Using also the proof of (4.1), we now have

$$(5.1) \qquad 1x = x, \quad 0x = aO = O \qquad (a \in \mathrm{R}, x \in V).$$

If $x \neq O$, then $x = cw$ with $c \neq 0$; the proof of (4.2) shows that:

(5.2) If $x \neq O$ then x generates V simply.

In particular, for $x \neq 0$, if $a \neq 0$ then $ax \neq O$; hence, recalling (5.1),

$$(5.3) \qquad ax = O \text{ iffif } a = 0 \text{ or } x = O \qquad (a \in \mathrm{R}, x \in V).$$

As before, the definition of addition is independent of the element $w \neq O$ used, and the associative, commutative and two distributive laws hold. Using (5.1) also shows that V is a vector space over R. Because of (P_2), $\dim(V) = 1$.

Recall that in a vector space, we may define *negatives* of elements and *subtraction* by setting

$$(5.4) \qquad -x = (-1)x, \quad x - y = x + (-y).$$

The usual laws for minus signs follow. For instance,

$$\begin{aligned} -(x - y) &= (-1)[x + (-1)y] = (-1)x + (-1)[(-1)y] \\ &= [(-1)(-1)]y + (-x) = 1y - x = y - x. \end{aligned}$$

DEFINITION 5B. A one-dimensional vector space V is *oriented* if a subset V^+ of V is designated as the set of *positive* elements of V; we assume that for some positive element w, aw is positive iffif $a > 0$. This is then true of any positive element.

The negatives of the positive elements form the set V^- of *negative* elements of V. Clearly each element of V is exactly one of the following: positive, negative, or zero (i.e. $=O$).

DEFINITION 5C. Let V be oriented. Define

$$x < y \text{ iffif } x + z = y \quad \text{for some } z \in V^+. \tag{5.5}$$

Now V is simply ordered; the proof is very similar to that of Section 4.

THEOREM 5D. *If V is oriented, then V^+, operated on by* R^+, *is a ray.*

If x is positive, then for $a>0$, ax is positive; thus R^+ operates on V. The rest of the proof is trivial.

DEFINITION 5E. A *biray* is an oriented one-dimensional vector space.

We now show that any two birays are isomorphic.

THEOREM 5F. *Let B and B' be birays, and suppose $w \in B^+$, $w' \in B'$. Then there is a unique mapping ϕ of B into B' such that $\phi(ax)=a\phi(x)$ ($x \in B$, $a \in \mathrm{R}$), and $\phi(w)=w'$;* (4.10) *holds. If $w' \neq O'$, then ϕ is one-one onto B'. If $w' \in B'^+$, then ϕ is an isomorphism onto B', which (therefore) preserves order.*

If $w'=O'$, the requirement on ϕ gives: $\phi(x)=O'$, all x. Otherwise, the proof of Theorem 4C applies, giving the present theorem.

COROLLARY 5G. *For each $w \in B^+$ there is a unique isomorphism of* R *onto B such that $\phi(1)=w$; then $\phi(a)=aw$.*

COROLLARY 5H. *The isomorphisms of B onto itself are in one-one correspondence with the numbers $\lambda \in \mathrm{R}^+$; they are given by $\phi_\lambda(x)=\lambda x$.*

See the proof of Corollary 4E.

6. Quantity structures. We now give our basic definition, telling us if a given model is a quantity structure S. The existence of such structures will be proved through a construction process, in Section 12.

As operators on S by powers, one might choose either Q or R; to leave open both (and possibly other) cases, we use the symbol R^*.

DEFINITION 6A. A *quantity structure* consists of a set S, which contains the real numbers R, a subset S^+ of S, a binary operation of multiplication in S, and a power operation on S^+ by a set R^* (either Q or R), subject to the following conditions:

Properties of multiplication:

(Q_1) Multiplication has its usual meaning in the subset R of S.

(Q_2) $1x=x$, $0x=0$. We define: $-x=(-1)x$.

(Q_3) Multiplication is associative and commutative.

Properties of subsets:

(Q_4) $R \cap S^+ = \mathrm{R}^+$.

(Q_5) For each $x \neq 0$, just one of x, $-x$ is in S^+.
(Q_6) If $x, y \in S^+$ then $xy \in S^+$.

Properties of powers ($x^\alpha \in S^+$ *if* $x \in S^+$, $\alpha \in R^*$):
(Q_7) a^α has its usual meaning for $a \in R^+$.
(Q_8) $x^1 = x$, $x^0 = 1$.
(Q_9) $(xy)^\alpha = x^\alpha y^\alpha$, $x^{\alpha+\beta} = x^\alpha x^\beta$, $(x^\alpha)^\beta = x^{\alpha\beta}$.

DEFINITION 6B. The elements of S^+ are called *positive*; their negatives form the set S^- of *negative* elements of S.

Example 6C. R, operating on itself by multiplication, and operating on its positive part by powers, is a quantity structure (which is contained in any other one with $R^* = R$).

We now prove some elementary properties of a quantity structure. We shall use letters $x, y, \cdots$ for elements of S (including R), $a, b, \cdots$ for numbers, and $\alpha, \beta, \cdots$ for elements of R^*.

Since $(-1)[(-1)x] = [(-1)(-1)]x = 1x = x$ and $[(-1)x]y = (-1)(xy)$,

$$-(-x) = x, \qquad (-x)y = -(xy), \qquad (-x)(-y) = xy. \tag{6.1}$$

Trichotomy: Each element of S is either positive, negative or zero, and is only one of these. For if $x \neq 0$, then by (Q_5), x is either positive or negative, but not both. By (Q_4), $0 \notin S^+$; and since $-0 = 0$, $0 \notin S^-$.

The *rule of signs* is: If x and y are both positive or both negative, then xy is positive; if one of x, y is positive and the other negative, then xy is negative. This follows from (6.1), (Q_5) and (Q_6).

Since $R \subset S$ and (Q_4) holds, a consequence of the rule of signs is:

$$\text{If } x \in S^+ \text{ and } a \in R, \text{ then } ax \in S^+ \text{ iffif } a \in R^+. \tag{6.2}$$

If $x, y \neq 0$, then the rule of signs gives $xy \neq 0$. Hence, using (Q_2) and (Q_3),

$$xy = 0 \text{ iffif } x = 0 \text{ or } y = 0. \tag{6.3}$$

Any element $x \neq 0$ has a *multiplicative inverse*, namely, x^{-1}, if x is positive, and $-(-x)^{-1}$ if x is negative. For if $x \in S^+$, then by (Q_8) and (Q_9), $xx^{-1} = x^1x^{-1} = x^{1-1} = x^0 = 1$, and if $x \in S^-$, then $-x \in S^+$, and $x(-(-x)^{-1}) = (-x)(-x)^{-1} = 1$. We may set $x^{-1} = -(-x)^{-1}$ if $x \in S^-$.

We prove the cancellation law:

$$\text{If } xz = yz \text{ and } z \neq 0 \text{ then } x = y. \tag{6.4}$$

For, taking u so that $zu = 1$, $xz = yz$ gives $xzu = yzu$, $x1 = y1$, $x = y$. (Note that we do not yet have addition in S, so we do not form $(x-y)z$.)

We may form *integral powers of any* $x \neq 0$. First, $x^2 = xx$, $x^3 = xxx$, etc. Next, we defined $x^{-1} = u$ above if $x \neq 0$; it is unique, by (6.4). Now set $x^{-n} = u^n$ for $n \in N$. It is easily verified that the properties in (Q_9) (with integral powers) continue to hold. For instance,

$$x^2x^{-4} = xxuuuu = xuuu = uu = x^{-2} = x^{2-4},$$
$$(x^{-3})^{-2} = (u^3)^{-2} = [(u^3)^{-1}]^2 = (x^3)^2 = x^{3\cdot 2} = x^{(-3)(-2)}.$$

7. Rays and birays in S. Given any $x \neq 0$ in the quantity structure S, set

$$(7.1) \qquad [x] = \{ax : a \in \mathrm{R}\}, \qquad [x]^+ = [x] \cap S^+.$$

By (Q_3), (P_1) of Section 4 holds with R in place of R^+. Now x generates $[x]$, and by (6.4), the generation is simple. Then, with addition in $[x]$ defined by (4.4), Theorem 5A shows that $[x]$ is a one-dimensional vector space. By (Q_5), we may set $y=x$ or $y=-x$, and have $y \in S^+$; by (6.2), $ay \in S^+$ iffif $a \in \mathrm{R}^+$. Thus $[x]$ *is a biray*, with $[x]^+$ as positive part; and by Theorem 5D, $[x]^+$ *is a ray*.

In particular, there is a *natural order* in $[x]$, and in particular, in $[x]^+$.

We now compare two birays in S. First note that any biray $[x]$ contains $0x=0$ as its zero element. We now prove:

$$(7.2) \qquad \text{For any } x, y \neq 0, \text{ either } [x] = [y] \text{ or } [x] \cap [y] = \{0\}.$$

For suppose $z \in [x] \cap [y]$, $z \neq 0$. Then we have $z = ax = by$ for some $a, b \neq 0$; hence $x = cz$, $y = dz$, and $y = dax$, $x = cby$, showing that $[y] = [x]$. We now see at once that (letting $\varnothing$ denote the null set) if $x \neq 0$, $y \neq 0$,

$$(7.3) \qquad \text{either } [x]^+ = [y]^+ \text{ or } [x]^+ \cap [y]^+ = \varnothing;$$

$$(7.4) \qquad [x]^+ = [y]^+ \text{ iffif } [x] = [y];$$

$$(7.5) \qquad [x] = [y] \text{ iffif } y \in [x].$$

We have two distributive laws, (4.4′) and (4.5), in each biray or ray, with $a, b \in \mathrm{R}$ or $a, b \in \mathrm{R}^+$. We give now a more general distributive law:

$$(7.6) \qquad z(x + y) = zx + zy \text{ if } x \text{ and } y \text{ are both in some } [w].$$

This is true if $z=0$; suppose $z \neq 0$. Say $x = aw$, $y = bw$. Now by (Q_3) and the definition of addition in $[w]$,

$$zx = z(aw) = a(zw), \qquad zy = b(zw),$$
$$z(x + y) = z[(a + b)w] = (a + b)(zw).$$

By (6.3), $zw \neq 0$, and thus the three elements shown are all in $[zw]$. The definition of addition in $[zw]$ now gives (7.6).

REMARK 7A. In any sum of terms, we will be careful to keep all the terms in one biray. Due to 0 being in all birays, other sums, for instance $x+(y-y)$, may have meaning; but note that $(x+y)-y$ in general is not defined. It would be possible to introduce any finite formal sum as a new sort of element, and add such elements componentwise; but this would not be useful for our purposes. (But see the end of Sect. 6 of **[5]**.)

8. The type space of a quantity structure. In mechanics, the physical dimension of length divided by the physical dimension of time gives the physical dimension of velocity. We shall give corresponding definitions in S.

DEFINITION 8A. In the quantity structure S, let each biray $[x]$ be considered as an element of a new structure $[S]$. We define multiplication and the power operation by R^* in $[S]$ as follows:

$$[x][y] = [xy], \qquad [x]^\alpha = [x^\alpha] \qquad (\text{taking } x \in S^+). \tag{8.1}$$

REMARKS. Henceforth we shall nearly always use elements x of S^+ to specify any biray $[x]$. Because of (7.4), we could use $[x]^+$ in place of $[x]$ in the definition.

We must show that the definitions (8.1) are independent of the elements x, y chosen. Suppose $[x]=[x']$, $[y]=[y']$ (all elements in S^+). Then we can write $x'=ax$, $y'=by$, and we have

$$x'y' = abxy \in [xy], \qquad x'^\alpha = a^\alpha x^\alpha \in [x^\alpha],$$

and hence, by (7.5), $[x'y']=[xy]$, $[x'^\alpha]=[x^\alpha]$.

THEOREM 8B. *$[S]$ is a multiplicative vector space over* R^*, *with identity element* $[1]$ *and inverses* $[x]^{-1}=[x^{-1}]$.

First, since $[1][x]=[1x]=[x]$ and $[x^{-1}][x]=[x^{-1}x]=[1]$, the last statements are true. Next, (Q_3) shows that multiplication in $[S]$ is associative and commutative. We must show that the operation by R^* on $[S]$ has the properties

$$(\Gamma\Delta)^\alpha = \Gamma^\alpha\Delta^\alpha, \qquad \Gamma^{\alpha+\beta} = \Gamma^\alpha\Gamma^\beta, \qquad \Gamma^{\alpha\beta} = (\Gamma^\alpha)^\beta, \qquad \Gamma^1 = \Gamma.$$

(With the additive notation, one recognizes the usual requirements for a vector space.) These properties follow at once from (Q_9) and (Q_8). For instance,

$$([x][y])^\alpha = [xy]^\alpha = [(xy)^\alpha] = [x^\alpha y^\alpha] = [x^\alpha][y^\alpha] = [x]^\alpha[y]^\alpha.$$

REMARK 8C. Going back to the actual birays, we have

$$[xy] = \{uv : u \in [x], v \in [y]\};$$

but $[x]^2 \neq \{u^2 : u \in [x]\}$, since u^2 is never negative.

9. Finitely generated structures. First we recall some definitions from the theory of vector spaces, written multiplicatively.

A set of elements $\Gamma_1, \cdots, \Gamma_n$ *generates*, or *spans*, $[S]$ iffif each element of $[S]$ can be written in the form $\Gamma_1^{\alpha_1} \cdots \Gamma_n^{\alpha_n}$ (with the α_i in R^*). The elements $\Gamma_1, \cdots, \Gamma_n$ are *independent* iffif

$$\Gamma_1^{\alpha_1} \cdots \Gamma_n^{\alpha_n} = [1] \quad \text{implies} \quad \alpha_i = 0 \quad (\text{all } i),$$

or equivalently,

$$\Gamma_1^{\alpha_1} \cdots \Gamma_n^{\alpha_n} = \Gamma_1^{\beta_1} \cdots \Gamma_n^{\beta_n} \quad \text{implies} \quad \alpha_i = \beta_i \quad (\text{all } i).$$

A *base* for $[S]$ is an independent generating set.

DEFINITION 9A. Given the quantity structure S, the elements $x_1, \cdots, x_n$ of S^+ *generate* S iffif $[x_1], \cdots, [x_n]$ generate $[S]$. Similarly for independent sets and bases.

We justify these definitions through the following theorems.

THEOREM 9B. *The elements $x_1, \cdots, x_n$ generate S iffif every $x \in S$ can be written in the form*

$$x = a x_1^{\alpha_1} \cdots x_n^{\alpha_n} \qquad (a \in \mathrm{R}; \alpha_1, \cdots, \alpha_n \in \mathrm{R}^*). \tag{9.1}$$

If $[x_1], \cdots, [x_n]$ generate $[S]$, then for any $x \neq 0$, $[x] \in [S]$, and we can write $[x] = [x_1]^{\alpha_1} \cdots [x_n]^{\alpha_n}$; thus $x \in [x_1^{\alpha_1} \cdots x_n^{\alpha_n}]$, and (9.1) is possible. The converse is clear.

THEOREM 9C. *The elements $x_1, \cdots, x_n$ of S^+ are independent iffif*

$$x_1^{\alpha_1} \cdots x_n^{\alpha_n} \in \mathrm{R} \quad \textit{implies} \quad \alpha_i = 0 \quad (\textit{all } i), \tag{9.2}$$

or equivalently, any expression of the form (9.1) *is unique, if $x \neq 0$.*

The first part of (9.2) gives $[x_1]^{\alpha_1} \cdots [x_n]^{\alpha_n} = [1]$; hence if the x_i are independent, all α_i are 0. The converse is clear. The last statement follows at once.

THEOREM 9D. *Given $x_1, \cdots, x_n \in S^+$, any $x \neq 0$ may be written uniquely in the form* (9.1) *iffif the x_i form a base for S.*

This follows from the last two theorems.

DEFINITION 9E. S is *finitely generated* if it has a (finite) generating set. Then it is a standard fact about vector spaces that S has a base, and all bases have the same number of elements. We call this number the *type dimension* dim (S) of S. We also recall from vector space theory:

THEOREM 9F. *Any independent set in a finitely generated S is part of a base.*

We now *compare two bases* $\{x_1, \cdots, x_n\}$, $\{y_1, \cdots, y_n\}$ in S. By Theorem 9D, we may write uniquely

$$y_i = a_i x_1^{\alpha_{i1}} \cdots x_n^{\alpha_{in}}, \qquad x_i = b_i y_1^{\beta_{i1}} \cdots y_n^{\beta_{in}} \qquad (i = 1, \cdots, n). \tag{9.3}$$

Substituting the first expression in the second gives

$$\begin{aligned} x_i &= b_i (a_1 x_1^{\alpha_{11}} \cdots x_n^{\alpha_{1n}})^{\beta_{i1}} \cdots (a_n x_1^{\alpha_{n1}} \cdots x_n^{\alpha_{nn}})^{\beta_{in}} \\ &= b_i a_1^{\beta_{i1}} \cdots a_n^{\beta_{in}} x_1^{\beta_{i1}\alpha_{11} + \cdots + \beta_{in}\alpha_{n1}} \cdots x_n^{\beta_{i1}\alpha_{1n} + \cdots + \beta_{in}\alpha_{nn}}; \end{aligned}$$

because of uniqueness, this gives (using the similar relation with the x_i and y_i interchanged)

$$\|\beta_{ij}\| = \|\alpha_{ij}\|^{-1}, \tag{9.4}$$

$$(9.5)\qquad a_i = b_1^{-\alpha_{i1}} \cdots b_n^{-\alpha_{in}}, \qquad b_i = a_1^{-\beta_{i1}} \cdots a_n^{-\beta_{in}} \qquad (i = 1, \cdots, n).$$

If an element is expressed in terms of the x_i, (9.3) gives at once its expression in terms of the y_i:

$$(9.6)\qquad cx_1^{\gamma_1} \cdots x_n^{\gamma_n} = cb_1^{\gamma_1} \cdots b_n^{\gamma_n} y_1^{\gamma_1\beta_{11}+\cdots+\gamma_n\beta_{n1}} \cdots y_n^{\gamma_1\beta_{1n}+\cdots+\gamma_n\beta_{nn}}.$$

Now let us consider a *type preserving* change of base: We require $[y_i] = [x_i]$ for each i. Then (9.3) reduces to

$$(9.7)\qquad y_i = \lambda_i x_i, \qquad x_i = \lambda_i^{-1} y_i \qquad (i = 1, \cdots, n);$$

the λ_i are positive, since the base elements are in S^+. Now (9.6) becomes

$$(9.8)\qquad cx_1^{\gamma_1} \cdots x_n^{\gamma_n} = c\lambda_1^{-\gamma_1} \cdots \lambda_n^{-\gamma_n} y_1^{\gamma_1} \cdots y_n^{\gamma_n}.$$

In the expression (9.1), we may call a the *coefficient* of x in terms of the given base. With (9.8), one finds the rule for finding the coefficient in terms of the new base from that in terms of the old. See, in this connection, Section 21 and the example of John's race in Section 18.

REMARK 9G. In the case of the trivial quantity structure, $S = R$, $[S]$ consists of a single element $[1]$, and $\dim(S) = 0$; a base is the null set.

10. Homomorphisms of quantity structures. We shall show, among other things, that two structures of the same type dimension are isomorphic.

DEFINITION 10A. A mapping ϕ of a quantity structure S into another, S', is a *homomorphism* if it preserves the structural properties, keeping R fixed. That is:

(H_1) If x is positive then $\phi(x)$ is positive.

(H_2) $\phi(xy) = \phi(x)\phi(y)$.

(H_3) $\phi(x^\alpha) = (\phi(x))^\alpha (x \in S^+)$.

(H_4) $\phi(a) = a$ $(a \in \mathrm{R})$.

Note that (H_2) and (H_4) give, for $x \in S$ and $a \in \mathrm{R}$,

$$(10.1)\qquad \phi(ax) = a\phi(x), \qquad \phi(0) = 0.$$

We shall consider now only finitely generated structures.

THEOREM 10B. *Let S and S' be structures with the same power operators* R^*, *and let $\{x_1, \cdots, x_n\}$ be a base in S. Then for each set of elements $x_1', \cdots, x_n'$ of S'^+ there is one and only one homomorphism ϕ of S into S' such that $\phi(x_i) = x_i'$ for each i; it is given by*

$$(10.2)\qquad \phi(ax_1^{\alpha_1} \cdots x_n^{\alpha_n}) = ax_1'^{\alpha_1} \cdots x_n'^{\alpha_n}.$$

Recalling Theorem 9D, (10.2) defines ϕ uniquely. The verification that ϕ is a homomorphism is immediate. For instance,

$$\phi((ax_1^{\alpha_1} \cdots x_n^{\alpha_n})^\gamma) = \phi(a^\gamma x_1^{\alpha_1\gamma} \cdots x_n^{\alpha_n\gamma}) = a^\gamma x_1'^{\alpha_1\gamma} \cdots x_n'^{\alpha_n\gamma}$$

$$= (ax_1^{\prime\alpha_1} \cdots x_n^{\prime\alpha_n})^\gamma = (\phi(ax_1^{\alpha_1} \cdots x_n^{\alpha_n}))^\gamma.$$

THEOREM 10C. *In the last theorem, ϕ is an isomorphism iffif $K' = \{x_1', \cdots, x_n'\}$ is independent; ϕ is onto S' iffif K' generates S'; ϕ is an isomorphism onto S' iffif K' is a base for S'.*

If K' is independent and $\phi(x) = \phi(y)$, then comparing the corresponding right hand sides of (10.2) and applying Theorem 9C shows that $x = y$, provided that $\phi(x) \neq 0$; in the contrary case, $x = y = 0$. The rest of the theorem is clear.

Recalling that a base consists of positive elements, the last theorem gives:

THEOREM 10D. *There is an isomorphism of S onto S' iffif* $\dim(S) = \dim(S')$; *any such isomorphism carries each base in S onto a base in S'.*

REMARK 10E. In Theorem 10B, take $S' = \text{R}$ and $x_i' = 1$ (all i); then ϕ gives the coefficient of an element in terms of the given base (see Sect. 9).

11. Similarities of S. Invariance under similarities is at the heart of dimensional analysis; see Section 22.

DEFINITION 11A. A *similarity* ϕ of the quantity structure S is an isomorphism of S onto itself which preserves type:

$$[\phi(x)] = [x]. \tag{11.1}$$

THEOREM 11B. *A homomorphism of S into itself satisfying* (11.1) *is a similarity. Also if $x \in S^-$ then $\phi(x) \in S^-$.*

If $x \in S^-$, then $-x \in S^+$, hence $\phi(-x) \in S^+$, by (H_1), and taking $a = -1$ in (10.1) shows that $\phi(x) = -\phi(-x) \in S^-$. Now suppose $\phi(x) = \phi(y) = u$. If $u \neq 0$, then $x, y \neq 0$, by (10.1), and $[x] = [\phi(x)] = [\phi(y)] = [y]$, so that $x = ay$ for some a; now

$$\phi(x) = \phi(ay) = a\phi(y) = a\phi(x), \qquad a = 1, \qquad x = y.$$

If $u = 0$, then $x = y = 0$ by what is already proved. Thus ϕ is an isomorphism. Finally, given x, $\phi(x) = ax$ for some $a > 0$; now $\phi((1/a)x) = x$, and ϕ is onto S.

THEOREM 11C. *Let $\{x_1, \cdots, x_n\}$ be a base for S. Then there is a one-one correspondence between ordered sets $\lambda = (\lambda_1, \cdots, \lambda_n)$ of positive numbers and similarities ϕ_λ of S, such that*

$$\phi_\lambda(x_i) = \lambda_i x_i \qquad (i = 1, \cdots, n); \tag{11.2}$$

we have

$$\phi_\lambda(x) = \lambda_1^{\alpha_1} \cdots \lambda_n^{\alpha_n} x \quad \textit{if } x \in [x_1^{\alpha_1} \cdots x_n^{\alpha_n}]. \tag{11.3}$$

First, given λ, $\{\lambda_1 x_1, \cdots, \lambda_n x_n\}$ is a base, and by Theorem 10C, the isomorphism ϕ_λ is determined. If $x = ax_1^{\alpha_1} \cdots x_n^{\alpha_n}$, then (10.2) gives

$$\phi_\lambda(x) = a(\lambda_1 x_1)^{\alpha_1} \cdots (\lambda_n x_n)^{\alpha_n},$$

and (11.3) follows. This shows that (11.1) holds; thus ϕ is a similarity. Clearly changing the λ_i changes the similarity. Finally, given any similarity ϕ, $\phi(x_i) \in [x_i] \cap S^+$, hence $\phi(x_i) = \lambda_i x_i$ for some $\lambda_i > 0$ (all i), and $\phi = \phi_\lambda$.

We now consider similarities which keep the elements of a substructure S' fixed.

DEFINITION 11D. A *substructure* S' of S is a subset of S containing R, which is closed under the operations of S. That is, if $x, y \in S'$, $a \in$ R, and $\alpha \in$ R*, then xy, ax and x^α are in S'. Thus S' is a quantity structure. By definition, $S'^+ = S' \cap S^+$.

If S' is a substructure of S, then $[S']$ is a subspace of $[S]$, and we may form the factor space $[S]$ mod $[S']$. Each $x \neq 0$ in S determines a biray $[x]$ in S, i.e. an element of $[S]$, and this has a coset

$$[x][S'] = \{[x]\Gamma : \Gamma \in [S']\} = \{[xy] : y \in S'^+\} \subset [S],$$

which is an element of $[S]$ mod $[S']$. (An attempt to define S mod S' brings in complications, since we must include R in a structure.)

Note that the minimal substructure of S is R; also $[S]$ mod $[\mathrm{R}]$ is just $[S]$ itself.

DEFINITION 11E. For any set $K = \{x_1, \cdots, x_n\}$ in S^+, set $K_{S'} = \{[x_1][S'], \cdots, [x_n][S']\}$ in $[S]$ mod $[S']$. We say K is *independent* mod S' if $K_{S'}$ is independent; K *generates* S mod S' if $K_{S'}$ generates $[S]$ mod $[S']$; and K is a *base* for S mod S' if $K_{S'}$ is a base for $[S]$ mod $[S']$.

THEOREM 11F. *Let S' be a substructure of S, and let $\{x_1, \cdots, x_m\}$ be a base for S mod S'. Then there is a one-one correspondence between ordered sets $\lambda = (\lambda_1, \cdots, \lambda_m)$ of positive numbers and similarities ϕ_λ of S which keep all elements of S' fixed, such that* (11.2) *holds. We have*

$$(11.4) \quad \phi_\lambda(x) = \lambda_1^{\alpha_1} \cdots \lambda_m^{\alpha_m} x \quad \textit{if } x \in [x_1^{\alpha_1} \cdots x_m^{\alpha_m} x'] \quad \textit{for some } x' \in S'.$$

Choose a base $\{y_1, \cdots, y_k\}$ for S'; then clearly the x_i and y_j form a base for S. Applying Theorem 11C to this base gives the present theorem.

12. Construction of S from a set of base elements. Suppose objects $x_1, \cdots, x_n$ are given; we wish to construct a quantity structure S from them, so that they will form a base for S. The elements of S will be the expressions

$$(12.1) \qquad a x_1^{\alpha_1} \cdots x_n^{\alpha_n} \qquad (a \in \mathrm{R}; \alpha_1, \cdots, \alpha_n \in \mathrm{R}^*),$$

with the identification

$$(12.2) \qquad 0 x_1^{\alpha_1} \cdots x_n^{\alpha_n} = 0 x_1^0 \cdots x_n^0 \qquad (\text{all } \alpha_1, \cdots, \alpha_n \in \mathrm{R}^*).$$

Certain of these elements we consider to be new names for elements of R:

$$(12.3) \qquad a\overset{0}{x_1}\cdots\overset{0}{x_n} = a \qquad (a \in \mathrm{R}).$$

The elements of S^+ are those in (12.1) with $a>0$. We define multiplication and powers by

$$(12.4) \qquad (a\overset{\alpha_1}{x_1}\cdots\overset{\alpha_n}{x_n})(b\overset{\beta_1}{x_1}\cdots\overset{\beta_n}{x_n}) = (ab)\overset{\alpha_1+\beta_1}{x_1}\cdots\overset{\alpha_n+\beta_n}{x_n},$$

$$(12.5) \qquad (a\overset{\alpha_1}{x_1}\cdots\overset{\alpha_n}{x_n})^\gamma = a^\gamma\overset{\alpha_1\gamma}{x_1}\cdots\overset{\alpha_n\gamma}{x_n} \qquad (a>0).$$

Note that if either element on the left of (12.4) has several names, then $a=0$ or $b=0$, hence $ab=0$, and the right hand side is well defined.

We now show that S is a quantity structure. Note first that

$$(12.6) \qquad (b\overset{0}{x_1}\cdots\overset{0}{x_n})(a\overset{\alpha_1}{x_1}\cdots\overset{\alpha_n}{x_n}) = (ba)\overset{\alpha_1}{x_1}\cdots\overset{\alpha_n}{x_n}.$$

Taking the α_i to be 0, (Q_1) follows. Using $b=1$ and $b=0$ in (12.6) gives (Q_2). Associativity and commutativity of multiplication in S follows directly from the same properties of multiplication in R and of addition in R^*. (Q_4) follows from the definition of S^+ and (12.3), and (Q_5) is proved with the help of (12.6), using $b=-1$. (Q_6) is clear.

Taking the α_i to be 0 in (12.5) gives (Q_7). Since (for $a>0$) $a^1=a$, $\alpha 1=\alpha$, and $a^0=1$, $\alpha 0=0$, (Q_8) holds. The proof of (Q_9) is straightforward. For instance, representing xy by (12.4), we have

$$(xv)^\gamma = (ab\overset{\alpha_1+\beta_1}{x_1}\cdots\overset{\alpha_n+\beta_n}{x_n})^\gamma = (ab)^\gamma\overset{(\alpha_1+\beta_1)\gamma}{x_1}\cdots\overset{(\alpha_n+\beta_n)\gamma}{x_n},$$

$$x^\gamma y^\gamma = (a^\gamma\overset{\alpha_1\gamma}{x_1}\cdots\overset{\alpha_n\gamma}{x_n})(b^\gamma\overset{\beta_1\gamma}{x_1}\cdots\overset{\beta_n\gamma}{x_n}) = (a^\gamma b^\gamma)\overset{\alpha_1\gamma+\beta_1\gamma}{x_1}\cdots\overset{\alpha_n\gamma+\beta_n\gamma}{x_n},$$

and these are equal.

We now introduce further shorthand notations: In (12.1), any x_i with the exponent 0 may be omitted. This agrees with (12.3). Also, if $a=1$ we may omit this symbol (unless all x_i have been omitted). For instance,

$$x_i \text{ means } 1\overset{0}{x_1}\cdots\overset{1}{x_i}\cdots\overset{0}{x_n}.$$

Now x_i to the power α is just $\overset{\alpha}{x_i}$, for

$$(1\overset{0}{x_1}\cdots\overset{1}{x_i}\cdots\overset{0}{x_n})^\alpha = 1\overset{0}{x_1}\cdots\overset{\alpha}{x_i}\cdots\overset{0}{x_n}.$$

We may now interpret the expressions (12.1) as products of powers. For, taking $n=3$ and x, y, z as base elements for short, we have

$$a(x)^\alpha(y)^\beta(z)^\gamma = (ax^0y^0z^0)(1x^\alpha y^0z^0)(1x^0y^\beta z^0)(1x^0y^0z^\gamma)$$
$$= ax^{0+\alpha+0+0}\ldots = ax^\alpha y^\beta z^\gamma.$$

In particular, the x_i are now elements of S. Moreover, comparing (12.1) with (9.1) and Theorem 9D shows that the x_i form a base for S.

13. Construction of S from base rays. Let the rays $\Gamma_1, \cdots, \Gamma_n$ be given (according to Definition 4B); we shall construct the quantity structure S from these. (One could equally well use birays in place of rays.) We could choose $x_i \in \Gamma_i$ and apply the definition of the last section; but we prefer to give an intrinsic definition. It is: The elements of S are the expressions

$$(13.1) \qquad ax_1^{\alpha_1} \cdots x_n^{\alpha_n} \ (a \in \mathrm{R};\ x_i \in \Gamma_i,\ \alpha_i \subset \mathrm{R}^*, \text{ for all } i);$$

we make the following identifications. For any positive numbers $\lambda_1, \cdots, \lambda_n$,

$$(13.2) \qquad a(\lambda_1 x_1)^{\alpha_1} \cdots (\lambda_n x_n)^{\alpha_n} = (a\lambda_1^{\alpha_1} \cdots \lambda_n^{\alpha_n})x_1^{\alpha_1} \cdots x_n^{\alpha_n};$$

we also make the identification (12.2), and introduce the abbreviation (12.3). We let the expression (13.1) be in S^+ iffif $a > 0$, and introduce multiplication and powers by (12.4) and (12.5).

To show that these definitions make sense and result in a quantity structure, we proceed as follows: Choose fixed elements $z_i \in \Gamma_i$, and define S_0 in terms of these, by Section 12. We now interpret the expressions shown above as products of powers in S_0. Say $x_i = c_i z_i$ $(i = 1, \cdots, n)$; then (13.2) becomes

$$a(\lambda_1 c_1 z_1)^{\alpha_1} \cdots (\lambda_n c_n z_n)^{\alpha_n} = (a\lambda_1^{\alpha_1} \cdots \lambda_n^{\alpha_n})(c_1 z_1)^{\alpha_1} \cdots (c_n z_n)^{\alpha_n},$$

which we recognize as a true equation in S_0. Thus each element of S is put into correspondence with a unique element of S_0. Since no identification other than (13.2) and (12.2) have been made, the elements of S may be identified with those of S_0. Now the definitions of positive elements, multiplication and powers agree in S and in S_0; thus S is precisely S_0. Therefore S is a quantity structure; moreover, if the definition of Section 12 is used to construct S, with chosen $x_i \in \Gamma_i$, it is independent of the choice of the x_i.

Since the x_i (in the Γ_i) form a base for S, the Γ_i form a set of base rays for S; equivalently, they form a base for $[S]$.

Here are three particular cases:

(a) No base rays at all: S consists of all $a (a \in \mathrm{R})$; here, $S = \mathrm{R}$.

(b) There is one base ray, say L. The elements of S are the expressions $al^{\alpha} (a \in \mathrm{R}, l \in L, \alpha \in \mathrm{R}^*)$. For instance, if L represents lengths, then L^2 represents areas, L^3 represents volumes, etc.

(c) There are three base rays: M, L and T. Interpreting these as representing mass, length and time, we have the basic quantity structure in Newtonian mechanics.

14. Structure of S^+. With a fixed base $x_1, \cdots, x_n$ for S, we may write the elements of S^+ in the form

$$(14.1) \quad e^{\alpha} x_1^{\alpha_1} \cdots x_n^{\alpha_n} = (\alpha, \alpha_1, \cdots, \alpha_n) \qquad (\alpha \in \mathrm{R};\ \alpha_1, \cdots, \alpha_n \in \mathrm{R}^*).$$

The operations in S^+ are given by

(14.2) $(\alpha, \alpha_1, \cdots, \alpha_n)(\beta, \beta_1, \cdots, \beta_n) = (\alpha + \beta, \alpha_1 + \beta_1, \cdots, \alpha_n + \beta_n),$

(14.3) $$(\alpha, \alpha_1, \cdots, \alpha_n)^\gamma = (\gamma\alpha, \gamma\alpha_1, \cdots, \gamma\alpha_n).$$

Thus, taking R^* to be R, S^+ is isomorphic with the vector space R^{n+1} over R. Note that the isomorphism depends on the choice of a base; the different isomorphisms differ by similarities in S. Also replacing e by some other $e' > 1$ would change the isomorphism.

The rays in S^+ are composed of the elements $(\alpha, \bar{\alpha}_1, \cdots, \bar{\alpha}_n)$ with fixed $\bar{\alpha}_i$. In this form, the definition of addition does not appear naturally; it is introduced through the definition

(14.4) $$(\alpha, \bar{\alpha}_1, \cdots) + (\beta, \bar{\alpha}_1, \cdots) = (\log(e^\alpha + e^\beta), \bar{\alpha}_1, \cdots).$$

15. Algebra in a quantity structure. We shall show how parts of elementary algebra fit into quantity structures; we consider simultaneous linear equations, quadratic equations, and roots of polynomials. *We now allow any symbol to denote elements of S.*

First we introduce division: Recalling that y^{-1} is defined if $y \neq 0$ (see the end of Section 6), we set

(15.1) $$x \div y = x/y = xy^{-1} \qquad (x, y \in S, y \neq 0).$$

Then

(15.2) $$x/y \in [x][y]^{-1}.$$

The usual rules follow (assuming certain elements to be nonzero):

(15.3) $$\frac{x}{y} \cdot \frac{z}{w} = (xy^{-1})(zw^{-1}) = (xz)(yw)^{-1} = \frac{xz}{yw},$$

(15.4) $$\frac{x}{y} \div \frac{z}{w} = (xy^{-1})(zw^{-1})^{-1} = xy^{-1}z^{-1}w = \frac{xw}{yz}.$$

Also, assuming x and y positive if α is not integral, since $(y^\beta)^\gamma = y^{\beta\gamma} = (y^\gamma)^\beta$,

(15.5) $$(x/y)^\alpha = (xy^{-1})^\alpha = x^\alpha(y^{-1})^\alpha = x^\alpha(y^\alpha)^{-1} = x^\alpha/y^\alpha.$$

Let us consider a triple of equations in three unknowns. Suppose we have birays $\Gamma_1, \Gamma_2, \Gamma_3, \Delta_1, \Delta_2, \Delta_3, \Delta$ in S, such that

(15.6) $$\Delta_1\Gamma_1 = \Delta_2\Gamma_2 = \Delta_3\Gamma_3 = \Delta.$$

Then the linear equations

(15.7) $$a_{i1}x_1 + a_{i2}x_2 + a_{i3}x_3 = c_i \qquad (i = 1, 2, 3)$$

have meaning, if we require

(15.8) $$a_{ij} \in \Delta_j, \qquad x_j \in \Gamma_j, \qquad c_i \in \Delta.$$

If the equations are true, then multiplying them by the elements

$$a_{22}a_{33} - a_{23}a_{32}, \qquad -a_{12}a_{33} + a_{13}a_{32}, \qquad a_{12}a_{23} - a_{13}a_{22}$$

of $\Delta_2\Delta_3$ respectively and adding gives

$$Dx_1 = D_1, \tag{15.9}$$

where

$$D = \begin{vmatrix} a_{11} & a_{12} & a_{13} \\ a_{21} & a_{22} & a_{23} \\ a_{31} & a_{32} & a_{33} \end{vmatrix} \in \Delta_1\Delta_2\Delta_3, \qquad D_1 = \begin{vmatrix} c_1 & a_{12} & a_{13} \\ c_2 & a_{22} & a_{23} \\ c_3 & a_{32} & a_{33} \end{vmatrix} \in \Delta\Delta_2\Delta_3; \tag{15.10}$$

similarly for x_2 and x_3. Thus if $D \neq 0$, we must have

$$x_1 = D_1/D \in (\Delta\Delta_2\Delta_3)(\Delta_1\Delta_2\Delta_3)^{-1} = \Delta\Delta_1^{-1} = \Gamma_1, \tag{15.11}$$

etc. Conversely, these x_i satisfy (15.7).

Next we consider quadratic equations.

LEMMA 15A. *If $x \neq 0$, then $x^2 \in S^+$, $1/x^2 = x^{-2} \in S^+$.*

This follows from the rule of signs, Section 6.

DEFINITION 15B. For $x \in S^+$, $\sqrt{x} = x^{1/2}$. (Recall that $x^\alpha \in S^+$, by definition, if α is not an integer.)

THEOREM 15C. *For $a \in S^+$, $x^2 = a$ has just two solutions, $\sqrt{a}$ and $-\sqrt{a}$.*

For if $x = \sqrt{a}$ and $y^2 = a$ also, then $y \in S^+$ implies $(y^2)^{1/2} = (x^2)^{1/2}$, $y^1 = x^1$, $y = x$, while if $y \in S^-$, then $-y \in S^+$ and $(-y)^2 = x^2$, and hence $-y = x$. Alternatively,

$$[y] = [y^2]^{1/2} = [x^2]^{1/2} = [x],$$

so that we may write $y^2 - x^2 = (y-x)(y+x)$, and (6.3) gives $y - x = 0$ or $y + x = 0$.

We now solve the general quadratic equation:

$$P(x) = ax^2 + bx + c = 0, \tag{15.12}$$

where

$$a \neq 0, \qquad a \in \Delta, \qquad b \in \Delta\Gamma, \qquad c \in \Delta\Gamma^2. \tag{15.13}$$

THEOREM 15D. *If $D = b^2 - 4ac \in S^-$, then* (15.12) *has no solutions. If $D = 0$, then the only solution is $x = -b/2a \in \Gamma$. If $D \in S^+$, then there are just two solutions, namely,*

$$\frac{1}{2a}(-b + \sqrt{D}), \qquad \frac{1}{2a}(-b - \sqrt{D}) \in \Gamma. \tag{15.14}$$

For

$$\frac{1}{a} P(x) = \left(x + \frac{b}{2a}\right)^2 - \frac{1}{4a^2}(b^2 - 4ac),$$

and the facts follow from Lemma 15A and Theorem 15C.

We end with a factorization theorem.

THEOREM 15E. *Suppose we define*

$$P(x) = x^n + a_1x^{n-1} + \cdots + a_{n-1}x + a_n, \tag{15.15}$$

where

$$x \in \Gamma, \quad a_i \in \Gamma^i \quad (i = 1, \cdots, n). \tag{15.16}$$

Suppose $x_1 \in \Gamma$ *and* $P(x_1) = 0$. *Then there is a polynomial*

$$Q(x) = x^{n-1} + b_1x^{n-2} + \cdots + b_{n-1} \qquad (x \in \Gamma, b_i \in \Gamma^i \textit{ for all } i),$$

such that $P(x) = (x - x_1)Q(x)$ $(x \in \Gamma)$.

This follows at once from the equations

$$P(x) - P(x_1) = (x^n - x_1^n) + a_1(x^{n-1} - x_1^{n-1}) + \cdots + a_{n-1}(x - x_1),$$
$$x^i - x_1^i = (x - x_1)(x^{i-1} + x_1x^{i-2} + \cdots + x_1^{i-2}x + x_1^{i-1}).$$

16. Calculus in a quantity structure. Even more than with algebra, it is commonly felt that calculus in actual fact can only deal with real numbers. This is a mistake; all the ordinary processes can be used in a quantity structure S. We indicate the main features here.

Functions. Let Γ and Δ be birays in S; we consider functions from Γ to Δ. In actual practice, the domain of the function will often be restricted to some subset of Γ. For our purpose, it is not necessary to go into this matter here.

Recall that algebraic operations carry us into new birays; this is of course true of physical quantities in general. There are certain important "elementary functions," the exponential, logarithmic and trigonometric functions for instance; these are from R to R. (See Sect. 19 for an example of the use of the exponential function.)

Absolute values. For any $x \in S$, define $|x|$ to be whichever of x, $-x$ is not in S^-; in particular, $|0| = 0$. By Theorem 15C, $|x| = \sqrt{x^2}$. The rule of signs shows that $|xy| = |x||y|$. Recalling that there is a natural order in any biray, the usual proof gives

$$|x + y| \leqq |x| + |y| \qquad (x \text{ and } y \text{ in the same biray}).$$

It is important to keep in mind that $|x|$ *is an element of* $[x]$, not a real number, in general; in fact, $|x| \in [x]^+$ if $x \neq 0$.

Limits. If $x_1, x_2, \cdots$ is a sequence of points on the biray Γ, we define $\lim x_i$ to be that element x of Γ (if it exists) such that for each $\gamma \in \Gamma^+$ there is an integer

N such that

$$|x - x_i| < \gamma \quad \text{if } i > N.$$

If the limit exists, it is unique. The usual rules are true; for instance: If $x_i \in \Gamma$, $y_i \in \Delta$, and $\lim x_i = x$, $\lim y_i = y$, then $\lim x_i y_i = xy \in \Gamma\Delta$. The usual proofs may be interpreted directly in S.

Given $f: \Gamma \to \Delta$, $\lim f(x)$ as $x \to x_0$ is defined similarly: It is $y \in \Delta$ iffif for each $\delta \in \Delta^+$ there is a $\gamma \in \Gamma^+$ such that

$$|y - f(x)| < \delta \quad \text{if } |x - x_0| < \gamma, \qquad x \neq x_0.$$

Continuous functions are defined as usual. A continuous function in a bounded closed interval of a biray takes on its minimum and maximum values, and all values between.

Differentiation. Let L and T be birays in S, and set $V = LT^{-1} = L/T$. We use this notation to suggest length, time and velocity. Let $f: T \to L$ be given. Then f is *differentiable* at $t_0 \in T$ iffif

$$\lim_{t \to t_0} \frac{f(t) - f(t_0)}{t - t_0} \in V$$

exists; if so, this limit is the *derivative* of f at t_0. The elementary theorems of differentiation, the differentiation of polynomials (compare Sect. 15), and so on, are found as usual.

Rolle's Theorem and the Theorem of the Mean are proved in the normal manner. As a corollary, if a function has the derivative 0 in an interval, it is constant in the interval. *Expansion in an infinite series* may be used.

Integration. Let $f: T \to V$ be continuous; suppose $VT = L$. We define

$$I(t', t'') = \int_{t'}^{t''} f(t)dt \in L$$

as follows. Given any subdivision $\mathfrak{S}$ of the interval (t', t'') in T, by the points $t_0 = t'$, $t_1, \cdots, t_n = t''$, set

$$F(\mathfrak{S}) = \sum_{i=1}^{n} f(t_{i-1})(t_i - t_{i-1});$$

then $F(\mathfrak{S}) \in L$. One shows as usual that there is an element $l \in L$ such that for any $\lambda \in L^+$, $|F(\mathfrak{S}) - l| < \lambda$ if the largest interval of $\mathfrak{S}$ is small enough. This l is defined to be the integral.

The "Fundamental Theorem of the calculus" holds, etc.

17. Calculus in 3-space. Here, we must put quantity structures S into relation with vectors in a vector space (the space of translations of 3-space). This is done by letting the *components* of vectors be elements of S. Letting u, v be vectors and p, q be elements of S, we may use the following relations:

$$p(u + v) = pu + pv, \qquad p(qv) = (pq)v, \qquad 1v = v.$$

If the coefficients of two vectors are in the same biray of S, we may add them; in particular,

$$pv + qv = (p + q)v \qquad (p, q \text{ in the biray } \Gamma).$$

We have here a structure that is a generalization of a vector space over a field, and of a module over a ring. (Since we shall not use this structure, we shall not attempt a full definition here.)

We may form tensors, and in particular, differential forms. We give one example.

Consider a *flow of fluid*. At the point P and time t, the velocity $v(P, t)$ and density $\rho(P, t)$ are defined. Suppose σ is a surface, with a given orientation across it; we wish to find a formula for the rate of flow of fluid through σ (in mass per unit time) at the time t. This is

$$I(\sigma, t) = \iint_\sigma \rho(P, t)v(P, t)\cdot d\sigma.$$

Note that, with the birays M, L and T of mass, length and time, we have, for coefficients,

$$\rho \in ML^{-3}, \qquad v \in LT^{-1}, \qquad d\sigma \in L^2,$$
$$I(\sigma, t) \in (ML^{-3})(LT^{-1})L^2 = MT^{-1}.$$

We remark that in relativity theory, going over to 4-dimensional space-time, one loses the distinction between the birays L and T. Still, an observer (under no acceleration) may observe the velocity of light and obtain the relation (2.1); he distinguishes L and T here.

If one wishes to put Maxwell's equations into intrinsic form, using quantity structures, there are other problems. For instance, one commonly differentiates between E and H, one being "axial" and the other "polar." With proper interpretations, this is not necessary. See, in this connection, Flanders, [**3**].

CHAPTER II. APPLICATIONS

In the first sections of this chapter we give some familiar examples of equations involving geometric or physical quantities. They may be read mostly without reference to Chapter I. When interpreted in terms of quantity structures, their meaning is clear and precise.

18. Some elementary problems. Here we use only simple algebra in a quantity structure.

Areas of rectangles. If we know that the area of a rectangle is the product of the lengths of the sides, then if these lengths are 2 ft and 5 ft respectively, the area is

$$A = 2\text{ ft} \times 5\text{ ft} = 2 \times 5 \times \text{ft} \times \text{ft} = 10\text{ ft}^2.$$

If we know the area is 10 ft² and one side is 2 ft, then the other side is

$$l = 10\text{ ft}^2 \div 2\text{ ft} = 10 \times 2^{-1} \times \text{ft}^{2-1} = 5\text{ ft}.$$

Velocities. John ran a quarter of a mile in two and a quarter minutes. What was his average speed, in ft per sec? Knowing that this is, by definition, the distance divided by the time, we have

$$v = (\tfrac{1}{4}\text{ mile})(2\tfrac{1}{4}\text{ min})^{-1} = (\tfrac{1}{4} \times 5280\text{ ft})(2\tfrac{1}{4} \times 60\text{ sec})^{-1}$$
$$= \tfrac{88}{9}\text{ ft sec}^{-1} = 9.78\text{ ft sec}^{-1}\text{ approximately.}$$

Here, we are working in a quantity structure constructed from the rays L for length and T for time; the speed lies in the ray LT^{-1}.

Flight of the ball. A ball is thrown up with the velocity 40 ft per sec. When is it 20 ft. high? We neglect friction of the air, and assume it is acted on by gravity only. Then, from elementary physics, we must solve

$$h = v_0 t - \tfrac{1}{2} g t^2 = 20\text{ ft},$$

with $v_0 = 40$ ft sec^{-1}, $g = 32$ ft sec^{-2}. Inserting gives

$$(40\text{ ft sec}^{-1})t - (16\text{ ft sec}^{-2})t^2 = 20\text{ ft},$$

or multiplying by sec²/4 ft,

$$4t^2 - 10\text{ sec } t + 5\text{ sec}^2 = 0.$$

The discriminant is (see Theorem 15D) $(10\text{ sec})^2 - 4 \times 4 \times 5\text{ sec}^2 = 20\text{ sec}^2$, which is in $T^2 \cap S^+$; hence, by Theorem 15D, there are two solutions (given by the usual formula):

$$t = \frac{5 - \sqrt{5}}{4}\text{ sec} \quad \text{or} \quad t = \frac{5 + \sqrt{5}}{4}\text{ sec}.$$

19. Some problems using calculus. For a *volume* problem, see Section 20.

A *pendulum* consists of a point mass m at the end of a string of length l. Using the arc length s and the angle $\theta = s/l$, the component of gravity along s is $-g \sin \theta$, or, keeping the amplitude small, $-g\theta$. By Newton's law of motion, $md^2s/dt^2 = -mg\theta$, or

$$\frac{d^2\theta}{dt^2} + \frac{g}{l}\theta = 0.$$

Note that

$$t \in T; \qquad l, s \in L; \qquad \theta \in \mathrm{R}; \qquad g \in LT^{-2};$$

and (see Sect. 16) the equation to be solved has its terms on the biray T^{-2}. The solution is

$$\theta = \alpha \sin(\sqrt{g/l}\, t - \beta),$$

with unknown constants α, $\beta \in R$. We have $\sqrt{g/l}\, t \in R$.

Consider the *growth of bacteria*, according to the law

$$\frac{du}{dt} = ku.$$

Here we use a quantity structure constructed from T and a ray Γ of quantities of bacteria. Since $du/dt \in \Gamma T^{-1}$, we must have $k \in T^{-1}$. The solution is

$$u = u_0 e^{kt}; \qquad u, u_0 \in \Gamma, \qquad kt \in R.$$

20. Proportionality. The volume V of a cylinder is proportional to the height h and to the square of the radius r of the base; it follows that $V = chr^2$ for some constant c. We give here a general theorem of this nature, first without using a quantity structure. We consider three independent variables for simplicity.

THEOREM 20A. *Let* Γ, Γ_1, Γ_2, Γ_3 *be rays, and let*

$$f(x_1, x_2, x_3) \qquad (x_i \in \Gamma_i;\ \textit{values of } f \textit{ in } \Gamma)$$

be a given function. Suppose there are numbers α_1, α_2, α_3 *such that, for any* $\lambda > 0$,

$$f(\lambda x_1, x_2, x_3) = \lambda^{\alpha_1} f(x_1, x_2, x_3), \tag{20.1}$$

and similarly for the second and third variables. Then

$$f(\lambda_1 x_1, \lambda_2 x_2, \lambda_3 x_3) = \lambda_1^{\alpha_1} \lambda_2^{\alpha_2} \lambda_3^{\alpha_3} f(x_1, x_2, x_3). \tag{20.2}$$

For

$$\begin{aligned} f(\lambda_1 x_1, \lambda_2 x_2, \lambda_3 x_3) &= \lambda_1^{\alpha_1} f(x_1, \lambda_2 x_2, \lambda_3 x_3) = \lambda_1^{\alpha_1} \lambda_2^{\alpha_2} f(x_1, x_2, \lambda_3 x_3) \\ &= \lambda_1^{\alpha_1} \lambda_2^{\alpha_2} \lambda_3^{\alpha_3} f(x_1, x_2, x_3). \end{aligned}$$

REMARK 20B. We can weaken the hypothesis in (20.1). We show, in fact, that if (for all $\lambda \in R^+$ and all x in a given ray)

$$f(\lambda x) = \phi(\lambda) f(x), \qquad \phi \text{ continuous},$$

then $\phi(\lambda) = \lambda^\alpha$, α being defined by $\phi(e) = e^\alpha$. For, first of all,

$$\phi(\lambda\mu) f(x) = f((\lambda\mu)x) = f(\lambda(\mu x)) = \phi(\lambda) f(\mu x) = \phi(\lambda)\phi(\mu) f(x),$$

and hence $\phi(\lambda\mu) = \phi(\lambda)\phi(\mu)$, by (4.3). Next, with n terms in the products,

$$e^\alpha = \phi(e) = \phi(e^{1/n} \cdots e^{1/n}) = \phi(e^{1/n}) \cdots \phi(e^{1/n}),$$

and hence $\phi(e^{1/n}) = (e^\alpha)^{1/n}$. It follows that $\phi(e^{k/n}) = (e^\alpha)^{k/n}$, and since ϕ is continuous, $\phi(e^u) = (e^\alpha)^u = (e^u)^\alpha$, if $u > 0$. Considering $\phi(1 \cdot 1)$ and $\phi(\lambda\lambda^{-1})$ gives the remaining cases.

The result (20.2) is expressed in terms of real variables λ_1, λ_2, λ_3. We now suppose the rays lie in a quantity structure, and express the result directly in this structure. We allow also subsidiary variables, in the form of a parameter u.

THEOREM 20C. *Let* $\Gamma, \Gamma_1, \cdots, \Gamma_r$ *be rays in a quantity structure S, let U be a set, and let* $f: \Gamma_1 \times \cdots \times \Gamma_r \times U \to \Gamma$ *be given. Suppose there are numbers* $\alpha_1, \cdots, \alpha_r$ *such that for each* $\lambda > 0$,

$$(20.3) \qquad f(x_1, \cdots, \lambda x_i, \cdots, x_r, u) = \lambda^{\alpha_i} f(x_1, \cdots, x_i, \cdots, x_r, u).$$

Then there is a uniquely defined function $F: U \to \Delta = \Gamma\Gamma_1^{-\alpha_1} \cdots \Gamma_r^{-\alpha_r}$ *such that*

$$(20.4) \qquad f(x_1, \cdots, x_r, u) = x_1^{\alpha_1} \cdots x_r^{\alpha_r} F(u).$$

If U is not used, $F(u)$ *is replaced by a constant* $c \in \Delta$.

REMARK. In the applications, we often have $\Delta = R^+$.

To show this, choose $z_i \in \Gamma_i$ for each i, and set

$$(20.5) \qquad F(u) = z_1^{-\alpha_1} \cdots z_r^{-\alpha_r} f(z_1, \cdots, z_r, u).$$

Now given $x_1, \cdots, x_r$ and u, there are positive numbers $\lambda_1, \cdots, \lambda_r$ such that $x_i = \lambda_i z_i$ for each i. Taking out the λ_i from $f(\lambda_1 x_1, \cdots)$ as before gives the result. Uniqueness is clear.

As noted in the case of the cylinder, the theorem does not give the actual constant. Let us find the constant, in the case of the *volume V of a pyramid* of base area A and height h; this will illustrate the use of calculus. Let the cross section at the distance h' from the vertex have the area $A(h')$. In the quantity structure constructed from the ray L representing length, we have: $h, h' \in L$; $A, A(h') \in L^2$; $V \in L^3$. From simple geometrical hypotheses, we then have $A(h') = kh'^2$ for some $k \in R^+$. In particular, $A = kh^2$; hence $k = A/h^2$, and

$$V = \int_0^h A(h')dh' = Ah^{-2}\int_0^h h'^2 dh' = \tfrac{1}{3}Ah^{-2}h'^3\Big|_0^h = \tfrac{1}{3}Ah.$$

21. Dropping and changing units. In working problems in mechanics, one may choose to express all quantities in terms of grams, centimeters and seconds. In the corresponding quantity structure S, this means the following: With the base gr, cm, sec (see Sect. 9), all quantities may be expressed in the form $a\,\mathrm{gr}^\alpha \mathrm{cm}^\beta \mathrm{sec}^\gamma$; the expression is unique if $a \neq 0$. Now if there are considerable computations to be made, one commonly drops the units, retaining only the coefficient a. At the end, one puts the units back in.

In terms of S, this means the following: By Remark 10E, there is a unique homomorphism ϕ of S into R such that $\phi(\mathrm{gr}) = \phi(\mathrm{cm}) = \phi(\mathrm{sec}) = 1$; the image of any element is its coefficient. Since ϕ is a homomorphism, it preserves the operations of addition, multiplication, and powers of positive elements. When a numerical answer is obtained, i.e. we have the coefficient a of the desired quantity, we simply put the proper units back in (since we are supposed to know the physical dimension of the final quantity).

If we wish to change units, this means that we must find the coefficients in terms of a new base. If the new base elements are of the same physical dimension as the old, i.e. (9.7) holds, the change in the coefficients is given by (9.8); in the general case, we must use (9.6).

In elementary work, it is very doubtful if units should be omitted, especially if units are to be changed. For instance, let us give the answer in the problem of John's race, in Section 18, as follows: "John ran the distance $\frac{1}{4}$ in the time $2\frac{1}{4}$, in terms of miles and minutes; his average speed was $\frac{88}{9}$, in terms of feet and seconds." The student will find it conceptually difficult to understand this, and more so to derive it. How does he remember in which direction the formulas for changing units work? Keeping in the units, all such difficulties disappear.

22. Dimensional analysis. In Newton's laws of motion, there is no particular constant of mass, length or time, or any combination of these appearing. When expressed in terms of a quantity structure S, the laws are invariant under transformations of S which leave all properties of S unchanged; that is, the laws are invariant under similarities of S. A common way of expressing this is: The laws do not depend on the choice of units in which they are expressed. (In S, one needs no units to express the laws.)

Suppose, in a given situation, a quantity q depends on other quantities $p_1, \cdots, p_s$: $q=f(p_1, \cdots, p_s)$, as a direct consequence of the laws of mechanics. Then a similarity ϕ of S, changing each p_i into $\phi(p_i)$, changes q into $\phi(q)$. Knowing this, one derives from the "Pi Theorem" considerable information about the function f. This is the method of "dimensional analysis," of great use in parts of mechanics, especially in hydrodynamics. A classical book on this subject is that of Bridgman, [1]. A more modern treatment, with a very detailed study of applications, will be found in Sedov, [4].

A great deal has been written on dimensional analysis, much of it of a rather vague and even philosophical nature: what a "dimension" really is, etc. Here, the advantages of using a mathematical model are clear; one can see precisely what physical assumptions are used, which require the function considered to be invariant under similarities; then the form of the function deduced is a mathematical consequence.

An important point to note is that there is not one model fixed for all considerations; one may choose a particular quantity structure for a particular application. We illustrate this in later sections.

23. The Pi Theorem. This appellation comes from the use of certain products, namely those in (23.3). We suppose we have a quantity structure S; actually, only the positive part, S^+, will be used.

DEFINITION 23A. Let $\Gamma, \Gamma_1, \cdots, \Gamma_s$ be rays in S^+. The function

$$f: \Gamma_1 \times \cdots \times \Gamma_s \to \Gamma$$

is *homogeneous* if it is invariant under similarities ϕ of S; that is,

$$f(\phi(x_1), \cdots, \phi(x_s)) = \phi(f(x_1, \cdots, x_s)). \tag{23.1}$$

If a substructure S' of S is given, we say f is *homogeneous* mod S' if it is invariant under similarities which keep all elements of S' fixed (see Sect. 11).

Note that "homogeneous mod R" is the same as "homogeneous."

LEMMA 23B. *Let the above f be homogeneous (or homogeneous* mod S'). *Then* Γ *is in the substructure generated by* $\Gamma_1, \cdots, \Gamma_s$ *(or by these* Γ_i *and* S').

Suppose not. Then we may choose a base $\{z_1, \cdots, z_n\}$ for S such that $z_1, \cdots, z_{n-1}$ generate a substructure containing the Γ_i and S', and $\Gamma = [z_n]$. By Theorem 11C, there is a similarity ϕ of S such that $\phi(z_i) = z_i$ for $i < n$, and $\phi(z_n) = 2z_n$. Now $\phi(x_i) = x_i$, $x_i \in \Gamma_i$ (all i), and ϕ leaves S' fixed; hence

$$f(x_1, \cdots, x_n) = f(\phi x_1, \cdots, \phi x_n) = \phi f(x_1, \cdots, x_n) = 2f(x_1, \cdots, x_n),$$

a contradiction (since all elements are in S^+).

We now give the Pi Theorem. To allow for subsidiary variables, we suppose a set U given; this may be omitted.

THEOREM 23C. *Let* $\Gamma, \Gamma_1, \cdots, \Gamma_s$ *be rays of* S^+, *let* S' *be a substructure, and suppose the function* $f: \Gamma_1 \times \cdots \times \Gamma_s \times U \to \Gamma$ *is homogeneous* (mod S') *for each fixed* $u \in U$. *Take a maximal subset* $\Gamma_1, \cdots, \Gamma_r$ *of the* Γ_i *(after renumbering) which are independent* (mod S'); *then (see the lemma) for some* Γ' *and* Γ_j' $(j > r)$ *in* S', *we may write*

$$\Gamma = \Gamma_1^{\beta_1} \cdots \Gamma_r^{\beta_r} \Gamma', \qquad \Gamma_j = \Gamma_1^{\alpha_{j1}} \cdots \Gamma_r^{\alpha_{jr}} \Gamma_j' \qquad (j > r). \tag{23.2}$$

Then there is a function $F: \Gamma_{r+1}' \times \cdots \times \Gamma_s' \times U \to \Gamma'$ *with the following property. Given* $x_i \in \Gamma_i$ *(all i) and* $u \in U$, *if we set*

$$\rho_j = x_j x_1^{-\alpha_{j1}} \cdots x_r^{-\alpha_{jr}} \in \Gamma_j' \qquad (j > r), \tag{23.3}$$

then

$$f(x_1, \cdots, x_s, u) = x_1^{\beta_1} \cdots x_r^{\beta_r} F(\rho_{r+1}, \cdots, \rho_s, u). \tag{23.4}$$

REMARK 23D. If S' is not used, then the rays Γ', Γ_j' are replaced by R^+ (and hence may be omitted from (23.2)); then the ρ_j and F have real values. If $r = s$, i.e. the Γ_i are independent (mod S'), then $F = F(u)$, and the theorem (without S') reduces to Theorem 20C, as is easily seen; if, moreover, there is no U, then F is replaced by a constant (in Γ' or in R^+).

REMARK 23E. In many applications, the function f will not be defined throughout the rays, but only on certain convex subsets; then the similarities used must be reasonably near the identity. We shall not study this type of question here.

To prove the theorem, choose elements $z_i \in \Gamma_i$ $(i = 1, \cdots, r)$, and set

$$z_j = z_1^{\alpha_{j1}} \cdots z_r^{\alpha_{jr}} \in \Gamma_1^{\alpha_{j1}} \cdots \Gamma_r^{\alpha_{jr}} = \Gamma_j \Gamma_j'^{-1} \qquad (j > r). \tag{23.5}$$

Since $\rho_j z_j \in \Gamma_j$ if $\rho_j \in \Gamma_j'$, we may define F by

$$(23.6)\qquad F(\rho_{r+1}, \cdots, \rho_s, u) = z_1^{-\beta_1} \cdots z_r^{-\beta_r} f(z_1, \cdots, z_r, \rho_{r+1}z_{r+1}, \cdots, \rho_s z_s, u).$$

Now given the x_i and u and hence the ρ_j we may write

$$(23.7)\qquad x_i = \lambda_i z_i, \qquad \lambda_i \in \mathrm{R}^+ \qquad (i \leqq r).$$

We may find a base for S consisting of $z_1, \cdots, z_r$, some elements forming a base for S' (if S' is used), and some further elements perhaps. Applying Theorem 11C shows that there is a similarity ϕ of S such that

$$\phi(x) = x \text{ (all } x \in S'), \qquad \phi(z_i) = \lambda_i z_i = x_i \qquad (i \leqq r).$$

Applying (11.3) to $\rho_j z_j \in \Gamma_j$ gives, using (23.2) and (23.3),

$$\phi(\rho_j z_j) = \lambda_1^{\alpha_{j1}} \cdots \lambda_r^{\alpha_{jr}} \rho_j z_j = \rho_j \lambda_1^{\alpha_{j1}} \cdots \lambda_r^{\alpha_{jr}} z_1^{\alpha_{j1}} \cdots z_r^{\alpha_{jr}}$$
$$= \rho_j x_1^{\alpha_{j1}} \cdots x_r^{\alpha_{jr}} = x_j \qquad (j > r).$$

Therefore, using (11.3) for ϕf in Γ and (23.6),

$$\begin{aligned} f(x_1, \cdots, x_s, u) &= f(\phi z_1, \cdots, \phi(\rho_{r+1}z_{r+1}), \cdots, u) \\ &= \phi f(z_1, \cdots, \rho_{r+1}z_{r+1}, \cdots, u) \\ &= \lambda_1^{\beta_1} \cdots \lambda_r^{\beta_r} z_1^{\beta_1} \cdots z_r^{\beta_r} F(\rho_{r+1}, \cdots, \rho_s, u), \end{aligned}$$

and using (23.7) gives (23.4).

24. An elementary area problem. One may construct different quantity structures to study a single problem; that one in which we have invariance under the largest class of similarities gives the most information. We illustrate this with the problem of finding the area A of a rectangle whose sides are of lengths x_1 and x_2: $A = f(x_1, x_2)$.

First, suppose we have no knowledge about area. Then we use $S = \mathrm{R}$ (thus assigning real numbers to all quantities). In the Pi Theorem, $\Gamma = \Gamma_1 = \Gamma_2 = \mathrm{R}^+$; $r = 0$ (each ray is dependent). The only similarity of S is the identity. Given $x_i \in \Gamma_i$ $(i = 1, 2)$, (23.3) reads $\rho_j = x_j$ $(j = 1, 2)$, and (23.4) gives $f(x_1, x_2) = F(x_1, x_2)$, an unknown function.

Next, try calling the "dimension" of each side L, and the "dimension" of A, L^2. Then $\Gamma_1 = \Gamma_2 = L$, $\Gamma = L^2$, and $r = 1$, $s = 2$. Now (23.3) reads $\rho_2 = x_2 x_1^{-1}$, and (23.4) gives

$$f(x_1, x_2) = x_1^2 F(x_2/x_1).$$

The assumption about similarities, in this case, is that multiplying all lengths by the same constant multiplies the area by that constant squared.

Finally, construct S from the rays Γ_1 and Γ_2, and take A in $\Gamma_1\Gamma_2$; then $r = s = 2$. The assumption about the homogeneity of f may be expressed as follows:

$$f(\lambda x_1, x_2) = \lambda f(x_1, x_2) = f(x_1, \lambda x_2),$$

i.e. the area is proportional to each side separately. Both the Pi Theorem and Theorem 20C give $f(x_1, x_2) = cx_1x_2$.

REMARK 24 A. With the last S above, if we have no knowledge about area, we may set $S' = S$, which allows only the trivial similarity; for the second case, in which Γ_1, and Γ_2 are dependent, we may let S' be generated by $\Gamma_2 \Gamma_1^{-1}$; for the last case, use $S' = \mathrm{R}$. The same results are obtained.

25. A falling body problem. If a ball is thrown up with the velocity v, when does it land on the ground (neglecting air resistance)? As in Section 18, we may solve $vt - 16$ ft sec$^{-2}t^2 = 0$ with $t > 0$, giving $t = cv$, with $c = \sec^2/16$ ft. (Hence if $v = 40$ ft/sec, then $t = 2.5$ sec.)

Now let us try dimensional analysis on the problem. We first assume known that the acceleration g of gravity enters in, but that it is a constant:

$$g = 32 \text{ ft sec}^{-2} \in LT^{-2}.$$

Assuming mass does not enter in, we construct S from L and T; we look for $t = f(v)$. Let g generate the substructure S'; we use only similarities keeping S' fixed. With the notations of Section 23, we have $r = s = 1$, and

$$\Gamma_1 = [v] = LT^{-1}, \qquad \Gamma = [t] = T = \Gamma_1^{\beta}\Gamma' = (LT^{-1})^{\beta}(LT^{-2})^{\gamma}$$

for some β and γ. This gives $\beta + \gamma = 0$, $-\beta - 2\gamma = 1$, and hence $\beta = 1$, $\gamma = -1$. The physical hypothesis is that f is homogeneous mod S'; assuming this, (23.4) gives $t = f(v) = cv$.

Now, travelling to other planets, we find that g is not constant; hence we look for a function $t = f(v, g)$. In this case, $r = s = 2$, and there is no S'; we have

$$\Gamma_1 = LT^{-1}, \qquad \Gamma_2 = LT^{-2}, \qquad \Gamma = \Gamma_1^{1}\Gamma_2^{-1};$$

thus the expression for Γ is the same as before, with $\Gamma^{-\frac{1}{2}}$ replacing Γ'. Again assuming homogeneity, (23.4) now gives $t = f(v, g) = cv/g$.

26. Pendulum problems. What is the period P of a pendulum of mass m and length l? We assume that the motion is periodic (an experimental fact), and include the amplitude α as a subsidiary variable. Let us first treat g as constant; we use the quantity structure generated by M, L and T, and let S' be generated by g. We wish $P = f(m, l, \alpha)$; we have $r = s = 2$. Now

$$\Gamma_1 = M, \qquad \Gamma_2 = L, \qquad \Gamma = \Gamma_1^{\beta}\Gamma_2^{\beta'}\Gamma'^{\gamma} = M^{\beta}L^{\beta'}(LT^{-2})^{\gamma};$$

hence $\beta = 0$, $\beta' = 1/2$, $\gamma = -1/2$, and hence, assuming f homogeneous mod S',

$$P = \sqrt{l}\,F(\alpha). \tag{26.1}$$

Now consider g as variable and look for $f(m, l, g, \alpha)$. Here, $\gamma = -1/2$ enters in, and we have $P = \sqrt{l/g}\,F(\alpha)$.

REMARK. The same analysis, and hence the same formula (with a different $F(\alpha)$) applies if we assume the pendulum is a rod of length l.

Now consider a more complex pendulum problem. We assume the pendulum consists of a rod, suspended at an interior point, with the length l below and the length l' above this point; we look for $P=f(l, l', g, \alpha)$ (omitting m). Here, $r=2$, $s=3$, and we take

$$\Gamma_1 = [l] = L, \qquad \Gamma_2 = [g] = LT^{-2}, \qquad \Gamma_3 = [l'] = L,$$

and $\Gamma=\Gamma_1^{1/2}\Gamma_2^{-1/2}$ as before. Now (23.3) becomes $\rho_3=l'/l$, and hence

$$P = \sqrt{l/g}\, F(l'/l, \alpha). \tag{26.2}$$

In place of l and l', we could have used for instance $l+l'$ and l'; the same analysis and formula would have resulted.

Note that if l' is small, and it seems a reasonable assumption that it may therefore be neglected, then it drops out of the above, and the preceding formula results. This illustrates the fact that in the applications, one must choose the model and its use to fit the individual case.

27. Satellites. First consider the motion of a pair of masses m_1 and m_2 about their common center of mass; we assume their distance apart r is constant. We recall the law of gravitation:

$$\text{force} = -\gamma m_1 m_2/r^2.$$

Hence

$$[\gamma] = [\text{force}][r^2][m_1m_2]^{-1} = (MLT^{-2})L^2M^{-2} = M^{-1}L^3T^{-2}. \tag{27.1}$$

We look for the period $P=f(m_1, m_2, r, \gamma)$; we could omit γ, and keep S', generated by $[\gamma]$, constant. Taking m_1, r, γ as independent, we use $\rho=m_2/m_1$, and

$$[P] = T = M^{\beta}L^{\beta'}(M^{-1}L^3T^{-2})^{\beta''},$$

giving $\beta=-1/2$, $\beta'=3/2$, $\beta''=-1/2$. The conclusion is

$$P = \sqrt{\frac{r^3}{m_1\gamma}}\, F\left(\frac{m_2}{m_1}\right).$$

Now assume that m_2 is small compared with m_1; then we may suppose that f is independent of m_2. In the case of a satellite about the earth, not far away from the earth, we should take account of a new variable, the radius of the earth. If we suppose that the force of gravity is the same as if all the mass of the earth were concentrated at the center, and we let r be the distance from the satellite to this center, then we may drop m_2 from the preceding analysis, giving (if γ is omitted) $P=c\sqrt{r^3/m_1}$.

As an application of this, note that if we know the mass and radius of the earth and the period of a satellite about the earth, and measure this period about the moon at a certain distance from the moon, we can deduce the mass of the moon from the above formula.

28. Water flowing from a reservoir. We assume there is a V-shaped opening of angle α through which the water runs out; we wish to find the rate r of flow of water in volume per unit time, as a function of the height h above the bottom of the V. We look for $r=f(h, g, \alpha)$ (we of course could take g as constant). Now

$$[r] = L^3T^{-1}, \qquad [h] = L, \qquad [g] = LT^{-2}, \qquad [r] = [h]^{5/2}[g]^{1/2}.$$

Now assume that water is a "perfect fluid." This means, for us, that f is homogeneous. Then the Pi Theorem gives

$$r = \sqrt{h^5 g}\, F(\alpha).$$

Actually, the formula is fairly accurate; thus the physical assumption is borne out.

29. Torque on a wire. We illustrate here a combination of methods, and the separation of two different kinds of length into different types (physical dimensions). First, consider a very short rod, of base A and length l. With the base fixed, its top is pulled a distance d (sideways) by the force F. It is an obvious physical assumption that F is proportional to A and to d, and inversely proportional to l; hence (Sect. 20)

$$F = \eta A d/l,$$

η being the rigidity of the substance. Take $d \in L$, $A \in L^2$, and $l \in L'$; then

$$[\eta] = [F][Ad]^{-1}[l] = [F]L^{-3}L'.$$

Now consider a rod (of the same substance) of radius r and length l, twisted through the (small) angle θ; we wish to find the torque τ needed. Clearly τ is proportional to θ (we could introduce a new dimension $[\theta]$ and obtain this; but this would render the physical assumption less clear); thus we now write

$$\tau = \theta f(r, l, \eta).$$

With $[\tau] = [F]L$, $[\theta] = [1]$, $[r] = L$, $[l] = L'$, and the same η, we have

$$[\tau] = [F]L = L^{\alpha}L'^{\beta}([F]L^{-3}L')^{\gamma};$$

hence $\gamma = 1$, $\beta = -1$, $\alpha = 4$, and we obtain

$$\tau = c\eta r^4\theta/l.$$

References

1. P. W. Bridgman, Dimensional analysis, Yale University Press, New Haven, 1922, 1931.
2. S. Drobot, On the foundations of dimensional analysis, Studia Math., 14 (1953) 84–99.
3. H. Flanders, Differential forms, Academic Press, New York-London, 1963.
4. L. I. Sedov, Similarity and dimensional methods in mechanics, Academic Press, New York-London, 1959.
5. H. Whitney, The Mathematics of Physical Quantities, Part I: Mathematical models for measurement, this MONTHLY, 75 (1968) 115–138.

1. LOGIC IN MATHEMATICS. Let us start with an example of reasoning, and consider it from two points of view :

(1) P or Q ; not P ; hence Q.

Here is one way of explaining this : A formal system of logic is set up ; it may contain "undefined terms," rules of procedure, and so on. One now derives various formulas and methods of carrying out proofs, and finds (1). In contrast, a young child might meet the reasoning as follows : Daddy comes home, and says to Johnny, "Which hand ?" Johnny tries the left hand, and finds nothing. He concludes that there is a present in Daddy's right hand.

Should Johnny be told that his reasoning is unjustified ? He knows the validity of his reasoning perfectly well ; the payoff is the present he gets. The "which hand" translates for him to "right or left," and the actual meaning of "or" here is best explained by the experiment carried out. In fact, the formal procedure here would be unjustified, unless one shows that the system of formal logic set up actually applies to this example.

In considering formal logic, perhaps we have gone to an extreme.It seems clear that it is good to insert in a text a chapter on logic, showing in a more informal manner the logic of propositions, quantifiers, and so on. Yet even here, in books containing such chapters, the later work is (fortunately) written so as to be essentially independent of the chapter. The basic difficulty is again that logic has been studied for itself instead of growing out of any direct need for it.

As a matter of fact, the "logic" studied in such texts is quite unsufficient for the need of mathematics. For instance, the reader might run into the following statement :

Logic : Fad or tool ?

Hassler WHITNEY **New-Jersey, U.S.A.**

Along with the tremendous growth of mathematics through the sciences has come the realization that the real worth of mathematics is not so much its ability to cope with complicated equations as its power in analyzing phenomena. It follows that in teaching mathematics, one should not merely give drill in carrying out certain operations ; one should study the basic ideas and workings of the subject. However, in trying to carry this out, for instance in the "New Math," textbook writers have tended to forget the manner in which one grasps and gets control of essential concepts. Indroducing beautiful abstract theories can wake the student to new realms of thought ; but when one tries to put them into conjunction with simple mathematical problems, the connection is apt not to take place.

If one really examines simple problems with care, the general concepts arise by themselves ; moreover, they are in a form already comprehended by the student. We shall illustrate some of the dangers of the too abstract approach, and some ways in which one may return to reality.

Revue périodique du Centre Belge de Pédagogie de la Mathématique

Periodiek tijdschrift van het Belgisch Centrum voor Metodiek van de Wiskunde

(2) If $x \neq 0$ then $\frac{x}{x} = 1$.

He looks up "If P then Q" to apply it here. He must determine the truth values of "$x = 0$" and of "$x/x = 1$". But "x/x" has no meaning if $x = 0$. Being somenwhat puzzled, he may try forming the "contrapositive" :

(3) If $\frac{x}{x} \neq 1$ then $x = 0$.

Who can blame him for giving up ?

A little later we shall see the real meaning of "If ... then... ."

2. QUANTIFIERS. Johnny has shown us that the logic of propositions need carry no fear. How about quantifiers ? Anyone who can understand the two following examples need not fear them either.

For any odd integer n, $n^2 - 1$ is divisible by 8. For $n^2 - 1 = (n - 1)(n + 1)$, and $n - 1$ and $n + 1$ are both even, and one of these is divisible by 4. Note that the full meaning of the proof is as follows : Suppose we are given any particular odd integer ; let us call it n . Using only the information that n is odd, we show that $n^2 - 1$ is divisible by 8. Then the proof may be applied to the integer 3 ; also to 19, to -7, in fact, to any odd integer ; so the statement holds in general.

An integer n is called composite if there exists an integer m such that $1 < m < n$ and n is divisible by m. Let us test to see if 15 is composite. We find the integer 3 with the desired property ; thus 15 is composite. Another proof : 5 has the desired property.

3. COMMUNICATION. Logic is the study of reasoning ; hence a chapter on logic should enable us to carry out reasoning correctly. The usual trouble is that one commonly defines "logic" in a general sense, and then supposes that studying certain particular rather formal items has taught us logic. The first problem is that of how one communicates ; however a theory may be set up, it must be communicated to the reader or listener. We shall consider only certain aspects of this problem.

A proof may be given in a very strict form ; the reader can follow each detailed step. But if he is to get a feeling of what is going on, there must be discussion also ; and this can be carried on in an informal manner. Thus there are two levels of communication, and it is very important to distinguish between them :

The formal level : One presents statements, with precise mathematical content. The reader must be able to understand exactly what is meant at each point. He may rephrase statements if he wishes, giving them different and perhaps simpler form, preserving the mathematical content. For instance, he may replace "but we recall that $x > 0$" by "known : $x > 0$."

The informal level : Here, one may use ordinary language, putting over ideas in whatever manner seems most appropriate. The student (and teacher) should be warned against trying to find the exact meaning of statements on this level. Of course much discussion (of an informal character) contains incidental statements with mathematical content ; thus the division into formal and informal is not always clear. But texts which mix up the manner of presentation so that the level of discussion is unclear make great trouble for the reader.

A typical word which is used sometimes formally and sometimes informally is "equation." For instance, the items displayed below are apt to be called "equations." Yet if their meanings are made clear, their meanings will probably differ considerably, and a single word will not apply to them all. In spite of this, a text is likely to decide to be "careful" at various points, and define an "equation". This implies that every use of the word will be formal ; and this is especially unfortunate if the teacher demands that the students apply a strict definition to a place where the definition does not hold.

A commonly used definition which is an utter confusion of the two levels is : "A ratio is a comparison of two quantities by division."

Thus one danger is that of faulty communication. Another is that of a lack of communication. From the manner in which some things are written down, the reader is supposed to figure out the full meanings attached. For instance certain equations are supposed to be automatically interpreted in particular ways. Here are some typical examples :

$3x - 5 = x + 3$:	an equation in x.
$y = x^2 - 2x + 15$:	defines y in terms of x.
$x + y = y + x$:	an identity.
$\frac{x}{a} + \frac{y}{b} = \frac{x + y}{a + b}$:	wrong.

Will the student feel confident in being able to rerpret similarly later equations ? Will he even be satisfied (better yet, should he be) with the interpretations given or only implied ?

4. THE USE OF SYMBOLS. A term that is sure to cause the student difficulties is that of "variable." A text usually tries to remove these difficulties by defining the term ; for

instance, "a variable is a letter that stands for any of a given domain." But the student worries about how is actually <u>used</u> ; and this seems often to be unrelated definition. And the text, in later considerations, may of be interpreted in different ways, in which definitions of given may or may not apply.

We illustrate this by considering the "solution of a equation" from different points of view.

(a) Tom has written down a certain number. He squares it, and subtracts his original number ; he tells us that he go 20. What can we find out about his original number ?

Here is a very clear problem ; let us go directly at it. A first basic principle is : To talk about something, give it a name (as though we knew what it is). Let us call Tom's number x . Then we are told that

$$(a_1) \qquad x^2 - x = 20.$$

Being clever, we reason as follows : Since $x^2 - x$ is 20, subtracting 20 will give 0 :

$$(a_2) \qquad x^2 - x - 20 = 0.$$

Next, if y, p, q are numbers (i.e. these letters are names for numbers), then

$$(y - p)(y + q) = y^2 - (p - q)y - pq.$$

This looks very much like (a_2) ; in fact, we can choose y to be x, and p = 5 , q = 4. Thus we find

$$(a_3) \qquad (x - 5)(x + 4) = x^2 - x - 20 = 0.$$

Now the product of the two numbers $x - 5$ and $x + 4$ is 0 ; this tells us that

(a.) $$x - 5 = 0 \quad \text{or} \quad x + 4 = 0,$$

or equivalently,

(a.) $$x = 5 \quad \text{or} \quad x = -4$$

Now we consider another problem, that looks much like the first :

(b) Solve the equation

(b.) $$x^2 - x = 20.$$

We should not confuse this with the preceding. For in (a.), "x" is a name for a perfectly definite number, which Tom wrote down ; such is not the case here. Thus the meaning of (b_1) must be explained. One way to phrase it is to say that "x" is variable, and we are to find the solution set of (b_1). (Here, "phrasing" may not mean "elucidating".)

An algebra book may "solve" (b_1) by the steps given in (a). For instance, subtracting 20 from both sides of (b_1) gives

(b.) $$x^2 - x - 20 = 0.$$

But how does one subtract a number from a side of an equation ? Looking up the definition of "equation" does not help him. Or, if x is a variable, is $x^2 - x$ something from which one may subtract a number ? The "definition" of variable does not help ! A proper use of language is of course possible, and one goes through the same steps, finally arriving at (b_5), looking like (a). But this also needs elucidation ; a final step might be :

(b_6) The solution set of (b_1) is $\{5, -4\}$.

Note that though the (b_i) have exactly the same expression as the (a_i), the meanings are quite different. In each (b_i), we could happily replace "x" by "u" for instance without altering the reasoning ; we can in (a_i) only if we made a separate statement as to the new letter.

One common way of phrasing the difference is : In each (a_i), "x" is a constant, while in each (b_i), "x" is a variable. But note that we did not need to introduce the notion "constant" to solve problem (a).

What is the best interpretation for studying the solution of equations ? Certainly one should start with the simplest interpretation, that in (a). Next we may note that we desire more general statements. For suppose Ann wrote down a number, and told us that after squaring it and subtracting the original number, she got 20. We did not solve this problem ; "x" stood for Tom's number, not Ann's. We really wish, for the second stage, to use (b), with the proper meaning so that it covers all such cases. Note, however, that we may carry out the work of solving (b) by going back to the original method :

Suppose x_0 is a number such that $x_0^2 - x_0 = 20$. Then we carry out the steps in (a), finding that x_0 is either 5 or -4, and we write down (b-). (We must also test, and note that $5^2 - 5 = 20$, and similarly for - 4.) Thus, with this explanation (and replacing x_0 by x) we may carry out the work in (b) with x being a "constant" instead of a "variable" (except in the original statement).

5. STATEMENTS, HYPOTHESES. Information is communicated through the assertion of statements. Let us take a typical statement, say (a_2). If it is of enough interest, should it not appear in an encyclopedia of mathematics ? Perhaps under

"some general facts" we might put

$$x^2 - x - 20 = 0.$$

The first trouble is, that this would not be understood. Seeing this would imply that we are dealing with numbers ; but what does "x" stand for ? Of course (a_2) is not to be understood all by itself, but in the light of what has already been explained.

To understand the assertion of a statement, the reader is supposed to know what structure is being dealt with, and what is assumed known about the particular items in the assertion. Such assumptions are called "premises" or "hypotheses." <u>An assertion has meaning only in terms of given preceding material</u>. Thus in (a_2), we know that "x" stands for a certain number chosen by Tom, and we know that (a_1) holds. We are not told whether x is supposed to be a natural number, an integer, or a real number, for instance ; in this regard we have not complete knowledge of the premises. (We are on pretty safe ground in dealing with real numbers ; if Tom does not consider negative numbers, we can then drop out the -4.)

We may phrase the meaning of an assertion as follows : <u>If</u> we are in the situation given at the moment of an assertion, <u>then</u> the assertion has meaning and is true.

We can now give a proper interpretation of any "if... then ..." form of statement : The "if" part is to be added to the hypotheses then in force ; under these conditions, the "then" part has meaning and is true.

For example, in (2), we add "$x \neq 0$" to any previous material ; then "$x/x = 1$" has meaning and is true. (Note that (3) would still probably not be understood). We remark that when the two parts of an "if... then..." statement both have meaning <u>in advance</u>, then the truth table method may be used to determine the truth of the statement.

Since the meaning of an assertion depends on the previous material, the information conveyed does also. Here is an illustration : After "solving" Tom's problem, we might think :

(a*) If $x > 0$ then $x = 5$.

Now this looks quite amazing : If we learn that a number is positive, we then know that it is 5. Of course this is not <u>any</u> number ; it is Tom's number. Knowing this, "$x > 0$" and "$x = 5$" convey the same information.

We note a general problem of communication. Since we must know at any point in a text just what hypotheses are in force, we must also know when hypotheses (and meanings of symbols) are dropped. For instance, in (2), the hypothesis $x \neq 0$ is probably supposed to last only through the displayed equation ; the symbol "x" may then be given another meaning if we are through with the argument.

6. <u>SOME GENERAL PROBLEMS OF EXPOSITION.</u> By this time it should be clear that the writing of a good text is very difficult ; the reader must be able to ascertain easily what the level of the discussion is, and if formal, just what he is supposed to know at that moment. Moreover, the reader will be expected to refer back and forth to different parts of the text ; the whole must form a consistent unit. Along with this is the problem of introducing concepts in a meaningful manner, bringing the reader into a good train of thought (through informal discussion).

We give some samples of difficulties of a general nature confronting the reader.

The reader is told that addition is a binary operation Later, in discussing subtraction, it is mentioned that subtraction is the inverse of addition. How can a binary operation have an inverse ?

To help understand the difficulty, here is a good topic for classroom discussion. Is subtraction commutative ? Jane says no ; for 6 - 4 and 4 - 6 are different. Bill says yes ; for if you first subtract 2 and then subtract 5, you get the same answer as if you first subtract 5 and then subtract 2. The students themselves should be able to unravel the problem, if they are urged to listen to each other and find the points of divergence. (The question itself may be given different interpretations).

In a "proof by contradiction", one introduces a new premise, and then shows that this cannot be ; the premise is thereby disproved (assuming the previous premises). During the proof, since things are contradictory, we are in a world of unreality. Here is an example.

Theorem. The square root of 2 is not rational. Proof Suppose it were. Say $\sqrt{2} = a/b$; we may assume that the fraction is written in its lowest terms, i.e. b has the smallest possible value. Squaring gives $2 = a^2/b^2$, and hence $a^2 = 2\,b^2$. Thus a^2 is even, and it follows that a is even. Hence we may write $a = 2\,c$. Now $4\,c^2 = 2\,b^2$, and $b^2 = 2\,c^2$; thus $\sqrt{2} = b/c$, contrary to the assumption about lowest terms.

The average student will follow the details, then memorize the proof ; he is then satisfied. But the brighter student may think : We have showed that there are no such numbers as a,b ; then how can we actually work with them ? How can we say that a is even, if it is not a number at all ? How do we square it ? The whole thing is quite mystifying.

Here is a way to help him. First, see what we can definitely say we have proved. It is :

Lemma : If a and b are natural numbers, then $a^2 \neq 2 b^2$.

Now we can say : There are no natural numbers a,b such that $\sqrt{2} = a/b$; that is, $\sqrt{2}$ is not rational.

We have replaced the first proof by one better understood. But the question remains : how about the meaning of the first proof ? The answer, for us, is simple : With the premise that we are dealing with the real number system, and with the additional premise that $\sqrt{2}$ is rational, we derive a contradiction ; thus the additional premise does not hold. In the course of the proof, the assertion that a is even has no meaning by itself ; it has meaning only in the light of all the premises. At that point, a is assumed to be a number with certain properties. After the proof is finished, when the premise that $\sqrt{2} = a/b$ is dropped, "a" no longer has any meaning.

As a last illustration, we show how presenting some lovely and harmless formal material may lead to confusion. It is stated that two sets A and B are equal if and only if each element of one is also an element of the other. (The text book writer is very familiar with this, and has planned carefully where to insert it). What is the reader confronted with He has two sets. He takes an element of A, and sees if it is in the other set B. Apparently the sets are different, without trying any test. And anyway, if they were the same, what is the point of taking an element of a set and seeing if it is in that very set ? Perhaps "equal" has some special meaning that has not really been presented ?

The basic difficulty is that material has been presented out of the blue, without reference to any application. Let us start from the other direction : Let A be the set of even numbers between 0 and 10, and let B be the set of numbers whose distance from 5 is either 1 or 3. are the sets the same ? Now it is clear that the statement has meaning in reference to two <u>expressions</u> of sets.

7. <u>A LAST WORD TO THE TEACHER.</u> Your students are mostly as bright and intelligent as they sometimes seem ; the authors of your texts are also, but they may not have succeeded in a very difficult problem. Your students can help you discover inconsistencies and difficulties in the text, and in your own statements. This is fine ; between you, further understanding will be achieved, and the book will be set right in many ways.

*

Comment on the Division of the Plane by Lines

Hassler Whitney

The purpose of this note is to illustrate what I consider a basic principle in mathematical research: If a problem is difficult, don't be afraid to get your hands dirty. That is, make up various examples, simple and complex, and test out what happens, in a rather complete manner.

The example I consider is: How many regions are formed by a set of straight lines in the plane? I refer to the recent paper of John E. Wetzel [1].

To follow the suggested principle, draw a number of lines in the plane, and actually look at the counting process. The question arises at once: In what order shall I count the regions? There are plenty of choices; so look for some simple choice.

If the "plane" is a large sheet of paper on a table, you might first count the ones furthest up, then move downwards. You could stretch a stick across and move it downwards, counting from left to right if several regions are touched at the same time. Or to avoid hiding parts of the figure, move a stretched string downwards. Call this a "sweep line"; we have come naturally to the "ingenious notion" in Wetzel's paper (p. 654). If the diagram is really large, and you are in the middle, you could first count those regions furthest away; the sweep line is replaced by a shrinking circle. A difficulty here (counting some regions twice) is apt to lead you to flatten the circles, thus bringing you to the sweep line.

My most important message here is toward the educational process: You learn math best by *doing* it; especially, by *needing* it in some situation. Especially in early grades, "dirtying your hands" becomes "act out the story." In this way, there is a much better chance that schoolchildren will regain some of the extraordinary rate of learning that they exhibited before entering school.

Reference

[1] John E. Wetzel, *On the division of the plane by lines*, this MONTHLY, 85 (1978), 647–656.

Some of these remarks were presented in an address to the American Mathematical Society at Brown University, August 10, 1978.

Institute for Advanced Study
Princeton, NJ 08540.

PERMISSIONS
Volume II

Birkhäuser Boston thanks the original publishers of the papers of Hassler Whitney for granting permission to reprint the papers in this collection.

[9] Reprinted from *AMS Bulletin* **38**, ©1932 by the American Mathematical Society.

[11] Reprinted from *AMS Transactions* **35**, ©1933 by the American Mathematical Society.

[15] Reprinted from *Annals of Mathematics* **34**, ©1933 by Princeton University Press.

[16] Reprinted from *Annals of Mathematics* **34**, ©1933 by Princeton University Press.

[28] Reprinted from *Annals of Mathematics* **37**, ©1936 by Princeton University Press.

[29] Reprinted from *Annals of Mathematics* **37**, ©1936 by Princeton University Press.

[30] Reprinted from *Compositio Mathematica* **4**, ©1937 by Kluwer Academic Publishers.

[35] Reprinted from *Annals of Mathematics* **38**, ©1937 by Princeton University Press.

[36] Reprinted from *AMS Bulletin* **43**, ©1937 by the American Mathematical Society.

[39] Reprinted from *Annals of Mathematics* **39**, ©1938 by Princeton University Press.

[41] Reprinted from *Proceedings of the National Academy of Science* **26**, ©1940 by permission of the author.

[42] Reprinted from *Proceedings of the National Academy of Science* **26**, ©1940 by permission of the author.

[43] Reprinted from *AMS Bulletin* **47**, ©1941 by the American Mathematical Society.

[48] Reprinted from *AMS Bulletin* **50**, ©1944 by the American Mathematical Society.

GPSR Compliance

The European Union's (EU) General Product Safety Regulation (GPSR) is a set of rules that requires consumer products to be safe and our obligations to ensure this.

If you have any concerns about our products, you can contact us on ProductSafety@springernature.com

In case Publisher is established outside the EU, the EU authorized representative is:

Springer Nature Customer Service Center GmbH
Europaplatz 3
69115 Heidelberg, Germany

Batch number: 10578397

Printed by Printforce, the Netherlands